N	0	1	2	3	4	5	6	7	8	9
0		000	301	477	602	699	778	845	903	954
1	000	041	079	114	146	176	204	230	255	279
2	301	322	342	362	380	398	415	431	447	462
3	477	491	505	519	531	544	556	568	580	591
4	602	613	623	633	643	653	663	672	681	690
5	699	708	716	724	732	740	748	756	763	771
6	778	785	792	799	806	813	820	826	833	839
7	845	851	857	863	869	875	881	886	892	898
8	903	908	914	919	924	929	934	940	944	949
9	954	959	964	968	973	978	982	987	991	996
10	000	004	009	013	017	021	025	029	033	037
11	041	045	049	053	057	061	064	068	072	076
12	079	083	086	090	093	097	100	104	107	111
13	114	117	121	124	127	130	134	137	140	143
14	146	149	152	155	158	161	164	167	170	173
15	176	179	182	185	188	190	193	196	199	201
16	204	207	210	212	215	217	220	223	225	228
17	230	233	236	238	241	243	246	248	250	253
18	255	258	260	262	265	267	270	272	274	276
19	279	281	283	286	288	290	292	294	297	299
20	301	303	305	308	310	312	314	316	318	320
21	322	324	326	328	330	332	334	336	338	340
22	342	344	346	348	350	352	354	356	358	360
23	362	364	365	367	369	371	373	375	377	378
24	380	382	384	386	387	389	391	393	394	396
25	398	400	401	403	405	407	408	410	412	413
26	415	417	418	420	422	423	425	427	428	430
27	431	433	435	436	438	439	441	442	444	446
28	447	449	450	452	453	455	456	458	459	461
29	462	464	465	467	468	470	471	473	474	476
30	477	479	480	481	483	484	486	487	489	490
31	491	493	494	496	497	498	500	501	502	504
32	505	507	508	509	511	512	513	515	516	517
33	519	520	521	522	524	525	526	528	529	530
34	531	533	534	535	537	538	539	540	542	543
35	544	545	547	548	549	550	551	553	554	555
36	556	558	559	560	561	562	563	565	566	567
37	568	569	571	572	573	574	575	576	577	579
38	580	581	582	583	584	585	587	588	589	590
39	591	592	593	594	596	597	598	599	600	601
40	602	603	604	605	606	607	609	610	611	612
41	613	614	615	616	617	618	619	620	621	622
42	623	624	625	626	627	628	629	630	631	632
43	633	634	635	636	637	638	639	640	641	642
44	643	644	645	646	647	648	649	650	651	652
45	653	654	655	656	657	658	659	660	661	662
46	663	664	665	666	667	667	668	669	670	671
47	672	673	674	675	676	677	678	679	679	680
48	681	682	683	684	685	686	687	688	688	689
49	690	691	692	693	694	695	695	696	697	698
N	0	1	2	3	4	5	6	7	8	9

N	0	1	2	3	4	5	6	7	8	9
50	699	700	701	702	702	703	704	705	706	707
51	708	708	709	710	711	712	713	713	714	715
52	716	717	718	718	719	720	721	722	723	723
53	724	725	726	727	728	728	729	730	731	732
54	732	733	734	735	736	736	737	738	739	740
55	740	741	742	743	744	744	745	746	747	747
56	748	749	750	751	751	752	753	754	754	755
57	756	757	757	758	759	760	760	761	762	763
58	763	764	765	766	766	767	768	769	769	770
59	771	772	772	773	774	775	775	776	777	777
60	778	779	780	780	781	782	782	783	784	785
61	785	786	787	787	788	789	790	790	791	792
62	792	793	794	794	795	796	797	797	798	799
63	799	800	801	801	802	803	803	804	805	806
64	806	807	808	808	809	810	810	811	812	812
65	813	814	814	815	816	816	817	818	818	819
66	820	820	821	822	822	823	823	824	825	825
67	826	827	827	828	829	829	830	831	831	832
68	833	833	834	834	835	836	836	837	838	838
69	839	839	840	841	841	842	843	843	844	844
70	845	846	846	847	848	848	849	849	850	851
71	851	852	852	853	854	854	855	856	856	857
72	857	858	859	859	860	860	861	862	862	863
73	863	864	865	865	866	866	867	867	868	869
74	869	870	870	871	872	872	873	873	874	874
75	875	876	876	877	877	878	879	879	880	880
76	881	881	882	883	883	884	884	885	885	886
77	886	887	888	888	889	889	890	890	891	892
78	892	893	893	894	894	895	895	896	897	897
79	898	898	899	899	900	900	901	901	902	903
80	903	904	904	905	905	906	906	907	907	908
81	908	909	910	910	911	911	912	912	913	913
82	914	914	915	915	916	916	917	918	918	919
83	919	920	920	921	921	922	922	923	923	924
84	924	925	925	926	926	927	927	928	928	929
85	929	930	930	931	931	932	932	933	933	934
86	934	935	936	936	937	937	938	938	939	939
87	940	940	941	941	942	942	942	943	943	944
88	944	945	945	946	946	947	947	948	948	949
89	949	950	950	951	951	952	952	953	953	954
90	954	955	955	956	956	957	957	958	958	959
91	959	960	960	960	961	961	962	962	963	963
92	964	964	965	965	966	966	967	967	968	968
93	968	969	969	970	970	971	971	972	972	973
94	973	974	974	975	975	975	976	976	977	977
95	978	978	979	979	980	980	980	981	981	982
96	982	983	983	984	984	985	985	985	986	986
97	987	987	988	988	989	989	989	990	990	991
98	991	992	992	993	993	993	994	994	995	995
99	996	996	997	997	997	998	998	999	999	000
N	0	1	2	3	4	5	6	7	8	9

SEVENTH EDITION

BASIC MATHEMATICS FOR ELECTRONICS

Nelson M. Cooke
Late President
Cooke Engineering Company

Herbert F. R. Adams
Former Chief Electronics Instructor
British Columbia Institute of Technology

Peter B. Dell
Late Chief Electronics Instructor
British Columbia Institute of Technology

T. Adair Moore
Former Electronics Instructor, Idaho State University
School of Vocational-Technical Education

GLENCOE

McGraw-Hill

New York, New York   Columbus, Ohio   Mission Hills, California   Peoria, Illinois

Basic mathematics for electronics. — 7th ed. / Nelson M. Cooke . . .
[et al.]
p. cm.
Includes index.
ISBN 0-02-800853-7
1. Electric engineering—Mathematics. 2. Electronics—
-Mathematics. I. Cooke, Nelson Magor.
TK153.B37 1991
510′.246213—dc20 91–20919
CIP

Imprint 1995

Send all inquiries to:
McGraw-Hill
936 Eastwind Drive
Westerville, OH 43081

ISBN 0-02-800853-7

Printed in the United States of America

4 5 6 7 8 9 10 11 12 13 14 15 RRDC/LP 03 02 01 00 99 98 97 96 95

CONTENTS

CHAPTER 4 ALGEBRA—MULTIPLICATION AND DIVISION 35

CHAPTER 5 EQUATIONS 56

CHAPTER 6 POWERS OF 10 74

CHAPTER 7 UNITS AND DIMENSIONS 90

CHAPTER 8 OHM'S LAW—SERIES CIRCUITS 115

CHAPTER 9 RESISTANCE—WIRE SIZES 142

CHAPTER 10 SPECIAL PRODUCTS AND FACTORING 156

CHAPTER 11 ALGEBRAIC FRACTIONS 180

CHAPTER 18 DETERMINANTS 309

CHAPTER 19 BATTERIES 328

CHAPTER 20 EXPONENTS AND RADICALS 335

CHAPTER 21 QUADRATIC EQUATIONS 358

CHAPTER 22 NETWORK SIMPLIFICATION 387

CHAPTER **27** TRIGONOMETRIC IDENTITIES AND EQUATIONS 468

CHAPTER **28** ELEMENTARY PLANE VECTORS 480

CHAPTER **29** PERIODIC FUNCTIONS 493

CHAPTER **30** ALTERNATING CURRENTS—FUNDAMENTAL IDEAS 507

CHAPTER **31** PHASOR ALGEBRA 521

CHAPTER 32 ALTERNATING CURRENTS—SERIES CIRCUITS 532

CHAPTER 33 ALTERNATING CURRENTS—PARALLEL CIRCUITS 566

CHAPTER 34 LOGARITHMS 592

CHAPTER 35 APPLICATIONS OF LOGARITHMS 617

PREFACE

This seventh edition of *Basic Mathematics for Electronics* continues a long tradition of providing mathematical concepts for students in electronics. This tradition began in 1942 with *Mathematics for Electricians and Radiomen* by Nelson M. Cooke, and has continued through many editions as *Basic Mathematics for Electronics*.

In the 1940s no one dreamed of the sophistication of electronic equipment that lay ahead. Since that first printing, *Basic Mathematics for Electronics* has gone through six editions, and an added edition with an introduction to calculus, with modifications matching the growth of electronics as a special study. This seventh edition stands as a continuing monument to Nelson Cooke. Well over half a million young men and women around the world have sharpened their mathematical skills in the book that continues to be known as "Cooke's Math." It also stands as a memorial to Peter B. Dell, who made valuable contributions to later editions.

This book can be used by students in formal classrooms of "regular" school or continuing education and those engaged in self-study. It provides a clear mathematical understanding of electricity and electronics. The material is in "block" form: algebra, trigonometry, logarithms, and computer mathematics. Thus students may concentrate on one topic to completion, or they may interleave their studies so that, after the initial chapters, different parts of the book are studied in parallel fashion. Sections on practical applications may be delayed until the theory or lab work has been covered—this is a book of practical mathematics, not a book of electronics theory.

We assume that students have a foundation of high school mathematics. Students who do not, or who have been away from it, may wish to review basic concepts.

A word about the International System of Units. Students should take advantage of the exercises in this book that require them to convert traditional units into metric units. The aim must be to become proficient in the incoming system, not in the outgoing.

As in previous metric editions, we continue to draw your attention to the different *symbols* of "pure" metric notation of measurement (*A* for ampere) and the circuit diagram and algebraic *symbols* (*I* for electric current). Students should learn to use the two sets of symbols correctly. The language of metric measurement and its symbolic usage are new to many students as well as to old practitioners. Teachers and students alike must learn to think and solve problems using the metric language.

Since computers are playing a greater role in education, computer-related information has been included in the margins, adjacent to the appropriate discussions in the text. We hope that these notes will allow the student greater versatility in using alternative methods of solving problems. However, these notes are in no way intended to replace a course in computer programming.

Many teachers and students have contributed to this revision and enlargement as we continue the effort to make this a useful text and a valuable reference book. Special thanks are due to those who have written to McGraw-Hill with comments and suggestions, including Robert A. Ciuffetti, Dennis A. Burden, Paul E. Grove, George Bulen, Royce M. Melvin, Anthony Ferrari, and Thomas M. Henderson. The authors also wish to acknowledge the helpful recommendations for the seventh edition from Professor A. E. Hall, Central Texas College and Dr. Kincheon H. Bailey, Jr., Wake Technical College. Improvements are always welcome, addressed to the publisher.

Herbert F. R. Adams

CHAPTER 1

INTRODUCTION

In the legions of textbooks on the subject of mathematics, all the basic principles contained here have been expounded in admirable fashion. However, students of electricity, radio, electronics, and computers have need for a course in mathematics that is directly concerned with application to electric and electronic circuits. This book is intended to provide those students with a sound mathematical background as well as further their understanding of basic circuitry.

1–1 MATHEMATICS—A LANGUAGE

The study of mathematics may be likened to the study of a language. In fact, mathematics *is* a language, the language of number and size. Just as the rules of grammar must be studied in order to master English, so must certain concepts, definitions, rules, terms, and words be learned in the pursuit of mathematical knowledge. These form the vocabulary or structure of the language. The more a language is studied and used, the greater becomes the vocabulary; the more mathematics is studied and applied, the greater its usefulness becomes.

There is one marked difference, however, between the study of a language and the study of mathematics. A language is based on words, phrases, expressions, and usages that have been brought together through the ages in more or less haphazard fashion according to the customs of the times. Mathematics is built upon the firm foundation of sound logic and orderly reasoning and progresses smoothly, step by step, from the simplest numerical processes to the most complicated and advanced applications, each step along the way resting squarely upon the steps which have been taken before. This makes mathematics the fascinating subject that it is.

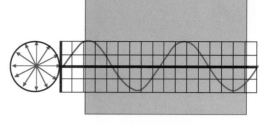

1–2 MATHEMATICS—A TOOL

As the builder works with a square and compasses, so the engineer employs mathematics. A thorough grounding in mathematics is essential to proficiency in any of the numerous branches of engineering. In no other branch is this more apparent than in the study of electrical and electronic subjects, for most of our basic ideas of electrical phenomena are based upon mathematical reasoning and stated in mathematical terms. This is a fortunate circumstance, for it enables us to build a structure of electrical knowledge with precision, assembling and expressing the components in clear and concise mathematical terms and arranging the whole in logical order. Without mathematical assistance, technicians must be content with the long and painful process of accumulating bits of information, details of experience, etc., and they may never achieve a thorough understanding of the field in which they live and work.

1–3 MATHEMATICS—A TEACHER

In addition to laying a foundation for technical knowledge and assisting in the practical application of knowledge already possessed, mathematics offers unlimited advantages in respect to mental training. The solution of a problem, no matter how simple, demands logical thinking for it to be possible to state the facts of the problem in mathematical terms and then proceed with the solution. Continued study in this orderly manner will increase your mental capacity and enable you to solve more difficult problems, understand more complicated engineering principles, and cope more successfully with the everyday problems of life.

1–4 METHODS OF STUDY

Before beginning detailed study of this text, you should carefully analyze it, in its entirety, in order to form a mental outline of its content, scope, and arrangement. You should make another preliminary survey of each individual chapter before attempting detailed study of the subject matter. After the detailed study, you should, before proceeding to new material, work problems until all principles are fixed firmly in your mind.

In working problems, the same general procedure is recommended. First, analyze a problem in order to determine the best method of solution. Then state the problem in mathematical terms by utilizing the principles that are applicable. If you make but little progress, it is probable you have not completely mastered the principles explained in the text, and a review is in order.

The authors are firm believers in the use of a workbook, preferably in the form of a loose-leaf notebook, which contains all the problems you have worked, together with the numerous notes made while studying the text. Such a book is an invaluable aid for purposes of review. The habit of jotting down notes while reading or studying should be cultivated. Such notes in your own words will provide a better understanding of a concept.

1–5 RATE OF PROGRESS

Home-study students should guard against too rapid progress. There is a tendency, especially in studying a chapter whose contents are familiar or easy to comprehend, to hurry on to the next chapter. Hasty reading may cause the loss of the meaning that a particular section or paragraph is intended to convey. Proficiency in mathematics depends upon thorough understanding of each step as it is encountered so that it can be used to master the step which follows.

1–6 IMPORTANCE OF PROBLEMS

Full advantage should be taken of the many problems distributed through the text. There is no approach to a full and complete understanding of any branch of mathematics other than the solution of numerous problems. Application of what has been learned from the text to practical problems in which you are primarily interested will not only help with the subject matter of the problem but also serve the purpose of fixing in mind the mathematical principles involved.

In general, the arrangement of problems is such that the most difficult problems appear at the end of each group. It is apparent that the working of the simpler problems first will tend to make the more difficult ones easier to solve. The home-study student is, therefore, urged to work all problems in the order given. At times, this may appear to be useless, and you may have the desire to proceed to more interesting things; but time spent in working problems will amply repay you in giving you a depth of understanding to be obtained in no other manner. This does not mean that progress should cease if a particular problem appears to be impossible to solve. Return to such a problem when your mind is fresh, or mark it for solution during a review period.

1–7 ILLUSTRATIVE EXAMPLES

Each of the illustrative examples in this book is intended to make clear some important principle or method of solution. The subject matter of the examples will be more thoroughly assimilated if, after careful analysis of the problem set forth, you make an independent solution and compare the method and results with the illustrative example.

1–8 REVIEW

Too much stress cannot be placed upon the necessity for frequent and thorough review. Points that have been missed in the original study of the text will often stand out clearly upon careful review. A review of each chapter before proceeding with the next is recommended.

1–9 SECTION REFERENCES

Throughout this book you will be referred to earlier sections for review or to bring to attention similar material pertaining to the subject under discussion. For convenience and ready reference, two sets of information are printed at the tops of the pages. On the left-hand page is the chapter number and title. On the right-hand page is the section number and heading of the last section or problem set on that page. Thus, wherever you open the book, these numbers show the section or sections covered on the pages in view. For example, Sec. 4–10 is easily found by leafing through the book while noting the inclusive numbers.

1–10 ABBREVIATIONS

Every profession and every technology has its own jargon—the particular words and phrases which describe the phenomena with which it deals. Electronics is particularly noteworthy in this respect, with inductance, capacitance, resistance, impedance, and frequency leading a host of others. Each phenomenon must be measured and described in understandable units so that other workers in the field will be able to understand exactly what is involved. After establishing such a vocabulary and list of units, the next logical development is a system of abbreviations—shorthand symbols which everyone will recognize as standing for the units and dimensions of the technology. For many years there was no single agreed-upon list of electronics abbreviations, and most of us had to be able to recognize several variations as acceptable abbreviations of the same term. For instance, A, a, amp, Amp, amps, and Amps were all used to represent *amperes,* depending upon the teacher, the author, and the publisher involved.

Even today, the exhaustive list of standard abbreviations recommended by the Institute of Electrical and Electronics Engineers is not wholly acceptable to all branches of the industry, and local variations and established forms continue to be used.

1–11 SIGNIFICANT FIGURES

The resistors, capacitors, and other devices used in electronic circuitry are often manufactured to convenient tolerances; 5%, 10%, and 20% are the most common. Accordingly, it is meaningless to calculate a resistance value to many decimal places, or to many "significant figures," when the circuit is to be constructed with a standard off-the-shelf resistor made to, say, ±10% accuracy. (Obviously a shunt to be made by hand may well be accurate to 0.5%, and then this argument would not apply.)

A calculator, whether of the hand-held variety or part of a desk-top computer, can be relied upon to give a satisfactory answer (three significant figures) to most of the problems at the level of study in this text. Answers computed by logarithms or by long multiplication or division will disagree with calculator answers and with each other—if they are taken to enough decimal places. It is safe for our purposes to assume that 1×10^{-10} or thereabouts on the calculator display is equivalent to zero. Of course, the accuracy of the answer to any mathematical problem is a function of the correctness of the computing device. Many students are tempted to believe that a nine- or ten-place answer *must* be more correct than a five- or six-place answer, without examining the manufacturer's details regarding the correctness of the final one or two places.

There are occasions, of course, when three significant figures may not be sufficient: accountants and auditors will want your financial calculations to be correct to the nearest cent, even when thousands of dollars are involved; the FCC will not be satisfied with a carrier frequency correct to only three significant figures; logarithms and trigonometric functions are given to four places, or five, or ten, and the answers achieved will reflect the accuracy of the tables used; angles greater than 90° must be converted into equivalents less than 90° for purposes of calculations, and they should not be rounded off prior to conversion. All the answers in this text reflect these notions, and you are accordingly encouraged to start using a good calculator early in your career. (See the section on calculators before purchasing one.) In this edition, we have included notes of special interest to the computer user, who must learn to use a few operational symbols and techniques that are very different from the customary ones which are used on calculators. Watch for these computer notes. They appear in boxes at appropriate locations in the text. Watch for the computer symbol.

1–12 METRICATION

Since the publication of the third edition of this book, both Canada and the United States have taken lengthy strides toward the adoption of the metric system of units. The metric system has undergone many refinements during its lifetime, from cgs through mks to mksa. Now, in the most logical refinement of all, the International System of Metric Units, known universally as SI (for Système International d'Unités), is being phased into the North American measurement scene. Since the SI metric units of electricity are those with which most of the users of this book are already familiar (amperes, volts, ohms, hertz, etc.), it only remains for us to learn to think in terms of meters for length and kilograms for mass to become proficient in the basic electronics requirements of metric units.

The SI units which you must know in order to study this textbook are introduced in Chap. 7, which was completely rewritten for the fourth edition.

1–13 USE OF CALCULATORS AND COMPUTERS

This text has been designed to develop *basic* mathematics, which involves the techniques of analysis and synthesis in the solution of problems. The book does include notes for the user of calculators and computers, and we hope that students will benefit from these notes as they use their own computing devices. However, there will be times in every technician's life when there is no calculator at hand. Accordingly, we prefer that at least the basic problems of every chapter be solved without a calculating device, which later may be used to check the longhand computation.

In the classroom, of course, the instructor will decide whether students may use electronic aids.

Before you purchase a calculator or computer, please read and consider the suggestions in the following sections. The requirements listed for calculators apply equally to computers—only the operating symbols and techniques are different.

1–14 ELECTRONIC AIDS TO MATHEMATICS FOR ELECTRONICS

Perhaps you have seen a sliderule—once the latest and best way to perform mathematical calculations quickly to an acceptable level of precision. This *analog* computer helped to eliminate reliance on extensive tables of logarithms and trigonometric functions (sometimes to 15 places of decimals or more). Just at the time that superb decitrig sliderules were having an impact on engineering calculations, new *digital* calculators were being developed. Suddenly, students who balked at paying $35.00 for a good slide rule were lining up to pay as much as $700.00 for a hand-held "pocket" calculator. Now, less than a generation later, many students own *personal computers* as well as calculators. Some calculators almost seem to be as powerful as computers.

But most electronic aids to computation are still limited to digital treatment of numerical problems. It is up to you to perform the algebra that leads to numerical solutions. Nevertheless, you should seriously consider the two major sections of this chapter *before* you invest in your personal electronic aids to mathematics for electronics.

1–15 THE CALCULATOR

Because of the rapid proliferation of hand-held electronic calculators, many students and, unfortunately, some teachers have decided that all their mathematical work is at an end—just punch a few keys, and the calculator will give them the answer. They have failed to realize that the calculator is able to perform the donkey work of calculation very quickly, but with no guarantees. All calculator work is subject to the law of GIGO: garbage in, garbage out. If you do not understand the principles involved in the mathematical processes, you will not be able to instruct your calculator correctly or interpret the results which it presents to you. It is still your mathematical skill that counts. And when—Heaven forbid!—your calculator batteries run down or your calculator itself is out of reach, all calculation depends upon your personal skill. Study these notes carefully. They will give you a solid foundation for estimating and calculating—and for instructing your calculator. They will also give you a solid foundation for choosing a calculator. The skills you develop from the use of this text will enable you to use it effectively.

1–16 POCKET CALCULATORS

Advances in integrated-circuit technology have enabled manufacturers to produce reasonably priced hand-held calculators capable of performing extremely complicated calculations very rapidly. Of course, the cheaper calculators are designed to perform only basic arithmetic, with perhaps square roots or percentages thrown in. Naturally, calculators equipped to deal with the additional complications with which people in the various fields of electronics have to deal are considerably more expensive.

It is assumed that you have or soon will have a suitable calculator and, further, that you will refer as often as necessary to the instruction manual or manuals that came with it. Users of pocket calculators should not overlook the importance of the *ideas* of powers of 10, even if they do not immediately feel the urgency to use the techniques extensively. Most pocket calculators give only six to ten places in their readouts, and that is insufficient for dealing with the wide range of units—picowatts to megohms and terahertz, at least—with which we are involved.

1–17 SELECTING YOUR CALCULATOR

Do not be in a rush to buy a low-priced calculator simply because you cannot afford a better one now. On the other hand, do not buy the most expensive model because it has ''everything.'' Take time to discuss with your teachers, and with senior students if possible, which features are essential and which others may be worth serious consideration. For your greatest satisfaction, try more than one *brand,* more than one *type,* and more than one *model* before you buy. Look over the manuals and handbooks that are provided by the manufacturers; the quality of the software may be a guide to the quality of the product. Does it work to eight or nine (or even more) significant figures internally while only displaying two or three significant figures? How accurate are the eighth or ninth significant figures? Are they worth an argument with someone who has a different brand, or even a different model of the same brand?

In the absence of other advice, and to achieve the greatest satisfaction in solving electronics circuit problems, you should choose a calculator which provides, as a minimum, the following features:

- Arithmetic. $+$, $-$, $\times$, $\div$, $\frac{1}{x}$, x^2, $\sqrt{x}$, y^x or x^y
- Logarithms. $\log x$, $\ln x$, 10^x, e^x
- Ease of entering and reading powers of 10 and selecting number of significant figures
- Decimal point. Fixed and floating, preferably selectable and easy to convert to scientific or engineering notation
- Angle selection. Degrees and radians, if not also gons (grads)
- Trigonometry. $\sin x$, $\cos x$, $\tan x$, $\arcsin x$, $\arccos x$, $\arctan x$ (or $\sin^{-1}x$, $\cos^{-1}x$, $\tan^{-1}x$)
- Memory. One or more (preferably more) memory registers, and preferably with an automatic memory stack (or "scratchpad memory")
- Special keys. π to a suitable number of places
- Special functions, →P, →R, polar and rectangular conversions
- Radix conversions to and from denary, binary, octal, and hexadecimal

In addition to those basic requirements, you should consider whether you will also find it worthwhile to pay more to obtain some useful extras:

- Metric conversions. Check these against tables; do they give Imperial gallons, U.S. dry gallons, or U.S. wet gallons?
- Statistical calculations. $\bar{x}$, s, $\Sigma +$, $\Sigma -$, %, Δ%

There are as many other extras available as manufacturers think you can be persuaded to pay for.

- Programming. When you are solving the same *form* of problem, such as evaluating determinants, only the numbers of which change, it can be very convenient to be able to enter a program to direct the calculator's mathematical processes. Should you consider a *keystroke* programming feature or a system of insertable *magnetic cards*? Or should you save money and key each step of the problem each time?
- Printout. Very often it is sufficient to report the *answer* to a problem. At other times, in order to check your work or justify an answer, it may be desirable to show *every step* of an involved calculation. Would it be worth the extra cost to you to have a calculator which can print out the entire process as well as deliver an answer? (How much is a roll of printout paper for the calculator you are considering?)

Another matter for serious consideration, *before* you buy any calculator, is the *power supply*. Does the calculator which appeals to you operate on batteries alone, or can it also operate, through an ac adapter, on 115 V? Are the batteries rechargeable? In an emergency, can you fit in ordinary cells? What is the manufacturer's estimate of the operating life of a new battery? If the batteries are rechargeable, does recharging take place if you are operating the calculator off the line? Should you consider buying a spare battery pack to be left charging while you are at school or work? Does it operate on sun or artificial light?

Ask about service, especially warranty service. Are parts and service available locally? Is there a minimum guaranteed turnaround time for service at some distant depot? A maximum time? Will the manufacturer still be in business when you require parts or service?

Check the readout. Can you read it in bright sunlight? Will you ever want to? Is an LED readout easier to see than LCD? Is blue easier to see than red? Is the angle of readout suitable in your normal working position?

Give serious thought to the method of notation which will best satisfy all your needs. Some calculators use an input system, called "direct algebraic notation," which uses this keying program:

KEY 2
KEY +
KEY 3
KEY =
Read 5

Others use a system called "reverse Polish notation (RPN)":

KEY 2
KEY ENTER
KEY 3
KEY +
Read 5

Do not commit yourself to either system until you have tried both. The authors suggest that RPN has advantages which may well be worth the extra few minutes spent in becoming familiar with it.

1–18 MATHEMATICAL TABLES

Since this edition does not contain several pages of tables of logarithms and trigonometric functions, it is essential that your calculator be able to provide at least the features listed above. In addition, it is up to you to know the condition of the power supply so your calculator will not let you down. For emergencies, you will find three-place tables inside the front covers. From time to time you should attempt solutions by using those tables so you will be able to use them when the need arises.

1–19 ACCURACY OF CALCULATORS

Different brands of calculators are programmed in different ways. Some manufacturers indicate the accuracy* of their readouts; others will make no mention of how correct their readouts are, regardless of the precision† they indicate. Consider, for example, π:

$$\pi = 3.141\ 592\ 65 \qquad \text{to } nine \text{ significant figures}$$

*Accuracy is defined as the conformity of an indicated value to an accepted standard or true value. The difference between an observed value and the true value is called the inaccuracy, or error. The smaller the error, the less the inaccuracy and the greater the accuracy.

†Precision is defined as the quality of being sharply defined or stated. A measure of the precision of a quantity is the number of distinguishable alternatives from which the quality was selected, which is sometimes indicated by the number of significant digits the quantity contains. For example, a quartz watch can indicate very *precisely* the elapsed time in seconds, but it may indicate very *inaccurately* the time of day because of a faulty setting.

Many teachers have mistakenly taught their students that $\pi = \frac{22}{7}$, although that is merely the simplest common fraction whose value is closely approximate to that of π. $\frac{22}{7} = 3.142\ 9$; that is, it is good enough to *three* significant figures. However, consider $\frac{355}{113}$ as a more accurate approximation of π. Check your own calculator. Does it have a selector to display π automatically? (You will be using π often.) If not, to what accuracy should you be working? Will you be using 20% components? 10%? 5%? 1%? In the past, electronics students learned that a 10-inch (in) slide rule could be read to 0.1%, which would guarantee three significant figures—adequate for all practical purposes. (Standard switchboard meters, for example, are seldom correct to within 3%.)

The answers to problems in this book should be taken to three significant figures (see Sec. 6–1) unless otherwise specified. That will eliminate almost all arguments between users of different brands and models of calculator. We suggest that you round off a display of five or more figures to three. That will meet almost all circuit requirements unless it is obvious from the problem that greater precision is required. There will certainly be times when three figures are inadequate; you must, for example, keep the frequency of your transmitter to a closer precision than three figures.

In your later studies in this book you will study binary fractions. 0.2 denary becomes 0.001100110 binary—accurate to about 2%. To achieve better accuracy for such number system conversions could require 12 to 15 digits!

1–20 THE COMPUTER

Soon after calculators entered the market, many opportunists got into the business. As fast as they got in—or faster—many of them went out of business. Purchasers who wanted service discovered that there was no service. It became cheaper in many cases to replace rather than repair a calculator. Will the computer market repeat this experience?

1–21 COMPUTER FACILITIES

Integrated circuits have crammed into cases the size of a typewriter, the memory capacity and calculating ability that once required a large building and thousands of electronic tubes. Even with transistors, computers were bulky and expensive. Modern designs permit a single package of *hardware* to use, and be controlled by, a great variety of *software* so that a single relatively inexpensive device will provide a variety of services to individual members of an entire family:

- Filing. Accessible memory of mailing lists, recipes, statistics
- Word processing. Typing that is easy to correct, alter, edit, and move around
- Financial service. Bookkeeping, spreadsheets, even income tax forms
- Calculation. From simple arithmetic to calculus, including all the functions necessary to basic mathematics for electronics
- Games. Not just checkers and chess, but a wide variety of skill- and reaction-testing entertainment
- Music. Such systems as MIDI that permit a keyboard instrument to control other keyboards, turn programming into sound, and sound into graphic symbols
- Graphics. In black and white or polychrome, orthogonal drafting, or freehand art

- Adaptability. If it can be done with paper and pencil, there is, or soon will be, software that will allow you to do it on a computer
- Printing. A wide variety of printer types will allow you to reproduce what you have put on the screen of your computer. (Printing nonstandard symbols may require a costly printer.)

You should make a point of studying the manuals that accompany the computers you are considering. Look for the amount of memory content required by the software you are interested in. Talk to computer users, instructors, and senior students before investing in a computer that may not be satisfactory for very long.

1–22 MEMORY

A modern computer contains within itself a limited amount of memory. Entries in a catalog of software will often indicate the number of *bits* or *bytes* of memory that are required to successfully run and operate a program. You will have to determine whether you will eventually want to be able to add additional floppy discs or that much larger capacity of a hard disc drive.

1–23 PERSONAL? MAINFRAME? COMPATIBILITY?

Before you buy a computer of your own, read the literature in order to determine what kind of computer to consider. A mainframe computer, suitable for a large business, could be far too costly for a student. But a business mainframe might accept and support a particular personal computer that can be interconnected. Will your family's business or a potential employer be able to accept your personal computer as a temporary workstation, connected as part of a local area network?

Should you consider buying a *compatible* computer that will allegedly respond to all the software developed to run, and run on, a more expensive ''brand-name'' computer? Talk to more than one dealer or user before you commit yourself.

1–24 COMPUTERS WITH THIS BOOK

Obviously, your use of a computer as a calculating device to accompany this book will primarily involve the more common routines of calculation for calculators (listed previously), with some additional application to drawing (and, perhaps, solving) linear and quadratic graphs. You must make sure that your computer, with its related software, will allow you to perform all the arithmetic, logarithms, and trigonometry that a good hand-held calculator can do. Discuss your probable needs with your instructors and senior students. Don't forget to ask what kinds of aids will be admitted to the examination room!

1–25 DIALECTS

Different computer manufacturers and software developers have different ideas about computer language. One program will require you to specify CLG whenever you want common logarithms, that is, logs to the base 10. Another computer will only perform if you demand LGT. Still another brand might respond to either command, and perhaps to one or two others as well. Before you can use your particular computer, you must consult the instruction manuals that come with it, and learn how to command it correctly. Keep those manuals close at hand!

CHAPTER 2

ALGEBRA— GENERAL NUMBERS

In general, arithmetic consists of the operations of addition, subtraction, multiplication, and division of a type of numbers represented by the digits 0, 1, 2, 3, . . . , 9. By using the above operations or combinations of them, we are able to solve many problems. However, a knowledge of mathematics limited to arithmetic is inadequate and a severe handicap to anyone interested in acquiring an understanding of electric circuits. Proficiency in performing even the most simple operations of algebra will enable you to solve problems and determine relations that would be impossible with arithmetic alone.

2–1 THE GENERAL NUMBER

Algebra may be thought of as a continuation of arithmetic in which letters and symbols are used to represent definite quantities whose actual values may or may not be known. For example, in electrical and electronics texts, it is customary to represent current by the letter I or i, voltage by V or v, resistance by R or r, etc. The base of a triangle is often represented by b, and the altitude may be specified as a. Such letters or symbols used for representing quantities in a general way are known as *general numbers* or *literal numbers*.

The importance of the general-number idea cannot be overemphasized. Although the various laws and facts concerning electricity can be expressed in English, they are more concisely and compactly expressed in mathematical form in terms of general numbers. As an example, Ohm's law states, in part, that the current in a certain part of a circuit is proportional to the potential difference (voltage) across that part of the circuit and inversely proportional to the resistance of that part. This same statement, in mathematical terms, is

$$I = \frac{V}{R}$$

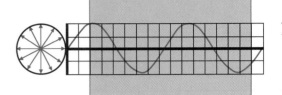

where I represents the current, V is the potential difference, and R is the resistance. Such an expression is known as a *formula*.

Although expressing various laws and relationships of science as formulas gives us a more compact form of notation, that is not the real value of formulas. As you attain proficiency in algebra, the value of general formulas will become more apparent. Our studies of algebra will consist mainly of learning how to add, subtract, multiply, divide, and solve general algebraic expressions, or formulas, in order to attain a better understanding of the fundamentals of electricity and related fields.

2–2 SIGNS OF OPERATION

In algebra the signs of operation $+$, $-$, $\times$, ( ), /, and $\div$ have the same meanings as in arithmetic. The sign $\times$ is generally omitted between literal numbers. For example, $I \times R$ is written IR and means that I is to be multiplied by R. Similarly, $2\pi fL$[†] means 2 times π times f times L. There are times when the symbol $\cdot$ is used to denote multiplication. Thus $I \times R$, $I \cdot R$, $(I)(R)$, and IR all mean I times R.

MULTIPLICATION
Most computers use * to represent multiplication. * must be used every time multiplication is required:
5*abc* will be input
as 5**a***b***c*

DIVISION
Division is indicated on most computers by a shilling bar:
$\frac{a}{c}$ is input
as *a*/*c*

2–3 THE ORDER OF SIGNS OF OPERATION

In performing a series of different operations, we will follow convention and perform first the multiplications and divisions, then the additions and subtractions. Thus,

$$16 \div 4 + 8 + 4 \times 5 - 3 = 4 + 8 + 20 - 3 = 29$$

Any parenthetical expressions are evaluated first.

The scope of operational symbols in computer usage is not automatic by definition. Computers require specific instructions, with parentheses in order to eliminate any possibility of confusion.

$\frac{a + bc}{2d + z}$
is input as
(*a* + (*b***c*))/(2**d* + *z*)

[†]See Table 4 in the Appendix for the Greek alphabet.

2–4 ALGEBRAIC EXPRESSIONS

An *algebraic expression* is one that expresses or represents a number by the signs and symbols of algebra. A *numerical algebraic expression* is one consisting entirely of signs and numerals. A *literal algebraic expression* is one containing general numbers or letters. For example, $8 - (6 + 2)$ is a numerical algebraic expression and I^2R is a literal algebraic expression.

2–5 THE PRODUCT

As in arithmetic, a *product* is the result obtained by multiplying two or more numbers. Thus, 12 is the product of 6×2.

2–6 THE FACTOR

If two or more numbers are multiplied together, each of them or the product of any combination of them is called a *factor* of the product. For example, in the product $2xy$, 2, x, y, $2x$, $2y$, and xy are all factors of $2xy$.

2–7 COEFFICIENTS

Any factor of a product is known as the *coefficient* of the product of the remaining factors. In the foregoing example, 2 is the coefficient of xy, x is the coefficient of $2y$, y is the coefficient of $2x$, etc. It is common practice to speak of the numerical part of an expression as the *coefficient* or as the *numerical coefficient*. If an expression contains no numerical coefficient, 1 is understood to be the numerical coefficient. Thus, $1abc$ is the same as abc.

2–8 PRIMES AND SUBSCRIPTS

When, for example, two resistances are being compared in a formula or it is desirable to make a distinction between them, they may be represented by R_1 and R_2 or R_a and R_b. The small numbers or letters written at the right of and below the R's are called *subscripts*. They are generally used to denote different values of the same units.

R_1 and R_2 are read "R sub one" and "R sub two" or simply "R one" and "R two."

Care must be used in distinguishing between subscripts and exponents. Thus V^2 is an indicated operation that means $V \cdot V$, whereas V_2 is used to distinguish one quantity from another of the same kind.

Primes and *seconds*, instead of subscripts, are often used to denote quantities. Thus one current might be denoted by I' and another by I''. The first is read "I prime" and the latter is read "I second" or "I double prime." I' resembles I^1 (I to the first power), but in general this causes little confusion.

Most computer readouts and printouts are not capable of delivering characters one-half line above or below the regular printed line. Accordingly, where we prefer to use superscript or subscript identifiers, your computer will print *R*1, *R*2, *R*3, etc. You must take care to read these as identifiers, and not as coefficients. *R*3 is the third resistance under discussion, *not R* multiplied by 3.

2–9 EVALUATION

To *evaluate* an algebraic expression is to find its numerical value. In Sec. 2–1, it was stated that in algebra certain signs and symbols are used to represent definite quantities. Also, in Sec. 2–4, an algebraic expression was defined as one that represents a number by the signs and symbols of algebra. We can find the numerical, or definite, value of an algebraic expression only when we know the values of the letters in the expression.

EXAMPLE 1 Find the value of $2ir$ if $i = 5$ and $r = 11$.

SOLUTION

$$2ir = 2 \times 5 \times 11 = 110$$

EXAMPLE 2 Evaluate the expression $23V - 3ir$ if $V = 10$, $i = 3$, and $r = 22$.

SOLUTION

$$\begin{aligned} 23V - 3ir &= 23 \times 10 - 3 \times 3 \times 22 \\ &= 230 - 198 = 32 \end{aligned}$$

EXAMPLE 3 Find the value of $\frac{V}{R} - 3I$ if $V = 230$, $R = 5$, and $I = 8$.

SOLUTION

$$\frac{V}{R} - 3I = \frac{230}{5} - 3 \times 8 = 46 - 24 = 22$$

When evaluating algebraic expressions on your computer, you will probably use a command such as LET: LET $x = 5$ will instruct your computer to make this substitution. (And don't forget to END your sequence of LETs)

```
LET i = 5
LET r = 11
END
RUN 2*i*r
read 110
```

PROBLEMS 2–1

Gross = 144?

NOTE *The accuracy of answers to numerical computations is, in general, that obtained with an electronic calculator rounded off to three significant figures.*

1. (*a*) What does the expression $(15)(x)$ mean?
   (*b*) What is the meaning of $12 \cdot i$?
   (*c*) What does $0.05R$ mean?
2. What is the value of:
   (*a*) $5i$ when $i = 7$ amperes (A)?
   (*b*) $4Z$ when $Z = 16$ ohms (Ω)?
   (*c*) $16V$ when $V = 110$ volts (V)?
3. One electrolytic capacitor costs \$3.45.
   (*a*) What will one gross of capacitors cost?
   (*b*) What will n capacitors cost?

4. One dozen resistors cost a total of \$2.04.
   (*a*) What is the cost of each resistor?
   (*b*) What is the cost of p resistors?
5. The current in a certain circuit is $25I$ A. What is the current if it is reduced to one-half its original value?
6. There are three resistances, of which the second is twice the first and one-sixth the third. If R represents the first resistance, what expressions describe the other two?
7. There are four capacitances, of which the second is two-thirds the first, the third is six times the second, and the fourth is twelve times the third. If C represents the first capacitance, in picofarads (pF), what expressions describe the other three?
8. If $P = 3$, $X = 5$, and $\psi = 12$, evaluate:
   (*a*) $P + \psi$
   (*b*) $\psi + X - P$
   (*c*) $\dfrac{\psi}{X}$
   (*d*) $\dfrac{X - P}{\psi}$
   (*e*) $\dfrac{P + \psi}{X}$
9. Write the expression which will represent each of the following:
   (*a*) A resistance which is R Ω greater than 16 Ω.
   (*b*) A voltage which is 220 V more than v V.
   (*c*) A current which is I A less than i A.
10. A circuit has a resistance of 125 Ω. Express a resistance which is R Ω less than six times this resistance.
11. An inductance L_1 exceeds another inductance L_2 by 115 millihenrys (mH). Express the inductance L_2 in terms of L_1.
12. When two capacitors C_1 and C_2 are connected in series, the resultant capacitance C_s of the combination is expressed by the formula

$$C_s = \frac{C_1C_2}{C_1 + C_2}$$

   What is the resultant capacitance if:
   (*a*) 5 pF is connected in series with 15 pF?
   (*b*) 150 pF is connected in series with 475 pF?
13. The current in any part of a circuit is given by $I = \dfrac{V}{R}$, in which I is the current in amperes through that part, V is the electromotive force (emf) in volts across that part, and R is the resistance in ohms of that part. What will be the current through a circuit with:
   (*a*) An emf of 220 V and a resistance of 5 Ω?
   (*b*) An emf of 50 V and a resistance of 200 Ω?
14. The time interval between the transmission of a radar pulse and the reception of the pulse's echo off a target is $t = \dfrac{2R}{c}$ seconds (s), where t is the time interval in seconds, R is the range in kilometers (km), and c is the speed of light, at which radio waves travel. [c = 300 000 kilometers per

second (km/s).] What is the time between the transmission of a pulse and the reception of its echo from a target at a distance of 124 km?

15. The relation of $t = \frac{2R}{c}$ in Prob. 14 is applicable to the transmission of sound in air and water. Owing to lower speeds of transmission, R is usually expressed in meters (m), and c is usually expressed in meters per second (m/s). (In air, $c \simeq 335$ m/s, and in salt water, $c \simeq 1460$ m/s. The sign $\simeq$ means "is approximately equal to.")
   (*a*) What is the time between the transmission of a short pulse of sound through air and the reception of its echo at a distance of 500 m?
   (*b*) What time will elapse if the sound pulse is transmitted through seawater at the same distance?
16. The relationship between the wavelength λ of a wave, the frequency f in hertz (Hz, or cycles per second), and the speed c at which the wave is propagated is $\lambda = \frac{c}{f}$. If λ is expressed in meters, then c must be expressed in meters per second; that is, λ and c must be expressed in the same units of length.
   (*a*) What is the wavelength in kilometers of a radio wave having a frequency of 980 kilohertz (kHz)? (980 kHz = 980 000 Hz; c = 300 000 km/s = 300 000 000 m/s.)
   (*b*) What is the wavelength in meters of a radio wave having a frequency of 121.5 megahertz (MHz)? (121.5 MHz = 121 500 000 Hz; c = 300 000 km/s = 300 000 000 m/s.)
17. The distance between a dipole antenna and its reflector is usually one-fifth of a wavelength. What will be this spacing for a signal at 480 MHz in meters?

2–10 EXPONENTS

To express "x is to be taken as a factor four times," we could write $xxxx$, but the general agreement is to write x^4 instead.

An *exponent,* or *power,* is a number written at the right of and above a second number to indicate how many times the second number is to be taken as a factor. The number to be multiplied by itself is called the *base*.

Thus, I^2 is read "I square" or "I second power" and means that I is to be taken twice as a factor; e^3 is read "e cube" or "e third power" and means that e is to be taken as a factor three times. Likewise, 5^4 is read "5 fourth power" and means that 5 is to be taken as a factor four times; thus,

$$5^4 = 5 \times 5 \times 5 \times 5 = 625$$

When no exponent, or power, is indicated, the exponent is understood to be 1. Thus, x is the same as x^1.

Since your computer probably reads and prints on only one line, it is necessary to use special symbols to denote exponentiation. Various computer dialects use ↑, **, ∧, or ∩ to indicate raising to a power. For example, x^y might be entered in a computer as

x ↑ y
or x ** y
or x ∧ y
or x ∩ y

2–11 THE RADICAL SIGN

The radical sign $\sqrt{\ }$ has the same meaning in algebra as in arithmetic; $\sqrt{e}$ means the square root of e, $\sqrt[3]{x}$ means the cube root of x, $\sqrt[4]{i}$ means the fourth root of i, etc. The small number in the angle of a radical sign, like the 4 in $\sqrt[4]{i}$, is known as the *index* of the root.

> Your computer keyboard probably does not have a radical sign. Some computer dialects will respond to SQR or SQRT. To indicate square roots:
> $\sqrt{64}$ will be set SQR64.
>
> Alternatively, to find $\sqrt[x]{a}$, use $a^{\frac{1}{x}}$. In Section 20–4 we will justify:
> $\sqrt{a} = a^{1/2} = a^{0.5}$
> $= a{*}{*}0.5$, etc.

2–12 TERMS

A *term* is an expression containing literal and/or numerical parts which are not separated by plus or minus signs. Terms may be parts of larger expressions in which the terms are separated by plus or minus signs. $3V^2$, IR, and $-2v$ are all terms of the expression $3V^2 + IR - 2v$.

Although the value of a term depends upon the values of the literal factors of the term, it is customary to refer to a term whose sign is plus as a *positive term*. Likewise, we refer to a term whose sign is minus as a *negative term*.

Terms having the same literal parts are called *like terms* or *similar terms*. $2a^2bx$, $-a^2bx$, $18a^2bx$, and $-4a^2bx$ are like terms in a^2bx.

Terms that are not alike in their literal parts are called *unlike terms* or *dissimilar terms*. $5xy$, $6ac$, $9I^2R$, and VI are *unlike terms*.

An algebraic expression consisting of but one term is known as a *monomial*.

A *polynomial*, or *multinomial*, is an algebraic expression consisting of two or more terms.

A *binomial* is a polynomial of two terms. Some examples: $v + ir$, $a - 2b$, and $2x^2y + xyz^2$ are binomials.

A *trinomial* is a polynomial consisting of three terms. For example, $2a + 3b - c$, $IR + 3v - V^2$, and $8ab^3c + 3d + 2xy$ are trinomials.

PROBLEMS 2–2

1. If $a = 4$, $b = 2$, and $c = 5$, evaluate the following:
   (a) $2abc$
   (b) $5a^2b + 3c$
   (c) $a^2b^2c^2$
   (d) $12ac^2 - 2b^2$
   (e) $\sqrt{4a^2b^2}$
   (f) $5\sqrt{9b^2c^2} + 3a^2$
2. If $V = 110$, $I = 6$, and $R = 25$, evaluate the following:
   (a) $5VI$
   (b) VI^2R
   (c) $I^2R + \dfrac{12V^2}{R}$
   (d) $\dfrac{25I^3R^2}{6IR} - \sqrt{\dfrac{100V^2}{R}}$
   (e) $\dfrac{36V^2IR}{I^3R} - 3R^2$
3. State which of the following are monomials, binomials, and trinomials:
   (a) $\dfrac{Q}{L}$
   (b) I^2R
   (c) $2\pi fL$
   (d) $a + jb$
   (e) $\theta + \phi + 90°$
   (f) $V_o - V_i$
   (g) $R + \dfrac{V}{X} + \dfrac{M}{\mu}$
   (h) $m^2 + 2mx + x^2$
   (i) $\dfrac{8(\pi L^2)}{20r_e'}$
   (j) $R_1 + R_2 + R_3$

4. In Probs. 1, 2, and 3, state which expressions are polynomials.
5. Write the following statements in algebraic symbols:
   (*a*) I is equal to V divided by R.
   (*b*) V is equal to I times R.
   (*c*) P is equal to R times the square of I.
   (*d*) R_1 is equal to the sum of R_2 and R_3.
   (*e*) K is equal to M divided by the square root of the product of L_1 and L_2.
   (*f*) R_p is equal to the product of R_1 and R_2 divided by their sum.
   (*g*) The meter multiplier N is equal to the meter resistance R_m divided by the shunt resistance R_s, all plus 1.
6. The approximate inductance of a single-layer air-core coil, such as used in the tuning circuits of radio receivers, can be calculated by the formula

$$L = \frac{2.54r^2n^2}{9r + 10l} \qquad \text{microhenrys } (\mu\text{H})$$

   where L = inductance, μH
   r = radius of winding, centimeters (cm)
   n = number of turns of wire in winding
   l = length of coil, cm

   What is the inductance of a coil that is 5 cm in diameter and 12 cm long and has 280 turns of wire?
7. The winding in Prob. 6 is removed from the coil form and smaller wire is substituted, so that, in the same length of coil, the number of turns is tripled. What is the inductance?
8. The power in any part of an electric circuit is given by the formula

$$P = I^2R \qquad \text{watts (W)}$$

   where P = power, W
   I = current, A
   R = resistance, Ω

   Find the power expended when:
   (*a*) The current is 0.25 A and the resistance is 10 000 Ω.
   (*b*) The current is 30 A and the resistance is 0.5 Ω.
9. In Prob. 8, if the resistance is kept constant, what happens to the power if the current is (*a*) doubled, (*b*) increased by a factor of 4, (*c*) reduced by a factor of 3?
10. The power in any part of an electric circuit is also given by the formula

$$P = \frac{V^2}{R} \qquad \text{W}$$

   where P = power expended, W
   V = electromotive force, V
   R = resistance, Ω

   What happens to the power if:
   (*a*) The voltage is doubled and the resistance is unchanged?
   (*b*) The voltage is halved and the resistance is unchanged?

(*c*) The resistance is doubled and the voltage is unchanged?
(*d*) The resistance is halved and the voltage is unchanged?

2–13 ORDER OF OPERATIONS

In Sec. 2–3 we indicated the standard order of *signs* of operation. To govern your work for the rest of your studies in mathematics, here is the sequence to follow for all mathematical operations:

1. Raise all numbers to indicated powers, including fractional powers (roots).
2. Evaluate any logarithms or trigonometric functions.
3. Evaluate within signs of grouping (including fraction bars), following the order of signs of operation.
4. Adjust for negation when removing signs of grouping.
5. Follow the order of signs of operation as you evaluate.
6. Perform any NOT operations.
   Perform any AND operations.
   Perform any OR operations.

Refer to this sequence periodically, and when you are faced with new topics in your studies.

2–14 SELF TESTS

All chapters (except the first) end with a self test that is very similar to the chapter test which your instructor will provide.

Your instructor will advise you if calculators may be used for the self tests and the subsequent chapter tests, and may require you to show all your work towards solutions.

You should strive for 100% mastery of these tests. If your algebraic answers are not identical to the answers in the back of the book, they *may* still be correct. Different is not necessarily wrong. Check your work carefully. If you do not detect any error, check for variations in sequence and location of plus and minus signs. Discuss your differences with your instructor. (Numerical answers may have been rounded off to three significant figures.)

Students who believe they know all the work in any future chapter may wish to use the self test as a form of *gating examination*. A result of 100% could indicate that you probably have 100% mastery of the topics tested, and you might be able to progress to the work of the next chapter with confidence.

Answers to the self tests are found in the Appendix.

SELF TEST

1. Write the expression that represents a potential difference that is 12 V greater than x V.
2. The wavelength λ of a signal is found from $\lambda = \frac{c}{f}$. If the speed of light c is 300 000 000 m/s, what is the wavelength of an electronic signal whose frequency is 500 000 000 Hz?
3. A capacitance C_1 is 50 microfarads (μF) greater than another capacitance C_2 and 20 μF less than a third capacitance C_3. Write the expression that describes C_3 in terms of C_2.

4. What is the value of $\frac{1}{L}$ when $L = 0.100$ henrys (H)?
5. What is the value of $\frac{V^2}{R}$ when $V = 6$ V and $R = 1000\ \Omega$?
6. Write the expression that describes the number of twenty-five cent coins that can be exchanged for D one-dollar bills.
7. The power absorbed by a circuit whose resistance is R ohms (Ω) when a current I amperes (A) flows through it is found from $P = I^2R$ watts (W). How much power is absorbed by a 1200-Ω circuit when a current of 0.000 025 A is flowing?
8. Which of the following expressions is a trinomial?
   (*a*) $5\ abc$
   (*b*) $a + bc$
   (*c*) $3a + 2b + 4c$
9. The inductance L henrys of a coil when subjected to a signal frequency is found from $L = \frac{X_L}{2\pi f}$, where X_L is the inductive reactance in ohms. What is the inductance of a coil that has a reactance of 2513 Ω when it is operating at 400 Hz?
10. Write the following statement in algebraic symbols: The average value V_{av} of a maximum applied voltage V_{max} is double the maximum value multiplied by a fraction whose value is unity divided by pi (π).

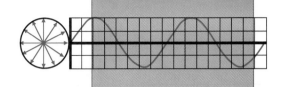

CHAPTER 3

ALGEBRA—ADDITION AND SUBTRACTION

The problems of arithmetic deal with positive numbers only. A *positive number* may be defined as any number greater than zero. Accepting this definition, we know that when such numbers are added, multiplied, and divided, the results are always positive. Such is the case in subtraction if a number is subtracted from a larger one. However, if we attempt to subtract a number from a smaller one, arithmetic furnishes us with neither a rule for carrying out this operation nor a meaning for the result.

3–1 NEGATIVE NUMBERS

Limiting our knowledge of mathematics to positive numbers would place us under a severe handicap, for there are many instances when it becomes necessary to deal with numbers that are called negative. Often, a negative number is defined as a number less than zero. Numerous examples of the uses of negative numbers could be cited. For example, zero degrees on the Celsius thermometer has been chosen as the temperature of melting ice—commonly referred to as freezing temperature. Now, everyone knows that in some climates it gets much colder than "freezing." Such temperatures are referred to as so many "degrees below zero." How shall we state, in the language of mathematics, a temperature of "10 degrees below zero"? Ten degrees above zero would be written 10°. Because 0° is the reference point, it is logical to assume that 10° below zero would be written as $-10°$, which, for our purposes, makes it a negative number.

Therefore, we see that a definition making a negative number less than zero is not completely correct. A negative number is some quantity away from a reference point in one direction (the defined negative direction), whereas the same positive quantity is simply the same quantity in the opposite direction (the defined positive direction).

Negative numbers are prefixed with the minus sign. Thus, negative 2 is written -2, negative $3ac$ is written $-3ac$ etc. If no sign precedes it, a number is assumed to be positive.

3–2 PRACTICAL NEED FOR NEGATIVE NUMBERS

The need for negative numbers often arises in the consideration of voltages or currents in electric and electronic circuits. It is common practice to select the ground, or earth, as a point of zero potential. This does not mean, however, that there can be no potentials below ground, or zero, potential. Consider the case of the three wire feeders connected as shown in Fig. 3–1.

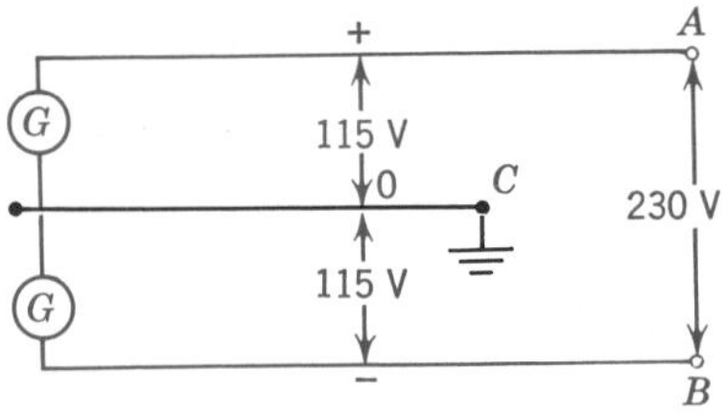

FIG. 3–1 Two 115-V generators connected in series with neutral wire grounded.

The generators *G*, which maintain a voltage of 115 V each, are connected in series so that their voltages add to give a voltage of 230 V across points *A* and *B*, and the neutral wire is grounded at *C*. Since *C* is at ground, or zero, potential, point *A* is 115 V positive with respect to *C* and point *B* is 115 V negative with respect to *C*. Therefore, the voltage at *A* with respect to ground, or zero, potential could be denoted as 115 V and the voltage at *B* with respect to ground could be denoted as -115 V.

Similar conditions exist in semiconductor circuits, as illustrated by the schematic diagram symbol of a 2N5457 N-channel junction field-effect transistor (NJFET) in Fig. 3–2. The drain current indicated by the arrow flows through the source resistor R_S and creates a potential of 4 V across it. Since the source is N-type material and the gate is P-type material, a *back-biased* PN junction is produced. The result is that the gate voltage, with respect to the source terminal, is negative. When measured with respect to ground or common, the source voltage is $+4$ V and the gate is at 0 V. The gate voltage, when measured with respect to the source terminal, is -4 V. This is usually written as $V_{GS} = -4$ V.

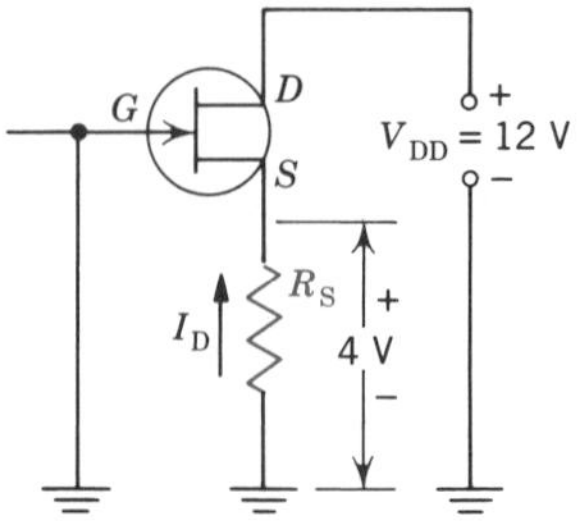

FIG. 3–2 The gate G is negative with respect to source S.

Your computer probably does not recognize our argument that -10 is greater than $+9$, but in the opposite direction.

$-x$ is treated as though x is smaller than zero: $-x < 0$

$+x$ is treated as though x is larger than zero: $+x > 0$

3–3 THE MATHEMATICAL NEED FOR NEGATIVE NUMBERS

From a purely mathematical viewpoint the need for negative numbers can be seen from the following series of operations in which we subtract successively larger numbers from 5:

$$\begin{array}{ccccc} 5 & 5 & 5 & 5 & 5 \\ \underline{0} & \underline{1} & \underline{2} & \underline{3} & \underline{4} \\ 5 & 4 & 3 & 2 & 1 \end{array}$$

$$\begin{array}{cccc} 5 & 5 & 5 & 5 \\ \underline{5} & \underline{6} & \underline{7} & \underline{8} \\ 0 & -1 & -2 & -3 \end{array}$$

The above subtractions result in the remainders becoming less until zero is reached. When the remainder becomes less than zero, the fact is indicated by placing the negative sign before the remainder. This is one reason for defining a negative number as a number less than zero. Mathematically, the definition is correct if we consider only the signs that precede the numbers.

You must not lose sight of the fact, however, that as far as magnitude, or size, is concerned, a negative number may represent a larger absolute value than some positive number. *The positive and negative signs simply denote reference from zero*. For example, if some point in an electric circuit is 1000 V negative with respect to ground, you can say so by writing −1000 V. But if you make good contact with your body between that point and ground, your chances of being electrocuted are just as good as if that point were positive 1000 V with respect to ground—and you wrote it +1000 V! In this case, *how much* is far more important than a matter of sign preceding the number. Similarly, −1000 V is greater than +500 V, but of different polarity. −$10 000 is greater than +$6000, except that it is owed, rather than owned.

3–4 THE ABSOLUTE VALUE OF A NUMBER

The numerical, or absolute, value of a number is the value of the number without regard to sign. Thus, the absolute values of numbers such as −1, +4, −6, and +3 are 1, 4, 6, and 3, respectively. Note that different numbers, such as −9 and +9, may have the same absolute value. To specify the absolute value of a number, such as Z, we write $|Z|$. This is often referred to as "the modulus of Z," or simply, "mod Z."

3–5 ADDITION OF POSITIVE AND NEGATIVE NUMBERS

Positive and negative numbers can be represented graphically as in Fig. 3–3. Positive numbers are shown as being directed toward the right of zero, which is the reference point, whereas negative numbers are directed toward the left.

Such a scale of numbers can be used to illustrate both addition and subtraction as performed in arithmetic. Thus, in adding 3 to 4, we can begin at 3 and count 4 units to the right to obtain the sum 7. Or, because these are positive numbers directed toward the right, we could draw them to scale, place them end

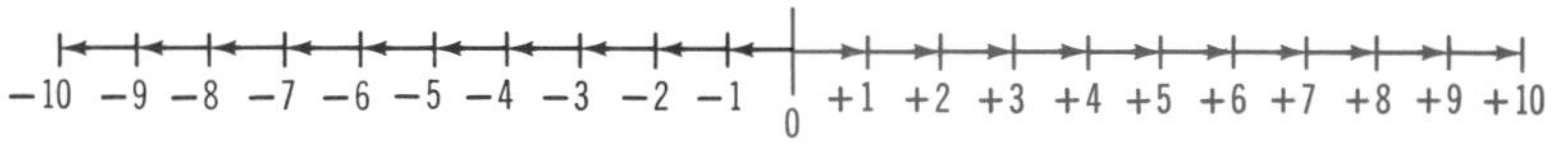

FIG. 3–3 Graphical representation of numbers from −10 to +10.

to end, and measure their total length to obtain a length of 7 units in the positive direction. This is illustrated in Fig. 3–4.

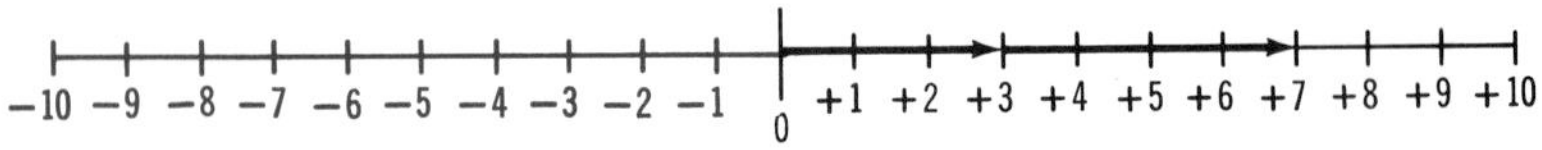

FIG. 3–4 Graphical addition of 3 and 4 to obtain 7.

In like manner, −2 and −3 can be added to obtain −5 as shown in Fig. 3–5.

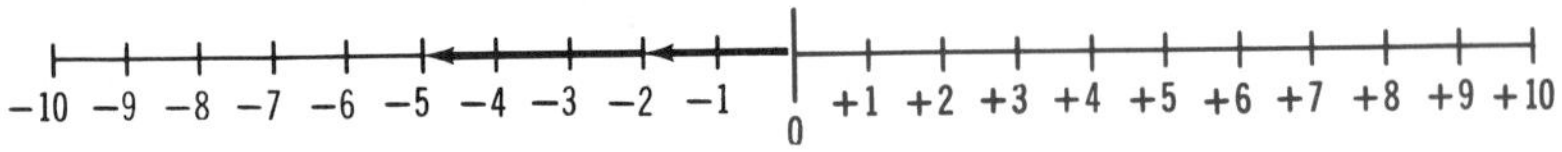

FIG. 3–5 Illustrating the addition of −2 and −3. The result is −5.

Note that adding −3 and −2 is the same as adding −2 and −3 as in the foregoing example. The sum −5 is obtained, as shown in Fig. 3–6.

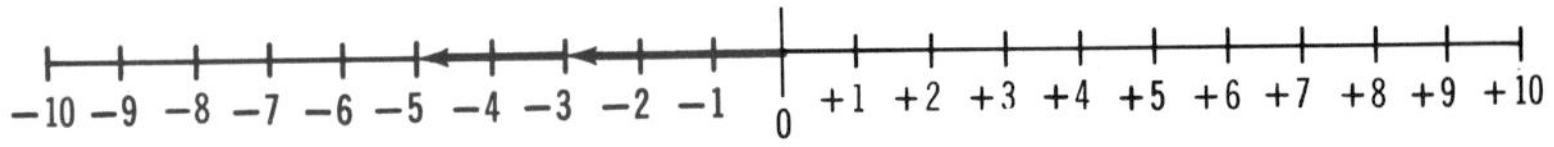

FIG. 3–6 Adding −3 and −2 is the same as adding −2 and −3. Each result is −5.

Suppose we want to add +6 and −10. We could accomplish this on the scale by first counting 6 units to the right and from *that* point counting 10 units to the left. In so doing, we would end up at −4, which is the sum of +6 and −10. Similarly, we could have started by first counting 10 units to the left, from zero, and from that point counting 6 units to the right for the +6. Again we would have arrived at −4.

Adding +6 and −10 can be accomplished graphically as in Fig. 3–7. The +6 is drawn to scale, and then the tail of the −10 is joined with the head of +6. The head of the −10 is then on −4. As would be expected, the same result is obtained by first drawing in the −10 and then the +6.

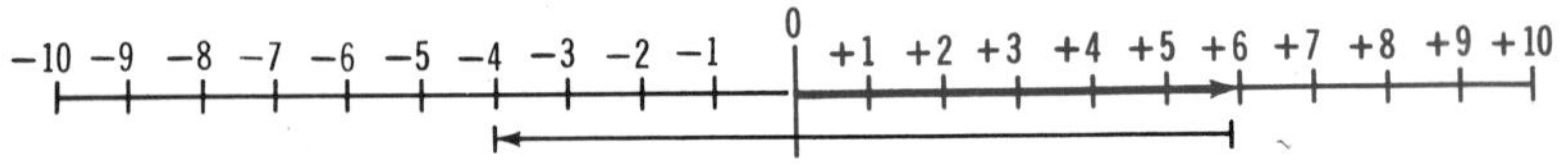

FIG. 3–7 Graphical addition of +6 and −10.

The following examples can be checked graphically in order to verify their correctness:

$$\begin{array}{r} +8 \\ +4 \\ \hline +12 \end{array} \qquad \begin{array}{r} +9 \\ -3 \\ \hline +6 \end{array} \qquad \begin{array}{r} +6 \\ -9 \\ \hline -3 \end{array} \qquad \begin{array}{r} -5 \\ +2 \\ \hline -3 \end{array} \qquad \begin{array}{r} -7 \\ +9 \\ \hline +2 \end{array} \qquad \begin{array}{r} -17 \\ -14 \\ \hline -31 \end{array}$$

Consideration of the above examples enables us to establish the following rule:

RULES

1. To add two or more numbers with like signs, find the sum of their absolute values and prefix this sum with the common sign.
2. To add a positive number to a negative number, find the difference of their absolute values and prefix to the result the sign of the number that has the greater absolute value.

When three or more algebraic numbers that differ in signs are to be added, find the sum of the positive numbers and then the sum of the negative numbers. Add these sums algebraically, and use Rule 2 to obtain the total algebraic sum.

The *algebraic sum* of two or more numbers is the result obtained by adding the numbers according to the preceding rules. Hereafter, the word "add" will mean "find the algebraic sum."

PROBLEMS 3–1

Add:

1. $\begin{array}{r} 22 \\ 46 \\ \hline \end{array}$
2. $\begin{array}{r} 36 \\ -18 \\ \hline \end{array}$
3. $\begin{array}{r} -95 \\ 61 \\ \hline \end{array}$
4. $\begin{array}{r} -18 \\ -47 \\ \hline \end{array}$
5. $\begin{array}{r} 124 \\ -96 \\ \hline \end{array}$
6. $\begin{array}{r} 165 \\ -572 \\ \hline \end{array}$
7. $\begin{array}{r} -121 \\ -615 \\ \hline \end{array}$
8. $\begin{array}{r} 0.0007 \\ -0.0052 \\ \hline \end{array}$
9. $\begin{array}{r} 416.82 \\ -2.91 \\ 15.27 \\ \hline \end{array}$
10. $\begin{array}{r} -97.63 \\ 5.74 \\ -26.32 \\ \hline \end{array}$
11. $\begin{array}{r} 6\frac{3}{4} \\ -2\frac{1}{4} \\ \hline \end{array}$
12. $\begin{array}{r} -6\frac{1}{4} \\ 2\frac{1}{8} \\ \hline \end{array}$
13. $\begin{array}{r} -8\frac{1}{32} \\ -3\frac{5}{16} \\ \hline \end{array}$
14. $\begin{array}{r} -\frac{5}{8} \\ 3\frac{1}{4} \\ \hline \end{array}$
15. $\begin{array}{r} \frac{1}{3} \\ -\frac{1}{5} \\ \hline \end{array}$

3–6 THE SUBTRACTION OF POSITIVE AND NEGATIVE NUMBERS

We may think of subtraction as the process of determining what number must be added to a given number in order to produce another given number. Thus, when we subtract 5 from 9 and get 4, we have found that 4 must be added to 5 in order to obtain 9. From this it is seen that subtraction is the inverse of addition.

EXAMPLE 1 $(+5) - (+2) = ?$

SOLUTION

In this example the question is asked, "What number added to $+2$ will give $+5$?" Using the scale of Fig. 3–8, start at $+2$ and count to the right (positive direction), until you reach $+5$. This requires three units. Then, the difference is $+3$, or $(+5) - (+2) = +3$.

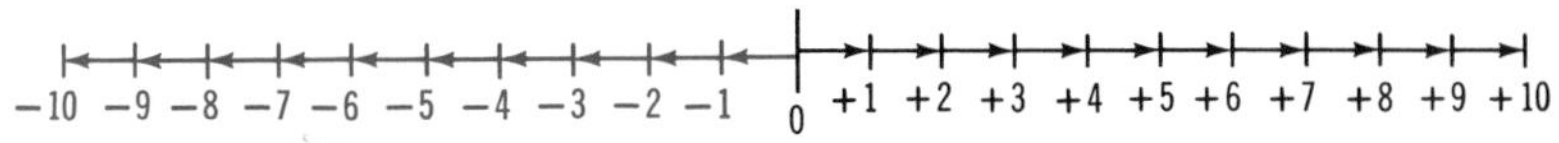

FIG. 3–8 Scale for graphical subtraction of positive and negative numbers.

EXAMPLE 2 $(+5) - (-2) = ?$

SOLUTION

In this example the question is asked, "What number added to -2 will give $+5$?" Using the scale, start at -2 and count the number of units to $+5$. This requires seven units, and because it was necessary to count in the positive direction, the difference is now shown as $+7$, or $(+5) - (-2) = +7$.

EXAMPLE 3 $(-5) - (+2) = ?$

SOLUTION

In this example the question is, "What number added to $+2$ will give -5?" Again using the scale, we start at $+2$ and count the number of units to -5. This requires seven units, but because it was necessary to count in the negative direction, the difference is -7, or now shown as $(-5) - (+2) = -7$.

EXAMPLE 4 $(-5) - (-2) = ?$

SOLUTION

Here the question is, "What number added to -2 will give -5?" Using the scale, we start at -2 and count the number of units to -5. This requires three units in the negative direction. This yields $(-5) - (-2) = -3$.

Summing up Examples 1 to 4, we have the following subtractions:

$$\begin{array}{r} +5 \\ \underline{+2} \\ +3 \end{array} \qquad \begin{array}{r} +5 \\ \underline{-2} \\ +7 \end{array} \qquad \begin{array}{r} -5 \\ \underline{+2} \\ -7 \end{array} \qquad \begin{array}{r} -5 \\ \underline{-2} \\ -3 \end{array}$$

A study of the foregoing subtractions illustrates the following principles:

1. Subtracting a positive number is equivalent to adding a negative number of the same absolute value.
2. Subtracting a negative number is equivalent to adding a positive number of the same absolute value.

These principles can be used for the purpose of establishing the following rule:

RULE

To subtract one number from another, change the sign of the subtrahend and add algebraically.

As in arithmetic, the number to be subtracted is called the *subtrahend*. The number from which the subtrahend is subtracted is called the *minuend*. The result is called the *remainder* or *difference*.

$$\begin{array}{lcr} \text{Minuend} & = & -642 \\ \text{Subtrahend} & = & 403 \\ \hline \text{Remainder} & = & -1045 \end{array}$$

PROBLEMS 3–2

Subtract the second line from the first:

1. $\begin{array}{r} 96 \\ \underline{33} \end{array}$

2. $\begin{array}{r} 25 \\ \underline{-96} \end{array}$

3. $\begin{array}{r} -237 \\ \underline{-416} \end{array}$

4. $\begin{array}{r} -125 \\ \underline{252} \end{array}$

5. $\begin{array}{r} 827 \\ \underline{-418} \end{array}$

6. $\begin{array}{r} 0.005\ 17 \\ \underline{0.083\ 29} \end{array}$

7. $\begin{array}{r} -6.13 \\ \underline{-8.29} \end{array}$

8. $\begin{array}{r} 2\frac{7}{16} \\ \underline{-1\frac{5}{8}} \end{array}$

9. $\begin{array}{r} -6\frac{5}{16} \\ \underline{-1\frac{3}{8}} \end{array}$

10. $\begin{array}{r} -\frac{1}{2} \\ \underline{\frac{5}{8}} \end{array}$

11. How many degrees must the temperature rise to change from (*a*) +6° to +73°, (*b*) −12° to +14°, and (*c*) −273° to −114°?
12. How many degrees must the temperature fall to change from (*a*) +212° to +32°, (*b*) +55° to −16°, and (*c*) −6° to −42°?
13. What amount of money is required to change an account from a debit of \$347.50 to a credit of \$225.95?

14. A certain point in a circuit is 570 V negative with respect to ground. Another point in the same circuit is 115 V positive with respect to ground. What is the potential difference between the two points?

15. In Fig. 3–2 what is the potential difference between drain D and source S?

3–7 ADDITION AND SUBTRACTION OF LIKE TERMS

In arithmetic, it is never possible to add unlike quantities. For example, we should not add inches and gallons and expect to obtain a sensible answer. Neither should we attempt to add volts and amperes, kilohertz and microfarads, ohms and watts, etc. So it goes on through algebra—we can never add quantities unless they are expressed in the same units.

The addition of two like terms such as $6VI + 12VI = 18VI$ can be checked by substituting numbers for the literal factors. Thus, if $V = 1$ and $I = 2$,

$$\begin{array}{rcrcrcr} 6VI & = & 6 \times 1 \times 2 & = & 6 \times 2 & = & 12 \\ 12VI & = & 12 \times 1 \times 2 & = & 12 \times 2 & = & 24 \\ \hline 18VI & = & 18 \times 1 \times 2 & = & 18 \times 2 & = & 36 \end{array}$$

From the foregoing, it is apparent that like terms may be added or subtracted by adding or subtracting their coefficients.

The addition or subtraction of unlike terms cannot be carried out but can only be indicated, because the unlike literal factors may stand for entirely different quantities.

EXAMPLE 5 Addition of like terms:

$$\begin{array}{r} -3i^2r \\ 8i^2r \\ \hline 5i^2r \end{array} \qquad \begin{array}{r} -16IR \\ 14IR \\ -3IR \\ \hline -5IR \end{array} \qquad \begin{array}{r} 13jIX \\ -20jIX \\ -32jIX \\ \hline -39jIX \end{array}$$

EXAMPLE 6 Subtraction of like terms:

$$\begin{array}{r} -8e_1 \\ 3e_1 \\ \hline -11e_1 \end{array} \qquad \begin{array}{r} 6iZ \\ -13iZ \\ \hline 19iZ \end{array} \qquad \begin{array}{r} -28L^2R \\ -29L^2R \\ \hline L^2R \end{array}$$

EXAMPLE 7 Addition of unlike terms:

$$\begin{array}{l} 3v \\ -3IX \\ 4V \\ \hline 3v - 3IX + 4V \end{array} \qquad \begin{array}{l} -3r \\ 4R \\ -16R_t \\ \hline 4R - 3r - 16R_t \end{array}$$

$$\begin{array}{l} 3VI \\ 10I^2R \\ -46W \\ \hline 3VI + 10I^2R - 46W \end{array}$$

3–8 ADDITION AND SUBTRACTION OF POLYNOMIALS

Polynomials are added or subtracted by arranging like terms in the same column and then combining terms in each column, as with monomials.

EXAMPLE 8 Addition of polynomials:

$$\begin{array}{rrr} -3ab & +\ 6cd & +\ x^2y \\ 14ab & & -\ 5x^2y \\ ab & -\ 3cd & \\ \hline 12ab & +\ 3cd & -\ 4x^2y \end{array} \qquad \begin{array}{rrr} 6V & +\ 3RI & -\ 8IZ \\ & RI & -\ 2IZ \\ -7V & & +\ 3IZ \\ \hline -V & +\ 4RI & -\ 7IZ \end{array}$$

EXAMPLE 9 Subtraction of polynomials:

$$\begin{array}{rrr} 3mn & +\ 16pq & -\ xy^2 \\ -9mn & & +\ 7xy^2 \\ \hline 12mn & +\ 16pq & -\ 8xy^2 \end{array} \qquad \begin{array}{rrr} 11R & +\ 4x & \\ 15R & & -\ 18Z \\ \hline -4R & +\ 4x & +\ 18Z \end{array}$$

PROBLEMS 3–3

Add (Problems 1 through 16):

1. $5i,\ 2i,\ -11i,\ 7i$
2. $3i^2r,\ 12i^2r,\ -32i^2r,\ 5i^2r$
3. $27IZ,\ 165IZ,\ -64IZ,\ -32IZ,\ 16IZ$
4. $65IR,\ -8.7IR,\ IR,\ -16.6IR,\ 15.2IR$
5. $2p + 12P,\ -3p - 18P$
6. $8jX,\ 26jX,\ -30jX,\ 18R,\ -5jX,\ 12R$
7. $18I,\ 5\frac{V}{R},\ -6I,\ -3\frac{V}{R},\ -24I,\ 16\frac{V}{R}$
8. $12\Omega,\ 2\omega,\ -16\omega,\ 4\Omega$
9. $30\alpha + 25\beta,\ 15\beta - 46\alpha,\ -27\alpha + 18\beta$
10. $2X,\ 12R,\ -15L,\ -6Q,\ 16X,\ 3R,\ 6L,\ 7Q$
11. $$\begin{array}{r} 6i^2r + 8W - 6vi + 32w \\ -3i^2r + 3W + 8vi + 18w \\ \underline{24i^2r - \ \ W - 5vi - \ \ \ w} \end{array}$$
12. $$\begin{array}{r} 25IX - 16IZ + 3IR \\ 14IZ + \ 2IX - \ \ IR \\ \underline{8IR + \ 4IX - 3IZ} \end{array}$$
13. $$\begin{array}{r} 16V_1 - 8V_2 + \ 4V_3 \\ 9V_1 + 2V_2 - 12V_3 \\ \underline{-14V_1 + 3V_2 - \ 6V_3} \end{array}$$

14. $2.15vi + 1.64\frac{v^2}{r} - 3.82i^2r, 0.57\frac{v^2}{r} + 1.94i^2r$

15. $\frac{1}{4}\pi ft, \ -3\pi Z, \ -\frac{2}{3}\pi ft, \frac{3}{16}\pi ft, \frac{7}{8}\pi Z$

16. $47IR + 3IZ$ to $-15IR - 4IZ$

17. From $35\phi + 7\theta$ subtract $22\phi - 4\theta$.

18. From $17.2\omega L + 5X_C - 13.2Z$ subtract $4.5\omega L - 3.2X_C + 5.6Z$.

19. From $\left(14.6I^2R + 3.7VI - 27\frac{V^2}{R}\right) + \left(3.1\ I^2R - 2.1VI + 18\frac{V^2}{R}\right)$ subtract $\left(-4.2VI + 22.4\ I^2R - 12\frac{V^2}{R}\right)$

20. Subtract $9.5X_C + \frac{3.26}{\omega C}$ from the sum of $-8.7X_C + \frac{2.46}{\omega C}$ and $-4.6X_C - \frac{1.98}{\omega C}$.

21. Take $1.25IR + 0.64IX - 2.81IZ$ from $-0.06IR + 0.23IX + 1.09IZ$.

22. How much more than $5V_g - 2iR$ is $3V_g + 6iR$?

23. What must be added to $11.93\mu + 4.7\lambda$ to obtain $22.86\mu - 9.4\lambda$?

24. What must be subtracted from $16.2a - 3.3b + 2.8c$ to obtain $8.1b + 1.7a - 2.6c$?

3–9 SIGNS OF GROUPING

Often it is necessary to express or group together quantities that are to be affected by the same operation. Also, it is desirable to be able to represent that two or more terms are to be considered as one quantity.

In order to meet the above requirements, signs of grouping have been adopted. These signs are the *parentheses* ( ), the *brackets* [ ], the *braces* { }, and the *vinculum* ____ . The first three are placed around the terms to be grouped as in $(V - IR)$, $[a + 3b]$, and $\{x^2 + 4y\}$. All have the same meaning: that the enclosed terms are to be considered as one quantity.

Thus, $16 - (12 - 5)$ means that the quantity $(12 - 5)$ is to be subtracted from 16. That is, 5 is to be subtracted from 12, and then the remainder 7 is to be subtracted from 16 to give a final remainder of 9. In like manner, $V - (IR + v)$ means that the sum of $(IR + v)$ is to be subtracted from V. (See Sec. 2–13.)

Carefully note that the sign preceding a sign of grouping, as the minus sign between V and $(IR + v)$ above, is a sign of *operation* and does not denote that $(IR + v)$ is a negative quantity.

The vinculum is used mainly with radical signs and fractions, as in

$$\sqrt{7245} \quad \text{and} \quad \frac{a + b}{x - y}$$

In the latter case the vinculum denotes the division of $a + b$ by $x - y$, in addition to grouping the terms in the numerator and denominator. When studying later chapters, you will avoid many mistakes by remembering that *the vinculum is a sign of grouping*.

In working problems involving signs of grouping, the operations within the signs of grouping should be performed first.

> When multiple parentheses (or other signs of grouping) are involved, the computer processes the *deepest*, or innermost parenthesis first.
>
> The expression
>
> 5[3 + 6(4 × 7 − 3)]
>
> is input as
>
> 5*(3 + 6*((4*7) − 3))
>
> and is worked out first as
>
> 5*(3 + 6*(28 − 3))
>
> then as
>
> 5*(3 + 6*(25))
>
> then as
>
> 5*(3 + 150)
>
> then as
>
> 5*(153)
>
> and finally reads out
>
> 765

EXAMPLE 10 $a + (b + c) = ?$

SOLUTION

This means, "What result will be obtained when the sum of $b + c$ is added to a?" Because both b and c are denoted as positive, it follows that we can write

$$a + (b + c) = a + b + c$$

because it makes no difference in which order we add.

EXAMPLE 11 $a + (b - c) = ?$

SOLUTION

This means, "What result will be obtained when the difference of $b - c$ is added to a?" Again, because it makes no difference in which order we add, we can write

$$a + (b - c) = a + b - c$$

EXAMPLE 12 $a - (b + c) = ?$

SOLUTION

Here the sum of $b + c$ is to be subtracted from a. This is the same as if we first subtract b from a and from the remainder subtract c. Therefore,

$$a - (b + c) = a - b - c$$

Because this is subtraction, we could change the signs and add algebraically, remembering that b and c are denoted as positive, as shown below:

$$\begin{array}{r} a \qquad\quad \\ b + c \\ \hline a - b - c \end{array}$$

EXAMPLE 13 $a - (-b - c) = ?$

SOLUTION

This means that the quantity $-b - c$ is to be subtracted from a. Performing this subtraction, we obtain

$$\begin{array}{r} a \\ -\ b - c \\ \hline a + b + c \end{array}$$

Therefore, $a - (-b - c) = a + b + c$

A study of Examples 10 to 13 enables us to state the following:

RULES

1. Parentheses or other signs of grouping preceded by a plus sign can be removed without any other change.
2. To remove parentheses or other signs of grouping preceded by a minus sign, change the sign of every term within the sign of grouping.

Although not apparent in the examples, another rule can be added as follows:

3. If parentheses or other signs of grouping occur one within another, remove the inner grouping first.

EXAMPLES

$$(x + y) + (2x - 3y) = x + y + 2x - 3y = 3x - 2y$$

$$3a - (2b + c) - a = 3a - 2b - c - a = 2a - 2b - c$$

$$10x - (-3x - 4y) + 2y = 10x + 3x + 4y + 2y = 13x + 6y$$

$$\begin{aligned} x - [2x + 3y - (3x - y) - 4x] &= x - [2x + 3y - 3x + y - 4x] \\ &= x - 2x - 3y + 3x - y + 4x \\ &= 6x - 4y \end{aligned}$$

PROBLEMS 3–4

Simplify by removing the signs of grouping and combining the similar terms:

1. $(a - 5y - 7) - (a + 2y - 9)$
2. $(5\lambda + 3\theta) - (-4\lambda + 5\theta + 6)$
3. $(2R + 4Z + 12X) - (-5R + 4Z + 15X - 7)$
4. $6I^2R + [-5VI + (-I^2R - 3VI) - 7VI] + 5$
5. $9W - [4w - (2VI - \overline{3w + 4W + 6VI})]$
6. $2A - \{5B - [2C - (A + 7B)]\}$
7. $6x - \{x + y - [z + x + y - (x + y + z) - 4x] - 3y\}$
8. $-\{-\theta - [\phi + \omega - 2\phi - (\omega + \phi) - \phi] - 2\theta - 3\omega\}$

9. $5x - [-3x - (-2y + 5z) - (10x - \overline{3y - 4z})]$
10. $5.4R - 2.6Z - \overline{1.5IX - 7R} -$
    $[4.6Z - (3X_C - 5.7IX) - 4.32R + 27]$

3–10 INSERTING SIGNS OF GROUPING

To enclose terms within signs of grouping preceded by a plus sign, rewrite the terms without changing their signs.

EXAMPLE 14

$$a + b - c + d = a + (b - c + d)$$

To enclose terms within signs of grouping preceded by a minus sign, rewrite the terms and change the signs of the terms enclosed.

EXAMPLE 15

$$a + b - c + d = a + b - (c - d)$$

No difficulty need be encountered when inserting signs of grouping because, by removing the signs of grouping from the result, the original expression should be obtained.

EXAMPLE 16

$$\begin{aligned} x - 3y + z &= x - (3y - z) \\ &= x - 3y + z \end{aligned}$$

PROBLEMS 3–5

1. Enclose the last three terms of each of the following expressions in parentheses preceded by a plus sign:
   (*a*) $2L_1 + 5L_2 - 4L_3 + L_T$
   (*b*) $2w + 12x - 3y + z$
   (*c*) $8\theta + 2\phi - 6\psi - 10\Omega$
   (*d*) $2R - 11\frac{V}{I} + 5\frac{P}{I^2} - 3X$
   (*e*) $10R_T - 3R_1 + 6R_2 - 5R_3 + 4R_4$
2. Enclose the last three terms of each of the following expressions in parentheses preceded by a minus sign:
   (*a*) $5w + 6x - 3y + 4z$
   (*b*) $2X_L + 7X_C - 5R_1 + 4Z + 3R_2$
   (*c*) $2I_1 + 7I_2 - 4I_3 - 7I_4$
   (*d*) $0.004IR - 2V + 1.6\frac{P}{I} - 0.006IZ$
   (*e*) $25\frac{V^2}{R} + W - 4I^2Z - 12VI - 18I^2R$
3. Write the amount by which Z is less than $\sqrt{R^2+X^2}$.
4. The sum of two currents is 526 milliamperes (mA). The larger of the two currents is i mA. What is the smaller?

5. The difference between two voltages is 38.2 V. The smaller voltage is v V. What is the greater?
6. Write the amount by which X_L exceeds $\frac{1}{2\pi fC}$.
7. What is the larger part of Z if $\sqrt{r^2+x^2}$ is the smaller part?
8. Write the amount by which Q is greater than $X - 2\pi fL$.
9. Write the amount by which V exceeds $IR + \frac{P}{I}$.
10. The difference between two numbers is 19.6. If the larger number is β, what is the smaller?
11. Write the smaller part of X_L if $2\pi fL$ is the larger part.
12. The difference between two numbers is X^2 and the larger of the two is Z^2. Write the relationship which describes R^2, which is the smaller of the two.

SELF TEST

1. Add: (256) and (-27)
2. Add: (-418) and (-16)
3. Subtract: (-31) from (-80)
4. A guy wire is fastened 4 m below the top of a mast, and a horizontal antenna is mounted 1.6 m above the guy fastening. How high is the top of the mast above the antenna?
5. Your bank account is overdrawn by \$112.85. You want to have the bank honor a new check for \$92.95 and also pay their \$10.00 charge for the overdraft. In order to have a new balance of \$422.50, how much must you deposit?
6. Add: $12P - 3\frac{V^2}{R} + 2IR$, $8P - 6IR + 21\frac{V^2}{R}$, and $11IR - 16P - 13\frac{V^2}{R}$.
7. From $12R + 19X_L + 3X_C$ subtract $3R - 12X_C + 16X_L$.
8. Simplify:
   $2M + 3P - 5R - [M - 2P + 3R - (2M - \overline{4P - 12R})]$
9. What must be subtracted from $12X^2 + 18Y^2 + 24Z^2$ to achieve $9X^2+24Y^2+44Z^2$?
10. How much less than $5a - 3b + 2c$ is $-2a - b - 5c$?

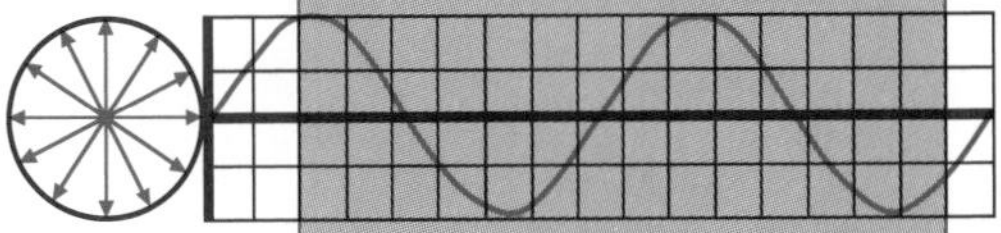

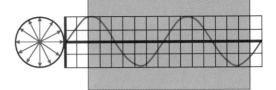

CHAPTER 4

ALGEBRA—MULTIPLICATION AND DIVISION

Multiplication is often defined as the *process of repeated addition*. Thus, 2×3 may be thought of as adding 2 three times, or $2 + 2 + 2 = 6$.

Considering multiplication as a shortened form of addition is not satisfactory, however, when the multiplier is a fraction. For example, it would not be sensible to say that $5 \times \frac{2}{7}$ was adding 5 two-sevenths of a time. This problem could be rewritten as $\frac{2}{7} \times 5$, which would be the same as adding $\frac{2}{7}$ five times. But this is only a temporary help, for if two fractions are to be multiplied together, as $\frac{3}{4} \times \frac{5}{6}$, the original definition of multiplication will not apply. However, the definition has been extended to include such cases, and the product of $5 \times \frac{2}{7}$ is taken to mean 5 multiplied by 2 and this product divided by 7; that is, by $5 \times \frac{2}{7}$ is meant $\dfrac{5 \times 2}{7}$. Also,

$$\frac{3}{4} \times \frac{5}{6} = \frac{3 \times 5}{4 \times 6} = \frac{15}{24}$$

4–1 MULTIPLICATION OF POSITIVE AND NEGATIVE NUMBERS

Because we are now dealing with both positive and negative numbers, it becomes necessary to determine what sign the product will have when combinations of these numbers are multiplied.

When only two numbers are to be multiplied, there can be but four possible combinations of signs, as follows:

(1) $(+2) \times (+3) = ?$
(2) $(-2) \times (+3) = ?$
(3) $(+2) \times (-3) = ?$
(4) $(-2) \times (-3) = ?$

Combination (1) means that $+2$ is to be added three times:

$$(+2) + (+2) + (+2) = +6$$

or

$$(+2) \times (+3) = +6$$

In the same manner, combination (2) means that -2 is to be added three times:

$$(-2) + (-2) + (-2) = -6$$

or

$$(-2) \times (+3) = -6$$

Combination (3) means that $+2$ is to be subtracted three times:

$$-(+2) - (+2) - (+2) = -6$$

or

$$(+2) \times (-3) = -6$$

Note that this is the same as subtracting 6 once, -6 being thus obtained. Combination (4) means that -2 is to be subtracted three times:

$$-(-2) - (-2) - (-2) = +6$$

or

$$(-2) \times (-3) = +6$$

This may be considered to be the same as subtracting -6 once; and because subtracting -6 once is the same as adding $+6$, we obtain $+6$ as above.

From the foregoing we have these rules:

RULES

1. The product of two numbers having like signs is positive.
2. The product of two numbers having unlike signs is negative.
3. If more than two factors are multiplied, Rules 1 and 2 are to be used successively.
4. The product of an even number of negative factors is positive. The product of an odd number of negative factors is negative.

These rules can be summarized in general terms as follows:

- Rule 1 $(+a)(+b) = +ab$
- Rule 1 $(-a)(-b) = +ab$
- Rule 2 $(+a)(-b) = -ab$
- Rule 2 $(-a)(+b) = -ab$
- Rule 3 $(-a)(+b)(-c) = +abc$
- Rule 4 $(-a)(-b)(-c)(-d) = +abcd$
- Rule 4 $(-a)(-b)(-c) = -abc$

PROBLEMS 4–1

Find the products of the following factors:

1. $8, 3$
2. $6, -5$
3. $-2.7, -8.2$
4. $-8.6, 3.8, -2.9$
5. $\frac{5}{16}, -\frac{3}{8}, \frac{1}{4}$
6. $\frac{2}{3}, -\frac{3}{4}, -\frac{7}{8}$
7. $-0.0386, -0.0002, -1.8, -0.06$
8. $1500, -0.03, 150, -0.001$
9. a, c, ω^2
10. $q, -r, -s, t$
11. $2\pi f, L_1, L_2$
12. $\theta^2, \phi^2, \lambda^2$
13. $\frac{1}{2}, \frac{1}{\pi}, \frac{1}{f}, \frac{1}{C_p}$
14. $\frac{1}{a}, -\frac{1}{b}, -\frac{1}{c}, -\frac{1}{d}$
15. $\alpha, \frac{1}{\psi}, -\frac{1}{\mu}, \varepsilon$

4–2 GRAPHICAL REPRESENTATION

Our system of representing numbers is a graphical one, as previously illustrated in Fig. 3–3. It might be well at this time to consider certain facts regarding multiplication.

When a number is multiplied by any other number except 1, we can think of the operation as having changed the absolute value of the multiplicand. Thus, 3 in × 4 becomes 12 in, 6 A × 3 becomes 18 A, etc. Such multiplications could be represented graphically by simply extending the multiplicand the proper amount, as shown in Fig. 4–1.

The multiplication of a negative number by a positive number is shown in Fig. 4–2.

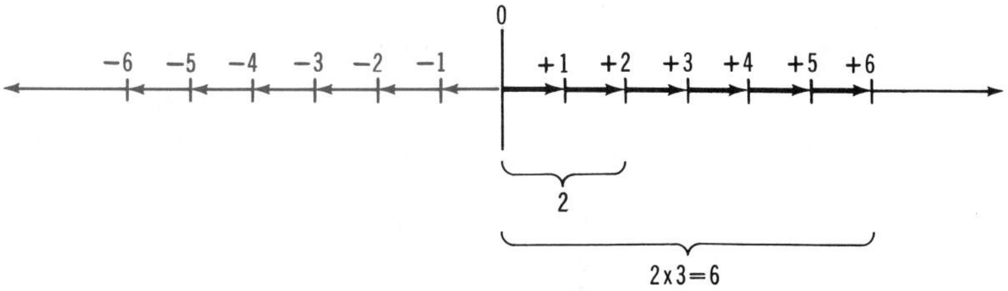

FIG. 4–1 Representation of the multiplication $2 \times 3 = 6$.

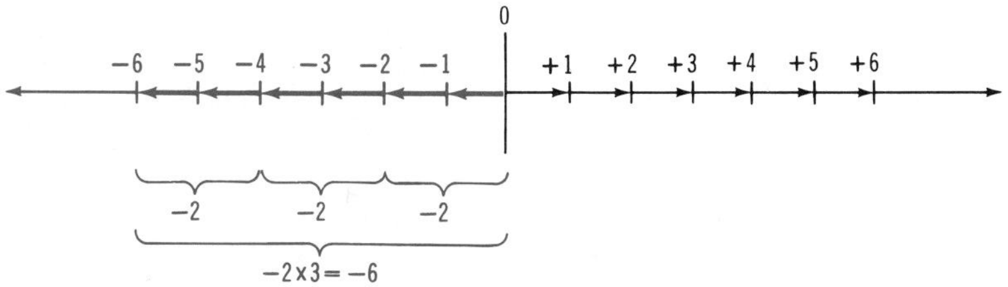

FIG. 4–2 Representation of the multiplication $-2 \times 3 = -6$.

From these examples, it is evident that a positive multiplier simply changes the absolute value, or magnitude, of the number being multiplied. What happens if the multiplier is negative? As an example, consider $2 \times (-3) = -6$. How will this be represented graphically?

Now, $2 \times (-3) = -6$ is the same as

$$2 \times (+3) \times (-1) = -6$$

Therefore, let us first multiply 2×3 to obtain $+6$ and represent it as shown in Fig. 4–1. We must multiply by -1 to complete the problem and in so doing should obtain -6, but -6 must be represented as a number six units in length and directed toward the left, as illustrated in Fig. 4–2. We therefore agree that multiplication by -1 causes counterclockwise rotation of a number in a direction that will be exactly opposite from its original direction. This is illustrated in Fig. 4–3.

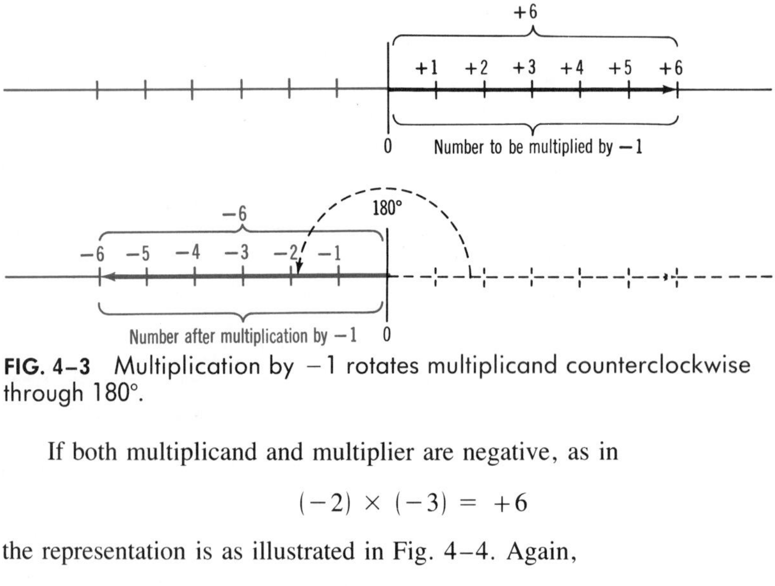

FIG. 4–3 Multiplication by -1 rotates multiplicand counterclockwise through 180°.

If both multiplicand and multiplier are negative, as in

$$(-2) \times (-3) = +6$$

the representation is as illustrated in Fig. 4–4. Again,

$$(-2) \times (-3) = +6$$

is the same as

$$(-2) \times (+3) \times (-1) = +6$$

The product has an absolute value of 6, and at the same time there has been rotation to $+6$ because of multiplication by -1.

The foregoing representations are applicable to division also, since the law of signs is the same as in multiplication.

The important thing to bear in mind is that multiplication or division by -1 causes counterclockwise rotation of a number to a direction exactly opposite the original direction. The number -1, when used as a multiplier or divisor, should be considered as an *operator* for the purpose of rotation. You should clearly understand this concept; you will encounter it later.

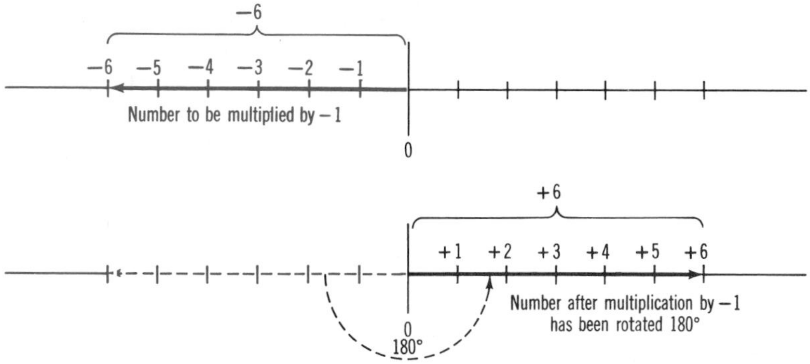

FIG. 4–4 Illustration of −6 rotated counterclockwise through 180° to become +6 owing to multiplication by −1.

4–3 LAW OF EXPONENTS IN MULTIPLICATION

As explained in Sec. 2–10, an exponent indicates how many times a number is to be taken as a factor. Thus $x^4 = x \cdot x \cdot x \cdot x$, $a^3 = a \cdot a \cdot a$, etc.

Because $\quad x^4 = x \cdot x \cdot x \cdot x$

and $\quad x^3 = x \cdot x \cdot x$

then $\quad x^4 \cdot x^3 = x \cdot x \cdot x \cdot x \cdot x \cdot x \cdot x = x^7$

or $\quad x^4 \cdot x^3 = x^{4+3} = x^7$

Thus, we have the rule:

RULE

To find the product of two or more powers having the same base, add the exponents.

EXAMPLES

$$a^3 \cdot a^2 = a^{3+2} = a^5$$
$$x^4 \cdot x^4 = x^{4+4} = x^8$$
$$6^2 \cdot 6^3 \cdot 6^5 \cdot = 6^{2+3+5} = 6^{10}$$
$$a^2 \cdot b^3 \cdot b^3 \cdot a^5 = a^{2+5} \cdot b^{3+3} = a^7b^6$$
$$e \cdot e^3 = e^{1+3} = e^4$$
$$3^2 \cdot 3^4 = 3^{2+4} = 3^6$$
$$e^a \cdot e^b = e^{a+b}$$

From the foregoing examples, it is seen that the law of exponents can be expressed in the well-known general form

$$a^m \cdot a^n = a^{m+n}$$

where $a \neq 0$ and m and n are literal numbers and may represent any number of factors.

4–4 MULTIPLICATION OF MONOMIALS

RULES

1. Find the product of the numerical coefficients and give it the proper sign, plus or minus, according to the rules for multiplication (Sec. 4–1).
2. Multiply this numerical product by the product of the literal factors. Use the law of exponents as applicable.

EXAMPLE 1 Multiply $3a^2b$ by $4ab^3$.

SOLUTION

$$\begin{aligned}(3a^2b)(4ab^3) &= +(3 \cdot 4) \cdot a^{2+1} \cdot b^{1+3} \\ &= 12a^3b^4\end{aligned}$$

EXAMPLE 2 Multiply $-6x^3y^2$ by $3xy^2$.

SOLUTION

$$\begin{aligned}(-6x^3y^2)(3xy^2) &= -(6 \cdot 3) \cdot x^{3+1} \cdot y^{2+2} \\ &= -18x^4y^4\end{aligned}$$

EXAMPLE 3 Multiply $-5e^2x^4y$ by $-3e^2x^2p$.

SOLUTION

$$\begin{aligned}(-5e^2x^4y)(-3e^2x^2p) &= +(5 \cdot 3)e^{2+2} \cdot p \cdot x^{4+2} \cdot y \\ &= 15e^4px^6y\end{aligned}$$

PROBLEMS 4–2

Multiply:

1. $a^5 \cdot a^3$
2. $-b^3 \cdot b^5$
3. $x^2 \cdot x^3 \cdot -x^4$
4. $-L \cdot L^2 \cdot -Y^5$
5. $(3y^2)(4y^4)$
6. $(12\theta)(-2\phi^2)$
7. $(4p)(3x^2)(-6p^3x)$
8. $(-5\mu)^2$
9. $(am^n)(bm^p)$
10. $(13b^x)(-2b^{a+y})$
11. $(2p)^3$
12. $(-5\lambda^2)^3$
13. $(-3a^2b^3cd^2)(-2abc^2d^5)$
14. $\left(\frac{1}{4}a^3\right)\left(-\frac{2}{3}ab^2\right)$
15. $\left(\frac{3}{16}X_L\right)\left(\frac{2}{3}M\right)(-2\pi)$
16. $(14a^2b^3cd)\left(-\frac{2}{7}ab^2de\right)$
17. $(0.2x^2y)(2.4a^2d)(-0.06xy)(a)$
18. $\left(\frac{5}{16}\theta\phi\right)\left(-\frac{3}{25}\mu\theta\right)\left(-\frac{24}{27}\theta^2\omega\right)$
19. $(r^3)^4$
20. $(3p^q)^r$

4–5 MULTIPLICATION OF POLYNOMIALS BY MONOMIALS

Another method of graphically representing the product of two numbers is as shown in Fig. 4–5. The product $5 \times 6 = 30$ is shown as a rectangle whose sides are 5 and 6 units in length; therefore, the rectangle contains 30 square units.

Similarly, the product of $5(6 + 9)$ can be represented as illustrated in Fig. 4–6.

Thus,

$$\begin{aligned} 5(6 + 9) &= 5 \times 15 \\ &= 75 \end{aligned}$$

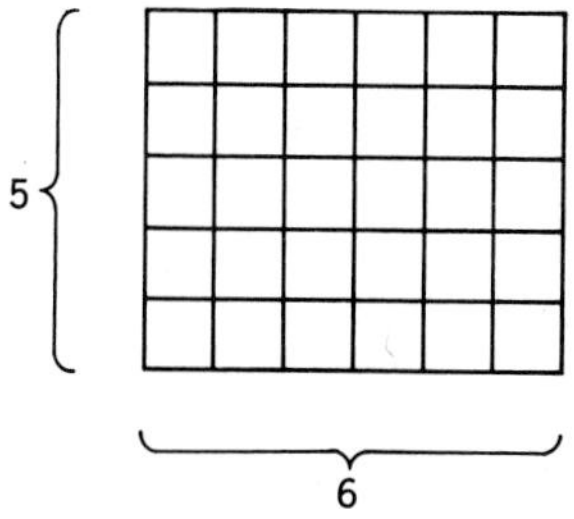

FIG. 4–5 Graphical representation of the multiplication $5 \times 6 = 30$.

Also,

$$\begin{aligned} 5(6 + 9) &= (5 \times 6) + (5 \times 9) \\ &= 30 + 45 \\ &= 75 \end{aligned}$$

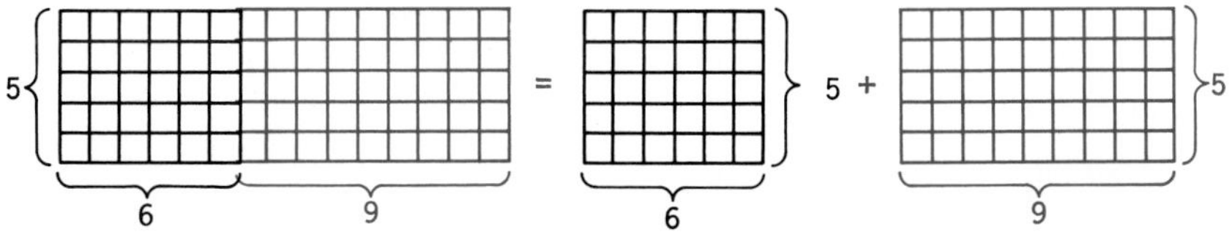

FIG. 4–6 Graphical representation of the multiplication $5(6 + 9) = 5 \times 6 + 5 \times 9 = 75$.

In like manner the product

$$a(c + d) = ac + ad$$

can be illustrated as in Fig. 4–7.

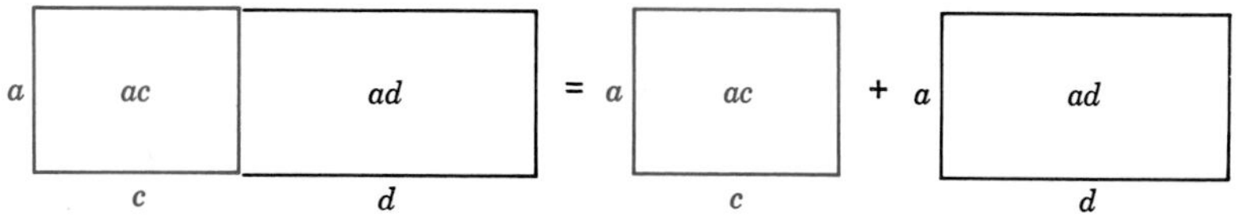

FIG. 4–7 Illustration of the product $a(c + d) = ac + ad$.

From the foregoing, you can show that

$$\begin{aligned} 3(4 + 2) &= 3 \times 4 + 3 \times 2 = 12 + 6 = 18 \\ 4(5 + 3 + 4) &= 4 \times 5 + 4 \times 3 + 4 \times 4 \\ &= 20 + 12 + 16 = 48 \\ x(y + z) &= xy + xz \\ p(q + r + s) &= pq + pr + ps \end{aligned}$$

Note that, in all cases, each term of the polynomial (the terms enclosed in parentheses) is multiplied by the monomial. From these examples, we develop the following rule:

RULE

To multiply a polynomial by a monomial, multiply each term of the polynomial by the monomial and write in succession the resulting terms with their proper signs.

EXAMPLE 4 $3x(3x^2y - 4xy^2 + 6y^3) = ?$

SOLUTION

Multiplicand $= 3x^2y - 4xy^2 + 6y^3$
Multiplier $= 3x$
Product $= 9x^3y - 12x^2y^2 + 18xy^3$

EXAMPLE 5 $-2ac(-10a^3 + 4a^2b - 5ab^2c + 7bc^2) = ?$

SOLUTION

Multiplicand $= -10a^3 + 4a^2b - 5ab^2c + 7bc^2$
Multiplier $= -2ac$
Product $= 20a^4c - 8a^3bc + 10a^2b^2c^2 - 14abc^3$

EXAMPLE 6 Simplify $5(2e - 3) - 3(e + 4)$.

SOLUTION

First multiply $5(2e - 3)$ and $3(e + 4)$, and then subtract the second result from the first, thus:

$$\begin{aligned} 5(2e - 3) - 3(e + 4) &= (10e - 15) - (3e + 12) \\ &= 10e - 15 - 3e - 12 \\ &= 7e - 27 \end{aligned}$$

PROBLEMS 4–3

Multiply:

1. $4x + 3y$ by 5
2. $2a + 3$ by $3a$
3. $2Y_1 + 4Y_2$ by $2Z^2$
4. $3.2c - jd$ by d^2
5. $\mu^3 + 2\lambda - 3\phi$ by 6π
6. $2\alpha^3 - 3\alpha^2 + 4\alpha$ by -5α
7. $2\alpha^2\beta - 5\theta\phi - 4\alpha\theta + 6$ by $6\beta\phi$
8. $6x^2y - 2x^2y^2 - 4xy^2$ by $0.5xy$
9. $-12m^2p - 5mp^2 + 4p^3$ by $-3mp$
10. $6\omega^2L_1 - 12\omega^2M + 3\omega^2L_2$ by $-5\omega L_1L_2$

11. $\frac{1}{2}I^2R - \frac{1}{4}I^2R^2 - \frac{1}{3}iZ$ by $\frac{2}{3}iIZ$

12. $8ab + 4ab^2 + 4$ by $-\frac{1}{4}ab^2$

13. $\frac{I^2R}{4} - \frac{i^2r}{2} + \frac{P}{6}$ by $12IP$

14. $5\mu^2k^2 + 3\eta k - 2\mu\eta^2$ by $-3\theta\omega$

15. $0.025V^3Z^2 + 0.05VZ^4 - 1.67Z^5$ by $6.28IZ$

Simplify:

16. $3ars(-4ar^2 + 2rs - 6as^2)$
17. $2(5x - 8y) - 5(7x + 3y)$
18. $\mu(\alpha - j\beta) + \mu(\alpha + j\beta)$
19. $\theta(\theta^2 + \phi) - \phi(\theta + \phi^2)$
20. $2x(3y^2 - a^2) - 5x(2y^2 - 3a^2)$
21. $0.5R_1(6I + 5R_1R_2 - IR^2) - 3I(0.5R_1 - IR_2 + 2R_1R^2)$
22. $\frac{1}{2}\gamma\beta(4\gamma^2\beta - 2\gamma\beta^2 - 10\gamma^3 + 5\beta^3)$
23. $3M\left(\frac{L^2}{4} + \frac{Ll}{8} - \frac{l^2}{12}\right)$
24. $5\theta(2\theta^2 + 3\theta\phi - 6\phi^2) - 3(6\theta^3 - 2\theta^2\phi - 7\theta\phi^2)$
25. $0.25I\left(\frac{R}{5} - \frac{R_1}{10} + \frac{3R_2}{5}\right) + 1.5(0.05IR - 0.375IR_2)$
26. $\frac{1}{2}\lambda\mu(6\lambda^2\mu - 5\lambda\mu^2 + 12\lambda - 4\mu)$
27. $4a(3a^2 + 2ab - 3b^2) - 2a(6a^2 + 4ab - 6b^2)$
28. $5\gamma^2\left(\frac{\gamma\lambda}{3} - \frac{\beta\lambda^2}{5} - \frac{\theta\beta^2}{10}\right) + 6\beta^2\left(\frac{\beta\lambda}{2} + \frac{\gamma\lambda}{5} - \frac{\gamma^2\theta}{12}\right)$
29. $6\left(\frac{s}{3} - \frac{s}{2} + \frac{2s}{3}\right) + 8\left(\frac{s}{4} + \frac{s}{2} + \frac{3s}{4}\right)$
30. $0.8P^2(3R_1 + 5R_2 - 2R_3) - 0.4P^2(0.6R_1 - 2R_2 - 0.06R_3)$

4–6 MULTIPLICATION OF A POLYNOMIAL BY A POLYNOMIAL

It is apparent that

$$(3 + 4)(6 - 3) = 7 \times 3 = 21$$

The above multiplication can also be accomplished in the following manner:

$$\begin{aligned}(3+4)(6-3) &= 3(6-3)+4(6-3)\\ &= (18-9)+(24-12)\\ &= 9+12\\ &= 21\end{aligned}$$

Similarly,

$$\begin{aligned}(2a-3b)(a+5b) &= 2a(a+5b)-3b(a+5b)\\ &= (2a^2+10ab)-(3ab+15b^2)\\ &= 2a^2+10ab-3ab-15b^2\\ &= 2a^2+7ab-15b^2\end{aligned}$$

From the foregoing, we have the following:

RULE

To multiply polynomials, multiply every term of the multiplicand by each term of the multiplier and add the partial products.

EXAMPLE 7 Multiply $2i - 3$ by $i + 2$.

SOLUTION

$$\begin{array}{ll}\text{Multiplicand} & = 2i - 3\\ \text{Multiplier} & = \underline{i + 2}\\ i \text{ times } (2i-3) & = 2i^2 - 3i\\ 2 \text{ times } (2i-3) & = \underline{ 4i - 6}\\ \text{Adding, product} & = 2i^2 + i - 6\end{array}$$

EXAMPLE 8 Multiply $a^2 - 3ab + 2b^2$ by $2a^2 - 3b^2$.

SOLUTION

$$\begin{array}{ll}\text{Multiplicand} & = a^2 - 3ab + 2b^2\\ \text{Multiplier} & = \underline{2a^2 - 3b^2}\\ \quad 2a^2 \text{ times} & \\ (a^2 - 3ab + 2b^2) & = 2a^4 - 6a^3b + 4a^2b^2\\ \quad -3b^2 \text{ times} & \\ (a^2 - 3ab + 2b^2) & = \underline{ - 3a^2b^2 + 9ab^3 - 6b^4}\\ \text{Adding, product} & = 2a^4 - 6a^3b + a^2b^2 + 9ab^3 - 6b^4\end{array}$$

Products obtained by multiplication can be tested by substituting any convenient numerical values for the literal numbers. It is not good practice to substitute the numbers 1 and 2. If there are exponents, then the use of 1 will not be a proof of correct work, for 1 to any power is still 1. Similarly, if addition should be involved, the use of 2 could give an incorrect indication, because $2 + 2 = 4$ and $2 \times 2 = 4$.

EXAMPLE 9 Multiply $a^2 - 4ab - b^2$ by $a + b$, and test by letting $a = 3$ and $b = 4$.

SOLUTION

$$\begin{array}{lcr} a^2 - 4ab - b^2 & = 9 - 48 - 16 = & -55 \\ \underline{a + b} & = 3 + 4 \quad = & \underline{7} \\ a^3 - 4a^2b - ab^2 & & -385 \\ \underline{\quad a^2b - 4ab^2 - b^3} & & \\ a^3 - 3a^2b - 5ab^2 - b^3 & = 27 - 108 - 240 - 64 = & -385 \end{array}$$

PROBLEMS 4–4

Multiply

1. $\alpha + 1$ by $\alpha + 1$
2. $\alpha + 1$ by $\alpha - 1$
3. $\alpha - 1$ by $\alpha - 1$
4. $\beta + 3$ by $\beta + 3$
5. $\beta + 3$ by $\beta - 3$
6. $\beta - 3$ by $\beta - 3$
7. $x + 4$ by $x + 3$
8. $L_1 - 5$ by $L_1 - 3$
9. $p - 12$ by $p + 4$
10. $a + 5$ by $a - 5$

NOTE *Parentheses or other signs of grouping are often used to indicate a product. Thus,* (ir + v)(2ir − 3v) *means* ir + v *multiplied by* 2ir − 3v. *Perform the indicated operations:*

11. $(y + 5)(y + 2)$
12. $(3X + Z)(5X + Z)$
13. $(x + 5y)(3x - 4y)$
14. $(ax + bx)(cx + dx)$
15. $(3N + P)(2N - 5P)$
16. $(9VI - 3I^2R)(5VI - 2I^2R)$
17. $(2\theta + 3\phi)(3\theta + 2\phi)$
18. $(1.5\psi + 0.5\phi)(2\psi + 1.75\phi)$
19. $(V - 2I)(2V - 5I)$
20. $\left(\frac{1}{3}m - \frac{1}{2}q\right)\left(\frac{3}{4}m + \frac{5}{6}q\right)$
21. $(3a^2 + a - 2)(2a + 5)$
22. $(6L_1^2 - 4L_1 - 5)(L_1 + 2)$
23. $(R + r)(2R^2 - 4Rr + 2r^2)$
24. $(x + y)(x + y)(x + y)$
25. $(a + b)(a - b)(a - b)$
26. $(p - q)(p - q)(p - q)$
27. $(\alpha - \beta)(\alpha + \beta)(\alpha - \beta)$
28. $(IR + P)(I^2R^2 - 2IRP + P^2)$
29. $(a^2 + 2ab + b^2)(a + b)$
30. $(a + 1)^2$
31. $(x + y)^2$
32. $(x - y)^2$
33. $(M - N)^2$
34. $(2\theta\phi + \psi + 1)^2$
35. $(3x + 3y)^3$
36. $(3\alpha + 7)(\alpha - 5) + (2\alpha - 3)(4\alpha - 1)$
37. $5(2VI + 1)(5VI - 3) - 2(3VI - 1)(VI + 3)$
38. $6(2X - 3Y + Z)(3X + 2Y - Z) - 2(X + 6Y + 2Z)(3X - Y - 2Z)$
39. $3a(2a + b - 1)^2 - 2a(a + 2b + 1)^2$
40. $3\theta^2(5\omega - \lambda + \theta)^2 - \theta^2(\omega + 7\lambda - 2\theta)^2$

4–7 DIVISION

The division of algebraic expressions requires the development of certain rules and new methods in connection with operations involving exponents. However, if you have mastered the processes of the preceding sections, algebraic division will be an easy subject.

For the purpose of review the following definitions are given:

1. The *dividend* is a number, or quantity, that is to be divided.
2. The *divisor* is a number by which a number, or quantity, is to be divided.
3. The *quotient* is the result obtained by division. That is,

$$\frac{\text{Dividend}}{\text{Divisor}} = \text{Quotient}$$

4–8 DIVISION OF POSITIVE AND NEGATIVE NUMBERS

Because division is the inverse of multiplication, the methods of the latter will serve as an aid in developing methods for division. For example,

because $6 \times 4 = 24$
then $24 \div 6 = 4$
and $24 \div 4 = 6$

These relations can be used in applying the rules for multiplication to division.

All the possible cases can be represented as follows:

$$(+24) \div (+6) = ?$$
$$(-24) \div (+6) = ?$$
$$(+24) \div (-6) = ?$$
$$(-24) \div (-6) = ?$$

Because division is the inverse of multiplication, we apply the rules for multiplication of positive and negative numbers and obtain the following:

$(+24) \div (+6) = +4$ because $(+4) \times (+6) = +24$
$(-24) \div (+6) = -4$ because $(-4) \times (+6) = -24$
$(+24) \div (-6) = -4$ because $(-4) \times (-6) = +24$
$(-24) \div (-6) = +4$ because $(+4) \times (-6) = -24$

Therefore, we have the following:

RULE

To divide positive and negative numbers,

1. If dividend and divisor have like signs, the quotient is positive.
2. If dividend and divisor have unlike signs, the quotient is negative.

PROBLEMS 4–5

Divide the first number by the second in Probs. 1 to 10:

1. 36, 4
2. −28, 7
3. −54, −3
4. −3.6, 120
5. $-\frac{2}{3}, \frac{1}{2}$

6. $\dfrac{21}{64}, \dfrac{7}{16}$
7. $2\pi fC, -1$
8. $1, 2\pi fC$
9. $V \times 10^8, L_v$
10. $\omega L, Q$

Supply the missing divisors:

11. $\dfrac{-33}{?} = \dfrac{1}{3}$
12. $\dfrac{16}{?} = -2$
13. $\dfrac{80}{?} = \dfrac{1}{4}$
14. $-\dfrac{27}{?} = -\dfrac{1}{3}$
15. $\dfrac{-\frac{15}{16}}{?} = \dfrac{3}{8}$

4–9 THE LAW OF EXPONENTS IN DIVISION

By previous definition of an exponent (Sec. 2–10),

$$x^6 = x \cdot x \cdot x \cdot x \cdot x \cdot x$$

and

$$x^3 = x \cdot x \cdot x$$

Then

$$x^6 \div x^3 = \frac{x^6}{x^3} = \frac{\cancel{x} \cdot \cancel{x} \cdot \cancel{x} \cdot x \cdot x \cdot x}{\cancel{x} \cdot \cancel{x} \cdot \cancel{x}} = x^3$$

This result is obtained by canceling common factors in numerator and denominator. The above could be expressed as

$$x^6 \div x^3 = \frac{x^6}{x^3} = x^{6-3} = x^3$$

In like manner,

$$\frac{a^7}{a^3} = a^{7-3} = a^4$$

From the foregoing, it is seen that the law of exponents can be expressed in the general form

$$a^m \div a^n = \frac{a^m}{a^n} = a^{m-n}$$

where $a \neq 0$ and m and n are general numbers.

4–10 THE ZERO EXPONENT

Any number, except zero, divided by itself results in a quotient of 1. Thus,

$$\frac{6}{6} = 1$$

Also,

$$\frac{a^3}{a^3} = 1$$

Therefore,

$$\frac{a^3}{a^3} = a^{3-3} = a^0 = 1$$

Then, in the general form,

$$\frac{a^m}{a^n} = a^{m-n}$$

If $\quad m = n$

then $\quad m - n = 0$

and $\quad \dfrac{a^m}{a^n} = a^{m-n} = a^0 = 1$

The foregoing leads to the definition that

Any base, except zero, affected by zero exponent is equal to 1.

Thus, a^0, x^0, y^0, 3^0, 4^0, etc., all equal 1.

4–11 THE NEGATIVE EXPONENT

If the law of exponents in division is to apply to all cases, it must apply when n is greater than m. Thus,

$$\frac{a^2}{a^5} = \frac{\not{a} \cdot \not{a}}{\not{a} \cdot \not{a} \cdot a \cdot a \cdot a} = \frac{1}{a^3}$$

or $\quad \dfrac{a^2}{a^5} = a^{2-5} = a^{-3}$

Therefore, $\quad a^{-3} = \dfrac{1}{a^3}$

Also, $\quad a^{-n} = \dfrac{1}{a^n}$

This leads to the definition that

Any base affected by a negative exponent is the same as 1 divided by that same base but affected by a positive exponent of the same absolute value as the negative exponent.

EXAMPLES

$$x^{-4} = \frac{1}{x^4}$$

$$2^{-2} = \frac{1}{2^2} = \frac{1}{4}$$

$$3^{-3} = \frac{1}{3^3} = \frac{1}{27}$$

$$\frac{4^3}{4^5} = \frac{4 \times 4 \times 4}{4 \times 4 \times 4 \times 4 \times 4} = \frac{1}{4 \times 4} = \frac{1}{4^2} = 4^{-2}$$

or $\quad \dfrac{4^3}{4^5} = 4^{3-5} = 4^{-2}$

It follows, from the consideration of negative exponents, that

Any *factor* of an algebraic term may be transferred from numerator to denominator, or vice versa, by changing the sign of the *exponent* of the factor; the sign of the *factor* does not change.

EXAMPLE 10

$$3a^2x^3 = \frac{3a^2}{x^{-3}} = \frac{3}{a^{-2}x^{-3}} = \frac{3x^3}{a^{-2}}$$

4–12 DIVISION OF ONE MONOMIAL BY ANOTHER

RULE

To divide one monomial by another,

1. Find the quotient of the absolute values of the numerical coefficients and affix the proper sign according to the rules for division of positive and negative numbers (Sec. 4–8).
2. Determine the literal coefficients with their proper exponents, and write them after the numerical coefficient found in 1 above.

EXAMPLE 11 Divide $-12a^3x^4y$ by $4a^2x^2y$.

SOLUTION

$$\frac{-12a^3x^4y}{4a^2x^2y} = -3ax^2$$

EXAMPLE 12 Divide $-7a^2b^4c$ by $-14ab^2c^3$. Express the quotient with positive exponents.

SOLUTION

$$\frac{-7a^2b^4c}{-14ab^2c^3} = \frac{ab^2}{2c^2}$$

EXAMPLE 13 Divide $15a^{-2}b^2c^3d^{-4}$ by $-5a^2bc^{-1}d^{-2}$. Express the quotient with positive exponents.

SOLUTION

$$\frac{15a^{-2}b^2c^3d^{-4}}{-5a^2bc^{-1}d^{-2}} = -\frac{3bc^4}{a^4d^2}$$

Division can be checked by substituting convenient numerical values for the literal factors or by multiplying the divisor by the quotient, the product of which should result in the dividend.

PROBLEMS 4–6

Divide:

1. $36\alpha^4\beta^6$ by $9\alpha^2\beta^2$
2. $-48a^6y^3z^4$ by $-6a^3yz^2$
3. $27\theta^4\phi^5\psi^6$ by $-9\theta^3\phi^2\psi^5$
4. $-96\omega^4L^8M^2$ by $-24\omega^2L^6M^2$
5. $95X_L{}^3Z^2$ by $-57X_L{}^2Z$
6. $-10^3a^{12}b^9c^{14}d^7$ by $10a^5b^5c^4d^4$

7. $-27x^6y^4z^{10}$ by $3x^2yz^9$

8. $6\ell^3m^2p$ into $-30\ell^5m^6p^3$

9. $-\frac{7}{16}m^4n^5p^2$ into $-\frac{21}{64}m^5n^7p^3$

10. $-\frac{5}{8}x^4y^{12}z^8$ into $-\frac{25}{48}x^3y^{14}z^6$

11. $\dfrac{18c^9d^2e^3}{-2c^8d^2e^3}$

12. $\dfrac{48\theta^4\phi^3\alpha}{12\theta^6\phi^2\alpha^3}$

13. $\dfrac{72\delta^5\varepsilon^6\eta^8}{-12\delta^2\varepsilon^4\eta^5}$

14. $\dfrac{45V^4E^6(IR)^8}{-0.5VE^2(IR)^3}$

15. $\dfrac{-3\alpha^{-3}\beta^4\lambda^{-4}\delta^2}{-24\alpha^{-2}\beta^{-3}\lambda^2\delta^{-2}}$

16. $\dfrac{-21rs^2t^{-4}u^6}{-63r^2s^{-2}t^{-3}u^2}$

17. $\dfrac{13\phi^2\theta^{-6}\psi\Omega^{-1}}{-52\phi^{-4}\theta^6\psi^2\Omega^2}$

18. $\dfrac{\frac{7}{16}x^3y^4\alpha^{-3}}{\frac{3}{8}x^{-2}y^3\alpha^{-1}}$

19. $\dfrac{250\alpha^8\beta^5\gamma^{-4}}{0.05\alpha^6\beta^{-2}\gamma^{-2}}$

20. $\dfrac{-0.000\,256I^4R^3Z}{0.016I^{-2}R^{-1}Z^3}$

4–13 DIVISION OF A POLYNOMIAL BY A MONOMIAL

Because $2 \times 8 = 16$

then $\dfrac{16}{2} = 8$

Also, because $3(a + 4) = 3a + 12$

then $\dfrac{3a + 12}{3} = a + 4$

Similarly, because $3I(2R + 3r) = 6IR + 9Ir$

then $\dfrac{6IR + 9Ir}{3I} = 2R + 3r$

From the foregoing we have the following:

RULE

To divide a polynomial by a monomial,

1. Divide each term of the dividend by the divisor.
2. Unite the results with the proper signs obtained by the division.

EXAMPLE 14 Divide $8a^2b^3c - 12a^3b^2c^2 + 4a^2b^2c$ by $4a^2b^2c$.

SOLUTION

$$\frac{8a^2b^3c - 12a^3b^2c^2 + 4a^2b^2c}{4a^2b^2c} = 2b - 3ac + 1$$

EXAMPLE 15 Divide $-27x^3y^2z^5 + 3x^4y^2z^4 - 9x^4y^3z^5$ by $-3x^3y^2z^4$.

SOLUTION

$$\frac{-27x^3y^2z^5 + 3x^4y^2z^4 - 9x^4y^3z^5}{-3x^3y^2z^4} = 9z - x + 3xyz$$

PROBLEMS 4–7

Divide:

1. $12x + 8y$ by 2
2. $18\theta - 12\phi$ by 6
3. $99m^2 - 45n^2$ by 9
4. $35x^6 - 28x^4 + 14x^2$ by $7x^2$
5. $24R_1 + 36R_2 - 60R_3$ by 6
6. $9\Delta^6 - 15\Delta^5 - 21\Delta^2$ by 3Δ
7. $0.025\mu^4\pi^2 + 50\mu^2\pi^4$ by $5\mu\pi^3$
8. $8.1\alpha^3\beta^2\gamma + 7.2\alpha^2\beta\gamma^3 - 3.6\alpha\beta\gamma$ by $0.09\alpha^2\beta^2\gamma^2$
9. $\frac{1}{2}m^2$ into $\frac{3}{20}m^5 - \frac{7}{10}m^3 - \frac{3}{5}m$
10. $-\frac{3}{4}I^2R$ into $\frac{5}{16}I^4R^2 - \frac{3}{8}I^2R + \frac{3}{10}I^{-2}R^{-1} + \frac{3}{4}I^{-4}R^{-2}$
11. $\dfrac{65xyz + 91x^2yz^2 - 39x^3yz^3 - 78x^5y^5z}{13xyz}$
12. $X^2(V + v) + x^2(V + v)$ by $V + v$
13. $8(\theta + \phi)^2 - 16(\theta + \phi)^4 + 12(\theta + \phi)^6$ by $4(\theta + \phi)$
14. $\alpha(\theta^2 + \phi^2)^2 - \beta(\theta^2 + \phi^2)^2$ by $-(\theta^2 + \phi^2)^2$
15. $8\pi(VI + P)^4 - 32\pi(VI + P)^2 + 96\pi(VI + P)$ by $16\pi(VI + P)^2$
16. $\dfrac{6I^2(R + r)(R - r) + 10I^4(R + r)^2(R - r)^2 - 12I^8(R + r)^4(R - r)^4}{-2I(R + r)(R - r)}$
17. $\dfrac{5I\left(\omega L - \frac{1}{\omega C}\right) - 10I^3\left(\omega L - \frac{1}{\omega C}\right)^3 - 25I^5\left(\omega L - \frac{1}{\omega C}\right)^5}{5I^2\left(\omega L - \frac{1}{\omega C}\right)^2}$
18. $\dfrac{(6\alpha + 5)(\alpha + 3) - (4\alpha + 2)(\alpha + 3)^2}{\alpha + 3}$

19. $\dfrac{-36\omega(\theta + \phi)(\theta - \phi) + 12\omega(\theta + \phi)^2(\theta - \phi)^2 - 24\omega(\theta + \phi)^3(\theta - \phi)^3}{4\omega(\theta + \phi)^2}$

20. $\dfrac{54V^2(R + R_1)(r + r_1) + 36V^4(R + R_1)^2(r + r_1)^2 - 108V^6(R + R_1)^3(r + r_1)^3}{9V(R + R_1)(r + r_1)}$

4–14 DIVISION OF ONE POLYNOMIAL BY ANOTHER

RULE

To divide one polynomial by another,

1. Arrange the dividend and divisor in ascending or descending powers of some common literal factor.
2. Divide the first term of the dividend by the first term of the divisor, and write the result as the first term of the quotient.
3. Multiply the entire divisor by the first term of the quotient; write the product under the proper terms of the dividend; and subtract the product from the dividend.
4. Consider the remainder a new dividend, and repeat 1, 2, and 3 until there is no remainder or until there is a remainder that cannot be divided by the divisor.

EXAMPLE 16 Divide $x^2 + 5x + 6$ by $x + 2$.

SOLUTION

Write the divisor and dividend in the usual positions for long division and eliminate the terms of the dividend, one by one:

$$\begin{array}{r|l} & x + 3 \\ \hline x + 2 & x^2 + 5x + 6 \\ & x^2 + 2x \\ \hline & 3x + 6 \\ & 3x + 6 \\ \hline \end{array}$$

x, the first term of the divisor, divides into x^2, the first term of the dividend, x times. Therefore, x is written as the first term of the quotient. The product of the first term of the quotient and the divisor $x^2 + 2x$ is then written under like terms in the dividend and subtracted. The first term of the remainder then serves as a new dividend, and the process of division is continued.

This result can be checked by multiplying the divisor by the quotient.

$$\begin{array}{rl} \text{Divisor} = & x + 2 \\ \text{Quotient} = & x + 3 \\ \hline & x^2 + 2x \\ & 3x + 6 \\ \hline \text{Dividend} = & x^2 + 5x + 6 \end{array}$$

EXAMPLE 17 Divide $a^2b^2 + a^4 + b^4$ by $-ab + b^2 + a^2$.

SOLUTION

First arrange the dividend and divisor according to step 1 of the rule. Because there are no a^3b or ab^3 terms, allowance is made by supplying 0 terms. Thus,

$$\begin{array}{r|l} & a^2 + ab + b^2 \\ \hline a^2 - ab + b^2 & a^4 + 0 + a^2b^2 + 0 + b^4 \\ & \underline{a^4 - a^3b + a^2b^2} \\ & \quad a^3b \\ & \quad \underline{a^3b - a^2b^2 + ab^3} \\ & \qquad a^2b^2 - ab^3 + b^4 \\ & \qquad \underline{a^2b^2 - ab^3 + b^4} \end{array}$$

EXAMPLE 18 Divide $4 + x^4 + 3x^2$ by $x^2 - 2$.

SOLUTION

$$\begin{array}{r|l} & x^2 + 5 \\ \hline x^2 - 2 & x^4 + 3x^2 + 4 \\ & \underline{x^4 - 2x^2} \\ & \quad 5x^2 + 4 \\ & \quad \underline{5x^2 - 10} \\ & \qquad\quad 14 = \text{remainder} \end{array}$$

The result is written $x^2 + 5 + \dfrac{14}{x^2 - 2}$

which is as it would be written in an arithmetical division that did not divide out evenly.

PROBLEMS 4–8

Divide:

1. $a^2 + 2a + 1$ by $a + 1$
2. $12m^2 + 10m - 12$ by $3m - 2$
3. $21x^2 + 22x - 8$ by $7x - 2$
4. $12Q^2 + 23Q + 10$ by $3Q + 2$
5. $8E^2 - 22E + 12$ by $2E - 4$
6. $\dfrac{15\theta^2 - 31\theta\varepsilon + 10\varepsilon^2}{3\theta - 5\varepsilon}$
7. $3R^3 + 9R^2 - 7R - 4RZ - 12Z - 21$ by $R + 3$
8. $2\phi^3 + 2\phi\omega^2 + 2\phi^2\omega + 2\omega^3$ by $\phi + \omega$
9. $E^3 + 5E^2 - 11E + 5$ by $E - 1$
10. $20k^3 - 23k^2\lambda - 18k\lambda^2 + 18\lambda^3$ by $4k - 3\lambda$

11. $E^2 - e^2$ by $E - e$
12. $E^3 - e^3$ by $E - e$
13. $V^4 - v^4$ by $V - v$
14. $L^3 + L^2IX + LI^2X^2 + I^3X^3$ by $L + IX$
15. $V^4 - I^4R^4$ by $V^2 - I^2R^2$
16. $\theta^5 + \phi^5$ by $\theta + \phi$
17. $X^6 - Y^6$ by $X + Y$
18. $\alpha^6 + \beta^6$ by $\alpha^4 - \alpha^2\beta^2 + \beta^4$
19. $x^3 + 3x^2y + 3xy^2 + y^3$ by $x + y$
20. $P^4 - P^3Q - 4P^2Q^2 - PQ^3 + Q^4$ by $P + Q$
21. $6R_2^3 - R_2^2 - 14R_2 + 3$ by $3R_2^2 + 4R_2 - 1$
22. $1 + 2m^4 + 4m^2 - m^3 + 7m$ by $3 + m^2 - m$
23. $24x^4 - 11x + 230x^2 + 51 - 124x^3$ by $2x^2 + 3 - 5x$
24. $\frac{1}{8}\theta^3 - \frac{9}{4}\theta^2\phi + \frac{27}{2}\theta\phi^2 - 27\phi^3$ by $\frac{1}{2}\theta - 3\phi$
25. $6R^2 - \frac{5}{6}R - \frac{1}{6}$ by $2R - \frac{1}{2}$
26. $n^3 - \frac{9}{5}n^2 - \frac{9}{25}n - \frac{27}{125}$ by $n - \frac{3}{5}$
27. $36x^2 + \frac{1}{9}y^2 + \frac{1}{4} - 4xy - 6x + \frac{1}{3}y$ by $6x - \frac{1}{3}y - \frac{1}{2}$
28. $\frac{1}{27}K^3 - \frac{1}{12}K^2 + \frac{1}{16}K - \frac{1}{64}$ by $\frac{1}{3}K - \frac{1}{4}$
29. $\frac{3}{2}L_1^2 - L_1 - \frac{8}{3}$ into $\frac{9}{16}L_1^4 - \frac{3}{4}L_1^3 - \frac{7}{4}L_1^2 + \frac{4}{3}L_1 + \frac{16}{9}$
30. $R_1^7 + \left(\frac{V}{I}\right)^7$ by $R_1 + \frac{V}{I}$

SELF TEST

1. Multiply 6 by -7 -42
2. Multiply (-2π) by $(-fL)$ $2\pi fL$
3. Multiply $\dfrac{3}{16}$, $-\dfrac{7}{30}$, and $\dfrac{4}{21}$ $\dfrac{1}{120}$

 $15I^2R - 10PIV^2$

4. Multiply $3I^2R - 2PV$ by $5IV$
5. Multiply $(\theta + 2)$ by $(\theta - 7)$ θ^2
6. Divide (-35) by (5) -7
7. Divide $45\lambda^2\Omega$ by $-9\lambda\Omega^2$
8. Divide:
 $$24P^3(L + X)(R + C) - 32P^2(L + X)(R + C) + 48P(L + X)(R + C)$$
 by $8P(X + L)(C + R)$
9. Divide $\theta^4 + \theta^2\phi^2 + \theta\phi^3 - \theta^3\phi - \phi^4$ by $\theta - \phi$
10. Divide $-10Y^4 - 31XY^3 + 29X^2Y^2 + 4X^3Y - 4X^4$ by $2X - 5Y$

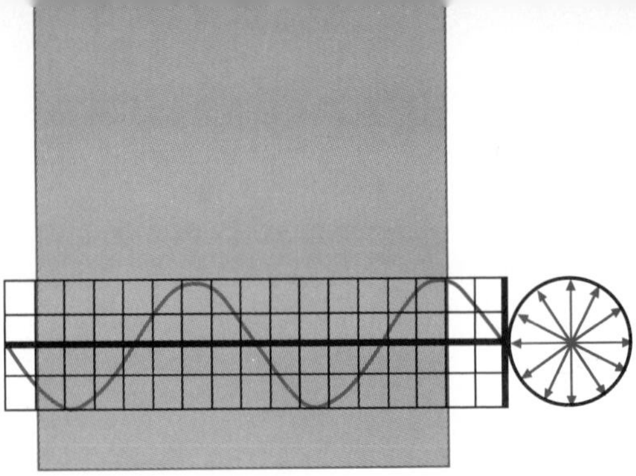

CHAPTER 5

EQUATIONS

In the preceding chapters, considerable time has been spent in the study of the fundamental operations of algebra. These fundamentals will be of little value unless they can be put to practical use in the solution of problems. This is accomplished by use of the equation, the most valuable tool in mathematics.

5–1 DEFINITIONS

An *equation* is a mathematical statement that two numbers, or quantities, are equal. The *equality sign* (=) is used to separate the two equal quantities. The terms to the left of the equality sign are known as the *left member* of the equation, and the terms to the right are known as the *right member* of the equation. For example, in the equation

$$3E + 4 = 2E + 6$$

$3E + 4$ is the left member and is equal to $2E + 6$, which is the right member.

An *identical equation*, or *identity*, is an equation whose members are equal for all values of the literal numbers contained in the equation. The equation

$$4I(r + R) = 4Ir + 4IR$$

is an identity because if $I = 2$, $r = 3$, and $R = 1$, then

$$4I(r + R) = 4 \cdot 2(3 + 1) = 32$$

Also,

$$\begin{aligned} 4Ir + 4IR &= 4 \cdot 2 \cdot 3 + 4 \cdot 2 \cdot 1 \\ &= 24 + 8 = 32 \end{aligned}$$

Any other values of I, r, and R substituted in the equation will produce equal numerical results in the two members of the equation.

An equation is said to be *satisfied* if, when numerical values are substituted for the literal numbers, the equation becomes an identity. Thus, the equation

$$ir - iR = 3r - 3R$$

is satisfied by $i = 3$, because when we substitute this value in the equation, we obtain

$$3r - 3R = 3r - 3R$$

which is an identity.

A *conditional equation* is one consisting of one or more literal numbers that is not satisfied by all values of the literal numbers. Thus, the equation

$$e + 3 = 7$$

is not satisfied by any value of e except $e = 4$.

To *solve* an equation is to find the value or values of the unknown number that will satisfy the equation. This value is called the *root* of the equation. Thus, if

$$i + 6 = 14$$

the equation becomes an identity only when i is 8, and therefore 8 is the root of the equation.

5–2 AXIOMS

An *axiom* is a truth, or fact, that is self-evident and needs no formal proof. The various methods of solving equations are derived from the following axioms:

1. If equal numbers are added to equal numbers, the sums are equal.

EXAMPLE 1

If		$x = x$
then		$x + 2 = x + 2$
because, if	$x = 4,$	$4 + 2 = 4 + 2$
or		$6 = 6$

Therefore, *the same number can be added to both members of an equation without destroying the equality.*

2. If equal numbers are subtracted from equal numbers, the remainders are equal.

EXAMPLE 2

If		$x = x$
then		$x - 2 = x - 2$
because, if	$x = 4,$	$4 - 2 = 4 - 2$
or		$2 = 2$

Therefore, *the same number can be subtracted from both members of an equation without destroying the equality.*

3. If equal numbers are multiplied by equal numbers, their products are equal.

EXAMPLE 3 If $x = x$
then $3x = 3x$
because, if $x = 4$, $3 \cdot 4 = 3 \cdot 4$
or $12 = 12$

Therefore, *both members of an equation can be multiplied by the same number without destroying the equality.*

4. If equal numbers are divided by equal numbers, their quotients are equal.

EXAMPLE 4 If $x = x$
then $\frac{x}{2} = \frac{x}{2}$
because, if $x = 4$, $\frac{4}{2} = \frac{4}{2}$
or $2 = 2$

Therefore, *both members of an equation can be divided by the same number without destroying the equality, except that division by zero is not allowed.*

5. Numbers that are equal to the same number or equal numbers are equal to each other.

EXAMPLE 5 If $a = x$ and $b = x$
then $a = b$
because, if $x = 4$, $a = 4$
and $b = 4$

Therefore, *an equal quantity can be substituted for any term of an equation without destroying the equality.*

6. Like powers of equal numbers are equal.

EXAMPLE 6 If $x = x$
then $x^3 = x^3$
because, if $x = 4$, $4^3 = 4^3$
or $64 = 64$

Therefore, *both members of an equation can be raised to the same power without destroying the equality.*

7. Like roots of equal numbers are equal.

EXAMPLE 7 If $x = x$
then $\sqrt{x} = \sqrt{x}$
because, if $x = 4$, $\sqrt{4} = \sqrt{4}$
or $2 = 2$

Therefore, *like roots of both members of an equation can be extracted without destroying the equality.*

5–3 NOTATION

In order to shorten the *explanations* of the solutions of various equations, we shall employ the letters **A, S, M,** and **D** for "add," "subtract," "multiply," and "divide," respectively. Thus,

- **A:** 6 will mean "add 6 to both members of the equation."
- **S:** $-6x$ will mean "subtract $-6x$ from both members of the equation."
- **M:** $-3a$ will mean "multiply both members of the equation by the factor $-3a$"
- **D:** 2 will mean "divide both members of the equation by 2."

5–4 THE SOLUTION OF EQUATIONS

A considerable amount of time and drill must be spent in order to become proficient in the solution of equations. It is in this branch of mathematics that you will find you must be familiar with the more elementary parts of algebra.

Some of the methods used in the solutions are very easy—so easy, in fact, that there is a tendency to employ them mechanically. This is all very well, but you should not become so mechanical that you forget the reason for performing certain operations.

We shall begin the solution of equations with very easy cases and attempt to build up general methods of procedure for all equations as we proceed to the more difficult problems.

If you are studying equations for the first time, you are urged to study the following examples carefully until you thoroughly understand the methods and the reasons behind them.

EXAMPLE 8 Find the value of x if $x - 3 = 2$.

SOLUTION

In this equation, it is seen by inspection that x must be equal to 5. However, to make the solution by the methods of algebra, proceed as follows:

Given	$x - 3 = 2$	
A: 3,	$x = 2 + 3$	(Axiom 1)
Collecting terms,	$x = 5$	

EXAMPLE 9 Solve for e if $e + 4 = 12$.

SOLUTION

Given	$e + 4 = 12$	
S: 4,	$e = 12 - 4$	(Axiom 2)
Collecting terms,	$e = 8$	

EXAMPLE 10 Solve for i if $3i + 5 = 20$.

SOLUTION

Given	$3i + 5 = 20$	
S: 5,	$3i = 20 - 5$	(Axiom 2)
Collecting terms,	$3i = 15$	
D: 3,	$i = 5$	(Axiom 4)

EXAMPLE 11 Solve for r if $40r - 10 = 15r + 90$.

SOLUTION

Given	$40r - 10 = 15r + 90$	
S: $15r$,	$40r - 10 - 15r = 90$	(Axiom 2)
A: 10,	$40r - 15r = 90 + 10$	(Axiom 1)
Collecting terms,	$25r = 100$	
D: 25	$r = 4$	(Axiom 4)

From the foregoing examples, it will be noted that adding or subtracting a term from both members of an equation is equivalent to *transposing* that number from one member to the other and changing its sign. This fact leads to the following rule:

RULE

A *term* can be transposed from one member of an equation to the other provided that its sign is changed.

By transposing all terms containing the unknown to the left member and all others to the right member and then collecting terms and dividing both members by the numerical coefficient of the unknown, the equation has been solved for the value of the unknown.

5–5 CANCELING TERMS IN AN EQUATION

EXAMPLE 12 Solve for x if $x + y = z + y$.

SOLUTION

Given	$x + y = z + y$	
S: y,	$x = z$	(Axiom 2)

The term y in both members of the given equation does not appear in the next equation as the result of subtraction. The result is the same as if the term were dropped from both members. This fact leads to the following rule:

RULE

If the same *term* preceded by the same sign occurs in both members of an equation, it can be canceled.

5–6 CHANGING SIGNS IN AN EQUATION

EXAMPLE 13 Solve for x if $8 - x = 3$.

SOLUTION

Given	$8 - x = 3$	
S: 8,	$-x = 3 - 8$	(Axiom 2)
M: -1	$x = -3 + 8$	(Axiom 3)
Collecting terms,	$x = 5$	

Note that multiplication by -1 has the effect of changing the signs of all terms. This gives the following rule:

RULE

The signs of all the *terms* of an equation can be changed without destroying the equality.

Although the foregoing rules involving mechanical methods are valuable, you should not lose sight of the fact that they are all derived from fundamentals, or axioms, as outlined in Sec. 5–2.

5–7 CHECKING THE SOLUTION

If there is any doubt that the value of the unknown is correct, the solution can be checked by substituting the value of the unknown in the original equation. If the two members reduce to an identity, the value of the unknown is correct.

EXAMPLE 14 Solve and test $3i + 14 + 2i = i + 26$.

SOLUTION

Given	$3i + 14 + 2i = i + 26$
Transposing,	$3i + 2i - i = 26 - 14$
Collecting terms,	$4i = 12$
D: 4,	$i = 3$

Test by substituting $i = 3$ in given equation.

CHECK

$$(3 \cdot 3) + 14 + (2 \cdot 3) = 3 + 26$$
$$9 + 14 + 6 = 3 + 26$$
$$29 = 29$$

PROBLEMS 5–1

Solve for the unknown in the following equations:

1. $4x - 8 = 8$ 8+8=16 x
2. $5\theta - 2 = 4\theta + 2$
3. $a - 8 = 4a - 26$
4. $3L - 2 = 2L + 1$
5. $6\theta - 27 = 2\theta - 3$
6. $27\lambda - 2 = 5\lambda + 6$
7. $11\pi - 22 = 4\pi + 13$
8. $5M + 2 = 3 + 4M$
9. $21 - 15\omega L = -8\omega L - 7$
10. $35Q + 16 = 61 + 20Q$
11. $8\alpha - 5(4\alpha + 3) = -3 - 4(2\alpha - 7)$
12. $3(\lambda - 2) - 10(\lambda - 6) = 5$
13. $5 + 2(V - 3) = 3(2V - 3) - 24$
14. $3(L - 3) - 2(L - 4) = 2(L - 4)$
15. $0 = 18 - 4Q + 27 + 9Q - 3 + 16Q$
16. $25R_1 - 19 - [3 - (4R_1 - 15)] - 3R_1 + (6R_1 + 21) = 0$
17. $19 - 5I(4I + 1) = 40 - 10I(2I - 1)$
18. $(\phi + 5)(\phi - 4) + 4\phi^2 = (5\phi + 3)(\phi - 4) + 2(\phi - 4) + 64$
19. $6(\beta - 1)(\beta - 2) - 4(\beta + 2)(\beta + 1) = 2(\beta + 1)(\beta - 1) - 24$
20. $18 - 3Z(2Z + 1) - [3 - 2(Z + 2)(Z - 3)]$
    $= 18 - 6Z - 4(Z - 5)(Z + 2)$

5–8 FORMING AND SOLVING EQUATIONS

As previously stated, we are continually trying to express certain laws and relations in the language of mathematics.

EXAMPLES

Words	Algebraic Symbols
The sum of the voltages V and v	$V + v$
The difference between resistances R and R_1	$R - R_1$
The excess of current I_1 over current I_2	$I_1 - I_2$
The number of centimeters in l meters	$100l$
The number of cents in d dollars	$100d$
The voltage V is equal to the product of the current I and the resistance R	$V = IR$

The solution of most problems consists in writing an equation that connects various observed data with known facts. This, then, is nothing more than translating from ordinary English, or words, into the symbolic language of mathematics. In relatively simple problems the translation can be made directly, almost word by word, into algebraic symbols (Examples 15 and 16).

It is almost impossible to lay down a set of rules for the solution of general problems, for they could not be made applicable to all cases. However, no rules will be needed if you thoroughly understand what is to be translated into the language of mathematics from the wording or facts of the problem at hand. The following outline will serve as a guide:

1. Read the problem carefully so that you understand every fact in it and recognize the relationships between the facts.
2. Determine what is to be found (the unknown quantity), and denote it by some letter. If there are two unknowns, try to represent one of them in terms of the other. If there are more than two unknowns, try to represent all but one of them in terms of that one.
3. Find two expressions which, according to the facts of the problem, represent the same quantity, and set them equal to each other. You can then solve the resulting equation for the unknown.

EXAMPLE 15 Five times a certain voltage diminished by 3,
$5 \quad \times \quad V \quad - \quad 3$
gives the same result as the voltage increased by 125.
$= \quad V \quad + \quad 125$

That is $$5V - 3 = V + 125$$
or $$V = 32\ \text{V}$$

EXAMPLE 16 What number increased by 42 is equal to 110?
$x \quad + \quad 42 \quad = \quad 110$

That is, $$x + 42 = 110$$
or $$x = 68$$

CHECK

$68 + 42 = 110$

PROBLEMS 5–2

1. The sum of two voltages is p V. One voltage is 120 V. What is the other?
2. The difference between two resistances is 68.8 Ω. One resistance is R Ω. What is the other?
3. How great a distance d will you travel in t hours (h) at r kilometers per hour (km/h)?
4. What is the fraction f whose numerator n is 8 less than its denominator?
5. An electronics timer has a guarantee of x years. We have been using it for u years. For how many years longer will the guarantee apply?

6. An oscilloscope is guaranteed for ℓ years, and it has been in service for s months. How much longer is it covered by the guarantee?
7. At what speed must a missile be traveling to cover D km in m min?
8. From what number must 8 be subtracted in order that the remainder is 27?
9. If a certain voltage is doubled and the result is diminished by 30, the remainder is 410 V. What is the voltage?
10. The volume of a parts container is v cubic centimeters (cm^3). Express the height in centimeters if the width is w cm and the length is l cm.
11. Write algebraically that the current is equal to the voltage divided by the resistance.
12. Write algebraically that the power dissipated by a resistor is equal to the square of the current multiplied by the resistance.
13. A stockroom is twice as long as it is wide, and its perimeter is 96 m. Find its length and width.
14. A multimeter and an oscilloscope together cost \$574. The oscilloscope costs \$356 more than the meter. Find the cost of (*a*) the oscilloscope and (*b*) the multimeter.
15. Find the lengths of three sides of a triangle whose perimeter is 180 m if the longest side is one and one-third times the second side, which is one and one-half times the shortest side.
16. The sum of the three angles in any triangle is 180°. The smallest angle in a given triangle is one-half the second angle and 52° smaller than the largest angle. How many degrees does each angle contain?
17. Write algebraically that the square on the hypotenuse h of a right triangle is equal to the sum of the squares on the other two sides, which are identified as a and b.
18. The sum of two consecutive numbers is 31. What are the numbers?
19. The sum of three consecutive numbers is 192. What are the numbers?
20. Write algebraically that the product of the impressed emf V and the resultant current I in a circuit is equivalent to the square of the emf divided by R, the resistance in the circuit.

5–9 LITERAL EQUATIONS—FORMULAS

A *formula* is a rule, or law, generally pertaining to some scientific relationship expressed as an equation by means of letters, symbols, and constant terms.

EXAMPLE 17 The area A of a rectangle is equal to the product of its base b and its altitude h. This statement written as a formula is

$$A = bh$$

EXAMPLE 18 The power P expended in an electric circuit is equal to its current I squared times the resistance R of the circuit. Stated as a formula

$$P = I^2R$$

The ability to handle formulas is of the utmost importance. The usual formula expresses one quantity in terms of other quantities, and it is often desirable to solve for *any* quantity contained in a formula. This is readily accomplished by using the knowledge gained in solving equations.

EXAMPLE 19 The voltage V across a part of a circuit is given by the current I through that part of the circuit times the resistance R of that part. That is,

$$V = IR$$

Suppose V and I are given but it is desired to find R.

Given	$V = IR$	
D: I,	$\frac{V}{I} = R$	(Axiom 4)
or	$R = \frac{V}{I}$	

Similarly, if we wanted to solve for I,

Given	$V = IR$	
D: R,	$\frac{V}{R} = I$	(Axiom 4)
or	$I = \frac{V}{R}$	

EXAMPLE 20 Solve for I if $v = V - IR$.

SOLUTION

Given	$v = V - IR$	
Transposing,	$IR = V - v$	
D: R,	$I = \frac{V - v}{R}$	(Axiom 4)

EXAMPLE 21 Solve for C if $X_C = \frac{1}{2\pi fC}$

SOLUTION

Given	$X_C = \frac{1}{2\pi fC}$	
D: X_C,	$1 = \frac{1}{2\pi fCX_C}$	(Axiom 4)
M: C,	$C = \frac{1}{2\pi fX_C}$	(Axiom 3)

It will be noted from the foregoing examples that if the numerator of a member of an equation contains but one term, any *factor* of that term may be transferred to the denominator of the other member as a *factor*. In like manner if the denominator of a member of an equation contains but one term, any *factor* of that term may be transferred to the numerator of the other member as a *factor*. These mechanical transformations simply make use of Axioms 3 and 4, and you should not lose sight of the real reasons behind them.

PROBLEMS 5–3

Given:	Solve for:
1. $C = \frac{Q}{V}$	Q, V
2. $V = IZ$	I, Z
3. $Z^2 = R^2 + X^2$	R^2 and X^2
4. $P = I^2R$	I^2 and R
5. $K = \frac{Rm}{L}$	L, R, and m
6. $R_t = R_1 + R_2 + R_3$	R_1, R_2, and R_3
7. $\lambda = \frac{v}{f}$	f and v
8. $C = 2\pi r$	r and π
9. $Q = \frac{\omega L}{R}$	L, R, and ω
10. $X_C = \frac{1}{2\pi fC}$	C and f
11. $f = \frac{1}{2\pi CX_C}$	C and X_C
12. $S = 2\pi rh$	r and h
13. $H = \frac{\phi}{A}$	ϕ and A
14. $N_s = \frac{V_sN_p}{V_p}$	N_p, V_s, and V_p
15. $BLv = V(10^8)$	B, L, and V
16. $A = \frac{T - ph}{2}$	T and h
17. $V_s = \frac{V_pI_p}{I_s}$	I_p
18. $H = \frac{F}{Li}$	F, L, and i

19. $I = \dfrac{V - v}{R}$	R, V, and v
20. $\mu = g_m r_p$	g_m
21. $t = \dfrac{\theta}{\omega}$	θ and ω
22. $h = \dfrac{V^2}{2g}$	g and V^2
23. $V_0 = 2V - V_t$	V and V_t
24. $n = \dfrac{\omega}{2\pi}$	ω
25. $3A = 4\pi r^3$	A
26. $\mu = \dfrac{B^2Al}{8\omega}$	l, ω, and A
27. $C = \dfrac{F(R - r)}{Z_t}$	Z_t, F, R, and r
28. $r = \dfrac{F}{4\pi^2 n^2 m}$	m and F
29. $R_L = \dfrac{V_b - v_b}{i}$	V_b, v_b and i
30. $t = \dfrac{T(C - F)}{C}$	T and F
31. $R = \dfrac{\rho l}{d^2}$	l, ρ, and d^2
32. $X = \dfrac{R}{PF}$	R and PF
33. $C = \dfrac{0.0884KA(n - 1)}{d}$	A and n
34. $M = k\sqrt{L_1 L_2}$	k
35. $Z_r = \dfrac{L}{RC}$	L, C, and R
36. $F = \dfrac{vI}{2kT_g}$	T_g
37. $\alpha = \dfrac{\eta\beta}{\gamma\omega}$	ω
38. $\dfrac{P_{so}}{P_{no}} = \dfrac{P_{si}}{P_{ni}}$	P_{no}
39. $\rho = \dfrac{Qe}{hv}$	Q

40. $V_b = \dfrac{V_B + V_{pt}}{W}$ $\quad V_{pt}$

41. $V_2 = V_3(1 - \omega^2 LC_2)$ $\quad C_2$

42. $4a = \dfrac{h + 2b}{v}$ $\quad b$

43. $Q = I_p p + I_n n$ $\quad I_n$

44. $g_m = (G_o - G)(1 + n)$ $\quad G_o$

45. $R_1 = \dfrac{1}{\omega_{01} C} - R_2$ $\quad \omega_{01}$

NOTE *When solving numerical problems which involve the solution of formulas, always solve the formula algebraically for the wanted factor before substituting the numerical values. This procedure permits you to check your work more easily. The reason is that the letters retain their identity through the various algebraic procedures, whereas once numbers are added, multiplied, etc., their identity is lost and your audit becomes more difficult.*

46. The power P in watts in any part of an electric circuit is given by

$$P = \frac{V^2}{R} \quad W$$

in which V is the emf applied to that part of the circuit and R is the resistance of that part. What is the resistance of a circuit in which 114 W is expended at an emf of 117 V?

47. The voltage drop V across any part of a circuit can be computed by the formula $V = IZ$ V, where I is the current in amperes through that part of the circuit and Z is the impedance in ohms of that part. Give the impedance of a circuit in which a voltage drop of 460 V is produced by a current flow of 0.115 A.

48. To find the frequency f of an alternator in hertz (Hz), that is, cycles per second, the number of pairs of poles P is multiplied by the speed of the armature S in revolutions per second (rev/s), $f = PS$. A tachometer connected to the rotor of a 60-Hz alternator reads 3600 revolutions per minute (rev/min). How many poles has the alternator?

49. For radio waves, the relationship between frequency f in megahertz and wavelength λ in meters is expressed by the formula

$$f = \frac{3 \times 10^2}{\lambda} \text{ MHz}$$

What is the wavelength of a radio wave at 60 MHz?

50. The length of a broadband dipole L_{fD} used for television reception can be computed by the formula $L_{fD} = \dfrac{141.3}{f}$ m, where f is the frequency in megahertz. The folded dipole shown in Fig. 5–1 is 0.7976 m long. For what frequency was it constructed?

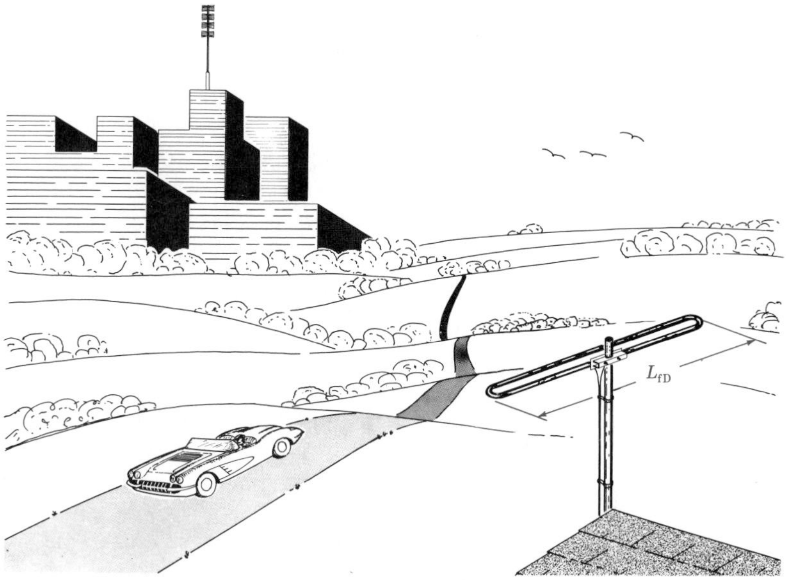

FIG. 5–1 Folded dipole of Prob. 50.

5–10 RATIO AND PROPORTION

Because proportions are special forms of equations, it is expedient to look now at the twin subjects ratio and proportion.

A ratio is a comparison of two things expressed in one of two ways: first, the "old-fashioned" method, *a:b,* pronounced "*a* is to *b*"; and second, as found in newer books, $\frac{a}{b}$. If the ratio of *x* to *y* is 1 to 4, or $\frac{1}{4}$, then *x* is one-quarter of *y*. Alternatively, *y* is four times as great as *x*.

EXAMPLE 22 Write the ratio of 25 cents (¢) to $3.00.

SOLUTION

25¢ to $3.00 may be written simply as 25¢:$3.00, but this does not tell us much. It is more helpful to convert both quantities to the same units:

$$\frac{25¢}{\$3.00} = \frac{25¢}{300¢} = \frac{1}{12}$$

Note that the two parts being compared are given the same units, in this case cents. Therefore, when the simplification is performed, not only the numbers but also the units are canceled. Thus a *true ratio* is a "pure," or dimensionless, number. Note also that a ratio may be an integer, that is, a fraction whose denominator is 1.

PROBLEMS 5–4

Write as a fraction the ratio of:

1. 12 mm to 60 mm
2. 12 square meters (m^2) to 18 m^2

3. 18 000 to 12 000
4. \$5.00 to 25¢
5. Write two different sets of numbers in the ratio 2:7.
6. Write two different sets of numbers in the ratio 0.125:1.
7. A recipe for ceramic insulators calls for 8 parts of type *A* clay to 24 parts of type *B* by mass. What is the ratio of type *A* to type *B*?
8. In Prob. 7, what is the ratio of the weight of type *A* to the weight of the total mixture?
9. The mechanical advantage (MA) of any machine is the ratio of load moved to effort applied. What is the MA of a system in which 32 kilograms (kg) of effort just starts motion of a 960-kg load?
10. In a certain alloy, 75% of the material is copper and 15% is zinc. What is the ratio of zinc to copper?

Just as ratios compare two things, so proportions are equalities of pairs of ratios.

When we draw a map to scale, the proportions on the map should equal those on the ground. If the scale is 1 cm to 10 km, then a trip which is 3 cm on the map must be 30 km on the ground. The proportion here is $\frac{1\text{ cm}}{3\text{ cm}} = \frac{10\text{ km}}{30\text{ km}}$ and, since the units cancel, our true proportion is an equality of two pure numbers. We could also write this proportion as $\frac{1\text{ cm}}{10\text{ km}} = \frac{3\text{ cm}}{30\text{ km}}$. Note that it is essential that the units on one side of the proportion be equal to those on the other side. This provides one good way of checking your work. If you perform a wrong operation such as multiplying instead of dividing, you will find that your units will reveal an error. The solution may read "cm/km = cm · km," and such an imbalance of units is a sure indication that you have made an error.

The usual purpose of proportions is to solve for one part when the other three parts are known.

EXAMPLE 23 Given the proportion $\frac{18}{a} = \frac{6}{5}$, solve for a.

SOLUTION

Obeying the usual rules of equations,

$$18 \times 5 = 6 \times a$$
$$a = 15$$

In the older form of writing ratios and proportions, $\frac{a}{b} = \frac{c}{d}$ would be written $a:b::c:d$ and pronounced "a is to b as c is to d." The elements on the outsides of the proportion were called the "extremes," and those in the middle the "means." Based on these definitions, you can prove the old law of proportions:

RULE

In a proportion, the product of the means equals the product of the extremes.

PROBLEMS 5–5

Find the missing term in each of the following proportions:

1. $\frac{3}{8} = \frac{?}{16}$
2. $\frac{3}{7} = \frac{?}{84}$
3. $\frac{?}{4} = \frac{60}{3}$
4. $\frac{80}{?} = \frac{60}{12}$
5. $\frac{X}{25} = \frac{70}{350}$
6. $\frac{q}{900} = \frac{0.2}{36}$
7. $\frac{3}{V} = \frac{8}{12}$
8. $\frac{0.6}{1.2} = \frac{0.4}{d}$
9. $\frac{0.007}{0.200} = \frac{Q}{0.04}$
10. $\frac{16}{Z} = \frac{8}{4}$

5–11 VARIATION AND PROPORTIONALITY

Often, in the study of electronics, you will hear such expressions as "the current is proportional to the voltage and inversely proportional to the resistance" and "the force is jointly proportional to the charges and inversely proportional to the square of the distance between them."

Sometimes an equivalent expression is used: "the current varies directly as the voltage," etc.

Two forms may be used to express mathematically the words "the current varies as the voltage." The first uses the symbol of proportionality: $I \propto V$. The second substitutes for the symbol $\propto$ the equivalent "$= k$," where k is the "konstant" of proportionality: $I = kV$. Other symbols such as b, c, n, etc., also are used as constants.

Similarly, the expression "the current is inversely proportional to the resistance" may be written $I \propto \frac{1}{R}$ or $I = k\frac{1}{R}$ or simply $I = \frac{k}{R}$.

"Jointly proportional" means "proportional to the product," so that "the force is jointly proportional to the masses" may be written $F \propto m_1m_2$ or $F = km_1m_2$.

Often, past experience, tables, and measurements, as well as calculations, may reveal the value of the constant of proportionality. For example, we know that the circumference of a circle is proportional to the radius. We may write this $C \propto R$ or $C = kR$. However, from previous knowledge, we can replace the general constant k by the known constant of proportionality 2π, and we can write $C = 2\pi R$.

EXAMPLE 24 If a varies directly as ρ and if $a = 8$ when $\rho = 4$, what will be the value of a when $\rho = 7$?

SOLUTION

$a \propto \rho = k\rho$. We know that $8 = k4$, from which $k = 2$. Substitute this value of k into the second condition:

$$a = k \times 7 = 2 \times 7 = 14$$

PROBLEMS 5–6

Write the following expressions in "proportionality" form and in "equation" form:

1. The current I varies directly as the voltage V.
2. The cost C varies directly as the weight W.
3. The capacitance C varies directly as the area A.
4. The reactance X_L varies jointly with the frequency f and the inductance L.
5. The capacitive reactance X_C varies inversely as the capacitance C.
6. The resistance varies directly as the length l and inversely as the cross-sectional area A.
7. The period T of vibration of a reed is directly proportional to the square root of the length l.
8. The volume of a sphere V is proportional to the cube of the radius r.
9. The pressure P on a gas varies inversely as the volume V.
10. The ratio of similar volumes V_1 and V_2 is proportional to the cube of the ratio of corresponding lengths l_1 and l_2.
11. The illumination L of an object varies inversely as the square of the distance d from the source of light.
12. If the current I varies directly as the voltage V and if $I = 1.2$ A when $V = 60$ V, what will be the value of I when $V = 85$ V?
13. In a certain varistor the current is proportional to the square of the voltage. If $I = 0.003$ A when $V = 90$ V, what voltage will produce a current of 2.25 A?
14. The resistance R of a wire varies directly as the wire length l and inversely as the square of the wire diameter d. If $R = 5.21\ \Omega$ when $l = 1$ km and if $d = 2.05$ mm, what will be the resistance of a 500-m length of wire 2.588 mm in diameter?
15. The load that a beam of given thickness can carry safely is directly proportional to the beam's width and inversely proportional to its length. If a beam 10 m long and 50 mm wide can support 9000 kg, what is the load that could be supported by a beam of identical thickness 25 m long and 75 mm wide?

SELF TEST

1. Solve for m: $5m = 2m + 9$
2. Solve for x: $6 + x(x + 5) = 4 + (x + 6)(x + 7)$
3. Solve for t: $3.5t - 1.84 = 1.6t + 1.22$
4. Solve for R:
   $(R + 3)(R - 5) + 5R^2 = (R + 2)(R - 5) + (R - 3)(5R - 7)$
5. Write the algebraic statement that the resistance at a second temperature R_2 is found by multiplying the resistance at 0°C R_0 by the sum of unity and the product of the temperature coefficient α and the change in temperature Δt.
6. A storeroom is half as wide as it is long. Its perimeter is 60 m. What are the dimensions of the storeroom?
7. Three times the sum of two consecutive numbers is 99. What are the numbers?
8. Solve for t_2: $\dfrac{R_1}{R_2} = \dfrac{t_1 + 234.5}{t_2 + 234.5}$
9. Solve for the missing term: $\dfrac{8}{24a} = \dfrac{?}{36a^2}$
10. The ratio of "similar areas" is equal to the square of the ratio of corresponding sides. If the area of Figure B is sixteen times the area of Figure A, how long is the longest side of Figure B compared to the longest side of Figure A?

CHAPTER 6

POWERS OF 10

Very few people enjoy performing numerical computations simply for the joy of "figuring." Practical people want concrete answers; therefore, they should use whatever tools or devices are available to help arrive at those answers with a minimum expenditure of time and effort. Engineers would normally simplify their calculations, especially when they are dealing with large numbers. Such a simplification device is the use of powers of 10.

6–1 SIGNIFICANT FIGURES

In mathematics, a number is generally considered as being exact. For example, 220 would mean 220.0000, etc., for as many added zeros as desired. A meter reading, however, is always an *approximation*. We might read 220 V on a certain switchboard type of voltmeter, whereas a precision instrument might show that voltage to be 220.3 V and a series of precise measurements might show it to be 220.36 V. It should be noted that the position of the decimal point does not determine the accuracy of a number. For example, 115 V, 0.115 kV, and 115 000 mV are of identical value and are equally accurate.

Any number representing a measurement, or the amount of some quantity, expresses the accuracy of the measurement. The figures required are known as *significant figures*.

The *significant figures* of any number are the figures 1, 2, 3, 4, . . . , 9, in addition to such ciphers, or zeros, as may occur between them or to their right or as may have been retained in properly rounding them off. Thus, the number of significant figures is an indication of *precision,* not *accuracy.*

EXAMPLES

0.002 36	is correct to *three* significant figures.
3.141 59	is correct to *six* significant figures.
980 000.0	is correct to *seven* significant figures.
24.	is correct to *two* significant figures.
24.0	is correct to *three* significant figures.
0.025 00	is correct to *four* significant figures.

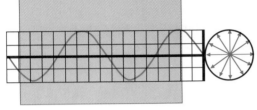

PROBLEMS 6–1

To how many significant figures have the following numbers been expressed?

1. 5.246 37
2. 0.000 003 14
3. 500 000
4. 23.0055
5. 7.00
6. 1
7. 0.000 01
8. 6.28
9. 0.000 025 38
10. 2 726.375

6–2 ROUNDED NUMBERS

A number is *rounded off* by dropping one or more figures at its right. If the last figure dropped is 6 or more, we increase the last figure retained by 1. Thus 3867 would be rounded off to 3870, 3900, or 4000. If the last figure dropped is 4 or less, we leave the last figure retained as it is. Thus 5134 would be rounded off to 5130, 5100, or 5000. If the last figure dropped is 5, add 1 if it will make the last figure retained *even;* otherwise, do not. Thus, $55.7\not{5} = 55.8$, but $67.6\not{5} = 67.6$.

> Your hand-held calculator probably always rounds up if the dropped figure is 5 or more. Thus, 67.65 becomes 67.7.

> Your computer probably always rounds *down:* INT(3.87) will probably yield 3, *not* 4
>
> Negative numbers will also probably round "down" to larger negative whole numbers. INT(−3.76) will probably yield −4

6–3 DECIMALS

An important consideration arises in making computations involving decimals. Electrical engineers and particularly electronics engineers are, unfortunately, required to handle cumbersome numbers ranging from extremely small fractions of electrical units to very large numbers, as represented by radio frequencies. The fact that these wide limits of numbers are encountered in the same problem does not simplify matters. This situation is becoming more complicated owing to the trend to the higher microwave frequencies with attendant smaller fractions of units represented by integrated circuit components.

The problem of properly placing the decimal point and thus reducing unnecessary work presents little difficulty to the person who has a working knowledge of the powers of 10.

Of course, using a calculator will place the decimal point accurately, leaving no room for confusion. More on calculators may be found in Chap. 1.

> Some computer dialects use E to represent multiplication by powers of 10. For example:
> 2.57E + 3 = 2.57×10^3
> 2.57E − 3 = 2.57×10^{-3}

6–4 POWERS OF 10

The powers of 10 are sometimes termed the "engineer's shorthand" or "scientist's shorthand." A thorough knowledge of the powers of 10 and the ability to apply the theory of exponents will greatly assist you in determining an approximation.

There was a time when the terms engineering notation and scientific notation were used interchangeably. Recently, however, a distinction has been made:

- *Scientific notation* describes the rewriting of any number as a number between 1 and 10 times the appropriate power of 10:

$$12\,345 = 1.234\,5 \times 10^4$$

- *Engineering notation* describes the more specialized technique of rewriting any number so as to use the third powers of 10, for which there are SI prefixes. This technique calls for the use of numbers between 0.1 and 1000 times the appropriate third power of 10. For example,

$$12\,345 = 12.345 \times 10^3$$

  If a unit of measure were involved, the 10^3 would be replaced by the prefix *kilo*.

Some calculators will permit you to FIX a readout to either SCI or ENG notation, and a choice of decimal places will be displayed. If a calculator is not used for computation, the powers of 10 enable one to work all problems by using convenient whole numbers. Either method offers a convenient way to obtain a final answer with the decimal point in its proper place.

Some of the multiples of 10 may be represented as shown in Table 6–1. From the table it is seen that any decimal fraction may be written as a whole

TABLE 6–1

Number		Power of 10		Expressed in English
0.000 001	=	10^{-6}	=	ten to the negative *sixth* power
0.000 01	=	10^{-5}	=	ten to the negative *fifth* power
0.000 1	=	10^{-4}	=	ten to the negative *fourth* power
0.001	=	10^{-3}	=	ten to the negative *third* power
0.01	=	10^{-2}	=	ten to the negative *second* power
0.1	=	10^{-1}	=	ten to the negative *first* power
1	=	10^0	=	ten to the *zero* power
10	=	10^1	=	ten to the *first* power
100	=	10^2	=	ten to the *second* power
1 000	=	10^3	=	ten to the *third* power
10 000	=	10^4	=	ten to the *fourth* power
100 000	=	10^5	=	ten to the *fifth* power
1 000 000	=	10^6	=	ten to the *sixth* power

number times some negative power of 10. This may be expressed by the following rule.

RULE

To express a decimal fraction as a whole number times a power of 10, move the decimal point to the right and count the number of places to the original point. The number of places counted is the proper negative power of 10.

EXAMPLES

$$0.006\,87 = 6.87 \times 10^{-3}$$
$$0.000\,048\,2 = 4.82 \times 10^{-5}$$
$$0.346 = 34.6 \times 10^{-2}$$
$$0.086\,43 = 86.43 \times 10^{-3}$$

Also, it is seen that any large number can be expressed as some smaller number times the proper power of 10. This can be expressed by the following rule.

RULE

To express a large number as a smaller number times a power of 10, move the decimal point to the left and count the number of places to the original decimal point. The number of places counted will give the proper positive power of 10.

EXAMPLES

$$435 = 4.35 \times 10^{2}$$
$$964\,000 = 96.4 \times 10^{4}$$
$$6\,835.2 = 6.835\,2 \times 10^{3}$$
$$5723 = 5.723 \times 10^{3}$$

PROBLEMS 6–2

Express the following numbers to three significant figures and write them in scientific notation as numbers between 1 and 10 times the proper power of 10:

1. 275 000
2. 13.6
3. 7133
4. 0.0963
5. 0.000 000 005 125
6. 8 743 000
7. 0.367
8. 59 235
9. 250×10^{-3}
10. $0.000\,086 \times 10^{6}$
11. $0.000\,399 \times 10^{8}$
12. $0.000\,399\,5 \times 10^{8}$
13. 259×10^{-4}
14. 0.031 415 9
15. 276 492.536 24
16. $1\,254\,325 \times 10^{-12}$
17. 0.000 000 107 52

18. $0.000\ 008\ 145\ 73 \times 10^{12}$
19. $2.000\ 915$
20. $0.000\ 055\ 55 \times 10^{-3}$
21. Rewrite each of your answers to Probs. 1 through 20 in engineering notation, using values between 0.1 and 1000 times the appropriate third power of 10.

6–5 MULTIPLICATION WITH POWERS OF 10

In Sec. 4–3 the law of exponents in multiplication was expressed in the general form

$$a^m \cdot a^n = a^{m+n} \qquad \text{(where } a \neq 0\text{)}$$

This law is directly applicable to the powers of 10.

EXAMPLE 1 Multiply 1000 by 100 000.

SOLUTION

$$1000 = 10^3$$

and $$100\ 000 = 10^5$$

then $$1000 \times 100\ 000 = 10^3 \times 10^5 = 10^{3+5} = 10^8$$

EXAMPLE 2 Multiply 0.000 001 by 0.001.

SOLUTION

$$0.000\ 001 = 10^{-6}$$

and $$0.001 = 10^{-3}$$

then $$0.000\ 001 \times 0.001 = 10^{-6} \times 10^{-3} = 10^{-6+(-3)} = 10^{-6-3} = 10^{-9}$$

EXAMPLE 3 Multiply 23 000 by 7000.

SOLUTION

$$23\ 000 = 2.3 \times 10^4$$

and $$7000 = 7 \times 10^3$$

then $$23\ 000 \times 7000 = 2.3 \times 10^4 \times 7 \times 10^3 = 2.3 \times 7 \times 10^7 = 16.1 \times 10^7\text{, or } 161\ 000\ 000$$

EXAMPLE 4 Multiply 0.000 037 by 600.

SOLUTION

$$0.000\ 037 \times 600 = 3.7 \times 10^{-5} \times 6 \times 10^2 = 3.7 \times 6 \times 10^{-3} = 22.2 \times 10^{-3}\text{, or } 0.0222$$

EXAMPLE 5 Multiply 72 000 × 0.000 025 × 4600.

SOLUTION

$$72\,000 \times 0.000\,025 \times 4600 = 7.2 \times 10^4 \times 2.5 \times 10^{-5} \times 4.6 \times 10^3$$
$$= 7.2 \times 2.5 \times 4.6 \times 10^2$$
$$= 82.8 \times 10^2, \text{ or } 8280$$

You will find that by expressing all numbers as numbers between 1 and 10 times the proper power of 10, the determination of the proper place for the decimal point will become a matter of inspection.

Study the instruction manual for your calculator until you can enter powers of 10 and interpret the readout. Can you FIX the number of places and choose SCI or ENG readouts?

PROBLEMS 6–3

Multiply the following. Although not all factors are expressed to three significant figures, express answers to three significant figures as numbers between 1 and 10 times the proper power of 10.

1. $1\,000 \times 0.001 \times 0.000\,01$
2. $0.000\,01 \times 10^5 \times 100$
3. 0.0006×360
4. $0.000\,25 \times 16 \times 10^{-4} \times 20 \times 10^5$
5. $0.000\,008\,4 \times 0.005 \times 0.000\,17$
6. $35\,000\,000 \times 680 \times 10^{-9} \times 5.5 \times 10^{-5}$
7. $8.62 \times 10^{12} \times 172\,000 \times 0.000\,027 \times 10^{-3}$
8. $500 \times 10^{-6} \times 782 \times 10^4 \times 0.000\,037 \times 10^{-8}$
9. $5\,960\,000 \times 0.000\,888 \times 604 \times 10^{-5}$
10. $2.846 \times 10^3 \times 0.009\,438 \times 10^6 \times 0.6848 \times 10^4$

The alternating-current (ac) inductive reactance of a circuit or an inductor is given by

$$X_L = 2\pi f L \qquad \Omega$$

where X_L = inductive reactance, Ω
f = frequency of alternating current, Hz
L = inductance of circuit or inductor, henrys (H)

Compute the inductive reactance when:

11. f = 60 Hz, L = 0.015 H
12. f = 1000 Hz, L = 0.015 H
13. f = 1 000 000 Hz, L = 0.015 H
14. f = 60 Hz, L = 1.5 H
15. f = 100 000 Hz, L = 0.000 026 5 H

6–6 DIVISION WITH POWERS OF 10

The law of exponents in division (Secs. 4–9 to 4–11) can be summed up in the following general form:

$$\frac{a^m}{a^n} = a^{m-n} \qquad \text{(where } a \neq 0\text{)}$$

EXAMPLE 6

$$\frac{10^5}{10^3} = 10^{5-3} = 10^2$$

or

$$\frac{10^5}{10^3} = 10^5 \times 10^{-3} = 10^2$$

EXAMPLE 7

$$\frac{72\ 000}{0.0008} = \frac{72 \times 10^3}{8 \times 10^{-4}}$$
$$= \frac{72}{8} \times 10^{3+4}$$
$$= 9 \times 10^7$$

or

$$\frac{72\ 000}{0.0008} = \frac{72 \times 10^3}{8 \times 10^{-4}}$$
$$\frac{72\ 000}{0.0008} = \frac{72}{8} \times 10^3 \times 10^4$$
$$= 9 \times 10^7$$

EXAMPLE 8

$$\frac{169 \times 10^5}{13 \times 10^5} = \frac{169}{13} \times 10^{5-5}$$
$$= 13 \times 10^0$$
$$= 13 \times 1 = 13$$

or

$$\frac{169 \times 10^5}{13 \times 10^5} = 13$$

It is apparent that powers of 10 which are factors that have the same exponents in numerator and denominator can be canceled. Also, you will note that powers of 10 which are factors can be transferred at will from denominator to numerator, or vice versa, if the sign of the exponent is changed when the transfer is made (Sec. 4–11).

6–7 APPROXIMATIONS

Multiplying 37 by 26 is very close to multiplying 40 by 25. The approximation 1000 is ‘‘within the order’’ of the actual product, 962. Usually, approximations which are within reason may be arrived at, and they serve as a guide to what the actual answer should be.

Such approximations should be made quickly before undertaking the exact calculations. The ‘‘order’’ of the calculated answer should be of the ‘‘order’’ of the approximation. If you expect an answer of the order of 1000 and you actually come up with 940 or 1050, the answer is probably correct. If, however, you arrive at an answer of 9.62, you should suspect that you have lost a factor of 10^2 somewhere, and you should check out your calculations. Although approximations will not guarantee the correctness of the calculated answer, they will reveal possible errors.

6–8 COMBINED MULTIPLICATION AND DIVISION

Combined multiplication and division is most conveniently accomplished by alternately multiplying and dividing until the problem is completed.

EXAMPLE 9 Simplify $\dfrac{0.000\,644 \times 96\,000 \times 3300}{161\,000 \times 0.000\,001\,20}$.

SOLUTION

First convert all numbers in the problem to numbers between 1 and 10 times their proper power of 10, thus:

$$\frac{6.44 \times 10^{-4} \times 9.6 \times 10^4 \times 3.3 \times 10^3}{1.61 \times 10^5 \times 1.2 \times 10^{-6}} = \frac{6.44 \times 9.6 \times 3.3 \times 10^4}{1.61 \times 1.2}$$

The problem as now written consists of multiplication and division of simple numbers. The answer approximates to

$$\frac{6 \times 10 \times 3 \times 10^4}{2 \times 1} = 90 \times 10^4$$

If the remainder of the problem is computed by calculator, then the answer 1056 can easily be adjusted to read 105.6×10^4, or 1.056×10^6. If the problem is solved without the aid of a calculator, there are no small decimals and no cumbersome large numbers to handle.

To solve this problem manually, first simplify the equation.

$$\frac{6.44 \times 9.6 \times 3.3 \times 10^4}{1.61 \times 1.2} = \frac{6.44 \times 8 \times 3.3 \times 10^4}{1.61}$$

Then find the products of the numerator and the denominator, respectively.

$$6.44 \times 8 \times 3.3 \times 10^4 = 170.016 \times 10^4$$
$$1.61 = 1.61$$

Finally, divide the numerator by the denominator.

$$\frac{170.016 \times 10^4}{1.61} = 105.6 \times 10^4$$

If we desire to express the answer in powers of 10, we would write 1.056×10^6, but written out without the power of 10, it would be 1 056 000.

6–9 RECIPROCALS

In electrical and electronics problems, many of the formulas that are used involve reciprocals. Some examples are

$$\frac{1}{R_t} = \frac{1}{R_1} + \frac{1}{R_2}$$

$$X_C = \frac{1}{2\pi fC}$$

$$f = \frac{1}{2\pi\sqrt{LC}}$$

The *reciprocal* of a number is 1 divided by that number. Finding a reciprocal presents no difficulty if the powers of 10 are used properly.

Also, many calculators offer the convenience of a $\frac{1}{x}$ key.

EXAMPLE 10 Simplify $\frac{1}{40\,000 \times 0.000\,25 \times 125 \times 10^{-6}}$.

SOLUTION

First convert all numbers in the denominator to numbers between 1 and 10 times their proper powers of 10, thus:

$$\frac{1}{4 \times 10^4 \times 2.5 \times 10^{-4} \times 1.25 \times 10^{-4}} = \frac{10^4}{4 \times 2.5 \times 1.25}$$

Multiplying the factors of the denominator results in

$$\frac{10^4}{12.5}$$

Instead of writing out the numerator as 10 000 and then dividing by 12.5, we could write the numerator as two factors in order better to divide mentally. That is, we can write the problem as

$$\frac{10^2 \times 10^2}{12.5} \quad \text{or} \quad \frac{100}{12.5} \times 10^2 = 8 \times 10^2$$

This method is of particular advantage because of the ease of estimating the final result.

If the final result is a decimal fraction, rewriting the numerator into two factors allows fixing the decimal point with the least effort.

EXAMPLE 11 Simplify $\frac{1}{625 \times 10^4 \times 2000 \times 64\,000}$.

SOLUTION

First convert all numbers in the denominator to numbers between 1 and 10 times their proper powers of 10, thus:

$$\frac{1}{6.25 \times 10^6 \times 2 \times 10^3 \times 6.4 \times 10^4} = \frac{10^{-13}}{6.25 \times 2 \times 6.4}$$

Multiplying the factors in the denominator results in

$$\frac{10^{-13}}{80}$$

Instead of writing out the numerator as 0.000 000 000 000 1 and dividing it by 80, we write the numerator as two factors in order better to divide mentally:

$$\frac{10^{2} \times 10^{-15}}{80} \quad \text{or} \quad \frac{100}{80} \times 10^{-15} = 1.25 \times 10^{-15}$$

If the value of the denominator product were over 100 and less than 1000, we would break up the numerator so that one of the factors would be 10^3, or 1000, and so on. This method will always result in a final quotient of a number between 1 and 10 times the proper power of 10.

PROBLEMS 6–4

Perform the indicated operations. Round off the figures in the results, if necessary, and express answers to three significant figures as a number between 1 and 10 times the proper power of 10:

1. $\dfrac{0.000\,75}{5000}$
2. $\dfrac{10}{0.000\,125 \times 80\,000}$
3. $\dfrac{0.6043}{5763}$
4. $\dfrac{420 \times 0.036}{0.0090}$
5. $\dfrac{0.375 \times 467 \times 10^{-9}}{531\,000}$
6. $\dfrac{1}{6.28 \times 452\,000 \times 0.000\,155}$
7. $\dfrac{2804 \times 74.23}{0.000\,900\,6 \times 0.008\,040}$
8. $\dfrac{1000}{248\,000 \times 5630 \times 10^{-3} \times 0.000\,090\,3 \times 10^{2}}$
9. $\dfrac{1 \times 10^{6}}{6.28 \times 10^{3} \times 2500 \times 10^{3} \times 0.25 \times 10^{-6}}$
10. $\dfrac{1}{6.28 \times 400 \times 10^{6} \times 50 \times 10^{-12}}$
11. $\dfrac{220 \times 188 \times 3.24}{1.97 \times 10^{2} \times 4.66 \times 9.33 \times 10^{4}}$
12. $\dfrac{65.3 \times 10^{-6} \times 504 \times 10^{6} \times 12\,700}{312 \times 10^{6} \times 0.007 \times 6.82}$

The ac capacitive reactance of a circuit, or capacitor, is given by the formula

$$X_C = \frac{1}{2\pi fC} \quad \Omega$$

where X_C = capacitive reactance, Ω
f = frequency of the alternating current, Hz
C = capacitance of the circuit, or capacitor, farads (F)

Compute the capacitive reactances when:

13. f = 60 Hz, C = 0.000 004 F
14. f = 28 000 000 Hz, C = 0.000 000 000 025 F
15. f = 132 000 000 000 Hz, C = 0.000 000 025 F

6–10 THE POWER OF A POWER

It becomes necessary, in order to work a variety of problems utilizing the powers of 10, to consider a few new definitions concerning the laws of exponents before we study them in algebra. This, however, should present no difficulty.

In finding the power of a power the exponents are multiplied. That is, in general,

$$(a^m)^n = a^{mn} \qquad (\text{where } a \neq 0)$$

EXAMPLE 12

$$100^3 = 100 \times 100 \times 100 = 1\ 000\ 000 = 10^6$$

or

$$100^3 = 10^2 \times 10^2 \times 10^2 = 10^6$$

then

$$100^3 = (10^2)^3 = 10^{2 \times 3} = 10^6$$

Numbers can be factored when raised to a power in order to reduce the labor in obtaining the correct number of significant figures, or properly fixing the decimal point.

EXAMPLE 13

$$\begin{aligned} 19\ 000^3 &= (1.9 \times 10^4)^3 \\ &= 1.9^3 \times 10^{4 \times 3} \\ &= 6.859 \times 10^{12} \end{aligned}$$

EXAMPLE 14

$$\begin{aligned} 0.000\ 007\ 5^2 &= (7.5 \times 10^{-6})^2 \\ &= 7.5^2 \times 10^{(-6) \times 2} \\ &= 56.25 \times 10^{-12} \\ &= 5.625 \times 10^{-11} \end{aligned}$$

In Example 13, 19 000 was factored into 1.9×10^4 in order to allow an easy mental check. Because 1.9 is nearly 2 and $2^3 = 8$, it is apparent that the result of cubing 1.9 must be 6.859, not 0.6859 or 68.59.

In Example 14, the 0.000 007 5 was factored for the same reason. We know that $7^2 = 49$; therefore, the result of squaring 7.5 must be 56.25, not 0.5625 or 5.625.

If your calculator offers x^y or y^x, review your instruction manual and practice raising to powers on your calculator.

6–11 THE POWER OF A PRODUCT

The power of a product is the same as the product of the powers of the factors. That is, in general,

$$(abc)^m = a^m b^m c^m$$

EXAMPLE 15

$$(10^5 \times 10^3)^3 = 10^{5 \times 3} \times 10^{3 \times 3}$$
$$= 10^{15} \times 10^9 = 10^{24}$$

or

$$(10^5 \times 10^3)^3 = (10^8)^3 = 10^{8 \times 3} = 10^{24}$$

6–12 THE POWER OF A FRACTION

The power of a fraction equals the power of the numerator divided by the power of the denominator. That is,

$$\left(\frac{a}{b}\right)^m = \frac{a^m}{b^m}$$

EXAMPLE 16

$$\left(\frac{10^5}{10^3}\right)^2 = \frac{10^{5 \times 2}}{10^{3 \times 2}} = \frac{10^{10}}{10^6} = 10^4$$

The above can be solved by first clearing the exponents inside the parentheses and then raising to the required power. Thus,

$$\left(\frac{10^5}{10^3}\right)^2 = (10^{5-3})^2 = (10^2)^2 = 10^4$$

6–13 THE ROOT OF A POWER

The root of a power in exponents is given by

$$\sqrt[n]{a^m} = a^{m \div n} \qquad \text{(where } a \text{ and } n \neq 0\text{)}$$

EXAMPLE 17

$$\sqrt{25 \times 10^8} = \sqrt{25} \times \sqrt{10^8} = 5 \times 10^{8 \div 2} = 5 \times 10^4$$

EXAMPLE 18

$$\sqrt[3]{125 \times 10^6} = \sqrt[3]{125} \times \sqrt[3]{10^6} = 5 \times 10^{6 \div 3} = 5 \times 10^2$$

In the general case when m is evenly divisible by n, the process of extracting roots is comparatively simple. When m is not evenly divisible by n, the result obtained by extracting the root is a fractional power.

EXAMPLE 19

$$\sqrt{10^5} = 10^{5 \div 2} = 10^{\frac{5}{2}}, \text{ or } 10^{2.5}$$

Such fractional exponents are encountered in various phases of engineering mathematics and are conveniently solved by the use of logarithms. However, in using the powers of 10, the fractional exponent is cumbersome for obtaining a final answer. It becomes necessary, therefore, to devise some means of extracting a root whereby an integer can be obtained as an exponent in the final result. The simplest way is to express the number, the root of which is desired, as some number times a power of 10 that is evenly divisible by the index of the required root. As an example, suppose it is desired to extract the square root of 400 000. Though it is true that

$$\begin{aligned}\sqrt{400\ 000} &= \sqrt{4 \times 10^5} \\ &= \sqrt{4} \times \sqrt{10^5} \\ &= 2 \times 10^{2.5}\end{aligned}$$

we have a fractional exponent that is not readily reduced to actual figures. However, if we express the number differently, we obtain an integer as an exponent. Thus,

$$\begin{aligned}\sqrt{400\ 000} &= \sqrt{40 \times 10^4} \\ &= \sqrt{40} \times \sqrt{10^4} \\ &= 6.32 \times 10^2\end{aligned}$$

It will be noted that there are a number of ways of expressing the above square root, such as

$$\sqrt{400\ 000} = \sqrt{0.4 \times 10^6}$$

or

$$\sqrt{4000 \times 10^2}$$

or

$$\sqrt{0.004 \times 10^8}$$

All are equally correct, but you should try to write the problem in a form that will allow a rough mental approximation in order that the decimal may be properly placed with respect to the significant figures.

PROBLEMS 6–5

Perform the indicated operations. When answers do not come out in round numbers, express them to three significant figures.

1. $(10^2)^3$
2. $(10^{-4})^3$
3. $(10^3 \times 10^4)^2$
4. $(4 \times 10^{-4})^2$
5. $(5 \times 10^3)^4$
6. $(3 \times 10^{-2})^3$
7. $(2 \times 10^4 \times 8 \times 10^{-5})^2$
8. $\left(\dfrac{32 \times 10^3}{8 \times 10^4}\right)^2$
9. $\sqrt{0.0625 \times 0.0004}$
10. $\sqrt{0.000\ 36 \times 0.009}$
11. $\sqrt{49 \times 10^2 \times 36 \times 10^{-2}}$
12. $\sqrt[3]{27 \times 10^{-3} \times 8 \times 10^{12}}$
13. $\dfrac{1}{6.28\sqrt{250 \times 10^{-3} \times 10^{-9}}}$
14. $\left(\dfrac{63 \times 10^6 \times 460 \times 10^{-12}}{5.1 \times 10^{-6}}\right)^2$

The resonant frequency of a circuit is given by the formula

$$f = \frac{1}{2\pi\sqrt{LC}} \qquad \text{Hz}$$

where f = resonant frequency, Hz
L = inductance of circuit, H
C = capacitance of circuit, F

Compute the resonant frequencies when:

15. $L = 0.000\ 045$ H, $C = 0.000\ 000\ 000\ 250$ F
16. $L = 0.000\ 018$ H, $C = 100 \times 10^{-12}$ F
17. $L = 5 \times 10^{-6}$ H, $C = 50 \times 10^{-12}$ F
18. $L = 0.000\ 23$ H, $C = 0.000\ 000\ 000\ 5$ F
19. $L = 70.4 \times 10^{-6}$ H, $C = 250 \times 10^{-12}$ F
20. $L = 40$ H, $C = 7 \times 10^{-6}$ F

6–14 ADDITION AND SUBTRACTION WITH POWERS OF 10

Sometimes it becomes necessary, when making calculations, to perform additions and subtractions with powers of 10. These operations present no difficulties if you remember that you are dealing with the addition and subtraction of like terms as described in Sec. 3–7. For example, you would not write $3x^2 + 5x^3 = 8x^5$, because $3x^2$ and $5x^3$ are unlike quantities. Similarly, you would not write

$$3 \times 10^2 + 5 \times 10^3 = 8 \times 10^5$$

because 3×10^2 and 5×10^3 also are unlike quantities.

The foregoing addition of

$$3 \times 10^2 + 5 \times 10^3$$

can be performed by either of two methods. You can convert the numbers so that no powers of 10 are involved and write $300 + 5000 = 5300$. Also, you can rewrite the terms to be added so that like powers of 10 are added, such as

$$3 \times 10^2 + 50 \times 10^2 = 53 \times 10^2$$

or

$$0.3 \times 10^3 + 5 \times 10^3 = 5.3 \times 10^3$$

This is the same as adding like terms.

EXAMPLE 20 Add 8.3×10^4 and 3.6×10^2.

SOLUTION

$$\begin{aligned} 8.3 \times 10^4 &= 83\ 000 = 830 \quad \times 10^2 \\ 3.6 \times 10^2 &= \underline{\quad\ 360} = \underline{\quad 3.6 \times 10^2} \\ &\quad\ 83\ 360 = 833.6 \times 10^2 \\ &\qquad\qquad\ = 8.336 \times 10^4 \end{aligned}$$

PROBLEMS 6–6

Perform the indicated operations. Express all answers (*a*) in ordinary form and (*b*) to three significant figures as numbers between 1 and 10 times the proper power of 10.

1. $5 \times 10^3 + 3 \times 10^2$
2. $25 \times 10^6 + 3.4 \times 10^3$
3. $1.73 \times 10^{12} + 2.46 \times 10^{12}$
4. $2 \times 10^3 + 4 \times 10^{-1}$
5. $186 \times 10^6 - 28 \times 10^{-3}$
6. The total capacitance C_p of two capacitors C_1 and C_2 in parallel is found from the formula

$$C_p = C_1 + C_2$$

What is the total capacitance of a circuit consisting of 200 pF in parallel with 50 nF?

SELF TEST

1. To how many significant figures has the following number been written?

0.001 250

2. Write in scientific notation: 17 625.
3. Write in engineering notation to three significant figures: 27 889.
4. Evaluate. Answer in engineering notation to three significant figures:

$$\frac{0.159 \times 22 \times 10^{-6} \times 400 \times 10^3}{1.37 \times 10^{-3} \times 0.588}$$

5. Evaluate. Answer in engineering notation to three significant figures:

$$\frac{175\,000 \times 0.000\,012\,868 \times 250 \times 10^{-5}}{2.86 \times 10^{-3} \times 0.008\,113}$$

6. What is the reciprocal of 0.0625?
7. If $R_X = R_L \dfrac{R_2}{R_1 + R_2}$, what is R_X if

$R_L = 12\,000\ \Omega$
$R_1 = 300\ \Omega$
$R_2 = 1\,800\ \Omega$

8. Evaluate. Answer in engineering notation to three significant figures:

$$(1.07 \times 10^3)^2$$

9. If $f = \dfrac{1}{2\pi\sqrt{LC}}$, find f when $L = 15 \times 10^{-3}$ Hz and $C = 50 \times 10^{-9}$ F.

10. $R_A = \dfrac{R_1 R_3}{R_1 + R_2 + R_3}$. What is R_A if $R_1 = 1200\ \Omega$, $R_2 = 4600\ \Omega$, and $R_3 = 33\ 000\ \Omega$?

CHAPTER 7

UNITS AND DIMENSIONS

As previously stated, the solution of every practical problem, when a concrete answer is desired, eventually reduces to an arithmetical computation; that is, the answer reduces to some *number*. In order for this answer, or number, to have a concrete meaning, it must be expressed in some *unit*. For example, if you were told that the resistance of a circuit is 16, the information would have no meaning unless you knew to what unit the 16 referred.

From the foregoing it is apparent that the expression for the magnitude of any physical quantity must consist of two parts. The first part, which is a number, specifies "how much"; the second part specifies the unit of measurement, or "what," as, for example, in 16 Ω, 20 A, or 100 m, the Ω, A, or m.

It is necessary, therefore, before beginning the study of circuits, to define a few of the more common electrical and dimensional units used in electrical and electronics engineering.

7–1 SYSTEMS OF MEASUREMENT

Over the years, the systems by which we have made measurements have changed considerably. We do not often now deal with grains of corn or the length of a man's forearm. Occasionally the civil engineer surveying an antenna site will talk about "chains" when we would have said "hundreds of meters." We in electronics are primarily concerned with four specific fields of measurement: distance, mass, time, and electric charge. All the electrical quantities are fundamentally related to these measurements, as you will discover if you study "higher mathematics."

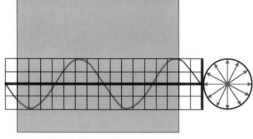

Generally speaking, North Americans have used two main systems for measuring some quantities, whereas the units of other quantities are the same in both systems. One of these systems is the so-called English fps (foot-pound-second) or "traditional" system, which was widely, almost exclusively, used by engineers in English-speaking countries until very recently. The other is the metric mks (meter-kilogram-second) system, which has grown in importance over the last century. The most modern refinement of the mks system is called SI, for Système International d'Unités. It is used by well over 90% of the world's population, and it has become even more widely used in the last few years with the conversion of Great Britain, India, Australia, Canada, and, as metric legislation is passed, a growing number of industries in the United States to SI. You should note that in both the U.S. Customary system and the SI several of the units are the same: seconds, volts, ohms, and amperes, especially.*

7–2 THE ENGLISH SYSTEM

The English system, developed over many centuries, contains many quite arbitrary relationships between the units and no systematic correlations. It is being superseded by the very logical SI metric units (Sec. 7–3). The small list given here shows only a very few of the many conversions which have been developed over the years.

12 inches (in) = 1 foot (ft)
3 feet (ft) = 1 yard (yd)
5280 feet (ft) = 1 statute mile (mi)
16 ounces (oz) = 1 pound (lb)
2000 pounds (lb) = 1 ton

7–3 THE SI METRIC SYSTEM

The metric system is a relatively newer, more orderly system related originally to the measurement of the earth itself. It uses decimal relationships throughout, rather than the arbitrary, hard-to-memorize conversions of the English system. The metric system started out as the centimeter-gram-second (cgs) system. Later, the cgs units were modified; the basic defined units were then the meter, the kilogram, and the second, and the name of the system was changed to mks. Later still, the ampere was added in order to elevate electrical units to the "physical" ones, and the name changed again to mksa.

The most recent development, the SI, is the result of many years of concentrated international cooperation and agreement, and has been published by the General Conference on Weights and Measures and the International Organization for Standardization (ISO). This system defines seven base measuring units, but it goes much farther in including other specific details. Altogether, SI involves:

- Seven base measuring units
- Two "supplementary" units
- An added collection of related units that can be defined in terms of the base and supplementary units

*A quite thorough introduction to the SI metric system was written by Mr. Adams and published by McGraw-Hill in a revised edition in June 1974. It is entitled *SI Metric Units: An Introduction*. Students and teachers with no background in the metric system at all may find it useful.

- An orderly use of decimal calculations, with powers of 10 notation, and special word prefixes representing numbers
- An international system of symbols and notation
- A comprehensive system of national and international standards

Some of the SI units have little direct meaning to us in electronics and are described below only for the sake of providing you with a complete list. Others are daily necessities, and you will find them used repeatedly throughout this book.

7–4 THE SEVEN BASE SI UNITS

LENGTH/METER The meter was originally defined to be 1×10^{-7} of the length of the line of longitude passing through Paris from the equator to the North Pole. The newest and most exact definition by the ISO is that the meter is the distance traveled in a vacuum by light in $\frac{1}{299\ 792\ 458}$ s. The denominator is, of course, the speed of light in meters per second.

$$1 \text{ meter (m)} = 39.370\ 079 \text{ inches (in)}$$

MASS/KILOGRAM The kilogram is simply defined to be the mass of a special cylinder of platinum-iridium alloy which is in the safekeeping of the International Bureau of Weights and Measures. This cylinder is called the *International Prototype Kilogram.*

$$1 \text{ kilogram (kg)} = 2.204\ 622\ 6 \text{ pounds (lb)}$$

TIME/SECOND The second is specifically defined as the duration of 9 192 631 770 periods of the radiation corresponding to the transition between the two hyperfine levels of the ground state of the atom of cesium 133. There are other special definitions of time based on the sun, stars, and moon, but the definition given here is one which can be duplicated in laboratories of Bureaus of Standards anywhere.

ELECTRIC CURRENT/AMPERE That intensity of electric current known as an ampere is defined as the constant current which, if maintained in two straight parallel conductors of infinite length and of negligible cross section in a vacuum exactly one meter apart, will produce a force between the conductors of 2×10^{-7} newton (N) per meter length of wire. The circuit symbol* for current is I, and the unit symbol for amperes is A.

TEMPERATURE/KELVIN The defined SI unit of temperature is the kelvin. The freezing point of pure water is 273.15 K (*not* °K). Ordinary temperature readings will be made on the Celsius scale, on which the freezing point of pure water is 0°C and on which the boiling point is 100°C (= 373.15 K).

*Circuit symbols appear in circuit or schematic diagrams, and they are used as ''quantity'' symbols in algebraic formulas. They are always printed in *italic* type. Unit symbols (not abbreviations) are used for units of measure. They are always printed in roman type (not slanting).

LUMINOUS INTENSITY/CANDELA The candela is the luminous intensity, in a given direction, of a source which emits a monochromatic yellow-green radiation of 540×10^{12} Hz, and whose energetic intensity in that direction is $\frac{1}{683}$ watts per steradian. (We will not concern ourselves with illumination in this book.)

MOLECULAR SUBSTANCE/MOLE The mole is the amount of substance (*not* mass) of a system which contains as many elementary entities (which must be specified: electrons, atoms, molecules) as there are atoms in 0.012 kg of carbon 12. (We will not concern ourselves with this more or less pure science unit in this book.)

ANGLES/RADIANS AND STERADIANS In addition to the seven base units, there are two supplementary units for the measurement of angles: the radian for the measurement of plane angles and the steradian for the measurement of solid angles. We will study plane angles in Chap. 23.

7–5 THE ADDITIONAL DEFINED SI UNITS

In addition to the base or standard SI units, there are 19 other units which are used so often that they have been given special names. Many of them are important to us in electronics, and they are listed first in the descriptions which follow. Again, all 19 units are listed for the benefit of users of this book who are studying beyond the book's limitations.

ELECTRIC CHARGE/COULOMB The coulomb is defined as the ampere-second. A reverse definition is that one ampere is the current intensity when one coulomb (C) flows in a circuit for one second. The coulomb is also defined as $6.241\,96 \times 10^{18}$ electronic charges. The circuit symbol for charge is Q.

ELECTRIC POTENTIAL/VOLT The volt is the practical unit of electromotive force (emf), or potential difference. It is defined as the watt per ampere. (See watt below.) A more common understanding of the volt is that it is the potential difference which will drive a current of one ampere through a resistance of one ohm. The circuit symbols for voltage are V and v, and the unit symbol for volt is V.

ELECTRIC RESISTANCE/OHM The ohm is the practical unit of resistance. It is defined as the volt per ampere; that is, the ohm is the amount of resistance which limits the current flow to one ampere when the applied electromotive force (potential) is one volt. The circuit symbol for resistance is R, and the unit symbol for ohms is Ω.

ELECTRIC CONDUCTANCE/SIEMENS The siemens is the practical unit of conductance. It is the reciprocal of the ohm, since the conductance is the reciprocal of the resistance. The relationship between ohms and siemens is given by

$$G = \frac{1}{R} \text{ siemens}$$

If resistance is thought of as representing the difficulty with which an electric current is forced to flow through a circuit, conductivity may be thought of as the ease with which a current will pass through the same circuit.

A conductance of one siemens will permit a current flow of one ampere under an electrical pressure of one volt. The siemens is a new unit name honoring a pioneer in electricity. Formerly, the unit of conductance was called the mho. The circuit symbol for conductance is *G*, and the unit symbol for siemens is S.

ELECTRIC CAPACITANCE/FARAD The farad is the unit of capacitance. It is an ampere-second per volt. A circuit, or capacitor, is said to have a capacitance of one farad when a change of one volt per second across it produces a current of one ampere. The circuit symbol for capacitance is *C*, and the unit symbol for farad is F. Capacitance will be further discussed in Chap. 32.

ELECTRIC INDUCTANCE/HENRY The henry is the unit of inductance. It is a volt-second per ampere. A circuit, or inductor, is said to have a self-inductance of one henry when a counterelectromotive force of one volt is generated within it by a rate of change of current of one ampere per second. The circuit symbol for inductance is *L*, and the unit symbol for henry is H. Inductance will be further discussed in Chap. 32.

FREQUENCY/HERTZ The SI unit of frequency is the hertz, which was formerly called the cycle per second. Since *cycle* is not a unit of measure as such, it is sufficient to describe the hertz as the reciprocal of time.

MAGNETIC FLUX/WEBER Magnetic flux is fully described as the volt-second.

MAGNETIC FLUX DENSITY/TESLA Tesla is the special name given to the "density" relationship of webers per square meter.

LUMINOUS FLUX/LUMEN This SI unit relates the amount of radiant energy in terms of candelas of luminous intensity multiplied by the solid angle in steradians from which the radiant flux "flows."

ILLUMINATION/LUX This unit describes the lumens per square meter relationship of luminous flux.

ENERGY/JOULE The SI unit for energy of all forms—mechanical work, electric energy, heat quantity, etc.—is the joule. The joule (J) may be expressed in terms of newtons of force multiplied by the distance in meters through which the force moves in the direction of its application.

FORCE/NEWTON The newton is the SI unit describing joules per meter.

PRESSURE/PASCAL The pascal (Pa) is defined as newtons per square meter. It is a very small unit of measurement.

POWER/WATT The watt is the SI unit for power of all forms—electric, mechanical, and so on. It is defined as the energy in joules expended per unit of time in seconds. The circuit symbol for power is *P*, and the unit symbol for watt is W.

In direct-current circuits the power in watts is the product of the voltage and the current, or

$$P = VI \qquad \text{W}$$

The watthour is the unit of electric energy, and its abbreviation is Wh. It is the amount of energy delivered by a power of one watt over a period of one hour.

CUSTOMARY TEMPERATURE/DEGREE CELSIUS Ordinary (nonscientific) temperature measurements will be made on the Celsius scale, which is related to the Kelvin scale by °C = K − 273.15.

RADIOACTIVITY/BECQUEREL The activity of radionuclides is measured in becquerels.

$$1 \text{ Bq} = 1 \text{ ``activity'' per second}$$

ABSORBED DOSE OF RADIATION/GRAY The gray is the absorbed dose of radiation that is equal to one joule per kilogram.

DOSE EQUIVALENT OF RADIATION/SIEVERT The sievert is the dose equivalent of ionizing radiation that is equal to one joule per kilogram. One sievert is not the same as one gray.

7–6 SOME SI METRIC INTERRELATIONSHIPS

The fundamental relationships between metric units are decimal. The following equations show some of the simpler multiples and submultiples. You will be involved in many such conversions as you continue your studies in electronics.

$$1 \text{ millimeter (mm)} = \frac{1}{1000} \text{ meter} = 10^{-3} \text{ m}$$

$$1 \text{ centimeter (cm)} = \frac{1}{100} \text{ meter} = 10^{-2} \text{ m}$$

$$1 \text{ kilometer (km)} = 1000 \text{ meters} = 10^{3} \text{ m}$$

$$1 \text{ gram (g)} = \frac{1}{1000} \text{ kilogram (kg)} = 10^{-3} \text{ kg}$$

7–7 RELATIONS BETWEEN THE SYSTEMS

Since the metric system is based on a decimal plan and the English system is not, there is no one numerical factor or constant which can be used for the conversion of one system to the other. Table 5 in the Appendix contains some conversion factors, and a few approximate equivalents are given here for your convenience:

1 inch (in) = 2.540 centimeters (cm) (exactly)
1 foot (ft) = 0.3048 meter (m) (exactly)
1 meter (m) = 39.37 inches (in)
1 mile (mi) = 1.609 kilometers (km)
1 kilometer (km) = 0.6214 mile (mi)
1 kilogram (kg) = 2.205 pounds (lb)
1 pound (lb) = 0.4536 kilogram (kg)

If you are unfamiliar with the metric system, try to visualize these relationships for future convenience. What is the weight in kilograms of a loaf of bread in your community? What is the distance in kilometers from your home to your work? What is your height in centimeters?

The units of time (seconds) and of electricity are identical in the two systems, and we will now deal with them in more detail.

Since the SI units are becoming increasingly important as Canada and the United States convert to the metric system, you should make the habit of thinking in metric units. Do not keep translating metric quantities into the old English units.

7–8 FREQUENCY

A current which reverses itself at intervals is called an *alternating current*. When this current rises from zero value to maximum value and returns to zero and then increases to maximum value in the opposite direction and again returns to zero, it is said to have completed *one cycle*. The number of times this cycle is repeated in one second is known as the *frequency* of the alternating current. The frequency of the average house current is 60 cycles per second (cps), and that of radio waves may be as high as several hundred million cycles per second. Note that frequency involves our other main unit, time, by measuring the number of events per second. In both the English and SI systems,

60 seconds (s) = 1 minute (min)
60 minutes (min) = 1 hour (h)
24 hours (h) = 1 day (d)

The International Electrotechnical Commission (IEC), the International Organization for Standardization (ISO), and the Conférence Générale des Poids et Mesures (CGPM) have adopted the name *hertz* (Hz) as the unit of frequency.

1 hertz = 1 cycle per second

7–9 RANGES OF UNITS

As stated in Sec. 6–3, the fields of communication and electrical engineering embrace extremely wide ranges in values of the foregoing units. For example, at the input of a radio receiver, we deal in millionths of a volt, whereas the output circuit of a transmitter may develop hundreds of thousands of volts. An electric clock might consume a fraction of a watt, whereas the powerhouse furnishing this power probably has a capability of millions of watts.

Furthermore, two of these units, the henry and the farad, are very large units, especially the latter. The average radio receiver employs inductances

TABLE 7–1 DECIMAL MULTIPLIERS

Number		Power of 10		Expressed in English		Prefix†	Symbol
0.000 000 000 000 000 001	=	10^{-18}	=	ten to the negative *eighteenth* power	=	atto	a
0.000 000 000 000 001	=	10^{-15}	=	ten to the negative *fifteenth* power	=	femto	f
0.000 000 000 001	=	10^{-12}	=	ten to the negative *twelfth* power	=	pico	p
0.000 000 001	=	10^{-9}	=	ten to the negative *ninth* power	=	nano	n
0.000 001	=	10^{-6}	=	ten to the negative *sixth* power	=	micro	μ
0.001	=	10^{-3}	=	ten to the negative *third* power	=	milli	m
1	=	10^{0}	=	ten to the *zero* power	=	unit	
1 000	=	10^{3}	=	ten to the *third* power	=	kilo	k
1 000 000	=	10^{6}	=	ten to the *sixth* power	=	mega	M
1 000 000 000	=	10^{9}	=	ten to the *ninth* power	=	giga	G
1 000 000 000 000	=	10^{12}	=	ten to the *twelfth* power	=	tera	T
1 000 000 000 000 000	=	10^{15}	=	ten to the *fifteenth* power	=	peta	P
1 000 000 000 000 000 000	=	10^{18}	=	ten to the *eighteenth* power	=	exa	E

ranging from a few millionths of a henry, as represented by tuning inductance, to several henrys for power filters. The farad is so large that even the largest capacitors are rated in millionths of a farad. Smaller capacitors used in radio circuits are often rated in terms of so many millionths of one-millionth of a farad.

The use of some power of 10 is very convenient in converting to larger multiples or smaller fractions of the basic units, called *practical units*.

7–10 DECIMAL MULTIPLIERS

Some of the more common multipliers and their unit names are explained below, and all of the multipliers are listed in Table 7–1.

MILLIUNITS The milliunit is one-thousandth of a unit. Thus, it takes 1000 millivolts to equal 1 volt, 500 milliamperes to equal 0.5 ampere, etc. This unit is commonly used with volts, amperes, henrys, and watts. It is abbreviated m; thus, 10 mH = 10 millihenrys.* Mathematically, milli = 10^{-3}, and 1 mW = 10^{-3} W.

MICROUNITS The microunit is one-millionth of a unit. That is, it takes 1 000 000 microamperes to make 1 ampere, 2 000 000 microfarads to equal 2

*See Table 3 in the Appendix for abbreviations and unit symbols.

†Proposed, but not official at the time this book went to press are four additional prefixes:
zepto, symbol z: 10^{-21}
zetta, symbol Z: 10^{21}
yocto, symbol y: 10^{-24}
yotta, symbol Y: 10^{24}

farads, etc. This unit, abbreviated μ (greek letter mu), is commonly used with volts, amperes, ohms, siemens, henrys, and farads. Thus,

$$5\ \mu\text{F} = 5 \text{ microfarads}$$

Mathematically, micro $= 10^{-6}$, and 1 μs $= 10^{-6}$ s.

PICOUNITS The picounit is one-millionth of one-millionth of a unit; that is, 1 farad is equivalent to 1 000 000 000 000, or 10^{12}, picofarads. This unit is seldom used for anything other than farads. It is represented by p; thus, 250 pF = 250 picofarads. Mathematically, pico $= 10^{-12}$. Older texts use the micromicrounit, abbreviated μμ. Thus,

$$2\ \mu\mu\text{F} = 2 \text{ micromicrofarads} = 2 \text{ pF}$$

KILOUNITS The kilounit is 1000 basic units. Thus, a kilovolt is equivalent to 1000 volts. This unit is commonly used with hertz, volts, amperes, ohms, watts, and volt-amperes. It is abbreviated k. Thus, 35 kW means 35 kilowatts; 2000 hertz (Hz) = 2 kilohertz (kHz). Mathematically, kilo $= 10^3$.

MEGAUNITS The megaunit is 1 000 000 basic units. Thus, 1 megohm is equal to 1 000 000 ohms. This unit is used mainly with ohms and hertz. It is abbreviated M. Thus, 3 MHz = 3 megahertz. Mathematically, mega $= 10^6$.

7–11 PREFERRED DECIMAL MULTIPLIERS

Table 7–1 gives the prefix names and abbreviations for the *third powers of 10*. These are the preferred powers, and, therefore, the preferred prefixes. Almost every calculation in electronics will result in a quantity involving one of the third powers of 10 prefixes: *milli*watts, *micro*amperes, *mega*hertz.

In a few cases, prefixes are also used for other powers of 10. Table 7–2 lists these denigrated, or nonstandard, powers of 10.

TABLE 7–2 DENIGRATED POWERS OF TEN

Number	Power of 10	Expressed in English	Prefix	Abbreviation
100	$= 10^2$	= ten squared	= hecto	h
10	$= 10^1$	= ten	= deca	da
0.1	$= 10^{-1}$	= ten to the negative *first* power	= deci	d
0.01	$= 10^{-2}$	= ten to the negative *second* power	= centi	c

In order to use the preferred third powers of 10, quantities will normally be expressed as numbers between 0.1 and 1000.

In Examples 1 to 3, express numbers by using preferred third powers of 10 prefixes.

EXAMPLE 1 0.01 A

SOLUTION

$$0.01\ \text{A} = 1 \times 10^{-2}\ \text{A}$$

Rewriting to a third power,

$$1 \times 10^{-2} = 10 \times 10^{-3}$$

Therefore, $$0.01\ \text{A} = 10 \times 10^{-3}\ \text{A} = 10\ \text{mA}$$

EXAMPLE 2 1320 kHz

SOLUTION

$$\begin{aligned} 1320\ \text{kHz} &= 1.32 \times 10^{3}\ \text{kHz} \\ 1320\ \text{kHz} &= 1.32 \times 10^{3} \times 10^{3}\ \text{Hz} \\ &= 1.32 \times 10^{6}\ \text{Hz} \\ &= 1.32\ \text{MHz} \end{aligned}$$

EXAMPLE 3 0.872 H

SOLUTION

This is a perfectly good number and need not be changed. However, some people may prefer to rewrite it as 872 mH, which is equally good.

7–12 DECIMAL CONVERSION FACTORS

Often it becomes necessary to convert microamperes to milliamperes, gigahertz to kilohertz, megawatts to watts, and so on. The more common conversions in simplified form are listed in Table 7–3.

TABLE 7–3 CONVERSION FACTORS

Multiply	By	To Obtain
Picounits	10^{-6}	Microunits
Picounits	10^{-9}	Milliunits
Picounits	10^{-12}	Units
Microunits	10^{6}	Picounits
Microunits	10^{-3}	Milliunits
Microunits	10^{-6}	Units
Milliunits	10^{9}	Picounits
Milliunits	10^{3}	Microunits
Milliunits	10^{-3}	Units
Units	10^{12}	Picounits
Units	10^{6}	Microunits
Units	10^{3}	Milliunits
Units	10^{-3}	Kilounits
Units	10^{-6}	Megaunits
Kilounits	10^{3}	Units
Kilounits	10^{-3}	Megaunits
Megaunits	10^{6}	Units
Megaunits	10^{3}	Kilounits

EXAMPLE 4 Convert 8 μF to farads.

SOLUTION

$$8\ \mu\text{F} = 8 \times 10^{-6}\ \text{F}$$

EXAMPLE 5 Convert 250 mA to amperes.

SOLUTION

$$\begin{aligned} 250\ \text{mA} &= 250 \times 10^{-3}\ \text{A} \\ &= 2.50 \times 10^{-1}\ \text{A} \end{aligned}$$

or

$$= 0.250\ \text{A}$$

EXAMPLE 6 Convert 1500 W to kilowatts.

SOLUTION

$$1500\ \text{W} = 1500 \times 10^{-3}\ \text{kW}$$

or

$$= 1.5\ \text{kW}$$

EXAMPLE 7 Convert 200 000 Ω to megohms.

SOLUTION

$$200\ 000\ \Omega = 200\ 000 \times 10^{-6}\ \text{M}\Omega = 0.2\ \text{M}\Omega$$

EXAMPLE 8 Convert 2500 kHz to megahertz.

SOLUTION

$$2500\ \text{kHz} = 2500 \times 10^{-3}\ \text{MHz} = 2.500\ \text{MHz}$$

EXAMPLE 9 Convert 0.000 450 S to microsiemens.

SOLUTION

$$0.000\ 450\ \text{S} = 0.000\ 450 \times 10^{6}\ \mu\text{S}$$

or

$$= 450\ \mu\text{S}$$

EXAMPLE 10 Convert 5 μs to seconds.

SOLUTION

$$5\ \mu\text{s} = 5 \times 10^{-6}\ \text{s}$$

PROBLEMS 7–1

Express answers to three significant figures as numbers between 1 and 10 times the proper power of 10:

1. 2700 V = (*a*) _____ mV = (*b*) _____ μV = (*c*) _____ kV
2. 6.85 A = (*a*) _____ mA = (*b*) _____ μA
3. 5.25 V = (*a*) _____ kV = (*b*) _____ μV = (*c*) _____ mV
4. 125 mA = (*a*) _____ μA = (*b*) _____ A
5. 3900 Ω = (*a*) _____ kΩ = (*b*) _____ MΩ = (*c*) _____ S

6. 50 μF = (*a*) _____ F = (*b*) _____ pF
7. 50 000 pF = (*a*) _____ F = (*b*) _____ μF
8. 16.5 mH = (*a*) _____ H = (*b*) _____ μH
9. 868 W = (*a*) _____ kW = (*b*) _____ mW = (*c*) _____ μW
10. 25.3 s = (*a*) _____ ms = (*b*) _____ μs
11. 1320 kHz = (*a*) _____ MHz = (*b*) _____ Hz
12. 47 kΩ = (*a*) _____ Ω = (*b*) _____ MΩ = (*c*) _____ S
13. 400 mW = (*a*) _____ W = (*b*) _____ kW
14. 220 μH = (*a*) _____ mH = (*b*) _____ H
15. 15 kHz = (*a*) _____ MHz = (*b*) _____ Hz
16. 8 μs = (*a*) _____ ms = (*b*) _____ s = (*c*) _____ ns
17. 0.055 A = (*a*) _____ μA = (*b*) _____ mA
18. 325 kV = (*a*) _____ V = (*b*) _____ MV
19. 5.6 MΩ = (*a*) _____ Ω = (*b*) _____ kΩ
20. 3.7 kWh = (*a*) _____ Wh = (*b*) _____ mWh
21. 3350 mH = (*a*) _____ μH = (*b*) _____ H
22. 506 MHz = (*a*) _____ kHz = (*b*) _____ Hz
23. 0.000 50 μF = (*a*) _____ pF = (*b*) _____ F
24. 1500 ms = (*a*) _____ μs = (*b*) _____ s = (*c*) _____ ns
25. 2.5 S = (*a*) _____ μS = (*b*) _____ Ω
26. 5000 μS = (*a*) _____ S = (*b*) _____ Ω
27. 1880 μA = (*a*) _____ mA = (*b*) _____ A
28. 0.15 kV = (*a*) _____ V = (*b*) _____ mV
29. 280 MW = (*a*) _____ W = (*b*) _____ kW
30. 980 000 Hz = (*a*) _____ kHz = (*b*) _____ MHz

7–13 INTERSYSTEM CONVERSIONS

In the early sections of this chapter we briefly reviewed the two systems with which we most often deal, and we listed some common conversion factors. Some books of tables give hundreds of such interrelationships, and you will meet them as you continue your studies.

You must realize that, without the units, your calculations are incomplete. When measurements are added, subtracted, multiplied, or divided, then the units pertaining to those measurements must also take part in the calculations.

EXAMPLE 11 Add 6 V and 12 V.

SOLUTION

$$6\ \text{V} + 12\ \text{V} = 18\ \text{V}$$

EXAMPLE 12 Add 3 m and 75 cm.

SOLUTION

In the metric system, the values of the prefixes represent decimal multipliers:

(*a*) 75 cm $= \frac{75}{100}$ m $=$ 0.75 m. Adding:

$$\begin{array}{rcr} 3\text{ m} & = & 3\ \ \ \ \text{m} \\ +75\text{ cm} & = & +0.75\text{ m} \\ \hline & & 3.75\text{ m} \end{array}$$

(*b*) Alternatively, 3 m $=$ 300 cm. Adding:

$$\begin{array}{rcr} 3\text{ m} & = & 300\text{ cm} \\ +75\text{ cm} & = & +\ \ 75\text{ cm} \\ \hline & & 375\text{ cm} \end{array}$$

Either answer is correct, but one form may be more acceptable than the other under some circumstances. The same person might properly describe the height of a child as 112 cm and later refer to a folded dipole antenna as 1.12 m long.

EXAMPLE 13 What is the speed of an object that traverses 30 m in 2 s?

SOLUTION

Speed is given in units of distance per unit of time. In this case, the speed is

$$\frac{30\text{ m}}{2\text{ s}} = 15\,\frac{\text{m}}{\text{s}} \qquad \text{(usually written m/s*)}$$

EXAMPLE 14 What is the area of a room 12 m long and 18 m wide?

SOLUTION

Areas are given in square measure:

$$(12\text{ m})(18\text{ m}) = 216\text{ square meters }(\text{m}^2)^\dagger$$

EXAMPLE 15

$$3\ \Omega + 6\ \Omega = 9\ \Omega$$
$$230\text{ V} - 115\text{ V} = 115\text{ V}$$

EXAMPLE 16

$$2\text{ m} \times 4\text{ m} = 2 \times 4 \times \text{m} \times \text{m} = 8\text{ m}^2$$
$$0.7\text{ m} \times 1.6\text{ m} = 0.7 \times 1.6 \times \text{m} \times \text{m} = 1.12\text{ m}^2$$
$$20\text{ cm} \times 1.2\text{ m} = 0.2\text{ m} \times 1.2\text{ m} = 0.24\text{ m}^2$$
$$3\text{ m} \times 5\text{ m} \times 2\text{ m} = 3 \times 5 \times 2 \times \text{m} \times \text{m} \times \text{m}$$
$$= 30\text{ m}^3$$

---

*m/s (*a shilling fraction*) has exactly the same meaning as $\frac{\text{m}}{\text{s}}$ (*a built-up fraction*); the *only* difference is in the manner of printing.

†It is generally preferred that areas be written in the form 8 m^2 rather than 8 sq m.

$$6\text{ m} \times 10\text{ m} = 6 \times 10 \times \text{meters} \times \text{meters}$$
$$= 60\text{ meters}^2 = 60\text{ m}^2$$

In a ratio between identical units, such as $\frac{60\text{ m}}{12\text{ m}}$, the unit symbols cancel and the result of the division is a "pure" number with no dimension.

EXAMPLE 17

$$\frac{60\text{ m}}{12\text{ m}} = \frac{60\,\cancel{\text{m}}}{12\,\cancel{\text{m}}} = 5$$

When quantities having different units are multiplied or divided, the result must express the operation.

EXAMPLE 18

$$4\text{ m} \times 5\text{ kg} = 4 \times 5 \times \text{m} \times \text{kg}$$
$$= 20\text{ kg} \cdot \text{m}$$

EXAMPLE 19

$$\frac{30\text{ m}}{10\text{ s}} = \frac{30}{10}\frac{\text{m}}{\text{s}} = 3\,\frac{\text{m}}{\text{s}}$$
$$= 3\text{ m/s} = 3\text{ m} \cdot \text{s}^{-1}$$

EXAMPLE 20

$$\frac{45\ \Omega}{15\text{ m}} = \frac{45}{15}\frac{\Omega}{\text{m}} = 3\ \Omega/\text{m}$$

In Example 19 note that m/s is read as "meters per second," and in Example 20 note that Ω/m is read as "ohms per meter." Per means *divided by.*

Thus some of the equivalent lengths stated in Sec. 7–7 can be expressed as follows:

- There are 2.540 cm/in.
- There is 0.3048 m/ft.
- There are 1.609 km/mi.
- There are 39.37 in/m.
- There is 0.6214 mi/km.

Using relations in forms such that units are treated mathematically as literal factors facilitates conversions and assures that results will be obtained with correct units.

EXAMPLE 21 Convert 3 in to centimeters.

SOLUTION

$$3\text{ in} \times 2.54\,\frac{\text{cm}}{\text{in}} = 3 \times 2.54 \cdot \cancel{\text{in}} \cdot \frac{\text{cm}}{\cancel{\text{in}}}$$
$$= 7.62\text{ cm}$$

EXAMPLE 22 How many meters are there in 236 ft?

SOLUTION

$$236\text{ ft} \times 0.3048\,\frac{\text{m}}{\text{ft}} = 236 \times 0.3048 \cdot \cancel{\text{ft}} \cdot \frac{\text{m}}{\cancel{\text{ft}}}$$
$$= 71.93\text{ m}$$

EXAMPLE 23 A certain resistance wire has a resistance of 3 Ω/m. What is the resistance of 6 m of this wire?

SOLUTION

$$3\,\frac{\Omega}{\text{m}} \times 6\ \text{m} = 3 \times 6 \cdot \frac{\Omega}{\cancel{\text{m}}} \cdot \cancel{\text{m}} = 18\ \Omega$$

EXAMPLE 24 Convert 1500 kHz to hertz.

SOLUTION

There are 10^3 Hz per kilohertz, that is, $10^3\,\frac{\text{Hz}}{\text{kHz}}$. Then

$$\begin{aligned} 1500\ \text{kHz} &= 1500\,\frac{\text{kcycles}}{\text{s}} \times 10^3\,\frac{\text{cycles}}{\text{kcycle}} \\ &= 1500 \times 10^3\,\frac{\cancel{\text{kcycles}}}{\text{s}} \cdot \frac{\text{cycles}}{\cancel{\text{kcycle}}} \\ &= 1.5 \times 10^6\ \text{cycles/s} \\ &= 1.5 \times 10^6\ \text{Hz} \end{aligned}$$

EXAMPLE 25 The wavelength λ of a radio wave in meters, the frequency f of the wave in hertz, and the velocity of propagation c in meters per second are related to one another by the formula

$$\lambda = \frac{c}{f}$$

or

$$\lambda = \frac{3 \times 10^8}{f}\quad \text{m}$$

Derive a formula for wavelength expressed in feet.

SOLUTION

Since there are 3.28 ft/m, this factor must be applied to express λ in feet. Thus,

$$\begin{aligned} \lambda &= \frac{3 \times 10^8}{f}\ \text{m} \times 3.28\,\frac{\text{ft}}{\text{m}} \\ &= \frac{3 \times 3.28 \times 10^8}{f} \cdot \text{m} \cdot \frac{\text{ft}}{\text{m}} \\ &= \frac{9.84 \times 10^8}{f}\ \text{ft*} \end{aligned}$$

*Some users of this book will be interested in the 1972 report issued by the National Bureau of Standards, which gives the value of c as 299 792.4562 km/s ± 1.1 m/s. 3×10^8 is a sufficiently accurate approximation for the correct solution to all the problems in this book, and, indeed, for most of the problems ever to be solved by the majority of electronics technicians anywhere.

EXAMPLE 26 By using the formula $\lambda = (3 \times 10^8)/f$ m, derive a formula for wavelength in meters when the frequency is expressed in megahertz.

SOLUTION

In the above formula f is expressed in hertz and it is desired to express the frequency in megahertz. Since

$$\text{MHz} = \text{Hz} \times 10^6$$

this is substituted for f in the formula. Thus,

$$\lambda = \frac{3 \times 10^8}{f \times 10^6}\ \text{m} = \frac{3 \times 10^2}{f}\ \text{m} = \frac{300}{f}\ \text{m}$$

PROBLEMS 7–2

1.	7 ft	= (*a*) ____	in	= (*b*) ____	cm	= (*c*) ____	mm
2.	3800 mm	= (*a*) ____	km	= (*b*) ____	ft	= (*c*) ____	yd
3.	1.76 m	= (*a*) ____	in	= (*b*) ____	cm	= (*c*) ____	yd
4.	21 250 ft	= (*a*) ____	km	= (*b*) ____	mi	= (*c*) ____	cm
5.	8025 yd	= (*a*) ____	mi	= (*b*) ____	m	= (*c*) ____	km
6.	60 mi/h	= (*a*) ____	m/s				

7. The radius of No. 14 wire is 32 thousandths of an inch. What is its diameter in millimeters?
8. Radio waves are often referred to by wavelength instead of frequency. The wavelength of waves at a frequency of 3000 MHz is 10 cm. What is that wavelength in millimeters?
9. Magnetic measurements that used to be made in cgs units, maxwells (Mx) and gausses (Gs), are now made in SI units, webers (Wb) and teslas (T). A flux of 1 Mx = 1 line of magnetism, and a flux density of 1 Gs = 1 Mx/cm^2. Today, a flux of 1 Wb = 10^8 Mx, and a flux density of 1 T = 1 Wb/m^2. What is 1 Gs in terms of teslas, in engineering notation?
10. The capacitance of a power line is measured at 4.98 nF/km. What is the capacitance per meter?
11. A transmission line 300 m long is found to have an attenuation loss of 0.18 decibel (dB). What is the attenuation in decibels per hundred meters?
12. A twisted-pair transmission line 200 m long has a loss of 42 dB. What is the loss in decibels per centimeter?
13. The high-frequency resistance of No. 10 copper wire was measured some years ago by using a 10-ft length of wire. At 100 MHz, the resistance was found to be 0.980 Ω. What was the resistance in ohms per centimeter at the same frequency?
14. The speed of free electrons in random motion is approximately 100 000 m/s. What is this speed in miles per hour?
15. The speed of electrons "drifting" in an electric current flow is about 0.2 cm/s. What is this speed in inches per minute?

7–14 PRACTICAL CONSIDERATIONS

In Secs. 6–3 and 7–9 and in several instances through the use of examples and problems, attempts have been made to emphasize that extremely wide ranges in values of units are encountered in electrical and electronics computations. This has been done in order to impress you with the necessity of exercising care in making computations if you are to obtain accurate results. For example, in computing inductive reactances, the frequency may be in megahertz and the inductance in microhenrys. In radar and other applications we are concerned with the velocity of propagation of radio waves (3×10^8 m/s) and with time intervals in microseconds. This is equally true in television reception, particularly as it relates to the production of duplicate images. As an example, Fig. 7–1 illustrates how a television receiver can receive a picture signal from

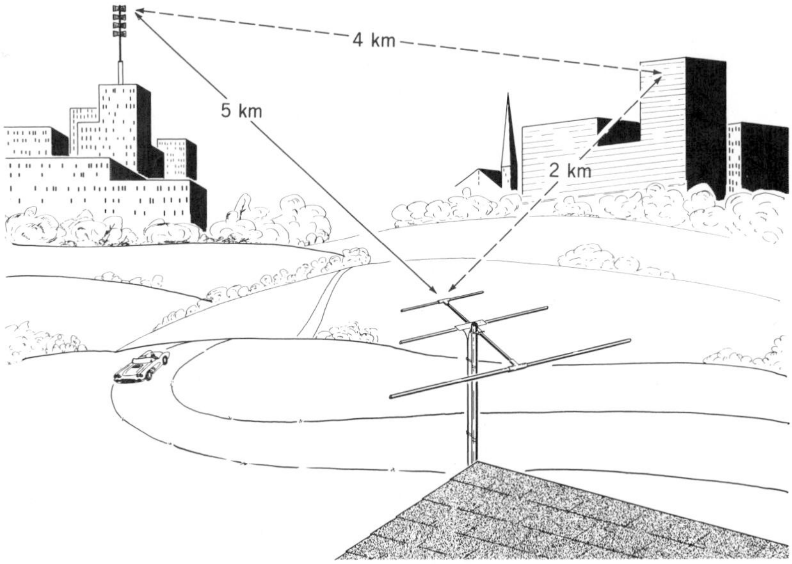

FIG. 7–1 Antenna receiving picture signal via two paths.

a transmitting station by different paths. The direct wave is received from the transmitter along one path, while the other signal arrives at the receiving antenna via a path 1 km longer than the direct path as a result of being reflected. Because the velocity of radio waves is 3×10^8 m/s, the reflected signal arrives at the receiver $1/(3 \times 10^5)$ s, or about 3.3 μs, later than the signal received via the direct path between transmitter and receiver. Since the electron scanning beam will scan one horizontal line in approximately 55 μs, on a picture 50 cm wide the beam scans about 1 cm in 1.1 μs. Therefore, the reflected signal arriving 3.3 μs late will produce a second picture 3 cm to the right in the direction of scanning as shown in Fig. 7–2. This duplicate image produced by the reflected wave is called a *ghost*.

FIG. 7–2 Television ghost. (*Courtesy of Radio Corporation of America*)

7–15 SIGNIFICANT FIGURES

The subjects of accuracy and significant figures were discussed in Sec. 6–1. Now that we have some idea of the various units used in electrical and electronics problems, two questions arise:

1. To how many significant figures should an answer be expressed?
2. How can we definitely show that an answer is correct to just so many significant figures?

The answer to the first question is comparatively easy. No answer can be more accurate than the figures, or data, used in the problem. It is safe to assume that the values of the average circuit components and calibrations of meters that we use in our everyday work are not known beyond three significant figures. Therefore, in the future we will round off long answers and express them to three significant figures. The exception will be when it is necessary to carry figures out in order to demonstrate or obey some fact or law, very carefully.

The second question brings up some interesting points. As an example, suppose we have a resistance of 500 000 Ω and we want to write this value so that it will be apparent to anyone that the figure 500 000 is correct to three significant figures. We can do so by writing

$$500 \times 10^3\ \Omega \quad \text{or} \quad 500\ \text{k}\Omega$$
$$50.0 \times 10^4\ \Omega$$
$$5.00 \times 10^5\ \Omega \quad \text{or} \quad 0.500\ \text{M}\Omega\text{, etc.}$$

Any one of these expressions definitely shows that the resistance is correct to three significant figures. Similarly, suppose we had measured the capacitance of a capacitor to be 3500 pF. How can we specify that the figure 3500 is correct to three significant figures? Again we can do so by writing

$$350 \times 10^1\ \text{pF}$$
$$35.0 \times 10^2\ \text{pF}$$
$$3.50 \times 10^3\ \text{pF} \quad \text{or} \quad 3.50\ \text{nF, etc.}$$

As in the preceding example, there are definitely three figures in the first factor that show the degree of accuracy.

7–16 CALCULATIONS WITH UNITS

In Sec. 7–13 we emphasized the necessity of keeping track of the units involved when performing calculations. The necessity becomes even more apparent when decimal multipliers of basic units are involved or when you are unsure how to proceed with a solution involving units of different measurements such as decibels and meters, ohms and meters, and hours and kilometers.

As long as your calculations are made in basic units, which are directly related, you will have no difficulty. For example, you know that

$$\text{Ohms} = \frac{\text{volts}}{\text{amperes}}$$

and

$$\text{Ohms} \neq \frac{\text{volts}}{\text{milliamperes}}$$

The milliamperes must be converted to amperes in order to keep the basic relationship in units. Therefore,

$$\text{Ohms} = \text{the number of } \frac{\text{volts}}{\text{milliamperes} \times 10^{-3}}$$

Of course, you could make up your own formulas for special cases and write, for example,

$$\text{Ohms} = \text{the number of } \frac{\text{volts} \times 10^{3}}{\text{milliamperes}}$$

but the task would be endless. Some frequently used formulas are derived for convenience, and you will derive some of them in Problems 7–3. However, when performing calculations, you will never go wrong if you first convert to basic units.

EXAMPLE 27 The voltage across a circuit is 250 V, and the current is 5 mA. What is the resistance of the circuit?

SOLUTION

Since ohms $= \frac{\text{volts}}{\text{amperes}}$, it is necessary to convert the current of 5 mA into amperes before calculating:

$$R = \frac{V}{I} = \frac{250}{5 \times 10^{-3}}$$
$$= 50 \times 10^{3}\ \Omega$$
$$= 50\ \text{k}\Omega$$

EXAMPLE 28 A current of 150 μA flows through a resistance of 30 kΩ. What is the voltage across the resistance?

SOLUTION

Since the current is in microunits and the resistance is in kilounits, both must be converted into basic units (amperes and ohms) before calculating:

$$\text{Volts} = \text{amperes} \times \text{ohms}$$

or

$$\begin{aligned} V &= I \times R \\ &= (150 \times 10^{-6})(30 \times 10^{3}) \\ &= 4.5 \text{ V} \end{aligned}$$

You will encounter cases in which you may be unsure how to proceed, particularly when you deal with units of differing measurements such as Ω/m, μF/km, m/s, kg/m², and dB/100 m. Keeping track of your units and handling them as literal numbers will ensure a correct numerical answer expressed in the proper units.

EXAMPLE 29 How long will it take to travel 225 km at an average speed of 45 km/h?

SOLUTION

Here we have kilometers and kilometers per hour, and we know the answer must be expressed in hours. Also, we know that

$$\text{Distance} = \text{speed} \times \text{time}$$

or

$$\text{Time} = \frac{\text{distance}}{\text{speed}}$$

That is

$$\text{h} = \frac{\text{km}}{\frac{\text{km}}{\text{h}}}$$

$$= \cancel{\text{km}} \cdot \frac{\text{h}}{\cancel{\text{km}}} = \text{h}$$

Knowing that the answer will be expressed in the proper unit, we can complete the calculation:

$$\begin{aligned} \text{Time} &= \frac{225 \text{ km}}{45 \frac{\text{km}}{\text{h}}} \\ &= \frac{225}{45} \text{ km} \cdot \frac{\text{h}}{\text{km}} \\ &= 5 \text{ h} \end{aligned}$$

EXAMPLE 30 A 3-km roll of No. 10 wire is measured and found to have a resistance of 9.81 Ω. What is the resistance of 100 m of this wire?

SOLUTION

The resistance must be expressed in ohms. Since the measurement was $\frac{9.81\ \Omega}{3000 \text{ m}}$,

$$\frac{9.81}{3000} \frac{\Omega}{\text{m}} = 3.27 \times 10^{-3} \frac{\Omega}{\text{m}}$$

Then the resistance of 100 m of this wire is

$$3.27 \times 10^{-3} \frac{\Omega}{\cancel{m}} \times 100 \cancel{m} = 0.327\ \Omega$$

This could be written as 0.327 Ω/100 m.

In the problems which follow, you will be asked to make conversions to accommodate readings in units which do not exactly fit the formulas relating the dimensions, as in Example 27, in which 5 mA had to be converted into amperes before proceeding. You will also be asked to convert the basic or classic formulas to adjust for units other than the basic ones. When both of these are asked for in one problem, follow this rule:

RULE

Adjust the units in which the measurements were made so that they will agree with the units for which the formula was developed. Then convert to other units as required.

PROBLEMS 7–3

1. The capacitive reactance of a circuit, or a capacitor, is given by the formula

$$X_C = \frac{1}{2\pi fC} \qquad \Omega$$

where X_C = capacitive reactance, Ω
f = frequency, Hz
C = capacitance of circuit, or capacitor, F

Show that $$X_C = \frac{159 \times 10^3}{fC} \qquad \Omega$$

when f = frequency, MHz
C = capacitance, pF

2. Referring to Prob. 1, what is the capacitive reactance of a capacitor of 0.000 02 μF at a frequency of 3200 MHz?
3. The inductive reactance of a circuit, or an inductor, is given by the formula

$$X_L = 2\pi fL \qquad \Omega$$

where X_L = inductive reactance, Ω
f = frequency, Hz
L = inductance of circuit, or inductor, H
Derive a formula for X_L
when f = frequency, MHz
L = inductance, μH

4. Referring to Prob. 3, an amplifier coil has an inductance of 15 μH. What is its inductive reactance at 1.8 MHz?

5. The resonant frequency of any circuit is given by the formula

$$f = \frac{1}{2\pi\sqrt{LC}} \qquad \text{Hz}$$

where f = frequency, Hz
L = inductance of circuit, H
C = capacitance of circuit, F
Derive a formula expressing f in megahertz
when L = inductance, μH
C = capacitance, pF

6. Referring to Prob. 5, what is the resonant frequency of a circuit with an inductance of 0.3 μH and a capacitance of 22 pF?

7. In copper conductors used in transmission lines, the depth of penetration of high-frequency currents is given by the formula

$$\delta = \frac{2.61 \times 10^{-3}}{\sqrt{f}} \qquad \text{in}$$

where f = frequency, MHz
Derive a formula for current penetration depth in centimeters when f is the frequency in megahertz.

8. Referring to Prob. 7, to what depth in millimeters will a current at 600 kHz penetrate a copper conductor?

9. The high-frequency resistance of a round copper wire or of round copper tubing was found in an old handbook to be

$$R_{ac} = 9.98 \times 10^{-4} \frac{\sqrt{f}}{d} \qquad \Omega/\text{ft}$$

where R_{ac} = high-frequency resistance, Ω/ft
f = frequency, MHz
d = outside diameter of conductor, in
Derive a formula for R_{ac} in ohms per centimeter when f is given in megahertz and d is given in centimeters.

10. Referring to Prob. 9, No. 40 wire has a diameter of 0.079 mm. What is the resistance per centimeter of the wire at a frequency of 120 MHz?

11. Use the formula in Example 25 to show that $\lambda = \dfrac{3 \times 10^4}{f}$ cm when f is given in megahertz.

12. Use the formula in Example 25 to derive a formula for wavelength (λ) in centimeters when f is given in kilohertz.

13. The midfrequency of television channel 4 is 69 MHz. Using the formula derived in Prob. 12, what is the length of one wavelength in centimeters?

14. The great majority of television receiving antennas consist of various combinations of dipoles. A dipole antenna is one that is approximately one-half wavelength long (0.5λ), as illustrated in Fig. 7–3. The actual length is slightly less than a half wave owing to "end effect" caused by the capac-

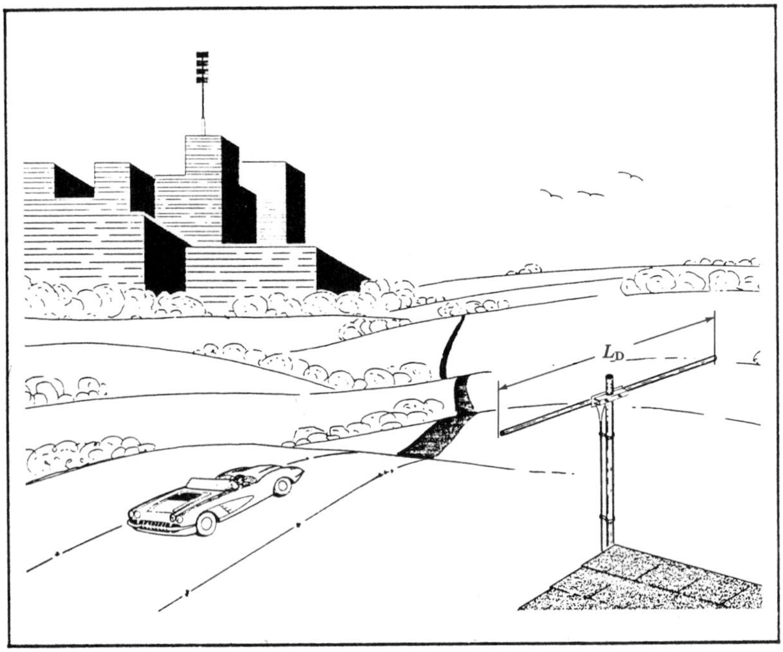

FIG. 7–3 Dipole antenna.

itance of the antenna, and it has been determined that dipoles used for television reception should be approximately 6% shorter than one-half wavelength. Use the formula derived in Prob. 12 to derive a formula for the length of a dipole antenna in centimeters when the frequency is in megahertz.

15. The midfrequency of television channel 13 is 213 MHz. Using the formula derived in Prob. 14, what length would you make a receiving antenna for this channel?

16. If a wire approximately one-half wavelength long is placed behind a dipole antenna, it acts as a reflector and increases the directivity of the antenna. This results in the reception of stronger signals when the dipole and the reflector are pointed at the transmitting station as illustrated in Fig. 7–4. For best results, the reflector should be 5% longer than the dipole. Referring to the formula for the length of a dipole derived in Prob. 14, derive a formula for the length of a reflector in centimeters when f is in megahertz.

17. The distance between a dipole and its reflector should be approximately one-fifth of one wavelength (0.2λ) as shown in Fig. 7–4. Referring to previously derived formulas, compute the following dimensions for the midfrequency of television channel 10, which is 195 MHz: (*a*) length of dipole, (*b*) length of reflector, and (*c*) spacing between dipole and reflector.

18. The directivity of a dipole-reflector combination, as shown in Fig. 7–4, can be increased by the addition of a conductor in front of the dipole as illustrated in Fig. 7–5. This wire or tube, which is known as a director, is

FIG. 7–4 Dipole antenna with reflector.

FIG. 7–5 Dipole antenna with reflector and director.

usually placed one-tenth wavelength (0.1 λ) from the dipole, and it should be about 5% shorter than the dipole. Derive a formula for the length of a director in centimeters when f is in megahertz.

19. Referring to Fig. 7–5, compute the following dimensions for the midfrequency of television channel 8, which is 183 MHz: (*a*) length of dipole, (*b*) length of reflector, (*c*) length of director, (*d*) spacing between dipole and reflector, and (*e*) spacing between dipole and director.
20. Ohm's law may be stated in the form $V = IR$, where V is measured in volts, I in amperes, and R in ohms. What voltage will appear across a resistor measuring 680 MΩ when a current of 0.250 μA flows through the resistor?

SELF TEST

1. Convert 1760 mA to amperes.
2. Convert 18 000 V to millivolts.
3. Convert 277 mA to microamperes.
4. What is the reciprocal of 600 kHz?
5. Convert a capacitance of 9.27 nF/mi to microfarads per meter.
6. $\lambda = \frac{c}{f}$. What is the wavelength of a "microwave" signal of 12 GHz?
7. Convert a speed of 800 ft/s to kilometers per hour.
8. The speed of electrons drifting in a current is approximately 4.5×10^{-3} mi/h. What is this speed in millimeters per second?
9. $$R_{ac} = 9.98 \times 10^{-4} \frac{\sqrt{f}}{d} \ \Omega/\text{ft}$$

   where f = frequency, MHz
   d = outside diameter of conductor, in.

   Derive a formula for R_{ac} in ohms per meter when f is given in gigahertz and d is given in millimeters.
10. The spacing between a dipole antenna and its reflector is given by $d = 0.2\ \lambda$. If the separation is 600 mm, what is the frequency for which the antenna is best suited?

CHAPTER 8

OHM'S LAW—SERIES CIRCUITS

Ohm's law for the electric circuit is the foundation of electric circuit analysis and is therefore of fundamental importance. The various relations of Ohm's law are easily learned and are readily applied to practical circuits. A thorough knowledge of these relations and their applications is essential for understanding electric circuits.

This chapter is concerned with the study of Ohm's law in dc series circuits as applied to *parts* of a circuit. For this reason, the internal resistance of a source of voltage, such as a generator or a battery, and the resistance of the wires connecting the parts of a circuit are not discussed in this chapter.

8–1 THE ELECTRIC CIRCUIT

An electric circuit consists of a source of voltage connected by conductors to the apparatus that is to use the electric energy.

An electric current will flow between two points in a conductor when a difference of potential exists across those points. The most generally accepted concept of an electric current is that it consists of a motion, or flow, of electrons from the negative toward a more positive point in a circuit. The force that causes the motion of electrons is called an *electromotive force,* a *potential difference,* or a *voltage,* and the opposition to the motion is called *resistance*.

The basic theories of electrical phenomena and the methods of producing currents are not within the scope of this book. You will find them adequately treated in the great majority of textbooks on the subject.

8–2 OHM'S LAW

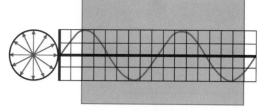

Ohm's law for the electric circuit, reduced to plain terms, states the relation that exists among voltage, current, and resistance. One way of stating this relation is as follows: The voltage across any *part* of a circuit is proportional to the product of the current through that *part* of the circuit and the resistance of that *part* of the circuit. Stated as a formula, the foregoing is expressed as

$$V = IR \qquad \text{V} \tag{1}$$

where V = voltage, or potential difference, V
I = current, A
R = resistance, Ω

If any two factors are known, the third can be found by solving Eq. (1). Thus,

$$I = \frac{V}{R} \qquad \text{A} \tag{2}$$

and

$$R = \frac{V}{I} \qquad \Omega \tag{3}$$

8–3 METHODS OF SOLUTION

The general outline for working problems given in Sec. 5–8 is applicable to the solution of circuit problems. In addition, a neat, simplified diagram of the circuit should be drawn for each problem. The diagram should be labeled with all the known values of the circuit such as voltage, current, and resistances. In this manner the circuit and problem can be visualized and understood. Solving a problem by making purely mechanical substitutions in the proper formulas is not conducive to gaining a complete understanding of any problem.

EXAMPLE 1 What will be the current through a resistance of 150 Ω if the applied voltage across the resistance is 117 V?

SOLUTION

The circuit is represented in Figs. 8–1 and 8–2.

Given $V = 117 \text{ V} \qquad R = 150\ \Omega$

$I = ?$

$$I = \frac{V}{R} = \frac{117}{150} = 0.780 \text{ A}$$

EXAMPLE 2 A voltmeter connected across a resistance reads 220 V, and an ammeter connected in series with the resistance reads 2.60 A. What is the value of the resistance?

SOLUTION

The circuit is represented in Fig. 8–3.

Given $V = 220 \text{ V} \qquad I = 2.60 \text{ A}$

$R = ?$

$$R = \frac{V}{I} = \frac{220}{2.60} = 84.6\ \Omega$$

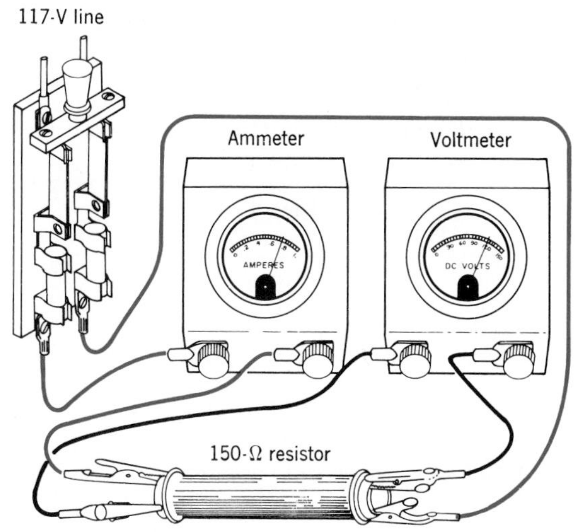

FIG. 8–1 Sketch of the circuit of Example 1, showing how the parts are connected to form the circuit.

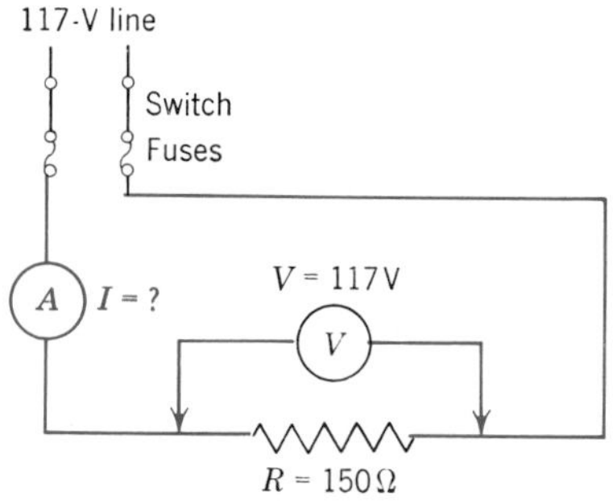

FIG. 8–2 Schematic circuit diagram of Example 1.

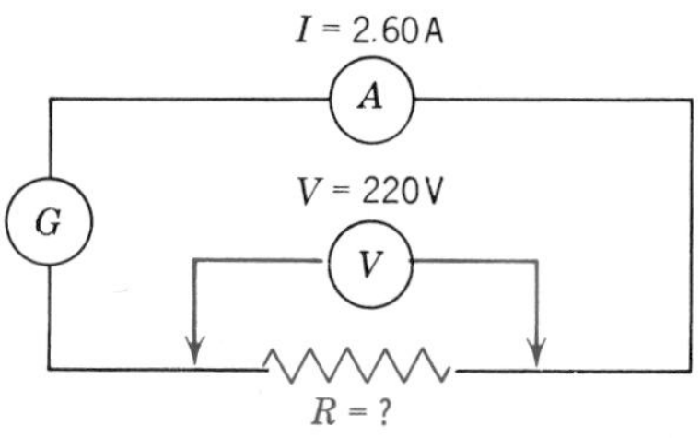

FIG. 8–3 Circuit of Example 2.

EXAMPLE 3 A current of 140 A flows through a resistance of 450 Ω. What should be the reading of a voltmeter connected across the resistance?

SOLUTION

The diagram of the circuit is shown in Fig. 8–4.

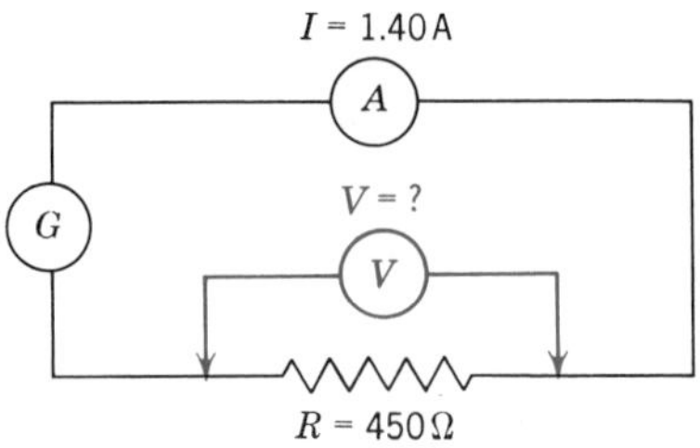

FIG. 8–4 Circuit of Example 3.

Given

$$I = 1.40\text{ A} \qquad R = 450\ \Omega$$
$$V = ?$$
$$V = IR = 1.40 \times 450 = 630\text{ V}$$

EXAMPLE 4 A measurement shows a potential difference of 63.0 μV across a resistance of 300 Ω. How much current is flowing through the resistance?

SOLUTION

The circuit is represented in Fig. 8–5.

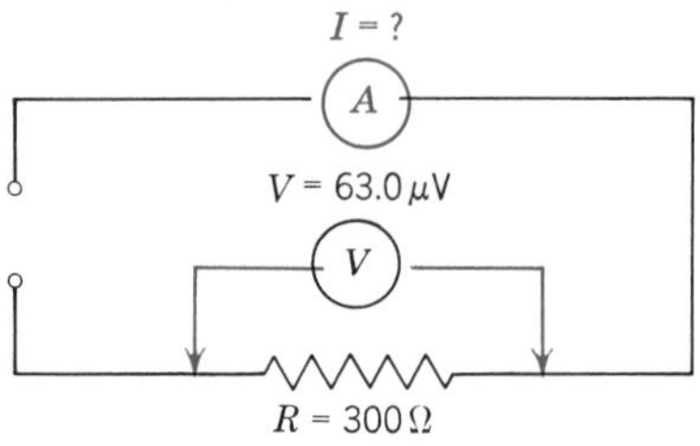

FIG. 8–5 Circuit of Example 4.

Given

$$V = 63.0\ \mu\text{V} = 6.3 \times 10^{-5}\text{ V} \qquad R = 300\ \Omega$$
$$I = ?$$
$$I = \frac{V}{R} = \frac{6.3 \times 10^{-5}}{300} = \frac{6.3 \times 10^{-7}}{3.00}$$
$$= 2.1 \times 10^{-7}\text{ A}$$

or

$$I = 0.21\ \mu\text{A} = 210\text{ mA}$$

EXAMPLE 5 There is a current of 8.60 mA through a resistance of 500 Ω. What voltage exists across the resistance?

SOLUTION

The circuit is represented in Fig. 8–6.

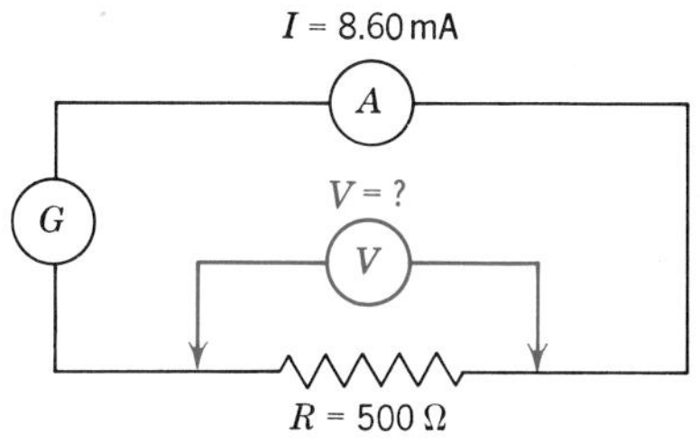

FIG. 8–6 Circuit of Example 5.

Given $I = 8.60 \text{ mA} = 8.60 \times 10^{-3} \text{ A} \qquad R = 500\ \Omega$

$V = ?$

$$
\begin{aligned}
V &= IR = 8.60 \times 10^{-3} \times 500 \\
&= 8.60 \times 10^{-3} \times 5 \times 10^{2} \\
&= 8.60 \times 5 \times 10^{-1} = 4.30 \text{ V}
\end{aligned}
$$

Carefully note, as illustrated in Examples 4 and 5, that the equations expressing Ohm's law are in units, that is, volts, amperes, and ohms.

PROBLEMS 8–1

1. What is the current through a 47.0-Ω resistor if a potential of 115 V is applied across the resistor?
2. A certain soldering iron draws 217 mA from a 115-V line. What is the resistance of the heating unit of the soldering iron?
3. What current will exist when an emf of 110 V is impressed across a 5.6-Ω resistor?
4. A milliammeter that is connected in series with a 10-kΩ resistor reads 8.0 mA. What is the voltage across the resistor?
5. A microvoltmeter connected across a 5-kΩ resistor reads 110 μV. What is the current through the resistor?
6. What voltage is required to cause a current flow of 6.2 mA through a resistance of 7.1 kΩ?
7. A certain milliammeter, with a scale of 0 to 1.0 mA, has a resistance of 28 Ω. If this milliammeter is connected directly across a 110-V line, what is the current through the meter? What conclusion do you draw?
8. The current through a 4.7-kΩ resistor is 5.1 mA. What should a voltmeter read when it is connected across the resistor?
9. The cold resistance of a carbon filament lamp is 222 Ω, and the hot resistance is 178 Ω. What is the current (*a*) when the lamp is switched across a 117-V line and (*b*) when constant operating temperature is reached?
10. A type 1N4455 semiconductor PN diode drops 0.7 V across its anode to cathode when connected in series with a 1-kΩ resistor and a 10-V direct-current source.
    (*a*) Determine the current passed by the diode.
    (*b*) Determine the resistance of the diode while it is conducting.

(*c*) What would be the resistance of the diode if the source voltage were increased to 17.5 V?

(*d*) If the answers to parts (*b*) and (*c*) are not the same, what reasons can you give?

8–4 POWER

In specifying the rating of electrical equipment, it is customary to state not only the voltage at which the equipment was designed to operate but also the rate at which the equipment produces or consumes electric energy.

The rate of producing or consuming energy is called *power,* and electric power is measured in watts or kilowatts. Thus, your study lamp may be rated 100 W at 120 V and a generator may be rated 2000 kW at 440 V.

Electric motors were formerly rated in terms of the mechanical energy output, measured in *horsepower,* which they could develop. The conversion from electric energy to this older unit of mechanical energy is given by the relation

$$746 \text{ W} = 1 \text{ horsepower (hp)}$$

With the advent of metrication, the SI unit of power will be used more and more, and the *watt,* or *joule per second,* will eliminate the horsepower rating.

$$1 \text{ watt (W)} = 1 \text{ J/s}$$

Because users of this book will undoubtedly be called upon to handle older motors rated in horsepower, a number of problems involving this older unit have been retained in this edition.

8–5 THE WATT

Energy is expended at a rate of one wattsecond (Ws or W·s)* (Joule, J) every second when one volt causes a current of one ampere to flow. In this case, we say that the power represented when one volt causes one ampere to flow is one watt. This relation is expressed as

$$P = VI \qquad \text{W} \tag{4}$$

This is a useful equation when the voltage and current are known.

Because, by Ohm's law, $V = IR,$ this value of V can be substituted in Eq. (4). Thus,

$$P = (IR)I$$

or

$$P = I^2R \qquad \text{W} \tag{5}$$

This is a useful equation when the current and resistance are known.

*The use of the center dot in the symbols for wattsecond and watthour is preferred in general physics relationships. However, it is customarily omitted in electricity and electronics usage.

By substituting the value of I of Eq. (2) in Eq. (4),

$$P = V\frac{V}{R}$$

or

$$P = \frac{V^2}{R} \quad \text{W} \tag{6}$$

This is a useful equation when the voltage and resistance are known.

WATTHOURS – KILOWATTHOURS The consumer of electric energy pays for the amount of energy used by his electrical equipment. This is measured by instruments known as *watthour* or *kilowatthour meters*. These meters record the amount of energy taken by the consumer.

Electric energy is sold at so much per kilowatthour (kWh). One watthour of energy is consumed when one watt of power continues in action for one hour. Similarly, 1 kWh is consumed when the power is 1000 W and the action continues for 1 h or when a 100-W rate persists for 10 h, etc. Thus, the amount of energy consumed is the product of the power and the time. Perhaps in time kilowatthour meters will be replaced by megajoule meters.

8–6 LOSSES

The study of the various forms in which energy may occur and the transformation of one kind of energy into another has led to the important principle known as the principle of the *conservation of energy*. Briefly, this states that energy can never be created or destroyed. It can be transformed from one form to another, but the total amount remains unchanged. Thus, an electric motor converts electric energy into mechanical energy, the incandescent lamp changes electric energy into heat energy, the loudspeaker converts electric energy into sound energy, the generator converts mechanical energy into electric energy, etc. In each instance the transformation from one type of energy to another is not accomplished with 100% efficiency because some energy is converted into heat and does no useful work as far as that particular conversion is concerned.

Resistance in a circuit may serve a number of useful purposes, but unless it has been specifically designed for heating or dissipation purposes, the energy transformed in the resistance generally serves no useful purpose.

8–7 EFFICIENCY

Because all electrical equipment contains resistance, some heat always develops when current flows. Unless the equipment is to be used for producing heat, the heat due to the resistance of the equipment represents wasted energy. No electrical equipment or other machine is capable of converting energy received into useful work without some loss.

The power that is furnished a machine is called the machine's *input,* and the power received from a machine is called its *output*. The efficiency of a machine is equal to the ratio of the output to the input. That is,

$$\text{Efficiency} = \frac{\text{output}}{\text{input}} \tag{7}$$

It is evident that the efficiency, as given in Eq. (7), is always a decimal, that is, a number less than 1. If we require the efficiency stated as a percentage, all that is required is to multiply Eq. (7) by 100%. Naturally, in Eq. (7), the output and input must be expressed in the same units. Hence, if the output is expressed in kilowatts, then the input must be expressed in kilowatts; if the output is expressed in horsepower, then the input must be expressed in horsepower; etc.

EXAMPLE 6 A voltage of 110 V across a resistor causes a current of 5 A to flow through the resistor. How much power is expended in the resistor?

SOLUTION 1

The circuit is represented in Fig. 8–7.

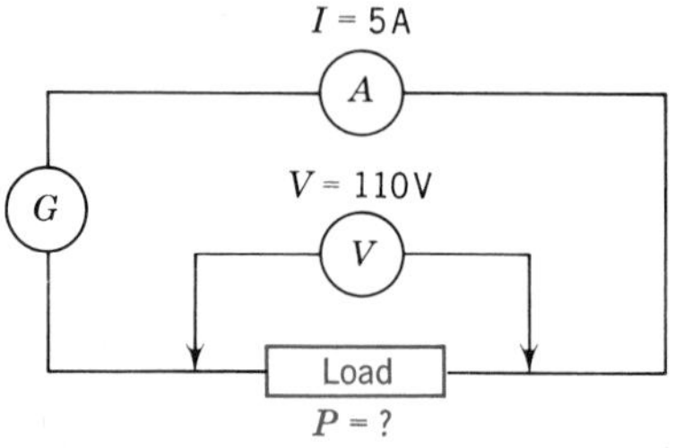

FIG. 8–7 Circuit of Example 6.

Given $V = 110 \text{ V} \quad I = 5 \text{ A} \quad P = ?$

Using Eq. (4), $P = VI = 110 \times 5 = 550 \text{ W}$

SOLUTION 2

Find the value of the resistance and use it to solve for *P*. Thus, using Eq. (3),

$$R = \frac{V}{I} = \frac{110}{5} = 22 \ \Omega$$

Using Eq. (5),

$$P = I^2R = 5^2 \times 22 = 5 \times 5 \times 22 = 550 \text{ W}$$

SOLUTION 3

Using Eq. (6),

$$P = \frac{V^2}{R} = \frac{110^2}{22} = \frac{110 \times 110}{22} = 550 \text{ W}$$

Solving a problem by two methods serves as an excellent check on the results, for there is then little chance of making the same error twice, as happens too often when the same method of solution is repeated.

EXAMPLE 7 A current of 2.5 A flows through a resistance of 40 Ω.
(*a*) How much power is expended in the resistor?
(*b*) What is the potential difference across the resistor?

SOLUTION 1

The circuit is represented in Fig. 8–8.

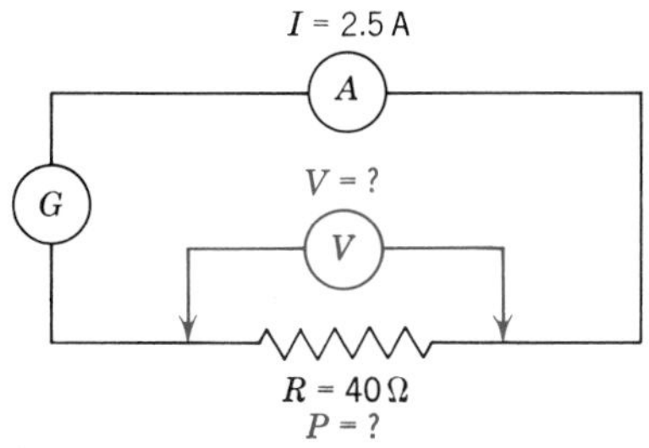

FIG. 8–8 Circuit of Example 7.

Given $I = 2.5\ \text{A} \quad R = 40\ \Omega \quad P = ? \quad V = ?$

(*a*) $P = I^2R = 2.5^2 \times 40$

$= 2.5 \times 2.5 \times 40 = 250\ \text{W}$

(*b*) $V = IR = 2.5 \times 40 = 100\ \text{V}$

SOLUTION 2

(*a*) Find *V*, as above, and use it to solve for *P*. Thus,

$$P = \frac{V^2}{R} = \frac{100^2}{40} = \frac{100 \times 100}{40} = 250\ \text{W}$$

or $P = VI = 100 \times 2.5 = 250\ \text{W}$

EXAMPLE 8 A voltage of 1.732 V is applied across a 500-Ω resistor.
(*a*) How much power is expended in the resistor?
(*b*) What is the current through the resistor?

SOLUTION

A diagram of the circuit is shown in Fig. 8–9.

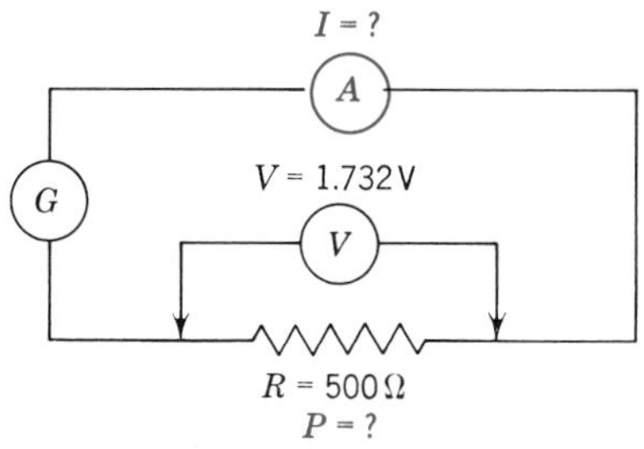

FIG. 8–9 Circuit of Example 8.

Given $V = 1.732\ \text{V} \qquad R = 500\ \Omega$

$P = ? \qquad I = ?$

(*a*)
$$P = \frac{V^2}{R} = \frac{1.732^2}{500} = \frac{1.732^2}{5 \times 10^2}$$
$$= \frac{1.732^2}{5} \times 10^{-2} = 0.006\ \text{W}$$

or $P = 6\ \text{mW}$

(*b*)
$$I = \frac{V}{R} = \frac{1.732}{500} = \frac{1.732}{5} \times 10^{-2}$$
$$= 0.346 \times 10^{-2}\ \text{A}$$

or $I = 3.46\ \text{mA}$

Check the foregoing solution for power by using an alternative method.

EXAMPLE 9 (*a*) What is the hot resistance of a 100-W 110-V lamp?
(*b*) How much current does the lamp take?
(*c*) At 4¢/kWh, how much does it cost to operate this lamp for 24 h?

SOLUTION 1

The circuit is represented in Fig. 8–10.

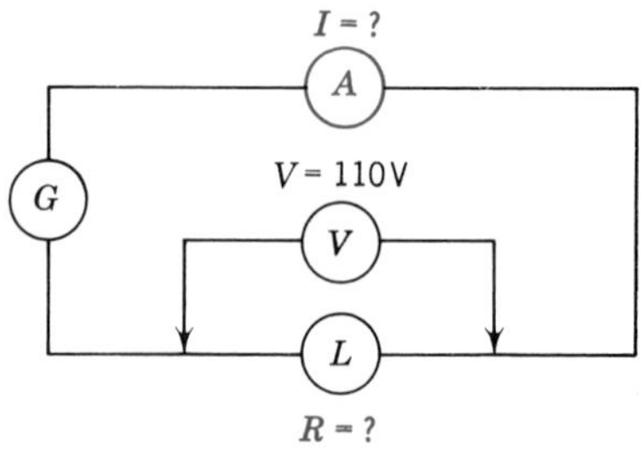

FIG. 8–10 Circuit of Example 9.

Given $P = 100\ \text{W} \qquad V = 110\ \text{V}$

(*a*) Because the power and voltage are known and the resistance is unknown, an equation that contains these three must be used. Thus,

$$P = \frac{V^2}{R}$$

hence
$$R = \frac{V^2}{P} = \frac{110^2}{100} = 121\ \Omega$$

(*b*) $$I = \frac{V}{R} = \frac{110}{121} = 0.909 \text{ A}$$

(*c*) If the lamp is lighted for 24 h, it will consume

$$100 \times 24 = 2400 \text{ Wh} = 2.40 \text{ kWh}$$

At 4¢/kWh the cost will be

$$2.4 \times 4 = 9.6¢$$

SOLUTION 2

The current may be found first by making use of the relation

$$P = VI$$

which results in $I = \frac{P}{V} = \frac{100}{110} = 0.909 \text{ A}$

The resistance can now be determined by

$$R = \frac{V}{I} = \frac{110}{0.909} = 121 \ \Omega$$

The solution can be checked by

$$P = I^2R = 0.909^2 \times 121 = 100 \text{ W}$$

which is the power rating of the lamp as given in the example. The cost is computed as before.

EXAMPLE 10 A motor delivering 6.50 mechanical horsepower is drawing 26.5 A from a 220-V line.

(*a*) How much electric power is the motor taking from the line?
(*b*) What is the efficiency of the motor?
(*c*) If power costs 3¢/kWh, how much does it cost to run the motor for 8 h?

SOLUTION

A diagram of the circuit is shown in Fig. 8–11.

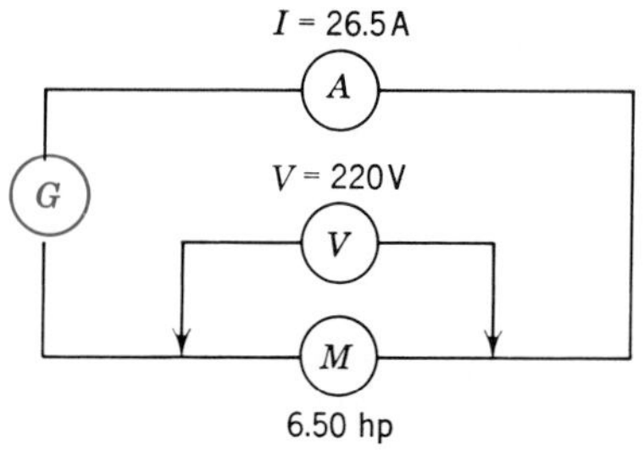

FIG. 8–11 Circuit of Example 10.

Given $V = 220\ \text{V} \qquad I = 26.5\ \text{A}$

and mechanical horsepower

$$\begin{aligned} P &= 6.5\ \text{hp} = 6.5 \times 746 \\ &= 4850\ \text{W} = 4.85\ \text{kW} \end{aligned}$$

(*a*) The power taken by the motor is

$$\begin{aligned} P = VI = 220 \times 26.5 &= 5830\ \text{W} \\ &= 5.83\ \text{kW} \end{aligned}$$

$$(b)\ \text{Efficiency} = \frac{\text{output}}{\text{input}} = \frac{4.85}{5.83} = 0.832 = 83.2\%$$

(*c*) Because the motor consumes 5.83 kW, in 8 h it will take

$$5.83 \times 8 = 46.6\ \text{kWh}$$

At 3¢/kWh, the cost will be $46.6 \times 0.03 = \$1.40$

NOTE *The cost was computed in two steps for the purpose of illustrating the solution. When you have become familiar with the method, the cost should be computed in one step. Thus,*

$$\text{Cost} = 5.83 \times 8 \times 0.03 = \$1.40$$

From the foregoing examples, it will be noted that computations involving power consist mainly in the applications of Ohm's law. Little trouble will be encountered if each problem is given careful thought and the systematic procedure previously outlined is followed in finding the solution.

PROBLEMS 8–2

1. 12 hp = (*a*) ______ W = (*b*) ______ kW
2. 32.6 kW = (*a*) ______ W = (*b*) ______ hp
3. What current is drawn by a 25-W soldering iron that is connected to a 115-V line?
4. How much power is expended in a 390-Ω resistor drawing 2.1 A?
5. What is the electric horsepower of a generator which delivers a current of 35.4 A at 240 V?
6. A voltmeter connected across a 15-kΩ resistor reads 80.5 V. How much power is being expended in the resistor?
7. A diesel engine is rated at 2200 hp. What is its electrical rating in kilowatts?
8. An ammeter is connected in the circuit of a 440-V motor. When the motor is running, the ammeter reads 2.27 A. How much power is being absorbed from the line?
9. The resistance of a certain ammeter is 0.064 Ω. Determine the power expended in the meter when it reads 1.6 A.

10. The resistance of a certain voltmeter is 300 kΩ. Determine the power expended in the voltmeter when it is connected across a 220-V line.

11. A type 2N5458 JFET used as an audio amplifier is self-biased by a 47-kΩ source resistor. A voltmeter connected across the resistor indicates 4.7 V dc.
   (*a*) What is the current through the resistor?
   (*b*) What is the continuous power radiated by the resistor while operating under these conditions?

12. A type 2N5459 JFET is operating with a 2.2-kΩ source bias resistor through which flows a current of 9 mA.
   (*a*) How much power is being expended by the resistor?
   (*b*) What is the voltage across the resistor?
   (*c*) What is the voltage at the gate with respect to the source terminal?

13. An emf of 80 μV is applied across a 680-Ω resistor.
   (*a*) How much power is expended in the resistor?
   (*b*) How much current will flow through the resistor?

14. A 1-kΩ resistor in the emitter circuit of a 2N1414 transistor produces a voltage of 6 V between ground and emitter.
   (*a*) What is the emitter current?
   (*b*) What is the power loss in this bias resistor?

15. A radar antenna motor is delivering 10 hp. A kilowattmeter that measures the power taken by the motor reads 8.24 kW.
   (*a*) What is the efficiency of the motor?
   (*b*) At 2.5¢/kWh, how much would it cost to run the motor continuously for 5 days?

16. A 440-V 10-hp forced-draft fan motor has an efficiency of 80%.
   (*a*) How many kilowatts does it consume?
   (*b*) How much current does it draw from the line?
   (*c*) At 2.5¢/kWh, how much would it cost to run this motor continuously for 1 week?

17. A generator which is 80% efficient delivers 50 A at 220 V. What must be the output of the diesel engine which drives the generator?

18. 23.9 kW is required to operate a 25-hp forced-draft fan motor.
   (*a*) What is its efficiency?
   (*b*) How much power is lost in the motor?

19. A generator delivers 80 A at 220 V with an efficiency of 88%. How much power is lost in the generator?

20. A 230-V $7\frac{1}{2}$-hp motor, which has an efficiency of 85%, is driving a radio transmitter 2-kV generator which has an efficiency of 80%. The motor is running fully loaded.
   (*a*) How much power does the motor take from the line?
   (*b*) How much current does the motor draw?
   (*c*) How much power will the generator deliver?
   (*d*) How much current will the generator deliver?
   (*e*) What is the overall efficiency; that is, what is the efficiency from motor input to generator output?

8–8 RESISTANCES IN SERIES

So far, our studies of the electric circuit have taken into consideration but one electrical component in the circuit, excluding the source of voltage. This is all very well for the purpose of becoming familiar with simple Ohm's law for power relations. However, practical circuits consist of more than one piece of equipment as far as circuit computations are concerned.

In a *series circuit* the various components comprising the circuit are so connected that the current, starting from the voltage source, must flow through each circuit component, in turn, before returning to the other side of the source.

There are three important facts concerning series circuits that must be borne in mind in order to understand thoroughly the action of such circuits and to facilitate their solution. In a series circuit:

1. The total voltage is equal to the sum of the voltages across the different parts of the circuit.
2. The current in any part of the circuit is the same.
3. The total resistance of the circuit is equal to the sum of the resistances of the different parts.

Point 1 is practically self-evident. If the sum of all the potential differences (voltage) around the circuit were not equal to the applied voltage, there would be some voltage left over which would cause an increase in current. This increase in current would continue until it caused enough voltage across some resistance just to balance the applied voltage. Hence,

$$V_t = V_1 + V_2 + V_3 + \ldots \tag{8}$$

Point 2 is self-evident, for the circuit components are so connected that the current must flow through each part in turn and there are no other paths back to the source.

To some, point 3 might not be self-evident. However, because it is agreed that the current I in Figs. 8–12 and 8–13 flows through all resistors, Eq. (8) can be used to demonstrate the truth of point 3. Thus, by dividing each member of Eq. (8) by I, we have

$$\frac{V_t}{I} = \frac{V_1 + V_2 + V_3}{I} + \ldots$$

or

$$\frac{V_t}{I} = \frac{V_1}{I} + \frac{V_2}{I} + \frac{V_3}{I} + \ldots$$

and by substituting R for $\frac{V}{I}$, we have

$$R_t = R_1 + R_2 + R_3 + \ldots \tag{9}$$

NOTE V_t *and* R_t *are used to denote total voltage and total resistance, respectively.*

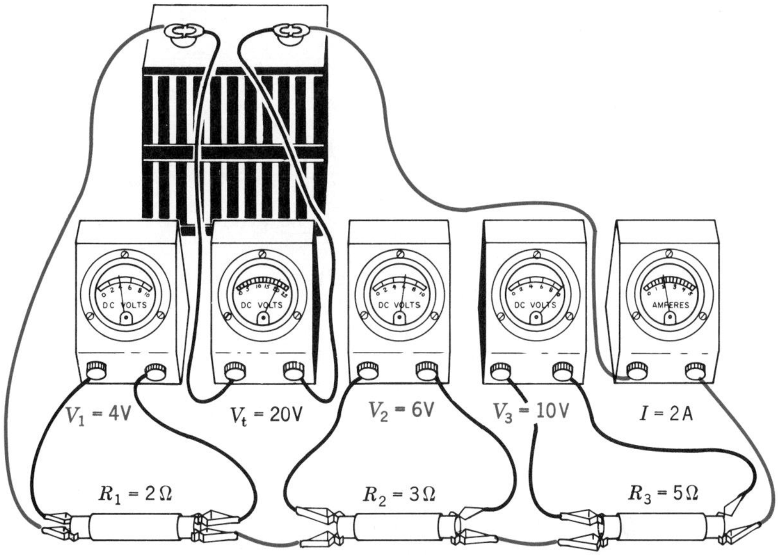

FIG. 8–12 Three resistors connected in series with a voltmeter connected across each resistor. The sum of the voltages across the resistors is equal to the battery voltage.

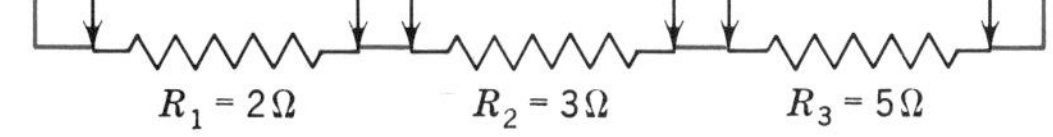

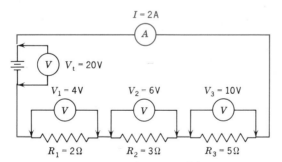

FIG. 8–13 Schematic diagram of the circuit represented in Fig. 8–12.

EXAMPLE 11 Three resistors $R_1 = 30\ \Omega$, $R_2 = 160\ \Omega$, and $R_3 = 40\ \Omega$ are connected in series across a generator. A voltmeter connected across R_2 reads 80 V. What is the voltage of the generator?

SOLUTION

Figure 8–14 is a diagram of the circuit.

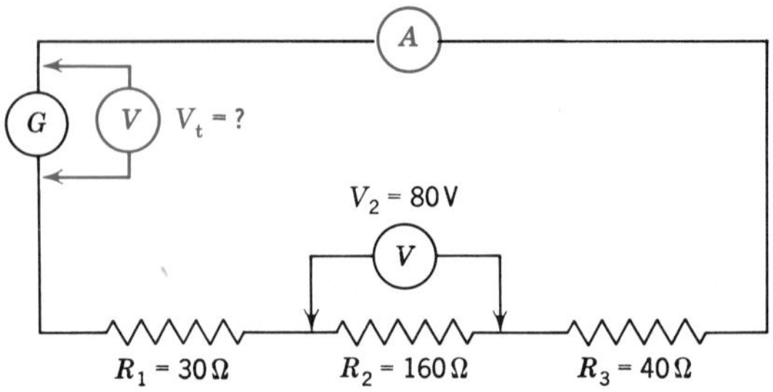

FIG. 8–14 Circuit of Example 11.

$$I = \frac{V_2}{R_2} = \frac{80}{160} = 0.5 \text{ A}$$
$$R_t = R_1 + R_2 + R_3$$
$$= 30 + 160 + 40 = 230\ \Omega$$
$$V_t = IR_t = 0.5 \times 230 = 115 \text{ V}$$

EXAMPLE 12 A 300-Ω relay must be operated from a 120-V line. How much resistance must be added in series with the relay coil to limit the current through it to 250 mA?

SOLUTION 1

The circuit is represented in Fig. 8–15. For a current of 250 mA in a 120-V circuit, the total resistance must be

$$R_t = \frac{V}{I} = \frac{120}{0.250} = 480\ \Omega$$

Because the relay coil has a resistance of 300 Ω, the resistance to be added is

$$R_x = R_t - R_c = 480 - 300 = 180\ \Omega$$

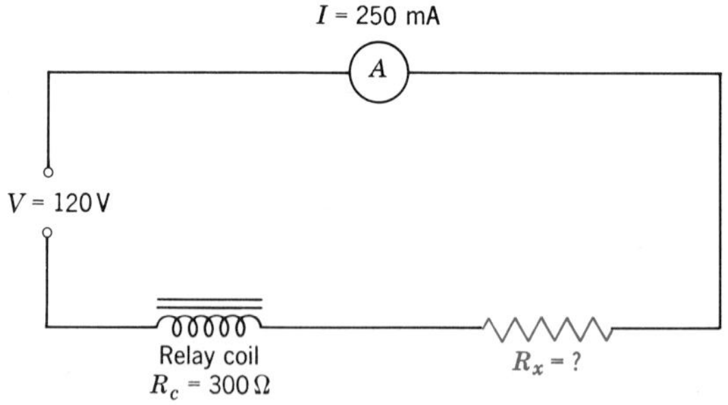

FIG. 8–15 Circuit of Example 12.

SOLUTION 2

For 0.250 A to flow through the relay coil, the voltage across the coil must be

$$V_c = IR_c = 0.250 \times 300 = 75 \text{ V}$$

Because the line voltage is 120 V, the voltage across the added resistance must be

$$V_x = V_t - V_c = 120 - 75 = 45 \text{ V}$$

Then the value of resistance to be added is

$$R_x = \frac{V_x}{I} = \frac{45}{0.250} = 180 \ \Omega$$

EXAMPLE 13 Three resistors $R_1 = 20\ \Omega$, $R_2 = 50\ \Omega$, and $R_3 = 30\ \Omega$ are connected in series across a generator. The current through the circuit is 2.5 A.
(*a*) What is the generator voltage?
(*b*) What is the voltage across each resistor?
(*c*) How much power is expended in each resistor?
(*d*) What is the total power expended?

SOLUTION
The circuit is represented in Fig. 8–16.

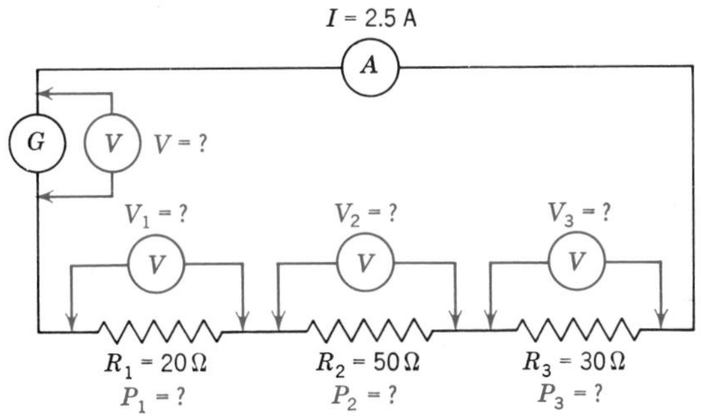

FIG. 8–16 Circuit of Example 13.

(*a*)

$$R_t = R_1 + R_2 + R_3 = 20 + 50 + 30 = 100\ \Omega$$

$$V = IR_t = 2.5 \times 100 = 250 \text{ V}$$

(*b*)

$$V_1 = IR_1 = 2.5 \times 20 = 50 \text{ V}$$
$$V_2 = IR_2 = 2.5 \times 50 = 125 \text{ V}$$
$$V_3 = IR_3 = 2.5 \times 30 = 75 \text{ V}$$

CHECK

$$V = V_1 + V_2 + V_3 = 50 + 125 + 75 = 250 \text{ V}$$

(*c*) Power in R_1, $P_1 = V_1 I = 50 \times 2.5 = 125$ W

CHECK

$$P_1 = I^2 R_1 = 2.5^2 \times 20 = 125 \text{ W}$$

Power in R_2, $P_2 = V_2 I = 125 \times 2.5 = 312.5$ W

CHECK

$$P_2 = I^2 R_2 = 2.5^2 \times 50 = 312.5 \text{ W}$$

Power in R_3, $P_3 = V_3 I = 75 \times 2.5 = 187.5$ W

CHECK

$$P_3 = I^2 R_3 = 2.5^2 \times 30 = 187.5 \text{ W}$$

(*d*) Total power, $P_t = P_1 + P_2 + P_3$

$$= 125 + 312.5 + 187.5 = 625 \text{ W}$$

CHECK

$$P_t = I^2 R_t = 2.5^2 \times 100 \times 625 \text{ W}$$

or

$$P_t = \frac{V^2}{R_t} = \frac{250^2}{100} = 625 \text{ W}$$

PROBLEMS 8–3

1. Three resistors, $R_1 = 470\ \Omega$, $R_2 = 220\ \Omega$, and $R_3 = 330\ \Omega$, are connected in series across 110 V.
   (*a*) What is the current through the circuit?
   (*b*) What is the voltage across R_2?
   (*c*) How much power is expended in R_1?
2. Three resistors, $R_1 = 1.8$ kΩ, $R_2 = 4.7$ kΩ, and $R_3 = 1.2$ kΩ are connected in series across 100 V.
   (*a*) What is the current through the circuit?
   (*b*) What is the voltage across each resistor?
3. A 120-V soldering iron which is rated at 40 W is to be used on a 220-V line.
   (*a*) How much resistance must be connected in series with the iron to limit the current to rated value?
   (*b*) If a standard resistor of 270 Ω is used in place of the calculated value, what minimum power rating must be specified for it?
   (*c*) If the standard resistor of (*b*) is used, what actual power will be delivered to the soldering iron?
4. Four identical 60-W lamps are connected in series across a 220-V line. The hot resistance of each lamp is 50 Ω.
   (*a*) What is the current through the lamps?
   (*b*) What is the voltage across each lamp?
   (*c*) What is the power dissipated by each lamp?
5. Three identical lamps are connected in series across a 440-V line. If the current through the lamps is 545 mA, what is the hot resistance of each lamp?
6. Three resistors, R_1, R_2, and R_3, are connected in series across a 470-V power supply. A voltmeter connected across R_1 reads 76 V. When connected across R_2, the voltmeter reads 51 V. R_3 is 150 kΩ.
   (*a*) What is the current through the circuit?
   (*b*) What is the value of R_1?

(*c*) What is the value of R_2?
(*d*) What is the wattage dissipated by each resistor?

7. Three resistors of 22, 15, and 68 Ω are connected in series across a 24-V source. If the current through the circuit is 222 mA, what is the resistance of the connecting wires and connections?
8. The visual readout system of a CPU terminal uses eight cathode-ray tubes. To conserve power drawn from the central control unit (CCU) power supply, the video monitors are connected in series directly across the 115-V line.

   Six of the monitors require 12.6-V filament voltage, and the remaining two require only 6.3 V. The filament current of all the monitors, to maintain *correct* operating temperature, is 210 mA. What value of series ballast resistor must be used to allow operation from the 115-V line?
9. Three resistors $R_1 = 4.7\ \Omega$, R_2, and R_3 are connected in series across a 100-V power supply, which delivers a current of 12.5 A. The voltage across R_3 is 18.75 V.
   (*a*) What is the value of R_3?
   (*b*) What is the value of R_2?
   (*c*) What is the total power drawn from the power supply?
10. Four resistors, $R_1 = 820\ \Omega$, $R_2 = 270\ \Omega$, $R_3 = 1.5\ k\Omega$, and $R_4 = 390\ \Omega$, are connected in series across a generator. The voltage appearing across R_3 is 504 V.
    (*a*) What is the generator voltage?
    (*b*) What is the power being dissipated by each resistor?

8–9 BIAS RESISTORS—FIELD-EFFECT TRANSISTORS

The field-effect transistor (FET) (Fig. 8–17) was introduced as a more suitable replacement for vacuum tubes than the bipolar junction transistor (BJT). The major reason was that the FET, like the vacuum tube, is a voltage-controlled device whereas the BJT is a current-operated device. Other reasons include the fact that the FET is normally ON, because it has only one major current carrier, whereas the BJT must have current flowing through all of the electrodes to set up the no-signal biasing conditions. When all the electrodes are passing current, the BJT has both electrons and holes (positive charges) passing, and recombining, among all three electrodes. The FET is controlled by the external circuitry, but the BJT's "fixed" parameters determine what external bias circuitry must be used (R_1 and R_2, Fig. 8–18, and R_B, Fig. 8–19). In other words, parameters of the FET do not control the external circuitry design to the extent that those of the BJT do.

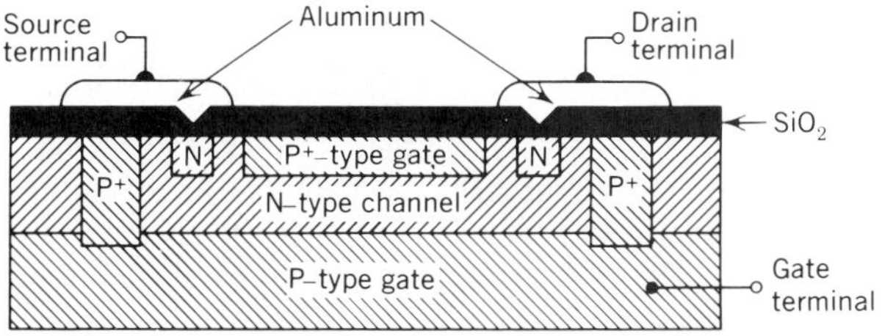

FIG. 8–17 (*a*) Block model of an N-channel junction FET.

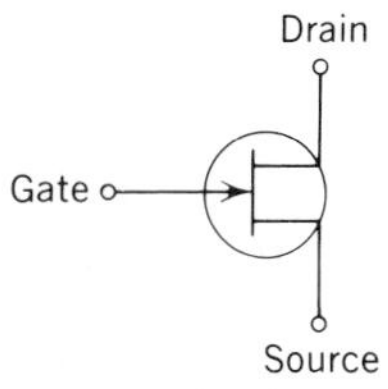

(*b*) Usual schematic symbol.

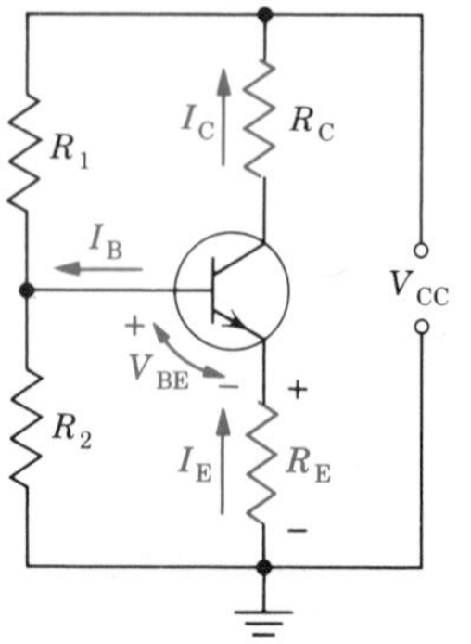

FIG. 8–18 Universal dc circuit.

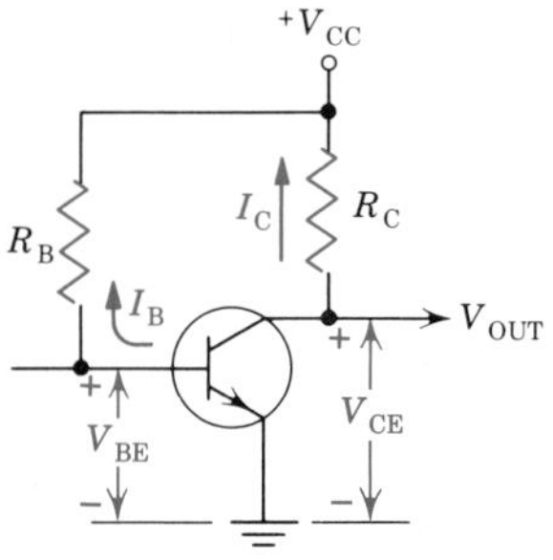

FIG. 8–19 Base bias of a common emitter BJT.

To clarify FET biasing, consider the schematic diagram of Fig. 8–20. The FET shown in the diagram is an N-channel junction FET or, simply, an NJFET. The N channel is indicated by the direction of the arrow of the gate terminal, which is similar to the arrow on the emitter electrode of the BJT. It indicates the type, NPN or PNP, of base material used. In Fig. 8–20 the arrow pointing in indicates that the material from source to drain is N-type semiconductor material (usually silicon).

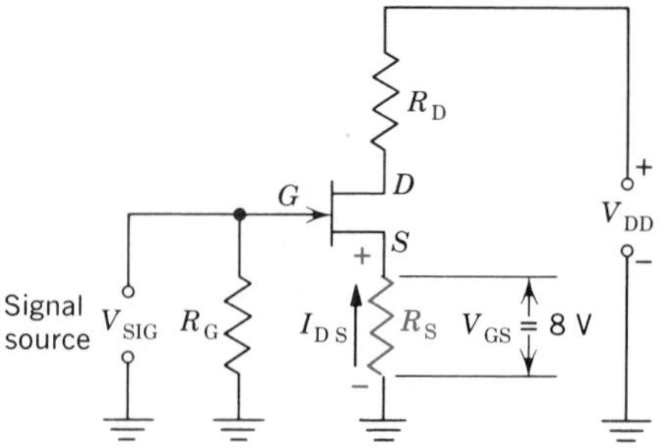

FIG. 8–20 Gate G is biased −8 V with respect to source S.

The FET of Fig. 8–20 indicates an N channel and a P-type gate; it is similar to any PN junction diode. The resistor R_S connected to the source terminal and ground passes a current I_{DS} from source to drain. This current will develop a voltage across the resistor R_S with polarity as shown. With a positive potential at the source terminal and, by connection, a negative potential at the gate terminal with respect to the source terminal, there now exists a reverse bias condition between the gate and source. Current flow through the junction is thereby prevented.

A further examination of Fig. 8–20 shows that there is a positive potential at the drain terminal which is much larger than the potential at the source terminal. Because of the difference, electrons will flow from the source to the drain and back to the supply V_{DD}. The gate terminal, while not passing current, will have a potential of −4 V when measured with respect to the source. We can say that the FET has a gate bias voltage V_{GS} of −4 V.

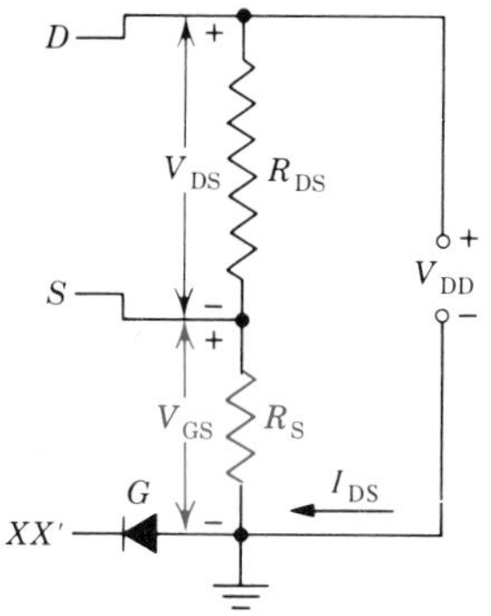

FIG. 8–21 Equivalent circuit of Fig. 8–20. Point *XX'* is connected into N-type bar between *D* and *S*.

The bias voltage just discussed is typical of a voltage across R_S. When the value of R_S is increased, the voltage at the gate will become more negative with respect to the source. It is usual to describe this type of biasing as either *source bias* or *self-bias*.

With the drain supply voltage V_{DD} maintaining the drain positive with respect to the source, electrons will flow from source to drain and introduce the current I_{DS}.

The schematic of Fig. 8–20 may be replaced by its equivalent dc circuit, as shown in Fig. 8–21. The purpose of the equivalent circuit is to illustrate that the N-type bar, when considered on its own, is nothing more than a piece of semiconductor material having a resistance r_{DS} which is made variable by the introduction of the P-type gate. By eliminating the external resistor R_D between the drain and the V_{DD} supply, we may show that the sum of the voltages V_{DS} and V_{R_S} must equal the supply V_{DD}.

$$V_{DS} + V_{R_S} = V_{DD}$$

You should notice that this formula does not take into account the voltage V_{GS}. That should not be too alarming, since V_{GS} provides bias, and the gate draws no current.

EXAMPLE 14 A 2N5459 JFET is to be operated as a class A audio amplifier. In that class of operation, the drain current I_D is 6 mA when the drain voltage V_{DS} is +8 V with respect to the source and the gate voltage V_{GS} is −4 V with respect to the source (achieved by voltage V_{RS}).

(*a*) What value of source-biasing resistor R_S is required?

(*b*) What power is dissipated by R_S?

(*c*) Ignoring the drain load resistor, what is the value of the drain voltage supply V_{DD}?

(*d*) What is the total power taken from the drain voltage supply?

SOLUTION

The circuit is shown schematically in Fig. 8–22.

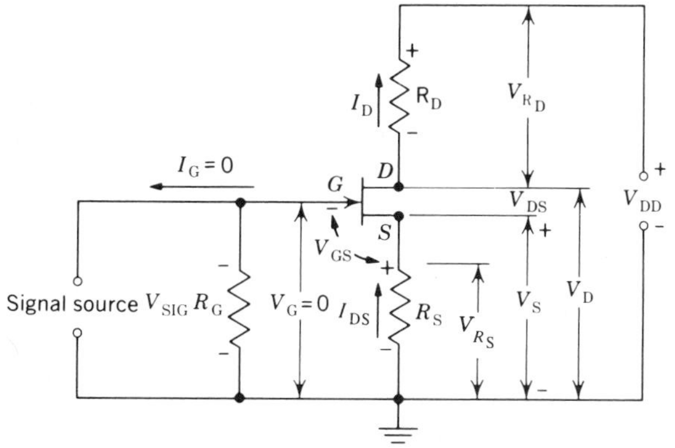

FIG. 8–22 Circuit of Example 14. NJFET using self/source bias. $V_S = V_{RS} = I_D R_S$. $V_G = 0$ V. $I_G = 0$ A.

(*a*)
$$R_S = \frac{V_S}{I_D} = \frac{4}{0.006} = \frac{4}{6 \times 10^{-3}}$$
$$= \frac{4}{6} \times 10^3 = 667\ \Omega$$

(*b*)
$$P_{R_S} = \frac{V_S{}^2}{R_S} = \frac{4^2}{667} = 24\text{ mW}$$

CHECK

$$P_{R_S} = I_D{}^2 R_S = (6 \times 10^{-3})^2 \times 667 = 24\text{ mW}$$

(*c*) $V_{DD} = V_{DS} + V_{RS} = 8 + 4 = 12$ V

(*d*) $P = I_D \times V_{DD} = 6 \times 10^{-3} \times 12 = 72$ mW

EXAMPLE 15 The FET of Example 14 is to work into a dc load resistance of $R_D = 2.2$ kΩ. What drain supply voltage V_{DD} will be required?

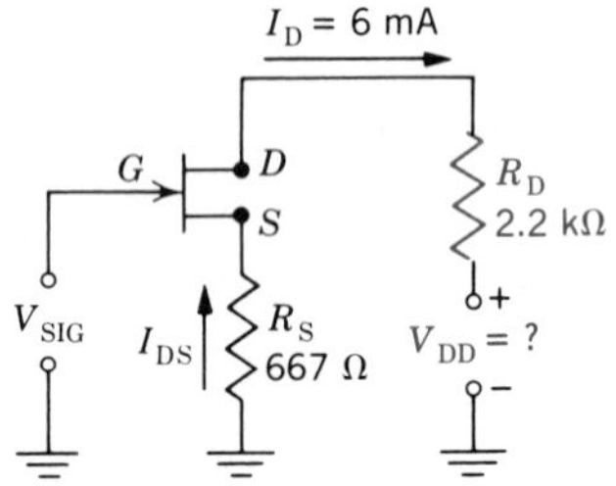

FIG. 8–23 Circuit of Example 15.

SOLUTION

The circuit is illustrated in Fig. 8–23. The voltage across the load resistor R_D is

$$V_{RD} = I_D R_D = 0.006 \times 2200 = 13.2\text{ V}$$
$$V_{DD} = I_D R_D + V_{DS} + V_{GS}$$
$$= 13.2 + 8 + 4$$
$$V_{DD} = 25.2\text{ V}$$

EXAMPLE 16 The FET type 2N5458 has the following characteristics:

$$V_{DD} = 25\text{ V},\ V_S = 6\text{ V, and } V_{DS} = 15\text{ V when } I_G = 0.$$

Use the circuit shown in Fig. 8–24 and determine values for (*a*) I_D, (*b*) R_S, (*c*) V_{GS}, and (*d*) the total current I_T.

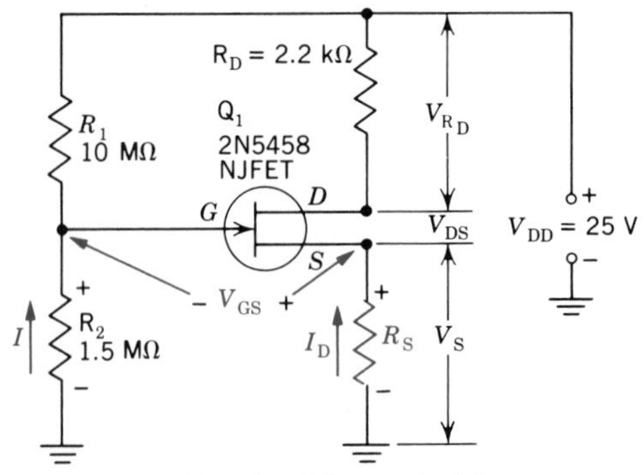

FIG. 8–24 Circuit of Example 16.

SOLUTION

$$V_D = V_{DD} - I_D R_D \quad \text{or} \quad V_D = V_{DS} + V_S$$
$$= 15 + 6$$
$$V_D = 21 \text{ V}$$

From this value for V_D we can determine that the voltage $I_D R_D$ is:

$$V_{DD} - V_D \quad \text{or} \quad 25 - 21 = 4 \text{ V}$$

From that voltage the value of I_D can easily be calculated:
(a) $I_D R_D = V_{R_D}$; therefore,

$$I_{DS} = \frac{4}{2.2 \times 10^3} = 1.8 \text{ mA}$$

(b) With $I_D = 1.8$ mA, $I_{DS} R_S = V_S$; therefore

$$R_S = \frac{V_S}{I_{DS}} = \frac{6}{1.8} \times 10^3 = 3.3 \text{ k}\Omega$$

(c) To determine the value of V_{GS}, we must first establish the value of the voltage across R_2. We can do so in either of two ways:

$$V_{DD} = I(R_1 + R_2)$$

Solve for current I necessary to produce IR_2. Alternatively,

$$V_{R_2} = \frac{R_2}{R_1 + R_2} \times V_{DD} = \frac{1.5 \times 10^6}{11.5 \times 10^6} \times 25$$
$$= 3.261 \text{ V}$$

The voltage V_{GS} is the difference between V_{R_2} and V_S; therefore,

$$V_{GS} = 3.261 - 6 = -2.739 \text{ V}$$

(d)

$$I_{R_2} = \frac{3.261}{1.5 \times 10^6} = 0.002 \text{ mA}$$

$$I_T = I_D + I_{R_2}$$
$$= 1.8 + 0.002 = 1.802 \text{ mA}$$

8–10 BIAS RESISTORS—TRANSISTORS

Proper operation of a transistor circuit requires that the emitter-base junction of the transistor be forward-biased and that the collector-base junction be reverse-biased, as shown in diagram form in Fig. 8–25. Whether discrete or in integrated circuits, all transistors require biasing, and biasing is generally achieved by means of resistors.

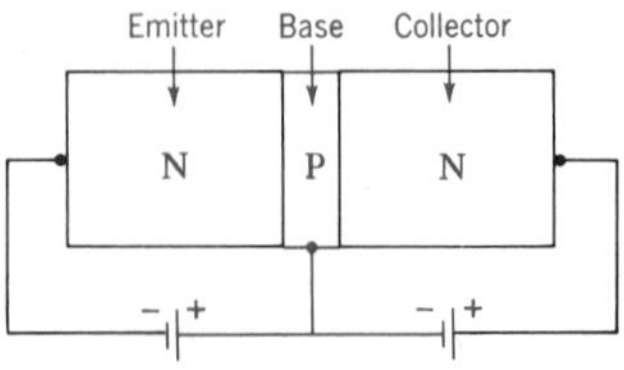

FIG. 8–25 NPN transistor biased for proper operation. The N-type emitter is forward-biased for low effective resistance, and the N-type collector is reverse-biased for high effective resistance.

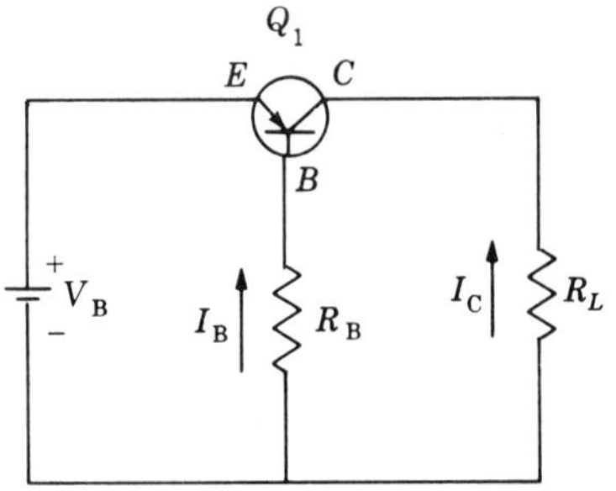

FIG. 8–26 Simple single-battery transistor biasing circuit of PNP transistor Q_1.

Sometimes the use of two different batteries is avoided by utilizing bias resistors, as in tube circuits. In addition, resistor values are chosen to limit current flows to acceptable levels. Figure 8–26 shows a simple circuit in which transistor Q_1 is supplied by a single battery V_B. The resistor in the base circuit R_B is chosen to regulate the base-emitter current I_B, and the output signal is taken across the load resistor R_L as the collector current I_C flows through it.

EXAMPLE 17 In Fig. 8–26, assuming that the voltage across the emitter-base junction is negligible, what must be the value of R_B if the base current must be limited to 80 μA? $V_B = 6$ V.

SOLUTION

$$R_B = \frac{V_B}{I_B} = \frac{6}{80 \times 10^{-6}} = 75\ \text{k}\Omega$$

When two batteries are used, as in Fig. 8–27, an analysis based upon constant-emitter-current bias reveals that

$$R_E = \frac{V_{EE}}{I_E}$$
$$I_C = \alpha I_E + I_{CO}$$
$$I_B = (1 - \alpha)I_E - I_{CO}$$

where I_{CO} = the very small leakage current in the collector circuit at room temperature
α = the current amplification factor under certain circuit arrangements; its value is usually slightly less than 1

EXAMPLE 18 In Fig. 8–27, the applied emf $V_{EE} = 12$ V and the specifications for transistor Q_1 indicate that the emitter current I_E should be limited to 10 mA. What value of resistor R_E should be chosen?

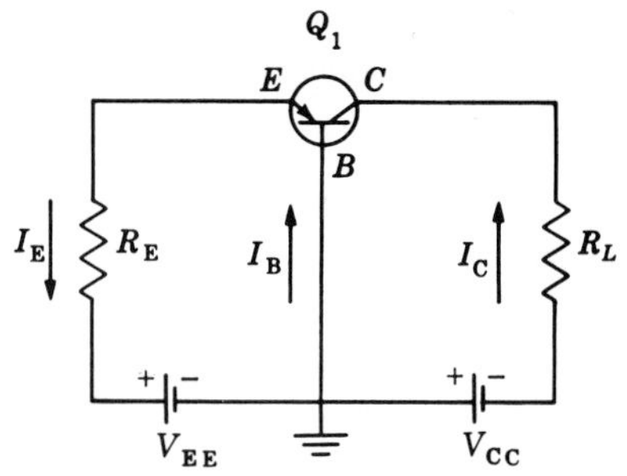

FIG. 8–27 PNP transistor Q_1 biased by means of two batteries, V_{EE} and V_{CC}.

SOLUTION

$$R_E = \frac{V_{EE}}{I_E} = \frac{12}{10 \times 10^{-3}} = 1.2\ \text{k}\Omega$$

EXAMPLE 19 For the circuit of Fig. 8–27, $V_{EE} = 12$ V, $I_E = 8$ mA, $\alpha = 0.95$, and $I_{CO} = 50\ \mu$A. Find (*a*) R_E, (*b*) I_C, and (*c*) I_B.

SOLUTION

(*a*)
$$R_E = \frac{V_{EE}}{I_E} = \frac{12}{0.008} = 1.5\ \text{k}\Omega$$

(*b*)
$$\begin{aligned} I_C &= \alpha I_E + I_{CO} \\ &= (0.95)(0.008) + 0.000\,050 \\ &= 7.65\ \text{mA} \end{aligned}$$

(*c*)
$$\begin{aligned} I_B &= (1 - 0.95)(0.008) - 0.000\,050 \\ &= 350\ \mu\text{A} \end{aligned}$$

PROBLEMS 8–4

1. The 2N5459 JFET is to be used as an audio amplifier operating class A. When the drain voltage V_D with respect to the source is 15 V, the drain current is 9 mA and the gate-to-source bias is −4.5 V.
   (*a*) What value of source resistor is required for this bias?
   (*b*) Disregarding any drain load resistance, what is the drain supply voltage V_{DD}?
2. The NPN transistor shown schematically in Fig. 8–18 is to be biased for operation as a common-emitter class A audio amplifier. The data book shows the following characteristics: $I_B = 0.1$ mA; forward current gain $\beta = 50$; $V_{BE} = 0.2$ V; $\beta I_B = I_C$; $R_C = 1.5\ \text{k}\Omega$; $R_1 = 9.8\ \text{k}\Omega$. If the voltage at the emitter measured with respect to ground is 1 V,
   (*a*) Determine the value of I_C and I_E when $V_{CC} = 12$ V.
   (*b*) Determine the values of R_2 and R_E.

(*c*) What will be the value of the current through R_2?
(*d*) What voltage will be measured at the collector with respect to ground?

3. The FET circuit shown in Fig. 8–24 is biased so that $V_S = 6$ V. $V_{DS} = 15$ V. With all other components and voltage supplies remaining the same,
(*a*) Determine the value of R_S.
(*b*) Determine the value of I_D.
(*c*) What will be the total current drawn from the 25-V source?
(*d*) How much power is drawn from the supply?

4. For the circuit used in Prob. 3 and the data given,
(*a*) Determine the voltage across R_1.
(*b*) Determine the voltage across R_2.
(*c*) What will be the value of V_{GS} when $V_S = 6$ V?
(*d*) Which resistor in this circuit will dissipate the most power?

5. A microphone preamplifier is wired as shown in Fig. 8–19. $R_C = 12$ kΩ, $I_B = 0.015$ mA, $V_{BE} = 0.4$ V, and $V_{CC} = 12$ V. If the transistor has a forward current gain $\beta = 50$, determine
(*a*) the value of I_C and I_E,
(*b*) the value of R_B,
(*c*) the magnitude and polarity of the voltage V_{CE}, and
(*d*) the power dissipated by the collector of the transistor.

6. In Fig. 8–26, assuming that the voltage across the emitter-base junction is negligible, what must be the value of R_B if the base current must be limited to 90 μA? $V_B = 6$ V.

7. It is desired to operate a transistor in grounded-base connection (Fig. 8–27) with a fixed bias of 6 V. The maximum current in the base circuit is 100 μA.
(*a*) What is the value of the resistor which will provide this voltage?
(*b*) What is the power which this resistor must radiate?

8. In the circuit of Fig. 8–27, emf $V_{EE} = 6$ V and the emitter current I_E should be limited to 8 mA. What value of resistor R_E should be chosen?

9. In the circuit of Fig. 8–27, what value should R_E be if $V_{EE} = 24$ V and I_E must be kept to 10 mA or less? Assume $V_{BE} = 0.6$ V.

10. In the circuit of Fig. 8–27, $V_{EE} = 12$ V and $V_C = 15$ V. $I_E = 10$ mA, $\alpha = 0.98$, and $I_{CO} = 75$ μA. Find (*a*) R_E, (*b*) I_C, and (*c*) I_B.

SELF TEST

1. What is the current through a hot 60-W lamp bulb operating on 117 V?
2. What is the resistance of a soldering iron that draws 250 mA from a 110-V source?
3. What voltage supply is necessary to drive a current of 22 mA through a resistance of 270 Ω?
4. A rectifier junction drops 1.2 V when connected in series with a 12-V power supply and a circuit that draws a current of 100 mA. What is the effective resistance of the junction?
5. What is the current through a stationary 0.2-Ω motor coil when a voltage of 440 V is applied directly across it?
6. A coiled-coil tungsten filament, cold before lit, draws 1.8 A when switched on to a household circuit of 120 V. What is the cold resistance of the filament?
7. A quarter-horsepower motor operating at 120 V draws a current of 1.8 A. What is the efficiency of the motor? 1 Hp = 746 W.
8. A workshop motor rated at 2 hp has a guaranteed efficiency of 95%. Under test, it draws 14.1 A from a 110-V source. Is it within its guarantee?
9. The only available soldering iron is rated 100 W, 110 V, but the only available voltage source is 220 V. How much resistance must be connected in series with the iron in order to use it safely?
10. Three resistors are connected in series. R_1 is rated 500 Ω, 0.5 W; R_2 is rated 1200 Ω, 2 W; and R_3 is rated 2000 Ω, 10 W. What is the greatest voltage that can be applied to the series circuit without exceeding any of their ratings?

CHAPTER 9

RESISTANCE—WIRE SIZES

The effects of resistance in series circuits were discussed in the preceding chapter. However, in order to prevent confusion while the more simple relations of Ohm's law were being discussed, the nature of resistance and the resistance of wires used for connecting sources of voltage with their respective loads were not mentioned.

In the consideration of practical circuits two important features must be taken into account: the resistance of the wires between the source of power and the electronic equipment that is to be furnished with power and the current-carrying capacity of these wires for a given temperature rise.

9–1 RESISTANCE

There is a wide variation in the ease (conductance) of current flow through different materials. No material is a perfect conductor, and the amount of opposition (resistance) to current flow within it is governed by the specific resistance and the length, cross-sectional area, and temperature of the material. Thus, for the same material and cross-sectional area, a long conductor will have a greater resistance than a shorter one. That is, *the resistance of a conductor of uniform cross-sectional area is directly proportional to the length of the conductor.* This is conveniently expressed as

$$\frac{R_1}{R_2} = \frac{L_1}{L_2} \qquad (1)$$

Where R_1 and R_2 are the resistances of conductors with lengths L_1 and L_2, respectively.

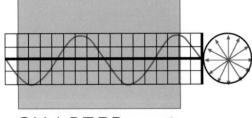

EXAMPLE 1 The resistance of No. 8 copper wire is 2.06 Ω/km. What is the resistance of 175 m of the wire?

SOLUTION

Given $R_1 = 2.06\ \Omega$, $L_1 = 1000$ m, and $L_2 = 175$ m, $R_2 = ?$ Upon solving Eq. (1) for R_2, we have

$$R_2 = \frac{R_1 L_2}{L_1} \frac{\Omega \cdot \text{m}}{\text{m}} = \frac{2.06 \times 175}{1000} \frac{\Omega \cdot \cancel{\text{m}}}{\cancel{\text{m}}} = 0.3605\ \Omega$$

For the same material and length, one conductor will have more resistance than another with a larger cross-sectional area. That is, *the resistance of a conductor is inversely proportional to the cross-sectional area of the conductor.* Expressed as an equation,

$$\frac{R_1}{R_2} = \frac{A_2}{A_1} \tag{2}$$

where R_1 and R_2 are the resistances of conductors with cross-sectional areas A_1 and A_2, respectively.

Because most wires are drawn round, Eq. (2) can be rearranged into a more convenient form. For example, let A_1 and A_2 represent the cross-sectional areas of two equal lengths of round wires with diameters d_1 and d_2, respectively. Because the area A of a circle of a diameter d is given by

$$A = \frac{\pi d^2}{4}$$

then $$A_1 = \frac{\pi d_1^2}{4} \quad \text{and} \quad A_2 = \frac{\pi d_2^2}{4}$$

Substituting in Eq. (2)

$$\frac{R_1}{R_2} = \frac{\frac{\pi d_2^2}{4}}{\frac{\pi d_1^2}{4}}$$

or $$\frac{R_1}{R_2} = \frac{d_2^2}{d_1^2} \tag{3}$$

Hence, the resistance of a round conductor varies inversely as the square of the diameter.

EXAMPLE 2 A rectangular conductor with a cross-sectional area of 1.04 square millimeters (mm^2) has a resistance of 0.075 Ω. What would be its resistance if its cross-sectional area were 2.08 mm^2?

SOLUTION

Given $R_1 = 0.075\ \Omega$, $A_1 = 1.04\ \text{mm}^2$, and $A_2 = 2.08\ \text{mm}^2$, $R_2 = ?$ Solving Eq. (2) for R_2,

$$R_2 = \frac{R_1 A_1}{A_2} \frac{\Omega \cdot \text{mm}^2}{\text{mm}^2}$$

$$= \frac{0.075 \times 1.04}{2.08} \frac{\Omega \cdot \cancel{\text{mm}^2}}{\cancel{\text{mm}^2}} = 0.0375\ \Omega$$

EXAMPLE 3 A round conductor with a diameter of 0.25 mm has a resistance of 8 Ω. What would be its resistance if its diameter were 0.5 mm?

SOLUTION

Given $d_1 = 0.25\ \text{mm}$, $R_1 = 8\ \Omega$, and $d_2 = 0.5\ \text{mm}$, $R_2 = ?$ Solving Eq. (3) for R_2,

$$R_2 = \frac{R_1 d_1^2}{d_2^2} \frac{\Omega \cdot \text{mm}^2}{\text{mm}^2} = \frac{8 \times 0.25^2}{0.5^2} \frac{\Omega \cdot \cancel{\text{mm}^2}}{\cancel{\text{mm}^2}} = 2\ \Omega$$

Hence, if the diameter is doubled, the cross-sectional area is increased four times and the resistance is reduced to one-quarter of its original value.

PROBLEMS 9–1

1. Number 16 copper wire has a resistance of 13.2 Ω/km.
   (*a*) What is the resistance of 500 m of this wire?
   (*b*) What is the resistance of 20 m of this wire?
2. Number 24 copper wire has a resistance of 84.2 Ω/km.
   (*a*) What is the resistance of 800 m of this wire?
   (*b*) What is the resistance of 1 m?
3. Using the information of Prob. 2, what is the resistance of a coil that has a mean diameter of 40 mm and is wound with 6280 turns of No. 24 copper wire?
4. The resistance of a 1-km run of No. 12 copper wire telephone line is measured and found to be 5.21 Ω.
   (*a*) What is the resistance per meter?
   (*b*) What is the resistance of a 720-m line?
   (*c*) What is the resistance of 2.2 km?
5. The telephone line of Prob. 4 is replaced with No. 10 copper wire, which has a resistance of 3.277 Ω/km. What is the resistance of the 720-m section?
6. A length of square conductor that is 0.5 cm on a side has a resistance of 0.0756 Ω. What will be the resistance of a similar length of 1.5-cm square conductor?
7. One kilometer of No. 6 wire, which has a diameter of 4.115 mm, has a resistance of 1.297 Ω. What is the resistance of 1 km of No. 2 wire whose diameter is 6.543 mm?

8. The resistance of 30 m of a specially drawn wire is found to be 32.1 Ω. A coil wound with identical wire has a measured resistance of 702 Ω. What is the length of wire in the coil?
9. It is desired to wind a milliammeter shunt having a resistance of 3.78 Ω, and No. 40 enameled copper wire with a resistance of 3.54 kΩ/km is available. What length of wire is required?
10. It is desired to wind a microammeter shunt having a resistance of 0.280 Ω, and No. 36 enameled copper wire with a resistance of 1.36 kΩ/km is available. What length of wire is required?

9–2 MICROHM-METER

Equations (1) and (2) from Sec. 9–1 can be combined to form the compound proportion

$$\frac{R_1}{R_2} = \frac{L_1}{L_2} \cdot \frac{A_2}{A_1}$$

Such ratios are extremely helpful in solving problems when sufficient information is available. However, as a statement regarding a single conductor, we fall back on the simple proportionality (Sec. 5–11)

$$R \propto \frac{L}{A}$$

When this proportionality (see Fig. 9–1) is written as an equation with a constant of proportionality, it becomes

$$R = \rho \frac{L}{A} \quad \Omega \tag{4}$$

where R = resistance of wire
ρ = specific resistance of material of which wire is made
L = length of wire
A = cross-sectional area of wire

From an algebraic rearrangement of Eq. (4), you can see that the units of ρ, the specific resistance, must relate to the units of L and A:

$$\rho = \frac{RA}{L}$$

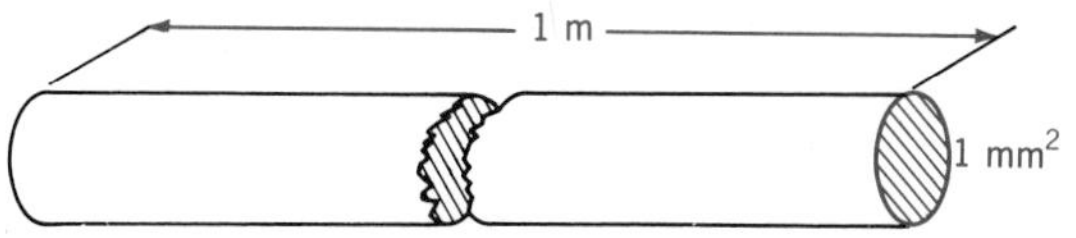

FIG. 9–1 Representation of 1 microhm-meter conductor.

In the older, or traditional, system of units, L was measured in feet and A was measured in circular mils (the circular-mil area was defined to be equal to the square of the diameter when the diameter was given in mils, or thousandths of an inch). This meant ρ was expressed in Ω-cmils/ft, commonly pronounced Ω/cmil-ft. (See Secs. 9–2 and 9–3 of the third edition of this book if you require more information about this superseded set of units.)

In the SI metric system, the unit of length is the meter and the unit of area is the square meter. Using those units, ρ would be expressed as

$$\rho = \frac{RA}{L} = \frac{\Omega \cdot \text{m}^2}{\text{m}} = \Omega \cdot \text{m}$$

Sometimes you will see tables of specific resistance giving values of ρ in ohm-meters. However, since more realistic sizes of wire will be given in square millimeters (see Appendix, Table 6), more often than not you will see practical values of ρ as:

$$\rho = \frac{RA}{L} = \frac{\Omega \cdot \text{mm}^2}{\text{m}} = \frac{\Omega \cdot \text{m}^2 \times 10^{-6}}{\text{m}}$$
$$= \Omega \cdot \text{m} \times 10^{-6} = \text{microhm-meters } (\mu\Omega \cdot \text{m})$$

Table 9–1 gives specific resistances of common conductive materials in microhm-meters.

Eq. (4) and its definition block may now be adjusted to read

$$R = \rho \frac{L}{A} \quad \Omega \tag{5}$$

where R = resistance of wire, Ω
ρ = specific resistance of wire, μΩ · m
L = length of wire, m
A = cross-sectional area of wire, mm^2

TABLE 9–1 SPECIFIC RESISTANCE AT 20°C

Material	μΩ · m
Aluminum	0.028 24
Brass	0.070 0
Constantan	0.490 0
Copper, hard drawn*	0.017 71
Gold	0.024 4
Iron	0.100 0
Lead	0.220 0
Mercury	0.957 8
Nickel	0.078 0
Silver	0.015 9
Tin	0.115 0
Zinc	0.058 0

*ASTM specifications list over 25 varieties of copper wire, rod, and tubing, with slight variations in specific resistance.
Use 0.017 71 μΩ·m as a reasonable *nominal* value for the problems in this book.

EXAMPLE 4 What is the resistance at 20°C of a copper wire that is 250 m long and 1.63 mm in diameter?

SOLUTION

Given $L = 250$ m, $d = 1.63$ mm ($r = 0.815$ mm), and also, from Table 9–1, $\rho = 0.017\ 71\ \mu\Omega \cdot \text{m}$, $R = ?$ Substituting in Eq. (4),

$$R = \frac{0.017\ 71 \times 250}{\pi(0.815)^2} = 2.122\ \Omega$$

EXAMPLE 5 The resistance of a conductor 1 km long and 2.05 mm in diameter is found to be 4.82 Ω at 20°C. What is the specific resistance of the wire?

SOLUTION

Given $L = 1000$ m, $d = 2.05$ mm ($r = 1.025$ mm), and $R = 4.82\ \Omega$, $\rho = ?$ Solving Eq. (5) for ρ,

$$\begin{aligned}\rho &= \frac{RA}{L} = \frac{4.82\ \pi(1.025)^2}{10^3} \qquad \mu\Omega \cdot \text{m} \\ &= 15.91 \times 10^{-3} \\ &= 0.015\ 91\ \mu\Omega \cdot \text{m}\end{aligned}$$

EXAMPLE 6 A roll of copper wire is found to have a resistance of 2.54 Ω at 20°C. The diameter of the wire is measured as 1.63 mm. How long is the wire?

SOLUTION

Solving Eq. (5) for L,

$$L = \frac{RA}{\rho} \qquad \text{m}$$

Substituting known values,

$$\begin{aligned}L &= \frac{2.54\pi(0.815)^2}{0.017\ 71} \ \text{m} \\ &= 299 \ \text{m}\end{aligned}$$

PROBLEMS 9–2

NOTE *In the following problems, consider that all the wire temperatures are 20°C.*

1. What is the resistance of a copper wire 120 m long and 0.643 mm in diameter?
2. With reference to Prob. 1, what is the resistance of an otherwise identical wire of aluminum?
3. With reference to Prob. 1, what is the resistance of an otherwise identical wire of iron?
4. What is the resistance of 300 m of copper wire with a diameter of 0.813 mm?

5. A special alloy wire 30 m long and 0.160 mm in diameter has a resistance of 56.7 Ω. What is the specific resistance of the alloy?
6. A constantan wire that has a specific resistance of 0.49 μΩ·m has a diameter of 0.254 mm and a length of 4.2 m. What is its resistance?
7. How many kilometers of copper wire 0.643 mm in diameter will it take to make 100 Ω of resistance?
8. What is the resistance of the wire in Prob. 7 in ohms per kilometer?
9. A coil of copper wire has a resistance of 2.38 Ω. If the diameter of the wire is 2.05 mm, find the length of the wire.
10. What is the resistance of 2 km of the wire in Prob. 9?

9–3 TEMPERATURE EFFECTS

In the preceding section the specific resistance of certain materials was given at a temperature of 20°C. The reason for stating the temperature is that the resistance of all pure metals increases with a rise in temperature. The results of experiments show that over ordinary temperature ranges this variation in resistance is directly proportional to the temperature. Hence, for each degree rise in temperature above some reference value, each ohm of resistance is increased by a constant amount α, called the *temperature coefficient of resistance*. The relation between temperature and resistance can be expressed by the equation

$$R_t = R_0(1 + \alpha \, \Delta t) \quad \Omega \tag{6}$$

where R_t = resistance at a temperature of t°C*
R_0 = resistance at 0°C
α = temperature coefficient of resistance at 0°C
Δt = change in temperature from 0°C

The temperature coefficient for copper is 0.004 27. That is, if a copper wire has a resistance of 1 Ω at 0°C, it will have a resistance of 1 + 0.004 27 = 1.004 27 Ω at 1°C. The value of the temperature coefficient for copper is essentially the same as that for most of the unalloyed metals such as gold, silver, aluminum, and lead.

The value of α varies with the temperature at which it is determined: 0.004 27 is valid only when the reference temperature is 0°C. If 20°C is taken as the reference temperature, the value of α changes to 0.003 93. Thus it is important to always relate to the reference temperature for which the value of α is specified. If the resistance is known at a certain temperature and you are required to determine the resistance corresponding to some other temperature, you *must* first calculate, as an intermediate step, the resistance for the reference temperature.

*°C stands for degrees Celsius.

EXAMPLE 7 The resistance of a coil of copper wire is 34 Ω at 15°C. What is the resistance at 70°C?

SOLUTION 1

Using $\alpha = 0.004\,27$ and relating to 0°C, $R_t = 34\ \Omega$ and $\Delta t = 15$ (15°C − 0°C).

$$R_t = R_0(1 + \alpha\,\Delta t)$$

Solving for R_0,
$$R_0 = \frac{R_t}{1 + \alpha\,\Delta t} = \frac{34}{1 + (0.004\,27)(15)} = \frac{34}{1.064\,05} = 31.95\ \Omega$$

Using R_0, solve for R_{70}:

$$R_{70} = 31.95(1 + 0.004\,27 \times 70) = 31.95(1.2989) = 41.5\ \Omega$$

SOLUTION 2

Using $\alpha = 0.003\,93$ and relating to 20°C, $R_t = 34\ \Omega$ and

$$\Delta t = -5 \qquad (15°\mathrm{C} - 20°\mathrm{C})$$

Solving for R_{20},
$$R_{20} = \frac{34}{1 + (0.003\,93)(-5)} = \frac{34}{1 - 0.019\,65} = \frac{34}{0.980\,35} = 34.68\ \Omega$$

Using R_{20}, solve for R_{70}:

$$R_{70} = 34.68(1 + 0.003\,93 \times 50) = 34.68(1.1965) = 41.5\ \Omega$$

A more convenient relation is derived by assuming that the proportionality between resistance and temperature extends linearly to the point where copper has a resistance of 0 Ω at a temperature of −234.5°C. This results in the ratio

$$\frac{R_2}{R_1} = \frac{234.5 + t_2}{234.5 + t_1} \tag{7}$$

where R_1 = resistance of copper in ohms at a temperature of t_1°C
R_2 = resistance of copper in ohms at a temperature of t_2°C

EXAMPLE 8 The resistance of a coil of copper wire is 34 Ω at 15°C. What is the coil's resistance at 70°C?

SOLUTION

Given $R_1 = 34\ \Omega$, $t_1 = 15°C$, and $t_2 = 70°C$. $R_2 = ?$ Solving Eq. (7) for R_2,

$$R_2 = \frac{234.5 + t_2}{234.5 + t_1} R_1$$

Substituting the known values,

$$R_2 = \frac{234.5 + 70}{234.5 + 15} \times 34 = 41.5\ \Omega$$

The specifications for electric machines generally include a provision that the temperature of the coils, etc., when the machines are operating under a specified load for a specified time, must not rise more than a certain number of degrees. Temperature rise can be computed by measuring the resistance of the coils at room temperature and again at the end of the test.

EXAMPLE 9 The field coils of a shunt motor have a resistance of 90 Ω at 20°C. After the motor was run for 3 h, the resistance of the field coils was 146 Ω. What was the temperature of the coils?

SOLUTION

Given $R_1 = 90\ \Omega$, $t_1 = 20°C$, $R_2 = 146\ \Omega$. $t_2 = ?$ Solving Eq. (7) for t_2,

$$t_2 = \frac{234.5 + t_1}{R_1} R_2 - 234.5$$

Substituting the known values,

$$t_2 = \frac{234.5 + 20}{90} \times 146 - 234.5$$
$$= 413 - 234.5 = 178.5°$$

The actual temperature rise is

$$t_2 - t_1 = 178.5° - 20° = 158.5°$$

PROBLEMS 9–3

1. The resistance of a 1-km coil of No. 16 copper wire at 20°C is 13.2 Ω. What will be the resistance at 0°C?
2. If the resistance of a copper coil is 4.13 Ω at 0°C, what will it be at 20°C?
3. What will be the resistance of the wire of Prob. 1 at 35°C?
4. The resistance of the primary winding of a transformer was 3.05 Ω at 20°C. After operation for 3 h, the resistance increased to 3.32 Ω. What was the final operating temperature?

5. The specifications for a high-power transformer included a provision that the transformer was to operate continuously under full load with the winding temperature not to exceed 55°C. The resistance of the primary coil was measured before the transformer was put on test, at 22°C, and found to be 52.7 Ω. After a day's test at rated load, the resistance was again measured and was found to be 60.0 Ω. Did the transformer meet the specifications?

9–4 WIRE MEASURE

Wire sizes are designated by numbers in a system known as the American wire gage (formerly Brown and Sharpe gage). These gage numbers, ranging from 0000, the largest size, to 40, the smallest size, are based on a constant ratio between successive gage numbers. The wire sizes and other pertinent data are listed in Table 6 in the Appendix.

Inspection of the wire table will reveal the progression formed by the wire sizes. As the sizes become smaller, every third gage number results in one-half the cross-sectional area and, therefore, double the resistance.

EXAMPLE 10 Number 10 wire has a cross-sectional area of 5.261 mm^2 and a resistance of 3.277 Ω/km. Three sizes smaller, No. 13 wire has an area of 2.63 mm^2 (almost exactly one-half of 5.261), and a resistance of 6.56 Ω (almost double 3.277). Similarly, half the resistance of No. 10 wire is provided by No. 7: 1.634 Ω/km, with an area of 10.55 mm^2 (double 5.261).

9–5 FACTORS GOVERNING WIRE SIZE IN PRACTICE

From an electrical viewpoint, three factors govern the selection of the size of wire to be used for transmitting current:

1. The safe current-carrying capacity of the wire
2. The power lost in the wire
3. The allowable voltage variation, or the voltage drop, in the wire

It must be remembered that the length of wire, for the purpose of computing wire resistance and its effects, is always twice the distance from the source of power to the load (outgoing and return leads).

EXAMPLE 11 A motor receives its power through No. 4 wire from a generator located at a distance of 1 km. The voltage across the motor is 220 V, and the current taken by the motor is 19.8 A. What is the terminal voltage of the generator?

SOLUTION

The circuit is represented in Fig. 9–2. Note that it consists of a simple series circuit which can be simplified to that of Fig. 9–3. The resistance of the 1 km of No. 4 wire from the generator to the motor is represented by R_0; reference to Table 6 shows it to be 0.8152 Ω. Similarly, the resistance from the motor back to the generator, which is represented by R_r, also is 0.8152 Ω. The voltage drop in each wire is

$$V = IR_0 = IR_r = 19.8 \times 0.8152 = 16.14 \text{ V}$$

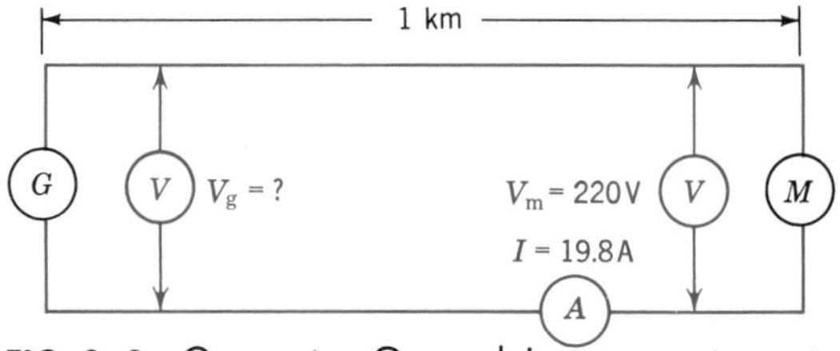

FIG. 9–2 Generator G supplying power to motor M at a distance of 1 km.

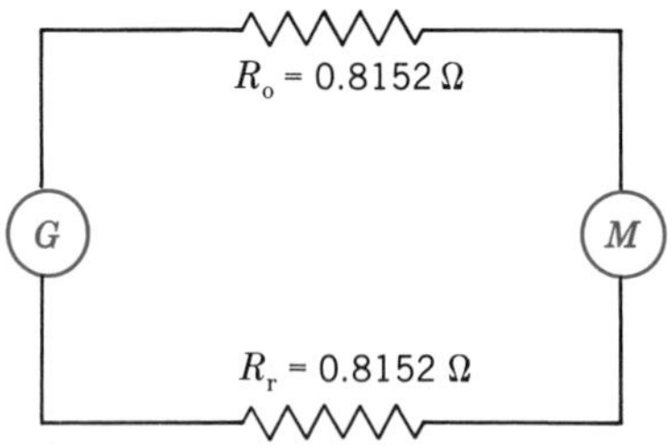

FIG. 9–3 Simplified form of circuit shown in Fig. 9–2.

Since the applied voltage must equal the sum of all the voltage drops around the circuit (Sec. 8–8), the terminal voltage of the generator is

$$V_g = 220 + 16.14 + 16.14 = 252.28 \text{ or } 252 \text{ V}$$

Since the resistance out R_0 is equal to the return resistance R_r, the foregoing solution is simplified by taking twice the actual wire distance for the length of wire that constitutes the resistance of the feeders. Therefore, the length of No. 4 wire between generator and motor is 2 km, which results in a line resistance R_L of

$$2 \times 0.8152 = 1.6304 \ \Omega$$

The circuit can be further simplified as shown in Fig. 9–4. Thus, the generator terminal voltage is

$$V_g = 220 + IR_L$$
$$= 220 + (19.8 \times 1.6304) = 252 \text{ V}$$

The power lost in the line is

$$P_L = I^2 R_L$$
$$= 19.8^2 \times 1.6304 = 639 \text{ W}$$

The power taken by the motor is

$$P_M = V_M I$$
$$= 220 \times 19.8$$
$$= 4356 \text{ W} = 4.356 \text{ kW}$$

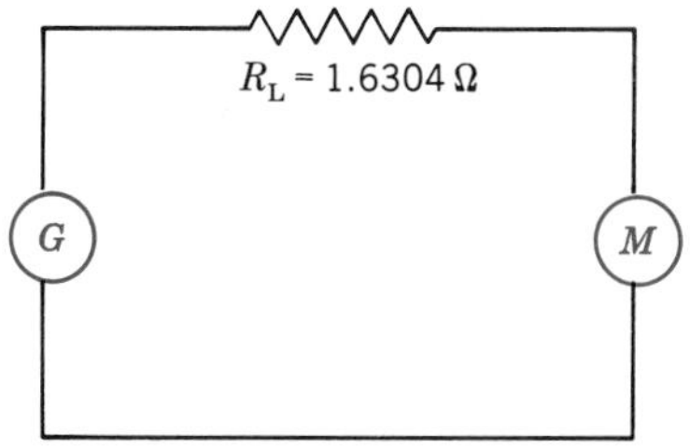

FIG. 9–4 Equivalent circuit of circuits shown in Figs. 9–2 and 9–3.

The power delivered by the generator is

$$P_G = P_L + P_M$$
$$= 639 + 4356 = 4995 \text{ W}$$

$$\text{Efficiency of transmission} = \frac{\text{power delivered to load}}{\text{power delivered by generator}}$$
$$= \frac{4356}{4995} = 0.872 = 87.2\% \qquad (8)$$

The efficiency of transmission is obtainable in terms of the generator terminal voltage V_G and the voltage across the load V_L. Because

$$\text{Power delivered to load} = V_L I$$

and

$$\text{Power delivered by generator} = V_G I$$

substituting in Eq. (8) gives us

$$\text{Efficiency of transmission} = \frac{V_L I}{V_G I}$$
$$= \frac{V_L}{V_G} \qquad (9)$$

and substituting the voltages in Eq. (9) gives us

$$\text{Efficiency of transmission} = \frac{220}{252}$$

$$= 0.873 = 87.3\%$$

PROBLEMS 9–4

NOTE *All wires in the following problems are of copper with characteristics as listed in Table 6 of the Appendix.*

1. (*a*) What is the resistance of 3200 m of No. 10 wire?
   (*b*) What is its weight?

2. (*a*) What is the resistance of 1.8 m of No. 8 wire?
   (*b*) What is its weight?
3. (*a*) What is the length of a 210-kg coil of No. 6 wire?
   (*b*) What is its resistance?
4. (*a*) What is the length of a 100-kg coil of No. 16 wire?
   (*b*) What is its resistance?
5. A telephone cable consisting of several pairs of No. 22 wire connects two cities 20 km apart. If a pair is short-circuited at one end, what will be the resistance of the loop thus formed?
6. A relay is to be wound with 1500 turns of No. 22 wire. The average diameter of a turn is 46 mm.
   (*a*) What will be the resistance?
   (*b*) What will be the weight of the coil?
7. Fifteen kilowatts of power is to be transmitted 200 m from a generator that maintains a constant terminal voltage of 240 V. If not over 5% line drop is allowed, what size wire must be used?
8. A generator with a constant brush potential of 230 V is feeding a motor 100 m away. The feeders are No. 6 wire, and the motor current is 27.7 A.
   (*a*) What would a voltmeter read if connected across the motor brushes?
   (*b*) What is the efficiency of transmission?
9. A motor requiring 34 A at 230 V is located 125 m from a generator that maintains a constant terminal voltage of 240 V.
   (*a*) What size wire must be used between generator and motor in order to supply the motor with rated current and voltage?
   (*b*) What will be the efficiency of transmission?
10. A 25-hp 230-V motor is to be installed 120 m from a generator that maintains a constant potential of 240 V.
    (*a*) If the motor is 84% efficient, what size wire should be used between motor and generator?
    (*b*) If the wire specified in (*a*) is used, what will be the motor voltage under rated load condition?

SELF TEST

1. No. 18 copper wire has a resistance of 21 Ω/km. What is the resistance of 420 m of this wire?
2. What is the specific resistance of the wire of Prob. 1 in microhm-meters? The cross-sectional area of No. 18 wire is 0.823 mm^2.
3. A 110-m coil of copper wire has a measured resistance of 0.145 Ω. What is the most probable gage size of the wire?
4. What is the mass of a 700-m coil of bare copper No. 18 wire?
5. Constantan wire has a specific resistance of 0.4900 μΩ·m. What will be the resistance of a 20-m coil of constantan wire that has a diameter of 0.203 mm?

6. The wire tables for Probs. 1 and 2 are based on a temperature of 20°C. What would be the resistance of a 1-km coil of No. 6 copper wire at 60°C? At 0°C, $\alpha = 0.004\ 27$.
7. The resistance of the primary winding of a transformer was measured as 2.48 Ω at 20°C. After operating continuously for several hours, the resistance has increased to a constant 3.22 Ω. The guarantee specifies that full-load temperature would not exceed 80°C. Does the transformer satisfy the guarantee?
8. A two-wire copper cable joins points 620 m apart. When the two wires are shorted at one end, the resistance of the resulting loop is measured as 2.56 Ω. What is the probable AWG size of the wire?
9. A generator that delivers a constant terminal voltage of 240 V must deliver 15 kW of power to a load 300 m away. If not more than 10% line drop can be tolerated, what size copper wire should be used?
10. If No. 0 copper wire is used for the line of Prob. 9, what is the voltage delivered to the load terminals?

CHAPTER 10

SPECIAL PRODUCTS AND FACTORING

In the study of arithmetic, it is necessary to memorize the multiplication tables as an aid to rapid computation. Similarly, in the study of algebra, certain forms of expressions occur so frequently that it is essential to be able to multiply, divide, or factor them by inspection.

10–1 FACTORING

To *factor* an algebraic expression means to find two or more expressions that, when multiplied, will result in the original expression.

EXAMPLE 1 $2 \times 3 \times 4 = 24$. Thus, 2, 3, and 4 are some of the factors of 24.

EXAMPLE 2 $b(x + y) = bx + by$. b and $(x + y)$ are the factors of $bx + by$.

EXAMPLE 3 $(x + 4)(x - 3) = x^2 + x - 12$. The quantities $(x + 4)$ and $(x - 3)$ are the factors of $x^2 + x - 12$.

10–2 PRIME NUMBERS

A number that has no factor other than itself and unity is known as a *prime number*. Thus, 3, 5, 13, x, and $(a + b)$ are prime numbers.

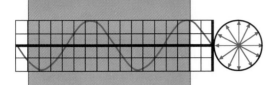

10–3 SQUARE OF A MONOMIAL

At this point you should review the law of exponents for multiplication in Sec. 4–3.

EXAMPLE 4

$$(2ab^2)^2 = (2ab^2)(2ab^2) = 4a^2b^4$$

EXAMPLE 5

$$(-3x^2y^3)^2 = (-3x^2y^3)(-3x^2y^3) = 9x^4y^6$$

By application of the rules for the multiplication of numbers having like signs and the law of exponents, we have the following rule:

RULE

To square a monomial, square the numerical coefficient, multiply this product by the literal factors of the monomial, and multiply the exponent of each letter by 2.

10–4 CUBE OF A MONOMIAL

EXAMPLE 6

$$(3a^2b)^3 = (3a^2b)(3a^2b)(3a^2b) = 27a^6b^3$$

EXAMPLE 7

$$\begin{aligned}(-2xy^3)^3 &= (-2xy^3)(-2xy^3)(-2xy^3) \\ &= -8x^3y^9\end{aligned}$$

Note that the cube of a *positive* number is always *positive* and that the cube of a *negative* number is always *negative*. Again, by application of the rules for the multiplication of positive and negative numbers and the law of exponents, we have the following rule:

RULE

To cube a monomial, cube the numerical coefficient, multiply this product by the literal factors of the monomial, multiply the exponent of each letter by 3, and affix the same sign as the monomial.

PROBLEMS 10–1

Find the values of the following indicated powers:

1. $(ac)^2$
2. $(\alpha\pi)^3$
3. $(i^2rX)^3$
4. $\pi\left(\frac{D}{2}\right)^2$
5. $(-3\theta\psi)^2$
6. $\left(\frac{4}{3}\pi R^3\right)^2$

7. $(-2\pi fL)^2$

8. $\left(3\frac{v}{i}\right)^2$

9. $(2\pi fC)^2$

10. $\left(-3\frac{ir}{v}\right)^3$

11. $-\left(\frac{1}{2\pi fC}\right)^2$

12. $-(-11X_L{}^2)^3$

13. $\left(-\frac{3M^2}{LC}\right)^3$

14. $\frac{(V_sN_p)^2}{V_p{}^3}$

15. $-\left(\frac{V2}{2g}\right)^3$

16. $\left(\frac{120f}{N}\right)^2$

17. $\left(\frac{3P}{5VI}\right)^3$

18. $-(2\pi fL)^3$

19. $-\left(\frac{4}{3}\pi R^3\right)^2$

20. $\left(\frac{5}{8}u^2v^3wx^4y^5\right)^3$

21. $\left(\frac{x^4y^6}{p^5}\right)^3$

10–5 SQUARE ROOT OF A MONOMIAL

The *square root* of an expression is one of two equal factors of the expression.

EXAMPLE 8 $\sqrt{3}$ is a number such that

$$\sqrt{3} \cdot \sqrt{3} = 3$$

EXAMPLE 9 $\sqrt{n}$ is a number such that

$$\sqrt{n} \cdot \sqrt{n} = n$$

Because $(+2)(+2) = +4$

and $(-2)(-2) = +4$

it is apparent that 4 has two square roots, $+2$ and -2. Similarly, 16 has two square roots, $+4$ and -4.

In general, every number has two square roots equal in magnitude, one positive and one negative. The positive root is known as the *principal root;* if no sign precedes the radical, the positive root is understood. (Your calculator gives only the principal root.) Thus, in practical numerical computations, the following is understood:

$$\sqrt{4} = +2$$

and $-\sqrt{4} = -2$

In dealing with literal numbers, the values of the various factors often are unknown. Therefore, when we extract a square root, we affix the double sign $\pm$ to denote "plus or minus."

EXAMPLE 10 Since $a^4 \cdot a^4 = a^8$ and $(-a^4)(-a^4) = a^8$,

then $\sqrt{a^8} = \pm a^4$

EXAMPLE 11 Since $x^2y^3 \cdot x^2y^3 = x^4y^6$ and $(-x^2y^3)(-x^2y^3) = x^4y^6$,

then $\sqrt{x^4y^6} = \pm x^2y^3$

From the foregoing examples, we formulate the following:

RULE

To extract the square root of a monomial, extract the square root of the numerical coefficient, divide the exponents of the letters by 2, and affix the $\pm$ sign.

EXAMPLE 12 $\sqrt{4a^4b^2} = \pm 2a^2b$

EXAMPLE 13 $\sqrt{\frac{1}{9}x^2y^6z^4} = \pm\frac{1}{3}xy^3z^2$

NOTE *A perfect monomial square is one that is positive and has a perfect square numerical coefficient and has only even numbers as exponents.*

10–6 CUBE ROOT OF A MONOMIAL

The *cube root* of a monomial is one of the three equal factors of the monomial.

Because $(+2)(+2)(+2) = 8$

then $\sqrt[3]{8} = 2$

Similarly, $(-2)(-2)(-2) = -8$

and $\sqrt[3]{-8} = -2$

From this it is evident that the cube root of a monomial has the same sign as the monomial itself.

Because $x^2y^3 \cdot x^2y^3 \cdot x^2y^3 = x^6y^9$

then $\sqrt[3]{x^6y^9} = x^2y^3$

The above results can be stated as follows:

RULE

To extract the cube root of a monomial, extract the cube root of the numerical coefficient, divide the exponents of the letters by 3, and affix the same sign as the monomial.

EXAMPLE 14 $\sqrt[3]{8x^6y^3z^{12}} = 2x^2yz^4$

EXAMPLE 15 $\sqrt[3]{-27a^3b^9c^6} = -3ab^3c^2$

NOTE *A perfect cube monomial has a positive or negative perfect cube numerical coefficient and exponents that are exactly divisible by 3.*

PROBLEMS 10–2

Find the value of the following:

1. $\sqrt{x^2}$
2. $\sqrt{a^4}$
3. $\sqrt{16\omega^2}$
4. $\sqrt{25^2}$
5. $\sqrt{(-\omega)^2}$
6. $\sqrt{100m^2n^{12}}$
7. $\sqrt{4\pi^2L^2C^2}$
8. $\sqrt[3]{81\alpha^6}$
9. $\sqrt[3]{8a^{12}}$
10. $\sqrt[3]{-125X^3}$
11. $\sqrt[3]{(-2)^6}$
12. $\sqrt{4\pi^2f^2L^2\times 10^2}$
13. $\sqrt{169m^4n^2p^6}$
14. $\sqrt[5]{32\lambda^5\psi^{10}}$
15. $\sqrt[3]{27\pi^6\theta^9\phi^{12}}$
16. $\sqrt{121x^{10}y^{12}z^6}$
17. $\sqrt{\dfrac{256\pi^2r^2x^4}{289z^6\phi^4}}$
18. $\sqrt{\dfrac{25m^4n^2p^8}{64a^4b^2c^6}}$
19. $-\sqrt{\dfrac{625r^6s^4t^8}{16x^6z^{10}}}$
20. $\sqrt[3]{\dfrac{-8\pi^3X_L{}^3}{27Z^6X_C{}^{12}}}$
21. $-\sqrt[3]{\dfrac{-64a^3\omega^6}{125x^6z^{12}}}$
22. $\sqrt{\dfrac{196h^2n^4p^6}{121a^2b^4c^2}}$
23. $\sqrt{\dfrac{25v^2t^2}{256a^8b^2x^2}}$
24. $\sqrt[3]{-\dfrac{1}{27}x^6y^{12}z^{15}}$

10–7 POLYNOMIALS WITH A COMMON MONOMIAL FACTOR

Type: $a(b + c + d) = ab + ac + ad$

RULE

To factor a polynomial whose terms contain a common monomial factor:

1. Determine by inspection the greatest common factor of its terms.
2. Divide the polynomial by this factor.
3. Write the quotient in parentheses preceded by the monomial factor.

EXAMPLE 16 Factor $3x^2 - 9xy^2$.

SOLUTION

The common monomial factor of both terms is $3x$.

$$\therefore\ 3x^2 - 9xy^2 = 3x(x - 3y^2)$$

EXAMPLE 17 Factor $2a - 6a^2b + 4ax - 10ay^3$.

SOLUTION

Each term contains the factor $2a$.

$$\therefore\ 2a - 6a^2b + 4ax - 10ay^3 = 2a(1 - 3ab + 2x - 5y^3)$$

EXAMPLE 18 Factor $14x^2yz^3 - 7xy^2z^2 + 35xz^5$.

SOLUTION

Each term contains the factor $7xz^2$.

$$\therefore\ 14x^2yz^3 - 7xy^2z^2 + 35xz^5 = 7xz^2(2xyz - y^2 + 5z^3)$$

PROBLEMS 10–3

Factor:

1. $3x + 6$
2. $\frac{1}{2}m + \frac{1}{4}q$
3. $6\alpha + \alpha\varepsilon + 4\alpha\phi$
4. $\frac{1}{4}x^3y - \frac{1}{3}xy^2 + \frac{1}{9}x^2y^2$
5. $16IZ - 12IR$
6. $12IV - 10RV$
7. $\frac{l^2m}{3} + \frac{l^3m^2}{8} - \frac{l^2m^3}{2}$
8. $4\omega^4X_L{}^4 - 12\omega X_L + 28\omega^2X_L{}^2$
9. $2a^3b^2c + 8a^2bc^3 + 12a^2b^2c^2$
10. $\frac{1}{4}I^2R^2Z^2 - \frac{1}{12}IRZ^2 + \frac{1}{16}I^2RZ$
11. $36\alpha^4\beta^3\omega^2 - 72\alpha^2\beta^2\omega^5 + 180\alpha^2\beta^5\omega^2$
12. $540\theta^3\lambda^2\phi + 810\theta^2\lambda\phi^3 - 1080\theta\lambda^3\phi^2$
13. $\frac{1}{27}I^2R^3 + \frac{1}{15}IR^2 - \frac{1}{3}I^3R$
14. $\frac{1}{36}X_L{}^3X_C{}^2 - \frac{1}{18}X_L{}^2X_C{}^3 + \frac{1}{72}X_L{}^2X_C{}^2$
15. $720\eta^4\theta^2\phi^3\omega + 1080\eta^2\theta^4\phi\omega^2 + 600\eta^3\theta^3\phi\omega^2 - 480\eta\theta^6\phi\omega$

10–8 SQUARE OF A BINOMIAL

Type: $(a + b)^2 = a^2 + 2ab + b^2$

The multiplication

$$\begin{array}{r} a + b \\ a + b \\ \hline a^2 + ab \\ + ab + b^2 \\ \hline a^2 + 2ab + b^2 \end{array}$$

results in the formula

$$(a + b)^2 = a^2 + 2ab + b^2$$

which can be expressed by the following rule:

RULE

To square the sum of two terms, square the first term, add twice the product of the two terms, and add the square of the second term.

EXAMPLE 19 Square $2b + 4cd$.

SOLUTION

$$\begin{aligned}(2b + 4cd)^2 &= (2b)^2 + 2(2b)(4cd) + (4cd)^2 \\ &= 4b^2 + 16bcd + 16c^2d^2\end{aligned}$$

EXAMPLE 20 Let x and y be represented by lengths. Then

$$(x + y)^2 = x^2 + 2xy + y^2$$

can be illustrated graphically as shown in Fig. 10–1.

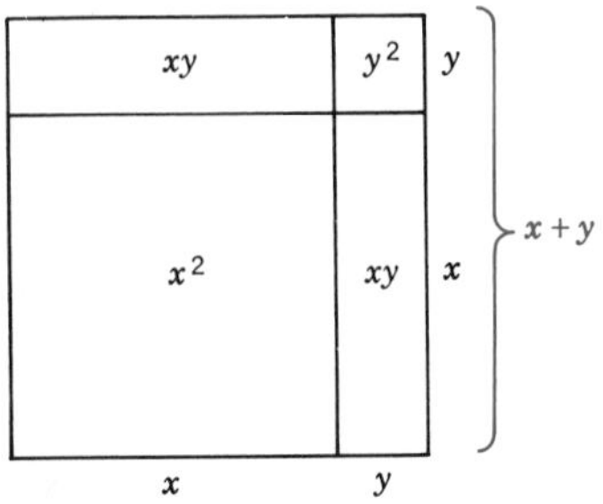

FIG. 10–1 Graphical illustration of $(x + y)^2 = x^2 + 2xy + y^2$.

The multiplication

$$\begin{array}{r} a - b \\ a - b \\ \hline a^2 - ab \quad\quad \\ - ab + b^2 \\ \hline a^2 - 2ab + b^2 \end{array}$$

results in the formula

$$(a - b)^2 = a^2 - 2ab + b^2$$

which can be expressed as follows:

RULE

To square the difference of two terms, square the first term, subtract twice the product of the two terms, and add the square of the second term.

EXAMPLE 21 Square $3a^2 - 5xy$.

SOLUTION

$$\begin{aligned}(3a^2 - 5xy)^2 &= (3a^2)^2 - 2(3a^2)(5xy) + (5xy)^2 \\ &= 9a^4 - 30a^2xy + 25x^2y^2\end{aligned}$$

EXAMPLE 22 Let x and y be represented by lengths. Then

$$(x - y)^2 = x^2 - 2xy + y^2$$

can be illustrated graphically as shown in Fig. 10–2. x^2 is the large square. The figure shows that the two rectangles taken from x^2 leave $(x - y)^2$. Since an amount y^2 is a part of one xy that has been subtracted from x^2 and is outside x^2, we must add it back in. Hence, we obtain

$$(x - y)^2 = x^2 - 2xy + y^2$$

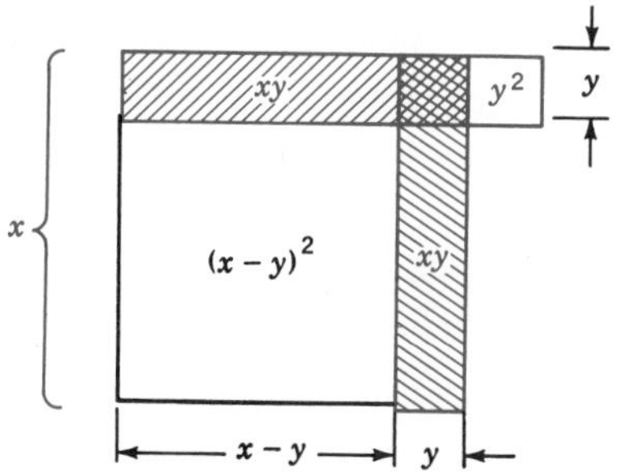

FIG. 10–2. Graphical illustration of $(x - y)^2 = x^2 - 2xy + y^2$.

Mentally, practice squaring sums and differences of binomials by following the foregoing rules. Proficiency in these and later methods will greatly reduce the labor in performing multiplications.

PROBLEMS 10–4

Mentally, square the following:

1. $x + 5$
2. $c + 7$
3. $C - w$
4. $R - 2$
5. $\phi + 12$
6. $m - 3$
7. $5V - v$
8. $2f + 5g$
9. $M - m$
10. $2\alpha - 3\beta$
11. $5\theta + 4\phi$
12. $2\lambda - 5\mu$
13. $9r_1 - 3r_2$
14. $m^2 + 6$
15. $1 + X_L^2$
16. $2\theta^2 - 13\phi$
17. $5a^2 - 2f^3$
18. $30 + 2$
19. $20 - 5$
20. $30 + 5$
21. $3\pi L^2 - 2\pi C^2$
22. $2\pi f L_1 - Z$
23. $1.5\theta^2 - 0.5\alpha$
24. $\frac{1}{2}R_1 + \frac{1}{4}R_2$
25. $\frac{3}{4}X^2 - \frac{1}{2}Z^2$
26. $\frac{1}{3}\phi^3\lambda + \frac{1}{2}\alpha^2$
27. $6\phi^2\omega - \frac{1}{4}\lambda^2$

Expand:

28. $(y + 2)^2$
29. $\left(y + \frac{1}{4}\right)^2$
30. $\left(\alpha + \frac{1}{3}\right)^2$
31. $\left(\frac{1}{2} - E\right)^2$
32. $\left(\mu - \frac{1}{12}\right)^2$
33. $(1 + x^5)^2$
34. $\left(X_1^2 + \frac{2}{3}\right)^2$
35. $\left(L^2 - \frac{7}{8}P\right)^2$
36. $\left(\frac{X}{2} + Y\right)^2$
37. $\left(\frac{a}{5} + \frac{b}{3}\right)^2$
38. $(2 + 7xy)^2$
39. $\left(R_1 - \frac{5}{8}R_2\right)^2$
40. $\left(2\phi + \frac{3}{4}\theta^2\right)^2$
41. Develop a graphical illustration of $(x + y)(x - y)$.

10–9 SQUARE ROOT OF A TRINOMIAL

In the preceding section, it was shown that

$$(a + b)^2 = a^2 + 2ab + b^2$$

and

$$(a - b)^2 = a^2 - 2ab + b^2$$

From these and other binomials that have been squared, it is evident that a trinomial is a perfect square if

1. Two terms are squares of monomials and are positive.
2. The other term is twice the product of the monomials and has affixed either a plus or a minus sign.

EXAMPLE 23 $x^2 + 2xy + y^2$ is a perfect trinomial square because x^2 and y^2 are the squares of the monomials x and y, respectively, and $2xy$ is twice the product of the monomials. Therefore,

$$x^2 + 2xy + y^2 = (x + y)^2$$

EXAMPLE 24 $4a^2 - 12ab + 9b^2$ is a perfect trinomial square because $4a^2$ and $9b^2$ are the squares of $2a$ and $3b$, respectively, and the other term is $-2(2a)(3b)$. Therefore,

$$4a^2 - 12ab + 9b^2 = (2a - 3b)^2$$

RULE

To extract the square root of a perfect trinomial square, extract the square root of the two perfect square monomials and connect them with the sign of the remaining term.

EXAMPLE 25 Supply the missing term in $x^4 + ? + 16$ so that the three terms will form a perfect trinomial square.

SOLUTION

The missing term is twice the product of the monomials whose squares result in the two known terms; that is, $2(x^2)(4) = 8x^2$. Hence,

$$x^4 + 8x^2 + 16 = (x^2 + 4)^2$$

EXAMPLE 26 Supply the missing term in $25a^2 + 30ab + ?$ so that the three terms will form a perfect trinomial square.

SOLUTION

The square root of the first term is $5a$. The missing term is the square of some number N such that $2(5a)(N) = 30ab$. Then by multiplying, we obtain $10aN = 30ab$, or $N = 3b$. Therefore,

$$25a^2 + 30ab + 9b^2 = (5a + 3b)^2$$

PROBLEMS 10–5

Supply the missing terms so that the three terms form perfect trinomial squares:

1. $a^2 + ? + 4$
2. $x^2 + ? + 9$
3. $u^2 - ? + 4$
4. $C^2 - ? + R^2$
5. $16w^2 + ? + l^2$
6. $25X_C^2 - ? + 4$
7. $25l^2 + ? + w^2$
8. $100L_1^2 + ? + 16M^2$
9. $4m^2 + ? + 9p^2$
10. $x^2 - 2xy + ?$
11. $a^2 + 2ab + ?$
12. $? + 70\theta\phi + 25\phi^2$
13. $? + 80pq + 100q^2$
14. $Z^2 + 12XZ + ?$
15. $\frac{1}{9}\theta^2\phi^2 - ? + \frac{1}{4}\omega^2$
16. $\frac{1}{16}\eta^4 - \frac{1}{4}\eta^2\theta + ?$
17. $? - \frac{\pi\phi}{3} + \frac{1}{4}\phi^2$
18. $\frac{R_1^2}{64} - \frac{1}{24}XR_1 + ?$

Extract the square roots of the following:

19. $E^2 + 6E + 9$
20. $p^2 - 12pq + 36q^2$
21. $9l^2 + 6lw + w^2$
22. $V^2 + 12VI + 36I^2$
23. $4\pi^4 r_1^2 + 28\pi^2 r_1 r_2 + 49r_2^2$
24. $64\omega^2\lambda^2 + 16\omega\lambda\Omega^2 + \Omega^4$
25. $\frac{9}{25}\pi^2R^4 + \frac{4}{5}\pi R^2 + \frac{4}{9}$
26. $4L_1^2 + \frac{12}{5}L_1L_2 + \frac{9}{25}L_2^2$
27. $\frac{10\phi\lambda}{21} + \frac{25\phi^2}{36} + \frac{4\lambda^2}{49}$
28. $-\frac{4Z^4M^2}{27} + \frac{4Z^8}{81} + \frac{M^4}{9}$

10–10 PRIME FACTORS OF AN EXPRESSION

In factoring a number, all of the number's prime factors should be obtained. After an expression is factored once, it may be possible to factor it again.

EXAMPLE 27 Find the prime factors of $12i^2r + 12iIr + 3I^2r$.

SOLUTION

$$\begin{aligned} 12i^2r + 12iIr + 3I^2r &= 3r(4i^2 + 4iI + I^2) \\ &= 3r(2i + I)(2i + I) \\ &= 3r(2i + I)^2 \end{aligned}$$

PROBLEMS 10–6

Find the prime factors of the following:

1. $2xy + 6yz$
2. $10abc + 35acd$
3. $5AC^2 + 10ACD + 5AD^2$
4. $5V^2i^2 - 10VIi^2R + 5I^2i^2R^2$

5. $18ir^2 + 60irz + 50iz^2$
6. $\frac{2V^4}{3r} + \frac{4V^2v^2}{r} + \frac{6v^4}{r}$
7. $\frac{24IR_1^2}{V} - \frac{48IR_1R_2}{V} + \frac{24IR_2^2}{V}$
8. $\frac{18R^2}{5\phi\omega} + \frac{6RfL}{5\phi\omega} + \frac{f^2L^2}{10\phi\omega}$
9. $\frac{5r\lambda^2}{16e} - \frac{5f^2r\lambda}{2e} + \frac{5f^4r}{e}$
10. $\frac{200Ff^2}{C} + \frac{480Ffx}{C} + \frac{288Fx^2}{C}$

10–11 PRODUCT OF THE SUM AND DIFFERENCE OF TWO NUMBERS

Type: $(a + b)(a - b) = a^2 - b^2$

The multiplication of the sum and difference of two general numbers, such as

$$\begin{array}{r} a + b \\ a - b \\ \hline a^2 + ab \qquad \\ - ab - b^2 \\ \hline a^2 \qquad\quad - b^2 \end{array}$$

results in the formula

$$(a + b)(a - b) = a^2 - b^2$$

which can be expressed by the following:

RULE

The product of the sum and difference of two numbers is equal to the difference of their squares.

EXAMPLE 28

$$(3x + 4y)(3x - 4y) = 9x^2 - 16y^2$$

EXAMPLE 29

$$(6ab^2 + 7c^3d)(6ab^2 - 7c^3d) = 36a^2b^4 - 49c^6d^2$$

PROBLEMS 10–7

Multiply by inspection:

1. $(x + 2)(x - 2)$
2. $(\lambda - 3)(\lambda + 3)$
3. $(R + Z)(R - Z)$
4. $(X_L + X_C)(X_L - X_C)$
5. $(2M - 3L)(2M + 3L)$
6. $(2\pi R_1 + 2\pi R_2)(2\pi R_1 - 2\pi R_2)$
7. $\left(\frac{3}{4}R_1 + Q\right)\left(\frac{3}{4}R_1 - Q\right)$
8. $\left(\frac{3}{4}\omega^2 - \frac{2}{5}\lambda\right)\left(\frac{3}{4}\omega^2 + \frac{2}{5}\lambda\right)$
9. $\left(\frac{2V^2}{R} + \frac{3I^2R}{P}\right)\left(\frac{2V^2}{R} - \frac{3I^2R}{P}\right)$
10. $\left(\frac{\theta^2}{2\phi} + \frac{3\alpha}{2\beta}\right)\left(\frac{\theta^2}{2\phi} - \frac{3\alpha}{2\beta}\right)$

10–12 FACTORING THE DIFFERENCE OF TWO SQUARES

RULE

To factor the difference of two squares, extract the square root of the two squares, add the roots for one factor, and subtract the second root from the first for the other factor.

EXAMPLE 30

$$x^2 - y^2 = (x + y)(x - y)$$

EXAMPLE 31

$$9a^2c^4 - 36d^6 = (3ac^2 + 6d^3)(3ac^2 - 6d^3)$$

PROBLEMS 10–8

Factor:

1. $I^2 - Z^2$
2. $Z^2 - R^2$
3. $9x^2 - 16y^2$
4. $9m^2 - 25p^2$
5. $\frac{1}{9} - x^2$
6. $\frac{\alpha^2}{\beta^2} - \frac{4\gamma^2}{9}$
7. $1 - 225\omega^2$
8. $\frac{1}{E_1{}^2} - \frac{1}{e^2}$
9. $64\pi^2\phi^4 - 1$
10. $\frac{1}{X_C{}^2} - \frac{V^2}{Q^2}$
11. $9c^2 - a^2 + 2ab - b^2$

Solution:

$$\begin{aligned} 9c^2 - a^2 + 2ab - b^2 &= 9c^2 - (a^2 - 2ab + b^2) \\ &= [3c + (a - b)][3c - (a - b)] \\ &= (3c + a - b)(3c - a + b) \end{aligned}$$

12. $(X^2 + 4XR + 4R^2) - Z^2$
13. $3x^2y^2 - 12s^2t^2 + 12z^2 - 12xyz$
14. $16I^2 - V^2 + \frac{14V}{X} - \frac{49}{X^2}$
15. $100acl - 144l^2 + 25a^2 + 100c^2l^2$

10–13 PRODUCT OF TWO BINOMIALS HAVING A COMMON TERM

Type: $(x + a)(x + b) = x^2 + (a + b)x + ab$

The multiplication

$$\begin{array}{r} x + a \\ x + b \\ \hline x^2 + ax \qquad\qquad \\ + bx + ab \\ \hline x^2 + ax + bx + ab \end{array}$$

when factored, results in $x^2 + (a + b)x + ab$.

This type of formula can be expressed as follows:

RULE

To obtain the product of two binomials having a common term, square the common term, multiply the common term by the algebraic sum of the second terms of the binomials, find the product of the second terms, and add the results.

EXAMPLE 32 Find the product of $x - 7$ and $x + 5$.

SOLUTION

$$\begin{aligned}(x - 7)(x + 5) &= x^2 + (-7 + 5)x + (-7)(+5)\\ &= x^2 - 2x - 35\end{aligned}$$

EXAMPLE 33

$$\begin{aligned}(ir + 3)(ir - 6) &= i^2r^2 + (+3 - 6)ir + (+3)(-6)\\ &= i^2r^2 - 3ir - 18\end{aligned}$$

Although the preceding examples have been written out in order to illustrate the method, the actual multiplication should be performed mentally. In Example 33, write the i^2r^2 term first. Then glance at the $+3$ and -6, note that their sum is -3 and their product is -18, and write down the complete product.

PROBLEMS 10–9

Mentally, multiply the following:

1. $(l + 5)(l + 2)$
2. $(\theta + 1)(\theta + 3)$
3. $(m + 1)(m - 2)$
4. $(\lambda - 4)(\lambda - 3)$
5. $(y + 5)(y + 7)$
6. $(2r + 3)(2r + 2)$
7. $(2V - 3)(2V + 1)$
8. $(5a - 3)(5a + 4)$
9. $(R_1 - 2)(R_1 - 5)$
10. $\left(\frac{1}{2}P + 2\right)\left(\frac{1}{2}P + 6\right)$
11. $(\alpha - 1)\left(\alpha - \frac{1}{4}\right)$
12. $(\lambda + 6)\left(\lambda - \frac{1}{3}\right)$
13. $\left(IR + \frac{1}{2}\right)\left(IR - \frac{1}{3}\right)$
14. $(2f + 12)\left(2f - \frac{1}{4}\right)$
15. $\left(\alpha + \frac{2}{3}\right)\left(\alpha + \frac{1}{3}\right)$
16. $\left(\frac{1}{R} + 3\right)\left(\frac{1}{R} - 1\right)$
17. $\left(\frac{1}{\sqrt{LC}} - f\right)\left(\frac{1}{\sqrt{LC}} - 3f\right)$
18. $\left(vt - \frac{1}{12}\right)\left(vt + \frac{1}{2}\right)$
19. $\left(\alpha\beta^2 + \frac{1}{10}\right)\left(\alpha\beta^2 + \frac{1}{5}\right)$
20. $\left(M + \frac{3L_1}{X}\right)\left(M - \frac{5L_1}{X}\right)$

10–14 FACTORING TRINOMIALS OF THE FORM $a^2 + ba + c$

A trinomial of the form $a^2 + ba + c$ can be factored if it is the product of two binomials having a common term.

RULE

To factor a trinomial of the form $a^2 + ba + c$, find two numbers whose sum is b and whose product is c. Add each of them to the square root of the first term for the factors.

EXAMPLE 34 Factor $a^2 + 7a + 12$.

SOLUTION

If this expression will factor, it will take the form

$$a^2 + 7a + 12 = (a + \quad)(a + \quad)$$

where the two blanks represent numbers whose product is 12 and whose sum is 7. The factors of 12 are

$$1 \times 12$$
$$2 \times 6$$
$$3 \times 4$$

The first two pairs will not do because the sum of neither pair is 7. The third pair gives the correct sum.

$$\therefore a^2 + 7a + 12 = (a + 3)(a + 4)$$

EXAMPLE 35 Factor $x^2 - 15x + 36$.

SOLUTION

Since the 36 is positive, its factors must bear the same sign; also, since -15 is negative, it follows that both factors must be negative. The factors of 36 are

1×36	4×9
2×18	6×6
3×12	

Inspection of these factors shows that 3 and 12 are the required numbers.

$$\therefore x^2 - 15x + 36 = (x - 3)(x - 12)$$

EXAMPLE 36 Factor $e^2 - e - 56$.

SOLUTION

Since we have -56, the two factors must have unlike signs. The sum of the factors must equal -1; therefore, the negative factor of -56 must have the greater absolute value. The factors of 56 are

1×56	4×14
2×28	7×8

Since the factors 7 and 8 differ in value by 1, we have

$$e^2 - e - 56 = (e + 7)(e - 8)$$

PROBLEMS 10–10

Factor:

1. $Q^2 + 3Q + 2$
2. $\phi^2 + 9\phi + 14$
3. $L^2 + 8L + 15$
4. $\alpha^2 - 9\alpha + 14$
5. $\beta^2 + 2\beta - 24$
6. $\theta^2 + 9\theta + 18$
7. $\theta^2 + 10\theta + 24$
8. $\omega^2 - 14\omega + 24$
9. $t^2 + 9t - 22$
10. $\alpha^2 t^2 - 13\alpha t + 36$
11. $q^2 + 4q - 12$
12. $g^2 - 22g - 75$
13. $\psi^2 + \psi - 12$
14. $I^2R^2 + 4VIR + 3V^2$
15. $\omega^2 - \omega f - 6f^2$
16. $\theta^4 - 4\theta^2\phi - 12\phi^2$
17. $\dfrac{1}{x^2} + \dfrac{3}{x} - 10$
18. $q^4 - \dfrac{1}{4}q^2 - \dfrac{1}{8}$
19. $I^2 + \dfrac{2IZ}{X} - \dfrac{24Z^2}{X^2}$
20. $v^4 i^4 + 2v^2 i^2 P - 3P^2$

10–15 PRODUCT OF ANY TWO BINOMIALS

Type: $(ax + b)(cx + d)$

Up to the present, if it was desired to multiply $5x - 2$ by $3x + 6$, we multiplied in the following manner:

$$\begin{array}{rrr} 5x & -\ 2 & \\ 3x & +\ 6 & \\ \hline 15x^2 & -\ 6x & \\ & +\ 30x & -\ 12 \\ \hline 15x^2 & +\ 24x & -\ 12 \end{array}$$

Note that $15x^2$ is the product of the first terms of the binomials and the last term is the product of the last terms of the binomials. Also, the middle term is the sum of the products obtained by multiplying the first term of each binomial by the second term of the other binomial.

The preceding example can be written in the following manner:

$$\begin{array}{ccc} 5x & - & 2 \\ & \nwarrow \ \nearrow & \\ & \swarrow \ \searrow & \\ 3x & + & 6 \\ \hline 15x^2 & +\ 24x & -\ 12 \end{array}$$

The middle term $(+24x)$ is the sum of *cross products* $(5x)(+6)$ and $(3x)(-2)$, which is obtained by multiplying the first term of each binomial by the second term of the other.

The usual method of obtaining this product is indicated by the following solution:

$$(5x - 2)(3x + 6) = 15x^2 + 24x - 12$$

RULE

For finding the product of any two binomials,

1. The first term of the product is the product of the first terms of the binomials.
2. The second term is the algebraic sum of the product of the two outer terms and the product of the two inner terms.
3. The third term is the product of the last terms of the binomials.

EXAMPLE 37 Find the product of $(4e + 7j)(2e - 3j)$.

SOLUTION

The only difficulty encountered in obtaining such products mentally is that of finding the second term.

$$(4e)(-3j) = -12ej$$
$$(7j)(2e) = 14ej$$
$$(-12ej) + (14ej) = +2ej$$
$$\therefore (4e + 7j)(2e - 3j) = 8e^2 + 2ej - 21j^2$$

EXAMPLE 38 Find the product $(7r^2 + 8Z)(8r^2 - 9Z)$.

SOLUTION

1. The first term of the product is $(7r^2)(8r^2) = 56r^4$.
2. Since $(7r^2)(-9Z) = -63r^2Z$ and $(8Z)(8r^2) = 64r^2Z$, the second term is $(-63r^2Z) + (64r^2Z) = +r^2Z$.
3. The third term is $(8Z)(-9Z) = 72Z^2$.

$$\therefore (7r^2 + 8Z)(8r^2 - 9Z) = 56r^4 + r^2Z - 72Z^2$$

By repeated drills you should acquire skill enough that you can readily obtain such products mentally. This type of product is frequently encountered in algebra, and the ability to multiply rapidly will save you much time.

PROBLEMS 10–11

Multiply:

1. $(\theta + 2)(\theta - 7)$
2. $(IZ - 5)(IZ + 3)$
3. $(2F + 1)(5F + 3)$
4. $(5\psi + 6)(2\psi + 2)$
5. $(4k - 2)(3k + 2)$
6. $(3A + 3)(2A - 4)$
7. $(2\omega + 7)(5\omega - 1)$
8. $(7\theta + 3)(3\theta + 7)$
9. $\left(\frac{1}{2}\omega + 8\right)\left(\frac{1}{2}\omega - 4\right)$

10. $\left(\frac{3}{\theta} - 6\right)\left(\frac{2}{\theta} - 12\right)$
11. $(3Y + IB)(2Y + 6IB)$
12. $(I - 18)(I + 6)$
13. $(3X - 20)(5X + 2)$
14. $(12M - 3)(3M - 12)$
15. $(15\theta - 2)(\theta - 5)$
16. $(4 + 5m)(5 - 4m)$
17. $(6 - 3\lambda)(9 - 5\lambda)$
18. $(4\theta + 6)(5\theta + 3)$
19. $(2a + 7b)(4a + 3b)$
20. $(3x + 7)(4x - 5)$
21. $(2a - 7t)(2a - 5t)$
22. $(\beta + 0.6)(\beta - 0.5)$
23. $(m + 0.7p)(m - 0.1p)$
24. $(IR - 0.9)(IR + 1)$
25. $\left(\frac{x}{8} + \frac{\lambda}{4}\right)(2x - 16\lambda)$
26. $\left(8\delta - \frac{2}{3\eta}\right)\left(9\delta - \frac{1}{2\eta}\right)$
27. $\left(5\alpha + \frac{1}{2\beta\theta}\right)\left(2\alpha + \frac{1}{3\beta\theta}\right)$
28. $\left(12\pi L - \frac{2}{3\pi C}\right)\left(10\pi L - \frac{2}{3\pi C}\right)$
29. $(0.2p - 0.7q)(0.8p - 0.3q)$
30. $\left(4m + \frac{3r}{p}\right)\left(6m - \frac{2r}{3p}\right)$

10–16 FACTORING TRINOMIALS OF THE TYPE $ax^2 + bx + c$

The method of factoring trinomials of the type $ax^2 + bx + c$ is best illustrated by examples.

EXAMPLE 39 Factor $3a^2 + 5a + 2$.

SOLUTION

It is apparent that the two factors are binomials and the product of the end terms must be $3a^2$ and 2. Therefore, the binomials to choose from are

$$(3a + 1)(a + 2)$$

and

$$(3a + 2)(a + 1)$$

However, the first factors when multiplied result in a product of $7a$ for the middle term. The second pair of factors when multiplied give a middle term of $5a$. Therefore,

$$3a^2 + 5a + 2 = (3a + 2)(a + 1)$$

EXAMPLE 40 Factor $6e^2 + 7e + 2$.

SOLUTION

Again, the end terms of the binomial factors must be so chosen that their products result in $6e^2$ and 2. Both last terms of the factors are of like signs, for the last term of the trinomial is positive. Also, both last terms of the factors must be positive, for the second term of the trinomial is positive. One of the several methods of arranging the work is as shown below. The tentative factors are arranged as if for multiplication:

Trial Factors	Products	
$(6e + 1)(e + 2)$	$= 6e^2 + 13e + 2$	Wrong
$(6e + 2)(e + 1)$	$= 6e^2 + 8e + 2$	Wrong
$(3e + 1)(2e + 2)$	$= 6e^2 + 8e + 2$	Wrong
$(3e + 2)(2e + 1)$	$= 6e^2 + 7e + 2$	Right

It is seen that any combination of the trial factors when multiplied results in the correct first and last term.

$$\therefore 6e^2 + 7e + 2 = (3e + 2)(2e + 1)$$

NOTE *This may seem to be a long process, but with practice, most of the factor trials can be tested mentally.*

EXAMPLE 41 Factor $12i^2 - 17i + 6$.

SOLUTION

The third term of this trinomial is $+6$; therefore, its factors must have like signs. Since the second term is negative, the cross products must be negative. Then it follows that both factors of 6 must be negative. Some of the combinations are as follows:

Trial Factors	Products	
$(2i - 3)(6i - 2)$	$= 12i^2 - 22i + 6$	Wrong
$(2i - 2)(6i - 3)$	$= 12i^2 - 18i + 6$	Wrong
$(3i - 3)(4i - 2)$	$= 12i^2 - 18i + 6$	Wrong
$(3i - 2)(4i - 3)$	$= 12i^2 - 17i + 6$	Right

$$\therefore 12i^2 - 17i + 6 = (3i - 2)(4i - 3)$$

EXAMPLE 42 Factor $8r^2 - 14r - 15$.

SOLUTION

The factors of -15 must have unlike signs. The signs of these factors must be so arranged that the cross product of greater absolute value is minus, because the middle term of the trinomial is negative.

Trial Factors	Products	
$(8r + 3)(r - 5)$	$= 8r^2 - 37r - 15$	Wrong
$(4r + 5)(2r - 3)$	$= 8r^2 - 2r - 15$	Wrong
$(4r + 3)(2r - 5)$	$= 8r^2 - 14r - 15$	Right

EXAMPLE 43 Factor $6R^2 - 7R - 20$.

NOTE *Many students prefer the following method to the trial-and-error method of the foregoing examples.*

SOLUTION

Multiply and divide the entire expression by the coefficient of R^2. The result is

$$\frac{36R^2 - 42R - 120}{6}$$

Take the square root of the first term, which is $6R$, and let that be some other letter such as x. Then, if

$$6R = x$$

by substituting the value of $6R$ in the above expression, we obtain

$$\frac{x^2 - 7x - 120}{6}$$

This results in an expression with a numerator easy to factor. Thus,

$$\frac{x^2 - 7x - 120}{6} = \frac{(x + 8)(x - 15)}{6}$$

Substituting $6R$ for x in the last expression, we obtain

$$\frac{(6R + 8)(6R - 15)}{6}$$

Factoring the numerator,

$$\frac{2(3R + 4)3(2R - 5)}{6}$$

Canceling, $$6R^2 - 7R - 20 = (3R + 4)(2R - 5)$$

NOTE *The denominator will* always *cancel out.*

EXAMPLE 44 Factor $4V^2 - 8VI - 21I^2$.

SOLUTION

Multiplying and dividing by the coefficient of V^2,

$$\frac{16V^2 - 32VI - 84I^2}{4}$$

Let the square root of the first term $4V = x$.

Then $\dfrac{x^2 - 8Ix - 84I^2}{4} = \dfrac{(x + 6I)(x - 14I)}{4}$

Substituting for x, $\dfrac{(4V + 6I)(4V - 14I)}{4}$

Factoring, $\dfrac{2(2V + 3I)2(2V - 7I)}{4}$

Canceling, $4V^2 - 8VI - 21I^2 = (2V + 3I)(2V - 7I)$

PROBLEMS 10–12

Factor:

1. $x^2 - 5x - 14$
2. $12\alpha^2 + 11\alpha + 2$
3. $8y^2 + 4y - 24$
4. $2R^2 - 11R + 12$
5. $10L^2 + L - 21$
6. $2\beta^2 - 9\beta + 10$
7. $14\mu^2 + 41\mu + 15$
8. $18L_1^2 + 31L_1 + 6$
9. $2\alpha^2 - \alpha\beta - 21\beta^2$
10. $10P^2 - 17PW + 3W^2$
11. $45y^2 - 8y - 21$
12. $20\lambda^2 - 22\lambda\phi + 6\phi^2$
13. $50\alpha^2 + 5\alpha\omega - 6\omega^2$
14. $10I^2 + 39IR + 14R^2$
15. $15Q^2 - 41MQ + 28M^2$
16. $42y^2z^2 + 11wyz - 20w^2$
17. $27l^2m^2 + 15lmw - 2w^2$
18. $2\mu^2 - 18\pi^2$
19. $6\psi^2 - 24\Omega^2$
20. $24m^4 - 43m^2p + 18p^2$
21. $15x^2 - 7\Delta x - 2\Delta^2$
22. $\alpha^2 - \dfrac{5\alpha}{6} + \dfrac{1}{6}$
23. $48\theta^2 + 5\theta + \dfrac{1}{8}$
24. $10Z^2 - \dfrac{3Z}{2} + \dfrac{1}{20}$
25. $0.32x^2 - 2$

10–17 SUMMARY

In this chapter, various cases of products and factoring have been treated separately in the different sections. Frequently, however, it becomes necessary to apply the principles underlying two or more cases to a single problem. It is very important, therefore, that you recognize the standard form for various types of algebraic expressions in order that you can apply the method of solution as needed. The standard forms are summarized in Table 10–1.

Problems 10–13 are included as a review of the entire chapter. If you can work all of them, you thoroughly understand the contents of this chapter. If not,

TABLE 10–1

General Type	Factors	Section
$ab + ac + ad$	$a(b + c + d)$	10–7
$a^2 + 2ab + b^2$	$(a + b)^2$	10–8
$a^2 - 2ab + b^2$	$(a - b)^2$	10–8
$a^2 - b^2$	$(a + b)(a - b)$	10–12
$a^2 + (b + c)a + bc$	$(a + b)(a + c)$	10–13
$acx^2 + (bc + ad)x + bd$	$(ax + b)(cx + d)$	10–15

a review of the doubtful parts is suggested, for a good working knowledge of special products and factoring makes it possible to do the following:

1. Multiply, divide, and factor very quickly in your head (mentally).
2. Find the solutions to problems which can be solved by (quick mental) factoring.

PROBLEMS 10–13

Find the value of the following:

1. $(-2\omega C)^2$
2. $(-5\alpha^2\beta^3\gamma)^3$
3. $\left(\dfrac{h^2k^3lm^2}{x^2y^2z^3}\right)^4$
4. $-\sqrt{100\alpha^2\beta^6\gamma^4}$
5. $\sqrt{\dfrac{144I^2R^2}{169F^2X_C{}^4}}$
6. $\sqrt[3]{-125a^9b^6c^{12}}$
7. $-\sqrt[3]{\dfrac{64p^3r^6}{125x^9y^{12}z^3}}$
8. $\sqrt{\dfrac{625I^4R^2P^2}{64V^2W^6}}$
9. $-\sqrt[3]{-216\theta^3\phi^6\omega^6}$
10. $125a\sqrt[3]{\dfrac{a^5x^6z^8}{a^2x^3z^2}}$

Factor:

11. $fL_1{}^2 - fL_2{}^2$
12. $\pi D_1 + \pi D_2 + \pi D_3$
13. $\dfrac{8v^2}{8r_1} + \dfrac{5v^2}{8r_2} - \dfrac{7v^2}{8r_3}$
14. $1.92C_1X - 0.12C_2X + 0.36C_3X$
15. $\dfrac{mp}{12} - \dfrac{5pr^2}{12} - \dfrac{7pt}{12}$
16. $\dfrac{5\pi d_1}{12} - \dfrac{5\pi d_2}{4} + \dfrac{5\pi d_3}{3}$

Mentally, find the products:

17. $(M + 8)^2$
18. $(2Q - 3P^2)^2$
19. $\left(7V^2 + \dfrac{5}{8}\right)^2$

20. $(0.2\alpha\lambda - 0.3\alpha\mu)^2$

21. $\left(\frac{5}{9}\beta - 3\lambda\right)^2$

22. $(0.3V + 0.5IR)^2$

Supply the missing term so that the three terms form a trinomial square:

23. $a^2 + ? + 16$

24. $25x^2 - ? + 4y^2$

25. $? + 36LM + 9L^2$

26. $9a^2 - 12ab + ?$

27. $x^2 + \frac{2xy}{3} + ?$

28. $? + \frac{3}{2}L^2M + \frac{9}{64}M^2$

Extract the square roots of the following:

29. $x^2 + 6x + 9$

30. $\alpha^2 + 8\alpha\beta + 16\beta^2$

31. $625L_1^2 + 700L_1X + 196X^2$

32. $Y^2 - \frac{2BY}{3} + \frac{B^2}{9}$

33. $\frac{I^2}{16} - \frac{Ii}{4} + \frac{i^2}{4}$

34. $\frac{4V^2}{25} - \frac{12VX}{5} + 9X^2$

Factor:

35. $8bY + 10BY$

36. $6mn - 10mp$

37. $3ir^2 + 18ir + 27i$

38. $10\pi^3dr_1^2 + 35\pi^2dr_1r_2 + 25\pi r_2^2$

39. $72pV^2 - 48pVIR + 8pI^2R^2$

40. $\frac{12P^2}{VI} - \frac{144PW}{VI} + \frac{432W^2}{VI}$

Find the products:

41. $(I + 5i)(I - 5i)$

42. $(2\alpha\gamma - 5\beta)(2\alpha\gamma + 5\beta)$

43. $(Y - 9)(Y + 9)$

44. $(8\theta + 7\phi)(8\theta - 7\phi)$

45. $\left(\frac{24V}{IR} + 2P\right)\left(\frac{24V}{IR} - 2P\right)$

46. $(0.3\varepsilon + 0.5\eta)(0.3\varepsilon - 0.5\eta)$

Factor:

47. $A^2 - 1$

48. $9 - g_o^2$

49. $9f^2R^2 - \frac{1}{4\pi^2C^2}$

50. $\frac{4}{25}\alpha^2\beta^4 - \frac{9}{16}\lambda^2$

51. $0.0025\psi^2 - 0.36\mu^2$

52. $0.01V^2 - 0.36I^2R^2$

Find the quotients:

53. $(a^2 - 4) \div (a - 2)$

54. $(4V^2 - I^2R^2) \div (2V - IR)$

55. $\left(\frac{1}{9}\alpha^2 - \frac{4}{49}\beta^2\right) \div \left(\frac{1}{3}\alpha - \frac{2}{7}\beta\right)$

56. $(0.04\alpha^2 - 0.09\beta^2) \div (0.2\alpha - 0.3\beta)$

57. $\left(\frac{9}{25}e^2 - \frac{16}{81}i^2r^2\right) \div \left(\frac{3}{5}e + \frac{4}{9}ir\right)$

58. $\dfrac{x^2 + 2xy + y^2 - 64}{x + y + 8}$

Find the products:

59. $(t + 5)(t - 6)$

60. $(5 - G)(3 - 2G)$

61. $(0.3X_L - 2)(X_L + 0.2)$

62. $(0.1Z + 0.6R)(0.3Z - R)$

63. $\left(p - \frac{1}{4}\right)\left(p + \frac{1}{7}\right)$

64. $(2\psi + 5)\left(\frac{\psi}{3} - \frac{1}{7}\right)$

65. $(2a + 5b)(5a - 3b)$

66. $(2\pi fL + X_C)(2\pi fL - 3X_C)$

67. $(2R - r)(0.3R + 0.2r)$

68. $(0.6x + 3y)(0.4x + 0.5y)$

69. $\left(4\phi + \frac{2\theta}{3}\right)\left(6\phi - \frac{\theta}{2}\right)$

70. $\left(\frac{2v}{3} - \frac{4s}{t}\right)\left(\frac{v}{2} - \frac{6s}{t}\right)$

Factor (remove any common factors first):

71. $15\alpha^2 + 23\alpha + 4$

72. $30X^2 - 2X - 4$

73. $\theta^2 - 5\theta + 6$

74. $m^2 - 0.4m - 0.05$

75. $y^2 - 3.2y + 0.6$

76. $A^2 - \dfrac{3A}{40} - \dfrac{1}{40}$

77. $20Y^2 - 7GY - 6G^2$

78. $3\alpha^2\beta^2\gamma^2 + \alpha\beta\gamma\Omega - 10\Omega^2$

79. $4X^2 + 0.3XZ - 0.1Z^2$

80. $g^2 + 0.1gh - 0.06h^2$

81. $\dfrac{X_C^2}{9} + \dfrac{2X_CZ}{3} + Z^2$

82. $a^2 + \dfrac{2a}{b} + \dfrac{1}{b^2}$

83. $15l^2 - 4lw - 3w^2$

84. $5\theta^2\omega - 5\phi^2\omega$

85. $5a^2 - 45$

86. $288f^2\lambda - \dfrac{2\lambda}{9}$

87. $\dfrac{3V^2}{2i} - \dfrac{18Vv}{i} + \dfrac{54v^2}{i}$

88. $\dfrac{x^3}{9Z} - \dfrac{x^2y}{6Z} + \dfrac{xy^2}{16Z}$

89. $\dfrac{a^2c}{2d} - \dfrac{145abc}{144d} + \dfrac{b^2c}{2d}$

90. $\dfrac{2VR_1^2}{3I} - \dfrac{1898VR_1R_2}{1350I} + \dfrac{2VR_2^2}{3I}$

SELF TEST

1. Evaluate: $(15\ P^3)^2$

2. Evaluate: $\sqrt{\dfrac{169R_1{}^2R_2{}^4}{81\alpha^4}}$

3. Factor: $15\pi R\lambda + 10\pi^2R\ \lambda^2 - 25\pi R^2\lambda^3$

4. Expand: $\left(\omega - \dfrac{7}{16}\right)^2$

5. Supply the missing term to form a perfect trinomial square:

$$9x^2 - 42xy + ?$$

6. What are the prime factors of:

$$18PX_L{}^2 + 60PX_LX_C + 50PX_C{}^2$$

7. Multiply: $\left(\dfrac{2\lambda}{3\pi} + \omega\right)\left(\dfrac{2\lambda}{3\pi} - \omega\right)$

8. Factor: $9a^2 - 30ab + 25b^2 - c^2$

9. Multiply: $\left(\theta + \dfrac{2\phi}{7}\right)\left(\theta - \dfrac{5\phi}{12}\right)$

10. Factor: $6l^4m^4 + 5l^2m^2r - 6r^2$

11. Multiply: $(5y + 2z)(3y - 4z)$

12. Factor: $\dfrac{6y^3}{5\alpha} - \dfrac{4y^2z}{\alpha} - \dfrac{16yz^2}{5\alpha}$

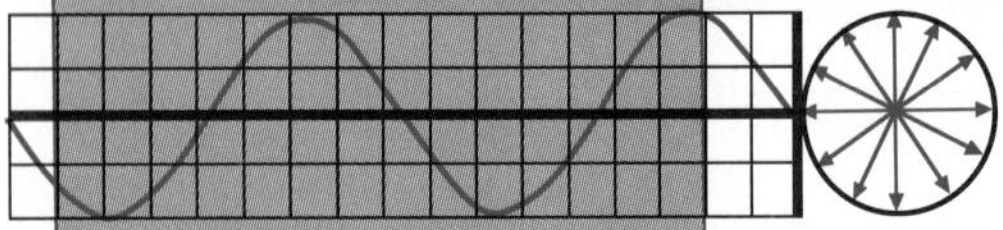

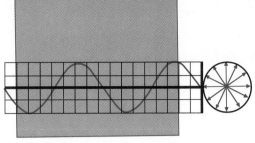

CHAPTER 11

ALGEBRAIC FRACTIONS

Algebraic fractions play an important role in mathematics, especially in equations for electric and electronic circuits.

At this time, if you feel you have not thoroughly mastered arithmetical fractions, you are urged to review them. Despite the current emphasis on metrication and the increasing use of decimal fractions in measurement, a good foundation in arithmetical fractions is essential, for every rule and operation pertaining to them is applicable to algebraic fractions. It is a fact that anyone who really knows arithmetical fractions rarely has trouble with algebraic fractions.

11–1 THE DEGREE OF A MONOMIAL

The degree of a monomial is determined by the number of literal factors the monomial has.

Thus, $6ab^2$ is a monomial of the third degree because $ab^2 = a \cdot b \cdot b$; $3mn$ is a monomial of the second degree. From these examples, it is seen that the degree of a monomial is the sum of the exponents of the literal factors (letters).

In such an expression as $5X^2Y^2Z$, we speak of the whole term as being of the fifth degree, X and Y as being of the second degree, and Z as being of the first degree.

The above definition for the degree of a monomial does not apply to letters in a denominator.

11–2 THE DEGREE OF A POLYNOMIAL

The degree of a polynomial is taken as the degree of the term of highest degree.

Thus, $3ab^2 - 4cd - d$ is a polynomial of the *third degree* and $6x^2y + 5xy^2 + x^2y^2$ is a polynomial of the *fourth degree*.

11–3 HIGHEST COMMON FACTOR

A factor of each of two or more expressions is a *common factor* of those expressions. For example, 2 is a common factor of 4 and 6; a^2 is a common factor of a^3, $(a^2 - a^2b)$, and $(a^2x^2 - a^2y)$.

The product of all the factors common to two or more numbers, or expressions, is called the *highest common factor*. That is, the highest common factor is the expression of highest degree that will divide each of the numbers, or expressions, without a remainder. It is commonly abbreviated HCF.

EXAMPLE 1 Find the HCF of

$$6a^2b^3(c + 1)(c + 3)^2$$

and

$$30a^3b^2(c - 2)(c + 3)$$

SOLUTION

6 is the greatest integer that will divide both expressions. The highest power of a that will divide both is a^2. The highest power of b that will divide both is b^2. The highest power of $(c + 3)$ that will divide both is $(c + 3)$. $(c + 1)$ and $(c - 2)$ will not divide both expressions.

$$\therefore 6a^2b^2(c + 3) = \text{HCF}$$

RULE

To determine the HCF:

1. Determine all the prime factors of each expression.
2. Take the common factors of all the expressions and give to each the lowest exponent it has in any of the expressions.
3. The HCF is the product of all the common factors as obtained in the second step.

EXAMPLE 2 Find the HCF of

$$50a^2b^3c(x + y)^3(x - y)^4$$

and

$$75a^2bc^2(x + y)^2(x - y)$$

SOLUTION

$$50a^2b^3c(x + y)^3(x - y)^4 = 2 \cdot 5 \cdot 5a^2b^3c(x + y)^3(x - y)^4$$
$$75a^2bc^2(x + y)^2(x - y) = 3 \cdot 5 \cdot 5a^2bc^2(x + y)^2(x - y)$$
$$\therefore \text{HCF} = 5^2a^2bc(x + y)^2(x - y)$$
$$= 25a^2bc(x + y)^2(x - y)$$

EXAMPLE 3 Find the HCF of

$$v^2 + vr \qquad v^2 + 2vr + r^2 \qquad \text{and} \qquad v^2 - r^2$$

SOLUTION

$$v^2 + vr = v(v + r)$$
$$v^2 + 2vr + r^2 = (v + r)^2$$
$$v^2 - r^2 = (v + r)(v - r)$$
$$\therefore \text{HCF} = v + r$$

PROBLEMS 11–1

Find the HCF of:

1. 42, 70
2. 60, 140, 220
3. $8\lambda^2\mu$, $24\lambda\mu\rho$, $48\lambda\mu^2\rho$
4. $10\alpha^2\beta$, $25\alpha^2\gamma$, $50\alpha^2\delta$
5. $0.5a^3b^2c$, $0.25a^2b^2c^2$, $0.1a^2bc^3$
6. $39x^4y^2z^3$, $78x^3y^3z^3$, $156x^2y^4z^3$
7. $63VI$, $189V^2I$, $21VI^2$
8. $18\alpha\beta^2\gamma^3$, $162\alpha^2\beta^3\gamma$, $220\alpha\beta^3\gamma^2$.
9. $X_L^2 - X_C^2$, $X_L^2 + X_LX_C$
10. $x^2 + 2xy + y^2$, $x^2 - y^2$
11. $\alpha^2 + 2\alpha + 1$, $\alpha^2 - 1$, $2\alpha^2 + 2\alpha$
12. $12\pi + 4\phi$, $9\pi^2 + 6\pi\phi + \phi^2$, $9\pi^2 - \phi^2$
13. $3P + \dfrac{6V^2}{R} + 3I^2R$, $PV + 2\dfrac{V^3}{R} + I^2RV$
14. $9I^2R^2 - 24VIR + 16V^2$, $3I^2R^2 - 10VIR + 8V^2$, $15I^2R^2 - 17VIR - 4V^2$
15. $10I^2 + 25\dfrac{VI}{R} + \dfrac{15V^2}{R^2}$, $30I + 45\dfrac{V}{R}$, $40I^2 + 40\dfrac{VI}{R} - 30\dfrac{V^2}{R^2}$

11–4 MULTIPLE

A number is a multiple of any one of its factors. For example, some of the multiples of 4 are 8, 16, 20, and 24. Similarly, some of the multiples of $a + b$ are $3(a + b)$, $a^2 + 2ab + b^2$, and $a^2 - b^2$. A *common multiple* of two or more numbers is a multiple of each of them. Thus, 45 is a common multiple of 1, 3, 5, 9, and 15.

11–5 LOWEST COMMON MULTIPLE

The smallest number that will contain each one of a set of factors is called the *lowest common multiple* of the factors. Thus, 48, 60, and 72 are all common multiples of 4 and 6, but the lowest common multiple of 4 and 6 is 12.

The lowest common multiple is abbreviated LCM.

EXAMPLE 4 Find the LCM of $6x^2y$, $9xy^2z$, and $30x^3y^3$.

SOLUTION

$$\begin{aligned}
6x^2y &= 2 \cdot 3 \cdot x \cdot x \cdot y \\
9xy^2z &= 3 \cdot 3 \cdot x \cdot y \cdot y \cdot z \\
30x^3y^3 &= 2 \cdot 3 \cdot 5 \cdot x \cdot x \cdot x \cdot y \cdot y \cdot y \\
\text{LCM} &= 2 \cdot 3 \cdot 3 \cdot 5 \cdot x \cdot x \cdot x \cdot y \cdot y \cdot y \cdot z
\end{aligned}$$

Because the LCM must contain *each* of the expressions, it must have 2, 3^2, and 5 as factors. Also, it must contain the literal factors of highest degree, or x^3y^3z.

$$\therefore \text{LCM} = 2 \cdot 3^2 \cdot 5 \cdot x^3y^3z = 90x^3y^3z$$

RULE

To determine the LCM of two or more expressions, determine all the prime factors of each expression. Find the product of all the different prime factors, taking each factor the greatest number of times it occurs in any one expression.

EXAMPLE 5 Find the LCM of

$$3a^3 + 6a^2b + 3ab^2$$
$$6a^4 - 12a^3b + 6a^2b^2$$
$$9a^3b - 9ab^3$$

SOLUTION

$$\begin{aligned} 3a^3 + 6a^2b + 3ab^2 &= 3a(a+b)^2 \\ 6a^4 - 12a^3b + 6a^2b^2 &= 2 \cdot 3 \cdot a^2(a-b)^2 \\ 9a^3b - 9ab^3 &= 3^2 \cdot ab(a+b)(a-b) \\ \therefore \text{LCM} &= 2 \cdot 3^2 \cdot a^2b(a+b)^2(a-b)^2 \\ &= 18a^2b(a+b)^2(a-b)^2 \end{aligned}$$

PROBLEMS 11–2

Find the LCM of the following:

1. 12, 30, 90
2. 14, 21, 154
3. 22, 66, 1815
4. ab^2c^3, a^2b^3c
5. $R^2L_1{}^3L_2{}^4M$, $RL_1{}^2L_2{}^2M^2Y$
6. $2\alpha^4\beta^2$, $10\alpha^2\beta^2\gamma^3$, $15\alpha^3\beta^3\gamma$
7. $5m^3n^2p^2$, $20m^2np$, $45mnp^4$
8. I^2, $3IR$, $17I^2R^2$
9. $a^2 - 4$, $a^2 + a - 6$, $2a^2 + 10a + 12$
10. $X^2 - 11X + 30$, $X^2 - 9X + 20$
11. $t^3 - 16t$, $t^2 - 8t + 16$
12. $10 + 35\lambda$, $4 - 49\lambda^2$, $20 - 140\lambda + 245\lambda^2$
13. $6\theta^2 + 7\theta - 3$, $44\theta^2 + 88\theta + 33$, $66\theta^2 + 11\theta - 11$

14. $4X_L^2 + 12X_LX_C + 8X_C^2$, $2X_L^2 + 10X_LX_C + 12X_C^2$, $X_L^2 + 4X_LX_C + 3X_C^2$

15. $8Q^2 - 38\frac{\omega LQ}{R} + 35\frac{\omega^2L^2}{R^2}$, $Q^2 - \frac{\omega^2L^2}{R^2}$, $2Q^2 - 9\frac{\omega LQ}{R} + 7\frac{\omega^2L^2}{R^2}$

11–6 DEFINITIONS

A fraction is an indicated division. Thus, we indicate 4 divided by 5 as $\frac{4}{5}$ (read four-fifths). Similarly, *X divided by Y* is written $\frac{X}{Y}$ (read X divided by Y or X over Y).

The quantity above the horizontal line is called the *numerator* and that below the line is called the *denominator* of the fraction. The numerator and denominator are often called the *terms* of the fraction.

11–7 OPERATIONS ON NUMERATOR AND DENOMINATOR

As in arithmetic, when fractions are to be simplified or affected by one of the four fundamental operations, we find it necessary to make frequent use of the following important principles:

1. The numerator and the denominator of a fraction can be multiplied by the same number or expression, except zero, without changing the value of the fraction.
2. The numerator and the denominator can be divided by the same number or expression, except zero, without changing the value of the fraction.

EXAMPLE 6

$$\frac{2}{3} = \frac{2 \times 3}{3 \times 3} = \frac{6}{9} = \frac{2}{3}$$

Also,

$$\frac{6}{9} = \frac{6 \div 3}{9 \div 3} = \frac{2}{3} = \frac{6}{9}$$

EXAMPLE 7

$$\frac{x}{y} = \frac{x \cdot a}{y \cdot a} = \frac{ax}{ay} = \frac{x}{y}$$

Also, $$\frac{ax \div a}{ay \div a} = \frac{x}{y} \quad \text{(where } a \neq 0\text{)}$$

No new principles are involved in performing these operations, for multiplying or dividing both numerator and denominator by the same number, except zero, is equivalent to multiplying or dividing the fraction by 1 in any form convenient for our use, such as

$$\frac{2}{2}, \frac{4}{4}, \frac{10}{10}$$

or

$$\frac{-1}{-1}$$

It will be noted that, in the foregoing principles, multiplication and division of numerator and denominator by zero are excluded. When any expression is multiplied by zero, the product is zero.

For example, $6 \times 0 = 0$. Therefore, if we multiplied both numerator and denominator of some fraction by zero, the result would be meaningless. Thus,

$$\frac{5}{6} \neq \frac{5 \times 0}{6 \times 0}$$

because

$$\frac{5 \times 0}{6 \times 0} = \frac{0}{0}$$

Division by zero is meaningless. Some people say that any number divided by zero results in a quotient of infinity, denoted by ∞. If we accept this, we immediately impose a severe restriction on operations with even simple equations. For example, let us assume for the moment that any number divided by zero *does* result in infinity. Then if

$$\frac{4}{0} = \infty$$

by following Axiom 3, we should be able to multiply both sides of this equation by 0. If so, we obtain

$$4 = \infty \cdot 0$$

which we know is not sensible. Obviously, there is a fallacy here; therefore, we shall simply say at this time that *division by zero is not a permissible operation.*

11–8 EQUIVALENT FRACTIONS

Examples 6 and 7 show that when a numerator and a denominator are multiplied or divided by the same number, except zero, we change the *form* but not the value of the given fraction. Therefore, two fractions having the same value but not the same form are called *equivalent fractions.*

PROBLEMS 11–3

Supply the missing terms:

1. $\dfrac{3}{5} = \dfrac{?}{20}$
2. $\dfrac{5}{32} = \dfrac{?}{128}$
3. $\dfrac{1}{R_1} = \dfrac{?}{R_1^2R_2}$
4. $\dfrac{2\theta}{9\phi} = \dfrac{?}{27\phi\omega}$

5. $\frac{3ab}{25c} = \frac{?}{75cd}$

6. $\frac{\alpha}{\alpha + 3} = \frac{?}{(\alpha + 3)(\alpha - 2)}$

7. $\frac{a + 1}{a - 3} = \frac{?}{a^2 - 9}$

8. $\frac{7 + \theta}{\theta - 1} = \frac{?}{\theta^2 - 1}$

9. $\frac{i + \alpha}{\alpha - 3\beta} = \frac{?}{6\alpha - 18\beta}$

10. $\frac{x - 2y}{2x + y} = \frac{?}{2x^2 + 5xy + 2y^2}$

11. Change the fraction $\frac{5}{8}$ into an equivalent fraction whose denominator is 64.

12. Change the fraction $\frac{7}{16}$ into an equivalent fraction whose denominator is 128.

13. Change the fraction $\frac{\omega L}{R}$ into an equivalent fraction whose denominator is $R^4 + 2R^3X + R^2X^2$

14. Change the fraction $\frac{L + 2}{L - 2}$ into an equivalent fraction whose denominator is $L^2 - 4$.

15. Change the fraction $\frac{Q}{VC + 1}$ into an equivalent fraction whose denominator is $2V^2C^2 - VC - 3$.

11–9 REDUCTION OF FRACTIONS TO THEIR LOWEST TERMS

If the numerator and denominator of a fraction have no common factor other than 1, the fraction is said to be in its lowest terms. Thus, the fractions $\frac{2}{3}$, $\frac{3}{5}$, $\frac{x}{y}$, and $\frac{x + y}{x - y}$ are in their lowest terms, for the numerator and denominator of each fraction have no common factor except 1.

The fractions $\frac{4}{6}$ and $\frac{3x}{9x^2}$ are not in their lowest terms, for $\frac{4}{6}$ can be reduced to $\frac{2}{3}$ if both numerator and denominator are divided by 2. Similarly, $\frac{3x}{9x^2}$ can be reduced to $\frac{1}{3x}$ by dividing both numerator and denominator by $3x$.

RULE

To reduce a fraction to its lowest terms, factor the numerator and denominator into prime factors and cancel the factors common to both.

Cancellation as used in the rule really means that we actually *divide* both terms of the fraction by the *common factors*. Then, to reduce a fraction to its lowest terms, it is only necessary to divide both numerator and denominator by the highest common factor, which leaves an equivalent fraction.

EXAMPLE 8 Reduce $\frac{27}{108}$ to lowest terms.

SOLUTION

$$\frac{27}{108} = \frac{\not{3} \cdot \not{3} \cdot \not{3}}{2 \cdot 2 \cdot \not{3} \cdot \not{3} \cdot \not{3}} = \frac{1}{4}$$

EXAMPLE 9 Reduce $\frac{24x^2yz^3}{42x^2yz^2}$ to lowest terms.

SOLUTION

$$\frac{24x^2yz^3}{42x^2yz^2} = \frac{\not{2} \cdot 2 \cdot 2 \cdot \not{3} \cdot x^2yz^3}{\not{2} \cdot \not{3} \cdot 7 \cdot x^2yz^2} = \frac{4z}{7}$$

Actually, the solution to Example 9 need not have been written out, for it can be seen by inspection that the HCF of both terms of the fraction is $6x^2yz^2$, which we divide into both terms in order to obtain the equivalent fraction $\frac{4z}{7}$.

Also, in reducing fractions, we may resort to direct cancellation as in arithmetic.

EXAMPLE 10 Reduce $\frac{x^2 - y^2}{x^3 - y^3}$ to lowest terms.

SOLUTION

$$\frac{x^2 - y^2}{x^3 - y^3} = \frac{(x + y)\cancel{(x - y)}}{\cancel{(x - y)}(x^2 + xy + y^2)}$$

$$= \frac{x + y}{x^2 + xy + y^2}$$

EXAMPLE 11 Reduce to lowest terms:

$$\frac{r^2 - R^2}{r^2 + 3rR + 2R^2}$$

SOLUTION

$$\frac{r^2 - R^2}{r^2 + 3rR + 2R^2} = \frac{\cancel{(r + R)}(r - R)}{(r + 2R)\cancel{(r + R)}} = \frac{r - R}{r + 2R}$$

PROBLEMS 11–4

Reduce to lowest terms:

1. $\frac{10}{48}$
2. $\frac{462}{2772}$
3. $\frac{27}{972}$
4. $\frac{30}{108}$
5. $\frac{x^3y^2}{x^2y^4}$
6. $\frac{3\theta^2\phi}{12\theta\phi^3}$

7. $\dfrac{125I^2R}{25IR^2}$

8. $\dfrac{32\theta\lambda^3\mu\phi^2}{80\theta^2\lambda\mu\phi^3}$

9. $\dfrac{\theta\phi}{\theta^2\phi - \phi^3}$

10. $\dfrac{7.5p + 0.5q}{2.5pq}$

11. $\dfrac{\alpha^2 - \beta^2}{\alpha^2 - 2\alpha\beta + \beta^2}$

12. $\dfrac{4m - 4n}{m^2 - n^2}$

13. $\dfrac{I^2 - R^2}{5I^2 + 10IR + 5R^2}$

14. $\dfrac{\alpha^2 + 3\alpha\beta - 10\beta^2}{2\alpha^2 + 11\alpha\beta + 5\beta^2}$

15. $\dfrac{\pi^2\omega^2 - 9\lambda^2\omega^2}{3\pi^2\omega - 8\pi\lambda\omega - 3\lambda^2\omega}$

11–10 SIGNS OF FRACTIONS

As stated in Sec. 11–6, a fraction is an indicated division or an indicated quotient. Heretofore, all our fractions have been positive, but now we must take into account three signs in working with an algebraic fraction: the sign of the numerator, the sign of the denominator, and the sign preceding the fraction. By the law of signs in division, we have

$$+\frac{+12}{+6} = +\frac{-12}{-6} = -\frac{+12}{-6} = -\frac{-12}{+6} = +2$$

or, in general,

$$+\frac{+a}{+b} = +\frac{-a}{-b} = -\frac{+a}{-b} = -\frac{-a}{+b}$$

Careful study of the above examples will show the truths of the following important principles:

1. The sign before either term of a fraction can be changed if the sign before the fraction is changed.
2. If the signs of both terms are changed, the sign before the fraction must not be changed.

That is, we can change *any two* of the three signs of a fraction without changing the value of the fraction.

It must be remembered that, when a term of a fraction is a polynomial, changing the sign of the term involves changing the sign of *each term of the polynomial*.

Changing the signs of both numerator and denominator, as mentioned in the second principle above, can be explained by considering both terms as multiplied or divided by -1, which, as previously explained, does not change the value of the fraction.

Multiplying (or dividing) a quantity by -1 twice does not change the value of the quantity. Hence, multiplying each of the two factors of a product by -1 does not change the value of the product. Thus,

$$\begin{aligned}(a - 4)(a - 8) &= (-a + 4)(-a + 8)\\ &= (4 - a)(8 - a)\end{aligned}$$

Also,

$$(a - b)(c - d)(e - f) = (b - a)(d - c)(e - f)$$

The validity of these illustrations should be checked by multiplication.

EXAMPLE 12 Change $-\frac{a}{b}$ to three equivalent fractions having different signs.

SOLUTION

$$-\frac{a}{b} = \frac{-a}{b} = \frac{a}{-b} = -\frac{-a}{-b}$$

EXAMPLE 13 Change $\frac{a - b}{c - d}$ to three equivalent fractions having different signs.

SOLUTION

$$\frac{a - b}{c - d} = \frac{-a + b}{-c + d} = -\frac{-a + b}{c - d} = -\frac{a - b}{-c + d}$$

EXAMPLE 14 Change $\frac{a - b}{c - d}$ to a fraction whose denominator is $d - c$.

SOLUTION

$$\frac{a - b}{c - d} = \frac{-a + b}{-c + d} = \frac{b - a}{d - c}$$

PROBLEMS 11–5

Express as fractions with positive numerators:

1. $-\frac{-5}{16}$
2. $\frac{-L_1L_2}{M - m}$
3. $\frac{-2\pi fL}{X_L - X_C}$
4. $-\frac{\sqrt{L_1L_2}}{\omega L}$
5. $\frac{-IR}{X - Z}$
6. $\frac{-\theta - \phi}{\rho - \tau}$

Express as fractions with positive denominators:

7. $\frac{GY}{-R - X}$
8. $\frac{\mu V_g}{-(R_p + R_L)}$

9. $\dfrac{\pi R^2}{-(A_1 - A_2)}$

10. $-\dfrac{\theta + \phi}{-2\lambda^2}$

Reduce to lowest terms:

11. $\dfrac{a - x}{x - a}$

12. $\dfrac{I - i}{-(i^2 - I^2)}$

13. $\dfrac{2\theta - \phi}{\phi^2 - \theta\phi - 2\theta^2}$

14. $\dfrac{x^2 - 2xy + y^2}{y^2 - 2yx + x^2}$

15. $\dfrac{2\omega - \omega^2 + 15}{5\omega^2 + 10 - 27\omega}$

16. $-\dfrac{4s^2t^2 + 3stv - v^2}{2s^2t^2 + stv - v^2}$

11–11 COMMON ERRORS IN WORKING WITH FRACTIONS

It has been demonstrated that a fraction may be reduced to lower terms by dividing both numerator and denominator by the same number (Sec. 11–9). Mistakes are often made by canceling parts of numerator and denominator that are not factors. For example,

$$\frac{5 + 2}{7 + 2} = \frac{7}{9}$$

Here is a case in which both terms of the fraction are polynomials and the terms, even if alike, can never be canceled. Thus,

$$\frac{5 + \cancel{2}}{7 + \cancel{2}} \neq \frac{5}{7}$$

because canceling terms has changed the value of the fraction. Similarly, it would be incorrect to cancel the x's in the fraction $\dfrac{6a - x}{6b - x}$, for the x's are not factors. At the same time, it is incorrect to cancel the 6's because, although they are factors of terms in the fraction, they are not factors of the complete numerator and denominator. Therefore, it is apparent that $\dfrac{6a - x}{6b - x}$ cannot be reduced to lower terms, for neither term (numerator or denominator) can be factored.

It is permissible to cancel x's in the fraction $\dfrac{6x}{ax + 5x}$, because each term of the denominator contains the common factor x. The denominator may be factored to give $\dfrac{6x}{x(a + 5)}$, the result being that x is a factor in both terms of the fraction. Note, however, that the single x in the numerator cancels both x's in the denominator.

Thus, we cannot remove, or cancel, like *terms* from the numerator and denominator of a fraction. Only like *factors* can be removed, or canceled.

Another important fact to be remembered is that adding the same number to or subtracting the same number from both numerator and denominator changes the value of the fraction. That is,

$$\frac{3}{4} \neq \frac{3+2}{4+2} \qquad \text{because the latter equals } \frac{5}{6}$$

Likewise,

$$\frac{3}{4} \neq \frac{3-2}{4-2} \qquad \text{because the latter equals } \frac{1}{2}$$

Similarly, squaring or extracting the same root of numerator and denominator results in a different value. For example,

$$\frac{3}{4} \neq \frac{3^2}{4^2} \qquad \text{because the latter equals } \frac{9}{16}$$

Likewise,

$$\frac{16}{25} \neq \frac{\sqrt{16}}{\sqrt{25}} \qquad \text{because the latter equals } \frac{4}{5}$$

Students sometimes thoughtlessly make the error of writing 0 (zero) as the result of the cancellation of all factors. For example,

$$\frac{4x^2y(a+b)}{4x^2y(a+b)} = 1, \textit{ not } 0$$

Another serious, although common, mistake is forgetting that the fraction bar, or vinculum, is a sign of grouping, so that $-\frac{x-y}{x}$ really means $-\left(\frac{x-y}{x}\right)$, or $-\left(\frac{x}{x}-\frac{y}{x}\right)$, or $-\left(1-\frac{y}{x}\right)$, and it does not reduce to $-(1-y)$.

Note that the *vinculum* is a sign of grouping and, when a minus sign precedes a fraction having a polynomial numerator, all the signs of the numerator must be changed in order to complete the process of subtraction.

Thus, $-\frac{x-y}{x}$ simplifies to $\frac{y}{x}-1$.

11–12 CHANGING MIXED EXPRESSIONS TO FRACTIONS

In arithmetic, an expression such as $3\frac{1}{3}$ is called a *mixed number;* $3\frac{1}{3}$ means $3 + \frac{1}{3}$. Similarly, in algebra, an expression such as $x + \dfrac{y}{z}$ is called a *mixed expression*. Because

$$4\frac{2}{3} = 4 + \frac{2}{3} = \frac{4}{1} + \frac{2}{3} = \frac{12}{3} + \frac{2}{3} = \frac{14}{3}$$

then,

$$x + \frac{y}{z} = \frac{x}{1} + \frac{y}{z} = \frac{xz}{z} + \frac{y}{z} = \frac{xz + y}{z}$$

Also,

$$3x^2 - 4x + \frac{3}{x^2 - 1} = \frac{3x^2}{1} - \frac{4x}{1} + \frac{3}{x^2 - 1}$$

$$= \frac{3x^2(x^2 - 1)}{x^2 - 1} - \frac{4x(x^2 - 1)}{x^2 - 1} + \frac{3}{x^2 - 1}$$

$$= \frac{3x^4 - 3x^2 - 4x^3 + 4x + 3}{x^2 - 1}$$

11–13 REDUCTION OF A FRACTION TO A MIXED EXPRESSION

As would be expected, reducing a fraction to a mixed expression is the reverse of changing a mixed expression to a fraction. That is, a fraction can be changed to a mixed expression by dividing the numerator by the denominator and adding to the quotient thus obtained the remainder, which is written as a fraction.

EXAMPLE 15 Change $\dfrac{12x^3 + 16x^2 - 8x - 3}{4x}$ to a mixed expression.

SOLUTION

Divide each term of the numerator by the denominator. Thus,

$$\frac{12x^3 + 16x^2 - 8x - 3}{4x} = 3x^2 + 4x - 2 - \frac{3}{4x}$$

EXAMPLE 16 Change $\dfrac{a^2 + 1}{a - 2}$ to a mixed expression.

SOLUTION

By division,

$$\begin{array}{r|l} & a \quad\quad + 2 \\ a - 2 & a^2 \quad\quad + 1 \\ & \underline{a^2 - 2a} \\ & 2a + 1 \\ & \underline{2a - 4} \\ & 5 \end{array}$$

$$\therefore \frac{a^2 + 1}{a - 2} = a + 2 + \frac{5}{a - 2}$$

PROBLEMS 11–6

Change the following mixed expressions to fractions:

1. $1\frac{5}{8}$
2. $3\frac{7}{8}$
3. $R + \frac{V}{I}$
4. $R - \frac{V}{I}$
5. $1 - \frac{3}{Q}$
6. $\frac{2}{A} - \frac{5}{A^2}$
7. $5 + \frac{3}{\beta + 2}$
8. $G + \frac{G}{2Z}$
9. $\omega L - 5 - \frac{L}{R}$
10. $5 + \frac{5x - 30}{x^2 - 2x}$
11. $\frac{9}{x^2} - \frac{14}{2x} - 2$
12. $4 - \frac{4}{c} - \frac{8}{c^2}$
13. $1 - \frac{4}{R} - \frac{21}{R^2}$
14. $\frac{a + b}{4} - \frac{a - b}{8}$
15. $1 - \frac{4\lambda + 1}{9\lambda^2 - 1}$
16. $\frac{x - 1}{2x} - \frac{x^2 - 1}{3x^2}$
17. $\frac{45}{\theta^2} + \frac{14}{\theta} - \frac{\theta + 1}{\theta - 1}$
18. $3 - \frac{3P(P - 5)}{P^2 - 25}$
19. $2\alpha^2 - 1 - \frac{4}{\alpha^2 - 3}$
20. $1 - \frac{50\omega\pi - 30\pi^2}{(5\omega - 3\pi)(3\omega + 5\pi)}$

Reduce the following fractions to mixed expressions:

21. $\frac{35}{18}$
22. $\frac{42}{5}$
23. $\frac{55}{16}$
24. $\frac{32\alpha^2 - 16\alpha + 4}{4\alpha}$
25. $\frac{x^3 + 12x^2 + 3x - 9}{x + 1}$
26. $\frac{G^2 + 6G + 8}{G - 1}$
27. $\frac{E^4 - e^4 - 1}{E + e}$
28. $\frac{6\phi^5 - \phi^4 + 4\phi^3 - 5\phi^2 - \phi + 20}{2\phi^2 - \phi + 3}$
29. $\frac{2x^3 + 2x^2 + x + 2}{x^2 + 1}$
30. $\frac{3x^3 - x^2y - 2xy^2 + 2y^3 - 2y^4}{x + y}$

11–14 REDUCTION TO THE LOWEST COMMON DENOMINATOR

The *lowest common denominator* (LCD) of two or more fractions is the lowest common multiple of their denominators.

EXAMPLE 17 Reduce $\frac{1}{3}$ and $\frac{3}{5}$ to their LCD.

SOLUTION

The LCM of 3 and 5 is 15. To change the denominator of $\frac{1}{3}$ to 15, we must multiply the 3 by 5 (15 ÷ 3). So that the value of the fraction will not be changed, we must also multiply the numerator by 5. Hence,

$$\frac{1}{3} = \frac{1}{3} \times \frac{5}{5} = \frac{5}{15}$$

For the second fraction, we must multiply the denominator by 3 in order to obtain a new denominator of 15 (15 ÷ 5). Again we must also multiply the numerator by 3 to maintain the original value of the fraction. Hence,

$$\frac{3}{5} = \frac{3}{5} \times \frac{3}{3} = \frac{9}{15}$$

EXAMPLE 18 Reduce $\dfrac{4a^2b}{3x^2y}$ and $\dfrac{6cd^2}{4xy^2}$ to their LCD.

SOLUTION

The LCM of the two denominators is $12x^2y^2$. This is the LCD.

For the first fraction the LCD is divided by the denominator. That is,

$$12x^2y^2 \div 3x^2y = 4y$$

Multiplying both numerator and denominator by $4y$, we have

$$\frac{4a^2b}{3x^2y} = \frac{4a^2b}{3x^2y} \cdot \frac{4y}{4y} = \frac{16a^2by}{12x^2y^2}$$

For the second fraction we follow the same procedure.

$$12x^2y^2 \div 4xy^2 = 3x$$

Multiplying both numerator and denominator by $3x$, we have

$$\frac{6cd^2}{4xy^2} = \frac{6cd^2}{4xy^2} \cdot \frac{3x}{3x} = \frac{18cd^2x}{12x^2y^2}$$

RULE

To reduce fractions to their LCD:

1. Factor each denominator into its prime factors and find the LCM of the denominators. This is the LCD.
2. For each fraction, divide the LCD by the denominator and multiply both numerator and denominator by the quotient thus obtained.

EXAMPLE 19 Reduce $\dfrac{3x}{x^2 - y^2}$ and $\dfrac{4y}{x^2 - xy - 2y^2}$ to their LCD.

SOLUTION

$$\frac{3x}{x^2 - y^2} = \frac{3x}{(x + y)(x - y)}$$

$$\frac{4y}{x^2 - xy - 2y^2} = \frac{4y}{(x + y)(x - 2y)}$$

The LCM of the two denominators, and therefore the LCD, is $(x + y)(x - y)(x - 2y)$.

For the first fraction, the LCD divided by the denominator is:

$$(x + y)(x - y)(x - 2y) \div (x + y)(x - y) = x - 2y$$

$$\therefore \frac{3x}{(x + y)(x - y)} = \frac{3x(x - 2y)}{(x + y)(x - y)(x - 2y)}$$

For the second fraction, the LCD divided by the denominator is:

$$(x + y)(x - y)(x - 2y) \div (x + y)(x - 2y) = x - y$$

$$\therefore \frac{4y}{(x + y)(x - 2y)} = \frac{4y(x - y)}{(x + y)(x - 2y)(x - y)}$$

To check the solution, the fractions having the LCD can be changed into the original fractions by cancellation.

PROBLEMS 11–7

Convert the following sets of fractions to equivalent sets having their LCD:

1. $\dfrac{1}{4}, \dfrac{2}{7}, \dfrac{3}{5}$
2. $\dfrac{3}{5}, \dfrac{8}{25}, \dfrac{16}{35}$
3. $\dfrac{1}{2}, \dfrac{7}{8}, \dfrac{9}{32}$
4. $\dfrac{1}{p}, \dfrac{1}{q}$
5. $\dfrac{V}{IR}, \dfrac{Z}{X}$
6. $\dfrac{1}{ir}, \dfrac{1}{\omega}, \dfrac{i}{\omega}$
7. $\dfrac{v}{r}, \dfrac{1}{ir}, pv$
8. $\dfrac{Q}{L_1}, \dfrac{1}{L_2}, \dfrac{\sqrt{L_1 L_2}}{M}$

9. $\dfrac{1}{a - b}, \dfrac{1}{a + b}$

10. $\dfrac{x}{y}, \dfrac{2x + y}{x - y}$

11. $\dfrac{7}{l - w}, \dfrac{5}{l + w}$

12. $\dfrac{3\phi}{1 - \phi^2}, \dfrac{2}{\phi + 1}, \dfrac{2}{1 - \phi}$

13. $\dfrac{\alpha}{\theta + \phi}, \dfrac{\beta}{\theta - \phi}, \dfrac{\alpha - \beta}{\phi - \theta}$

14. $\dfrac{1}{2M + 2}, \dfrac{5}{3M - 3}, \dfrac{3M - 1}{1 - M^2}$

15. $\dfrac{\pi^2 - \phi^2}{\pi\phi}, \dfrac{\pi\phi - \phi^2}{\pi\phi - \pi^2}$

16. $\dfrac{R + 3Z}{4R^2 + 12RZ + 8Z^2}, \dfrac{R + Z}{4R^2 + 20RZ + 24Z^2}, \dfrac{R + 2Z}{R^2 + 4RZ + 3Z^2}$

11–15 ADDITION AND SUBTRACTION OF FRACTIONS

The sum of two or more fractions having the same denominator is obtained by adding the numerators and writing the result over the common denominator.

EXAMPLE 20

$$\frac{2}{7} + \frac{1}{7} + \frac{5}{7} = \frac{2 + 1 + 5}{7} = \frac{8}{7}$$

EXAMPLE 21

$$\frac{3v}{R + r} + \frac{v}{R + r} + \frac{5v}{R + r} = \frac{3v + v + 5v}{R + r} = \frac{9v}{R + r}$$

To subtract two fractions having the same denominator, subtract the numerator of the subtrahend from the numerator of the minuend and write the result over their common denominator.

EXAMPLE 22

$$\frac{4}{5} - \frac{3}{5} = \frac{4 - 3}{5} = \frac{1}{5}$$

EXAMPLE 23

$$\frac{a}{x} - \frac{b}{x} = \frac{a - b}{x}$$

EXAMPLE 24

$$\frac{a}{x} - \frac{b - c}{x} = \frac{a - b + c}{x}$$

Note that *the vinculum is a sign of grouping* and that, when a minus sign precedes a fraction having a polynomial numerator, all the signs in the numerator must be changed in order to complete the process of subtraction.

We thus have the following rule:

RULE

To add or subtract fractions having unlike denominators:

1. Reduce them to equivalent fractions having their LCD.
2. Combine the numerators of these equivalent fractions, in parentheses, and give each the sign of the fraction. This is the numerator of the result.
3. The denominator of the result is the LCD.
4. Simplify the numerator by removing parentheses and combining terms.
5. Reduce the fraction to the lowest terms.

EXAMPLE 25 Simplify $\frac{a-5}{6x} - \frac{2a-5}{16x}$.

SOLUTION

$$\begin{aligned}\frac{a-5}{6x} - \frac{2a-5}{16x} &= \frac{8(a-5)}{48x} - \frac{3(2a-5)}{48x}\\ &= \frac{8(a-5) - 3(2a-5)}{48x}\\ &= \frac{8a - 40 - 6a + 15}{48x}\\ &= \frac{2a-25}{48x}\end{aligned}$$

CHECK

Let $a = 6$, $x = 1$.

$$\frac{a-5}{6x} = \frac{1}{6} \qquad \frac{2a-5}{16} = \frac{7}{16}$$

$$\frac{1}{6} - \frac{7}{16} = \frac{8-21}{48} = -\frac{13}{48}$$

Also,

$$\frac{2a-25}{48} = \frac{12-25}{48} = -\frac{13}{48}$$

Solution is correct.

EXAMPLE 26 Simplify $x^2 - xy + y^2 - \frac{2y^3}{x+y}$.

SOLUTION

$$\begin{aligned}x^2 - xy + y^2 - \frac{2y^3}{x+y} &= \frac{(x+y)x^2}{x+y} - \frac{(x+y)xy}{x+y} + \frac{(x+y)y^2}{x+y} - \frac{2y^3}{x+y}\\ &= \frac{x^3 + x^2y - x^2y - xy^2 + xy^2 + y^3 - 2y^3}{x+y}\\ &= \frac{x^3 - y^3}{x+y}\end{aligned}$$

PROBLEMS 11–8

Perform the following indicated additions and subtractions:

1. $\frac{1}{4} + \frac{3}{5} - \frac{2}{3}$
2. $\frac{7}{32} - \frac{5}{8} + \frac{3}{16}$
3. $\frac{7}{8} - \frac{5}{16} - \frac{1}{3}$
4. $\frac{2x}{3} - \frac{x}{5} + \frac{5x}{4}$
5. $\frac{5V}{3} + \frac{7V}{16} - \frac{3V}{8}$
6. $\frac{1}{N} - \frac{1}{n}$
7. $\frac{p}{q} - \frac{s}{t}$
8. $\frac{5l}{6m} - \frac{l}{5m} - \frac{3l}{24m}$
9. $\frac{\alpha}{M} - \frac{\beta}{N} - \frac{\gamma}{P}$
10. $\frac{4}{L_1} + \frac{3}{L_2} + \frac{7}{L_1L_2}$
11. $\frac{8}{V} - \frac{2}{IR} + \frac{5}{IRV}$
12. $\frac{3\alpha}{\phi\lambda} + \frac{2\phi}{\alpha\lambda} + \frac{6\lambda}{\alpha\phi}$
13. $\frac{3I - i}{2} + \frac{5I + 2i}{3}$
14. $\frac{a + 4}{7} - \frac{a - 1}{3}$
15. $\frac{3}{L_1 - L_2} + \frac{1}{L_1 + L_2}$
16. $\frac{5}{2\alpha + 6} - \frac{2}{\alpha + 3}$
17. $\frac{5}{L_1 - 2} - \frac{2}{L_1 + 6}$
18. $\frac{a}{c + d} + \frac{b}{c - d} - \frac{a - b}{d - c}$
19. $\frac{1}{2\theta + 2} - \frac{5}{3\theta - 3} + \frac{2\theta - 1}{1 - \theta^2}$
20. $\frac{8}{\alpha^2 - 9} - \frac{2}{\alpha^2 - 5\alpha + 6}$
21. $\frac{7}{Y^2 + 5Y} - \frac{3}{Y} + \frac{5}{Y - 5}$
22. $\frac{11R_1 - 2}{3R_1^2 - 3} - \frac{5R_1 + 1}{2R_1^2 - 2}$
23. $\frac{21}{14 - \pi} - \frac{35 - 2\pi^2}{\pi^2 - 11\pi - 42}$
24. $\frac{2L - 4M}{2L - 2M} - \frac{3M^2 - 3LM}{L^2 - 2LM + M^2}$
25. $\frac{x + y}{x - y} - \frac{x - y}{x + y} - \frac{2xy}{x^2 + 2xy + y^2}$
26. $\frac{2X_C}{2X_C + 3X_L} - \frac{3X_L}{2X_C - 3X_L} + \frac{8X_L^2}{4X_C^2 - 9X_L^2}$
27. $\frac{E - 1}{E^2 - 9E + 20} - \frac{E + 1}{E^2 - 11E + 30}$
28. $a + b - \frac{a^2 - b^2}{a - b} + 1$
29. $\frac{\omega^2 + 3\omega + 9}{\omega^2 - 3\omega + 9} - \frac{54}{\omega^3 + 27} - \frac{\omega - 3}{\omega + 3}$
30. $\frac{\theta + 3\pi}{4\theta^2 + 12\theta\pi + 8\pi^2} + \frac{\theta + 2\pi}{\theta^2 + 4\theta\pi + 3\pi^2} - \frac{\theta + \pi}{4\theta^2 + 20\theta\pi + 24\pi^2}$

11–16 MULTIPLICATION OF FRACTIONS

The methods of multiplication of fractions in algebra are identical with those in arithmetic.

The product of two or more fractions is the product of the numerators of the fractions divided by the product of the denominators.

EXAMPLE 27

$$\frac{2}{3} \times \frac{3}{5} = \frac{6}{15}$$

EXAMPLE 28

$$\frac{a}{b} \cdot \frac{x}{y} = \frac{ax}{by}$$

When a factor occurs one or more times in *any* numerator and in *any* denominator of the product of two or more fractions, it can be canceled the same number of times from both. This process results in the product of the given fractions in lower terms.

EXAMPLE 29 Multiply $\frac{6x^2y}{7b}$ by $\frac{21b^2c}{24xy^2}$.

SOLUTION

$$\frac{6x^2y}{7b} \cdot \frac{21b^2c}{24xy^2} = \frac{3bcx}{4y}$$

EXAMPLE 30 Simplify

$$\frac{2a^2 - ab - b^2}{a^2 + 2ab + b^2} \cdot \frac{a^2 - b^2}{4a^2 + 4ab + b^2}$$

SOLUTION

$$\frac{2a^2 - ab - b^2}{a^2 + 2ab + b^2} \cdot \frac{a^2 - b^2}{4a^2 + 4ab + b^2}$$

$$= \frac{(2a + b)(a - b)}{(a + b)(a + b)} \cdot \frac{(a + b)(a - b)}{(2a + b)(2a + b)}$$

$$= \frac{(a - b)(a - b)}{(a + b)(2a + b)} = \frac{a^2 - 2ab + b^2}{2a^2 + 3ab + b^2}$$

It is very important that you understand clearly what we are allowed to cancel in the numerators and the denominators. The *whole* of an expression is always canceled, *never one term*. For example, in the expression $\frac{8a}{a - 5}$, it is not permissible to cancel the a's and obtain $\frac{8}{-5}$. It must be remembered that the denominator $a - 5$ denotes *one quantity*. Because of the parentheses, we would not cancel the a's if the expression were written $\frac{8a}{(a - 5)}$. However, the parentheses are not needed, for the *vinculum, which is also a sign of grouping, serves the same purpose*. We will consider this again in the next chapter.

11–17 DIVISION OF FRACTIONS

As with multiplication, the methods of division of fractions in algebra are identical with those of arithmetic. Therefore, to divide by a fraction, invert the divisor fraction and proceed as in the multiplication of fractions.

EXAMPLE 31

$$\frac{5}{2} \div \frac{2}{3} = \frac{5}{2} \cdot \frac{3}{2} = \frac{15}{4}$$

EXAMPLE 32

$$\frac{ab^2}{xy} \div \frac{a^2b}{xy^2} = \frac{ab^2}{xy} \cdot \frac{xy^2}{a^2b} = \frac{by}{a}$$

EXAMPLE 33

$$\frac{x}{y} \div \left(a + \frac{b}{c}\right) = \frac{x}{y} \div \frac{ac + b}{c}$$

$$= \frac{x}{y} \cdot \frac{c}{ac + b} = \frac{cx}{y(ac + b)} = \frac{cx}{acy + by}$$

Students often ask why we must invert the divisor and multiply by the dividend in dividing fractions. As an example, suppose we have $\frac{a}{b} \div \frac{x}{y}$. The dividend is $\frac{a}{b}$, and the divisor is $\frac{x}{y}$. Now

$$\text{Quotient} \times \text{divisor} = \text{dividend}$$

Therefore, the quotient must be a number such that, when multiplied by $\frac{x}{y}$, it will give $\frac{a}{b}$ as a product. Then,

$$\left(\frac{a}{b} \cdot \frac{y}{x}\right) \cdot \frac{x}{y} = \frac{a}{b}$$

Hence, the quotient is $\frac{a}{b} \cdot \frac{y}{x}$, which is the dividend multiplied by the inverted divisor.

PROBLEMS 11–9

Simplify:

1. $\frac{3}{5} \times \frac{5}{8} \times \frac{64}{65}$
2. $\frac{6}{14} \times \frac{3}{15} \times \frac{35}{10}$
3. $\frac{1}{16} \times \frac{8}{35} \times \left(-\frac{7}{32}\right)$
4. $\frac{3}{5} \div \frac{15}{10}$
5. $\frac{5}{8} \div \frac{3}{16}$
6. $-\frac{2}{3}\left(-\frac{5}{16} \div \frac{15}{64}\right)$
7. $\frac{9\theta^3}{2\phi} \times \frac{10\phi^2}{27\theta^2}$
8. $2m\left(\frac{3r}{4m^2} \times \frac{15mr}{18}\right)$
9. $\frac{50\alpha^3\gamma^2\delta^3}{\beta^5\gamma\delta^2} \div \frac{25\alpha^2\gamma\delta^2}{\beta^3\delta}$

10. $\dfrac{2\pi fL}{5} \div \dfrac{2\pi fC}{10}$

11. $\dfrac{\omega L}{R} \div 2\pi fL$

12. $\left(\dfrac{x^2 + 3x}{x}\right) \div \dfrac{x + 3}{x}$

13. $\dfrac{4}{x - y} \div \dfrac{x^2 - y^2}{x^2 + 2xy + y^2}$

14. $\dfrac{4\theta^2 - 1}{\theta^3 - 16\theta} \div \dfrac{2\theta - 1}{\theta - 4}$

15. $\dfrac{9B^2 - Y^2}{4B^2G - 25G} \div \dfrac{21B + 7Y}{2BG - 5G}$

16. $\dfrac{I^2 - 4i^2}{Ii + 2i^2} \cdot \dfrac{2i}{I - 2i}$

17. $\dfrac{\theta^2 - 2\theta\phi + \phi^2}{3\theta - 3\phi} \times \dfrac{3\theta + 3\phi}{\rho^3 - 3\rho^2 + 2\rho} \times \dfrac{\rho^3 - \rho}{\theta^2\rho - \phi^2\rho}$

18. $\dfrac{F^2 + 2F + 1}{P^3 - PZ^2} \cdot \dfrac{P^2 - Z^2}{5F^3 + 10F^2 + 5F} \cdot \dfrac{F^2P - 10FP + 25P}{F^2 - 110F + 525}$

19. $\dfrac{(a^2 - 2a - 3)}{-12b^2} \times \dfrac{3ab^4 + 15b^4}{a^2b - 8a + ab - 8} \times \dfrac{6ab - 48}{a^2b + 2ab - 15b}$

20. $\dfrac{R^2 - r^2}{r^2 + Rr} \cdot \dfrac{R(R - r)}{(R - r)^2} \div \dfrac{R^2 - 3Rr + 2r^2}{Rr - 2r^2}$

21. $\dfrac{V^2 - 6V + 8}{V^2} \times \dfrac{5V^4 + 5V^3}{V^2 - 9V + 20} \div \dfrac{V^2 - V - 2}{2V^2 - 10V}$

22. $\dfrac{16I^4R^2 - 9}{4\left(I^2R + \dfrac{3}{4}\right)} \cdot \dfrac{I^4R^2 - 3I^2R - 28}{2I^4R^2 - 32} \div \dfrac{8I^4R^2 - 62I^2R + 42}{8I^2R - 32}$

23. $\left(4 - \dfrac{4}{c} - \dfrac{8}{c^2}\right)\left(\dfrac{3c^4 - 6c^3}{2c^2 - 2c - 4}\right)\left(\dfrac{2c^2 + 8c}{3c^3 + 6c^2 - 24c}\right)$

24. $\left(m - \dfrac{m^2}{m}\right)\left(\dfrac{m^2 - n^2}{m^2 + mn}\right)\left(\dfrac{m + n}{m^2 + mn}\right)$

25. $\left(\dfrac{5\phi^5 - 5\phi^4}{\phi^2 - \phi - 20}\right)\left(\dfrac{\phi^2 + 11\phi + 28}{5\phi - 5}\right) \div \left(\dfrac{\phi^4 + 9\phi^3 + 14\phi^2}{\phi^2 - 3\phi - 10}\right)$

26. $\left(\dfrac{I^2 + I - 6}{I^4 - 9I^2}\right)\left(I^2 + 4I + \dfrac{12I}{I - 3}\right) \div \left(\dfrac{I^2 - I - 2}{I^2 - 6I + 9}\right)$

27. $\left(\dfrac{\omega L + R}{2} + \dfrac{\omega L - R}{4}\right)\left(\dfrac{4}{9\omega^2L^2 + 6\omega LR + R^2}\right)$

28. $\left(\dfrac{45}{\theta^2} + \dfrac{14}{\theta} + 1\right)\left(\dfrac{3\theta^3 + 6\theta^2}{\theta^2 + 18\theta + 81}\right)\left(\dfrac{\theta^2 + 13\theta + 36}{\theta^2 + 9\theta + 20}\right)\left(\dfrac{1}{3\theta + 3}\right)$

29. $\left(\dfrac{3y^3 - 6y^2 - 45y}{3y^2 + 4y - 4}\right)\left(\dfrac{y^2 + 2y}{2y^2 + 7y + 3}\right)\left(\dfrac{6y^3 - y^2 - 2y}{y^2 - 7y + 10}\right)\left(\dfrac{1}{y} - \dfrac{2}{y^2}\right)$

30. $\left(\dfrac{1}{f^2} + \dfrac{2}{f} + 1\right)\left(\dfrac{f^3 - f^2}{f^2 - 5f - 6}\right)\left(2 - \dfrac{12f - 2}{f^2 - 1}\right)$

11–18 COMPLEX FRACTIONS

A *complex fraction* is one with one or more fractions in its numerator, denominator, or both. The name is an unfortunate one. There is nothing complex or intricate about such compounded fractions, as we shall see.

RULE

To simplify a complex fraction, reduce both numerator and denominator to simple fractions; then perform the indicated division.

EXAMPLE 34 Simplify $\dfrac{\frac{1}{3}+\frac{1}{5}}{4-\frac{1}{5}}$.

SOLUTION

$$\frac{\frac{1}{3}+\frac{1}{5}}{4-\frac{1}{5}} = \frac{\frac{5+3}{15}}{\frac{20-1}{5}}$$

$$= \frac{\frac{8}{15}}{\frac{19}{5}} = \frac{8}{15} \times \frac{5}{19} = \frac{8}{57}$$

EXAMPLE 35 Simplify $\dfrac{5-\frac{1}{a+1}}{3+\frac{2}{a+1}}$.

SOLUTION

$$\frac{5-\frac{1}{a+1}}{3+\frac{2}{a+1}} = \frac{\frac{5(a+1)-1}{a+1}}{\frac{3(a+1)+2}{a+1}}$$

$$= \frac{\frac{5a+4}{a+1}}{\frac{3a+5}{a+1}}$$

$$= \frac{5a+4}{a+1} \cdot \frac{a+1}{3a+5}$$

$$= \frac{5a+4}{3a+5}$$

NOTE *It is evident that if the same factor occurs in both numerators of a complex fraction, the factors can be canceled. Also, if a factor occurs in both denominators, it can be canceled. Thus,* (a + 1) *could have been canceled in Example 35 after the numerators and denominators were reduced from mixed expressions to simple fractions.*

EXAMPLE 36 Simplify $\dfrac{\dfrac{a}{b}+\dfrac{a+b}{a-b}}{\dfrac{a}{b}-\dfrac{a-b}{a+b}}$.

SOLUTION

$$\frac{\dfrac{a}{b}+\dfrac{a+b}{a-b}}{\dfrac{a}{b}-\dfrac{a-b}{a+b}} = \frac{\dfrac{a(a-b)+b(a+b)}{b(a-b)}}{\dfrac{a(a+b)-b(a-b)}{b(a+b)}}$$

$$= \frac{\dfrac{a^2-ab+ab+b^2}{b(a-b)}}{\dfrac{a^2+ab-ab+b^2}{b(a+b)}}$$

$$= \frac{\dfrac{a^2+b^2}{b(a-b)}}{\dfrac{a^2+b^2}{b(a+b)}} = \frac{a+b}{a-b}$$

PROBLEMS 11–10

Simplify:

1. $\dfrac{\dfrac{3}{5}-3}{2-\dfrac{2}{5}}$

2. $\dfrac{25-\left(\dfrac{2}{7}\right)^2}{5+\dfrac{2}{7}}$

3. $\dfrac{2+\dfrac{3M}{4P}}{M+\dfrac{8P}{3}}$

4. $\dfrac{R-\dfrac{1}{X}}{R+\dfrac{1}{X}}$

5. $\dfrac{I}{\dfrac{1}{fC_1}+\dfrac{1}{fC_2}}$

6. $\dfrac{\dfrac{i^2}{8}-8}{1+\dfrac{i}{8}}$

7. $\dfrac{Y}{Y-\dfrac{G}{B}}$

8. $\dfrac{5\theta+\dfrac{2\lambda}{5\phi}}{\dfrac{2\lambda}{5\theta}+5\phi}$

9. $\dfrac{\dfrac{E^2}{e^2}-1}{\dfrac{E^2+e^2}{2Ee}+1}$

10. $\dfrac{\dfrac{\lambda+\pi}{\lambda^2+\pi^2}-\dfrac{1}{\lambda+\pi}}{\dfrac{1}{\lambda+\pi}-\dfrac{\lambda}{\lambda^2+\pi^2}}$

11. $\dfrac{\dfrac{l}{l+w}}{1+\dfrac{w}{l-w}}$

12. $\dfrac{\omega+2-\dfrac{15}{\omega}}{1-\dfrac{8}{\omega}+\dfrac{15}{\omega^2}}$

13. $\dfrac{1-\dfrac{x+y}{x-y}}{1+\dfrac{x+y}{x-y}}$

14. $\dfrac{\dfrac{\theta}{\theta+\phi}-\dfrac{\theta}{\theta-\phi}}{\dfrac{\theta}{\theta+\phi}+\dfrac{\theta}{\theta-\phi}}$

15. $\dfrac{1}{v-\dfrac{v^2-1}{v+\dfrac{1}{v-1}}}$

16. $\dfrac{L_1}{Q-\dfrac{1}{Q+\dfrac{1}{Q}}}-\dfrac{L_1}{Q+\dfrac{1}{Q-\dfrac{1}{Q}}}$

SELF TEST

1. What is the HCF of:

   $ax^2 + ab, \quad ax + 2ab, \quad ax^2 + abx - 2ab^2$

2. What is the LCM of:

   $12\theta, \quad 8\phi, \quad 36\lambda$

3. Solve for the missing part:

$$\frac{3 + L}{5 - x} = \frac{21 + 10L + L^2}{?}$$

4. Reduce to lowest terms: $\dfrac{x^2 + xy - 2y^2}{x^2 + 5xy + 6y^2}$

5. Express as a fraction with a positive numerator:

$$-\frac{l - m}{p - q}$$

6. Change the following fraction to a mixed expression:

$$\frac{3R^3 + 5R^2V - RV^2 - 3V^3 + 3}{R + V}$$

7. Convert the following set of fractions to an equivalent set having its LCD:

$$\frac{a}{b}, \quad \frac{a + 1}{b + 1}, \quad \frac{a - 2}{b + 3}$$

8. Perform the indicated operations:

$$\frac{2}{\theta + \lambda} + \frac{5\theta}{\theta - \lambda} - \frac{2\lambda}{\theta^2 - \lambda^2}$$

9. Simplify:

$$\frac{a^2 - b^2}{4a^2 + 8ab + 3b^2} \times \frac{4a^2 - b^2}{2a^2 + ab - b^2} \div \frac{a^2 - b^2}{2a^2 - 2b^2}$$

10. Simplify: $\dfrac{5 + \dfrac{1}{x}}{5 - \dfrac{1}{x}}$

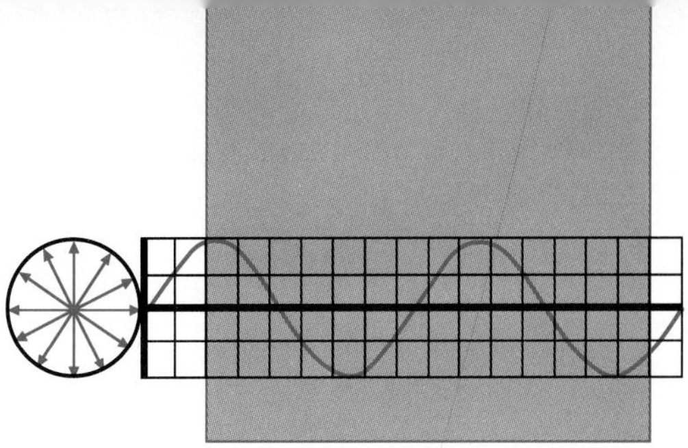

CHAPTER 12

FRACTIONAL EQUATIONS

An equation containing a fraction in which the unknown occurs in a denominator is called a *fractional equation*. Equations of this type are encountered in many problems involving electric and electronic circuits. Simple fractional equations, wherein the unknown appeared only as a factor, were studied in earlier chapters.

12–1 FRACTIONAL COEFFICIENTS

A number of problems lead to equations containing *fractional coefficients*. This type of equation is included in this chapter because the methods of solution apply to fractional equations also.

EXAMPLE 1

$$\frac{3x}{4} + \frac{3}{2} = \frac{5x}{8} \quad \text{and} \quad \frac{x}{2} + \frac{x}{3} = 5$$

are equations having fractional coefficients.

EXAMPLE 2

$$\frac{60}{x} - 3 = \frac{60}{4x} \quad \text{and} \quad \frac{x-2}{x} = \frac{4}{5}$$

are fractional equations.

You are familiar with the methods of solving simple equations that do not contain fractions. An equation involving fractions can be changed to an equation containing no fractions by canceling the denominators and then solved as heretofore. To accomplish this we have the following rule:

RULE

To solve an equation containing fractions:

1. First clear the equation of fractions by multiplying every term by the LCD of the whole equation. (This will permit canceling all denominators.)
2. Solve the resulting equation.

EXAMPLE 3 Given $\frac{5x}{12} - 13 = \frac{x}{18}$. Solve for x.

SOLUTION

Given $$\frac{5x}{12} - 13 = \frac{x}{18}$$

M: 36, the LCD, $$\frac{36 \cdot 5x}{12} - 36 \cdot 13 = \frac{36x}{18}$$

Canceling, $$\frac{\overset{3}{\cancel{36}} \cdot 5x}{\cancel{12}} - 36 \cdot 13 = \frac{\overset{2}{\cancel{36}}x}{\cancel{18}}$$

Simplifying, $$15x - 468 = 2x$$

Collecting terms, $$13x = 468$$

D: 13, $$x = 36$$

CHECK

Substitute 36 for x in the original equation:

$$\frac{5 \cdot 36}{12} - 13 = \frac{36}{18}$$

Clearing fractions, $$15 - 13 = 2$$
$$2 = 2$$

EXAMPLE 4 Given, $\frac{e - 4}{9} = \frac{e}{10}$. Solve for e.

SOLUTION

Given $$\frac{e - 4}{9} = \frac{e}{10}$$

M: 90, the LCD, $$\frac{90(e - 4)}{9} = \frac{90e}{10}$$

Canceling, $$\frac{\overset{10}{\cancel{90}}(e - 4)}{\cancel{9}} = \frac{\overset{9}{\cancel{90}}e}{\cancel{10}}$$

Simplifying, $$10(e - 4) = 9e$$

or $$10e - 40 = 9e$$

Collecting terms, $$10e - 9e = 40$$

or $$e = 40$$

CHECK

Substitute 40 for e in the original equation:

$$\frac{40 - 4}{9} = \frac{40}{10}$$

Clearing fractions, $$4 = 4$$

Note that when the fractions were cleared and the equation was written in simplified form in the above solution, the resulting equation was

$$10(e - 4) = 9e$$

which is equivalent to multiplying each member by the denominator of the other member and expressing the resulting equation with no denominators. This is called *cross multiplication*. You will see the justification of this if each member is expressed as a fraction having the LCD. Although the method is convenient, it must be remembered that *cross multiplication is permissible only when each term of an equation has the same denominator.*

PROBLEMS 12–1

Solve the following equations:

1. $\frac{\theta}{3} = \frac{\theta}{7} + 4$
2. $\frac{y}{4} = \frac{y}{8} + 3$
3. $\frac{2\gamma}{3} + \frac{\gamma}{4} = 5 + \frac{\gamma}{2}$
4. $I - \frac{1}{4} = \frac{2I}{5} - \frac{1}{16}$
5. $\frac{1 + V}{2} + \frac{2V - 3}{3} = 3$
6. $\frac{1}{3} + \frac{Z}{5} = \frac{Z}{3}$
7. $\frac{6 + 3\phi}{4} + \frac{12 - 2\phi}{15} = \frac{6\phi}{5} - \frac{37}{60}$
8. $\frac{F}{6} + \frac{F - 3}{18} = \frac{3 + 3F}{12}$

NOTE *If a fraction is negative, the sign of each term of the numerator must be changed after removing the denominator. (See Sec. 11–10.) Remember that* ***the vinculum is a sign of grouping.***

9. $\lambda + \frac{3\lambda - 5}{2} = 12 - \frac{2\lambda - 4}{3}$
10. $\frac{4I + 3}{5} - \frac{I - 5}{10} = \frac{I}{3}$

11. $\frac{\theta + 5}{5} - \frac{\theta - 7}{7} = 4$

12. $x - \frac{3 + 4x}{5} + \frac{2x - 3}{6} - \frac{5x - 4}{15} = 0$

13. $\frac{1}{15}(2Z - 9) - \frac{1}{3}(9Z - 6) = \frac{1}{5}(15Z + 95)$

14. $\frac{2}{3}(z + 1) - \frac{3}{4}(z + 2) = \frac{1}{6}(z + 1)$

15. $\frac{1}{4}\left(4 + \frac{3\omega}{2}\right) = \frac{1}{7}\left(2\omega - \frac{1}{3}\right) + \frac{31}{28}$

12–2 EQUATIONS CONTAINING DECIMALS

An equation containing decimals is readily solved by first clearing the equation of the decimals. This is accomplished by multiplying both members by a power of 10 that corresponds to the largest number of decimal places appearing in any term.

EXAMPLE 5 Solve $0.75 - 0.7a = 0.26$.

SOLUTION

Given	$0.75 - 0.7a = 0.26$
M: 100,	$75 - 70a = 26$
Collecting terms,	$70a = 49$
D: 70,	$a = 0.7$

CHECK

Substitute 0.7 for a in the original equation:

$$\begin{aligned} 0.75 - 0.7 \cdot 0.7 &= 0.26 \\ 0.75 - 0.49 &= 0.26 \\ 0.26 &= 0.26 \end{aligned}$$

If decimals occur in any denominator, multiply both numerator and denominator of the fraction by a power of 10 that will reduce the decimals to integers.

EXAMPLE 6 Solve $\frac{5m - 1.33}{0.02} - \frac{m}{0.05} = 1083.5$.

SOLUTION

Given $\frac{5m - 1.33}{0.02} - \frac{m}{0.05} = 1083.5$. Multiplying numerator and denominator of each fraction by 100,

$$\frac{500m - 133}{2} - \frac{100m}{5} = 1083.5$$

The equation is then solved and checked by the usual methods.

PROBLEMS 12–2

Solve the following equations:

1. $0.06I = 0.03$
2. $0.5V = 1.5$
3. $0.8\theta = 2 + 0.4\theta$
4. $0.375\lambda + 2 + 0.125\lambda = 0.625\lambda + 1.375$
5. $0.3r + 4 = 0.7r - 8$
6. $\phi + 2.6 - 0.2\phi = 1.4 + 0.3\phi$
7. $16.5 - 1.5(2R - 0.5) - 15.6 + 2.1(R + 0.3) = 0.03$
8. $0.018 - 0.004\alpha + 0.027 + 0.009\alpha - 0.003 + 0.016\alpha = 0$
9. $\dfrac{0.5b}{6} - \dfrac{0.2b - 0.5}{30} = \dfrac{0.3b + 0.3}{15}$
10. $\dfrac{0.5(\theta - 5)}{3.75} = \dfrac{0.3(\theta + 5)}{7.5} - \dfrac{0.2(3\theta - 2)}{5}$
11. $\dfrac{1.4\alpha - 6}{20} = \dfrac{0.4\alpha - 6}{4}$
12. $\dfrac{0.6x - 16}{2} - \dfrac{0.8x - 51}{3} + \dfrac{0.4x + 12}{6} - 0.18x = 1.4$
13. $\dfrac{0.4Y - 0.6}{0.06Y - 0.07} = \dfrac{2Y - 3}{0.3Y - 0.4}$
14. $\dfrac{0.2(\omega - 1)}{0.5(\omega + 5)} - \dfrac{0.3(1 - \omega)}{0.7(\omega + 5)} - \dfrac{29}{140} = 0$
15. $(0.7\alpha - 0.7)(0.2 + \alpha) = (1 - 1.4\alpha)(0.1 - 0.5\alpha)$

12–3 FRACTIONAL EQUATIONS

Fractional equations are solved in the same manner as equations containing fractional coefficients (Sec. 12–1). That is, every term of the equation must be multiplied by the LCD.

EXAMPLE 7 Solve $\dfrac{x + 2}{3x} - \dfrac{2x^2 + 3}{6x^2} = \dfrac{1}{2x}$.

SOLUTION

Given
$$\frac{x + 2}{3x} - \frac{2x^2 + 3}{6x^2} = \frac{1}{2x}$$

M: $6x^2$, the LCD,
$$\frac{6x^2(x + 2)}{3x} - \frac{6x^2(2x^2 + 3)}{6x^2} = \frac{6x^2}{2x}$$

Canceling,
$$\frac{\overset{2x}{\cancel{6x^2}}(x + 2)}{\cancel{3x}} - \frac{\cancel{6x^2}(2x^2 + 3)}{\cancel{6x^2}} = \frac{\overset{3x}{\cancel{6x^2}}}{\cancel{2x}}$$

Rewriting,
$$2x(x + 2) - (2x^2 + 3) = 3x$$

Simplifying, $$2x^2 + 4x - 2x^2 - 3 = 3x$$

Collecting terms, $$4x - 3x = 3$$

or $$x = 3$$

CHECK

Substituting 3 for x in the original equation,

$$\frac{3+2}{9} - \frac{18+3}{54} = \frac{1}{6}$$

That is, $$\frac{30}{54} - \frac{21}{54} = \frac{9}{54}$$

EXAMPLE 8 Solve $\dfrac{8a+2}{a-2} - \dfrac{2a-1}{3a-6} + \dfrac{3a+2}{5a-10} + 5 = 15.$

SOLUTION

Given $$\frac{8a+2}{a-2} - \frac{2a-1}{3a-6} + \frac{3a+2}{5a-10} + 5 = 15$$

Factoring denominators,

$$\frac{8a+2}{a-2} - \frac{2a-1}{3(a-2)} + \frac{3a+2}{5(a-2)} + 5 = 15$$

M: $15(a-2)$, the LCD,

$$\frac{15(a-2)(8a+2)}{a-2} - \frac{15(a-2)(2a-1)}{3(a-2)} + \frac{15(a-2)(3a+2)}{5(a-2)} + 15(a-2)(5) = 15(a-2)(15)$$

Canceling,

$$\frac{15\cancel{(a-2)}(8a+2)}{\cancel{a-2}} - \frac{\overset{5}{\cancel{15}}\cancel{(a-2)}(2a-1)}{\cancel{3}\cancel{(a-2)}} + \frac{\overset{3}{\cancel{15}}\cancel{(a-2)}(3a+2)}{\cancel{5}\cancel{(a-2)}} + 15(a-2)(5) = 15(a-2)(15)$$

Rewriting,

$$15(8a+2) - 5(2a-1) + 3(3a+2) + 15(a-2)(5) = 15(a-2)(15)$$

Simplifying,

$$120a + 30 - 10a + 5 + 9a + 6 + 75a - 150 = 225a - 450$$

Collecting terms,

$$\begin{aligned} 120a - 10a + 9a + 75a - 225a &= -30 - 5 - 6 + 150 - 450 \\ -31a &= -341 \\ a &= 11 \end{aligned}$$

Check the solution by the usual method.

PROBLEMS 12–3

Solve the following equations:

1. $\frac{1}{a} + \frac{2}{a} = 3 - \frac{3}{a}$
2. $\frac{2}{x} + \frac{5}{6} = 1.5$
3. $\frac{16}{m} - \frac{7}{m} - 2 = \frac{3}{m} - \frac{2}{m}$
4. $\frac{5}{2L} - \frac{1}{3} + \frac{7}{5L} + \frac{1}{60} = 1$
5. $\frac{5}{2} + \frac{1}{2L} = \frac{8}{L}$
6. $\frac{5}{3R} - \frac{17}{3R} + \frac{8}{R} = 2$
7. $\frac{12 - w}{w} - \frac{6}{w} = \frac{3}{w}$
8. $\frac{9}{14 + 2P} = \frac{1}{6 - 2P}$
9. $\frac{40 - \pi}{24\pi} + \frac{5}{6} - \frac{40 + \pi}{8\pi} = 0$
10. $\frac{10}{W} - 3 = \frac{2 - W}{W}$
11. $\frac{40 + v_0}{8v_0} - \frac{5}{6} - \frac{40 - v_0}{24v_0} = 0$
12. $\frac{6m - 17}{3m + 3} - \frac{2m - 5}{9 + m} = 0$
13. $\frac{6}{x - 1} - \frac{5}{1 - x} - \frac{8}{x - 1} + \frac{x}{1 - x} = 0$
14. $\frac{3}{5 + R} + \frac{R}{R + 2} = \frac{R + 4}{R + 5}$
15. $\frac{R}{R + 1} + 2 = \frac{3R}{R + 2}$
16. $\frac{5 + R}{5 - R} - \frac{16R}{25 - R^2} + \frac{5 - R}{5 + R} + 2 = 0$
17. $\frac{\omega + 3}{\omega - 8} - \frac{5 - \omega}{\omega + 1} = \frac{2\omega^2 - 2}{\omega^2 - 7\omega - 8}$
18. $\frac{2\phi + 7}{6\phi - 4} - \frac{17\phi + 7}{9\phi^2 - 4} - \frac{3\phi - 5}{9\phi + 6} = 0$
19. $\frac{4(9 - \theta)}{\theta^2 - 9} - \frac{3\theta - 2}{\theta + 3} - \frac{3\theta + 2}{3 - \theta} = 0$
20. $\frac{a - 7}{a + 2} - \frac{6}{a + 3} = \frac{a^2 - a - 42}{a^2 + 5a + 6}$
21. *A* can do a piece of work in 8 h, and *B* can do it in 6 h; how long will it take them to do it together?

Solution: Let n = number of hours it will take them to do it together. Now *A* does $\frac{1}{8}$ of the job in 1 h; therefore, *A* will do $\frac{n}{8}$ in n h. Also, *B* does $\frac{1}{6}$ of the job in 1 h; therefore, *B* will do $\frac{n}{6}$ in n h. Then they will do $\frac{n}{8} + \frac{n}{6}$ in n h.

The entire job will be completed in n h, which we may represent by $\frac{8}{8}$ or $\frac{6}{6}$ of itself, which is 1.

$$\therefore \frac{n}{8} + \frac{n}{6} = 1$$

M: 24, the LCD,

$$3n + 4n = 24$$
$$7n = 24$$
$$n = 3\tfrac{3}{7} \text{ h}$$

22. A technician can install a television transmission line in 6 h, and the technician's helper can do it in 10 h. In how many hours should they be able to do it if they work together?
23. A journeyman can make up a multichassis electronics kit in 8 h. An apprentice takes 14 h to do the same job. Assuming that they can work as efficiently together as they do separately, how long should it take the two of them to complete the kit?
24. A can do a piece of work in a days, and B can do it in b days. Derive a general formula for the number of days it would take both together to do the work.

Solution: Let x = number of days it will take both together.

Now A will do $\frac{x}{a}$ of the job in x days. Also, B will do $\frac{x}{b}$ of the job in x days.

Then $\frac{x}{a} + \frac{x}{b} = 1$

M: *ab,* $bx + ax = ab$

Factoring, $x(a + b) = ab$

D: $(a + b)$, $x = \frac{ab}{a + b}$

Alternate solution: Let x = number of days it will take both together. Then $\frac{1}{x}$ = part that both together can do in 1 day; $\frac{1}{a}$ = part that A alone can do in 1 day; and $\frac{1}{b}$ = part that B can do in 1 day.

Now, $\frac{1}{a} + \frac{1}{b} = \frac{1}{x}$

M: *abx,* $bx + ax = ab$

Factoring, $x(b + a) = ab$

D: $(a + b)$, $x = \frac{ab}{a + b}$

25. A can do a piece of work in a days, B in b days, and C in c days. Derive a general formula for the number of days it would take them to do the work together.
26. A tank can be filled by pipe A in 5 h and by pipe B in 8 h. The tank can be emptied by the drain pipe in 6 h. If all three pipes are opened at the same time, how long will it take to fill the tank?
27. Three circuits are operated on a storage battery. Circuit 1 completely discharges the battery in 24 h, circuit 2 in 18 h, and circuit 3 in 9 h. All circuits are connected in parallel. Circuit 1 is switched on at 8:00 A.M. and the other two circuits are switched on at 10:00 A.M. At what time will the battery be technically discharged?
28. A tank can be filled by one of two pipes in x h and by the other of the two in y h; it can be emptied by a drain pipe in z h. Derive a general formula for the number of hours required to fill the tank with all pipes open.

29. A bottle contains 1 liter (L) of a mixture of equal parts of acid and water. How much water must be added to make a mixture that will be one-tenth acid?

Solution: Let n = number of liters of water to be added.

$$\begin{aligned}
1\text{ L} &= \text{amount of original mixture}\\
\text{and}\quad 0.5\text{ L} &= \text{amount of acid}\\
\text{Hence}\quad n + 1 &= \text{amount of new one-tenth acid mixture}\\
\text{Now,}\quad \frac{1}{10} &= \frac{\text{amount of acid}}{\text{total mixture}}\\
\text{Then,}\quad \frac{1}{10} &= \frac{0.5}{n + 1}\\
n + 1 &= 5\\
n &= 4\text{ L of water to be added}
\end{aligned}$$

30. How much metal containing 25% copper must be added to 20 kg of pure copper to obtain an alloy having 50% copper?

Solution: Let x = desired amount of metal containing 25% copper.

$$\begin{aligned}
\text{Then}\quad 0.25x &= \text{amount of copper in this metal}\\
20 + 0.25x &= \text{amount of copper in mixture}\\
x + 20 &= \text{total weight of mixture}\\
0.5(x + 20) &= \text{amount of copper in mixture}\\
0.5x + 10 &= 20 + 0.25x\\
x &= 40\text{ kg}
\end{aligned}$$

31. How much 15% nickel alloy must be added to 10 kg of 40% nickel alloy to form a 20% nickel alloy?

32. A full radiator contains 50 L of a 30% mixture of antifreeze. How much antifreeze is required to obtain a 45% mixture?

Solution: The radiator now contains 50 L of 30% antifreeze = 15 L. We want it to contain 50 L of 45% antifreeze = 22.5 L. But to get the mixture we want, we must drain off some quantity of 30% mixture and replace it with 100% antifreeze. Let the volume replaced be x L:

$$\begin{aligned}
15 - 0.3x + x &= 22.5\\
x &= 10.7\text{ L}
\end{aligned}$$

33. A diesel engine driving a 100-kW generator for an isolated communications center has a 250-L cooling system which, during the summer, contains a 40% antifreeze solution. How much coolant must be drained off and replaced with pure antifreeze to produce an 85% antifreeze solution?

34. A fighter plane traveling at 900 km/h leaves its base at 9:00 A.M. in order to overtake a bomber which departed from the same base at 7:00 A.M. and is traveling at 475 km/h. How much time is required for the fighter to overtake the bomber?

35. The sum of two numbers is 420. When the larger is divided by the smaller, the quotient is 5. Find the numbers.

36. The numerator of a fraction is 54 greater than the denominator. When 9 is subtracted from each term, the quotient is 4. What is the value of the fraction?

37. The sum of three consecutive numbers is $4\frac{1}{2}$. Find the numbers.
38. A certain number, plus 23, is divided by the same number plus 12. The quotient is $\frac{4}{3}$. What is the number?
39. The perimeter of a stockroom is 160 m. The room is three times as long as it is wide. What are its dimensions?
40. A screened room is two-thirds as wide as it is long. If it had been 3 m wider and 3 m shorter, its area would have been 3 m^2 larger. What are its dimensions?

12–4 LITERAL EQUATIONS

Equations in which some or all of the numbers are replaced by letters are called *literal equations;* they were studied in Chap. 5. Having attained more knowledge of algebra, such as factoring and fractions, we are now ready to proceed with the solution of more difficult literal equations, or formulas. No new methods are involved in the actual solutions—we are prepared to solve a more complicated equation simply because we have available more tools with which to work. Again, we point out that the ability to solve formulas is of utmost importance.

EXAMPLE 9 Given $I = \dfrac{V}{R + r}$, solve for r.

SOLUTION

Given $$I = \frac{V}{R + r}$$

M: $(R + r)$, $$I(R + r) = V$$

Removing parentheses, $$IR + Ir = V$$

S: IR, $$Ir = V - IR$$

D: I, $$r = \frac{V - IR}{I}$$

EXAMPLE 10 Given $S = \dfrac{RL - a}{R - 1}$, solve for L.

SOLUTION

Given: $$\frac{RL - a}{R - 1} = S$$

M: $(R - 1)$, $$RL - a = S(R - 1)$$

A: a, $$RL = S(R - 1) + a$$

D: R, $$L = \frac{S(R - 1) + a}{R}$$

EXAMPLE 11 Given $\dfrac{a}{x - b} = \dfrac{2a}{x + b}$, solve for x.

SOLUTION

Given $$\frac{a}{x - b} = \frac{2a}{x + b}$$

M: $(x^2 - b^2)$, the LCD, $$\frac{(x^2 - b^2)a}{x - b} = \frac{(x^2 - b^2)2a}{x + b}$$

Canceling, $\dfrac{\overset{x+b}{\cancel{(x^2-b^2)}}a}{\cancel{x-b}} = \dfrac{\overset{x-b}{\cancel{(x^2-b^2)}}2a}{\cancel{x+b}}$

Rewriting,	$(x+b)a = (x-b)2a$
Removing parentheses,	$ax + ab = 2ax - 2ab$
Collecting terms,	$ax - 2ax = -2ab - ab$
or	$-ax = -3ab$
M: -1,	$ax = 3ab$
D: a,	$x = 3b$

NOTE *The last two steps can be combined into a single step by dividing* $-ax = -3ab$ *by* $-a$ *to obtain* $x = 3b$.

CHECK

Substitute $3b$ for x in the given equation:

$$\frac{a}{3b-b} = \frac{2a}{3b+b}$$

Simplifying, $\dfrac{a}{2b} = \dfrac{2a}{4b}$

or $\dfrac{a}{2b} = \dfrac{a}{2b}$

PROBLEMS 12–4

Given:	Solve for:
1. $A = \dfrac{h}{2}(a+b)$	h, b
2. $a = \dfrac{w - w_1}{w - w_2}$	w_1, w_2
3. $v = V_b - IR$	V_b, I
4. $R_{eq} = R(F-1)$	R, F
5. $L = \dfrac{V_3 - V_2}{\omega^2 C_2 V_3}$	C_2, V_2
6. $s = \dfrac{V_1 - R(I_1 + I_2)}{I_2}$	I_1, V_1, R
7. $R_t = R_o(1 + \alpha\,\Delta t)$	$\alpha, \Delta t$
8. $I_{\lambda_2} = \dfrac{V_{e_2} + V_\lambda - V_2}{R_b}$	V_λ, V_2
9. $r = \dfrac{vR}{V - v}$	v, R
10. $\mu = \dfrac{g_m}{g_m' - g_m}$	g_m', g_m

11. $R_1 = \dfrac{1}{\omega^2 C_1 C_2 R_3}$ — C_1

12. $\mu = \dfrac{2G_L + g_p - 2G_2}{G_2 - G_L}$ — G_2, G_L, g_p

13. $\beta = \dfrac{V_0 - I_0 R_0}{\mu V_0}$ — V_0

14. $\beta_m = \dfrac{m\pi a}{a + b}$ — a, b

15. $I = \dfrac{E_g - E_t}{R}$ — E_t, E_g

16. $I_n = \dfrac{\gamma I_p}{1 - \gamma}$ — γ, I_p

17. $Z_1 = \dfrac{Z_3(E - IZ_2)}{I(Z_2 + Z_3)}$ — Z_2, Z_3, $\dfrac{E}{I}$

18. $Z_0 = \dfrac{R_a R}{(\mu + 1)R + R_a}$ — R, R_a, μ

19. $A = \dfrac{V(R_x + R_y)}{R_y V_1 - R_x V}$ — R_x, R_y, $\dfrac{V}{V_1}$

20. $B_c = \dfrac{\pi\sqrt{2DFf_b}}{\sqrt{2D + F}}$ — D, F

21. $X_p = \dfrac{X_s^2 R + Z_{ab}^2(X_s + R)}{Z_{ab}^2}$ — R, Z_{ab}^2

22. $Z_1 = \dfrac{(\mu + 1)R_1 R + R_a(R_1 + R)}{R_a + R}$ — R_a, R

23. $R_1 = \dfrac{R_2(V_n - I_2 R)}{I_2 R - 2V_n}$ — V_n, $I_2 R$

24. $2C_2 R_3 = \sqrt{2} - C_1 R_1\left(\dfrac{R_2}{R_3} - 1\right)$ — R_1, R_2, C_1

25. $C = \dfrac{5}{9}(F - 32)$ — F

26. $r = \dfrac{\mu V_g - PR_p}{P}$ — P, μ

27. $V_1 = \dfrac{BI_0 R_0}{R + R_0}$ — R_0, R

28. $f_{out} = \dfrac{C_1 f_{in}}{C_1 + C_2}$ — f_{in}, C_1, C_2

29. $r = \dfrac{\mu m N}{Kg} - R_H$	R_H, K
30. $r_p = \dfrac{GR_{pg}}{g_m R_{pg} - G}$	G, g_m
31. $\alpha = \dfrac{Z_1 + Z_2 - R}{Z_1(1 - k) + Z_2}$	R, Z_1, k
32. $X = \dfrac{K}{(f_1 - f_2) - (f_0 - f_2)}$	f_1, K
33. $f = \dfrac{\alpha F_{12} F_s}{2(F_2 + F_s)}$	F_2, F_{12}
34. $\mu\beta = \dfrac{2N}{2L + N}$	L, N
35. $H_2 S = \left(\dfrac{1}{R_1}\right)\left(\dfrac{S}{S + \alpha}\right)$	α, S
36. $\dfrac{n_2 - n_1}{n_1} = \dfrac{-hv}{kT}$	n_1
37. $\dfrac{r_1}{r_1 + r_2} = \dfrac{r_3}{r_3 + r_4}$	r_1, r_4
38. $F = 1 + 2\left(\dfrac{T_s}{T_a}\right)\left(\dfrac{1}{X}\right)$	T_a, X, T_s
39. $C_v = \dfrac{C_0(f_c - fX)}{f_c}$	C_0, X
40. $\dfrac{C_3}{C_1 + C_2} = \dfrac{R_3}{\dfrac{1}{R_1} + \dfrac{1}{R_2}}$	C_2, R_1
41. $R_1 = \dfrac{R_0(E - I_2 R_2)}{I_2(R_0 + R_2)}$	I_2
42. $\alpha = 1 + \dfrac{1}{\mu_0}\left(1 + 1.5\dfrac{d_2}{d_1}\right)$	d_1
43. $\dfrac{V - v_0}{v_0} = \dfrac{R_2}{R_1}\left(\dfrac{i_1 + i_2}{i_1}\right)$	v_0, i_1
44. $Z_{am}{}^2 = R\dfrac{(X_p - X_s)Z_{ab}{}^2}{Z_{ab}{}^2 + X_s{}^2}$	$Z_{ab}{}^2$, X_p
45. $G = \dfrac{\mu_1 \mu_2}{(1 - \mu_1\beta_1)(1 - \mu_2\beta_2)}$	μ_1, β_1

46. $C_g = C_{gf} + C_{gp}\left(1 + \dfrac{\mu R_b}{r_p + r_b}\right)$ $\qquad R_b, C_{gp}$

47. $I_1 = \dfrac{\pi\lambda^2\gamma_1(2I_f + 1)}{\sigma_0(\gamma_1 + \gamma_f)} - \dfrac{1}{2}$ $\qquad I_f, \sigma_0$

48. $K_\varepsilon{}^2\left(1 + \dfrac{\tan^2 K_a}{\varepsilon_p{}^2}\right) = -a^2$ $\qquad \varepsilon_p{}^2$

49. $\lambda = \dfrac{\pi n' d_0(d_1 - d_0)}{1 - d_1}$ $\qquad d_1, n'$

50. $I_2 = \dfrac{VR_0}{R_1R_0 + R_1R_2 + R_2R_0}$ $\qquad VR_0, R_1$

51. $\alpha = \dfrac{Z_1R_2 + Z_2R_2 + ZZ_2}{ZZ_2 - RkZ_2}$ $\qquad R_2, Z, R$

52. $\dfrac{V_b - V_c}{\mu} = V_c + V_s\left(\dfrac{R_p}{R_1 + R_p}\right)$ $\qquad R_1, V_s$

53. $\dfrac{r_1}{r_1 + r_2} = \dfrac{r_3}{r_3 + r_4}$ $\qquad r_1, r_3, r_4$

54. $\dfrac{S^2}{N^2} = \dfrac{\alpha F}{2f\left(1 + \dfrac{F_s}{F_2}\right)}$ $\qquad F_s, F_2$

55. $G = \dfrac{V_{out}C_fC_{fg}}{C_{fg}Q - V_{out}C_f(C_d + C_{fg})}$ $\qquad V_{out}$

56. $T_m = \dfrac{T}{\dfrac{\omega_{32}kv_{12}T_m}{\omega_{21}hv_{12}} - 1}$ $\qquad T, h$

57. $\dfrac{P_L}{2p} = \dfrac{\omega\varepsilon_2 p_2(\tan\delta)}{2CN(p_1 + p_2)}$ $\qquad p_2$

58. $a_2 = \dfrac{FC}{(\Omega_1 - B)(\Omega_2 - B) + c^2}$ $\qquad \Omega_1$

59. $\dfrac{1}{R_p} = \dfrac{1}{R_1} + \dfrac{1}{R_2}$ $\qquad R_p, R_1, R_2$

60. $i_s = \dfrac{v}{L\left(S_s + \dfrac{R}{L}\right)}$ $\qquad L, R$

61. $\mu = \dfrac{V_0}{R_3}\left(\dfrac{R_a + R_3}{V - V_0}\right)$ $\qquad R_3, R_a, \dfrac{V_0}{V}$

62. $\dfrac{\omega_{01}L}{\dfrac{R_1R_2}{R_1 + R_2}} = 1$ $\quad L, R_1$

63. $M = \dfrac{k}{1 + \dfrac{N}{4\pi}k}H_0$ $\quad \pi, k$

64. $HS = \dfrac{\dfrac{1}{C}}{S + \dfrac{1}{R_cC}}$ $\quad C, R_c$

65. $b = \dfrac{d(X^2 + X'^2)}{(X + X')^2}$ $\quad d$

66. $(G_2)(p) = \dfrac{A(p + \omega_1)}{p + \omega - \dfrac{AC_2}{C_1 + C_2}p}$ $\quad C_2$

67. $\dfrac{V_0}{V} = \dfrac{\mu R_1 + R_a}{\mu R_1 + R_a + (R_s + R_1)\left(1 + \dfrac{R_a}{R_3}\right)}$ $\quad R_a, R_s, \mu$

68. $\dfrac{V_0}{V} = \dfrac{h_{fe} + 1 + \dfrac{h_{ie}}{R_B}}{h_{fe} + 1 + h_{ie}\left(\dfrac{1}{R_B} + \dfrac{1}{R_E}\right)}$ $\quad V, R_B$

69. $R_0 = \left(\dfrac{1}{1 + \mu\dfrac{R_1}{R_a}}\right)\left(\dfrac{1}{\dfrac{1}{R_s + R_1} + \dfrac{1}{R_2}}\right)$ $\quad R_a, R_2, \mu$

70. $R_{in} = R_E\left[\dfrac{h_{fe} + 1 + h_{ie}\left(\dfrac{1}{R_E} + \dfrac{1}{R_B}\right)}{1 + \dfrac{h_{ie}}{R_B}}\right]$ $\quad R_B, R_E$

71. $R_i = R_1\dfrac{\mu + R_a\left(\dfrac{1}{R_1} + \dfrac{1}{R_2} + \dfrac{1}{R_3}\right)}{1 + R_a\left(\dfrac{1}{R_1} + \dfrac{1}{R_2} + \dfrac{1}{R_3}\right)}$ $\quad R_a$

72. $MH = \dfrac{4\pi r^2}{T^2\left(1 + \dfrac{\alpha}{\dfrac{1}{2}\pi - \alpha}\right)}$ $\quad \alpha, \pi$

73. $$\frac{\alpha - \dfrac{\pi}{\alpha - \beta}}{\alpha + \dfrac{\pi}{\alpha - \beta}} - 1 = \frac{\alpha}{\beta} \qquad \pi, \beta$$

74. The force between two magnetic poles of strength S_1 and S_2 at a separation of d cm is

$$F = \frac{10S_1S_2}{d^2} \text{ micronewtons} \qquad (\mu\text{N})$$

When the poles are separated by a distance of 20 cm, a force of 100 μN exists between them. $S_2 = 50$ units. What is the value of S_1?

75. The force acting to close the air gap of a simple electromagnetic relay is

$$F = \frac{B^2A}{2\mu} \quad \text{N}$$

What will be the value of A, the cross-sectional area of the gap, in square meters, which will permit a flux density B of 48×10^3 webers per square meter (Wb/m²) to exert a force F of 128 N? μ, the permeability of air, is $4\pi \times 10^{-7}$ SI units.

76. When two impedances Z_1 and Z_2 are connected in parallel, the resultant joint impedance Z_p is

$$Z_p = \frac{Z_1Z_2}{Z_1 + Z_2}$$

Solve for Z_2.

77. Using the formula given in Prob. 76, what is the value of Z_2 when $Z_p = 5\ \Omega$ and $Z_1 = 12\ \Omega$?

78. $\dfrac{N_p}{N_s} = \dfrac{V_p}{V_s}$; $V_p = 100$, $V_s = 20$, $N_p = 400$. Find the value of N_s.

79. $\dfrac{V_1}{V_2} = \dfrac{R_1}{R_2}$; $V_1 = 20.8$ V, $V_2 = 41$ V, $R_1 = 34.8\ \Omega$. What is R_2?

80. Corresponding temperature readings in Fahrenheit degrees (°F) can be obtained from a Celsius thermometer by the use of the formula $F = \dfrac{9}{5}C + 32$, where C is the temperature in degrees Celsius. When the temperature is 77°F, what is the Celsius temperature?

81. Use the formula given in Prob. 80 to find the temperature at which the Fahrenheit and Celsius temperatures are equal, that is, at which $F = C$.

82. $L_t = L_0 + L_0\alpha t$. If $L_t = 20$, $\alpha = 8.33 \times 10^{-2}$, and $t = 24$, what is the value of L_0?

83. $R_t = R_0\,(1 + 0.0042(\Delta t))\ \Omega$. What is the resistance R_0 at 0°C, if, with a change in temperature Δt of 65°C, the resistance $R_t = 48\ \Omega$?

84. $P = \dfrac{LI^2}{2}$. The energy P stored in a circuit is 1250 joules (J). If the current $I = 2.5$ A, find the value of the coefficient of self-induction L.

85. When two capacitors C_1 and C_2 are connected in series, the resultant total capacitance can be computed by means of the equation

$$\frac{1}{C_s} = \frac{1}{C_1} + \frac{1}{C_2}$$

If $C_s = 5$ pF and $C_2 = 50$ pF, what is the value of C_1?

86. The joint conductance $\dfrac{1}{R_p}$ siemens of three resistances connected in parallel is expressed by

$$\frac{1}{R_p} = \frac{1}{R_1} + \frac{1}{R_2} + \frac{1}{R_3}$$

Solve for R_p.

87. A lens formula is $\dfrac{1}{f} = \dfrac{1}{p} + \dfrac{1}{q}$. What is the value of p when $q = 120$ and $f = 60$?

88. Use the lens equation that was given in Prob. 87 to find the image distance q when the focal length $f = 10$ cm and the object distance $p = 40$ cm.

89. $P = \dfrac{V^2}{R}$. (*a*) How is the value of P changed when V is doubled? (*b*) How is the value of P changed when R is doubled?

90. A source of emf consists of n cells in parallel, and each cell has an emf of V V and an internal resistance of r Ω. The current that flows through a load of R Ω is given by the relation

$$I = \frac{V}{R + \dfrac{r}{n}} \quad \text{A}$$

Solve for r and R.

91. Use the formula stated in Prob. 90 to find the value of R when $V = 1.2$ V, $r = 0.03$ Ω, $I = 2$ A, and $n = 4$ cells.

92. Use the formula stated in Prob. 90 to find n in terms of I and V when $R = 32$ Ω and $r = 0.1$ Ω.

93. A source of emf consists of n cells in series, and each cell has an emf of V V and an internal resistance of r Ω. The current flowing through a load of R Ω is given by the relation

$$I = \frac{nV}{R + nr} \quad \text{A}$$

Solve for R and n.

94. Use the formula stated in Prob. 93 to find the number of identical cells of internal resistance $r = 0.6\ \Omega$ each, if they provide an emf of $V = 2.1$ V each, when they drive a current $I = 2$ A through a load $R = 4.5\ \Omega$.

95. A signal voltage v_g is supplied to the gate of an FET which has a voltage gain expressed as the ratio of drain voltage v_d to gate signal voltage v_g. The gain can be expressed as follows:

$$\text{Gain} = \frac{i_d r_d}{r_s + \dfrac{1}{g_m}} = \frac{v_d}{v_g}$$

where g_m is the transconductance in siemens and r_s is the ac source resistance in ohms. Solve for r_s and g_m.

96. Use the formula in Prob. 95 to find the value of i_d if $r_s = 400\ \Omega$, $r_d = 5\ \text{k}\Omega$, and $g_m = 0.005$ S, producing a gain of 8.33.

97. Does $\dfrac{IR + V}{R} = \text{I} + \text{V}$? Explain your answer.

98. If $I = \dfrac{V}{R_1 + R_2 + R_3}$, does $R_3 = \dfrac{V}{R_1 + R_2 + I}$? Explain your answer.

99. $S = v_0 t + \frac{1}{2}gt^2$. What is the value of the initial velocity v_0 in terms of S, g, and t?

100. Using the formula stated in Prob. 99, what is the acceleration due to gravity g if the initial velocity $v_0 = 3$ m/s, $S = 520.5$ m, and $t = 10$ s?

101. A radiosonde is dropped from an airplane and falls freely until its parachute opens. Twelve seconds after its parachute opens, it has fallen an additional 1.2 km. What was its velocity when the parachute opened? Use your value of g from Prob. 100.

102. If

$$\frac{a}{b} = \frac{a - \dfrac{x}{a - b}}{a + \dfrac{x}{a - b}} - 1$$

what is the value of x when $b = 4.62$ and $a = 3$?

103. The transconductance g_m of a JFET can be found by dividing the incremental changes in signal drain current i_d by the signal gate voltage v_{gs}. That is,

$$g_m = \frac{\Delta i_d}{v_{gs}}$$

Solve for v_{gs} and write an expression to explain Δi_d.

104. Use the formula in Prob. 103 to determine the value of v_{gs} when $\Delta i_d = 7.5\ \mu A$ and $g_m = 7500\ \mu S$.

105. $I_C = \dfrac{V_{CC} - V_C}{R_C}$. What will be the value of V_{CC} when $I_C = 4.9$ mA and the collector voltage V_C measured with respect to ground is -10 V? R_C has a value of 5 kΩ. With respect to the data given and the calculated values, is this transistor type PNP or NPN?

106. $V = L\dfrac{I_1 - I_2}{t}$. What is the change in current when a voltage $V = 1.8$ kV is induced in an inductance $L = 6$ H in time $t = 0.2$ s?

107. $I = C\dfrac{V_1 - V_2}{t}$. What is the change of voltage which will produce a current flow of $I = 150$ A during the discharge of a 50-μF capacitor in 100 μs?

108. $R_a = \dfrac{R_1R_3}{R_1 + R_2 + R_3}$. Three resistances $R_1 = ?$, $R_2 = 3\ \Omega$, and $R_3 = 2.14\ \Omega$ are connected in delta to produce an equivalent Y-circuit branch $R_a = 0.6\ \Omega$. Find R_1.

109. In transistor parameters, $\beta = \dfrac{\alpha}{1 - \alpha}$. Solve for α in terms of β.

110. Using the formula stated in Prob. 109, what is α when $\beta = 284.7$?

SELF TEST

1. Solve for θ: $\dfrac{\theta + 2}{3} - \dfrac{\theta - 4}{2} = 2$

2. Solve for z: $\dfrac{5z + 0.3}{1.5} - \dfrac{3.5z - 0.6}{1.2} = 1.2$

3. Solve for ω: $\dfrac{\omega + 1}{3\omega} - \dfrac{4\omega^2 + 2}{12\omega^2} = \dfrac{1}{4\omega}$

4. Solve for ϕ: $\dfrac{3}{4 + \phi} = \dfrac{1}{12 - \phi}$

5. A beginning student can assemble a multichassis electronics kit by herself in 22 h. An experienced apprentice can do the same job alone in 16 h. How long should the student and apprentice be able to do the job working together?

6. A storage tank can be filled by one pipe in 16 h and by a second pipe in 12 h. The full tank can be drained completely in 30 h. At startup, all three pipes are opened simultaneously. How long will it take to fill the tank completely?

7. A 40-L radiator is full of 4% antifreeze. How much coolant must be removed and replaced by 100% antifreeze to result in a radiator full of 60% antifreeze?

8. Solve for t_2: $\frac{R_2}{R_1} = \frac{234.5 + t_2}{234.5 + t_1}$

9. Solve for L_2: $L_p = \frac{L_1 L_2}{L_1 + L_2}$

10. Solve for d_2: $\alpha = 1 + \frac{1}{\mu_o}\left(1 + 1.5\frac{d_2}{d_1}\right)$

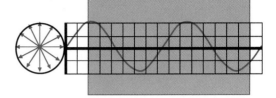

CHAPTER 13

OHM'S LAW—PARALLEL CIRCUITS

Most of the systems employed for the distribution of electric energy consist of parallel circuits; that is, a source of emf is connected to a pair of conductors, known as *feeders,* and various types of load are connected across the feeders. A simple distribution circuit consisting of a motor and a bank of five lamps is represented schematically in Fig. 13–1 and pictorially in Fig. 13–2. The motor

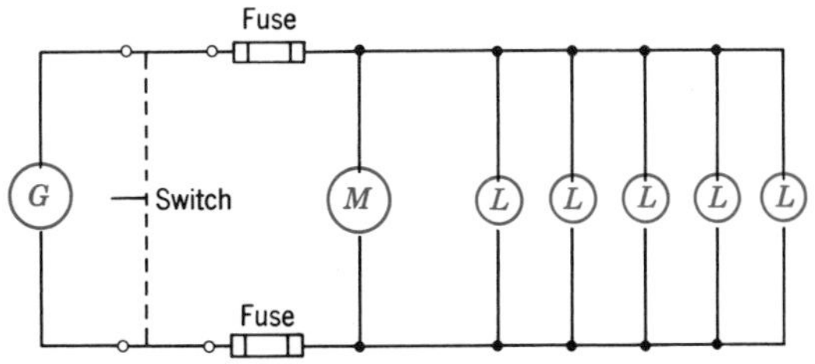

FIG. 13–1 Schematic diagram of a generator G connected to a motor *M* in parallel with a bank of five lamps *L*.

and the lamps are said to be in *parallel,* and it is evident that the current supplied by the generator divides between the motor and the lamps.

In this chapter you will analyze parallel circuits and solve parallel circuit problems. The solution of a parallel circuit generally consists in reducing the entire circuit to a single equivalent resistance which could replace the original circuit without any change in the supply voltage or current.

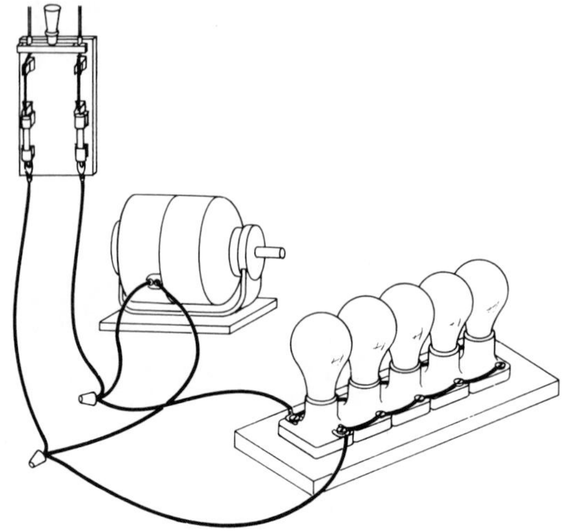

FIG. 13–2 Illustration of circuit shown in Fig. 13–1.

13–1 TWO RESISTANCES IN PARALLEL

The schematic diagram of Fig. 13–3 and the accompanying circuit shown in Fig. 13–4 represent two resistors R_1 and R_2 connected in parallel across a source of voltage V. An examination of the circuit arrangement brings out two important facts:

1. The same voltage exists across the two resistors.
2. The total current I_t delivered by the generator enters the paralleled resistors at junction a, divides between the resistors, and leaves the parallel circuit at junction b. Thus, the sum of the currents I_1 and I_2, which flow through R_1 and R_2, respectively, is equal to the total current I_t.

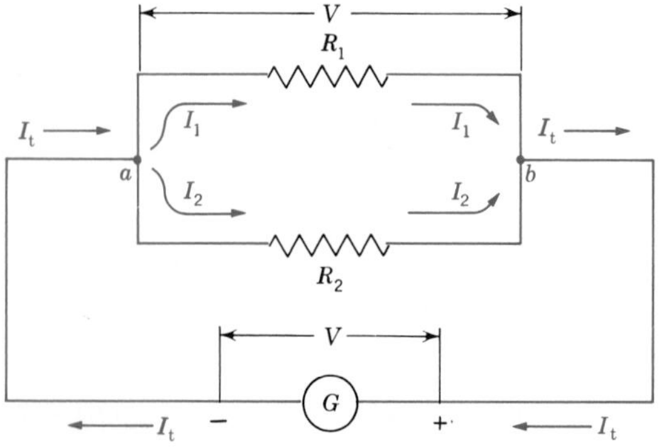

FIG. 13–3 Resistors R_1 and R_2 connected in parallel across generator G, which maintains a potential of V V.

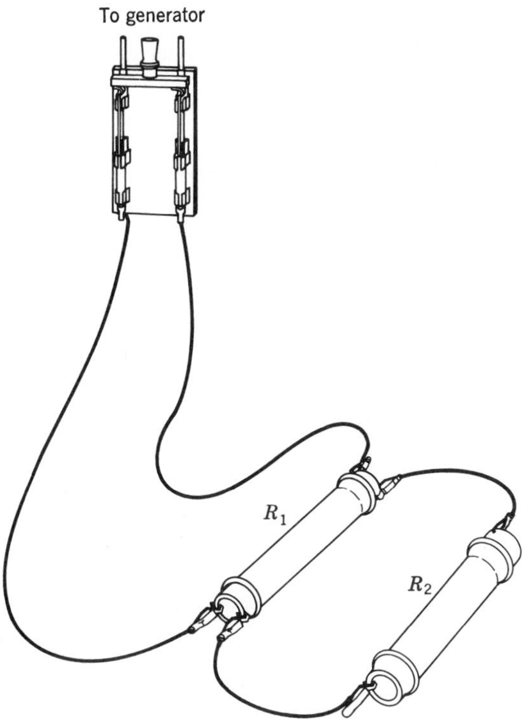

FIG. 13–4 Illustration of schematic circuit shown in Fig. 13–3.

By making use of these facts and applying Ohm's law, it is easy to derive equations that show how paralleled resistances combine. From 1 above,

$$I_1 = \frac{V}{R_1} \qquad I_2 = \frac{V}{R_2} \qquad \text{and} \qquad I_t = \frac{V}{R_p}$$

where R_p is the total resistance of R_1 and R_2, or the equivalent resistance of the parallel combination. From 2 above,

$$I_t = I_1 + I_2 \tag{1}$$

Substituting in Eq. (1) the value of the currents,

$$\frac{V}{R_p} = \frac{V}{R_1} + \frac{V}{R_2}$$

D: V,

$$\frac{1}{R_p} = \frac{1}{R_1} + \frac{1}{R_2} \tag{2}$$

Equation (2) states that the total conductance (Sec. 7–5) of the circuit is equal to the sum of the parallel conductances of R_1 and R_2; that is,

$$G_t = G_1 + G_2 \tag{3}$$

It is evident, therefore, that, when resistances are connected in parallel, each additional resistance represents another path (conductance) for current. Hence, increasing the number of resistances in parallel increases the total conductance of the circuit and thus decreases the equivalent resistance of the circuit.

> If your calculator has a $\frac{1}{x}$ key, you will find it extremely helpful when evaluating parallel circuits.

EXAMPLE 1 What is the total resistance of the circuit of Fig. 13–3 if $R_1 = 6\ \Omega$ and $R_2 = 12\ \Omega$?

SOLUTION 1

Given $R_1 = 6\ \Omega$ and $R_2 = 12\ \Omega$, $R_p = ?$

Substituting the known values in Eq. (2),

$$\frac{1}{R_p} = \frac{1}{6} + \frac{1}{12} = 0.1667 + 0.0833$$

or

$$\frac{1}{R_p} = 0.250$$

Solving for R_p,

$$R_p = \frac{1}{0.250} = 4.0\ \Omega$$

SOLUTION 2

A more convenient formula for the total resistance of two parallel resistances is obtained by solving Eq. (2) for R_p. Thus,

$$R_p = \frac{R_1 R_2}{R_1 + R_2} \qquad (4)$$

Hence, the total resistance of two resistances in parallel is equal to the product of the resistances divided by the sum.

Substituting the values of R_1 and R_2 in Eq. (4),

$$R_p = \frac{6 \times 12}{6 + 12} = \frac{72}{18} = 4.0\ \Omega$$

Thus, the paralleled resistors R_1 and R_2 are equivalent to a single resistance of 4.0 Ω. Note that the total resistance is *less* than either of the resistances in parallel.

EXAMPLE 2 (*a*) Give the total resistance of the circuit of Fig. 13–3 if $R_1 = 21\ \Omega$ and $R_2 = 15\ \Omega$. (*b*) If the generator supplies 12 V across points *a* and *b*, what is the generator (line) current?

SOLUTION 1

(*a*) $$R_p = \frac{R_1R_2}{R_1 + R_2} = \frac{21 \times 15}{21 + 15} = 8.75\ \Omega$$

(*b*) $$I_t = \frac{V}{R_t} = \frac{12}{8.75} = 1.371\ \text{A}$$

SOLUTION 2

Since 12 V exists across both resistors, the current through each can be found and the two currents can be added to obtain the total current. Thus,

Current through R_1, $I_1 = \frac{V}{R_1} = \frac{12}{21} = 0.571\ \text{A}$

Current through R_2, $I_2 = \frac{V}{R_2} = \frac{12}{15} = 0.8\ \text{A}$

Total current, $I_t = I_1 + I_2 = 0.571 + 0.8 = 1.371\ \text{A}$

Hence, $R_p = \frac{V}{I_t} = \frac{12}{1.371} = 8.75\ \Omega$

From the foregoing, it is evident that R_1 and R_2 could be replaced by a single resistor of 8.75 Ω, connected between *a* and *b*, and the generator would be working under the same load conditions. Also, it is apparent that when a current enters a junction of resistors connected in parallel, the current divides between the branches in inverse proportion to the resistances; that is, the greatest current flows through the least resistance.

EXAMPLE 3 In the circuit shown in Fig. 13–3, $R_1 = 25\ \Omega$, $V = 220$ V, and $I_t = 14.3$ A. What is the resistance of R_2?

SOLUTION 1

The current through R_1 is

$$I_1 = \frac{V}{R_1} = \frac{220}{25} = 8.8\ \text{A}$$

Since $$I_t = I_1 + I_2$$

the current through R_2 is

$$I_2 = I_t - I_1 = 14.3 - 8.8 = 5.5\ \text{A}$$

Then $$R_2 = \frac{V}{I_2} = \frac{220}{5.5} = 40\ \Omega$$

SOLUTION 2

$$R_p = \frac{V}{I_t} = \frac{220}{14.3} = 15.4\ \Omega$$

Solving Eq. (2) or (4) for R_2,

$$R_2 = \frac{R_1 R_p}{R_1 - R_p} = \frac{25 \times 15.4}{25 - 15.4} = 40\ \Omega$$

PROBLEMS 13–1

1. Two 470-Ω resistors are connected in parallel. What is the equivalent resistance?
2. Two resistors, one of 3600 Ω and the other of 6800 Ω, are connected in parallel. What is the equivalent resistance of the combination?
3. What is the total resistance of 22 kΩ in parallel with 47 kΩ?
4. What is the equivalent resistance of 33 kΩ in parallel with 8.2 kΩ?
5. What is the equivalent resistance of:
   (*a*) Two 1200-Ω resistors in parallel?
   (*b*) Two 8.2-kΩ resistors in parallel?
   (*c*) Two 18-Ω resistors in parallel?
6. State a general formula for the total resistance R_p of two equal resistances of R Ω connected in parallel.
7. In the circuit of Fig. 13–3, how much generator voltage would be required to deliver a total current of 1.11 A through a parallel combination of $R_1 = 180\ \Omega$ and $R_2 = 270\ \Omega$?
8. How much power would be absorbed by the 270-Ω resistor of Prob. 7?
9. In the circuit shown in Fig. 13–3, $I_t = 20.3$ mA, $V = 220$ V, and $R_1 = 12$ kΩ. What is the resistance of R_2?
10. How much power is dissipated by R_1 of Prob. 9?
11. How much total power is drawn from the generator of Prob. 9?
12. In the circuit of Fig. 13–3, $R_1 = 18$ kΩ and the current through R_2 is 14.71 mA. A total current $I_t = 70.27$ mA flows through the parallel combination. What is the resistance of R_2?
13. How much power is expended in R_2 of Prob. 12?
14. How much power is drawn from the generator of Prob. 12?
15. What is the generated voltage of Prob. 12?

13–2 THREE OR MORE RESISTANCES IN PARALLEL

The procedure for deriving a general equation for the total resistance of three or more resistances in parallel is the same as that of the preceding section. For example, Fig. 13–5 represents three resistors R_1, R_2, and R_3 connected in parallel across a source of voltage V. The total line current I_t splits at junction a into currents I_1, I_2, and I_3, which flow through R_1, R_2, and R_3, respectively. Then

$$I_1 = \frac{V}{R_1} \qquad I_2 = \frac{V}{R_2}$$

$$I_3 = \frac{V}{R_3} \qquad I_t = \frac{V}{R_p}$$

where R_p is the total resistance of the parallel combination.

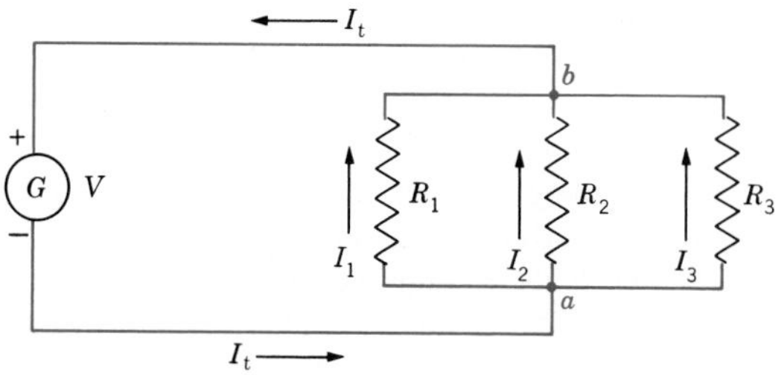

FIG. 13–5 Resistors R_1, R_2, and R_3 connected in parallel.

Since $$I = I_1 + I_2 + I_3$$

by substituting, $$\frac{V}{R_p} = \frac{V}{R_1} + \frac{V}{R_2} + \frac{V}{R_3}$$

D: V, $$\frac{1}{R_p} = \frac{1}{R_1} + \frac{1}{R_2} + \frac{1}{R_3} \qquad (5)$$

From Eq. (5), it is evident that the total conductance of the circuit is equal to the sum of the paralleled conductances of R_1, R_2, and R_3; that is,

$$G_p = G_1 + G_2 + G_3$$

In like manner, it can be demonstrated that the total resistance R_p of any number of resistances connected in parallel is

$$\frac{1}{R_p} = \frac{1}{R_1} + \frac{1}{R_2} + \frac{1}{R_3} + \frac{1}{R_4} + \frac{1}{R_5} + \ldots$$

Or, in terms of conductances,

$$G_p = G_1 + G_2 + G_3 + G_4 + G_5 + \ldots$$

EXAMPLE 4 What is the total resistance of the circuit of Fig. 13–5 if $R_1 = 5\ \Omega$, $R_2 = 10\ \Omega$, and $R_3 = 12.5\ \Omega$?

SOLUTION

Substituting the known values in Eq. (5),

$$\frac{1}{R_p} = \frac{1}{5} + \frac{1}{10} + \frac{1}{12.5}$$
$$= 0.2 + 0.1 + 0.08$$

or $$\frac{1}{R_p} = 0.38$$

Solving for R_p, $$R_p = \frac{1}{0.38} = 2.63\ \Omega$$

If Eq. (5) is solved for R_p, the result is

$$R_p = \frac{R_1R_2R_3}{R_1R_2 + R_1R_3 + R_2R_3} \qquad (6)$$

It is seen that Eq. (6) is somewhat cumbersome for computing the total resistance of three resistances connected in parallel. However, you should recognize such expressions for three or more resistances in parallel, for you will encounter them in the analysis of networks.

Finding the total resistance of any number of resistors in parallel is facilitated by arbitrarily assuming a voltage to exist across the parallel combination. The currents through the individual branches are added to obtain the total line current. The assumed voltage divided by the total current results in the total resistance of the combination.

In order to avoid decimal quantities, that is, currents of less than 1 A, the assumed voltage should be numerically greater than the highest resistance of any parallel branch.

EXAMPLE 5 Three resistances $R_1 = 10\ \Omega$, $R_2 = 15\ \Omega$, and $R_3 = 45\ \Omega$ are connected in parallel. Find their total resistance.

SOLUTION

Assume $V_a = 100$ V to exist across the combination.

Current through R_1, $I_1 = \frac{V_a}{R_1} = \frac{100}{10} = 10\text{ A}$

Current through R_2, $I_2 = \frac{V_a}{R_2} = \frac{100}{15} = 6.67\text{ A}$

Current through R_3, $I_3 = \frac{V_a}{R_3} = \frac{100}{45} = 2.22\text{ A}$

Total current, $I_t = 18.89\text{ A}$

Total resistance, $R_p = \frac{V_a}{I_t} = \frac{100}{18.89} = 5.3\ \Omega$

PROBLEMS 13–2

1. What is the equivalent resistance of 18 Ω, 27 Ω, and 47 Ω connected in parallel?
2. What is the total resistance of 180 Ω, 470 Ω, and 680 Ω connected in parallel?
3. Three resistors of 15 Ω, 470 Ω, and 6.8 Ω are connected in parallel. What is their total resistance?
4. Three resistors of 27 Ω, 270 Ω, and 2700 Ω are connected in parallel. Find the total resistance of the combination.
5. What is the equivalent resistance of 220 Ω, 150 Ω, 680 Ω, and 820 Ω connected in parallel?
6. Four resistors of 11 Ω, 16 Ω, 22 Ω, and 47 Ω are connected in parallel. What is the total resistance of the combination?
7. What is the total resistance of:
   (*a*) Three 1.8-kΩ resistors in parallel?
   (*b*) Four 12-kΩ resistors in parallel?

8. What is the total resistance of:
   (*a*) Three 100-kΩ resistors connected in parallel?
   (*b*) Four 100-kΩ resistors connected in parallel?
   (*c*) Five 100-kΩ resistors connected in parallel?
9. State a general formula for the resistance R_p of n equal resistances of R Ω each connected in parallel.
10. In the circuit of Fig. 13–5, the total current $I_t = 18.03$ A, $R_1 = 100$ Ω, $R_2 = 150$ Ω, and $V = 475$ V. What is the resistance of R_3?
11. If the values of Prob. 10 are used, what is the power delivered to the 150-Ω resistor?
12. What would be the resistance in Prob. 10 if the 150-Ω resistor were shorted out?
13. In the circuit of Fig. 13–5, $R_1 = 15$ Ω, $R_2 = 18$ Ω, $I_3 = 15.25$ A, and $V = 100$ V. Find (*a*) the value of R_3 to two significant figures and (*b*) the total power delivered to the circuit.
14. In the circuit of Fig. 13–5, $R_2 = 510$ Ω, $R_3 = 270$ Ω, $I_t = 4.38$ A, and $I_1 = 1.52$ A. Find the value of R_1 to two significant figures.
15. In the circuit of Fig. 13–5, $R_1 = R_2 = 2.2$ kΩ, and R_3 is disconnected. $I_t = 156$ mA. What must be the value of R_3 connected into the circuit to result in a total current of 0.50 A?
16. A 10-kΩ 100-W resistor, a 15-kΩ 50-W resistor, and a 100-kΩ 10-W resistor are connected in parallel.
    (*a*) What is the maximum voltage which may be applied without exceeding the rating of any resistor?
    (*b*) What is the total current drawn by the combination when the voltage of part (*a*) is applied?

13–3 COMPOUND CIRCUITS

The solution of circuits containing combinations of series and parallel branches generally consists in reducing the parallel branches to equivalent series circuits and combining them with the series branches. No set rules can be formulated for the solution of all types of such circuits, but from the examples that follow you will be able to build up your own methods of attack.

EXAMPLE 6 Find the total resistance of the circuit represented in Fig. 13–6.

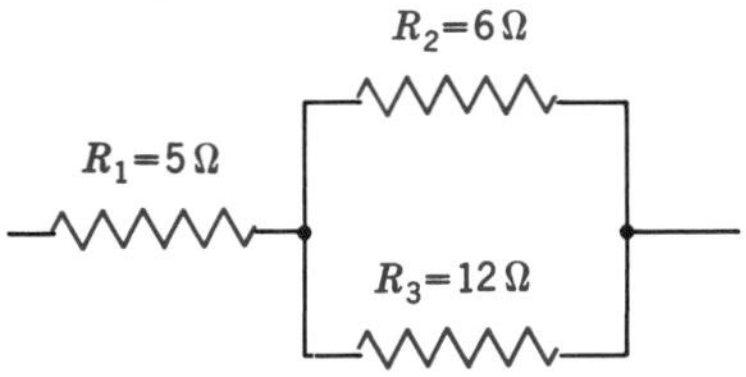

FIG. 13–6 Series-parallel circuit of Example 6.

SOLUTION

Note that the parallel branch of Fig. 13–6 is the circuit of Example 1. Since the equivalent series resistance of the parallel branch is

$$\frac{R_2R_3}{R_2 + R_3}$$

the circuit reduces to two resistances in series, the total resistance of which is

$$R_t = R_1 + \frac{R_2R_3}{R_2 + R_3} = 5 + \frac{6 \times 12}{6 + 12} = 9.0\ \Omega$$

> Calculator users may find it useful to try
>
> $$R_t = R_1 + \frac{1}{\frac{1}{R_2} + \frac{1}{R_3}}$$
>
> and use the $\frac{1}{x}$ key.

EXAMPLE 7 Find the total resistance of the circuit represented in Fig. 13–7.

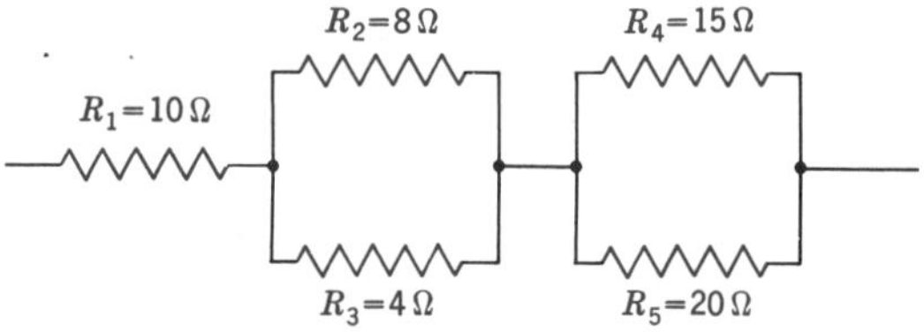

FIG. 13–7 Circuit of Example 7, consisting of one resistance in series with two parallel branches.

SOLUTION

The circuit of Fig. 13–7 is similar to that shown in Fig. 13–6, but with an additional parallel branch. By utilizing the expression for the total resistance of two resistances in parallel, the entire circuit reduces to three resistances in series, the total resistance of which is

$$\begin{aligned} R_t &= R_1 + \frac{R_2R_3}{R_2 + R_3} + \frac{R_4R_5}{R_4 + R_5} \\ &= 10 + \frac{8 \times 4}{8 + 4} + \frac{15 \times 20}{15 + 20} \\ &= 21.2\ \Omega \end{aligned}$$

EXAMPLE 8 Find the total resistance between points a and b in Fig. 13–8.

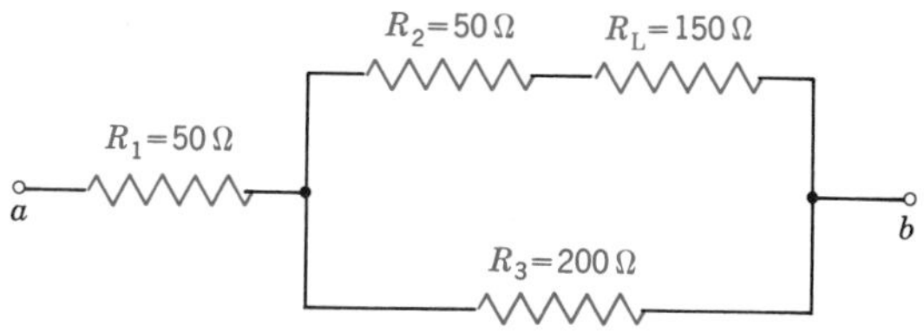

FIG. 13–8 Circuit of Example 8.

SOLUTION

Since R_2 and R_L are in series, they must be added before being combined with R_3. Again, by utilizing the expression for the total resistance of two resistances in parallel, the entire circuit reduces to two resistances in series. Thus, the total resistance is

$$\begin{aligned} R_t &= R_1 + \frac{R_3(R_2 + R_L)}{R_3 + (R_2 + R_L)} \\ &= 50 + \frac{200(50 + 150)}{200 + 50 + 150} \\ &= 150\ \Omega \end{aligned}$$

Note that the circuit of Fig. 13–8 is identical with that of Fig. 13–9. The latter is the customary method for representing T networks, often encountered in communication circuits, where R_L is the load or receiving resistance.

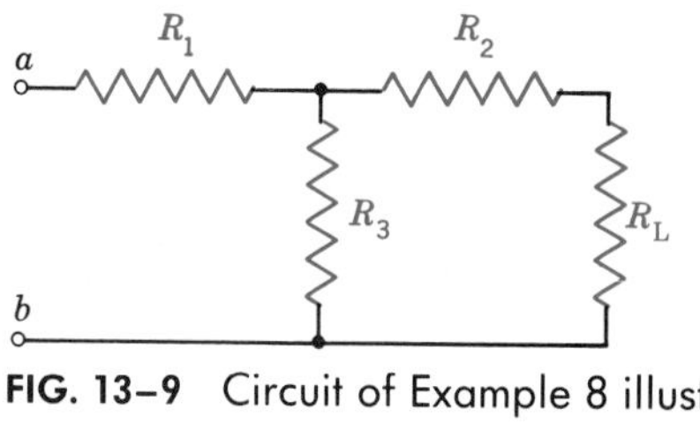

FIG. 13–9 Circuit of Example 8 illustrated in T-network form.

EXAMPLE 9 Find the resistance between points a and b in Fig. 13–10.

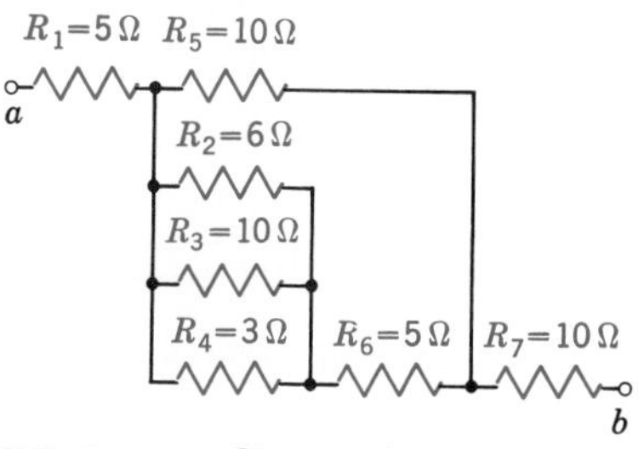

FIG. 13–10 Circuit of Example 9.

SOLUTION

In many instances a circuit diagram that *appears* to be complicated can be better understood and analyzed by redrawing it in a form which is more simplified. For example, Fig. 13–11 represents the circuit of Fig. 13–10.

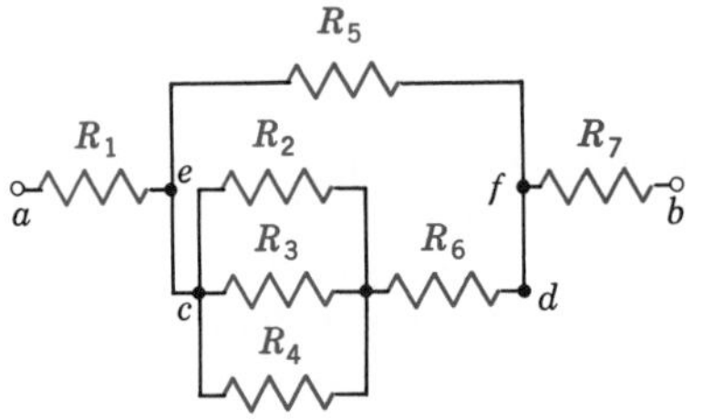

FIG. 13–11 Simplified circuit of Example 9.

First, find the equivalent series resistance of the parallel group formed by R_2, R_3, and R_4 and add it to R_6, which will result in the resistance R_{cd} between points c and d. Now, combine R_{cd} with R_5, which is in parallel, to give an equivalent series resistance R_{ef} between points e and f. The circuit is now reduced to an equivalence of R_1, R_{ef}, and R_7 in series, which are added to obtain the total resistance R_{ab} between points a and b. The total resistance of R_2, R_3, and R_4 is 1.67 Ω, which, when added to R_6, results in a resistance $R_{cd} = 6.67\ \Omega$ between c and d. The total series resistance R_{ef} between points e and f, formed by R_{cd} and R_5 in parallel, is 4.0 Ω. Therefore, the resistance R_{ab} between points a and b is

$$R_{ab} = R_1 + R_{ef} + R_7 = 19\ \Omega$$

PROBLEMS 13–3

1. In the circuit of Fig. 13–12, $R_1 = 360\ \Omega$, $R_2 = 470\ \Omega$, and $R_3 = 160\ \Omega$. $V_G = 120$ V. What is the total current I_t of the circuit?

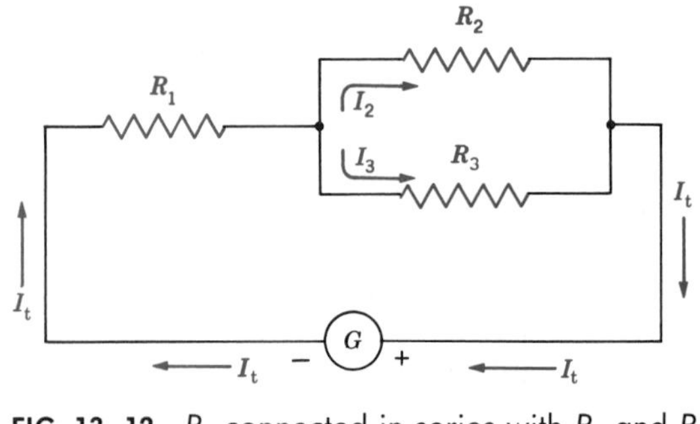

FIG. 13–12 R_1 connected in series with R_2 and R_3 in parallel.

2. In Prob. 1, how much power is expended in R_3?
3. In Prob. 1, if R_1 is short-circuited, how much power is expended in R_2?
4. In Prob. 1, what will be the total current I_t if R_2 is open-circuited?
5. In the circuit of Fig. 13–12, $R_1 = 62$ kΩ, $R_2 = 15$ kΩ, and $I_t = 3.26$ mA and the voltage across R_3 is 27.9 V. Find (*a*) V_G, (*b*) R_3, (*c*) R_t, (*d*) I_2, (*e*) I_3.
6. In Prob. 5, how much current will the generator supply if R_3 is short-circuited?
7. In the circuit of Fig. 13–12, $R_t = 4.979$ kΩ, $R_1 = 2.2$ kΩ, $V_G = 1000$ V, and $I_2 = 82.1$ mA. Find (*a*) voltage across R_1, (*b*) voltage across R_2, (*c*) resistance of R_2 to two significant figures, (*d*) resistance of R_3 to two significant figures, (*e*) total current I_t, (*f*) current through R_3, (*g*) total power expended in the circuit.
8. In Prob. 7, if R_1 is short-circuited, (*a*) how much power will be expended in R_2 and (*b*) what will be the current through R_3?
9. In the circuit of Fig. 13–9, R_1, R_2, and R_3 are 200-Ω resistors and $R_L = 470$ Ω. What is the effective resistance between points *a* and *b*?
10. In the circuit of Fig. 13–9, $R_1 = R_2 = R_3 = 300$ Ω and $R_L = 600$ Ω. What is the resistance between points *a* and *b*?
11. In the circuit of Fig. 13–9, $R_1 = R_2 = R_L = 500$ Ω and $R_3 = 1$ kΩ. What is the resistance between points *a* and *b*?
12. In the circuit of Fig. 13–13, $R_1 = R_2 = R_4 = R_5 = 10$ Ω and $R_3 = R_L = 600$ Ω. If a voltage of 30 V exists across R_L, what is the total current I_t?

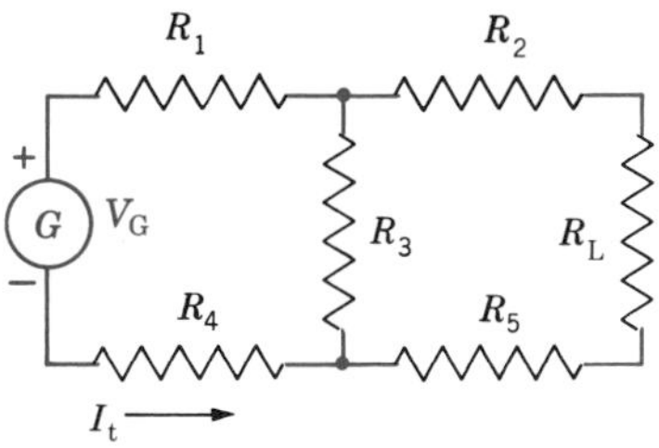

FIG. 13–13 Circuit of Prob. 12.

13. In the circuit of Fig. 13–14 (next page), the generator voltage $V_G = 3500$ V, $R_4 = 1.5$ kΩ, $R_2 = 6.8$ kΩ, $I_2 = 52.9$ mA, $R_3 = 2.7$ kΩ, and $I_t = 273$ mA. Find to two significant figures (*a*) resistance of R_1, (*b*) resistance of R_5, and (*c*) power expended in R_3.
14. In the circuit represented in Fig. 13–15 (next page), find the total current I_t.
15. If, in Fig. 13–15 (next page), points *a* and *b* are short-circuited, find the total power expended.

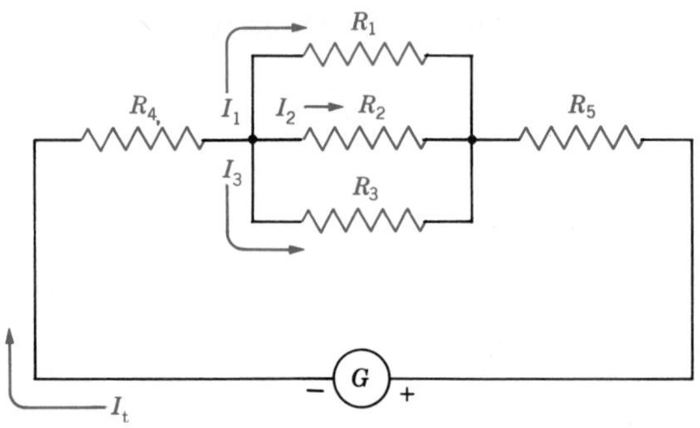

FIG. 13–14 Series-parallel circuit of Prob. 13.

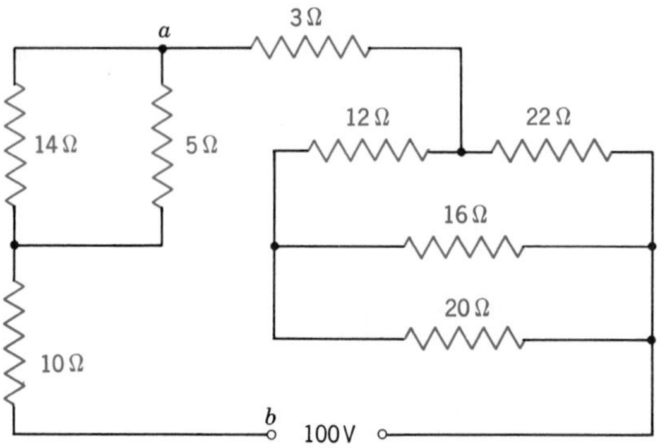

FIG. 13–15 Circuit of Prob. 14.

16. What is the total current I_t in the circuit shown in Fig. 13–16?
17. In the circuit of Prob. 16, what is the current flow through the 5-Ω resistor?
18. What would be the power expended in the circuit of Fig. 13–16 if points *a* and *b* were short-circuited?

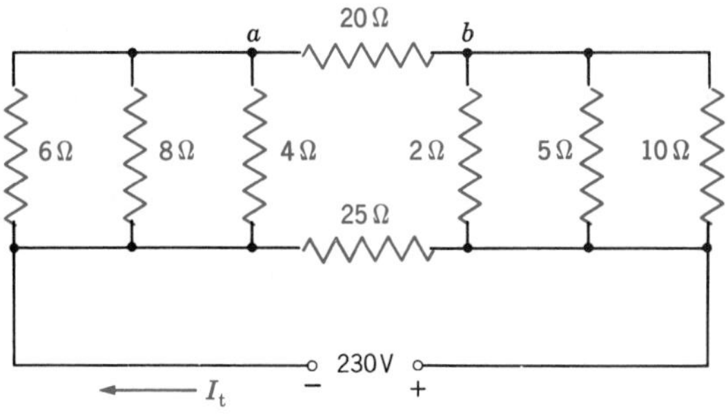

FIG. 13–16 Circuit of Prob. 16.

SELF TEST

1. You require a resistance of 920 Ω. What standard value resistor could you connect in parallel with a 1.2-kΩ resistor to come very close to the desired value?
2. Four resistances, 1800 Ω, 4700 Ω, 6800 Ω, and 120 kΩ, are connected in parallel. What is the resistance of the combination?
3. Three resistors are connected in parallel.
   R_1 is rated 12 kΩ, 50 W.
   R_2 is rated 68 kΩ, 12 W.
   R_3 is rated 120 kΩ, 10 W.
   What is the maximum voltage that may be safely applied without exceeding the rating of any resistor?

CHAPTER 14

METER CIRCUITS

Chapters 8 and 13 dealt with the study of Ohm's law as applied to series and parallel circuits, and in Chap. 9 consideration was given to the effects of resistance in current-carrying conductors. The principles and methods learned therein are applied in the present chapter to circuits relating to *dc instruments* used for servicing electrical, radio, and other electronic equipment.

14–1 DIRECT-CURRENT INSTRUMENTS—BASIC METER MOVEMENT

The most common measuring instruments used with electric and electronic circuits are the *voltmeter* and the *ammeter*. As the names imply, a voltmeter is an instrument used to measure voltage and an ammeter is a current-measuring instrument.

The great majority of meters used with direct currents employ the D'Arsonval movement illustrated in Fig. 14–1. This movement utilizes a coil of wire mounted on jeweled bearings between the pole pieces of a permanent magnet. When direct current flows through the coil, a magnetic field is set up around the coil, thereby producing a force which, in conjunction with the magnetic field of the permanent magnet, causes the coil to rotate from the no-current position. Since the arc of rotation is proportional to the amount of current passing through the coil, a pointer can be attached to the coil and the deflection of the pointer over a calibrated scale can be used to indicate values of current.

The *sensitivity* of a current-indicating meter is the amount of current necessary to cause full-scale deflection of the pointer. For example, an instrument of wide usage is the 0–1 milliammeter illustrated in Fig. 14–2. This meter has a sensitivity of 1 mA because, when a current of 1 mA flows through the meter, the pointer indicates full-scale deflection. This particular meter has an internal resistance of 55 Ω. Other meter movements have different sensitivities with various values of internal resistance.

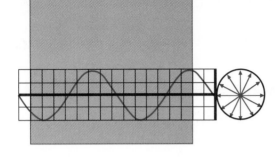

FIG. 14–1 D'Arsonval meter movement. (*Courtesy of Western Electrical Instrument Corporation*)

FIG. 14–2 0–1 milliammeter. (*Courtesy of Triplett Electrical Instrument Company*)

14–2 MULTIRANGE CURRENT METERS

Instead of utilizing a number of meters to make various current measurements, it is common practice to select a meter movement with sufficient sensitivity and, with the aid of one or more shunts, extend the range and therefore the usefulness of the meter. A shunt, in this application, is a resistor that is shunted (connected in parallel) across the meter coil as shown in Fig. 14–3.

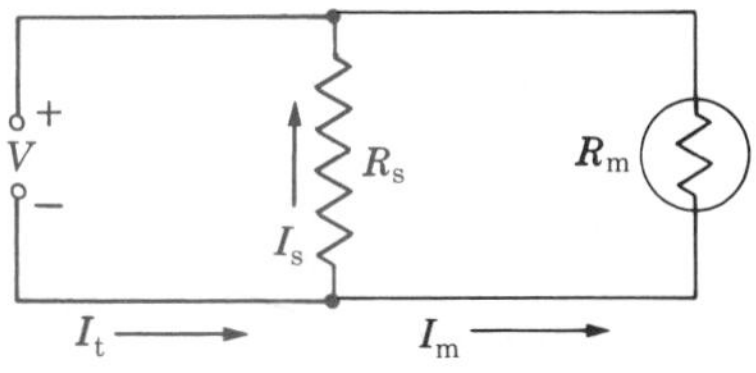

FIG. 14–3 Total current I_t consists of current I_s, which flows through shunt resistor R_s, and the meter current I_m, which flows through the coil of the meter. That is, $I_t = I_s + I_m$.

A meter such as illustrated in Fig. 14–2, with a resistance of 55 Ω, is connected to measure the circuit current of Fig. 14–4. In this condition the switch S is open and the meter indicates a full-scale deflection of 1 mA. In Fig. 14–5 the switch S is closed, thereby shunting the 55-Ω resistor R_s across the meter. Since the meter resistance and shunt resistance are equal, the circuit current I_t divides equally between them and the meter reads 0.5 mA.

In Fig. 14–4, with the switch open, the meter would indicate actual values of current. In Fig. 14–5, with the switch closed, circuit current would be obtained by multiplying the meter readings by a factor of 2 or by re-marking the scale as shown in Fig. 14–6.

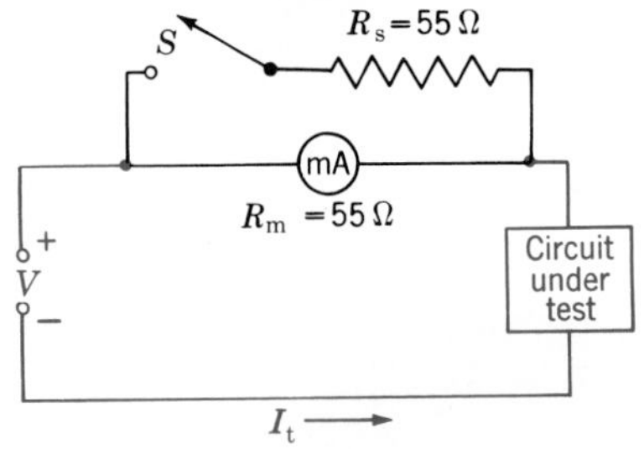

FIG. 14–4 Total current I_t flows through the milliammeter, which indicates a full-scale deflection of 1 mA.

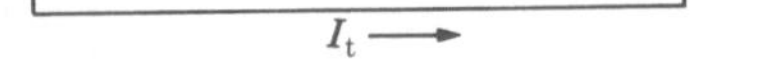

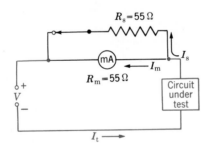

FIG. 14–5 Total current I_t divides equally between meter resistance R_m and shunt resistance R_s. $I_t = I_m + I_s = 1$ mA and $I_s = I_m = 0.5$ mA.

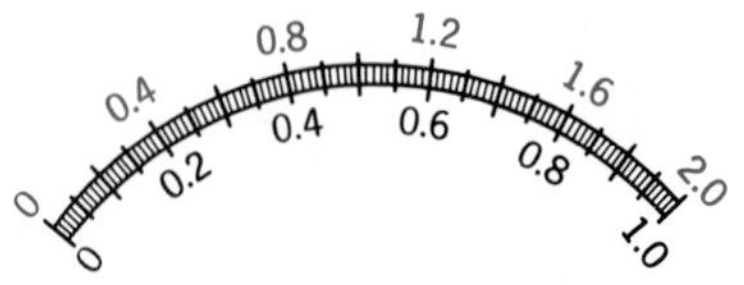

FIG. 14–6 Multirange meter scale.

EXAMPLE 1 A 0–1 milliammeter has an internal resistance of 70 Ω. Design a circuit that will allow this meter to be used as a multirange meter having the ranges 0–1, 0–10, and 0–100 mA and 0–1 A.

SOLUTION

The circuit is shown in Fig. 14–7. The switch S is used for range selection by switching in the proper shunt resistor. In its present position no shunt resistor

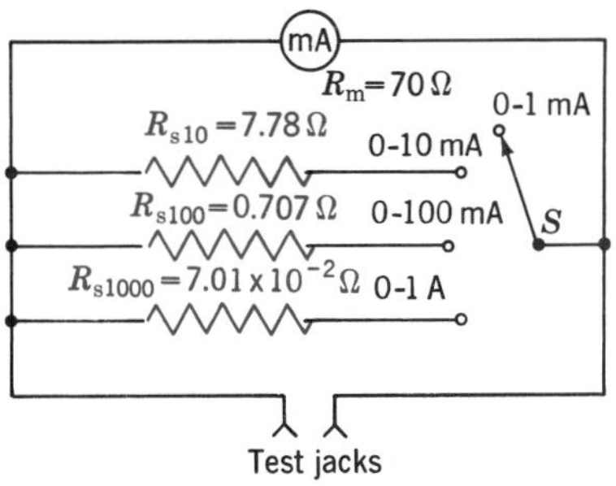

FIG. 14–7 Circuit for extending range of 0–1 milliammeter. Test leads from jacks are connected in series with circuit in which current is to be measured.

is used and therefore the meter is connected to measure within its basic range of 0–1 mA. At full-scale deflection the voltage across the meter will be

$$V_m = I_t R_m = 0.001 \times 70 = 7 \times 10^{-2} \text{ V}$$

Since whatever shunt resistor is in use will be in parallel with the resistance of the meter R_m, the same voltage will appear across the shunt resistance. That is,

$$V_m = V_s = 7 \times 10^{-2} \text{ V}$$

When the 0- to 10-mA range is used, the switch S will connect R_{s10} in parallel with the meter and therefore its internal resistance R_m. For full-scale deflection, 1 mA must flow through the meter coil, which leaves 9 mA to flow through R_{s10}. For this condition the value of R_{s10} must be

$$R_{s10} = \frac{V_s}{I_s} = \frac{7 \times 10^{-2}}{9 \times 10^{-3}} = 7.78 \ \Omega$$

Similarly, when the 0- to 100-mA range is placed in operation by switching to shunt resistor R_{s100}, full-scale deflection 1 mA still must flow through the meter coil, leaving 99 mA to flow through R_{s100}. Then,

$$R_{s100} = \frac{V_s}{I_s} = \frac{7 \times 10^{-2}}{99 \times 10^{-3}} = 0.707 \ \Omega$$

Likewise, when the 0- to 1-A (0- to 1000-mA) range is used, 999 mA must flow through the shunt resistor for full-scale deflection.

$$\therefore R_{s1000} = \frac{V_s}{I_s} = \frac{7 \times 10^{-2}}{999 \times 10^{-3}} = 0.0701\ \Omega$$

It will be noted that only basic Ohm's law was used in Example 1. This was done to emphasize the usefulness of the law. Also, special seldom-used formulas are difficult to remember and handbooks for ready reference are not always available on the job. Actually, you can find the resistance of a meter shunt by using your knowledge of current distribution in parallel circuits. For the 0- to 10-mA range of Example 1, the 70-Ω meter movement must carry 1 mA and the shunt resistor must carry 9 mA. Since the shunt carries nine times the meter current, the shunt resistance must be one-ninth the resistance of the meter, or $\frac{1}{9} \times 70 = 7.78\ \Omega$.

Similarly, for the range of 0 to 100 mA, the meter movement still must carry 1 mA, leaving 99 mA to flow through the shunt. Therefore, the resistance of the shunt will be one ninety-ninth of the resistance of the meter movement, or $\frac{1}{99} \times 70 = 0.707\ \Omega$.

Now that the principles of meter shunts are understood, it is left as an exercise for you to show that

$$R_s = \frac{R_m}{N - 1} \quad \Omega \tag{1}$$

where R_s = shunt resistance, Ω
R_m = meter resistance, Ω
N = ratio obtained by dividing new full-scale reading by basic full-scale reading, both readings in same units

The ratio N is known as the *multiplying power* of the shunt resistor, that is, the factor by which the basic meter scale is multiplied when the shunt resistor R_s is connected in parallel with the meter resistance R_m. From Eq. (1),

$$N = \frac{R_m}{R_s} + 1$$

EXAMPLE 2 By what factor must the scale readings be multiplied when a resistance of 100 Ω is connected across a meter movement of 400 Ω?

SOLUTION

$$N = \frac{R_m}{R_s} + 1 = \frac{400}{100} + 1 = 5$$

14–3 SHUNTING METHODS

Although mechanical details are not shown in Fig. 14–7, it is necessary to use a shorting switch in this type of circuit to avoid damage to the meter movement. When operation requires switching from one shunt to another, the new shunt must be connected before contact with the shunt in use is broken. If it is not, the entire circuit current will flow through the meter movement while the switch is moving from one contact to another.

By another method of switching, illustrated in Fig. 14–8, shunts are connected into the circuit by a two-pole rotary switch *S* which makes connections

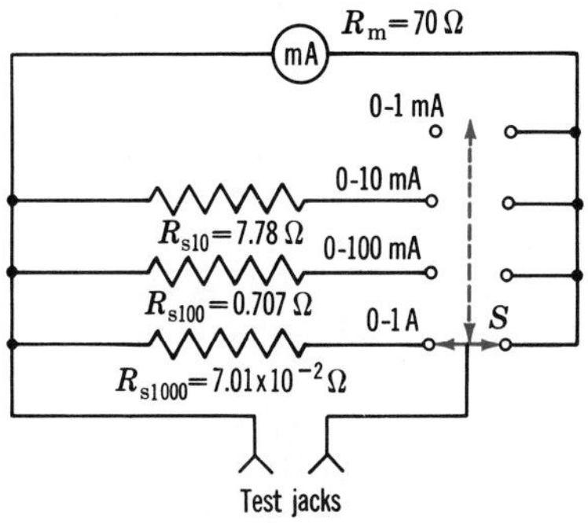

FIG. 14–8 Method of switching shunts.

between two sets of contacts. With this arrangement, the meter movement is protected by an open circuit when the operator switches from one shunt to another.

Still another method of employing shunts is shown in Fig. 14–9. This is known as the *Ayrton,* or *universal,* shunt. In addition to other advantages, it

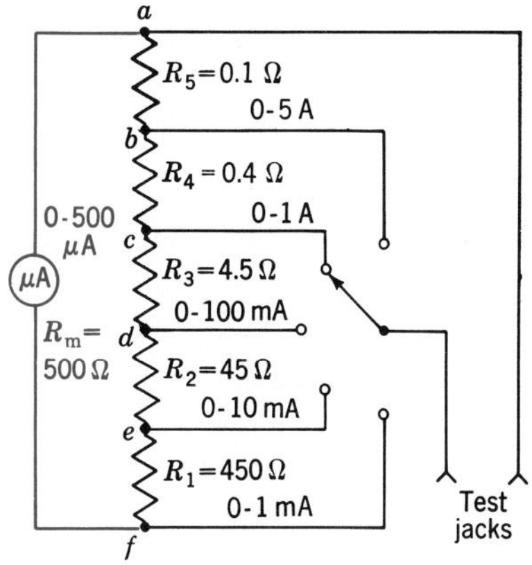

FIG. 14–9 Multicurrent test meter using universal shunt.

provides a safe and convenient method of switching from one range to another. The total shunt resistance, which is permanently connected across the meter, generally has the same resistance as the meter movement. The value of the resistance for each range shunt can be computed by dividing the total circuit resistance $R_{a-f} + R_m$ by the multiplying power *N*. This is demonstrated by the development which follows:

$$\begin{aligned}
R_{a-e}(I_t - I_m) &= (R_{e-f} + R_m)I_m \\
&= (R_{a-f} - R_{a-e} + R_m)I_m \\
R_{a-e}I_t - R_{a-e}I_m &= R_{a-f}I_m - R_{a-e}I_m + R_mI_m \\
R_{a-e}I_t &= R_{a-f}I_m + R_mI_m \\
R_{a-e} &= \frac{I_m}{I_t}(R_{a-f} + R_m) \\
R_{a-e} &= \frac{1}{N}(R_{a-f} + R_m) \qquad (2)
\end{aligned}$$

where R_{a-e} = portion of Ayrton shunt which is connected in shunt for the meter connection at point e. In Fig. 14–9, this is $R_{a-e} = R_2 + R_3 + R_4 + R_5$.
R_{a-f} = total Ayrton shunt (in Fig. 14–9, this is illustrated as $R_1 + R_2 + R_3 + R_4 + R_5$)
R_m = meter movement resistance
N = multiplier for switch setting (in Fig. 14–9, $N = 2$ for setting at f, $N = 20$ for setting at e, and so on)

For example, the 0–500 microammeter movement has a resistance R_m of 500 Ω and total shunt resistance R_{a-f} connected across the meter is 500 Ω. When the switch is on the 0- to 1-mA position, the multiplying power N is 2.

For the 0- to 10-mA range, N would be 20 because 10 mA is 20 times the original full scale of 0.5 mA. Therefore, the required shunt for this range is

$$R_{a-e} = \frac{R_{a-f} + R_m}{N} = \frac{500 + 500}{20} = 50\ \Omega$$

Since the entire shunt resistance is 500 Ω,

$$R_1 = R_{a-f} - R_{a-e} = 500 - 50 = 450\ \Omega$$

When the switch is connected to the 0- to 100-mA range, N becomes 200 and R_1 and R_2 in series (R_{d-f}) form the shunt. That is,

$$R_{a-d} = \frac{R_{a-f} + R_m}{N} = \frac{500 + 500}{200} = 5\ \Omega$$

NOTE $\quad R_{a-d} = \dfrac{2R_m}{N}$ when $R_{a-f} = R_m$

Since $\quad R_1 = 450\ \Omega$ and $R_{a-d} = 5\ \Omega$

then

$$\begin{aligned}
R_2 &= R_{a-f} - (R_1 + R_{a-d}) \\
&= 500 - (450 + 5) \\
&= 45\ \Omega
\end{aligned}$$

The values of the remaining shunts are computed in the same manner.

PROBLEMS 14–1

1. A 0–1 milliammeter has an internal resistance of 47 Ω. What shunt resistance is required to extend the meter range to 0–50 mA?
2. A meter movement with a sensitivity of 100 μA has an internal resistance of 1250 Ω. How much shunt resistance is required to result in a 0- to 10-mA range?
3. The meter in Prob. 1 is being used as a multicurrent instrument. The shunt for the 0- to 50-mA range is burned out, but a spool of No. 30 enamel-covered copper wire is on hand. How much of this wire is needed to wind a substitute shunt?
4. A 0–1 milliammeter has an internal resistance of 42 Ω. If this meter is shunted with a 0.4719-Ω resistor, by what must the meter readings be multiplied to obtain the correct values of current?
5. It is desired to use the milliammeter illustrated in Fig. 14–2 as a multicurrent meter. What values of shunts are required for the following ranges: (*a*) 0–10 mA, (*b*) 0–100 mA, (*c*) 0–1 A, (*d*) 0–10 A?
6. In the circuit of Fig. 14–10, the total shunt resistance is equal to the resistance of the meter movement. Find the values of R_1, R_2, R_3, R_4, and R_5.

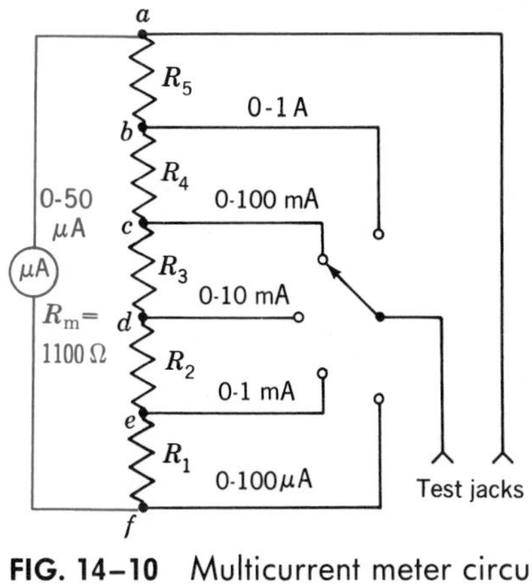

FIG. 14–10 Multicurrent meter circuit of Prob. 6.

7. A 0–1 milliammeter is available. Design an Ayrton shunt to permit it to be used for the following ranges: (*a*) 0–10 mA, (*b*) 0–100 mA, (*c*) 0–1 A, (*d*) 0–10 A. The meter resistance is 1500 Ω.

14–4 VOLTMETERS

In Fig. 14–11, a voltage of 1 V is impressed across a circuit consisting of a 0–1 milliammeter in series with a variable resistor. The resistor is so adjusted that the circuit is limited to 1 mA; therefore, the meter indicates a full-scale deflection, or a reading of 1 mA. If the resistor is unchanged and the voltage

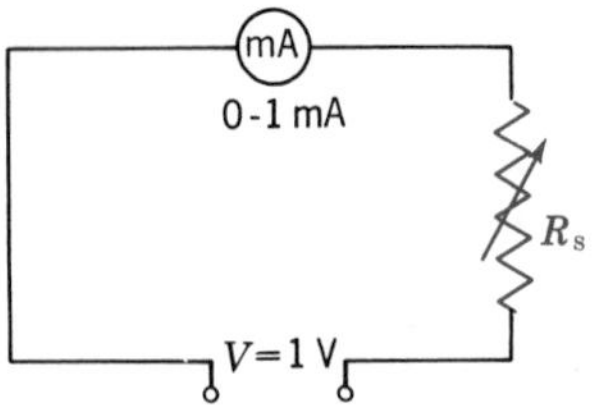

FIG. 14–11 Basic circuit of milliammeter used to indicate voltage.

is reduced to 0.5 V, then the circuit current will be reduced to one-half its original value and the meter will read 0.5 mA. Even though the meter deflection is the result of current, actually the meter can be used as a 0–1 voltmeter, indicating 1 V in the first instance and 0.5 V when the voltage is reduced.

Similarly, if the resistor is adjusted to a higher safe value so that the application of 150 V causes full-scale deflection, the instrument can be used as a 0–150 voltmeter. In that case voltage values will be obtained by multiplying the basic scale readings by a factor of 150 or by substituting a new scale as shown in Fig. 14–12.

FIG. 14–12 Panel voltmeter. (*Courtesy of Western Electrical Instrument Corporation*)

EXAMPLE 3 It is desired to use the milliammeter of Fig. 14–2 as a 0–10 voltmeter. What resistance R_{mp} must be connected in series with the instrument to accomplish this?

SOLUTION

The additional series resistance is called a *multiplier* resistance, and its value must be such that, when it is added to the resistance of the meter movement, the total resistance will limit the current through the instrument to 1 mA when 10 V is applied. The circuit is shown in Fig. 14–13. R_{mp} is the multiplier resistance, and R_m = 55 Ω is the resistance of the meter movement.

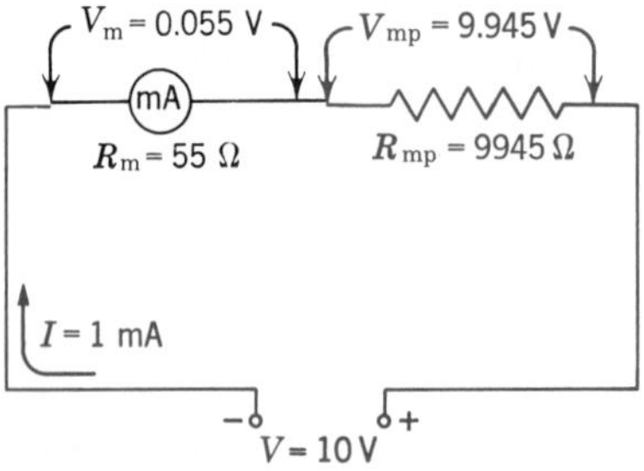

FIG. 14–13 Voltmeter circuit of Example 3.

If 10 V is to be applied across the two series resistances as shown in Fig. 14–13, in order to limit the current to 1 mA, 0.055 V must appear across the meter because

$$V_m = IR_m = 10^{-3} \times 55 = 0.055 \text{ V}$$

The remaining voltage, which is 10 − 0.055 = 9.945 V, must appear across R_{mp}. Accordingly,

$$R_{mp} = \frac{V_{mp}}{I} = \frac{9.945}{10^{-3}} = 9945 \ \Omega$$

If a 10 000-Ω resistor is used as a multiplier, with 10 V applied to the jacks, and if an observer could discern the difference, the voltage reading would be in error by only 0.05 V. (What percent error does this represent?)

EXAMPLE 4 A 0–50 microammeter, with a resistance of 1140 Ω, is to be used as a 0–100 voltmeter. What value of multiplier resistance is needed?

SOLUTION

For full-scale deflection the voltage across the meter must be limited to

$$V_m = IR_m = 50 \times 10^{-6} \times 1140$$
$$= 0.057 \text{ V}$$

The voltage across the multiplier is 100 − 0.057 = 99.943 V which results in

$$R_{mp} = \frac{V_{mp}}{I} = \frac{99.943}{50 \times 10^{-6}} = 1\ 998\ 860 \ \Omega$$

Naturally, a 2-MΩ resistor would be used.

14–5 VOLTMETER SENSITIVITY

The *sensitivity* of a voltmeter is expressed in the number of ohms in the multiplier for each volt of range. For example, the voltmeter of Example 3 has a range of 10 V and a multiplier of 10 000 Ω, which results in a sensitivity of 1000 Ω/V. The voltmeter of Example 4 has a sensitivity of 20 000 Ω/V.

14–6 VOLTMETER LOADING EFFECTS

The sensitivity of a voltmeter is a good indication of the meter's accuracy. This is particularly true when the voltages in the low-current circuits often encountered in electronic equipment are measured. For example, a 0–150 voltmeter with a sensitivity of 200 Ω/V would give excellent service, say as a power switchboard meter, at a reasonable cost. However, it would not be satisfactory for some other applications. In Fig. 14–14, two 60-kΩ resistors are connected in series across 120 V. In this condition, 60 V will appear across each resistor.

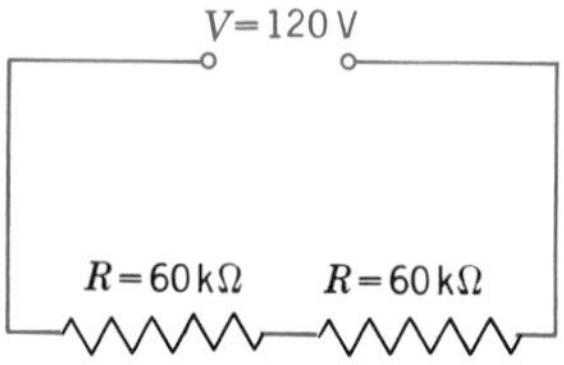

FIG. 14–14 The current through the resistors is 1 mA, and the voltage across each resistor if 60 V.

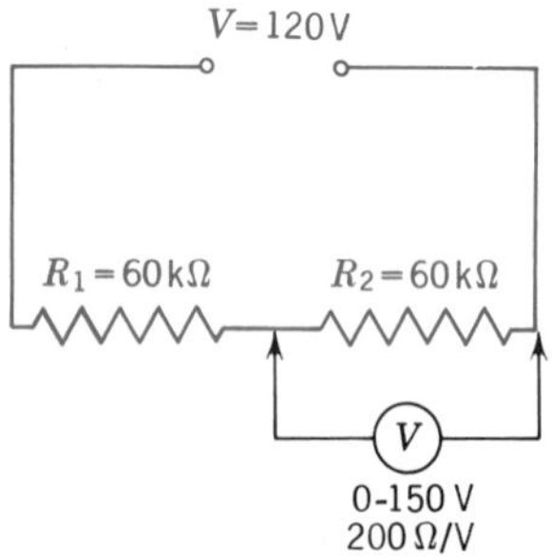

FIG. 14–15 A 30-kΩ voltmeter connected across R_2. Total circuit current is now 1.5 mA, and the voltage across R_2 is 30 V.

If the voltmeter is connected across R_2 as shown in Fig. 14–15, the joint resistance R_p of R_2 and R_{mp} becomes

$$R_p = \frac{R_2 R_{mp}}{R_2 + R_{mp}} = 20\ 000\ \Omega$$

The total resistance of the circuit is now

$$\begin{aligned} R_t &= R_1 + R_p = 60\ 000 + 20\ 000 \\ &= 80\ 000\ \Omega = 80\ \text{k}\Omega \end{aligned}$$

This results in a circuit current of

$$I_t = \frac{V}{R_t} = \frac{120}{80\ 000} = 1.5 \times 10^{-3}\ \text{A}$$

Therefore, the voltage existing across R_2 due to the shunting effect of the voltmeter is

$$\begin{aligned} V_p = I_t R_p &= 1.5 \times 10^{-3} \times 20\ 000 \\ &= 30\ \text{V} \end{aligned}$$

It is left as an exercise to show that if the voltmeter of Example 4 is used to measure the voltage across R_2, the reading will be 59.1 V.

14–7 MULTIRANGE VOLTMETERS

Using a single multiplier provides only one voltmeter range. Similar to the usage of current-measuring instruments, it has become practice to increase the usefulness of an instrument by selecting a meter movement of sufficient sensitivity and, with the use of several multipliers, use the instrument as a multirange voltmeter. Such an arrangement is shown in Fig. 14–16.

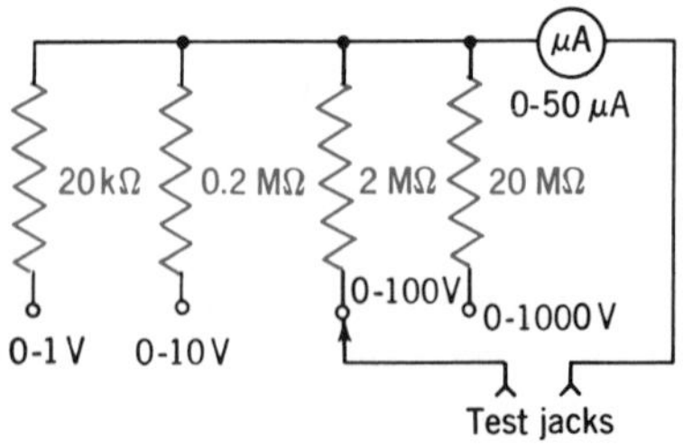

FIG. 14–16 A 0–50 microammeter used with multipliers for multirange voltmeter.

14–8 OHMMETERS

Owing to the fact that a change in the resistance of a circuit will cause a change in the current in that circuit, a current-measuring instrument can be calibrated to indicate values of resistance required for a given change in current. Such a calibrated instrument is called an *ohmmeter*.

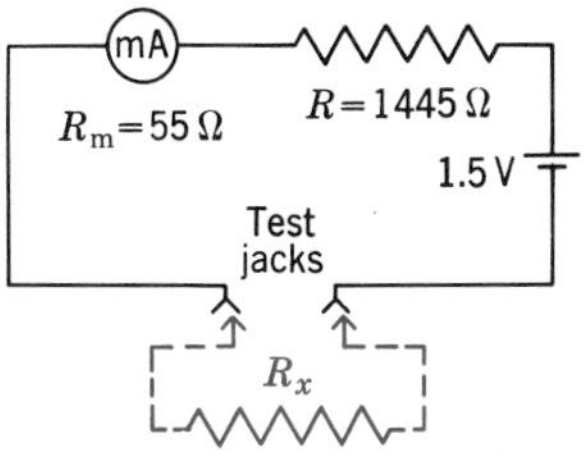

FIG. 14–17 A 0–1 milliammeter used in ohmmeter circuit.

In the schematic diagram shown in Fig. 14–17, the 0–1 milliammeter of Fig. 14–2 is connected in series with a 1.5-V battery and a resistance of 1445 Ω. Since the total resistance of the circuit is 1500 Ω, if the test jacks are short-circuited, the meter will read full scale. If the short circuit is removed and a resistance R_x of 1500 Ω is connected across the jacks, the meter will indicate half-scale deflection because now the total circuit resistance is 3000 Ω. Therefore, at full-scale deflection the meter scale could be marked 0 Ω of external circuit resistance, and at half scale it could be marked 1500 Ω. Similarly, other values of known resistance could be used to calibrate the scale throughout its range. Also, unknown resistances can be used to calibrate the scale by making use of the relation

$$R_x = R_c \frac{I_1 - I_2}{I_2} \quad \Omega \tag{3}$$

where R_x = unknown resistance, Ω
R_c = circuit resistance when test jacks are short-circuited, Ω
I_1 = current when test jacks are short-circuited, A
I_2 = current when R_x is connected in circuit, A

Use your knowledge of Ohm's law and Axiom 5 (Sec. 5–2) to derive Eq. (3).

As a provision for compensating for battery aging and maintaining calibration, variable resistors controlled from the instrument panel are connected in ohmmeter circuits by either of the methods illustrated in Figs. 14–18 and 14–19. In each case the test leads are short-circuited and the resistor control is adjusted until the meter reads full scale, or 0 Ω. An example of such a control is the "Ω ADJ" on the instrument as shown in Fig. 14–20.

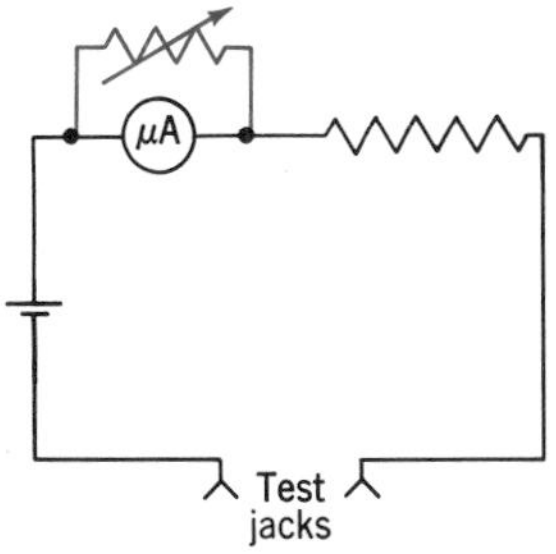

FIG. 14–18 Ohmmeter circuit with variable shunt resistance.

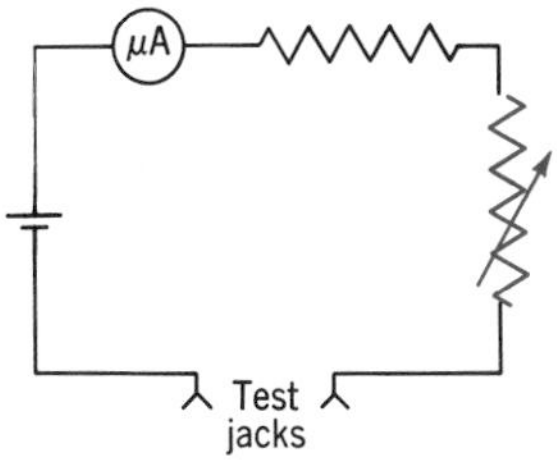

FIG. 14–19 Ohmmeter with variable series resistance.

FIG. 14–20 Multimeter; see the arrangement of shunts and multipliers on the selector switch. (*Courtesy of Triplett Electrical Instrument Company*)

Since zero resistance between the test jacks results in maximum current and larger values of resistance result in less current, certain types of ohmmeter scales are marked with numbers increasing from right to left as illustrated on the ohms scale in Fig. 14–20.

In practice, the use of the ordinary ohmmeter should be limited from about one-tenth of to ten times the center-scale resistances reading because of the small deflection changes at the ends of the scale. For this reason multirange ohmmeters are employed for changing midscale values, and the ranges generally are designed to multiply the basic scale by some power of 10.

14–9 MULTIMETERS

For the purposes of convenience and economy, meters combining the functions and desired ranges of ammeters, voltmeters, and ohmmeters are incorporated into one instrument called a multimeter, one type of which is illustrated in Fig. 14–20. If the test leads are plugged into the proper pin jacks and the rotary switch is switched to the proper function and range, the instrument can be utilized for several functions.

PROBLEMS 14–2

1. In the circuit of Fig. 14–21: (*a*) What voltages are across R_1 and R_2? (*b*) A 0–100 voltmeter with a sensitivity of 1000 Ω/V is connected across R_1. What is the reading of the voltmeter?

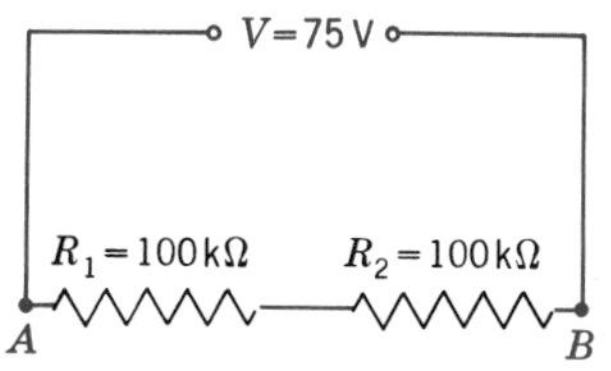

FIG. 14–21 Circuit of Probs. 1 and 2.

2. In the circuit of Fig. 14–21:
   (*a*) A 0–100 voltmeter with a sensitivity of 20 000 Ω/V is connected across R_1. What is the voltmeter reading?
   (*b*) What will the voltmeter read if connected across points *A* and *B*?
   (*c*) When the voltmeter is connected across points *A* and *B*, what current flows through R_2?
3. What are the values of the multiplier resistors R_1, R_2, R_3, and R_4 in Fig. 14–22?

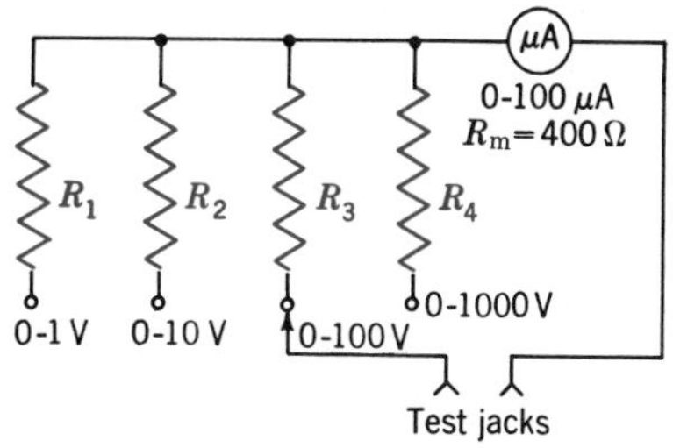

FIG. 14–22 Multirange voltmeter circuit of Prob. 3.

4. Refer to Eq. (1). Did you show that $R_s = \dfrac{R_m}{N - 1}\Omega$?
5. Refer to the end of Sec. 14–6. Did you show that the voltmeter reading will be 59.1 V?
6. Refer to Eq. (3). Did you show that $R_x = R_c\dfrac{I_1 - I_2}{I_2}$?

14–10 DIGITAL METERS

The D'Arsonval meter movements discussed in this chapter have been in service for at least three generations, and they will continue to find many applications for years to come. These analog devices have long service lives (or their successors have, after their users progress past the beginner stage), and they are superior to digital meters when it comes to monitoring varying values. However, when accuracy of reading, ease of reading, and many accurate repetitions of a reading for the same value of parameter are required, digital displays are preferable.

The various methods employed to convert analog signals to digital readouts involve mathematical concepts beyond the scope of this chapter. They require the use of graphs (Chap. 16) and time constants (Chap. 34), and they rely heavily on operational amplifiers and a variety of readout devices.

With a profusion of LSI and VLSI devices now on the market, digital meters which may be accurate to ±0.01%, ±1 digit are available. (You may find it interesting to investigate the meanings of the various ways in which meter accuracy may be stated.)

SELF TEST

1. A 0–1 milliammeter with an internal resistance of 45 Ω is to have its range extended to 0–50 mA. What shunt resistance is required?
2. The meter movement of Prob. 1 is to be used to indicate voltages up to 220 V. What value of resistance should be connected as a multiplier?
3. Design an Ayrton shunt to use the meter movement of Prob. 1 as a multi-range ammeter to measure currents of 0–2 mA, 0–50 mA, and 0–500 mA.
4. A voltmeter with a sensitivity of 10 000 Ω/V is set at 500 V to measure the voltage across resistor R_2 in Fig. 14–23. What is the reading of the voltmeter?

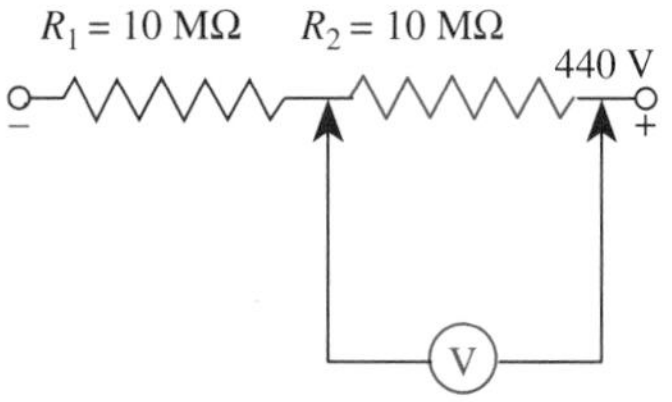

FIG. 14–23 Circuit for self test Prob. 4.

CHAPTER 15

DIVIDER CIRCUITS AND WHEATSTONE BRIDGES

In this chapter consideration is given to voltage and current divider circuits. Computations involving voltages and currents in these circuits are simply applications of Ohm's law to series and parallel circuits.

The source of power for radio and television receivers, amplifiers, and similar electronic equipment generally consists of a filtered direct voltage which has been obtained from a rectified alternating voltage. For reasons of economy and design considerations, rectifier power supplies are usually so designed that only the highest voltage desired is available at the output. In most applications, however, other voltages are needed. For example, cathode ray tubes require higher voltages than other circuit elements require. Screen grids require yet other voltages. Also, bias voltages are required. These voltages can be made available from single sources of voltage by the use of *voltage dividers*.

15–1 VOLTAGE DIVIDERS

That several values of voltage can exist around a circuit was first demonstrated in Sec. 8–8 and Figs. 8–12 and 8–13. A similar situation exists when tapped resistors, or resistors in series, are connected across the output of a power supply as illustrated in Fig. 15–1. This represents a simple *voltage divider.*

Since the resistors are of equal value, one-third of the 300-V output voltage will appear across each one. Therefore, since terminal D is at zero or ground potential, terminal C will be $+100$ V with respect to D, terminal B will be $+200$ V, and terminal A will be $+300$ V.

In addition to serving as a voltage divider, the total resistance connected across the output of a power supply generally serves as a *load resistor* and as a *bleeder*. The latter serves to "bleed off" the charge of the filter capacitors after the rectifier is turned off. As a compromise between output voltage reg-

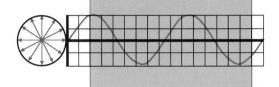

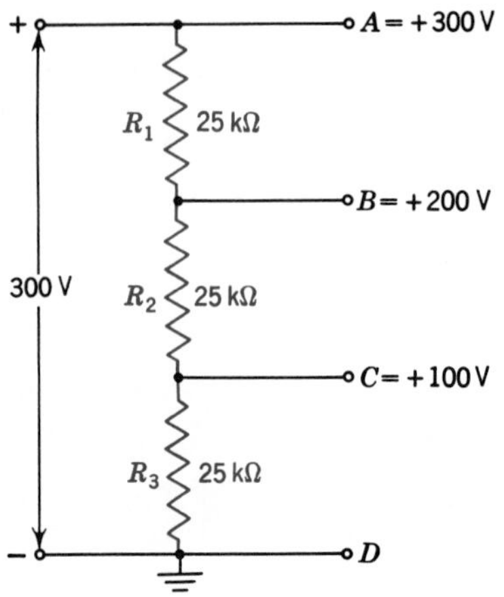

FIG. 15–1 Voltage divider consisting of three 25-kΩ resistors connected across 300-V power supply.

ulation and efficiency of operation, the total value of the voltage divider resistance is so designed that the bleeder current will be about 10% of the full-load current. The bleeder current in Fig. 15–1 with no loads connected to the various voltage divider terminals is

$$I = \frac{V}{R_1 + R_2 + R_3} = \frac{300}{75\,000}$$
$$= 4.00 \text{ mA}$$

The grounded point of a voltage divider is generally used as the reference point for circuit voltages supplied by the voltage divider. In Fig. 15–1, this is at grounded terminal D.

If the power supply output voltage is grounded at no other point, the voltage divider can be grounded at an intermediate point so as to obtain both positive and negative voltages. For example, if the voltage divider resistors of Fig. 15–1 are grounded as shown in Fig. 15–2, the voltage relations change. Terminal D is now -100 V with respect to ground, B is $+100$ V, and A is $+200$ V.

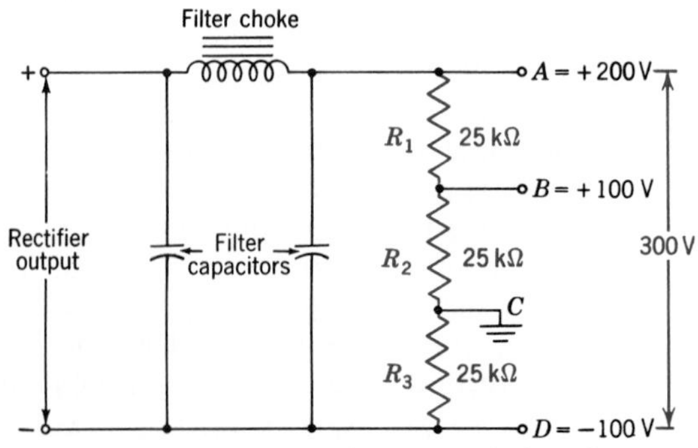

FIG. 15–2 Voltage divider grounded at C.

15–2 VOLTAGE DIVIDERS WITH LOADS

The voltage dividers of Figs. 15–1 and 15–2 have no loads connected to them; only the bleeder current of 4 mA flows through the voltage divider resistors. When loads are connected to the various terminals, the resulting additional currents must be taken into consideration because they affect the operating voltages. For example, assume a load of $R_4 = 50\ 000\ \Omega$ connected between terminals C and D of Fig. 15–1. Under these conditions, the resistance between terminals C and D is

$$R_{CD} = \frac{R_3 R_4}{R_3 + R_4} = \frac{25\ 000 \times 50\ 000}{25\ 000 + 50\ 000}$$
$$= 16\ 700\ \Omega = 16.7\ \text{k}\Omega$$

The total resistance between terminals A and D is

$$R_{AD} = R_1 + R_2 + R_{CD} = 66\ 700\ \Omega$$

resulting in a total current of

$$I_t = \frac{V}{R_{AD}} = 4.50\ \text{mA}$$

The voltage across terminals B and D is

$$V_{BD} = I_t R_{BD}$$
$$= 188\ \text{V} \qquad \text{(instead of 200 V)}$$

and across terminals C and D it is

$$V_{CD} = I_t R_{CD}$$
$$= 75\ \text{V} \qquad \text{(instead of 100 V)}$$

The circuit is shown in Fig. 15–3.

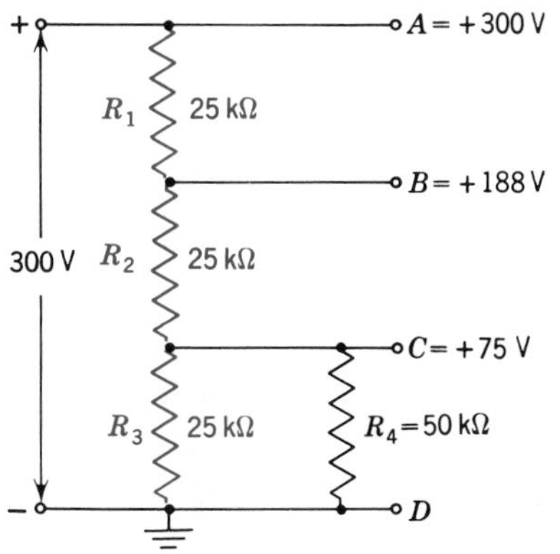

FIG. 15–3 Load of 50 kΩ connected across terminals C and D.

Show that, if an additional load of $R_5 = 50\ \text{k}\Omega$ is connected across terminals B and D, the terminal voltages would be as illustrated in Fig. 15–4.

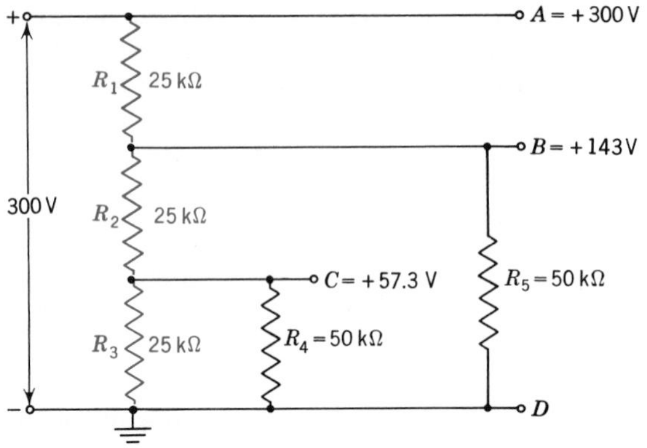

FIG. 15–4 Loads $R_4 = R_5 = 50\ \text{k}\Omega$ connected to voltage divider.

EXAMPLE 1 Design a voltage divider circuit for a 250-V power supply. The connected loads are 60 mA at 250 V and 40 mA at 150 V. Allow a 10% bleeder current.

SOLUTION

The circuit is shown in Fig. 15–5. The total load current is 100 mA; therefore, the bleeder current, which flows through R_2, is 10 mA. Since the voltage across R_2 is 150 V,

$$R_2 = \frac{150}{10 \times 10^{-3}} = 15\ 000\ \Omega = 15\ \text{k}\Omega$$

The current through R_1 is 40 + 10 = 50 mA, and the voltage across R_1 is 250 − 150 = 100 V. Then

$$R_1 = \frac{100}{50 \times 10^{-3}} = 2000\ \Omega$$

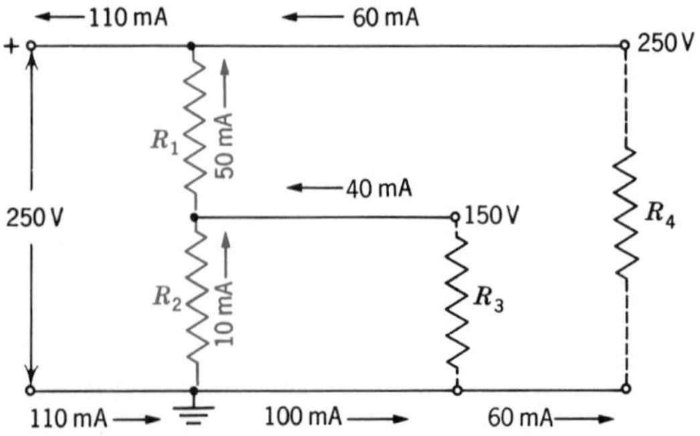

FIG. 15–5 Circuit of Example 1.

EXAMPLE 2 What are the values of the voltage divider resistors in Fig. 15–6 if the bleeder current is 10% of the total load current?

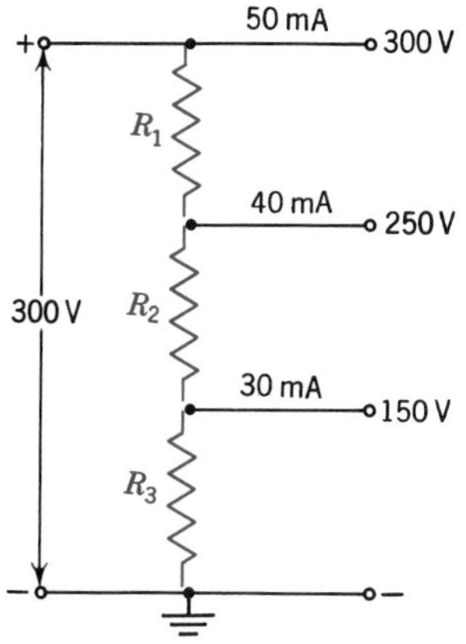

FIG. 15–6 Voltage divider of Example 2.

SOLUTION

The total load current I_L is

$$I_L = 50 + 40 + 30 = 120 \text{ mA}$$

The bleeder current is

$$I_B = 0.1 \times 120 = 12 \text{ mA}$$

The complete circuit is shown in Fig. 15–7. The voltage across R_3 is 150 V, and only the bleeder current of 12 mA flows through this resistor. Therefore,

$$R_3 = \frac{150}{12 \times 10^{-3}} = 12.5 \text{ k}\Omega$$

The 30-mA load current of the 150-V load terminal combines with the bleeder current of 12 mA for a total of 42 mA through R_2, across which is 100 V. Therefore,

$$R_2 = \frac{100}{42 \times 10^{-3}} = 2.38 \text{ k}\Omega$$

Similarly, 82 mA flows through R_1, across which is 50 V. Then

$$R_1 = \frac{50}{82 \times 10^{-3}} = 610 \ \Omega$$

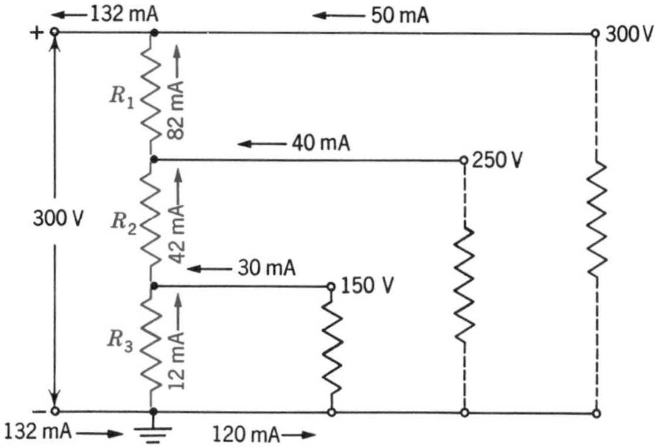

FIG. 15–7 Complete circuit of Example 2.

NOTE *Resistors* $R_1 = 610 \ \Omega$, $R_2 = 2380 \ \Omega$, *and* $R_3 = 12\ 500 \ \Omega$ *are not readily available commercially. Try substituting standard preferred values of* $R_1 = 560 \ \Omega$, $R_2 = 2.4 \text{ k}\Omega$, *and* $R_3 = 12 \text{ k}\Omega$ *for the computed values, and determine how this would affect the loads.*

EXAMPLE 3 Find the values of the voltage divider resistors of Fig. 15–8. The −50-V bias terminal draws no current, and the bleeder current is 10% of the total load current.

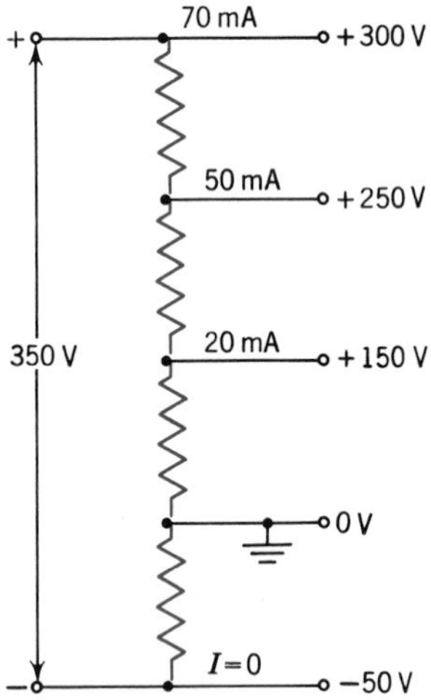

FIG. 15–8 Voltage divider of Example 3.

SOLUTION

The total load current I_L is

$$I_L = 70 + 50 + 20 = 140 \text{ mA}$$

The bleeder current is

$$I_B = 0.1 I_L = 0.1 \times 140 = 14 \text{ mA}$$

The complete circuit is illustrated in Fig. 15–9. There is a voltage of 50 V across R_4, and the total current of 154 mA flows through this resistor. Therefore

$$R_4 = \frac{50}{154 \times 10^{-3}} = 325 \ \Omega$$

Since R_3 carries only the bleeder current and the voltage across this resistor is 150 V,

$$R_3 = \frac{150}{14 \times 10^{-3}} = 10.7 \text{ k}\Omega$$

In like manner,

$$R_2 = \frac{100}{34 \times 10^{-3}} = 2.94 \text{ k}\Omega$$

and

$$R_1 = \frac{50}{84 \times 10^{-3}} = 595 \ \Omega$$

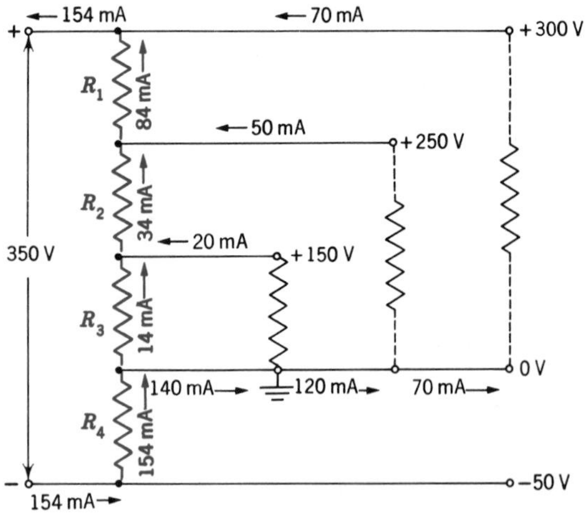

FIG. 15–9 Complete circuit of Example 3.

NOTE *As a problem, substitute the commercially available preferred values of* $R_1 = 620 \ \Omega$, $R_2 = 3 \text{ k}\Omega$, $R_3 = 11 \text{ k}\Omega$, *and* $R_4 = 300 \ \Omega$ *for the computed values, and determine how the loads would be affected.*

PROBLEMS 15–1

1. The vertical attenuator of an oscilloscope is illustrated in Fig. 15–10. With an input voltage of 60 V, what voltages appear between the switch positions and the input to the vertical amplifier?

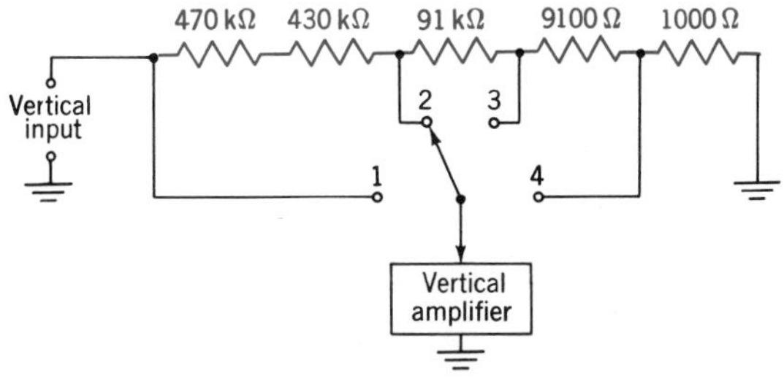

FIG. 15–10 Circuit of Prob. 1. **NOTE** *No current flows from the circuit.*

2. The horizontal hold control of a television receiver is illustrated in Fig. 15–11. What range of control voltage is available from the potentiometer to the horizontal hold control?

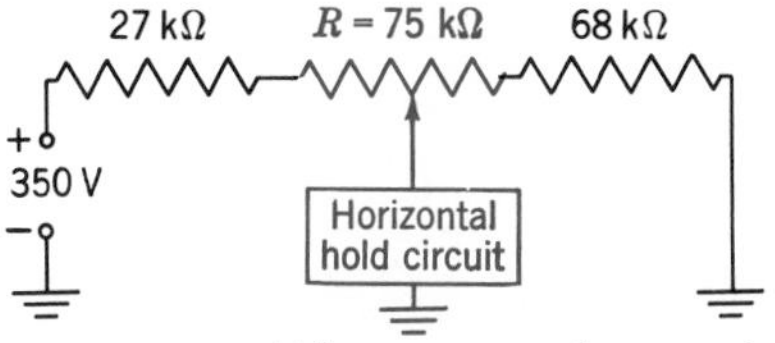

NOTE *The horizontal hold draws no current from the circuit.*

FIG. 15–11 What range of control voltage is available from the potentiometer to the horizontal hold control?

3. What is the power dissipated by each of the resistors and the potentiometer of Prob. 2?
4. Determine the values of the voltage divider resistors of Fig. 15–12 if a total of 180 mA is drawn from the power supply.
5. A voltage divider, similar to that of Fig. 15–12, is to be used to deliver the following services: 80 mA at 48 V, 60 mA at 24 V, and 60 mA at 12 V. What are the values of R_1, R_2, and R_3 if the bleeder current is 10% of the total load current?
6. What are the values of the voltage divider resistors of Fig. 15–13 if the bleeder current is 20 mA?

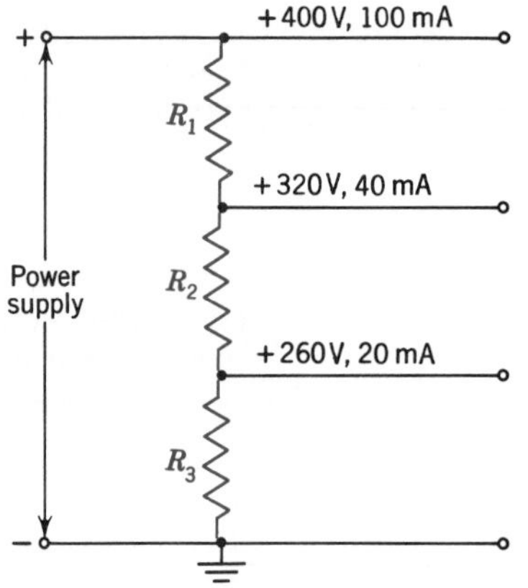

FIG. 15–12 Circuit of Probs. 4 and 5.

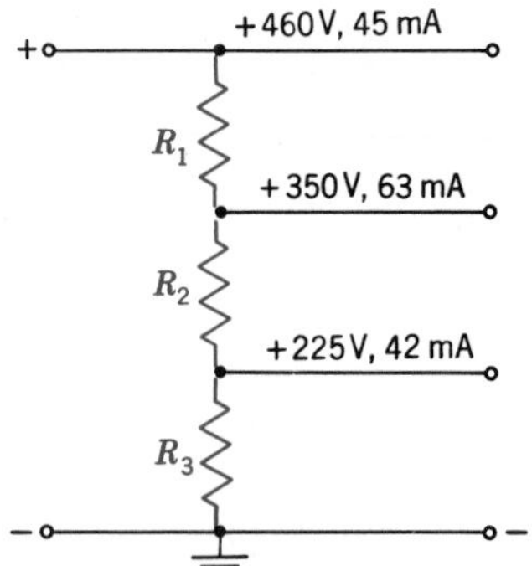

FIG. 15–13 Circuit of Probs. 6, 7, and 8.

7. What is the power dissipated by each of the resistors in Prob. 6?
8. What is the total power delivered by the voltage source in Prob. 6?
9. What are the values of the voltage divider resistors of Fig. 15–14 if the bleeder current is 10 mA?

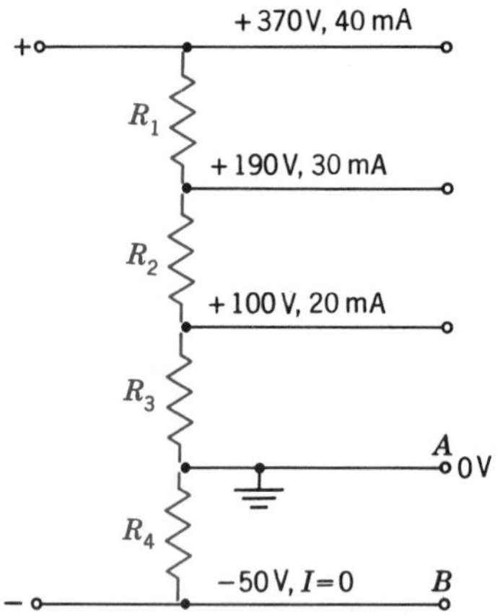

FIG. 15–14 Circuit of Probs. 9, 10, and 11.

10. What wattage ratings should be used for the resistors in Prob. 9?
11. What is the total power delivered by the voltage source of Prob. 9?
12. If the biasing resistor R_4 of Fig. 15–14 became open-circuited, what would be the voltage between terminals *A* and *B*?
13. Referring to Sec. 15–2, did you show that Fig. 15–4 is the result when Fig. 15–3 is changed by the addition of a 50-kΩ load?

14. Referring to Example 2, did you try substituting standard 5% preferred values into the voltage divider of Fig. 15–7?
15. Referring to Example 3, did you try substituting standard 5% preferred values into the voltage divider of Fig. 15–9?

15–3 CURRENT DIVIDERS

We have seen that voltage dividers are employed to develop voltage drops across series resistors. Each voltage drop is proportional to the resistance value related to the total series resistance. When resistors are connected in parallel, the voltage is the same across each, but the current is divided in *inverse* proportion. (See Sec. 13–1.)

EXAMPLE 4 In Fig. 15–15, since

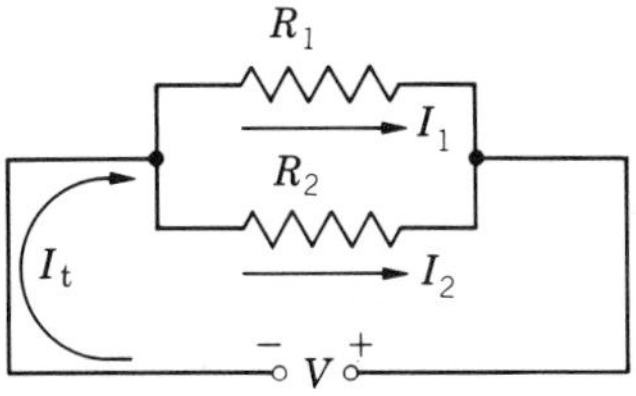

FIG. 15–15 Current divider circuit of Example 4.

$$V = \text{voltage drop across } R_1 = I_1R_1$$

and $$V = \text{voltage drop across } R_2 = I_2R_2$$

and $$I_t = I_1 + I_2$$

Therefore, $$I_1R_1 = I_2R_2$$

and $$I_1R_1 = (I_t - I_1)R_2 = I_tR_2 - I_1R_2$$

Collecting like terms, $$I_1R_1 + I_1R_2 = I_tR_2$$

$$I_1(R_1 + R_2) = I_tR_2$$

so that $$I_1 = I_t\left(\frac{R_2}{R_1 + R_2}\right) \tag{1}$$

Note carefully that the numerator in Eq. (1) is R_2 and not R_1. You should now prove to your own satisfaction that

$$I_2 = I_t\left(\frac{R_1}{R_1 + R_2}\right) \tag{2}$$

PROBLEMS 15–2

1. In Fig. 15–15, $R_1 = 1.2\ \text{k}\Omega$, $R_2 = 3.6\ \text{k}\Omega$, and $V = 25$ V. Find (*a*) I_1; (*b*) I_2.
2. In Fig. 15–15, $R_1 = 2.2\ \text{k}\Omega$ and $R_2 = 4.7\ \text{k}\Omega$. $V = 150$ V. Find (*a*) I_1; (*b*) I_2.
3. In Fig. 15–15, $I_t = 312$ mA, $R_1 = 560\ \Omega$, and $R_2 = 750\ \Omega$. Find (*a*) I_1; (*b*) I_2.
4. In Fig. 15–15, it is required that $I_1 = 25$ mA and $I_2 = 75$ mA. A resistance of 200 Ω has been established for R_1.
   (*a*) What value must be used for R_2?
   (*b*) What emf must be applied to achieve the required output?
5. In Fig. 15–15, the source of emf is replaced with a constant current source capable of delivering 5 A under all conditions. $R_1 = 100\ \text{k}\Omega$ is a shunt resistor permanently connected across the constant current source. $R_2 = 1.2\ \text{k}\Omega$ is a load driven by the shunted constant current source. Find I_2.

15–4 WHEATSTONE BRIDGE CIRCUITS

The accuracy of resistance measurements by the voltmeter-ammeter method is limited, mainly because of errors in the meters and the difficulty of reading the meters precisely. Probably the most widely used device for precise resistance measurement is the Wheatstone bridge, the circuit diagram of which is shown in Fig. 15–16.

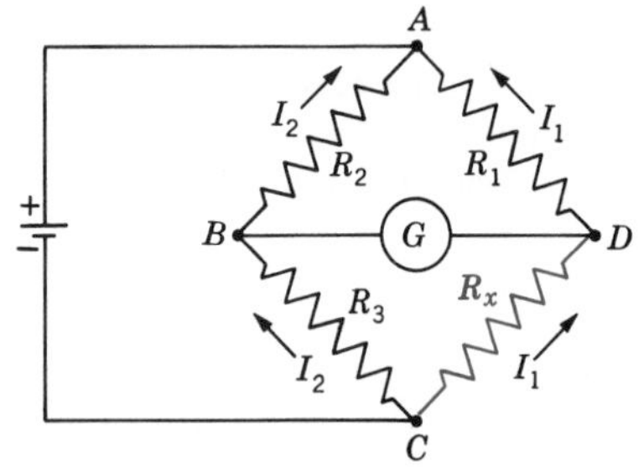

FIG. 15–16 Schematic diagram of Wheatstone bridge.

Resistors R_1, R_2, and R_3 are known values, and R_x is the resistance to be measured. In most bridges, R_1 and R_2 are adjustable in ratios of 1:1, 10:1, 100:1, etc., and R_3 is adjustable in small steps. In measuring a resistance, R_3 is adjusted until the galvanometer reads zero, and in this condition the bridge is said to be "balanced." Since the galvanometer reads zero, it is evident that the points B and D are exactly at the same potential; that is, the voltage drop from A to B is the same as from A to D. Expressed as an equation,

$$V_{AD} = V_{AB}$$

or

$$I_1R_1 = I_2R_2 \tag{3}$$

Similarly, the voltage drop across R_x must be equal to that across R_3; hence,

$$I_1R_x = I_2R_3 \tag{4}$$

Dividing Eq. (4) by Eq. (3),

$$\frac{I_1R_x}{I_1R_1} = \frac{I_2R_3}{I_2R_2}$$

$$\therefore \frac{R_x}{R_1} = \frac{R_3}{R_2} \tag{5}$$

Equation (5) is the fundamental equation of the Wheatstone bridge. By solving it for the only unknown, R_x, the value of the resistance under measurement can be computed.

Since the balance conditions of the bridge are not directly related to the voltage of the energy source, you may be tempted to think that the value of the voltage is immaterial. Obviously, for mere mathematical analysis, any convenient voltage may be assumed. In practice, however, the resistors used in Wheatstone bridges are very precise, and they are usually delicate. You must be sure that their power-dissipating capabilities are not exceeded by applying too high a voltage.

EXAMPLE 5 In the circuit illustrated in Fig. 15–16, $R_1 = 10\ \Omega$, $R_2 = 100\ \Omega$, and $R_3 = 13.9\ \Omega$. If the bridge is balanced, what is the value of the unknown resistance?

SOLUTION

Solving Eq. (5) for R_x,

$$R_x = \frac{R_1 R_3}{R_2}$$

Substituting the known values,

$$R_x = \frac{10 \times 13.9}{100} = 1.39\ \Omega$$

Locating the point at which a telephone cable or a long control line is grounded is simplified by the use of two circuits that are modifications of the Wheatstone bridge. These are the Murray loop and the Varley loop.

Figure 15–17 represents the method of locating the grounded point in a cable by using a Murray loop. A spare ungrounded cable is connected to the

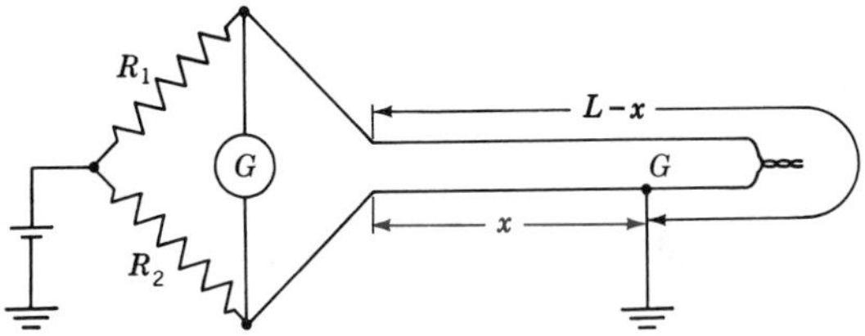

FIG. 15–17 Murray loop.

grounded cable at a convenient location beyond the grounded point G. This forms a loop of length L, one part of which is the distance x from the point of measurement to the grounded point G. The other part of the loop is then $L - x$. These two parts of the loop form a bridge with R_1 and R_2, which are adjusted until the galvanometer shows no deflection. Because this results in a balanced bridge circuit,

$$\frac{R_2}{R_1} = \frac{x}{L - x} \tag{6}$$

Solving for x,

$$x = \frac{R_2}{R_1 + R_2} L \tag{7}$$

EXAMPLE 6 A Murray loop is connected as in Fig. 15–17 to locate a ground in a cable between two cities 60 km apart. The lines forming the loop are identical. With the bridge balanced, $R_1 = 645\ \Omega$ and $R_2 = 476\ \Omega$. How far is the grounded point from the test end?

SOLUTION

Substituting the known values in Eq. (7),

$$x = \frac{476}{645 + 476} \times 2 \times 60 = 50.95 \text{ km}$$

If the two cables forming the loop are not the same size, the relations of Eq. (7) can be used to compute the resistance R_x of the grounded cable from the point of measurement to the grounded point. Then if R_L is the resistance of the entire loop,

$$R_x = \frac{R_2}{R_1 + R_2} R_L \tag{8}$$

EXAMPLE 7 A Murray loop is connected as shown in Fig. 15–17. The grounded cable is No. 19 wire, and wire of a different size is used to complete the loop. The resistance of the entire loop is 126 Ω, and when the bridge is balanced, $R_1 = 342\ \Omega$ and $R_2 = 217\ \Omega$. How far is the ground from the test end?

SOLUTION

Substituting the known values in Eq. (8),

$$R_x = \frac{217}{342 + 217} \times 126 = 48.9\ \Omega$$

Since No. 19 wire has a resistance of 24.6 Ω/km, 48.9 Ω represents 1.852 km of wire between the test end and the grounded point.

PROBLEMS 15–3

1. In the balanced Wheatstone bridge of Fig. 15–16, $R_1 = 0.05\ \Omega$, $R_2 = 1.2\ \Omega$, $R_3 = 46.5\ \Omega$. What is the value of the unknown resistance?
2. In the Wheatstone bridge of Fig. 15–16, the ratio of $R_2{:}R_3$ is 100:1. $R_1 = 4.52\ \Omega$. What is the unknown resistor?
3. In the Wheatstone bridge, the ratio of $R_1{:}R_2$ is 1000 and R_x is believed to be 520 Ω. At what setting of R_3 may a balance be expected?
4. A ground exists on one conductor of a lead-covered No. 19 pair. A Murray loop is used to locate the fault by connecting the pair together at the far end (Fig. 15–17). When the bridge circuit is balanced, $R_1 = 33.3\ \Omega$ and $R_2 = 21.7\ \Omega$. If the cable is 2.2 km long, how far from the test end is the cable grounded?
5. Several No. 8 wires run between two cities located 65 km apart. One wire becomes grounded, and a Murray loop is used in one city to locate the fault by connecting two of the wires in the other city. When the bridge is balanced, $R_1 = 716\ \Omega$ and $R_2 = 273\ \Omega$. How far from the test end is the wire grounded?

6. A No. 6 wire, which is known to be grounded, is made into a loop by connecting a wire of different size at its far end. The resistance of the loop thus formed is 5.62 Ω. When a Murray loop is connected and balanced, the value of R_1 is 16.8 Ω and that of R_2 is 36.2 Ω. How far from the test end does the ground exist?
7. As a research project, discover the details of the Varley loop and develop its equation, which is similar to that for the Murray loop.

SELF TEST

1. Determine the actual values and power dissipation of the resistors of a voltage divider that will provide the following services:
   + 36 V, 100 mA, + 18 V, 40 mA, + 12 V, 40 mA,
   + 6 V, 20 mA. Allow a bleeder current of 20 mA.
2. Two resistors, $R_1 = 1.2$ kΩ, and $R_2 = 3.6$ kΩ, are connected in parallel. If the total current flow into the parallel circuit is I_T, what percentage of I_T will flow through resistor R_1?
3. In the Wheatstone Bridge of Fig. 15–18, you are informed that R_x is probably 12 Ω. What value of R_1 should you set for a trial balance?
4. In Fig. 15–19, $R_1 = 330$ Ω, $R_2 = 180$ Ω, and $R_3 = 220$ Ω. The applied emf is 36 V. How much power is dissipated by R_3?
5. What is the total resistance of the circuit in Fig. 15–20?

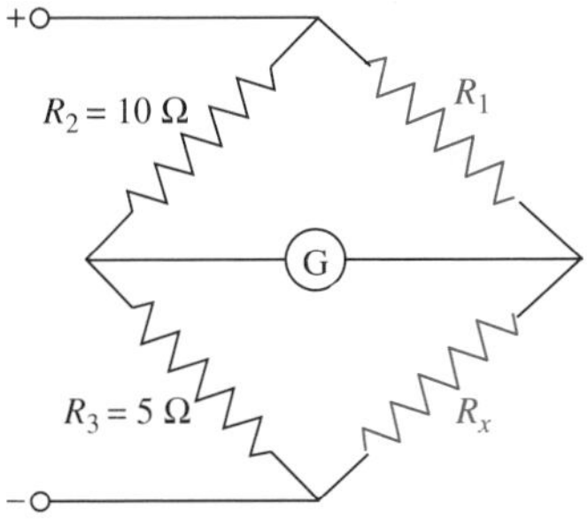

FIG. 15–18 Wheatstone Bridge circuit of self test Prob. 3.

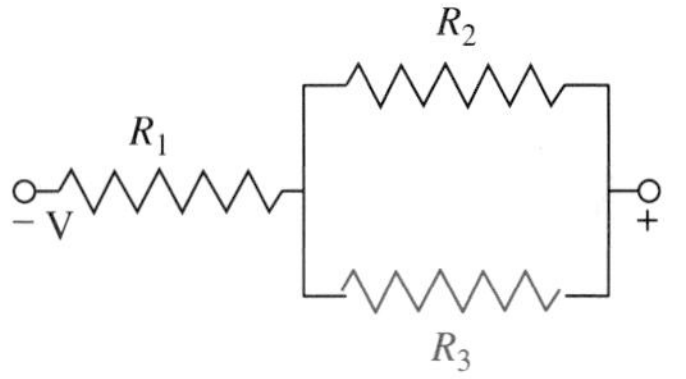

FIG. 15–19 Circuit for self test Prob. 4.

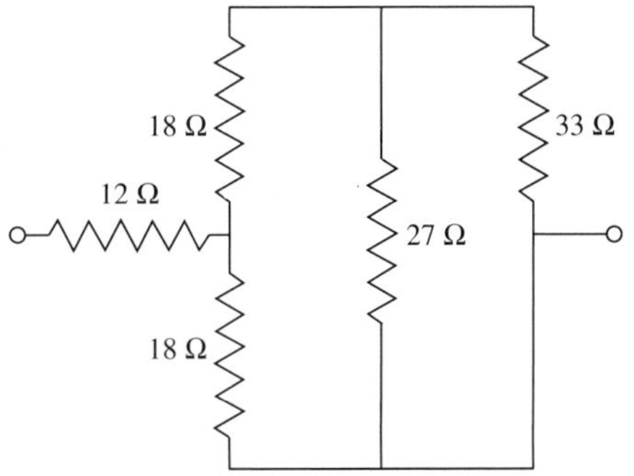

FIG. 15–20 Circuit for self test Prob. 5.

CHAPTER 16

GRAPHS

A graph is a pictorial representation of the relationship between two or more quantities. Everyone is familiar with various types of graphs or graphic charts. They are used extensively in magazines, newspapers, annual reports, and trade journals published for engineers, manufacturers, and others concerned with relative values. It is difficult to conceive how engineers and technicians could dispense with them.

We have already used simple graphic representations in Chap. 3, and here we will develop a few of the uses of straight-line graphs. In later chapters we will use graphs in working out the solutions of problems and in quickly presenting information in varied forms.

The notions presented here are fundamental to the use of all graph forms, and we are paving the way for some important and interesting topics which will follow in later chapters.

16–1 LOCATING POINTS ON A GRAPH

The accurate location of points is vital, and the manner of marking points can help or seriously hinder in arriving at a correct solution to a problem. One of the most common methods of locating a point is by using a large dot (Fig. 16–1). But this is the poorest form of location, and Fig. 16–1 illustrates why. Do you draw the line through the center, through the top, or through the bottom of a large dot? Can you be sure where the center is? The possibility of introducing errors is great, and you should study the variations of error illustrated in the various parts of Fig. 16–1.

A more acceptable way to mark a point is to use an X, with the intersection marking the spot, or else a circled dot, ⊙, with the tiny point marking the spot and the circle attracting your attention to it. These correct methods are illustrated in Fig. 16–2, and they should be used in all your graph-drawing practice.

A second important item to watch always is the placing of the points. If there is a choice, the points should be far apart, so that the line joining them

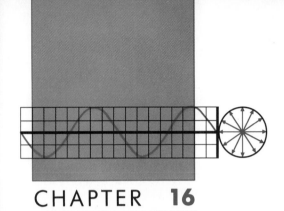

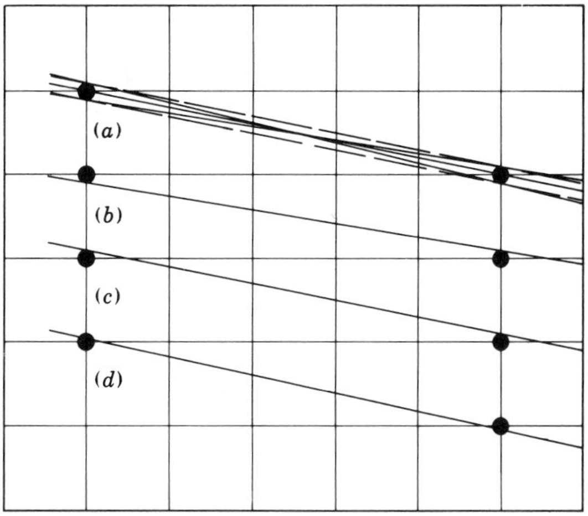

FIG. 16–1 Illustration of errors introduced by the use of large dots to locate points.
(*a*) Instead of a single fine line, a broad range of possibilities is presented.
(*b*) Shall we join the outside edges of the dots?
(*c*) Should we join top to top or bottom to bottom?
(*d*) Should we just pick a line that somehow touches both dots somewhere?

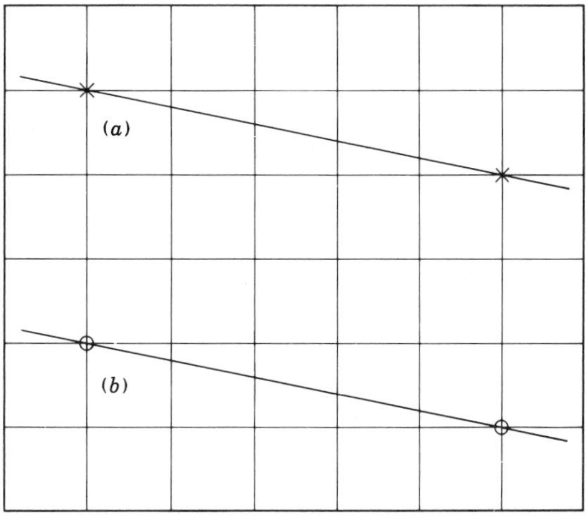

FIG. 16–2 Illustration of the correct method of locating points: After the line is drawn, the small point locations are still indicated, but only a single line can be drawn between the points.

spans the most important area of the graph. Thus, any error in locating the points themselves is minimized. If the points are located close together and an error is made in locating either one point or both points, then other useful locations "outside" the points plotted will be subject to greater error. This fault is illustrated in Fig. 16–3, in which the two circled dots have been plotted

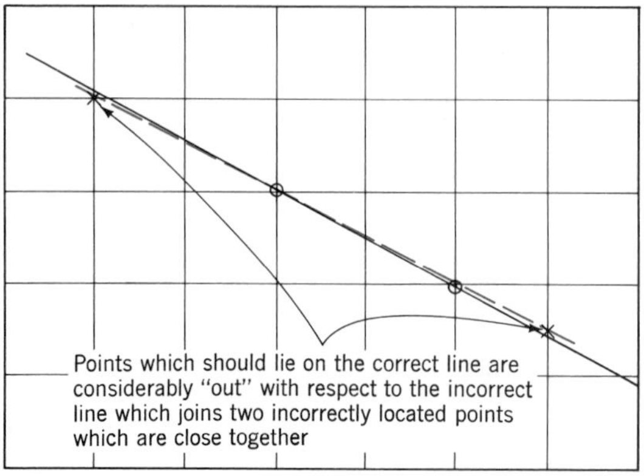

Points which should lie on the correct line are considerably "out" with respect to the incorrect line which joins two incorrectly located points which are close together

FIG. 16–3 Illustration of the error introduced when points are plotted close together. If the points are slightly incorrect then useful points "outside" the plotted area are even further off, and the error is enlarged.

slightly off their desired locations. The line joining them comes some distance away from the X points, which should lie on the line. In Fig. 16–4, the two circled dots are again plotted slightly off their desired locations, but since they are widely separated, the amount of error of intermediate points is less.

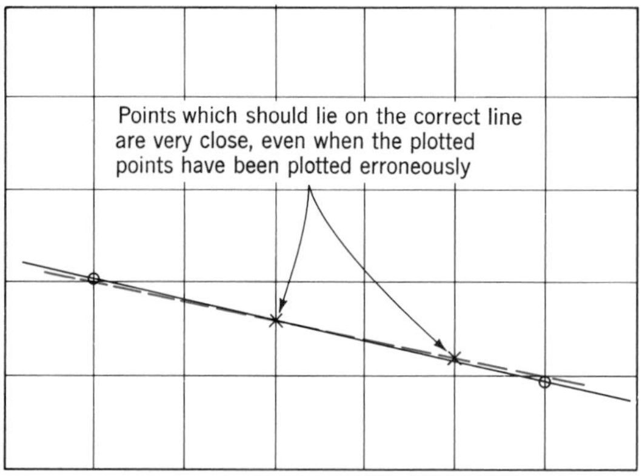

Points which should lie on the correct line are very close, even when the plotted points have been plotted erroneously

FIG. 16–4 Illustration of the reduction in error when incorrectly plotted points are far apart, so that the line joining them spans the working area of the graph. The error in locating each circled dot is the same as the error in Fig. 16–3, but the × locations are closer to the incorrect line which joins the plotted points.

16–2 SOLVING PROBLEMS BY MEANS OF GRAPHS

In many instances there arise problems involving relationships that, though readily solved by usual arithmetical or algebraic methods, are more clearly understood when solved graphically. It is also true that there are many problems which can be solved graphically with less labor than is required for the purely mathematical solutions. The following illustrative examples will show how some problems can be worked graphically.

Check your computer user's manual to determine whether you have graphing possibilities. Look for HCHAR OR VCHR. Can you call up different colors to identify different curves? Do you have a "mouse" that helps you to draw lines?

EXAMPLE 1 Steamship *A* sails from New York at 6 A.M., steaming at an average speed of 10 knots (kn). (A knot is a measure of speed and is one nautical mile per hour.) The same day, at 9 A.M., steamship *B* sails from New York, steering the same course as *A* but steaming at 15 kn. (*a*) How long will it take *B* to overtake *A*? (*b*) What will be the distance from New York at that time?

SOLUTION

Choose convenient scales on graph paper, and plot the distance in nautical miles (nmi) covered by each vessel against the time in hours, as shown in Fig. 16–5. This is conveniently accomplished by making a table like Table 16–1.

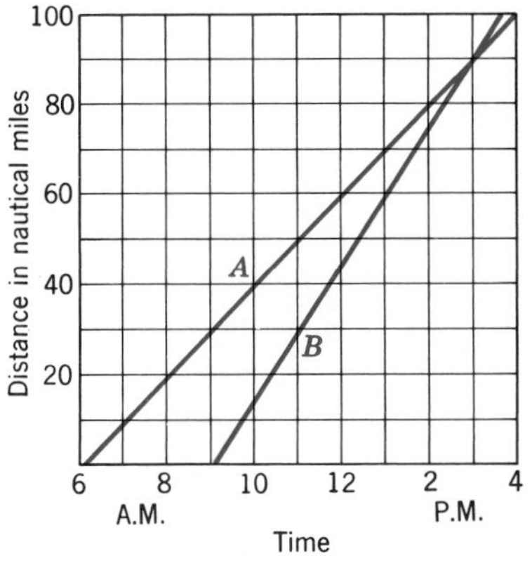

FIG. 16–5 Graph of Example 1.

TABLE 16–1

Time, o'clock	Distance Covered by *A*, nmi	Distance Covered by *B*, nmi
6 A.M.	0	0
8	20	0
10	40	15
12	60	45
2 P.M.	80	75
4	100	105

It will be noted that the graphs of the two distances intersect at 90 nmi, or at 3 P.M. This means the two ships will be 90 nmi from New York at 3 P.M. Because both are steering the same course, *B* will overtake *A* at this time and distance.

The graphic solution furnishes us with other information. For example, by measuring the vertical distance between the graphs, we can determine how far apart the ships will be at any time. Thus, at 11 A.M. the ships will be 20 nmi apart, at 1 P.M. they will be 10 nmi apart, etc.

EXAMPLE 2 Ship *A* is 200 nmi at sea, and ship *B* is in port. At 8 A.M., *A* starts toward the port, making a speed of 20 kn. At the same time, *B* leaves port at a speed of 30 kn to intercept *A*. After traveling 2 h, *B* is delayed for 1 h and 40 min at the lightship. *B* then continues on its course to intercept *A*. (*a*) At what time will the two ships meet? (*b*) How far will they be from port at that time?

SOLUTION

Fig. 16–6 is a graph showing the conditions of the problem. The graph is constructed as in Example 1. A table of distances against time is made up; a convenient scale is chosen; and the points are plotted and joined with a straight line.

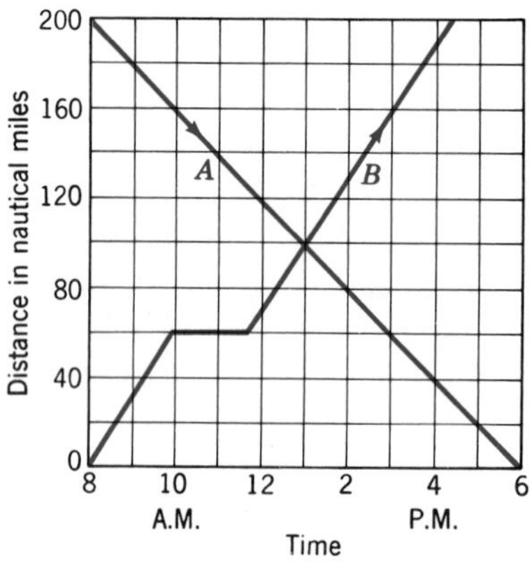

FIG. 16–6 Graph of Example 2.

The intersection of the graphs illustrates that the ships will meet 100 nmi from port at 1 P.M. Why is there a horizontal portion in the graph of *B*'s distance from port? If *A* and *B* continue their speeds and courses, at what time will *A* reach port? At what time will *B* arrive at *A*'s 10 A.M. position? What will be the distance between the ships at that time?

PROBLEMS 16–1

1. A circuit consists of a 10-Ω resistor R_c connected across a variable emf V_v. Plot the current *I* through the resistor against the voltage *V* across the resistor as V_v is varied in 10-V steps from 0 to 100 V. What conclusion do you draw from this graph?

2. A circuit consists of a 50-Ω resistor R_L connected across the variable emf V_v of Prob. 1. On the same graph sheet as your solution to Prob. 1, plot the current I through R_L against the voltage V as V_v is varied between 0 and 100 V. What conclusion do you draw from the pair of graphs?
3. The distance s covered by a moving object is equal to the product of the object's velocity v and the time t during which the object is moving; that is, $s = vt$. Plot the distance in kilometers traveled by an automobile averaging 55 km/h against time for every hour from 9 A.M. to 6 P.M. What conclusions do you draw from the graph?
4. A variable resistor R_v is connected across a generator which maintains a constant voltage V_c of 120 V. Plot the current I through the resistor as the resistance is varied in 5-Ω steps between 5 and 50 Ω. What conclusions do you draw from this graph?

Solve these problems graphically:

5. Train *A* leaves a city at 8 A.M. traveling at the rate of 80 km/h. Two hours later train *B* leaves the same city, on the same track, traveling at the rate of 120 km/h.
   (*a*) At what time does train *B* overtake *A*?
   (*b*) How far from the starting point will the trains be at the time of part (*a*)?
   (*c*) How far apart will the trains be 2 h after *B* starts?
6. Two people start toward each other from points 144 km apart, the first traveling at 96 km/h and the second at 64 km/h.
   (*a*) How long will it be before they meet?
   (*b*) How far will each have traveled when they meet?
   (*c*) How far apart will they be after 30 min of travel?
7. *A* owns a motor that consumes 10 kWh per day, and *B* owns a motor that consumes 30 kWh per day. Beginning on the first day of a 30-day month, *A*'s motor runs continuously. *B*'s motor runs for 1 day, is idle for 4 days, then runs for 2 days, is idle for 6 days, and then runs continuously for the rest of the month. On what days of the month will *A*'s and *B*'s power bills be the same?
8. The owner of a radio store decides to pay the salespeople according to either of two plans. The first plan provides for a fixed salary of $50 per week plus a commission of $3 for each radio sold. According to the second plan, a salesperson may take a straight commission of $5.50 for each radio sold. Determine at which point the second plan becomes more attractive for an energetic salesperson.
9. The exact conversion of Fahrenheit temperature readings to their Celsius equivalents is given by the formula

$$C = \frac{5}{9}(F - 32)$$

An approximation of this formula that is convenient for mental arithmetic, especially for people still "getting comfortable" with metric change is given by

$$C \simeq \frac{1}{2}(F - 30)$$

(*a*) On the same sheet of plain graph paper, draw graphs of both of these formulas. For the Celsius scale, plot the range between $-15°C$ and $+15°C$. For Fahrenheit, plot between $+5°F$ and $+55°F$.
(*b*) Algebraically determine the temperature at which the two formulas agree. Does your pair of graphs confirm this temperature?
(*c*) On either side of your exact curve, draw a cone representing Fahrenheit values, exact $\pm 10\%$. What is the range of your graph for which your approximate curve lies within this 10% cone?

16–3 COORDINATE NOTATION

Let us suppose you are standing on a street corner and a stranger asks you for directions to some prominent building. You tell the stranger to go four blocks east and five blocks north. By these directions, you have automatically made the street intersection a *point of reference,* or *origin,* from which distances are measured. From this point you could count distances to any point in the city, using the blocks as a unit of distance and pairs of directions (east, north, west, or south) for locating the various points.

To draw a graph, we had to use two lines of reference, or *axes*. These correspond to the streets meeting at right angles. Also, in fixing a point on a graph, it was necessary to locate that point by pairs of numbers. For example, when we plot distance against time, we need one number to represent the time and another number to represent the distance covered in that time.

So far, only positive numbers have been used for graphs. To restrict graphs to positive values would impose just as severe a handicap as if we were to restrict algebra to positive numbers. Accordingly, a system must be established for plotting pairs of numbers either or both of which may be positive or negative. In such a system, a sheet of squared paper is divided into four sections, or quadrants, by drawing two intersecting axes at right angles to each other. The point O, at the intersection of the axes, is called the *origin*. The horizontal axis is generally known as the *x axis,* and the vertical axis is called the *y axis*.

There is nothing new about measuring distances along the *x axis;* it is the basic system described in Sec. 3–5 and shown in Fig. 3–3. That is, we agree to regard distances along the *x axis* to the *right* of the origin as *positive* and those to the *left* as *negative*. Also, we consider distances along the *y axis* as *positive* if *above* the origin and *negative* if *below* the origin. In effect, we have simply added to our method of graphical representation as originally outlined in Fig. 3–3.

With this system of representation, which is called a system of *rectangular coordinates,* we are able to locate any pair of numbers regardless of the signs. Because this system was developed by the French mathematician René Descartes, you will often hear it referred to as the system of *Cartesian coordinates*.

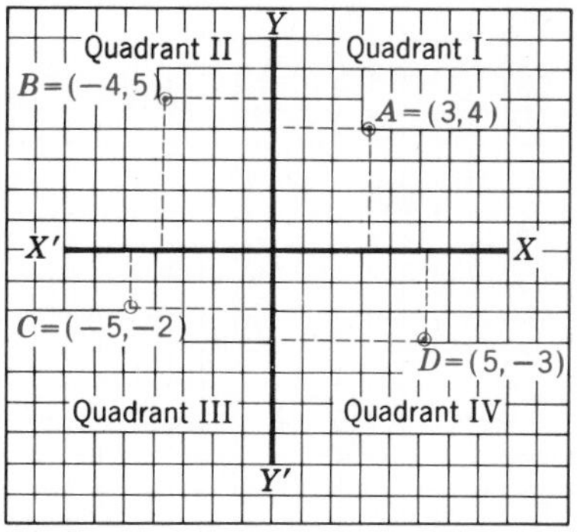

FIG. 16–7 System of rectangular coordinates.

EXAMPLE 3 Referring to Fig. 16–7,

- Point A is in the first quadrant. Its x value is $+3$, and its y value is $+4$.
- Point B is in the second quadrant. Its x value is -4, and its y value is $+5$.
- Point C is in the third quadrant. Its x value is -5, and its y value is -2.
- Point D is in the fourth quadrant. Its x value is $+5$, and its y value is -3.

Thus, every point on the surface of the paper corresponds to a pair of coordinate numbers that completely describe the point.

The two signed numbers that locate a point are called the *coordinates* of that point. The x value is called the *abscissa* of the point, and the y value is called the *ordinate* of the point.

In describing a point in terms of its coordinates, the abscissa is always stated first. Thus, to locate point A in Fig. 16–7, we write $A = (3, 4)$, meaning that, to locate point A, we count three divisions to the right of the origin along the x axis and up four divisions along the y axis. In like manner, we completely describe point B by writing $B = (-4, 5)$. Also,

$$C = (-5, -2) \text{ and } D = (5, -3)$$

PROBLEMS 16–2

1. On a map, which lines correspond to the x axis, latitude or longitude?
2. Plot the following points: (2, 3), (-6, -1), (3, -7), (0, -6), (0, 0), (-8, 0).
3. Plot the following points: (-1.5, 10), (-6.5, -7.5), (3.6, -4), (0, 2.5), (6.5, 8.5), (3.5, 0).
4. Using Fig. 16–8, give the coordinates of the points *A, B, C, D, E, F, G, H, I, J, K, L, M,* and *N*.
5. Plot the following points: $A = (-1, -2)$, $B = (5, -2)$, $C = (5, 4)$, $D = (-1, 4)$. Connect these points in succession. What kind of figure is *ABCD?* Draw the diagonals *DB* and *CA*. What are the coordinates of the point of intersection of the diagonals?

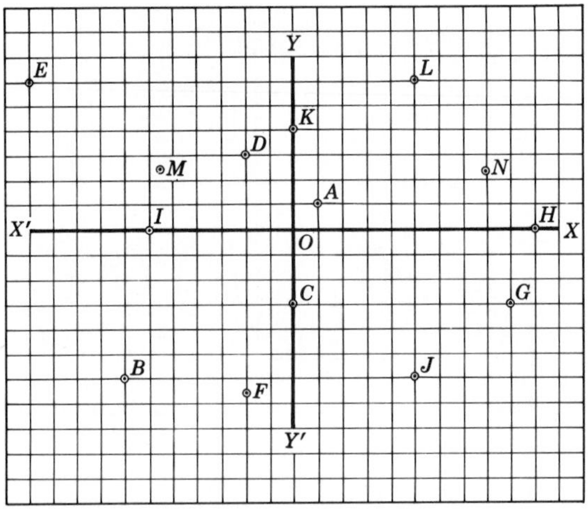

FIG. 16–8 Graph of Prob. 4.

16–4 GRAPHS OF LINEAR EQUATIONS

A relation between a pair of numbers, not necessarily connected with physical quantities such as those in foregoing exercises, can be expressed by a graph.

Consider the following problem: The sum of two numbers is equal to 5. What are the numbers? Immediately it is evident there is more than one pair of numbers that will fulfill the requirements of the problem. For example, if only positive numbers are considered, we have, by addition,

0	1	2	3	4	5
5	4	3	2	1	0
5	5	5	5	5	5

Similarly, if negative numbers are included, we can write

−1	−2	−3	−4	−5	−6
+6	+7	+8	+9	+10	+11
5	5	5	5	5	5

and so on, indefinitely.

Also, if fractions or decimals are considered, we have

1.5	−3.75	−1.63	−8.36
3.5	+8.75	+6.63	+13.36
5	5	5	5

and so on, indefinitely.

It follows that there is an infinite number of pairs of numbers whose sum is 5.

Let x represent any possible value of one of these numbers, and let y represent the corresponding value of the second number. Then

$$x + y = 5$$

For any value assigned to x, we can solve for the corresponding value of y. Thus, if $x = 1$, $y = 4$. Also, if $x = 2$, $y = 3$. Likewise, if $x = -4$, $y = 9$, because, by substituting -4 for x in the equation, we obtain

$$-4 + y = 5$$

or

$$y = 9$$

In this manner, there may be obtained an unlimited number of values for x and y that satisfy the equations, some of which are listed below.

If $x =$	-6	-4	-2	0	2	4	6	8	10
Then $y =$	11	9	7	5	3	1	-1	-3	-5
Coordinates of	A	B	C	D	E	F	G	H	I

With the tabulated pairs of numbers as coordinates, the points are plotted and connected in succession, as shown in Fig. 16–9. The line drawn through these points is called the *graph of the equation* $x + y = 5$.

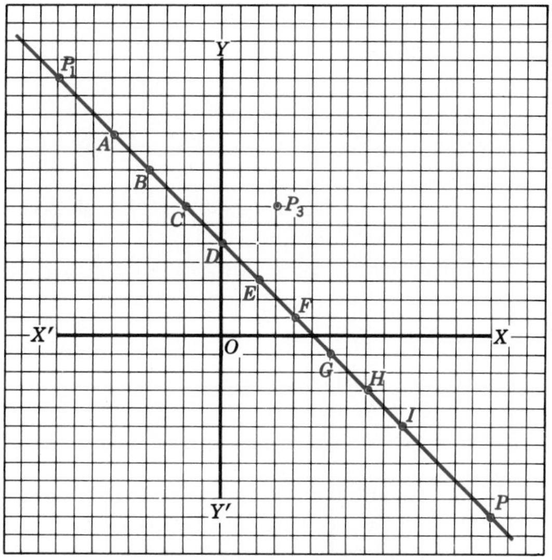

FIG. 16–9 Graph of the equation $x + y = 5$.

Regardless of what pairs of numbers (coordinates) are chosen from the graph, it will be found that each pair satisfies the equation. For example, the point P has coordinates $(15, -10)$; that is, $x = 15$ and $y = -10$. These numbers satisfy the equation because $15 - 10 = 5$. Likewise, the point P_1 has coordinates $(-9, 14)$ that also satisfy the equation because $-9 + 14 = 5$. The point P_3 has coordinates $(3, 7)$. This point is not on the line, nor do its coordinates satisfy the equation; for $3 + 7 \neq 5$. The straight line, or graph, can be extended in either direction, always passing through points whose coordinates satisfy the conditions of the equation. This is as would be expected, for there is an infinite number of pairs of numbers called *solutions* that, when added, are equal to 5.

PROBLEMS 16–3

1. Graph the equation $x - y = 8$ by tabulating and plotting five pairs of values for x and y that satisfy the equation. Can a straight line be drawn through these points? Plot the point (4, 4). Is it on the graph of the equation? Do the coordinates of this point satisfy the equation? From the graph, when $x = 0$, what is the value of y? When $y = 0$, what is the value of x? Do these pairs of values satisfy the equation?
2. Graph the equation $2x + 3y = 6$ by tabulating and plotting at least five pairs of values for x and y that satisfy the equation. Can a straight line be drawn through these points? Plot the point $(-15, 12)$. Is this point on the graph of the equation? Do the coordinates satisfy the equation? Plot the point $(10, -5)$. Is this point on the graph of the equation? Do the coordinates satisfy the equation? From the graph, when $x = 0$, what is the value of y? When $y = 0$, what is the value of x? Do these pairs of values satisfy the equation?

16–5 VARIABLES

When two variables, such as x and y, are so related that a change in x causes a change in y, then y is said to be a *function* of x. By assigning values to x and then solving for the value of y, we make x the *independent variable* and y the *dependent variable*.

The above definitions are applicable to all types of equations and physical relations. For example, in Fig. 16–5, distance is plotted against time. The distance covered by a body moving at a constant velocity is given by

$$s = vt$$

where s = distance
v = velocity
t = time

In this equation and therefore in the resulting graph, the distance is the dependent variable because it depends upon the amount of time. The time is the independent variable, and the velocity is a constant.

Similarly, in Prob. 1 of Problems 16–1, the formula $I = \frac{V}{R}$ is used to obtain values for plotting the graph. Here the resistance R is the constant, the voltage V is the independent variable, and the current I is the dependent variable.

In Prob. 4 of Problems 16–1, the same formula $I = \frac{V}{R}$ is used to obtain coordinates for the graph. Here the voltage V is a constant, the resistance R is the independent variable, and the current I is the dependent variable.

From these and other examples, it is evident, as will be illustrated in Sec. 16–6, that the graph of an equation having variables of the first degree is a straight line. This fact does not apply to variables in the denominator of a fraction as in the case above where R is a variable. $I = \frac{V}{R}$ is not an equation of the first degree as far as R is concerned because, by the law of exponents, $I = VR^{-1}$.

It is general practice to plot the independent variable along the horizontal, or x axis, and the dependent variable along the vertical, or y axis.

In plotting the graph of an equation, it is convenient to solve the equation for the dependent variable first. Values are then assigned to the independent variable in order to find the corresponding values of the dependent variable.

If an equation or formula contains more than two variables, we must, after choosing the dependent variable, decide which is to be the independent variable for each separate investigation, or graphing. For example, consider the formula

$$X_L = 2\pi f L$$

where X_L = inductive reactance of an inductor, Ω
f = frequency, Hz
L = inductance, H
2π = 6.28 . . .

In this case, we can vary either the frequency f or the inductance L in order to determine the effect upon the inductive reactance X_L, but we must not vary both at the same time. Either f must be fixed at some constant value and L varied or L must be fixed. A little thought will show the difficulty of plotting, on a plane, the variations X_L if f and L are varied simultaneously.

16–6 THE GRAPH-EQUATION RELATIONSHIPS

Each one of the equations that have been plotted is of the *first degree* (Sec. 11–1) and contains *two unknowns*. From the graphs the following important facts are obtained:

1. The graph of an equation of the first degree is a straight line.
2. The coordinates of every point on the graph satisfy the conditions of the equation.
3. The coordinates of every point not on the graph do not satisfy the conditions of the equation.

Because the graph of every equation of the first degree results in a straight line, as stated under 1 above, first-degree equations are called *linear equations*. Also, because such equations have an infinite number of solutions, they are called *indeterminate equations*.

As x changes in value in such an equation, the value of y also changes. Hence, x and y are called *variables*.

Now consider Fig. 16–9 on page 277, the graph of $x + y = 5$. This equation may be written in the form $y = -x + 5$. Here y is called the dependent variable, because its value depends upon the value of x; and x is called the independent variable, because we may assign to it any value we choose.

Notice in the graph first of all that the y intercept, the point where the curve cuts the y axis, is at the point $x = 0$, $y = 5$, and this value is revealed in the equation $y = -x + 5$ because, at the y axis, $x = 0$ and y then equals 5.

Second, note the slope of the line. For every step in the x direction (positive to the right), there is a downward (negative) step in the y direction. By definition, the slope of a line is the ratio of the change in the y values between two points to the corresponding change in x values between the same two points:

$$\text{Slope} = \frac{\Delta y}{\Delta x}$$

where the symbol Δ (Greek letter delta) means "the change in."

Figure 16–9 has been redrawn in Fig. 16–10 to show the changes in x and y between two arbitrarily selected points B and H. The slope of the graph equals

$$\frac{\Delta y}{\Delta x} = \frac{-12}{+12} = -1$$

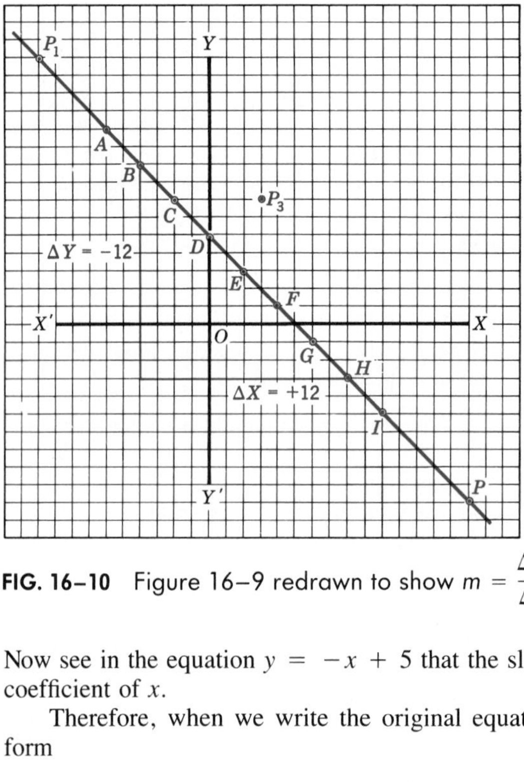

FIG. 16–10 Figure 16–9 redrawn to show $m = \dfrac{\Delta y}{\Delta x}$.

Now see in the equation $y = -x + 5$ that the slope, -1, is indicated in the coefficient of x.

Therefore, when we write the original equation $x + y = 5$ in standard form

$$y = -x + 5$$

the slope of the line is the coefficient of the x term and the y intercept is the constant term.

The general form of equation for a straight line is

$$y = mx + b$$

where y = dependent variable
x = independent variable
m = slope of the curve (straight line)
b = value of the y intercept

16–7 METHODS OF PLOTTING

To graph a linear equation of two variables,

1. Convert the equation to the standard form $y = mx + b$ to indicate quickly the values of the slope m and the y intercept b.
2. Choose a suitable value for x, substitute it into the standard form equation, and solve for the corresponding value of y. This results in one solution, or one set of coordinates.
3. Choose another value for x, and again solve for y. This second x value should be reasonably well spaced from the first (see Figs. 16–3 and 16–4).
4. Plot the two points whose coordinates were calculated in steps 2 and 3. Connect them with a fine straight line.
5. Check the resulting graph by solving for and plotting a third point. This third point must lie on the same straight line or its extension.

EXAMPLE 4 Graph the equation $2x - 5y = 10$.

SOLUTION

1. Rewrite the equation in the standard form: $y = \frac{2}{5}x - 2$.
2. Always plot first the value of y when $x = 0$. This value is immediately obtained from the "-2" of the equation, which shows the y intercept. This inspection results in a point, which we shall call A, whose coordinates are $(0, -2)$.
3. Now choose some value of x. Any value will serve, but one which cancels the denominator of the fractional coefficient will be the best choice. Let $x = 5$ and, by solving the equation, obtain $y = 0$. This gives the second point, B, at $(5, 0)$. (Sometimes it may be more convenient to choose, as the second point, the value of $y = 0$ and solve for x.)
4. Choose another value of x in order to solve for the third (check) point. Let $x = -10$. Then $y = -6$, and this gives point C at $(-10, -6)$.
5. Draw the line of the equation by joining the three points. The points and the finished graph are shown in Fig. 16–11.

When x was set equal to zero, the resulting point A had coordinates that located the point where the graph crossed the y axis. This point is called the *y intercept*. Likewise, when y was set equal to zero, the resulting point B had coordinates that located the point where the graph crossed the x axis. This point is called the *x intercept*. Not only are these easy methods of locating two points

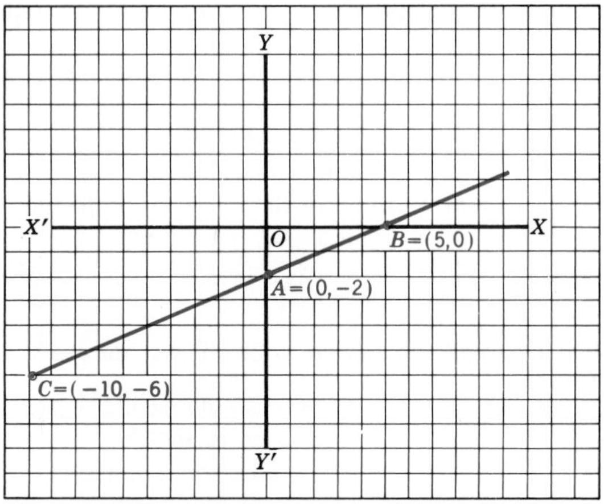

FIG. 16–11 Graph of the equation $2x - 5y = 10$.

with which to graph the equation but also these two points give us the exact location of the intercepts. The intercepts are important, as will be shown later.

The x intercept is often referred to as the *root* or *zero* of the equation.

An alternative method of plotting straight-line graphs is to use the information obtained from the standard form $y = mx + b$. If we locate the y intercept b immediately and then step over and up (or down) in accordance with the slope m, we can locate additional points. If, for example, $y = 2x + 9$, then the y intercept is at $+9$ and the slope is $+2{:}1$.

Follow the development of the graph in Fig. 16–12. First plot the y intercept, $+9$. Then step one unit in the positive x direction and two units in the positive y direction and plot the first point. Next, since $+\frac{2}{1} = \frac{-2}{-1}$, again starting at the y intercept, step one unit in the negative x direction and two units in the negative y direction and plot the second point. If these two points are too close together to be reliable, space them better by moving greater distances in the x and y directions while keeping the ratio $\frac{\Delta y}{\Delta x}$ equal to 2:1 ($= m$). Finally, join the two points so located with a straight line which passes through the third, or test, point, the y intercept.

PROBLEMS 16–4

Graph the following equations and determine the x and y intercepts:

1. $5x + 4y = 12$
2. $2x - y = 8$
3. $x - 3y = 3$
4. $2x + y = 9$

5. Plot the following equations on the same sheet of graph paper (same axes), and carefully study the results: (*a*) $x - y = -8$; (*b*) $x - y = -5$;

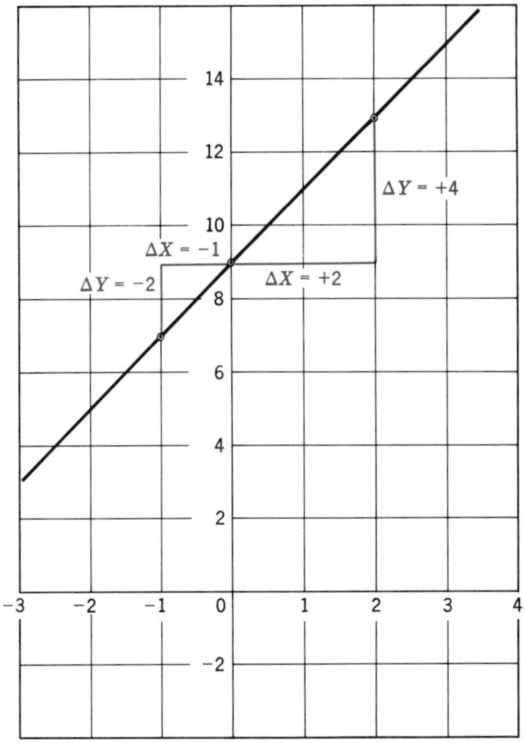

FIG. 16–12 Alternative method of plotting a straight line: First, locate y-intercept, given by the constant in the standard form equation. Second, step off Δx and Δy so that $\frac{\Delta y}{\Delta x} = m$, or slope (also given in the standard form equation), first in the $+x$ direction and then in the $-x$ direction.

(*c*) $x - y = 0$; (*d*) $x - y = 4$; (*e*) $x - y = 8$. Are the graphs parallel? Note that all left members of the given equations are identical. Solve each of these equations for y and write them in a column, thus:

$$
\begin{aligned}
(a)\ y &= x + 8 \\
(b)\ y &= x + 5 \\
(c)\ y &= x + 0 \\
(d)\ y &= x - 4 \\
(e)\ y &= x - 8
\end{aligned}
$$

In each equation, does the last term of the right member represent the y intercept?
When the equations are solved for y, as above, each coefficient of x is $+1$. All the graphs slant to the right because the coefficient of each x is positive. Each time an x increases one unit, note that the corresponding y increases one unit. That is because the coefficient of x in each equation is 1.

6. Plot the following equations on the same sheet of graph paper (same axes), and carefully study the results.
(*a*) $4x - 2y = -30$; (*b*) $4x - 2y = -16$; (*c*) $4x - 2y = 0$; (*d*) $4x - 2y = 12$
(*e*) $4x - 2y = 30$; (*f*) $8x - 4y = 60$.
Are all the graphs parallel? Again note that all left members are identical. Does the graph of Eq. (*f*) fall on that of Eq. (*e*)? Note that (*e*) and (*f*) are *identical equations*. Why?
Solve each of these equations, except (*f*), for y and write them in a column, thus:

$$\begin{aligned} (a)\ y &= 2x + 15 \\ (b)\ y &= 2x + 8 \\ (c)\ y &= 2x + 0 \\ (d)\ y &= 2x - 6 \\ (e)\ y &= 2x - 15 \end{aligned}$$

In each equation, does the last term of the right member represent the y intercept? When linear equations are written in this form, this last term is known as the *constant term*.
Are all the coefficients of the x's positive? That is why all the graphs slant upward to the right. Lines slanting in this manner are said to have *positive slopes*.
Each time an x increases or decreases one unit, note that y respectively increases or decreases two units. That is because the coefficient of each x is 2. If a graph has a *positive slope,* an increase or decrease in x always results in a corresponding increase or decrease in y. In these equations, each line has a slope of $+2$, the coefficient of each x.

7. Plot the following equations on the same set of axes: (*a*) $x + 2y = 18$; (*b*) $x + 2y = 10$; (*c*) $x + 2y = 0$; (*d*) $x + 2y = -14$; (*e*) $x + 2y = -22$; (*f*) $3x + 6y = -66$.
Are all the graphs parallel? How should you have known they would be parallel without plotting them?
Does the graph of (*f*) fall on that of (*e*)? How should you have known (*e*) and (*f*) would plot the same graph without actually plotting them? Solve each equation for y as in Probs. 5 and 6. Does the constant term denote the y intercept in each case? Is the coefficient of each x equal to $\frac{1}{2}$? The minus sign means that each graph has a *negative slope;* that is, the lines slant downward to the right. Thus, when x increases, y decreases, and vice versa. The $\frac{1}{2}$ slope means that, when x varies one unit, y is changed $\frac{1}{2}$ unit. Therefore, the variations of x and y are completely described by saying the slope is $-\frac{1}{2}$.

8. Plot the following equations on the same set of axes: (*a*) $x - 4y = 0$; (*b*) $x - 2y = 0$; (*c*) $x - y = 0$; (*d*) $2x - y = 0$; (*e*) $4x - y = 0$; (*f*) $4x + y = 0$; (*g*) $2x + y = 0$; (*h*) $x + y = 0$; (*i*) $x + 2y = 0$; (*j*) $x + 4y = 0$. Solve the equations for y, as before, and carefully analyze your results.

16–8 EQUATIONS DERIVED FROM GRAPHS

Often we obtain a set of readings relating two variables and want to know whether there is any definite relationship between the variables. This investigation makes use of both the graph showing the relationship and our understanding of the standard form of a straight-line equation.

$$y = mx + b$$

1. Plot the observed values carefully on a graph. If a straight-line relationship is indicated, draw it.
2. Sometimes one or more points appear to be off the trend. There may or may not be errors in these readings. For the present, we will *assume* that they are errors.
3. If the trend is a straight line, but some points are off, try to draw the line so that there is an equal number of floating points above and below the line. (Use a transparent straightedge.)
4. The straight-line result must now obey the law $y = mx + b$.

EXAMPLE 5 Given the following set of readings, draw the graph and determine the law relating the variables:

x	-2	2	4	6
y	-7	5	11	17

SOLUTION

First, plot the points as they have been given, and try them with a straightedge for a straight-line relationship. Since in Fig. 16–13 a straight line is indicated,

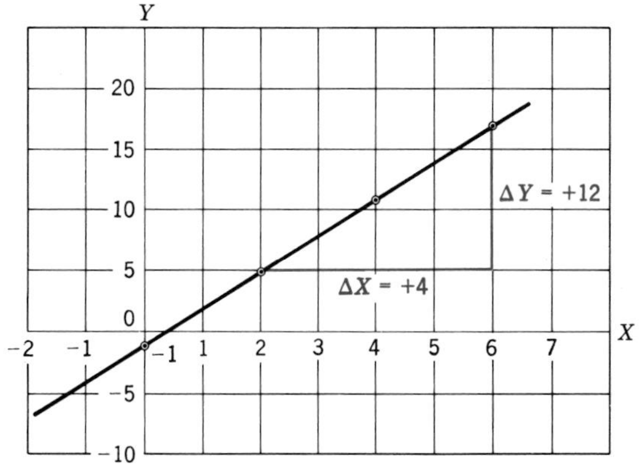

FIG. 16–13 Graph of Example 5.

draw the line that joins the points. The y intercept is seen to be -1. This gives the value of b in the standard form. Then, to determine the slope m, choose any two convenient points, reasonably spaced, say (2, 5) and (6, 17). The difference between the points in the y direction is $17 - 5 = 12$. The difference between the points in the x direction is $6 - 2 = 4$.

$$\text{Then the slope } m = \frac{\Delta y}{\Delta x} = \frac{17 - 5}{6 - 2} = \frac{+12}{+4}$$
$$= +3$$

and the relationship is $y = 3x - 1$.

EXAMPLE 6 Given the readings relating P and V, determine the law relating them:

P	-4	-2	2	6	10
V	17	11	-5	-22	-38

SOLUTION

Plot the points and test for a straight-line relationship. Because some of the points are not quite on the line, draw the straight line which will balance the floating points (Fig. 16–14). Now the V intercept is seen to be $+2$, and the

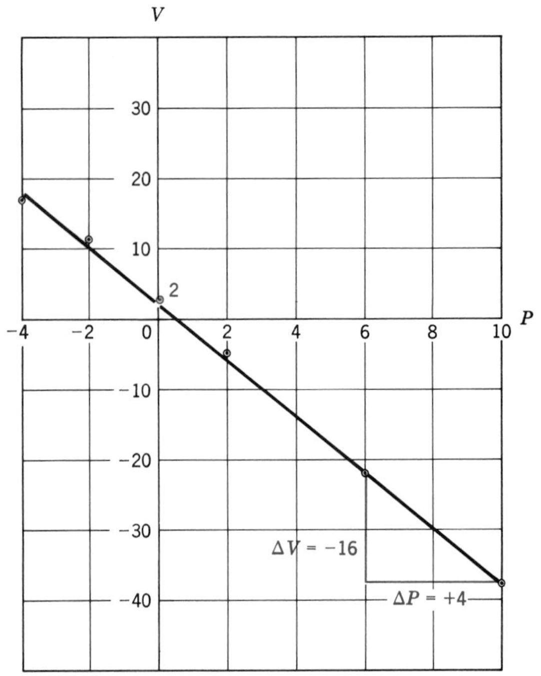

FIG. 16–14 Graph of Example 6.

equation relating P and V will be of the form $V = mP + 2$. To evaluate the slope m, choose any two convenient points on the line, and arrive at m:

$$m = \frac{\Delta V}{\Delta P} = \frac{-38 - (-22)}{10 - 6}$$
$$= \frac{-16}{+4} = -4$$

and the relationship is seen to be

$$V = -4P + 2$$

Referring to Examples 5 and 6, see how m may be found algebraically by realizing that $\Delta y = y_2 - y_1$, the difference of the values of y when going from point 1 to point 2, and $\Delta x = x_2 - x_1$, the difference of the values of x going from point 1 to point 2. Then

$$m = \frac{\Delta y}{\Delta x} = \frac{y_2 - y_1}{x_2 - x_1}$$

Always call your starting point 1 and your finishing point 2. That will yield the correct *sign* as well as the correct *value* of the slope.

PROBLEMS 16–5

1. What is the $y = mx + b$ form equation for the graph of Fig. 16–11?
2. A series of readings shows values of y for predetermined values of x:

x	5	10	15	20	25	30
y	100	200	300	400	500	600

   Plot values of y against values of x and determine the values of the constants m and b which connect x and y in the form $y = mx + b$.

3. A laboratory experiment relates x and y as follows:

x	−10	10	20	30	40	50	60
y	−0.4	1.2	2.0	2.8	3.6	4.4	5.2

   What is the equation, in the form $y = \alpha x + \theta$, which relates x and y?

4. The following is a series of readings relating s and t:

t	50	125	210	250	360	435
s	0.36	0.30	0.23	0.20	0.11	0.05

   Plot s against t and, assuming s and t are connected by a law of the form $s = u + qt$, find u and q.

5. The following is a set of laboratory readings relating R and T:

T	30	75	150	210	270	300	360	390	425	450
R	0.38	0.35	0.31	0.26	0.22	0.195	0.16	0.13	0.12	0.10

Plot the graph of R versus T and determine the formula which relates them.

6. A comparison of Celsius (C) and Fahrenheit (F) temperatures is given in the following table:

°C	0	10	38	60	100
°F	32	50	100	140	212

Plot °F against °C.

(*a*) Determine from the graph the relationship between the two temperature scales in the form $F = \theta C + \phi$.

(*b*) From the graph, what is the Fahrenheit equivalent of 25°C?

(*c*) From the graph, what is the Celsius equivalent of 165°F?

7. The readings of current flow I through a certain resistor as the emf V is changed are given in the following table:

V	10	20	30	40	50	60	70	80	90	100	V
I	0.2125	0.4255	0.638	0.851	1.062	1.278	1.49	1.702	1.915	2.125	A

(*a*) From the graph, what is the ratio $\dfrac{\text{change in voltage}}{\text{change in current}}\left(\dfrac{\Delta V}{\Delta I}\right)$?

(*b*) What is the ratio $\dfrac{\text{change in current}}{\text{change in voltage}}\left(\dfrac{\Delta I}{\Delta V}\right)$?

(*c*) From Ohm's law, what is the resistance of the resistor?

(*d*) What conclusions do you draw from your answers to questions (*a*), (*b*), and (*c*)?

8. The following is a series of readings of the avalanche breakdown of a zener diode:

V	−14.2	−14.4	−14.6	−14.7	−14.8	−14.9	−15	−15.1	−15.2	−15.3	−15.4	−15.5	V
I	0	0	0	−10	−18.9	−28.2	−37.4	−46.8	−56	−65.2	−74.6	−83.9	mA

Plot the graph of I versus V and determine:

(*a*) What $\dfrac{\Delta I}{\Delta V}$ is after the voltage goes more negative than 14.6 V.

(*b*) What the ratio is for voltages less negative than 14.6 V.

9. When the base current I_B of an NPN transistor is 0.01 mA, the readings of collector-to-emitter voltage V_{CE} produce the following set of readings for the collector current I_C. Plot I_C against V_{CE}.

V_{CE}	12.5	14.3	16.5	18.5	20.0	21.5	23.2	24.4	V
I_C	0.9	1.1	1.2	1.4	1.6	1.62	1.65	1.69	mA

(*a*) Over what range of voltages is I_C considered constant?
(*b*) What would be your interpretation of I_C if I_B were doubled?
(*c*) How do you interpret the current gain of the transistor over the tabulated voltages? Explain your observations.
(*d*) What is the value of beta $\left(\beta = \frac{I_C}{I_B}\right)$ over the tabulated spread of current? How do you explain the differences?

10. The readings of current versus applied voltage for a tunnel diode are as follows:

V_1	0.002	0.008	0.011	0.016	0.02	0.023	0.027	0.03	0.04	0.07	0.095	0.105	0.115
I_1	0.1	0.2	0.3	0.4	0.5	0.6	0.7	0.8	0.9	1.0	0.9	0.8	0.7

0.125	0.135	0.145	0.160	0.20	0.32	0.39	0.42	0.43	0.45	0.46	0.47	0.48	V
0.6	0.5	0.4	0.3	0.2	0.1	0.2	0.3	0.4	0.5	0.7	0.8	0.9	mA

(*a*) Draw the graph of I_1 versus V_1.
(*b*) Note specifically the range of voltages which makes the tunnel diode act like a negative resistance.
(*c*) Note the ranges of voltages which make the tunnel diode act like a positive resistance.

SELF TEST

1. Given the following table of *I-V* readings for two resistors:

V	0 V	10 V	20 V	30 V	40 V	50 V
I_1	0 A	1 A	2 A	3 A	4 A	5 A
I_2	0 mA	200 mA	400 mA	600 mA	800 mA	1 A

(*a*) Plot the graphs *I* against *V*, for the two resistors.
(*b*) Determine the resistance of each resistor, and label the graphs.
(*c*) Add and label the graphs of an ideal short circuit and an ideal open circuit.

2. A small sales-service business has been operating a 100-W display-window floodlamp 24 h per day. The owner is now considering a timer control that can be programmed to light up the window during important "traffic" times.

   Program A would turn on two 200-W floodlamps at 2 P.M. At 8 P.M. it would turn off one 200-W lamp and keep the other on until 10 P.M. At 10 P.M. it would switch off the 200-W lamp and switch on a 100-W lamp that would burn until 6 A.M., when it would switch off.

   Program B would operate two 200-W lamps from 2 P.M. until 6 P.M. Then it would operate a single 200-W lamp until midnight, when it would turn off the 200-W lamp and turn on a 60-W night light that would burn until 6 A.M., and then switch off.

   Plot a graph of energy in watthours against time to compare the two programs with the original 100-W lamp operating 24 h per day. The time axis should show a 24-h cycle beginning at 10 A.M.

3. Graph the equation $5y - 10x = 15$.

4. Plot the graph for the following set of readings and determine the law that relates them in the form $M = mP + b$.

P	−10	0	10	20	30	40	50	60	70
M	1.75	1.5	1.25	1.00	0.75	0.5	0.25	0	−0.25

CHAPTER 17

SIMULTANEOUS EQUATIONS

Many times in electronics we find several circuit conditions applying at the same time and therefore requiring interlocking solutions. Accordingly, the study of simultaneous equations and their most common methods of solution is a vital one for electronics technicians.

The subject of simultaneous equations also provides us with an excellent application of the linear graphs discussed in Chap. 16. This chapter leans heavily on the notions presented there, although, once the meaning of simultaneous solutions is understood, we can quickly move on to various algebraic methods of solution.

17–1 GRAPHICAL SOLUTION OF SIMULTANEOUS LINEAR EQUATIONS

The graphs of the equations

$$x + 2y = 12$$

and

$$3x - y = 1$$

are shown in Fig. 17–1. The point of intersection of the lines has the coordinates (2, 5); that is, the x value is 2 and the y value is 5. Now this point is on both of the graphs; it follows, therefore, that the x and y values should satisfy both equations. Substituting 2 for x and 5 for y in each equation results in the identities

$$2 + 10 = 12$$

and

$$6 - 5 = 1$$

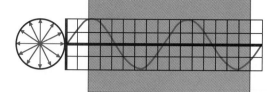

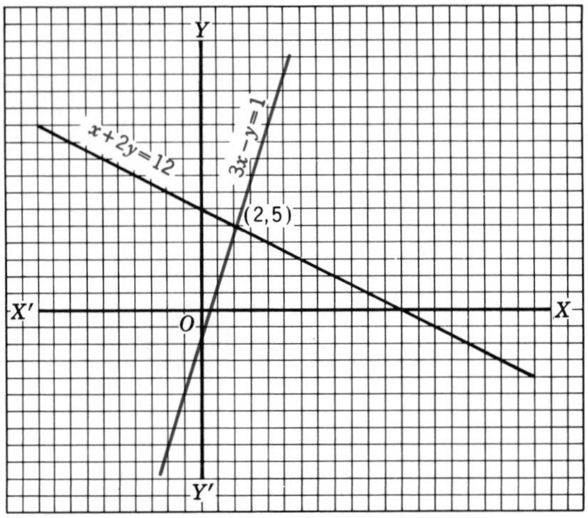

FIG. 17–1 Graph of the equations $x + 2y = 12$ and $3x - y = 1$

From this it is observed that, if the graphs of two linear equations intersect, they have one common set of values for the variables, or one common solution. Such equations are called *simultaneous linear equations*.

Because two straight lines can intersect at only one point, there can be only one common set of values or one common solution that satisfies both equations.

Two equations, each with two variables, are called *inconsistent equations* when their plotted lines are parallel to each other. Because parallel lines do not intersect, there is no common solution for two or more inconsistent equations.

Considerable care must be used in graphing equations, for a deviation in the graph of either equation will cause the intersection to be in the wrong place and hence will lead to an incorrect solution.

PROBLEMS 17–1

Solve the following pairs of equations graphically, and check your solutions by substituting them into each of the original equations:

1. $2x + y = 9$
   $4x - y = 6$
2. $6x - y = 25$
   $2x + 3y = 25$
3. $\alpha + 2\beta = -2$
   $15\alpha - 4\beta = 106$
4. $x + 2y = 26$
   $4x - y = 32$
5. $9E + 2I = 34$
   $6E + 5I = -14$
6. $l - 8m = 0$
   $l + m = 45$
7. $7\alpha + 3\beta = -23$
   $5\beta + 4\alpha = -23$
8. $8F - f = 0$
   $3f + 4F = 14$
9. $5 + L = 2M$
   $-20 + 3L = -4M$
10. $8P + 6Q = 9$
    $12P - 2Q = 8$

17–2 SOLUTION OF SIMULTANEOUS LINEAR EQUATIONS BY ADDITION AND SUBTRACTION

It has been shown in preceding sections that an unlimited number of pairs of values of variables satisfy one linear equation. Also, it can be determined graphically whether there is one pair of values, or solution, that will satisfy two given linear equations. The solution of two simultaneous linear equations can also be found by algebraic methods, as illustrated in the following examples:

EXAMPLE 1 Solve the equations $x + y = 6$ and $x - y = 2$.

SOLUTION

Given	$x + y = 6$	(a)
	$x - y = 2$	(b)
Add (a) and (b),	$2x = 8$	(c)
D: 2 in (c),	$x = 4$	
Substitute this value of x in (a),	$4 + y = 6$	
Collect terms,	$y = 2$	

The common solution for (a) and (b) is

$$x = 4 \qquad y = 2$$

CHECK

Substitute in (a),	$4 + 2 = 6$
Substitute in (b),	$4 - 2 = 2$

In Example 1 the coefficients of y in Eqs. (a) and (b) are the same except for sign. That being so, y can be *eliminated* by adding these equations, and the resulting sum is an equation in one unknown. This method of solution is called *elimination by addition*.

Because the coefficients of x are the same in Eqs. (a) and (b) of Example 1, x could have been eliminated by subtracting either equation from the other, and an equation containing only y as a variable would have been the result. This method of solution is called *elimination by subtraction*. The remaining variable x would have been solved for in the usual manner by substituting the value of y in either equation.

EXAMPLE 2 Solve the equations $3x - 4y = 13$ and $5x + 6y = 9$.

SOLUTION

Given	$3x - 4y = 13$	(a)
	$5x + 6y = 9$	(b)
M: 3 in (a),	$9x - 12y = 39$	(c)
M: 2 in (b),	$10x + 12y = 18$	(d)
Add (c) and (d),	$19x = 57$	(e)
D: 19 in (e),	$x = 3$	(f)
Substitute this value of x in (a),	$9 - 4y = 13$	(g)
Collect terms,	$-4y = 4$	(h)
D: -4 in (h),	$y = -1$	

The common solution for (a) and (b) is

$$x = 3 \qquad y = -1$$

CHECK

Substitute in (a),	$9 + 4 = 13$
Substitute in (b),	$15 - 6 = 9$

In Example 2 the coefficients of x and y in Eqs. (a) and (b) are not the same. The coefficients of y were made the same absolute value in the Eqs. (c) and (d) in order to eliminate y by the method of addition.

EXAMPLE 3 Solve the equations $4a - 3b = 27$ and $7a - 2b = 31$.

SOLUTION

Given	$4a - 3b = 27$	(a)
	$7a - 2b = 31$	(b)
M: 7 in (a),	$28a - 21b = 189$	(c)
M: 4 in (b),	$28a - 8b = 124$	(d)
Subtract (d) from (c),	$-13b = 65$	(e)
D: $-$ 13 in (e),	$b = -5$	(f)
Substitute this value of b in (a),	$4a + 15 = 27$	(g)
Collect terms,	$4a = 12$	(h)
D: 4 in (h),	$a = 3$	(i)

The common solution for (a) and (b) is

$$a = 3 \qquad b = -5$$

CHECK

Substitute the values of the variables (a) and (b) as usual.

In Example 3 the coefficients of a and b in Eqs. (a) and (b) are not the same. The coefficients of a were made the same absolute value in the Eqs. (c) and (d) in order to eliminate a by the method of subtraction.

RULE

To solve two simultaneous linear equations having two variables by the method of elimination by addition or subtraction:

1. If necessary, multiply each equation by a number that will make the coefficients of one of the variables of equal absolute value.
2. If these coefficients of equal absolute value have like signs, subtract one equation from the other; if they have unlike signs, add the equations.
3. Solve the resulting equation.
4. Substitute the value of the variable found in step 3 in one of the original equations, and then solve the resulting equation for the remaining variable.
5. Check the solution by substituting in both the original equations.

PROBLEMS 17–2

Solve for the unknowns by the method of addition and subtraction:

1. $2a - b = 4$
   $2a + 3b = 12$
2. $\alpha - 5\beta = 23$
   $2\alpha - 2\beta = 14$
3. $4x + 2 = -2z$
   $3x + 12 = 2z$
4. $5V + 6I = 7$
   $2V + 5I = -5$
5. $R_1 - 6R_2 = -2$
   $3R_1 + R_2 = 13$
6. $5\theta + 4\phi = 12$
   $\theta - 2\phi = 8$
7. $3p - 50 = -7q$
   $5p - 2q = 15$
8. $2\alpha - \beta = 3$
   $4\beta + 3\alpha = 10$
9. $5M + L = 13$
   $3M + 2L = 12$
10. $4p - 3q = 5$
    $9p - 8q = 0$
11. $3I_1 - 3I_2 = 24$
    $I_1 + 2I_2 = -1$
12. $3Z_1 + Z_2 = 14$
    $Z_1 + 2Z_2 = 13$
13. $3i + 5I = -9$
    $3I - 4i = -17$
14. $I + 3i = 25$
    $4i + I = 31$
15. $\theta + 5 = 2\phi$
    $3\theta - 20 = -4\phi$
16. $5\alpha + 1 = -3\beta$
    $7\beta + 3\alpha - 15 = 0$
17. $0.3V + 0.2v = -0.9$
    $0.5V = -1.9 - 0.3v$
18. $0.9X_L + 0.04X_C = 9.4$
    $0.05X_L + 2.5 = 0.3X_C$
19. $0.03I - 0.54 = -0.02i$
    $21 - i = I$
20. $0.4L + 1.6 = 0.9X$
    $0.7X + 0.2 = 0.6L$
21. Solve the problems of Problems 17–1 by the method of addition and subtraction, and confirm the answers obtained by the graphical method.

17–3 SOLUTION BY SUBSTITUTION

Another common method of solution is called *elimination by substitution.*

EXAMPLE 4 Solve the equations $16x - 3y = 10$ and $8x + 5y = 18$.

SOLUTION

Given
$$16x - 3y = 10 \qquad (a)$$
$$8x + 5y = 18 \qquad (b)$$

Solve (*a*) for x in terms of y,
$$x = \frac{10 + 3y}{16} \qquad (c)$$

Substitute this value of x in (*b*),
$$8\left(\frac{10 + 3y}{16}\right) + 5y = 18 \qquad (d)$$

M: 16 in (*d*),
$$8(10 + 3y) + 80y = 288 \qquad (e)$$
Expand (*e*),
$$80 + 24y + 80y = 288 \qquad (f)$$
Collect terms in (*f*),
$$104y = 208 \qquad (g)$$
D: 104 in (*g*),
$$y = 2 \qquad (h)$$

Substitute value of y in (*a*),	$16x - 6 = 10$	(*i*)
Collect terms in (*i*),	$16x = 16$	(*j*)
D: 16 in (*j*),	$x = 1$	(*k*)

CHECK

Usual method.

Not only is the method of substitution a very useful one, it also serves to emphasize that the values of the variables are the same in both equations. The method of solving by substitution can be stated as follows:

RULE

To solve by substitution:

1. Solve one of the equations for one of the variables in terms of the other variable.
2. Substitute the resultant value of the variable, found in step 1, in the remaining equation.
3. Solve the equation obtained in step 2 for the second variable.
4. In the simpler of the original equations, substitute the value of the variable found in step 3 and solve the resulting equation for the remaining unknown variable.

PROBLEMS 17–3

Solve by the method of substitution. (*Continued on p. 297.*)

1. $V + 4I = 14$
   $V - 4I = -2$
2. $a + 2b = 13$
   $3a + 15 = 3b$
3. $R_1 + R_2 = 8$
   $R_1 - R_2 = 2$
4. $\lambda + 8\mu = 122$
   $\lambda + \mu = 17$
5. $6B - 4G = 38$
   $4B + 2G = 44$
6. $5I_1 + 7I_2 = 74$
   $5I_2 - 7I_1 = 0$
7. $2X_L - 7 = 5X_C$
   $7X_L - 40 = 2X_C$
8. $3X_L + 20 = 8X_C$
   $3X_C - 44 = -8X_L$
9. $5I_1 - 39 = 3I_2$
   $3I_1 + 3 = 4I_2$
10. $3\lambda_1 + 11 = 4\lambda_2$
    $3\lambda_2 = 9 + 2\lambda_1$
11. $4L_1 + 2L_2 = 20$
    $2L_2 + 1 = 3L_1$
12. $18 - 6I_1 = 8I_2$
    $5I_1 + 4I_2 - 22 = 0$
13. $5Y - 42 = -3Z$
    $Z - 18 = -2Y$
14. $2\pi - 8 = \omega$
    $2\omega + 3\pi = 5$
15. $2R_1 + 4R_2 = 12$
    $5R_1 - 6R_2 = 78$
16. $4X_L - 9X_C = -16$
    $7X_C + 2 = 6X_L$
17. $0.6\theta + 1.7\phi = 3.5$
    $1.4\theta - 3.9 = 0.3\phi$
18. $0.6I + 0.8i = 2.6$
    $7.0 - 0.5I = -0.3i$
19. $1.2a - 2b = 1$
    $1.4a - 1.5b = 1.5$
20. $0.6L + 0.2M = 2040$
    $0.5L + 0.3M = 1860$

21. Solve Probs. 1 to 20 graphically, and confirm the answers obtained algebraically.

17–4 SOLUTION BY COMPARISON

In the method of solution by comparison we solve for the value of the same variable in each equation in terms of the other variable and place these values equal to each other. The result is an equation having only one unknown.

EXAMPLE 5 Solve the equations $x - 4y = 14$ and $4x + y = 5$.

SOLUTION

Given	$x - 4y = 14$	(*a*)
	$4x + y = 5$	(*b*)
Solve (*a*) for x in terms of y,	$x = 14 + 4y$	(*c*)
Solve (*b*) for x in terms of y,	$x = \dfrac{5 - y}{4}$	(*d*)
Equate values of x in (*c*) and (*d*),	$14 + 4y = \dfrac{5 - y}{4}$	(*e*)
M: 4 in (*e*),	$56 + 16y = 5 - y$	(*f*)
Collect terms in (*f*),	$17y = -51$	(*g*)
D: 17 in (*g*),	$y = -3$	
Substitute the value of y in (*a*),	$x + 12 = 14$	
Collect terms,	$x = 2$	

CHECK

Usual method.

PROBLEMS 17–4

Solve by the method of comparison:

1. $5Y + 3G = 8$
   $Y + G = 2$
2. $6I_1 - I_2 = 5$
   $I_1 + 2I_2 = 16$
3. $\varepsilon + 3\eta = 11$
   $4\varepsilon + 7\eta = 29$
4. $4E + 3e = 15$
   $2E + 11e = 36$
5. $3\alpha + 2\beta = 22$
   $2 + 3\alpha = \beta$
6. $5L_1 + 24 = 6L_2$
   $9L_2 - 22 = 4L_1$
7. $3X - 5Z = 35$
   $X + 2Z = 19$
8. $2M - 24Q = 0$
   $3M - 20Q = 16$
9. $0.5m + 0.2p = 3.10$
   $0.7m - 1.30 = 0.1p$
10. $2.8I - 2.7i = 19.9$
    $6 + 5i = 2.1I$

Solve Probs. 1 to 10 graphically and by the other algebraic methods.

17–5 FRACTIONAL FORM

Simultaneous linear equations having fractions with numerical denominators are readily solved by first clearing the fractions from the equations and then solving by means of any method which is considered to be most convenient.

EXAMPLE 6 Solve the equations $\frac{x}{4} + \frac{y}{3} = \frac{7}{12}$ and $\frac{x}{2} - \frac{y}{4} = \frac{1}{4}$.

SOLUTION

Given $$\frac{x}{4} + \frac{y}{3} = \frac{7}{12} \qquad (a)$$

$$\frac{x}{2} - \frac{y}{4} = \frac{1}{4} \qquad (b)$$

M: 12, the LCD, in (*a*), $3x + 4y = 7$

M: 4, the LCD, in (*b*), $2x - y = 1$

The resulting equations contain no fractions. Inspection of them shows that solution by addition is most convenient. The solution is

$$x = 1 \qquad y = 1$$

PROBLEMS 17–5

Solve the following sets of equations:

1. $\frac{x}{6} + \frac{3y}{6} = 6$
   $\frac{x}{3} - y = -4$

2. $\frac{x}{3} - \frac{6y}{10} = 2$
   $\frac{x}{3} - y = 4$

3. $\frac{5\lambda}{6} + \frac{7\mu}{2} = 3$
   $\frac{\lambda}{3} - \frac{5\mu}{2} = 9$

4. $2E - \frac{15e}{26} = 3\frac{1}{13}$
   $\frac{13E}{33} - \frac{8e}{99} = 1$

5. $V_1 + V_2 = 25$
   $\frac{V_1}{3} + \frac{V_2}{5} = 21$

6. $\frac{I}{3} + \frac{i}{5} = -\frac{1}{15}$
   $\frac{7i}{30} + \frac{I}{10} = \frac{1}{2}$

7. $\frac{G}{4} + \frac{B}{14} = \frac{1}{2} = \frac{3G}{6} + \frac{6B}{28}$

8. $\frac{Z_1 + 2Z_2}{24} - \frac{Z_2 - 5}{4} = \frac{Z_1 + Z_2 + 1}{36}$
   $\frac{Z_1 - 2}{12} = \frac{5 + Z_2}{3} - \frac{2Z_2 + 6}{6}$

9. $\frac{\theta - \phi}{4} + \frac{5\theta}{4} = \frac{\theta}{2} + \frac{7\phi}{4}$
   $\frac{\theta + \phi}{3} = 3\phi - 1.6$

10. $\frac{4I - i}{15} + \frac{1}{8} = 2i - \frac{12I}{5}$
    $i - I = \frac{3}{16}$

17–6 FRACTIONAL EQUATIONS

When variables occur in denominators, it is generally easier to solve without clearing the equations of fractions.

EXAMPLE 7 Solve the equations $\frac{5}{x} - \frac{6}{y} = -\frac{1}{2}$ and $\frac{2}{x} - \frac{3}{y} = -1$.

SOLUTION

Given
$$\frac{5}{x} - \frac{6}{y} = -\frac{1}{2} \quad (a)$$
$$\frac{2}{x} - \frac{3}{y} = -1 \quad (b)$$

M: 2 in (a),
$$\frac{10}{x} - \frac{12}{y} = -1 \quad (c)$$

M: 5 in (b),
$$\frac{10}{x} - \frac{15}{y} = -5 \quad (d)$$

Subtract (d) from (c),
$$\frac{3}{y} = 4$$
$$y = \tfrac{3}{4}$$

Substitute $\frac{3}{4}$ for y in (b),
$$\frac{2}{x} - 4 = -1$$

Collect terms,
$$\frac{2}{x} = 3$$
$$\therefore x = \tfrac{2}{3}$$

CHECK

Usual method.

> Alternative appearance:
>
> $$5\left(\frac{1}{x}\right) - 6\left(\frac{1}{y}\right) = -\frac{1}{2}$$
> $$2\left(\frac{1}{x}\right) - 3\left(\frac{1}{y}\right) = -1$$
>
> Solve for $\frac{1}{x}$ and $\frac{1}{y}$.
>
> $\frac{1}{y} = \frac{4}{3}$ and $\frac{1}{x} = \frac{3}{2}$
>
> and invert:
>
> $x = \frac{2}{3}$ and $y = \frac{3}{4}$

PROBLEMS 17–6

Solve the following sets of equations:

1. $\frac{9}{R_1} - \frac{5}{R_2} = 2$
   $\frac{3}{R_2} + \frac{5}{R_1} = 30$

2. $\frac{12}{V} - \frac{9}{v} = 3$
   $\frac{6}{V} + \frac{3}{v} = 4$

3. $\dfrac{4}{G-1} = \dfrac{3}{1-Y}$

$\dfrac{5}{2Y-5} - \dfrac{7}{2G-39} = 0$

4. $\dfrac{6}{p} = \dfrac{1}{3}$

$\dfrac{5}{q} - \dfrac{3}{p} = \dfrac{1}{4}$

5. $\dfrac{3}{\theta} + \dfrac{3}{\phi} = \dfrac{51}{60}$

$\dfrac{4}{\theta} - \dfrac{2}{\phi} = \dfrac{19}{30}$

6. $\dfrac{4}{a-1} = \dfrac{3}{1-b}$

$\dfrac{7}{2a-39} = \dfrac{5}{2b-5}$

7. $\dfrac{R_1 + 3R_2}{R_2 - R_1} = \dfrac{1}{3}$

$\dfrac{R_1}{4} - \dfrac{1}{6} = \dfrac{R_1 - 2R_2}{6}$

8. $\dfrac{\lambda + 3\pi}{7} + 1 = \pi$

$\dfrac{2}{\lambda} - \dfrac{4}{\pi} = 0$

9. $\dfrac{1}{I_1} + \dfrac{1}{I_2} = \dfrac{73}{15}$

$\dfrac{12}{I_1} - \dfrac{3}{I_2} = \dfrac{156}{15}$

10. $\dfrac{1}{\pi} + \dfrac{1}{\lambda} = 3\dfrac{31}{35}$

$\dfrac{1}{2\pi} + \dfrac{1}{4\lambda} = 1\dfrac{19}{35}$

17–7 LITERAL EQUATIONS IN TWO UNKNOWNS

The solution of literal simultaneous equations involves no new methods of solution. In general, it will be found that the addition or subtraction method will suffice for most cases.

EXAMPLE 8 Solve the equations $ax + by = c$ and $mx + ny = d$.

SOLUTION

Given
$$ax + by = c \qquad (a)$$
$$mx + ny = d \qquad (b)$$

First eliminate x.

M: m in (a), $$amx + bmy = cm \qquad (c)$$
M: a in (b), $$amx + any = ad \qquad (d)$$
Subtract (d) from (c), $$bmy - any = cm - ad \qquad (e)$$
Factor (e), $$y(bm - an) = cm - ad \qquad (f)$$
D: $(bm - an)$ in (f), $$y = \frac{cm - ad}{bm - an}$$

Now go back to (a) and (b), and eliminate y.

M: n in (a), $$anx + bny = cn \qquad (g)$$
M: b in (b), $$bmx + bny = bd \qquad (h)$$
Subtract (h) from (g), $$anx - bmx = cn - bd \qquad (i)$$
Factor (i), $$x(an - bm) = cn - bd \qquad (j)$$
D: $(an - bm)$ in (j), $$x = \frac{cn - bd}{an - bm}$$
$$= \frac{bd - cn}{bm - an}$$

EXAMPLE 9 Solve the equations

$$\frac{a}{x} + \frac{b}{y} = \frac{1}{xy} \quad \text{and} \quad \frac{c}{x} + \frac{d}{y} = \frac{1}{xy}$$

SOLUTION

Given

$$\frac{a}{x} + \frac{b}{y} = \frac{1}{xy} \qquad (a)$$

$$\frac{c}{x} + \frac{d}{y} = \frac{1}{xy} \qquad (b)$$

First eliminate y, although it makes no difference which variable is eliminated first.

M: xy, the LCD, in (a),	$ay + bx = 1$	(c)
M: xy, the LCD, in (b),	$cy + dx = 1$	(d)
M: c in (c),	$acy + bcx = c$	(e)
M: a in (d),	$acy + adx = a$	(f)
Subtract (f) from (e),	$bcx - adx = c - a$	(g)
Factor (g),	$x(bc - ad) = c - a$	(h)
D: $(bc - ad)$ in (h),	$x = \dfrac{c - a}{bc - ad}$	

Now go back to (a) and (b) to eliminate x, and find

$$y = \frac{b - d}{bc - ad}$$

PROBLEMS 17–7

Given:	Solve for
1. $a - 5\pi = b$ $a + b = \theta$	a and b
2. $5x + 2y = \alpha$ $2x - 7y = \beta$	x and y
3. $\dfrac{5Y}{b} - YG + \dfrac{3G}{a} = 0$ $YG - \dfrac{5G}{b} = \dfrac{3Y}{a}$	a and b
4. $4L_1 + 3L_2 = C$ $3L_1 - 2L_2 = C$	L_1 and L_2
5. $R_1 + R_2 = Z_1$ $IR_1 - iR_2 = Z_2$	R_1 and R_2
6. $5r + 3R = Z_1$ $3r + 7R = Z_2$	R and r
7. $0.25\ X_1 + 0.375X_2 = Z_1$ $0.03\ X_1 + 0.02X_2 = Z_2$	X_1 and X_2

8. $\dfrac{R_L}{4} + \dfrac{R_p}{3} = R_T$

$\dfrac{R_L}{2} - \dfrac{R_p}{4} = R_1$ $\qquad R_L$ and R_p

9. $\dfrac{1}{R_x} + \dfrac{3}{R_y} = \dfrac{1}{R_p}$

$\dfrac{5}{R_x} + \dfrac{1}{R_y} = \dfrac{1}{R_t}$ $\qquad R_x$ and R_y

10. $\frac{1}{3}(Z_1 - Z_2) = Z_1 - Z_2 - X_C$

$\frac{2}{5}Z_1 - Z_2 = 0$ $\qquad Z_1$ and Z_2

17–8 EQUATIONS CONTAINING THREE UNKNOWNS

In the preceding examples and problems, two equations were necessary to solve for two unknown variables. For problems involving three variables, three equations are necessary. The same methods of solution apply for problems involving three variables.

EXAMPLE 10 Solve the equations

$$2x + 3y + 5z = 0 \qquad (a)$$
$$6x - 2y - 3z = 3 \qquad (b)$$
$$8x - 5y - 6z = 1 \qquad (c)$$

SOLUTION

Choose a variable to be eliminated. Let it be x.

M: 3 in (a), $\qquad 6x + 9y + 15z = 0 \qquad (d)$

$\qquad 6x - 2y - 3z = 3 \qquad (b)$

Subtract (b) from (d), $\qquad 11y + 18z = -3 \qquad (e)$

M: 4 in (a), $\qquad 8x + 12y + 20z = 0 \qquad (f)$

$\qquad 8x - 5y - 6z = 1 \qquad (c)$

Subtract (c) from (f), $\qquad 17y + 26z = -1 \qquad (g)$

This gives Eqs. (e) and (g) in two variables y and z. Solving them, we obtain $y = 3$, $z = -2$.
Substitute these values into (a),

$$2x + 9 - 10 = 0 \qquad (h)$$

Collect terms, $\qquad 2x = 1 \qquad (i)$

D: 2 in (i), $\qquad x = \frac{1}{2}$

CHECK

Substitute the values of the variables in the equations.

PROBLEMS 17–8

Solve:

1. $10I_t - 4I_2 + I_1 = 5$
   $4I_t - I_2 + 3I_1 = 11$
   $3I_t + 6I_2 - 2I_1 = 9$
2. $I_1 - I_2 + I_3 = 6$
   $I_2 + I_3 + I_1 = 10$
   $I_1 - I_3 + I_2 = 0$
3. $I_1 + 2I_2 + I_3 = 8$
   $3I_1 - 3I_2 - 2I_3 = 9$
   $2I_1 + I_2 + 3I_3 = 43$
4. $a - 2b + c = 3$
   $a + b + 2c = 1$
   $2a - b + c = 2$
5. $\frac{1}{R_L} - \frac{1}{R_p} - \frac{1}{R_1} = \frac{31}{126}$
   $\frac{1}{R_L} + \frac{1}{R_p} - \frac{1}{R_1} = \frac{67}{126}$
   $\frac{1}{R_p} - \frac{1}{R_1} - \frac{1}{R_L} = \frac{-59}{126}$
6. $\frac{1}{a} - \frac{1}{b} - \frac{1}{c} = 1$
   $\frac{1}{b} - \frac{1}{a} - \frac{1}{c} = 1$
   $\frac{1}{c} - \frac{1}{a} - \frac{1}{b} = 1$
7. $0.1r - 0.1R + 0.6R_L = 4$
   $2r + 3R + 6R_L = 53$
   $\frac{3}{40}r + \frac{1}{20}R - \frac{1}{40}R_L = \frac{1}{20}$
8. $E_1 - E_2 - E_3 = \alpha$
   $E_3 - E_1 - E_2 = \beta$
   $E_2 - E_3 - E_1 = \gamma$
9. $a - 8 = b$
   $a - 2c = -4$
   $3c - 3b = 12$
10. $a + 5 = c$
    $7b = 3c - 1$
    $2b - a = c - 9$

17–9 METHODS OF SOLUTION OF PROBLEMS

It is convenient to solve a problem involving more than one unknown by setting up a system of simultaneous equations according to the statements of the problem.

EXAMPLE 11

When a certain number is increased by one-third of another number, the result is 23. When the second number is increased by one-half of the first number, the result is 29. What are the numbers?

SOLUTION

Let x = first number and y = second number.

Then
$$x + \frac{1}{3}y = 23 \qquad (a)$$

Also,
$$y + \frac{1}{2}x = 29 \qquad (b)$$

Solving the equations, we obtain x = 16, y = 21.

CHECK

When 16, the first number, is increased by one-third of 21, we have

$$16 + 7 = 23$$

When 21, the second number, is increased by one-half of 16, we have

$$21 + 8 = 29$$

EXAMPLE 12 Two airplanes start from Omaha at the same time. The plane traveling west has a speed 130 km/h faster than that of the plane traveling east. At the end of 4 h they are 2600 km apart. What is the speed of each plane?

SOLUTION

Let x = rate of plane flying west and y = rate of plane flying east.

Then $x - y = 130$ (*a*)

Since Rate × time = distance

then $4x$ = distance traveled by plane flying west

and $4y$ = distance traveled by plane flying east

hence, $4x + 4y = 2600$ (*b*)

Solving Eqs. (*a*) and (*b*), we obtain

$$x = 390 \text{ km/h}$$
$$y = 260 \text{ km/h}$$

CHECK

Substitute these values of x and y into the statements of the example.

Often it is possible to derive a formula from known data and thereby eliminate terms which are not desired or cannot be used conveniently in some investigation.

EXAMPLE 13 The effective voltage V of an alternating voltage is equal to 0.707 times its maximum value V_{max}. That is,

$$V = 0.707V_{max} \tag{1}$$

Also, the average value V_{av} is equal to 0.637 times the maximum value. That is,

$$V_{av} = 0.637V_{max} \tag{2}$$

It is desired to express the effective value V in terms of the average value V_{av}.

SOLUTION

V_{max} must be eliminated.

Solving Eq. (1) for V_{max}, $V_{max} = \dfrac{V}{0.707}$

Solving Eq. (2) for V_{max}, $V_{max} = \dfrac{V_{av}}{0.637}$

By Axiom 5, $\dfrac{V}{0.707} = \dfrac{V_{av}}{0.637}$

Solving for V, $V = 1.11V_{av}$ (3)

Equation (3) shows that the effective value of an alternating voltage is 1.11 times the average value of the voltage.

EXAMPLE 14 You know that in a dc circuit $P = VI$ and also that $P = I^2R$. Derive a formula for V in terms of I and R.

SOLUTION

It is evident that P must be eliminated. Because both equations are equal to P, we can equate them (Axiom 5) and obtain

$$VI = I^2R$$

D: I, $$V = IR \tag{4}$$

EXAMPLE 15 The quantity of electricity Q, in coulombs, in a capacitor is equal to the product of the capacitance C and the applied voltage V. That is,

$$Q = CV \tag{5}$$

The total voltage across capacitors C_a and C_b connected in series is $V = V_a + V_b$. Find C in terms of C_a and C_b.

SOLUTION

Solve for V, V_a, and V_b. Thus

$$V = \frac{Q}{C}$$

$$V_a = \frac{Q}{C_a}$$

and $$V_b = \frac{Q}{C_b}$$

Then, since $$V = V_a + V_b$$

By substitution $$\frac{Q}{C} = \frac{Q}{C_a} + \frac{Q}{C_b}$$

D: Q, $$\frac{1}{C} = \frac{1}{C_a} + \frac{1}{C_b}$$

M: CC_aC_b, the LCD, $$C_aC_b = CC_b + CC_a$$

Transposing, $$CC_a + CC_b = C_aC_b$$

D: $(C_a + C_b)$, $$C = \frac{C_aC_b}{C_a + C_b} \tag{6}$$

This is the formula for the resultant capacitance C of two capacitors C_a and C_b connected in series.

PROBLEMS 17–9

1. The sum of two voltages is V_s V, and their difference is V_d V. What are the voltages?
2. Find two numbers whose sum is 16 and whose difference is 4.

3. If 3 is subtracted from each term of a fraction, the value of the fraction becomes $\frac{1}{2}$, and if 3 is added to each term, the value of the fraction becomes $\frac{2}{3}$. What is the fraction?
4. In a right triangle, the acute angles are complementary (that is, they add up to 90°). What are the angles if their difference is 10°?
5. The difference between the two acute angles of a right triangle is $\alpha°$. Find the angles.
6. The sum of the three angles of any triangle is 180°. Find the three angles of a particular triangle if the smallest angle is one-third the middle angle and the largest is 5° larger than the middle one.
7. A TV repair technician goes to the parts dealer for an assortment of common resistors and capacitors. The sales clerk replies: "We have two such assortments: 25 resistors and 10 capacitors for \$7.25, or 50 of each for \$25.00. Both assortments come under the same discount schedule." "I'll take the larger selection," says the technician, "if you'll figure out the price of one resistor and one capacitor."
Help the sales clerk.
8. A takes one hour longer than B to walk 25 km, but if A's pace were doubled, then A would take 1.5 h less than B. Find their walking rates.
9. In 2 h, J drives 110 km less than A does in 3 h. In 6 h, J drives 120 km more than A does in 4 h. Find their average rates of driving.
10. $v = gt$ and $s = \frac{1}{2}gt^2$. Solve for v in terms of s and t.
11. $C = \dfrac{Q}{V}$ and $W = \dfrac{QV}{2}$. Solve for W in terms of C and Q.
12. $I = \dfrac{V}{R}$ and $P = I^2R$. If $P = 2.7$ kW and $V = 180$ V, find the current I and the resistance R.
13. $v = u + at$ and $s = \frac{1}{2}(u + v)t$. Find the distance s in terms of initial velocity u and acceleration a and time t.
14. Use the information of Prob. 13 to show that $v^2 = u^2 + 2as$.
15. Gain $= \dfrac{i_d r_d}{v_{gs}}$ and gain $= \dfrac{v_d}{i_d\left(r_s + \dfrac{1}{g_m}\right)}$.

    Solve for gain in terms of r_d and g_m when $r_s = 0$.
16. $R = 2D_L fL$ and $Q = \dfrac{2\pi fL}{R}$. Solve for D_L in terms of π and Q.
17. $R = \omega LQ$, $Q = \dfrac{\omega L}{r}$, and $\omega^2 = \dfrac{1}{LC}$. Solve for R in terms of L, C, and r.
18. $I = \dfrac{V}{R}$ and $I_1 = \dfrac{V}{R + R_1}$. Solve for R in terms of R_1, I, and I_1.

19. $Q = It$ coulombs (C), and $I = \dfrac{CV}{t}$ A. Solve for Q in terms of C and V.

20. Given $P = VI$ W, $I = \dfrac{V}{R}$ A, and $H = 0.24I^2Rt$ J, solve for H in terms of P and t.

21. Use the data of Prob. 20 to find H when $V = 50$ V over a time $t = 8$ s if the heater resistance $R = 450\ \Omega$.

22. Given $I_aR_a = I_bR_b$, $\dfrac{Q_a}{Q_b} = \dfrac{C_a}{C_b}$, $I_a = \dfrac{Q_a}{t}$, and $I_b = \dfrac{Q_b}{t}$, show that $R_aC_a = R_bC_b$.

23. $P = VI$ W, $I = \dfrac{V}{R}$ A, and $H = I^2Rt$ J. Solve for H in terms of P and t.

24. Use the data of Prob. 23 to find V_p when $\mu = 50$, Vg $= 5$ V, Ip $= 12.5$ mA, and $R_p = 10\text{k}\Omega$.

25. The three-Varley method of cable fault location yields the following relationships:

$$R_AV_1 + R_AR_Y = R_AV_2 - R_YR_B$$
$$R_AV_2 + R_AR_X = R_AV_3 - R_XR_B$$
$$R_AV_1 + R_AR_T = R_AV_3 - R_TR_B$$

Solve these three equations for R_Y, R_X, and R_T, and show that $R_T = R_Y + R_X$.

26. Given $V = I_x(R + R_x)$, $V = I_a(R + R_a)$, and $V = IR$. Show that

$$R_x = R_a \times \frac{\dfrac{I - I_x}{I_x}}{\dfrac{I - I_a}{I_a}}$$

27. Given three star-delta transformation equations:

$$R_a = \frac{R_1R_3}{R_1 + R_2 + R_3}$$
$$R_b = \frac{R_1R_2}{R_1 + R_2 + R_3}$$
$$R_c = \frac{R_2R_3}{R_1 + R_2 + R_3}$$

Solve for R_1, R_2, and R_3 in terms of R_a, R_b, and R_c.

28. If three resistors R_1, R_2, and R_3 are connected in parallel so that the total circuit current I_t is divided into I_1, I_2, and I_3, respectively, determine relationships similar to Eq. (1) in Chap. 15 for I_1, I_2, and I_3 in terms of R_1, R_2, R_3, and I_t.

SELF TEST

1. Solve the following pair of equations graphically:

$$2y = -3x + 4$$
$$6y = 7x - 4.2$$

2. Solve by addition and subtraction:

$$m - n = 5$$
$$2m + 2n = 14$$

3. Solve by substitution:

$$3R + 2Z = 5$$
$$5R - 3Z = 21$$

4. Solve by comparison:

$$2y + 3x = 4$$
$$6y - 7x = -4.2$$

5. Solve:

$$X + R = 10.5$$
$$\frac{X}{3} + \frac{R}{9} = 1.7$$

6. Solve:

$$\frac{3}{A} + \frac{8}{B} = 2.6$$
$$\frac{4}{A} + \frac{7}{B} = 2.55$$

7. Given $R_1 - 3R_2 = \alpha$ and $2R_1 + 4R_2 = \beta$, solve for R_1 and R_2.

8. Solve:

$$5I_1 + 4I_2 - 2I_3 = 13$$
$$6I_1 - 3I_2 - 8I_3 = 11$$
$$2I_1 + 2I_2 + 2I_3 = 18$$

9. Given the delta-star transformation equations:

$$R_1 = \frac{R_aR_b + R_bR_c + R_aR_c}{R_c}$$
$$R_2 = \frac{R_aR_b + R_bR_c + R_aR_c}{R_a}$$
$$R_3 = \frac{R_aR_b + R_bR_c + R_aR_c}{R_b}$$

Solve for R_a, R_b, and R_c, each in terms of R_1, R_2, and R_3.

CHAPTER 18

DETERMINANTS

In Chap. 17 we learned four methods of solving simultaneous equations of the second order, and we used some of those methods to solve equation sets of the third order. Indeed, some of the methods we learned are limited to solving simultaneous equations of the second order, while others may be used to solve third-, fourth-, fifth-, or even higher-order systems.

However, after about the third order, the method of repeated addition and subtraction, with its attendant multiplication, becomes tedious. In this chapter we shall investigate a ''mechanical'' method of solving simultaneous equations. This method, known as the method of determinants, is usually not introduced until students are well along in advanced mathematics, so we are not going to study all the fascinating developments which the whole subject of determinants may involve. (That would take a separate book of its own.) Instead, we are going to see how determinants may be put to work for us in order to simplify our solutions to simultaneous equations.

18–1 SECOND-ORDER DETERMINANTS

In Sec. 17–2, we learned how to solve pairs of simultaneous equations by the method of addition and subtraction. Let us apply this method to a pair of *general equations:*

$$a_1x + b_1y = c_1 \qquad (1)$$

$$a_2x + b_2y = c_2$$

where a_1, a_2, b_1, b_2, c_1, and c_2 represent any numbers, positive or negative, integers or fractions, or zero. Let us solve these general equations for x:

$$a_1x + b_1y = c_1 \qquad (a)$$

$$a_2x + b_2y = c_2 \qquad (b)$$

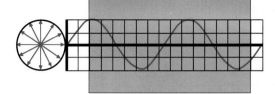

M: b_2 in (a), $\qquad\qquad a_1 b_2 x + b_1 b_2 y = b_2 c_1$ $\hfill (c)$

M: b_1 in (b), $\qquad\qquad a_2 b_1 x + b_1 b_2 y = b_1 c_2$ $\hfill (d)$

Subtract (d) from (c),

$$(a_1 b_2 - a_2 b_1)x = b_2 c_1 - b_1 c_2 \hfill (e)$$

Solve for x, $\qquad\qquad\qquad x = \dfrac{b_2 c_1 - b_1 c_2}{a_1 b_2 - a_2 b_1}$ $\hfill (2)$

It is left as an exercise for you to prove similarly that

$$y = \dfrac{a_1 c_2 - a_2 c_1}{a_1 b_2 - a_2 b_1} \hfill (3)$$

Observe that we have kept the literal factors in alphabetical order for convenience in checking.

Note several interesting facts about these two solutions:

1. Their denominators are identical, and they contain only the coefficients of x and y.
2. The numerator for the solution of y contains no y coefficients.
3. The numerator for the solution of x contains no x coefficients.

For a few minutes, let us consider just the denominator: $a_1 b_2 - a_2 b_1$. We are going to define a new, alternative method of writing this expression.

$$a_1 b_2 - a_2 b_1 = \begin{vmatrix} a_1 & b_1 \\ a_2 & b_2 \end{vmatrix}$$

This arrangement is called the *determinant* of the denominator. It is a mechanical statement made up of two horizontal *rows* and two vertical *columns* of two elements each, and it is a *second-order determinant*. Whenever this form appears, it is understood to mean $a_1 b_2 - a_2 b_1$. To obtain this evaluation of the determinant, we perform diagonal multiplication, first of all downward to the right to obtain

$$\begin{vmatrix} a_1 & b_1 \\ a_2 & b_2 \end{vmatrix}_{a_1 b_2} \qquad\quad \text{(this is, by definition, positive multiplication)}$$

and, second, we multiply upward to the right to obtain

$$\begin{vmatrix} a_1 & b_1 \\ a_2 & b_2 \end{vmatrix} - a_2 b_1 \qquad\quad \text{(this is, by definition, negative multiplication)}$$

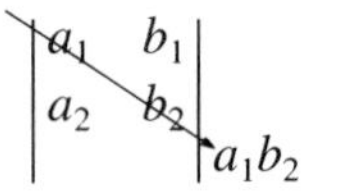

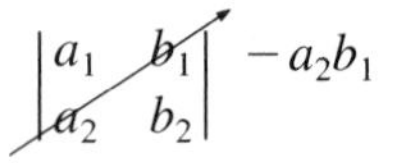

RULES

1. The diagonal multiplication in determinants derives its sign from the direction of the multiplication, and not primarily from any algebraic signs of the elements being multiplied.
2. After the individual steps of multiplication, with the appropriate sign of the multiplication affixed, the products are added algebraically to form the evaluation of the determinants.

EXAMPLE 1 Evaluate the determinant

$$\begin{vmatrix} -3 & 2 \\ 5 & 1 \end{vmatrix}$$

SOLUTION

Perform the signed diagonal multiplication:

$$+(-3)(1) - (5)(2) = -3 - 10 = -13$$

PROBLEMS 18–1

Evaluate the following determinants:

1. $\begin{vmatrix} 5 & 1 \\ 12 & 1 \end{vmatrix}$
2. $\begin{vmatrix} 2 & 6 \\ 15 & -9 \end{vmatrix}$
3. $\begin{vmatrix} 6 & -7 \\ 3 & 8 \end{vmatrix}$
4. $\begin{vmatrix} -2 & 3 \\ 7 & 15 \end{vmatrix}$
5. $\begin{vmatrix} 2 & -7 \\ 2 & -7 \end{vmatrix}$
6. $\begin{vmatrix} 3 & -2 \\ 1 & 2 \end{vmatrix}$
7. $\begin{vmatrix} 6 & 2 \\ 18 & -5 \end{vmatrix}$
8. $\begin{vmatrix} -3 & -7 \\ -4 & -2 \end{vmatrix}$
9. $\begin{vmatrix} 0.9 & 0.1 \\ 0.2 & 0.3 \end{vmatrix}$
10. $\begin{vmatrix} -0.06 & 0.02 \\ 0.05 & -1.6 \end{vmatrix}$
11. $\begin{vmatrix} a & b \\ a & b \end{vmatrix}$
12. $\begin{vmatrix} a & b \\ x & y \end{vmatrix}$
13. $\begin{vmatrix} b & a \\ y & x \end{vmatrix}$
14. $\begin{vmatrix} a & x \\ b & y \end{vmatrix}$
15. $\begin{vmatrix} b & y \\ a & x \end{vmatrix}$
16. $\begin{vmatrix} y & x \\ b & a \end{vmatrix}$

18–2 SOLUTION OF EQUATIONS

Consider Eqs. (2) and (3) to be the solutions for x and y in the general equations (1):

$$x = \frac{b_2c_1 - b_1c_2}{a_1b_2 - a_2b_1} \qquad y = \frac{a_1c_2 - a_2c_1}{a_1b_2 - a_2b_1}$$

Or, in determinant form:

$$x = \frac{\begin{vmatrix} b_2 & c_2 \\ b_1 & c_1 \end{vmatrix}}{\begin{vmatrix} a_1 & b_1 \\ a_2 & b_2 \end{vmatrix}} \tag{4}$$

$$y = \frac{\begin{vmatrix} a_1 & c_1 \\ a_2 & c_2 \end{vmatrix}}{\begin{vmatrix} a_1 & b_1 \\ a_2 & b_2 \end{vmatrix}} \tag{5}$$

Let us see how the determinant form may be developed directly from the original equations without performing the intervening additions and subtractions. Given the original equations:

$$\begin{aligned} a_1x + b_1y &= c_1 \\ a_2x + b_2y &= c_2 \end{aligned} \tag{1}$$

First, produce the determinant of the denominator by setting, in order, the coefficients of the unknowns:

$$\begin{vmatrix} a_1 & b_1 \\ a_2 & b_2 \end{vmatrix}$$

Second, by using the denominator determinant as a base, develop the determinant of the numerator of the solution for x by replacing the column of x coefficients by the corresponding column of constants (the right-hand sides of the equations). Then complete the new determinant by putting in the column of the y coefficients in its original position:

$$\begin{vmatrix} c_1 & b_1 \\ c_2 & b_2 \end{vmatrix}$$

Confirm that this determinant is identical in value with

$$\begin{vmatrix} b_2 & c_2 \\ b_1 & c_1 \end{vmatrix}$$

given as Eq. (4), but easier to develop automatically.

Third, still using the denominator determinant as a starting place, develop the determinant of the numerator of y by replacing the column of y coefficients by the column of constants and leaving the column of x coefficients in its original position:

$$\begin{vmatrix} a_1 & c_1 \\ a_2 & c_2 \end{vmatrix}$$

Last, put these three determinants together to form the full solution statements:

$$x = \frac{\begin{vmatrix} c_1 & b_1 \\ c_2 & b_2 \end{vmatrix}}{\begin{vmatrix} a_1 & b_1 \\ a_2 & b_2 \end{vmatrix}} \tag{4}$$

$$y = \frac{\begin{vmatrix} a_1 & c_1 \\ a_2 & c_2 \end{vmatrix}}{\begin{vmatrix} a_1 & b_1 \\ a_2 & b_2 \end{vmatrix}} \tag{5}$$

RULE

To solve two simultaneous equations having two variables by the method of determinants:

1. Form the denominator determinant by using the coefficients of the unknowns in their correct rows and columns.
2. Form the x numerator determinant by replacing the column of x coefficients in the denominator determinant by the column of constants.
3. Form the y numerator determinant by replacing the column of y coefficients in the denominator by the column of constants.
4. Combine the three determinants so formed to produce the pair of solution equations.

EXAMPLE 2 Solve the simultaneous equations

$$\begin{aligned} 3p + 2q &= 8 \\ 5p + q &= 11 \end{aligned}$$

SOLUTION

The denominator determinant is

$$\begin{vmatrix} 3 & 2 \\ 5 & 1 \end{vmatrix}$$

Using this determinant as a base, the determinant for the numerator of p must be $\begin{vmatrix} 8 & 2 \\ 11 & 1 \end{vmatrix}$ and the determinant for the numerator of q must be $\begin{vmatrix} 3 & 8 \\ 5 & 11 \end{vmatrix}$. Thus,

$$p = \frac{\begin{vmatrix} 8 & 2 \\ 11 & 1 \end{vmatrix}}{\begin{vmatrix} 3 & 2 \\ 5 & 1 \end{vmatrix}} \qquad \text{and} \qquad q = \frac{\begin{vmatrix} 3 & 8 \\ 5 & 11 \end{vmatrix}}{\begin{vmatrix} 3 & 2 \\ 5 & 1 \end{vmatrix}}$$

When evaluating determinants, *always* evaluate the denominator first. (The reason will be explained soon.) The value of the denominator is

$$+(3)(1) - (5)(2) = -7$$

The numerator of p has the value

$$+(8)(1) - (11)(2) = -14$$
$$p = \frac{-14}{-7} = 2$$

The numerator of q has the value $+(3)(11) - (5)(8) = -7$, and

$$q = \frac{-7}{-7} = 1$$

18–3 CONSISTENCY OF EQUATIONS

For solving systems of second-order simultaneous equations, there are three main possibilities:

1. The equations may represent straight lines which intersect. These are said to be *independent equations*. They are in no way related to each other except that the unknowns have similar symbols, A, b, x, θ, etc., and one pair of values constitutes the whole solution.
2. The equations may represent superimposed lines. These are said to be *dependent equations*. They are related to each other, and every solution of the one is also a solution of the other. There is an endless number of solutions.
3. The equations may represent parallel lines. These are said to be *inconsistent equations*. They differ only in the constant terms (the y intercepts), and there is no solution for one equation which satisfies the other.

The values of the denominator and the numerators quickly show us into which classification any system of simultaneous equations falls:

1. To be independent, the denominators may not equal zero.
2. To be dependent, the denominator is zero and the numerators equal zero.
3. To be inconsistent, the denominator is zero and at least one of the numerators does not equal zero.

This is why we evaluate the denominator first. If it is zero, there is no single set of values which will constitute the entire solution, and, in electronics problems, there is no use investigating further.

PROBLEMS 18–2

Solve these systems of simultaneous equations by using determinants:

1. $2x + y = 9$
 $4x - y = 6$

2. $3x + y = 20$
 $2x + 3y = 25$

3. $3I_1 - 4I_2 = 17$
 $I_1 + 3I_2 = -3$

4. $R_1 + 3R_2 = 29$
 $R_1 - 3R_2 = -7$

5. $I + i = 2$
 $4I + 3i = 11$

6. $3V + 2V_g = 1$
 $V_g + V = -2$

7. $3\alpha - 7\beta = 19$
 $2\alpha - \beta = 9$

8. $4X_C + 3X_L = 2.9$
 $30X_L = 17 - 8X_C$

9. $0.5R_1 + 0.2R_2 = 315$
 $0.6R_1 - 54 = 0.03R_2$

10. $Z_1 = 9300 - Z_2$
 $192 + 0.06Z_2 = 0.04Z_1$

18–4
THIRD-ORDER DETERMINANTS

When solving sets of three simultaneous equations, naturally, we arrive at third-order determinants consisting of three columns and three rows of three elements each, such as

$$\begin{vmatrix} 3 & 1 & 2 \\ 2 & 6 & 5 \\ 4 & 8 & 1 \end{vmatrix} \qquad \begin{vmatrix} a_1 & b_1 & c_1 \\ a_2 & b_2 & c_2 \\ a_3 & b_3 & c_3 \end{vmatrix}$$

Now, when we multiply on the diagonal, we find a slight complication. Multiplying the main diagonal is simple:

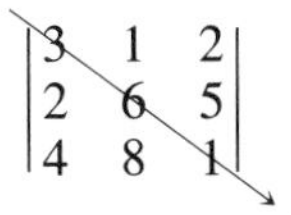

$$\begin{vmatrix} 3 & 1 & 2 \\ 2 & 6 & 5 \\ 4 & 8 & 1 \end{vmatrix} = +(3)(6)(1) = +18$$

but the next diagonal gets complicated:

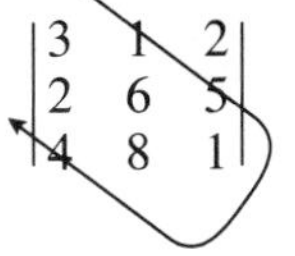

$$\begin{vmatrix} 3 & 1 & 2 \\ 2 & 6 & 5 \\ 4 & 8 & 1 \end{vmatrix} = +(1)(5)(4) = +20$$

and also the next:

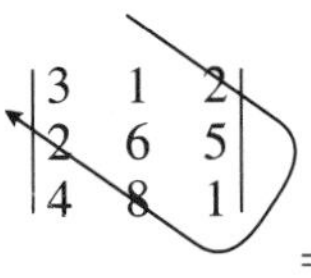

$$\begin{vmatrix} 3 & 1 & 2 \\ 2 & 6 & 5 \\ 4 & 8 & 1 \end{vmatrix} = +(2)(8)(2) = +32$$

And you can see that the negative diagonals will be just as complicated. So we devise a method of notation which gets around this complication and enables us to perform straight-line multiplication. First, we set down the determinant in its usual form, with three columns and three rows. Then, to the right of this determinant, we repeat the first two columns. This process straightens out the diagonals

$$\begin{vmatrix} 3 & 1 & 2 \\ 2 & 6 & 5 \\ 4 & 8 & 1 \end{vmatrix}\begin{matrix} 3 & 1 \\ 2 & 6 \\ 4 & 8 \end{matrix}$$

$$= -(1)(2)(1) = -2$$

$$= +(3)(6)(1) = +18$$

and we obtain, with a complete program of diagonal multiplication, the value of the determinant $= -100$.

EXAMPLE 3 Evaluate the determinant

$$\begin{vmatrix} 2 & -1 & 4 \\ 1 & 6 & 5 \\ 7 & -3 & -2 \end{vmatrix}$$

SOLUTION

Rewrite the determinant and repeat the first two columns outside to the right:

$$\begin{vmatrix} 2 & -1 & 4 \\ 1 & 6 & 5 \\ 7 & -3 & -2 \end{vmatrix}\begin{matrix} 2 & -1 \\ 1 & 6 \\ 7 & -3 \end{matrix}$$

Then perform the diagonal multiplication, signed, as for second-order determinants and obtain

$$-24 - 35 - 12 - 168 + 30 - 2 = -211$$

EXAMPLE 4 Solve the third-order set of simultaneous equations:

$$\begin{aligned} a + 2b + c &= 7 \\ 2a + b + 2c &= 2 \\ a + 3b + 4c &= 14 \end{aligned}$$

SOLUTION

First, write and evaluate the denominator determinant:

$$\begin{vmatrix} 1 & 2 & 1 \\ 2 & 1 & 2 \\ 1 & 3 & 4 \end{vmatrix}\begin{matrix} 1 & 2 \\ 2 & 1 \\ 1 & 3 \end{matrix} = -9$$

Second, develop the determinant for the numerator of a, replacing the column of a coefficients by the column of constants, and evaluate it:

$$\begin{vmatrix} 7 & 2 & 1 \\ 2 & 1 & 2 \\ 14 & 3 & 4 \end{vmatrix}\begin{matrix} 7 & 2 \\ 2 & 1 \\ 14 & 3 \end{matrix} = 18$$

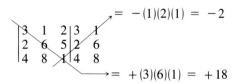

Third, combine the denominator and numerator to evaluate a:

$$a = \frac{18}{-9} = -2$$

You should immediately prove that $b = 4$ and $c = 1$.

PROBLEMS 18–3

Evaluate these third-order determinants:

1. $\begin{vmatrix} 1 & 1 & 1 \\ 3 & -1 & 2 \\ 5 & 2 & -1 \end{vmatrix}$

2. $\begin{vmatrix} 2 & 5 & 1 \\ 7 & 18 & 4 \\ -1 & -20 & -5 \end{vmatrix}$

3. $\begin{vmatrix} 4 & 3 & -2 \\ 5 & -1 & 3 \\ -3 & 2 & 2 \end{vmatrix}$

4. $\begin{vmatrix} -3 & -2 & 3 \\ 0 & -7 & 2 \\ 0 & 7 & -4 \end{vmatrix}$

5. $\begin{vmatrix} 6 & 4 & -3 \\ -5 & 2 & 6 \\ 1 & -3 & -8 \end{vmatrix}$

6. $\begin{vmatrix} 3 & 8 & 3.2 \\ 12 & 20 & 16.5 \\ -16 & -12 & -7.8 \end{vmatrix}$

Solve these simultaneous equations by using determinants:

7. $4I_1 + 3I_2 - 2I_3 = 15$
   $5I_1 - I_2 + 3I_3 = 17$
   $-3I_1 + 2I_2 + 2I_3 = 12$
8. $R_1 + R_2 + R_3 = 3$
   $5R_1 - 2R_2 + 6R_3 = 40$
   $-2R_1 + 3R_2 - 3R_3 = -25$
9. $6x + 4y + 3z = 43$
   $-5x + 2y + 6z = -2$
   $x - 3y - 8z = -11$
10. $3r + 5p - 2q = -3$
    $p + q = 4r$
    $3p - 7q + 2r = -42$
11. $3V + 8v + 12(IR) = 11.25$
    $6(IR) - 8V -- 2v = 1.00$
    $16v - 24(IR) + 9V = -6.75$
12. $12I_1 + 20I_2 + 10I_3 = 16.5$
    $8I_2 - 6I_3 + 3I_1 = 3.2$
    $20I_3 - 16I_1 - 12I_2 = -7.8$

18–5 MINORS

The method of diagonal multiplication works perfectly for both second- and third-order determinants. Unfortunately, it will not work for higher-order systems. Thus, if we are required to evaluate by determinants a fourth- or fifth-order set of equations such as might arise from the solution of a complicated circuit (see Chap. 22), we must work out another useful system. Since we can do this without difficulty, we will not try to prove the statement above. (Even many "higher mathematics" texts say simply: Do not use diagonal multiplication for fourth-order determinants or higher.)

This is how minors come about: Let us evaluate the general third-order determinant:

$$\begin{vmatrix} a_1 & b_1 & c_1 \\ a_2 & b_2 & c_2 \\ a_3 & b_3 & c_3 \end{vmatrix} \begin{matrix} a_1 & b_1 \\ a_2 & b_2 \\ a_3 & b_3 \end{matrix}$$

$$= a_1b_2c_3 + a_3b_1c_2 + a_2b_3c_1 - a_3b_2c_1 - a_1b_3c_2 - a_2b_1c_3 \qquad (6)$$

Consider the terms which involve the value a_1. These may be collected to yield $a_1(b_2c_3 - b_3c_2)$, which in turn could be written

$$a_1\begin{vmatrix} b_2 & c_2 \\ b_3 & c_3 \end{vmatrix}$$

where the new second-order determinant is called the *minor of the element* a_1.

We can develop this minor from the original third-order determinant by selecting the element a_1, crossing out the other elements of the row and column which contain a_1, and writing the minor with the elements remaining.

$$\begin{vmatrix} \textcircled{a_1} & \not{b_1} & \not{c_1} \\ \not{a_2} & b_2 & c_2 \\ \not{a_3} & b_3 & c_3 \end{vmatrix}$$

yields

$$\begin{vmatrix} b_2 & c_2 \\ b_3 & c_3 \end{vmatrix}$$

RULE

To find the *minor* of any element in a determinant, select the element, cross out the row and column containing that element, and write the lower-order determinant which contains all the other elements that remain.

Thus, in the third-order determinant of Eq. (6), the minor of the element b_3 is

$$\begin{vmatrix} a_1 & c_1 \\ a_2 & c_2 \end{vmatrix}$$

EXAMPLE 5 Evaluate the minor of 2 in the determinant

$$\begin{vmatrix} 1 & 4 & 0 \\ 3 & 1 & 5 \\ 5 & 6 & 2 \end{vmatrix}$$

SOLUTION

Striking out the elements in the row and column containing the 2 yields

$$\begin{vmatrix} 1 & 4 \\ 3 & 1 \end{vmatrix} = +1 - 12 = -11$$

PROBLEMS 18–4

Write and evaluate the *minors* of the indicated elements:

1. $\begin{vmatrix} 6 & 4 & -3 \\ -5 & 2 & 6 \\ 1 & -3 & \textcircled{8} \end{vmatrix}$

2. $\begin{vmatrix} 3 & 1 & -1 \\ \textcircled{8} & -2 & 2 \\ -13 & -3 & -1 \end{vmatrix}$

3. $\begin{vmatrix} 2 & 7 & 5 \\ \textcircled{-3} & -2 & 8 \\ 5 & 0 & 6 \end{vmatrix}$

4. $\begin{vmatrix} -3 & -7 & 16 \\ -8 & 2 & 84 \\ 2 & 3 & \textcircled{-26} \end{vmatrix}$

5. $\begin{vmatrix} 18 & 6 & -3 \\ 4 & 0 & 5 \\ 0 & \textcircled{-7} & -1 \end{vmatrix}$

6. $\begin{vmatrix} -8 & \textcircled{-13} & 10 \\ 0 & 2 & 5 \\ 2 & 10 & -20 \end{vmatrix}$

7. $\begin{vmatrix} 0 & 0 & 6 \\ 0 & 5 & 2 \\ -8 & \textcircled{4} & 0 \end{vmatrix}$

8.* $\begin{vmatrix} 3 & 6 & -3 & 2 \\ 2 & -2 & 2 & -1 \\ 5 & \textcircled{25} & 0 & 3 \\ 0 & 5 & -5 & 1 \end{vmatrix}$

▶ ***HINT** The minor of any element of a fourth-order determinant will be a third-order determinant which may itself be evaluated by the diagonal method or by second-step cofactors, which are discussed in the following section.

18–6 COFACTORS

A simple step converts the *minor* into a *cofactor*. When evaluating a complete determinant by the method of cofactors, we first find the minors of all the elements in any given row or column. Then we convert these minors into cofactors by assigning them algebraic signs according to this simple rule:

RULE

Each element of a determinant, regardless of its actual algebraic value, has a cofactor sign according to its place in the determinant. The signs are found by a checkerboard arrangement:

$$\begin{vmatrix} + & - & + \\ - & + & - \\ + & - & + \end{vmatrix}$$

The only thing to remember is to always start the upper left-hand corner (the element in row 1 and column 1) with a + sign. All the rest follows automatically, regardless of the number of elements in the determinant.

EXAMPLE 6 Evaluate the following determinant by means of cofactors:

$$\begin{vmatrix} 1 & 4 & 0 \\ -3 & 1 & 5 \\ 5 & 6 & -2 \end{vmatrix}$$

SOLUTION

Choose any convenient row or column, and, one after the other, set down the individual elements of that row or column, together with their minors:

$$4\begin{vmatrix} -3 & 5 \\ 5 & -2 \end{vmatrix} \qquad 1\begin{vmatrix} 1 & 0 \\ 5 & -2 \end{vmatrix} \qquad 6\begin{vmatrix} 1 & 0 \\ -3 & 5 \end{vmatrix}$$

Then assign the cofactor signs according to the checkerboard plan:

$$-4\begin{vmatrix} -3 & 5 \\ 5 & -2 \end{vmatrix} \qquad +1\begin{vmatrix} 1 & 0 \\ 5 & -2 \end{vmatrix} \qquad -6\begin{vmatrix} 1 & 0 \\ -3 & 5 \end{vmatrix}$$

Evaluate each minor, multiply its value by the element of which it is the minor, and add algebraically according to the cofactor signs and the actual algebraic sign of the multiplications:

$$-4(6 - 25) + 1(-2 - 0) - 6(5 - 0) = 76 - 2 - 30 = 44$$

You should immediately evaluate the same third-order determinant by the cofactors of the elements of each other row and column in turn. The answer must always be 44.

EXAMPLE 7 Solve this set of simultaneous equations by means of cofactors:

$$\begin{aligned} 2p + 10q + 5r &= 9 \\ -3p + 9q + 4r &= -3 \\ 7p - 6q - r &= 17 \end{aligned}$$

SOLUTION

Using the information now at hand, we may immediately set up the determinant form of solution:

$$p = \frac{\begin{vmatrix} 9 & 10 & 5 \\ -3 & 9 & 4 \\ 17 & -6 & -1 \end{vmatrix}}{\begin{vmatrix} 2 & 10 & 5 \\ -3 & 9 & 4 \\ 7 & -6 & -1 \end{vmatrix}} \qquad q = \frac{\begin{vmatrix} 2 & 9 & 5 \\ -3 & -3 & 4 \\ 7 & 17 & -1 \end{vmatrix}}{\begin{vmatrix} 2 & 10 & 5 \\ -3 & 9 & 4 \\ 7 & -6 & -1 \end{vmatrix}}$$

$$r = \frac{\begin{vmatrix} 2 & 10 & 9 \\ -3 & 9 & -3 \\ 7 & -6 & 17 \end{vmatrix}}{\begin{vmatrix} 2 & 10 & 5 \\ -3 & 9 & 4 \\ 7 & -6 & -1 \end{vmatrix}}$$

Always evaluate the denominator first. To solve by means of cofactors, we choose any row or column in the denominator determinant, evaluate the minors, and multiply by the elements, adding algebraically and using the checkerboard signs.

$$\begin{vmatrix} 2 & 10 & 5 \\ -3 & 9 & 4 \\ 7 & -6 & -1 \end{vmatrix}$$

$$\begin{aligned} &= -(-3)\begin{vmatrix} 10 & 5 \\ -6 & -1 \end{vmatrix} + (9)\begin{vmatrix} 2 & 5 \\ 7 & -1 \end{vmatrix} - (4)\begin{vmatrix} 2 & 10 \\ 7 & -6 \end{vmatrix} \\ &= 3(-10 + 30) + 9(-2 - 35) - 4(-12 - 70) \\ &= 60 - 333 + 328 \\ &= 55 \end{aligned}$$

Since the denominator is not zero, we should evaluate the numerators, in turn, of p, q, and r. The numerator of

$$p = \begin{vmatrix} 9 & 10 & 5 \\ -3 & 9 & 4 \\ 17 & -6 & -1 \end{vmatrix}$$

$$\begin{aligned} &= +(17)\begin{vmatrix} 10 & 5 \\ 9 & 4 \end{vmatrix} - (-6)\begin{vmatrix} 9 & 5 \\ -3 & 4 \end{vmatrix} + (-1)\begin{vmatrix} 9 & 10 \\ -3 & 9 \end{vmatrix} \\ &= +110 \end{aligned}$$

Therefore, $p = \dfrac{110}{55} = 2$. Now prove that $q = -1$ and $r = 3$.

18–7 USEFUL PROPERTIES OF DETERMINANTS

The evaluation of determinants by the methods of diagonal multiplication or cofactors will yield the correct answers if you keep close watch on your arithmetic and the algebraic signs of positive and negative diagonals or of the checkerboard cofactor signs. There are, however, a few very useful properties of determinants which will simplify the process of evaluation. These properties are described briefly below, and it is left to you to perform the diagonal multiplication or cofactor evaluation methods to confirm them immediately when you meet them.

1. When all the elements of any row (or column) are zero, the value of the determinant is zero:

$$\begin{vmatrix} a_1 & b_1 & 0 \\ a_2 & b_2 & 0 \\ a_3 & b_3 & 0 \end{vmatrix} = 0$$

EXAMPLE 8 Evaluate the determinant:

$$\begin{vmatrix} 2 & 4 & -3 \\ 0 & 0 & 0 \\ -4 & 6 & 1 \end{vmatrix}$$

SOLUTION

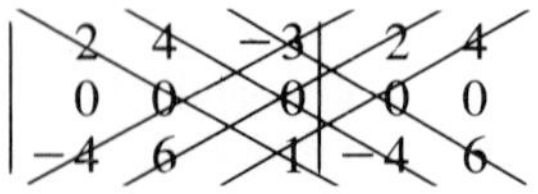

Each diagonal multiplication introduces a factor of zero. Therefore, each diagonal product is zero, and the value of the determinant is zero.

2. When all the elements to the right (or left) of the principal diagonal are zero, the value of the determinant is the product of the elements of the principal diagonal:

$$\begin{vmatrix} a_1 & 0 & 0 \\ a_2 & b_2 & 0 \\ a_3 & b_3 & c_3 \end{vmatrix} = a_1b_2c_3$$

(It is left to you to prove that this is true for fourth-order determinants also.)

EXAMPLE 9 Evaluate the determinant:

$$\begin{vmatrix} 3 & 8 & 5 \\ 0 & -2 & 7 \\ 0 & 0 & -5 \end{vmatrix}$$

SOLUTION

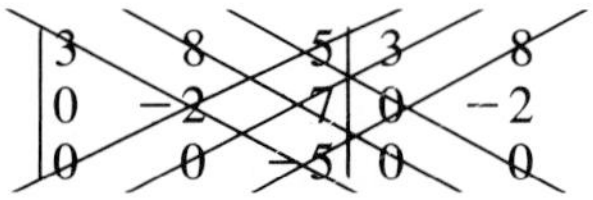

All of the diagonal multiplications except the first one, through the principal diagonal, are zero. Therefore, the value of the determinant is $(3)(-2)(-5) = 30$.

3. Interchanging all the rows and columns gives the identical result, both absolute value and algebraic sign:

$$\begin{vmatrix} a_1 & a_2 & a_3 \\ b_1 & b_2 & b_3 \\ c_1 & c_2 & c_3 \end{vmatrix} = \begin{vmatrix} a_1 & b_1 & c_1 \\ a_2 & b_2 & c_2 \\ a_3 & b_3 & c_3 \end{vmatrix}$$

4. Interchanging two rows (or columns) gives the same absolute value but the opposite algebraic sign:

$$\begin{vmatrix} c_1 & b_1 & a_1 \\ c_2 & b_2 & a_2 \\ c_3 & b_3 & a_3 \end{vmatrix} = - \begin{vmatrix} a_1 & b_1 & c_1 \\ a_2 & b_2 & c_2 \\ a_3 & b_3 & c_3 \end{vmatrix}$$

5. When the corresponding elements of any two rows (or columns) are identical or proportional, the value of the determinant is zero:

$$\begin{vmatrix} a_1 & ka_1 & c_1 \\ a_2 & ka_2 & c_2 \\ a_3 & ka_3 & c_3 \end{vmatrix} = 0 \qquad (k \text{ may } = 1)$$

EXAMPLE 10 Evaluate the determinant:

$$\begin{vmatrix} 3 & 5 & 6 \\ 2 & -1 & 4 \\ 7 & 4 & 14 \end{vmatrix}$$

SOLUTION

$$\begin{vmatrix} 3 & 5 & 6 \\ 2 & -1 & 4 \\ 7 & 4 & 14 \end{vmatrix} \begin{matrix} 3 & 5 \\ 2 & -1 \\ 7 & 4 \end{matrix}$$

Diagonal multiplication yields a zero value. Observation of the first and third columns shows that col 3 = 2 × col 1.

6. A common factor of any row (or column) may be factored out as a common factor of the whole determinant:

$$\begin{vmatrix} a_1 & b_1 & kc_1 \\ a_2 & b_2 & kc_2 \\ a_3 & b_3 & kc_3 \end{vmatrix} = k \begin{vmatrix} a_1 & b_1 & c_1 \\ a_2 & b_2 & c_2 \\ a_3 & b_3 & c_3 \end{vmatrix}$$

EXAMPLE 11 Evaluate the determinant:

$$\begin{vmatrix} 3 & 6 & 2 \\ -2 & 8 & 5 \\ 40 & 30 & -70 \end{vmatrix}$$

SOLUTION

$$\begin{vmatrix} 3 & 6 & 2 \\ -2 & 8 & 5 \\ 40 & 30 & -70 \end{vmatrix} = 10 \begin{vmatrix} 3 & 6 & 2 \\ -2 & 8 & 5 \\ 4 & 3 & -7 \end{vmatrix} \begin{matrix} 3 & 6 \\ -2 & 8 \\ 4 & 3 \end{matrix}$$

$$= 10(-253) = -2530$$

7. When the elements of any row (or column) are increased by a constant times the corresponding elements of any other row (or column), the value of the determinant is unchanged. (k may equal 1, −1, or any other positive or negative integer or fraction):

$$\begin{vmatrix} a_1 & b_1 & ka_1 + c_1 \\ a_2 & b_2 & ka_2 + c_2 \\ a_3 & b_3 & ka_3 + c_3 \end{vmatrix} = \begin{vmatrix} a_1 & b_1 & c_1 \\ a_2 & b_2 & c_2 \\ a_3 & b_3 & c_3 \end{vmatrix}$$

EXAMPLE 12 Evaluate the determinant:

$$\begin{vmatrix} 2 & 8 & 3 \\ 3 & 7 & -6 \\ -1 & 2 & 1 \end{vmatrix}$$

SOLUTION

If the spaces filled by the elements 8, 3, and −6 can be converted to zeros, the evaluation of the determinant will be the product of the elements of the principal axis. Or if any two spaces in any row or column can be adjusted to zero, the evaluation becomes a single element times its cofactor.

Using the principle introduced above, let us attempt to eliminate the element 3. We will multiply each element of the third row by -3 and add the result to the corresponding elements of the first row:

$$\begin{vmatrix} 2 & 8 & 3 \\ 3 & 7 & -6 \\ -1 & 2 & 1 \end{vmatrix} = \begin{vmatrix} 2 + (-3)(-1) & 8 + (-3)(2) & 3 + (-3)(1) \\ 3 & 7 & -6 \\ -1 & 2 & 1 \end{vmatrix}$$

$$= \begin{vmatrix} 5 & 2 & 0 \\ 3 & 7 & -6 \\ -1 & 2 & 1 \end{vmatrix}$$

Then, to eliminate the -6, we will multiply the third row by 6 and add the results to the second row:

$$\begin{vmatrix} 5 & 2 & 0 \\ 3 & 7 & -6 \\ -1 & 2 & 1 \end{vmatrix} = \begin{vmatrix} 5 & 2 & 0 \\ 3 + (6)(-1) & 7 + (6)(2) & -6 + (6)(1) \\ -1 & 2 & 1 \end{vmatrix}$$

$$= \begin{vmatrix} 5 & 2 & 0 \\ -3 & 19 & 0 \\ -1 & 2 & 1 \end{vmatrix}$$

This determinant may be evaluated by the product of the element 1 and its cofactor:

$$\begin{vmatrix} 5 & 2 & 0 \\ -3 & 19 & 0 \\ -1 & 2 & 1 \end{vmatrix} = +1 \begin{vmatrix} 5 & 2 \\ -3 & 19 \end{vmatrix}$$

$$= 95 + 6 = 101$$

You should test this solution by the diagonal multiplication of the original determinant. Alternatively, the simplification may continue by removal of the element 2 in the first row. If we add to the first row the product of $-\frac{2}{19}$ (second row),

$$\begin{vmatrix} 5 + \left(-\frac{2}{19}\right)(-3) & 2 + \left(-\frac{2}{19}\right)(19) & 0 + \left(-\frac{2}{19}\right)(0) \\ -3 & 19 & 0 \\ -1 & 2 & 1 \end{vmatrix}$$

$$= \begin{vmatrix} 5\frac{6}{19} & 0 & 0 \\ -3 & 19 & 0 \\ -1 & 2 & 1 \end{vmatrix}$$

Evaluation by the principal diagonal yields

$$\left(5\frac{6}{19}\right)(19)(1) = 101$$

With practice, the addition of a fraction in the form $-\frac{a_x}{a_y}a_y$ will reveal itself as a valuable tool.

8. When the elements of any row (or column) may be written as sums, the determinant may be written as the sum of two determinants with the rows (or columns) of the sum elements in their corresponding places:

$$\begin{vmatrix} a_1 & b_1 & p_1 + q_1 \\ a_2 & b_2 & p_2 + q_2 \\ a_3 & b_3 & p_3 + q_3 \end{vmatrix} = \begin{vmatrix} a_1 & b_1 & p_1 \\ a_2 & b_2 & p_2 \\ a_3 & b_3 & p_3 \end{vmatrix} + \begin{vmatrix} a_1 & b_1 & q_1 \\ a_2 & b_2 & q_2 \\ a_3 & b_3 & q_3 \end{vmatrix}$$

Now apply these fundamental properties of determinants in the solution of the following problems and problems like them in later chapters.

PROBLEMS 18–5

Evaluate the following determinants by means of the *cofactors* of the indicated rows or columns:

1. $\begin{vmatrix} 6 & 43 & -3 \\ -5 & -2 & 6 \\ 1 & -11 & -8 \end{vmatrix}$ Row 1

2. $\begin{vmatrix} 3 & 1 & -1 \\ 7 & -4 & -1 \\ 2 & -3 & 5 \end{vmatrix}$ Col 2

3. $\begin{vmatrix} 15 & 3 & -2 \\ 17 & -1 & 3 \\ 12 & 2 & 2 \end{vmatrix}$ Row 3

4. $\begin{vmatrix} -35 & 5 & 1 \\ 96 & 0 & -12 \\ 18 & -6 & 3 \end{vmatrix}$ Col 1

5. $\begin{vmatrix} 3 & 0 & 21.7 \\ 2 & 3 & 15.3 \\ 0 & -2 & 1.9 \end{vmatrix}$ Col 3

6. $\begin{vmatrix} 2 & 4 & 10 \\ -8 & -16 & -13 \\ 0 & 16 & 2 \end{vmatrix}$ Row 2

7. $\begin{vmatrix} 2 & 2 & 2 & 2 \\ -3 & -3 & 3 & -3 \\ 6 & 6 & -6 & -6 \\ 5 & -5 & 5 & 5 \end{vmatrix}$ Col 2

▶ **HINT** The cofactors of elements in a fourth-order determinant will themselves be third-order determinants which may be evaluated by diagonals or by cofactors.

8. $\begin{vmatrix} 2 & 16 & 12 & -10 & -2 \\ 5 & 2 & 2 & 3 & -9 \\ 11 & 0 & 0 & 5 & 4 \\ 5 & 0 & 2 & 15 & 4 \\ 0 & -4 & 10 & -8 & 0 \end{vmatrix}$ Row 3

Solve by using determinants and cofactors:

9. $\begin{aligned} 5I_1 + 2I_2 + 6I_3 &= 41 \\ 2I_1 + 3I_2 - 2I_3 &= 1 \\ -4I_1 - I_2 + 3I_3 &= 8 \end{aligned}$

10. $\begin{aligned} 2\theta + \phi - \lambda &= 3 \\ 3\theta - 2\phi + 2\lambda &= 8 \\ 4\theta - 3\phi - \lambda &= -13 \end{aligned}$

11. $\begin{aligned} 3\alpha + 2\beta - 5\gamma &= 10.2 \\ 4\alpha - 5\beta + 8\gamma &= -3.7 \\ \alpha + 2\beta - 3\gamma &= 8.6 \end{aligned}$

12. $\begin{aligned} 2I_1 + 3I_2 + 2I_3 &= -26 \\ -8I_1 + 2I_2 - 10I_3 &= 84 \\ -3I_1 - 7I_2 + 4I_3 &= 16 \end{aligned}$

13. $$\begin{aligned} 2I_1 + 6I_2 - 3I_3 - 3I_4 &= -23 \\ -5I_1 + 2I_2 - 2I_3 + 4I_4 &= -34 \\ 4I_1 - 5I_2 + 2I_3 + I_4 &= 31 \\ I_1 - 2I_2 + 4I_3 + 6I_4 &= 16 \end{aligned}$$

14. $$\begin{aligned} 2x + 4y + 10z &= 10 \\ -8x - 16y + 5z &= -13 \\ 16y - 20z &= 2 \end{aligned}$$

15. $$\begin{aligned} 2\alpha + 2\beta + 2\gamma + 2\delta &= 12 \\ -3\alpha - 3\beta + 3\gamma - 3\delta &= 18 \\ 6\alpha + 6\beta - 6\gamma - 6\delta &= 12 \\ 5\alpha - 5\beta + 5\gamma + 5\delta &= 20 \end{aligned}$$

16. $$\begin{aligned} 2.5I_1 + 1.7I_2 - 6.8I_3 - 7.2I_4 + 3.6I_5 &= 9.6480 \\ -5.7I_1 + 2.1I_2 + 1.9I_3 + 8.5I_4 - 1.6I_5 &= 31.5510 \\ 14.1I_1 - 3.9I_2 + 2.1I_3 - 6.6I_4 + 2.2I_5 &= -9.4820 \\ -11.2I_1 - 2.6I_2 - 3.3I_3 + 4.5I_4 - 0.6I_5 &= -61.1650 \\ 8.4I_1 + 1.1I_2 - 4.2I_3 - 2.7I_4 + 1.8I_5 &= 18.1180 \end{aligned}$$

17. Solve selected problems from Chap. 17 by means of determinants.
18. Use determinants for the solution of appropriate problems throughout the remainder of this book.
19. The vidicon tubes of a color TV camera receive light information that is transformed into color and luminance signals. The output of the three vidicon tubes is measured in the following proportions. Solve for the resulting red, green, and blue percentages:

$$\begin{aligned} R - 0.7G + 0.3B &= -0.08 \\ 0.6R + 0.2G + 0.2B &= 0.32 \\ 0.5R + 0.5G - B &= 0.335 \end{aligned}$$

If a white picture is Y, where Y is the *luminance signal* made up of the red, green, and blue components, show that:

$$Y = 0.3R + 0.59G + 0.11B$$

SELF TEST

1. Evaluate: $\begin{vmatrix} 3 & 4 \\ 7 & 1 \end{vmatrix}$

2. Solve by means of determinants:

$$\begin{aligned} 2y + 3x &= 4 \\ 6y &= 7x - 4.2 \end{aligned}$$

3. Evaluate: $\begin{vmatrix} 1 & 1 & -1 \\ 3 & -2 & 3 \\ 7 & 5 & -3 \end{vmatrix}$

4. Solve using determinants:

$$\begin{aligned} 5\alpha &= -4\beta + 2\gamma + 13 \\ 8\gamma &= 6\alpha - 3\beta - 11 \\ 2\beta &= 18 - 2\alpha - 2\gamma \end{aligned}$$

5. Evaluate by using the cofactors of column 1:

$$\begin{vmatrix} 1 & -11 & -8 \\ 6 & 43 & -3 \\ -5 & -2 & 6 \end{vmatrix}$$

6. Solve for p. Use determinants and cofactors:

$$\begin{aligned} 8p + q - 3r + 2s &= 37 \\ 4p - 3q + 2r - s &= 7 \\ -2p + 3q + 3r + 5s &= 5 \\ 3p + 2q + 2r + s &= 7 \end{aligned}$$

CHAPTER 19

BATTERIES

In preceding discussions of electric circuits, to avoid confusion, all sources of electromotive force have been considered to be sources of constant potential, and nothing has been said of their internal resistances. At the same time, no mention has been made of the actual sources of the emf. In this chapter we will consider both of these factors. First of all, electrical devices that produce electric energy, as well as those that consume energy, have a certain amount of internal resistance which materially affects their operation. The application of Ohm's law to the internal resistance of batteries is the feature topic of this chapter. And despite the prevalence of utility power supply, batteries are still useful, indeed necessary, sources of portable power. For this reason, the electronics technician should be aware of the problems which arise in the use of batteries.

19–1 ELECTROMOTIVE FORCE

A battery is a device which converts chemical energy into electric energy. Essentially, it consists of a cell, or several cells connected in series or parallel, conveniently packaged. The emf of the battery is the total voltage developed by the chemical action. However, not all of this total voltage is available for doing useful work in an external circuit, because some of it is needed to overcome the internal resistance of the battery itself. The voltage which is supplied to the external circuit is known as the terminal voltage; that is,

$$\text{Terminal voltage} = \text{emf} - \text{internal voltage drop}$$

19–2 BATTERIES

The word *battery* is taken to mean two or more *cells* connected to each other, although a single cell is often referred to as a battery.

Figure 19–1 represents a circuit by which the voltage existing across the cell can be read with the resistance connected across the battery or with the resistance disconnected from the circuit.

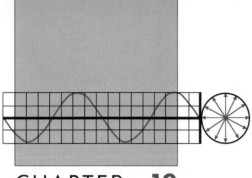

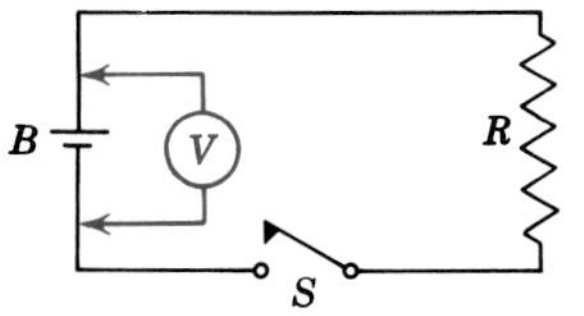

FIG. 19–1 High-resistance voltmeter used for measuring electromotive force of a cell.

The emf of a cell is the total amount of voltage developed by the cell. For all practical purposes the emf of a cell can be read with a high-resistance voltmeter connected across the cell when the cell is not supplying current to any other circuit, as when the switch *S*, Fig. 19–1, is open.

When a cell supplies current to an external circuit, as when the switch in Fig. 19–1 is closed, it will be found that the voltmeter no longer reads the open-circuit voltage (emf) of the cell. The reason is that part of the emf is used in forcing current through the resistance of the cell and the remainder is used in forcing current through the external circuit. Expressed as an equation,

$$V = V_t + Ir \tag{1}$$

where V is the emf of the cell or group of cells and V_t is the voltage measured across the terminals while forcing a current I through the internal resistance r. Since I also flows through the external circuit of resistance R, Eq. (1) can be written

$$V = IR + Ir$$

or

$$V = I(R + r) \tag{2}$$

EXAMPLE 1 A cell whose internal resistance is 0.15 Ω delivers 0.50 A to a resistance of 2.85 Ω. What is the emf of the cell?

SOLUTION

Given $r = 0.15\ \Omega$, $R = 2.85\ \Omega$, and $I = 0.50$ A.

From Eq. (2), $V = 0.50(2.85 + 0.15) = 1.5$ V

EXAMPLE 2 Figure 19–2 represents a cell with an emf of 1.2 V and an internal resistance r of 0.2 Ω connected to a resistance R of 5.8 Ω. How much current is in the circuit?

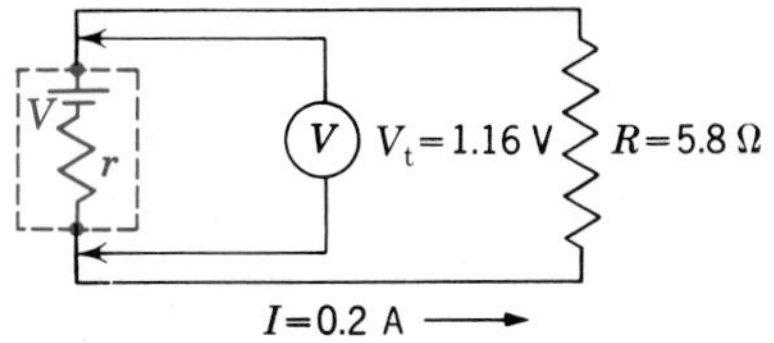

FIG. 19–2 Circuit of Example 2.

SOLUTION

Solving Eq. (2) for the current,

$$I = \frac{V}{R + r} \tag{3}$$

$$= \frac{1.2}{5.8 + 0.2} = 0.2 \text{ A}$$

Note the significance of Eq. (3). It says that the current in a circuit is proportional to the emf of the circuit and inversely proportional to the *total* resistance of the circuit. This is Ohm's law for the *complete circuit*.

EXAMPLE 3 A cell with an emf of 1.6 V delivers a current of 2 A to a circuit of 0.62 Ω. What is the internal resistance of the cell?

SOLUTION

Solving Eq. (2) for the internal resistance,

$$r = \frac{V - IR}{I} \tag{4}$$

$$= \frac{1.6 - 2 \times 0.62}{2} = 0.18\ \Omega$$

Therefore, the significance of Eq. (4) is that a voltage equal to $V - IR$ is sending the current I through the internal resistance r.

Since Eq. (4) can be rearranged to

$$r = \frac{V}{I} - R$$

and

$$\frac{V}{I} = R_t$$

Eq. (4) can be written

$$r = R_t - R$$

or

$$R_t = R + r \tag{5}$$

Equation (5) states simply that the resistance of the entire circuit is equal to the resistance of the external circuit plus the internal resistance of the source of the emf.

PROBLEMS 19–1

1. A battery taken off the shelf gives a voltmeter reading of 9 V. When connected across a 22-Ω circuit, it drives a current of 408 mA. What is its internal resistance?
2. A 24-cell battery measures 38.4 V on open circuit. If the total internal resistance is 6.24 Ω, how much current will flow through a 470-Ω circuit?
3. A 6-V battery drives a current of 590 mA through a 10-Ω load. What is the internal resistance of the battery?
4. With the circuit of Prob. 3, how much power is absorbed by the internal resistance of the battery?
5. With the circuit of Prob. 3, (*a*) how much power is delivered to the load and (*b*) what is the efficiency of the circuit?

19–3 CELLS IN SERIES

If n identical cells are connected in series, the emf of the combination will be n times the emf of each cell. Similarly, the total internal resistance of the circuit will be n times the internal resistance of each cell. By modifying Eq. (2), the expression for the current through an external resistance of R Ω is

$$I = \frac{nV}{R + nr} \tag{6}$$

EXAMPLE 4 Six cells, each having an emf of 2.1 V and an internal resistance of 0.1 Ω, are connected in series, and a resistance of 3.6 Ω is connected across the combination.
(*a*) What is the current in the circuit?
(*b*) What is the terminal voltage of the group?

SOLUTION

Figure 19–3 is a diagram of the circuit. The resistance *nr* represents the total internal resistance of all cells in series.

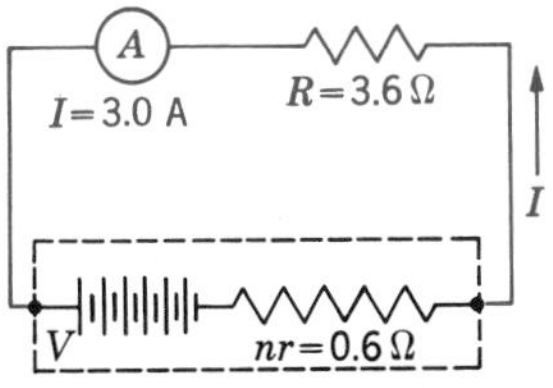

FIG. 19–3 Circuit of Example 4.

(*a*) $I = \dfrac{nV}{R + nr} = \dfrac{6 \times 2.1}{3.6 + 6 \times 0.1} = 3.0\ \text{A}$

(*b*) The terminal voltage of the group is equal to the total emf minus the voltage drop across the internal resistance. From Eq. (1),

$$\begin{aligned} V_t &= nV - Inr \\ &= 6 \times 2.1 - 3 \times 6 \times 0.1 \\ &= 10.8\ \text{V} \end{aligned}$$

Since the terminal voltage exists across the external circuit, a more simple relation is

$$V_t = IR = 3 \times 3.6 = 10.8\ \text{V}$$

19–4 CELLS IN PARALLEL

If *n* identical cells are connected in parallel, the emf of the group will be the same as the emf of one cell and the internal resistance of the group will be equal to the internal resistance of one cell divided by the number of cells in parallel, that is, to $\dfrac{r}{n}$. By modifying Eq. (2), the expression for the current through an external resistance of *R* Ω is

$$I = \frac{V}{R + \dfrac{r}{n}} \tag{7}$$

EXAMPLE 5 Three cells, each with an emf of 1.4 V and an internal resistance of 0.15 Ω, are connected in parallel, and a resistance of 1.35 Ω is connected across the group.
(*a*) What is the current in the circuit?
(*b*) What is the terminal voltage of the group?

SOLUTION

Figure 19–4 is a diagram of the circuit. The resistance $\dfrac{r}{n}$ represents the internal resistance of the group.

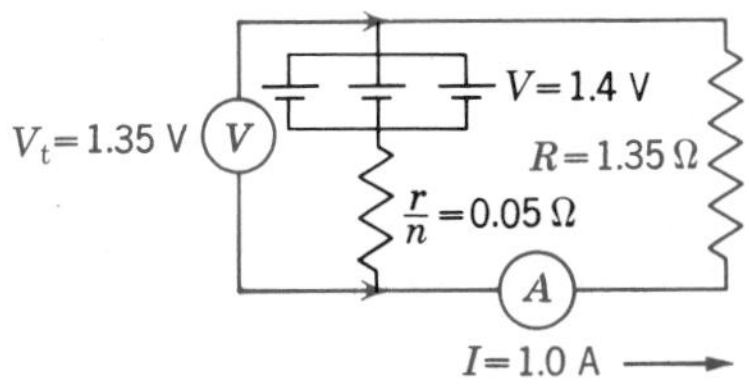

FIG. 19–4 Circuit of Example 5.

(*a*) $I = \dfrac{V}{R + \dfrac{r}{n}} = \dfrac{1.4}{1.35 + \dfrac{0.15}{3}} = 1.0\ \text{A}$

(*b*) $V_t = IR = 1.0 \times 1.35 = 1.35\ \text{V}$

PROBLEMS 19–2

1. The emf of a cell is 1.2 V, and the internal resistance of the cell is 0.08 Ω. When the cell supplies current to a load, the voltage drop across the internal resistance is 0.05 V.
   (*a*) What is the terminal voltage?
   (*b*) What is the current?
   (*c*) What is the connected load?
2. A cell whose emf is 1.4 V is supplying 1.5 A to a 0.733-Ω circuit.
   (*a*) What is the internal resistance of the cell?
   (*b*) How much power is lost in the cell?
3. A cell of emf 1.6 V develops a terminal voltage of 1.58 V when delivering 158 mA to an external circuit.
   (*a*) What is the internal resistance of the cell?
   (*b*) How much power is expended in the cell?
   (*c*) What is the resistance of the external circuit?
   (*d*) How much power is absorbed by the load circuit?
   (*e*) What is the efficiency of the power transfer?
4. A high-resistance voltmeter reads 2 V when connected across the terminals of an open-circuit cell. What will the meter read when a 4-A current is delivered to a 0.33-Ω load if the internal resistance of the cell is 0.17 Ω?
5. Using the data and results of Prob. 4, how much current results if the cell itself were short-circuited?
6. A cell with an emf of 2 V and an internal resistance of 0.1 Ω is connected to a load consisting of a variable resistor.
   (*a*) Plot the power delivered to the load as the load resistance is varied in 0.01-Ω steps from 0.05 to 0.15 Ω. What conclusion do you draw from this graph?
   (*b*) Plot the efficiency of power transfer over the same resistance range. What conclusion do you draw?
7. Six identical cells, each of emf 1.6 V and internal resistance 0.1 Ω are connected in series across a load resistor, and they deliver a circuit current of 1.0 A.
   (*a*) What is the resistance of the load?
   (*b*) How much power is absorbed by the battery?
   (*c*) What is the current if the battery were short-circuited?
8. If the cells in Prob. 7 are connected in parallel, how much power will be delivered to the load?
9. Ten cells of emf 1.5 V and internal resistance of 0.6 Ω each are connected in series across a load of 33 Ω.
   (*a*) What is the current in the circuit?
   (*b*) What will be the terminal voltage of the battery?
   (*c*) How much power will be delivered to the load?
10. If the cells of Prob. 9 are connected in parallel across the same load, how much current will there be?
11. Twelve identical cells are hooked up so that four groups of three cells each in series are connected in parallel as shown in Fig. 19–5. The emf of each

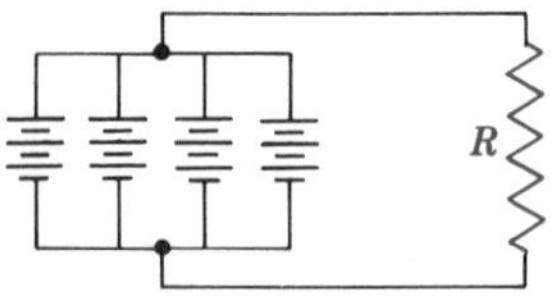

FIG. 19–5 Circuit of Prob. 11.

cell is 1.6 V, and each cell has an internal resistance of 0.2 Ω. If the load R is 0.85 Ω and the measured current flow through R is 4.8 A:
(*a*) What is the terminal voltage of the battery?
(*b*) What is the emf of each cell?
(*c*) How much power is expended in each cell?

12. The cells of Prob. 11 are so arranged that there are two-cells-per-series groups (six groups in parallel).
(*a*) How much power is dissipated in R?
(*b*) What is the current through each cell?

13. Each cell of a six-cell storage battery has an emf of 1.6 V and an internal resistance of 0.02 Ω. The battery is to be charged from a 14-V line.
(*a*) How much resistance must be connected in series with the battery to limit the charging current to 15 A?
(*b*) What would the current be if the battery were disconnected from the charging circuit and short-circuited?

14. Sixteen storage batteries of three cells each are to be charged in series from a 115-V line. Each cell has an emf of 2.1 V and an internal resistance of 0.02 Ω.
(*a*) How much resistance must be connected in series with the battery to limit the charging current to 10 A?
(*b*) How much power is dissipated in the entire circuit?
(*c*) How much power is dissipated in the series charging resistance?
(*d*) What would the current be if the batteries were disconnected from the charging circuit and short-circuited?

15. Six identical cells that are connected in series deliver 4 A to a circuit of 2.7 Ω. When two of the same cells are connected in parallel, they deliver 5 A to an external resistance of 0.375 Ω. What are the emf and internal resistance of each cell?
Solution:

Let V = emf of each cell
r = internal resistance of each cell
I = current in external circuit
R = resistance of external circuit

For the series connection,
$6V$ = emf of six cells in series
and
$6r$ = internal resistance of six cells in series

Substituting in Eq. (2),

$$6V = 4(2.7 + 6r) = 10.8 + 24r \qquad (a)$$

For the parallel connection,

V = emf of cells in parallel

and

$\frac{r}{2}$ = internal resistance of two cells in parallel

Substituting in Eq. (2),

$$V = 5\left(0.375 + \frac{r}{2}\right)$$

or

$$2V = 3.75 + 5r \qquad (b)$$

Solve Eqs. (*a*) and (*b*) simultaneously to obtain

$$V = 2.0 \text{ V}$$

and

$$r = 0.05\ \Omega$$

16. Ten identical cells connected in series send a current of 3 A through a 1-Ω circuit. When three of the cells are connected in parallel, they send a current of 6 A through an external resistance of 0.1 Ω. What are the emf and internal resistance of each cell?
17. Six cells connected in series send a current of 880 mA through a resistance of 10 Ω. When four of the cells are connected in parallel, they send 1.55 A through 1 Ω. What are the emf and internal resistance of each cell?
18. Twelve cells in series, each with an emf of 2.0 V, send a certain current through a 2.4-Ω circuit. The current is the same through a 0.24-Ω circuit when five of the cells are connected in parallel. What is the value of the current and what is the internal resistance of each cell?
19. A cell with an internal resistance of 0.035 Ω sends a 3-A current through an external circuit. Another cell, having the same emf but with an internal resistance of 0.385 Ω, sends a current of 2 A through the external circuit when substituted for the first cell. What is the emf of the cells and what is the resistance of the external circuit?
20. A cell sends a 20-A current through an external circuit of 0.06 Ω. When the resistance of the external circuit is increased to 3.98 Ω, the current is 0.4 A. What is the emf and the internal resistance of the cell?

SELF TEST

1. An apparently new battery gives an open-circuit voltage reading of 9 V. When connected across a 20-Ω test load, it drives a current of 446 mA. What is its internal resistance?
2. Twelve storage batteries of two cells in series each are connected in parallel. Each cell has an open-circuit emf of 2.2 V and an internal resistance of 0.015 Ω. How much current will the battery drive through a 40-Ω load?
3. A rechargeable battery undergoing a test drives a 4.26-A current through a test circuit of 2 Ω, but only 1.80 A through a second 5-Ω circuit. What is the emf and the internal resistance of the battery?

CHAPTER 20

EXPONENTS AND RADICALS

In earlier chapters, examples and problems have been limited to those containing exponents and roots that consisted of integers. In this chapter the study of exponents and radicals is extended to include new operations that will enable you to solve electrical formulas and equations of a type hitherto omitted. In addition, new ideas that will be of fundamental importance in your study of alternating currents are introduced.

20–1 FUNDAMENTAL LAWS OF EXPONENTS

As previously explained, if n is a positive integer, a^n means that a is to be taken as a factor n times. Thus, a^4 is defined as being a shortened form of notation for the product $a \cdot a \cdot a \cdot a$. The number a is called the *base,* and the number n is called the *exponent*.

For the purpose of review, the fundamental laws for the use of *positive-integer exponents* are listed below:

$$a^m \cdot a^n = a^{m+n} \qquad \text{(Sec. 4–3)} \qquad (1)$$

$$a^m \div a^n = a^{m-n} \quad (\text{when } n < m) \qquad \text{(Sec. 4–9)} \qquad (2)$$

$$= \frac{1}{a^{n-m}} \quad (\text{when } n > m) \qquad \text{(Sec. 4–11)}$$

$$(a^m)^n = a^{mn} \qquad \text{(Sec. 6–10)} \qquad (3)$$

$$(ab)^m = a^m b^m \qquad \text{(Sec. 6–11)} \qquad (4)$$

$$\left(\frac{a}{b}\right)^m = \frac{a^m}{b^m} \quad (b \neq 0) \qquad \text{(Sec. 6–12)} \qquad (5)$$

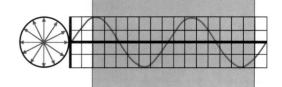

> Remember your computer cannot print superscripts. You must give it an exponentiation command.
> Instead of
>
> $a^m a^n$
>
> you will have to key in something like
>
> a**m*a**n
>
> or
>
> a ↑ m*a ↑ n

20–2 ZERO EXPONENT

If a^0 is to obey the law of exponents for multiplication as was stated under Eq. (1) of the preceding section, then

$$a^m \cdot a^0 = a^{m+0} = a^m$$

Also, if a^0 is to obey the law of exponents for division, then

$$\frac{a^m}{a^0} = a^{m-0} = a^m$$

Therefore, the zero power of any number, except zero, is defined as being equal to 1, for 1 is the only number that, when used to multiply another number, does not change the value of the multiplicand.

20–3 NEGATIVE EXPONENTS

If a^{-n} is to obey the multiplication law, then

$$\frac{a^n}{a^n} = a^{n-n} = a^0 = 1$$

In Sec. 4–11, it was shown that a *factor* can be transferred from one term of a fraction to the other if the sign of its exponent is changed, that is, from numerator to denominator, or vice versa.

PROBLEMS 20–1

By making use of the five fundamental laws of exponents, write the results of the indicated operations:

1. $x^5 \cdot x^2$
2. $\theta^2 \cdot \theta^4$
3. $\alpha^2 \cdot \alpha$
4. $\varepsilon^3 \cdot \varepsilon^6$
5. $p^l p^q$
6. $\theta^{3x} \cdot \theta^{2x}$
7. $V^\alpha \cdot V^\beta$
8. $w^{a+b} \cdot w^{a-b}$
9. $x^{12} \div x^7$
10. $\alpha^{5.3} \div \alpha^{2.7}$
11. $X^{5y} \div X^2$
12. $e^{\pi+2} \div e^3$
13. $\theta^{\alpha+\beta} \div \theta^{\alpha-\beta}$
14. $\psi^{\alpha+\beta} \div \psi^{\alpha-\gamma}$
15. $(I^2)^5$
16. $(x^5)^3$
17. $(x^2y^3)^3$
18. $(I^2Rt)^3$
19. $(a^n)^m$
20. $(a^4)^x$
21. $(-x^l y^m z^p)^4$
22. $(-x^a b^y)^3$
23. $\left(\frac{V}{R}\right)^2$

24. $\left(\dfrac{R_1R_2}{R_3}\right)^3$

25. $\left(\dfrac{2\pi fL}{R}\right)^2$

26. $\left(\dfrac{Z_1{}^2}{Z_3Z_4}\right)^2$

27. $\left(\dfrac{-X_C{}^2}{X_L}\right)^3$

28. $\left(\dfrac{\pi D^2}{4}\right)^4$

29. $\left(\dfrac{\gamma^{5x}}{\gamma^{x+2}}\right)^2$

30. $\left(\dfrac{a^{3\pi}}{a^{5\lambda}}\right)^{4\gamma}$

Express with all positive exponents:

31. $0.159f^{-1}C^{-1}$

32. $a^{-4}y^{-3}$

33. $m^{-n}p^{2q}$

34. $25X_L{}^{-4}X_C{}^{-2}$

35. $\theta^4\phi^{-3}\lambda^{-2x}$

36. $(\pi R^2)^{-2i}$

37. $\dfrac{a^{-3}b}{c^{-1}}$

38. $\left(\dfrac{Z_1Z_2}{Z_4}\right)^{-3}$

39. $\dfrac{3I^3R^{-2}}{12I^2r^{-3}}$

40. $\dfrac{\alpha^3}{2(4\beta\gamma)^{-2}}$

20–4 FRACTIONAL EXPONENTS

The meaning of a base affected by a fractional exponent is established by methods similar to those employed in determining meanings for zero or negative exponents. If we assume that Eq. (1) of Sec. 20–1 holds for fractional exponents, we should obtain, for example,

$$a^{\frac{1}{2}} \cdot a^{\frac{1}{2}} = a^{\frac{1}{2}+\frac{1}{2}} = a^1 = a$$

Also, $$a^{\frac{1}{3}} \cdot a^{\frac{1}{3}} \cdot a^{\frac{1}{3}} = a^{\frac{1}{3}+\frac{1}{3}+\frac{1}{3}} = a^1 = a$$

That is, $a^{\frac{1}{2}}$ is one of two equal factors of a, and $a^{\frac{1}{3}}$ is one of three equal factors of a. Therefore, $a^{\frac{1}{2}}$ is the square root of a, and $a^{\frac{1}{3}}$ is the cube root of a. Hence,

$$a^{\frac{1}{2}} = \sqrt{a}$$

and $$a^{\frac{1}{3}} = \sqrt[3]{a}$$

Likewise, $$a^{\frac{2}{3}} \cdot a^{\frac{2}{3}} \cdot a^{\frac{2}{3}} = a^{\frac{2}{3}+\frac{2}{3}+\frac{2}{3}} = a^{\frac{6}{3}} = a^2$$

Hence, $$(a^{\frac{2}{3}})^3 = a^2$$

or $$a^{\frac{2}{3}} = \sqrt[3]{a^2}$$

In a fractional exponent, the denominator denotes the root and the numerator denotes the power of the base.

In general, $$a^{\frac{m}{n}} = \sqrt[n]{a^m} = (\sqrt[n]{a})^m$$

EXAMPLE 1

$$a^{\frac{3}{5}} = \sqrt[5]{a^3}$$

EXAMPLE 2

$$(-8)^{\frac{1}{3}} = \sqrt[3]{-8} = -2$$

PROBLEMS 20–2

Find the value of:

1. $25^{\frac{1}{2}}$
2. $(-64)^{\frac{1}{3}}$
3. $81^{\frac{1}{4}}$
4. $-(-64)^{\frac{1}{6}}$
5. $(-27x^3y^{12}b^9c^3)^{\frac{1}{3}}$
6. $(L_1{}^4L_2{}^4)^{\frac{1}{2}}$
7. $(\omega^2L^2)^{\frac{3}{2}}$
8. $(X^2C^2)^{\frac{3}{2}}$
9. $\left(\dfrac{8x^6}{\pi^9}\right)^{\frac{2}{3}}$
10. $\left(\dfrac{r^{12}R^8}{16V^4}\right)^{\frac{3}{4}}$

Express with radical signs:

11. $12^{\frac{1}{2}}$
12. $15^{\frac{1}{2}}$
13. $(27Z)^{\frac{1}{3}}$
14. $6^{\frac{2}{3}}$
15. $\theta^{\frac{3}{4}}\lambda^{\frac{3}{4}}$
16. $x^{\frac{2}{3}}y^{\frac{3}{2}}$

Express with fractional exponents:

17. $\sqrt{x^5}$
18. $\sqrt[3]{M^2}$
19. $\sqrt[3]{81R^2}$
20. $\sqrt[3]{a^2b^4c^6}$
21. $\sqrt[3]{G^2B^6}$
22. $\alpha\sqrt[5]{\beta^2}$
23. $\sqrt[5]{\alpha^2\beta^2}$
24. $4L\sqrt{\omega^3}$
25. $2\pi\sqrt[3]{20f^5}$
26. $5\alpha^2\sqrt[5]{-32\alpha^3\beta^7}$

20–5 RADICAND

The meaning of the radical sign was explained in Sec. 2–11. The number under the radical sign is called the *radicand.*

20–6 SIMPLIFICATION OF RADICALS

The form in which a radical expression is written can be changed without altering the numerical value of the expression. Such a change is desirable for many reasons. For example, addition of several fractions containing different radicals in the denominators would be more difficult than addition with the radicals removed from the denominators. Similarly, it is left for you to show that

$$\frac{1}{\sqrt{3}} = \frac{\sqrt{3}}{3}$$

It is apparent that the value to several decimal places could be computed more easily from the second fraction than from the first.

Because we are chiefly concerned with radicals involving a square root, only that type will be considered.

20–7 REMOVING A FACTOR FROM THE RADICAND

Since, in general, $\sqrt{ab} = \sqrt{a} \cdot \sqrt{b}$, the following is evident:

RULE

A radicand can be separated into two factors one of which is the greatest perfect square it contains. The square root of this factor can then be written as the coefficient of a radical the other factor of which is the radicand.

EXAMPLE 3

$$\begin{aligned}\sqrt{27} &= \sqrt{9 \cdot 3}\\ &= \sqrt{9} \cdot \sqrt{3}\\ &= \pm 3\sqrt{3}\end{aligned}$$

EXAMPLE 4

$$\begin{aligned}\sqrt{8} &= \sqrt{4 \cdot 2}\\ &= \sqrt{4} \cdot \sqrt{2}\\ &= \pm 2\sqrt{2}\end{aligned}$$

EXAMPLE 5

$$\begin{aligned}\sqrt{75} &= \sqrt{25 \cdot 3}\\ &= \sqrt{25} \cdot \sqrt{3}\\ &= \pm 5\sqrt{3}\end{aligned}$$

EXAMPLE 6

$$\begin{aligned}\sqrt{200a^5b^3c^2d} &= \sqrt{100a^4b^2c^2} \cdot \sqrt{2abd}\\ &= \pm 10a^2bc\sqrt{2abd}\end{aligned}$$

PROBLEMS 20–3

Simplify by removing factors from the radicand:

1. $\sqrt{27}$
2. $\sqrt{32}$
3. $\sqrt{12}$
4. $\sqrt{56}$
5. $\sqrt{75}$
6. $\sqrt{24}$
7. $\sqrt{180}$
8. $\sqrt{28}$
9. $\sqrt{720}$
10. $\sqrt{27x^4}$
11. $\sqrt{18\alpha^2\beta^4}$
12. $\sqrt{99A^3D}$
13. $2\sqrt{225\omega^2L}$
14. $3\pi\sqrt{72r^3z^5\pi^3}$
15. $6\omega\sqrt{63f^4F^3T^5}$
16. $7x\sqrt{147xy^2z^3D^3}$
17. $(2A^3)\sqrt{162A^3B^5C^6}$
18. $8\sqrt{567X_L{}^2Z_1{}^4}$
19. $2\pi^2\sqrt{507\pi^4C^4L^4}$
20. $5\theta\sqrt{289\theta^5\lambda^7}$

20–8 SIMPLIFYING RADICALS CONTAINING FRACTIONS

Since $$\sqrt{\frac{4}{9}} = \pm\frac{2}{3}$$

and $$\frac{\sqrt{4}}{\sqrt{9}} = \pm\frac{2}{3}$$

then $$\sqrt{\frac{4}{9}} = \pm\frac{\sqrt{4}}{\sqrt{9}}$$

Also, $$\sqrt{\frac{16}{25}} = \pm\frac{4}{5}$$

and $$\frac{\sqrt{16}}{\sqrt{25}} = \pm\frac{4}{5}$$

then $$\sqrt{\frac{16}{25}} = \frac{\sqrt{16}}{\sqrt{25}}$$

Or, in general terms, $$\sqrt{\frac{a}{b}} = \frac{\sqrt{a}}{\sqrt{b}}$$

The above relation permits simplification of radicals containing fractions by removing the radical from the denominator. This process, by which the denominator is made a rational number, is called *rationalizing the denominator.*

RULE

To rationalize the denominator:

1. Multiply both numerator and denominator by a number that will make the resulting denominator a perfect square.
2. Simplify the resulting radical by removing factors from the radicands.

EXAMPLE 7

$$\begin{aligned}\sqrt{\frac{2}{5}} &= \sqrt{\frac{2}{5}\cdot\frac{5}{5}}\\ &= \sqrt{\frac{10}{25}}\\ &= \frac{\sqrt{10}}{\sqrt{25}}\\ &= \pm\frac{\sqrt{10}}{5}\end{aligned}$$

EXAMPLE 8

$$\sqrt{\frac{1}{2}} = \sqrt{\frac{1}{2}\cdot\frac{2}{2}} = \sqrt{\frac{2}{4}} = \frac{\sqrt{2}}{\sqrt{4}} = \pm\frac{\sqrt{2}}{2}$$

EXAMPLE 9

$$\frac{3}{\sqrt{6}} = \frac{3}{\sqrt{6}}\cdot\frac{\sqrt{6}}{\sqrt{6}} = \pm\frac{1}{2}\sqrt{6}$$

EXAMPLE 10

$$\sqrt{\frac{3a}{5x}} = \sqrt{\frac{3a}{5x}\cdot\frac{5x}{5x}} = \sqrt{\frac{15ax}{25x^2}} = \frac{\sqrt{15ax}}{\sqrt{25x^2}} = \pm\frac{1}{5x}\sqrt{15ax}$$

PROBLEMS 20–4

Simplify the following:

1. $\sqrt{\frac{1}{5}}$
2. $\sqrt{\frac{1}{13}}$
3. $\sqrt{\frac{2}{5}}$
4. $\sqrt{\frac{4}{5}}$
5. $\sqrt{\frac{3}{8}}$
6. $\sqrt{\frac{7}{15}}$
7. $\frac{16}{\sqrt{2}}$
8. $\frac{9}{\sqrt{3}}$
9. $\frac{1}{\sqrt{\lambda}}$
10. $\frac{21\sqrt{35}}{\sqrt{7}}$
11. $\sqrt{\frac{9}{16\theta}}$
12. $\sqrt{\frac{Q}{R}}$
13. $\alpha\sqrt{\frac{\omega}{\alpha}}$
14. $\pi\sqrt{\frac{X_L}{2\pi f L}}$
15. $\sqrt{\frac{\beta^2}{\theta}}$
16. $\frac{2F}{f_0}\sqrt{\frac{f_0}{F}}$
17. $\frac{\pi R^3}{4A}\sqrt{\frac{A}{\pi}}$
18. $\sqrt{\frac{E - e}{E + e}}$
19. $\sqrt{X_L^2 - \left(\frac{X_L}{4}\right)^2}$

20. $\sqrt{R^2 + \left(\frac{R}{3}\right)^2}$ 21. $\sqrt{Q^4 - \left(\frac{Q}{3}\right)^4}$

22. Use your calculator to confirm numerical answers to Probs. 1 through 11 above.

20–9 ADDITION AND SUBTRACTION OF RADICALS

Terms that are the same except in respect to their coefficients are called *similar terms*. Likewise, *similar radicals* are defined as radicals that have the same index and the same radicand and differ only in their coefficients. For example, $-2\sqrt{5}$, $3\sqrt{5}$, and $\sqrt{5}$ are similar radicals.

Similar radicals can be added or subtracted in the same way that similar terms are added or subtracted.

EXAMPLE 11

$$3\sqrt{6} - 4\sqrt{6} - \sqrt{6} + 8\sqrt{6} = 6\sqrt{6}$$

EXAMPLE 12

$$\begin{aligned}\sqrt{12} + \sqrt{27} &= 2\sqrt{3} + 3\sqrt{3}\\ &= 5\sqrt{3}\end{aligned}$$

Note that, in the simplification of radicals, the positive root is assumed.

EXAMPLE 13

$$\sqrt{48x} + \sqrt{\frac{x}{3}} + \sqrt{3x} = 4\sqrt{3x} + \frac{1}{3}\sqrt{3x} + \sqrt{3x} = \frac{16}{3}\sqrt{3x}$$

If the radicands are alike, then factors removed are assumed to be positive roots. If the radicands are not alike and cannot be reduced to a common radicand, then the radicals are dissimilar terms and addition and subtraction can only be indicated. Thus the following statement can be made:

RULE

To add or subtract radicals:

1. Reduce the radicals to their simplest form.
2. Combine similar radicals, and assume positive square roots of factors removed from the radicands.
3. Indicate addition or subtraction of dissimilar radicals.

PROBLEMS 20–5

Simplify:

1. $7\sqrt{2} - 4\sqrt{2}$
2. $5\sqrt{3} + 2\sqrt{12}$
3. $6\sqrt{3} - \sqrt{27}$
4. $\sqrt{847} - \sqrt{28}$
5. $a\sqrt{2} - b\sqrt{2} + c\sqrt{2}$
6. $\alpha\sqrt{2} + \beta\sqrt{8} - \gamma\sqrt{50}$
7. $2\sqrt{27} + 5\sqrt{48} - \sqrt{75}$
8. $2\sqrt{\frac{1}{3}} + \sqrt{\frac{1}{3}}$
9. $7\sqrt{5} - \frac{15}{\sqrt{5}} - 16\sqrt{\frac{5}{16}}$
10. $6\sqrt{27} + 5\sqrt{32}$
11. $6\sqrt{\frac{1}{18}} + 6\sqrt{\frac{1}{2}} + 3\sqrt{2}$

12. $\dfrac{R_1}{3} + \sqrt{\dfrac{16R_1^{\,2}}{3}}$

13. $\sqrt{\dfrac{4}{7}} - \sqrt{\dfrac{5}{35}}$

14. $\sqrt{\dfrac{\varepsilon + \eta}{\varepsilon - \eta}} + \sqrt{\dfrac{\varepsilon - \eta}{\varepsilon + \eta}}$

15. $\sqrt{\dfrac{\alpha}{27}} - \sqrt{\dfrac{\alpha}{54}}$

16. $\sqrt{\dfrac{7R^2}{16V}} + \sqrt{\dfrac{M^2V}{28}} - 4\sqrt{\dfrac{63}{16V}}$

20–10 MULTIPLICATION OF RADICALS

Obtaining the product of radicals is the inverse of removing a factor, as will be shown in the following examples:

EXAMPLE 14

$$\begin{aligned} 3\sqrt{3} \cdot 5\sqrt{4} &= 15\sqrt{3 \cdot 4} \\ &= 15 \cdot 2\sqrt{3} \\ &= 30\sqrt{3} \end{aligned}$$

EXAMPLE 15

$$\begin{aligned} 4\sqrt{3a} \cdot 2\sqrt{6a} &= 8\sqrt{3a \cdot 6a} \\ &= 8\sqrt{18a^2} \\ &= 8\sqrt{9 \cdot 2a^2} \\ &= 24a\sqrt{2} \end{aligned}$$

EXAMPLE 16 Multiply $3\sqrt{2} + 2\sqrt{3}$ by $4\sqrt{2} - 3\sqrt{3}$.

SOLUTION

$$\begin{array}{llll} 3\sqrt{2} & + 2\sqrt{3} & & \\ 4\sqrt{2} & - 3\sqrt{3} & & \\ \hline 24 & + 8\sqrt{6} & & \\ & - 9\sqrt{6} & - 18 & \\ \hline 24 & - \sqrt{6} & - 18 & = 6 - \sqrt{6} \end{array}$$

PROBLEMS 20–6

Perform the indicated operations:

1. $\sqrt{2} \cdot \sqrt{5}$
2. $\sqrt{8}\sqrt{2}$
3. $3\sqrt{15} \cdot \sqrt{3}$
4. $8\sqrt{5} \cdot 4\sqrt{15}$
5. $2\sqrt{5} \cdot 3\sqrt{8}$
6. $\sqrt{6} \cdot \sqrt{24}$
7. $\sqrt[3]{2} \cdot \sqrt[3]{32}$
8. $\sqrt{\dfrac{7}{16}} \cdot \sqrt{\dfrac{21}{3}}$
9. $\left(\sqrt{x^2 + R^2}\right)^2$
10. $\left(\varepsilon + \sqrt{3}\right)\left(\varepsilon - \sqrt{3}\right)$
11. $\left(\sqrt{X} - \sqrt{X - 5}\right)^2$
12. $\left(3 + \sqrt{5}\right)^2$
13. $\sqrt{(Z - Y)^2}$

14. $(2\sqrt{5} + 3\sqrt{2})(\sqrt{5} + 5\sqrt{2})$
15. $\sqrt{2\varepsilon} \cdot \sqrt{50\pi^2\varepsilon}$
16. $\sqrt{2(x^2 - 4x + 4)} \cdot \sqrt{\dfrac{8}{4x^2 + 16x + 16}}$
17. $(-2 - \sqrt{5})(3 - 5\sqrt{5})$
18. $(4 + 2\sqrt{3})(2 - \sqrt{3})$
19. $\left(\dfrac{36 - 9\sqrt{5}}{2}\right)\left(\dfrac{2\sqrt{5} + 8}{11}\right)$
20. $\dfrac{(\sqrt{\alpha} - \sqrt{\beta})(\alpha + 2\sqrt{\alpha\beta} + \beta)}{\alpha - \beta}$

20–11 DIVISION

An indicated root whose value is irrational but whose radicand is rational is called a *surd*. Thus, $\sqrt[3]{3}$, $\sqrt{2}$, $\sqrt[4]{5}$, $\sqrt{3}$, etc., are surds. If the indicated root is the square root, then the surd is called a *quadratic surd*. For example, $\sqrt{2}$, $\sqrt{5}$, $\sqrt{6}$, $\sqrt{15}$ are quadratic surds. Then, by extending the definition, such expressions as $3 + \sqrt{2}$ and $\sqrt{3} - 6$ are called *binomial quadratic surds*.

It is important that you become proficient in the multiplication and division of binomial quadratic surds. One method of solving ac circuits, which will be discussed later, makes wide use of these particular operations. Multiplication of such expressions was covered in the preceding section. However, a new method is necessary for division.

Consider the two expressions $a - \sqrt{b}$ and $a + \sqrt{b}$. They differ only in the sign between the terms. These expressions are *conjugates;* that is, $a - \sqrt{b}$ is called the conjugate of $a + \sqrt{b}$, and $a + \sqrt{b}$ is called the conjugate of $a - \sqrt{b}$. Remember this meaning of ''conjugate,'' for it is the same with reference to certain circuit characteristics.

To divide a number by a binomial quadratic surd, rationalize the divisor (denominator) by multiplying both dividend (numerator) and divisor by the conjugate of the divisor.

EXAMPLE 17

$$\frac{1}{3 + \sqrt{2}} = \frac{3 - \sqrt{2}}{(3 + \sqrt{2})(3 - \sqrt{2})} = \frac{3 - \sqrt{2}}{7}$$

EXAMPLE 18

$$\frac{1}{3\sqrt{3} - 1} = \frac{3\sqrt{3} + 1}{(3\sqrt{3} - 1)(3\sqrt{3} + 1)} = \frac{3\sqrt{3} + 1}{26}$$

EXAMPLE 19

$$\frac{3-\sqrt{2}}{4+\sqrt{2}}=\frac{(3-\sqrt{2})(4-\sqrt{2})}{(4+\sqrt{2})(4-\sqrt{2})}$$
$$=\frac{14-7\sqrt{2}}{14}$$
$$=\frac{2-\sqrt{2}}{2}$$

NOTE *In each of the foregoing examples the resulting denominator is a rational number. In general, the product of two conjugate surd expressions is a rational number. This important fact is widely used in the solution of ac problems.*

PROBLEMS 20–7

Perform the indicated division:

1. $\frac{2\sqrt{15}}{\sqrt{20}}$
2. $\frac{5}{5-\sqrt{3}}$
3. $\frac{36}{5+\sqrt{7}}$
4. $\frac{7}{3\sqrt{5}+2}$
5. $\frac{25}{5-5\sqrt{2}}$
6. $\frac{x+\sqrt{y}}{x-\sqrt{y}}$
7. $\frac{x-\sqrt{y}}{x+\sqrt{y}}$
8. $\frac{3-\sqrt{5}}{2+\sqrt{5}}$
9. $\frac{2+5\sqrt{2}}{5+5\sqrt{2}}$
10. $\frac{\sqrt{R}+\sqrt{Z}}{\sqrt{R}-\sqrt{Z}}$
11. $\frac{\sqrt{5}+3}{2+\sqrt{3}}$
12. $\frac{20+\mathrm{j}18}{3+\mathrm{j}5}$

▶ **HINT** Maintain order j18, j5, etc. and treat terms containing the symbol j as if they were radicals.

20–12 THE OPERATOR j

In our studies so far, we have met with several mathematical symbols which actually indicate *commands;* $+$, $-$, $\times$, $\div$, and $\sqrt{\ }$ are all symbols which actually tell us to perform some specific operation. In Sec. 3–5, for instance, we saw that the minus sign is equivalent to a rotation of a quantity through 180°, and, by definition, this rotation is in the positive, or counterclockwise, direction.

Now we must meet the operator j, which also provides a rotation, not of 180°, but of 90°. You have noticed that all the algebraic symbols used so far in this book are printed in *italic* (slanting) type. The operator j, however, is printed in roman (regular) type to distinguish it as an operator and to constantly remind the student that it is not just another algebraic symbol. The operator j is extremely useful in the solution of electronic circuits, and although the idea is simple and straightforward—*just rotate through 90° in a counterclockwise (ccw) direction*—it is essential that we understand exactly how to operate with it. In Fig. 20–1, the line *OA*, which lies on the *x* axis and is *a* units long, can

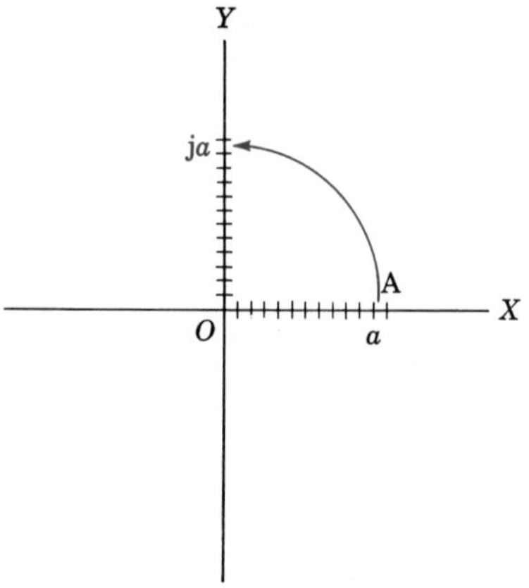

FIG. 20–1 Representation of a quantity affected by the operator j.

be operated on by the operator j to become *ja*, a line of the same length as before but now rotated ccw through 90° to lie on the *y* axis.

Note how the rotated quantity is described: first is given the symbol of the operator j, and then the quantity which has been operated upon, *a*. Thus, when *a* is "j'd", it becomes *ja*. This practice of placing the operator first draws attention to the fact that we are not dealing with some quantity j multiplied by some other quantity *a*, but that the j operator is operating on the quantity *a*. The algebraic symbol *ja* represents for us the geometric symbol of a line rotated through 90° in a counterclockwise direction.

Any quantity operated upon by j will rotate through 90° in a counterclockwise direction; similarly, any quantity operated upon by −j will rotate through 90° in a clockwise direction. (See Fig. 20–2.)

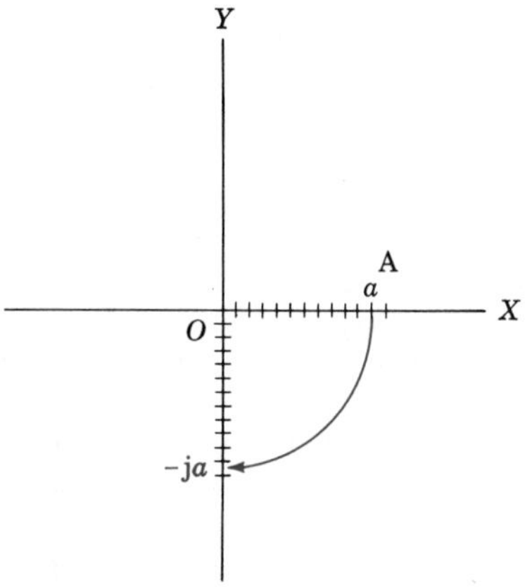

FIG. 20–2 Representation of a quantity affected by the operator −j.

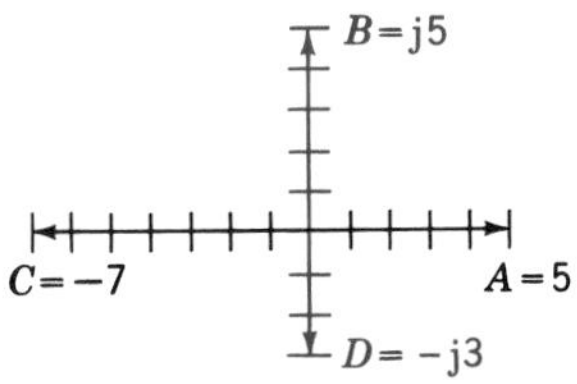

FIG. 20–3 Comparison of quantities $A = +5$, $B = +j5$, $C = -7$, and $D = -j3$.

Figure 20–3 relates four different quantities for purposes of review: $A = 5$, $B = j5$, $C = -7$, and $D = -j3$.

A quantity may be j-operated more than once. If we start with a quantity ja, as in Fig. 20–1, and j it again, we cause it to rotate through an additional 90° ccw, as shown in Fig. 20–4.

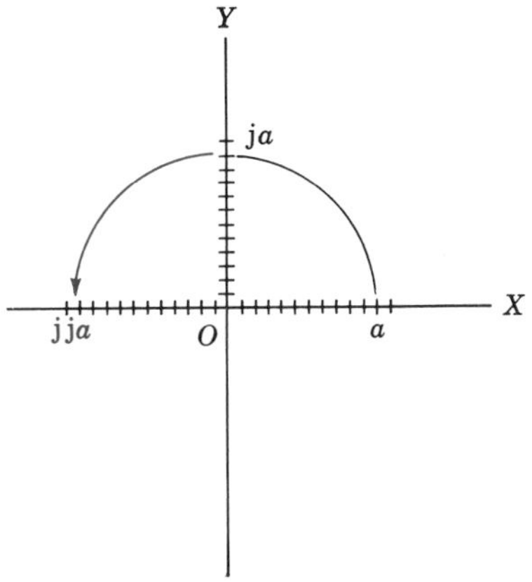

FIG. 20–4 Representation of repeated operation by j.

$j(ja)$ may be written jja, or, more simply, j^2a. Similarly, j^3 indicates that a quantity has been operated on three times in succession; that is, it has been rotated through 90° ccw three times in succession. Figures 20–5 and 20–6 indicate repeated rotations resulting from repeated operations by j and −j.

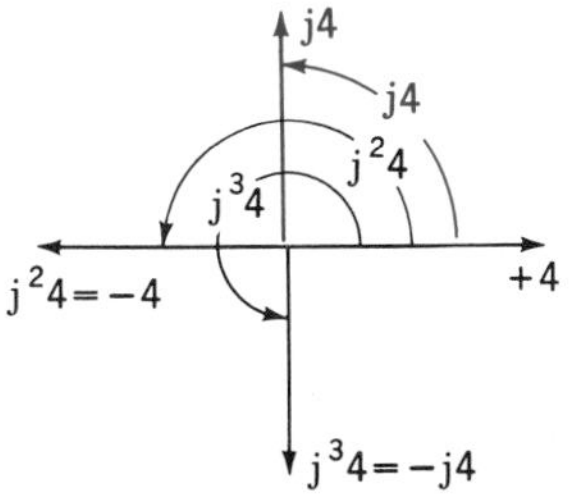

FIG. 20–5 Repeated rotation of numbers in counterclockwise direction.

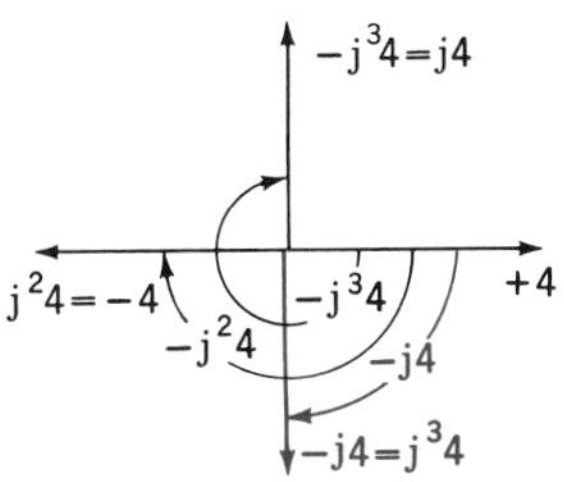

FIG. 20–6 Repeated rotation in clockwise direction.

Note, in passing, a very interesting point about j^2a: j-ing a twice in succession brings it to the same point as a single operation with a minus sign. From this graphic illustration, you can see that

$$j^2 = -1$$

and

$$j = \sqrt{-1}$$

This added relationship, $j = \sqrt{-1}$, is an extremely interesting one, because so far, in the removal of factors from radicands, all the radicands have been positive numbers. In Sec. 20–13, we will use the important relationship $j = \sqrt{-1}$ to factor negative radicands and to determine (or, at least represent) the square roots of negative numbers.

First, however, let us continue with the fascinating relationships exhibited by repeated operations with j. Since $j^2 = -1$, then j^3 must equal $j(-1)$, or $-j$, and j^4 must equal j^2j^2, that is, $(-1)(-1) = +1$. The truth of these statements can be justified by the following considerations:

$$\sqrt{-1} \cdot \sqrt{-1} = -1$$

That is,

$$j \cdot j = -1$$

$$\therefore j^2 = -1$$

Also, $\sqrt{-1} \cdot \sqrt{-1} \cdot \sqrt{-1} = -1 \cdot \sqrt{-1} = -j$

That is, $j \cdot j \cdot j = j^3$

$$\therefore j^3 = -j$$

Also, $\sqrt{-1} \cdot \sqrt{-1} \cdot \sqrt{-1} \cdot \sqrt{-1} = (\sqrt{-1} \cdot \sqrt{-1})(\sqrt{-1} \cdot \sqrt{-1})$

$$= (-1)(-1) = 1$$

That is, $j \cdot j \cdot j \cdot j = j^4$

$$\therefore j^4 = 1$$

Similarly, it can be shown that successive multiplication by each $+j$ rotates the number 90° in a counterclockwise direction.

If we consider successive multiplication by $-j$, we have

$$(-\sqrt{-1})(-\sqrt{-1}) = -1$$

That is, $(-j)(-j) = j^2$

$$\therefore (-j)^2 = -1$$

Also, $(-\sqrt{-1})(-\sqrt{-1})(-\sqrt{-1}) = (-1)(-\sqrt{-1}) = \sqrt{-1}$

That is, $(-j)(-j)(-j) = (j^2)(-j)$

$$= (-1)(-j) = j$$

$$\therefore (-j)^3 = j$$

To demonstrate that $(-j)^4 = 1$ and $\frac{1}{j} = -j$ is left as an exercise for you.

Note the convenience of the graphic method of representation of the j operations, Figs. 20–5 and 20–6. This method is an advantageous one because, if we can *visualize* a graph or diagram when we come up against certain types of numbers and equations, we often have a better understanding of the manner in which the quantities vary or are related.

One special note must be drawn to your attention: Long before the operator j was found to have practical application in electrical and electronics calculations, mathematicians used the symbol *i* to represent $\sqrt{-1}$. When electrical theory adopted the symbol *i* for instantaneous current flow in a circuit, we switched the mathematicians' *i* to j for our symbol of rotation through 90° ccw. Sometimes in your reading you will meet *i* instead of j, but you will know what it really means: "Rotate the quantity operated upon by 90° in a counterclockwise direction."

As a mathematical definition, j is sometimes referred to as the "complex operator," but, as we have seen, there is nothing particularly complex about j.

20–13 INDICATED SQUARE ROOTS OF NEGATIVE NUMBERS

So far, in the removal of factors from radicands, all the radicands have been positive numbers. Also, we have extracted the square roots of positive numbers only. How shall we proceed to factor negative radicands, and what is the meaning of the square root of a negative number?

According to our laws for multiplication, no number multiplied by itself or raised to any even power will produce a negative result. For example, what does $\sqrt{-25}$ mean when we know of no number that, when multiplied by itself, will produce -25?

The indicated square root of a negative number is known as an *imaginary number*. It is probable that this name was assigned before mathematicians could visualize such a number and that the word "imaginary" was originally used to distinguish such numbers from the so-called "real numbers" previously studied. In any event, calling such a number imaginary might be considered unfortunate, because in working with circuits such numbers become very real in the physical sense. If you accidentally touch a large capacitor that is highly charged, you are likely to be killed by the current driven by some of those "imaginary" volts. This will be discussed later.

To avoid the difficulty of operations with the indicated square roots of negative numbers, or imaginary numbers, it becomes necessary to introduce a new type of number. That is, we agree that every imaginary number can be expressed as the product of a positive number and $\sqrt{-1}$.

EXAMPLE 20

$$\begin{aligned}\sqrt{-25} &= \sqrt{(-1)25}\\ &= \sqrt{-1}\sqrt{25}\\ &= \sqrt{-1}\cdot 5\end{aligned}$$

As we saw in Sec. 20–12, $\sqrt{-1}$ may be represented by the operator j, and we may now rewrite $\sqrt{-1}\cdot 5$ as j5.

EXAMPLE 21

$$\begin{aligned}\sqrt{-16} &= \sqrt{(-1)16}\\ &= \sqrt{-1}\sqrt{16}\\ &= \sqrt{-1}\cdot 4\\ &= \mathrm{j}4\end{aligned}$$

EXAMPLE 22

$$\begin{aligned}\sqrt{-X^2} &= \sqrt{(-1)X^2}\\ &= \sqrt{-1}\sqrt{X^2}\\ &= \sqrt{-1}\cdot X\\ &= \mathrm{j}X\end{aligned}$$

EXAMPLE 23

$$\begin{aligned}-\sqrt{-4X^2} &= -\sqrt{(-1)4X^2}\\ &= -\sqrt{-1}\sqrt{4X^2}\\ &= -\sqrt{-1}\cdot 2X\\ &= -\mathrm{j}2X\end{aligned}$$

PROBLEMS 20–8

Express the following by using the operator j:

1. $\sqrt{-16}$
2. $\sqrt{-100}$
3. $\sqrt{-625}$
4. $\sqrt{-Y^2}$
5. $-\sqrt{-z^2}$
6. $-\sqrt{-49\omega^2}$
7. $\sqrt{-R^2Z^4}$
8. $\sqrt{\dfrac{-Q^4}{\omega^2L^2}}$

9. $-3\sqrt{-25}$
10. $2\sqrt{-48}$
11. $\sqrt{\dfrac{-25}{81}}$
12. $-\sqrt{\dfrac{169}{-\alpha^2}}$
13. $\sqrt{\dfrac{-75}{28}}$
14. $-\sqrt{-\lambda^2\pi}$
15. $-\sqrt{\dfrac{-V^2}{P}}$
16. Did you demonstrate that $(-j)^4 = 1$?
17. Did you demonstrate that $\dfrac{1}{j} = -j$?

20–14 COMPLEX NUMBERS

If a "real" number is united to an "imaginary" number by a plus or a minus sign, the expression thus obtained is called a *complex number*. Thus, $3 - j4$, $a + jb$, $R + jX$, etc., are complex numbers. At this time, we shall consider, not their graphical representation, but simply how to perform the four fundamental operations algebraically. Figure 20–7 shows the representation of the complex number $a + jb$.

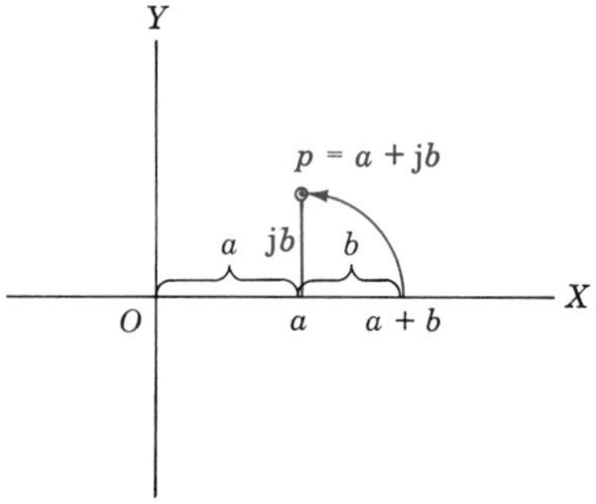

FIG. 20–7 Representation of a complex number $a + jb$. a lies in OX; b is rotated through 90° counterclockwise; the point p represents the "sum" of a and jb.

20–15 ADDITION AND SUBTRACTION OF COMPLEX NUMBERS

Combining a real number with an imaginary number cannot be accomplished by the usual methods of addition and subtraction; these processes can only be expressed. As an example—if we have the complex number $5 + j6$, this is as far as we can simplify it at this time. We should not attempt to add 5 and j6 arithmetically, for these two numbers are at right angles to each other, and such an operation would be meaningless. However, we *can* add and subtract complex numbers by treating them as ordinary binomials.

EXAMPLE 24 Add $3 + j7$ and $4 - j5$.

SOLUTION

$$\begin{array}{r} 3 + j7 \\ 4 - j5 \\ \hline 7 + j2 \end{array}$$

EXAMPLE 25 Subtract $-15 - j6$ from $-5 + j8$.

SOLUTION

$$\begin{array}{r} -5 + j8 \\ -15 - j6 \\ \hline 10 + j14 \end{array}$$

PROBLEMS 20–9

Find the indicated sums:

1. $\begin{array}{r} 4 + j9 \\ 3 + j7 \\ \hline \end{array}$
2. $\begin{array}{r} 12 + j7 \\ 16 + j7 \\ \hline \end{array}$
3. $\begin{array}{r} 30 + j7 \\ 18 - j4 \\ \hline \end{array}$
4. $\begin{array}{r} 96 - j22 \\ 32 - j5 \\ \hline \end{array}$
5. $\begin{array}{r} 28 - j11 \\ 115 + j5 \\ \hline \end{array}$
6. $\begin{array}{l} 32 \\ \ \ 5 + j6 \\ \hline \end{array}$
7. $\begin{array}{r} 20 + j3 \\ - j5 \\ \hline \end{array}$
8. $\begin{array}{l} 26 - j6 \\ 31 \\ \hline \end{array}$

9 to 16. Subtract the lower complex number from the upper in each of the above problems.

20–16 MULTIPLICATION OF COMPLEX NUMBERS

As in addition and subtraction, complex numbers are treated as ordinary binomials when multiplied. However, when writing the result, we must not forget that $j^2 = -1$.

EXAMPLE 26 Multiply $4 - j7$ by $8 + j2$.

SOLUTION

$$\begin{array}{rrr} 4 - j7 & & \\ 8 + j2 & & \\ \hline 32 - j56 & & \\ + j8 & - j^2 14 & \\ \hline 32 - j48 & - j^2 14 & \end{array}$$

Since $j^2 = -1$, the product is

$$\begin{aligned} 32 - j48 - (-1)(14) &= 32 - j48 + 14 \\ &= 46 - j48 \end{aligned}$$

EXAMPLE 27 Multiply $7 + j3$ by $6 + j2$.

SOLUTION

$$\begin{array}{rr} 7 + j3 & \\ 6 + j2 & \\ \hline 42 + j18 & \\ + j14 & + j^2 6 \\ \hline 42 + j32 & + j^2 6 = 36 + j32 \end{array}$$

20–17 DIVISION OF COMPLEX NUMBERS

As in the division of binomial quadratic surds, we simplify an indicated division by rationalizing the denominator in order to obtain a "real" number as divisor (Sec. 20–11). We do this by multiplying by the conjugate in the usual manner.

EXAMPLE 28

$$\frac{10}{1 + j2} = \frac{10(1 - j2)}{(1 + j2)(1 - j2)} = \frac{10(1 - j2)}{1 - j^2 4} = \frac{10(1 - j2)}{5} = 2(1 - j2)$$

EXAMPLE 29

$$\frac{5 + j6}{3 - j4} = \frac{(5 + j6)(3 + j4)}{(3 - j4)(3 + j4)} = \frac{15 + j38 + j^2 24}{9 - j^2 16} = \frac{-9 + j38}{25}$$

EXAMPLE 30

$$\frac{a + jb}{a - jb} = \frac{(a + jb)(a + jb)}{(a - jb)(a + jb)} = \frac{a^2 + j2ab + j^2 b^2}{a^2 - j^2 b^2} = \frac{a^2 + j2ab - b^2}{a^2 + b^2}$$

PROBLEMS 20–10

Find the indicated products:

1. $5(2 - j6)$
2. $(4 + j3)(5 + j6)$
3. $(a + jb)(a - jb)$
4. $(2 - j5)(7 - j9)$
5. $(\theta + j\phi)(\theta + j\phi)$
6. $(G - jB)(G + jL)$

Find the quotients:

7. $\frac{1}{1 + j1}$
8. $\frac{6}{1 - j2}$
9. $\frac{1 + j1}{1 - j1}$
10. $\frac{1 - j1}{1 + j1}$
11. $\frac{5}{5 + j5}$
12. $\frac{4 + j1}{5 - j3}$
13. $\frac{6}{6 - jx}$
14. $\frac{\theta + j\phi}{\theta - j\phi}$
15. $\frac{R + j\omega X}{R - j\omega X}$
16. $\frac{j3}{2 + j3}$
17. $\frac{j\phi}{\theta - j\phi}$
18. $\frac{1 + j\frac{\omega}{\omega_o}}{1 - j\frac{\omega}{\omega_o}}$
19. $\frac{R}{\frac{1}{j\omega C} + R + j\omega L}$

20. In an equation relating complex numbers, such as

$$a + jb = P - jQ$$

the in-phase quantities are equal ($a = P$) and the out-of-phase quantities are equal ($b = -Q$).

Write in the form $a + jb$: $\dfrac{(1 + j\omega\tau_1)(1 + j\omega\tau_2)}{\mu_o - \beta}$

20–18 RADICAL EQUATIONS

An equation in which the unknown occurs in a radicand is called an *irrational* or *radical equation*. To solve such an equation, arrange it so that the radical is the only term in one member of the equation. Then eliminate the radical by squaring both members of the equation.

EXAMPLE 31 Given $\sqrt{3x} = 6$; solve for x.

SOLUTION

$$\sqrt{3x} = 6$$

Squaring, $\quad 3x = 36$

D: 3, $\quad x = 12$

CHECK

Substituting 12 for x in the given equation,

$$\sqrt{3 \cdot 12} = 6$$
$$\sqrt{36} = 6$$
$$6 = 6$$

EXAMPLE 32 Given $\sqrt{2x + 3} = 7$; solve for x.

SOLUTION

$$\sqrt{2x + 3} = 7$$

Squaring, $\quad 2x + 3 = 49$

S: 3, $\quad 2x = 46$

D: 2, $\quad x = 23$

CHECK

$$\sqrt{2 \cdot 23 + 3} = 7$$
$$\sqrt{49} = 7$$
$$7 = 7$$

EXAMPLE 33 The time for one complete swing of a simple pendulum is given by

$$t = 2\pi\sqrt{\frac{L}{g}}$$

where t = time, s
L = length of pendulum
g = acceleration due to gravity
Solve for L.

SOLUTION

Given $$t = 2\pi\sqrt{\frac{L}{g}} \qquad (a)$$

Squaring (*a*), $$t^2 = 4\pi^2\frac{L}{g} \qquad (b)$$

M: g in (*b*), $$gt^2 = 4\pi^2 L \qquad (c)$$

D: t^2 in (*c*), $$g = \frac{4\pi^2 L}{t^2} \qquad (d)$$

Rewrite (*c*), $$4\pi^2 L = gt^2 \qquad (e)$$

D: $4\pi^2$ in (*e*), $$L = \frac{gt^2}{4\pi^2}$$

EXAMPLE 34 Given $V = I_p Z_p + j\omega M I_s$ and $I_s Z_s = -j\omega M I_p$. Show that

$$V = I_p\left[Z_p + \frac{(\omega M)^2}{Z_s}\right]$$

SOLUTION

Since I_s does not appear in the final equation, it must be eliminated. Solving the given equations for I_s,

$$I_s = \frac{V - I_p Z_p}{j\omega M} \qquad (a)$$

$$I_s = \frac{-j\omega M I_p}{Z_s} \qquad (b)$$

Equating the right members of (*a*) and (*b*),

$$\frac{V - I_p Z_p}{j\omega M} = \frac{-j\omega M I_p}{Z_s}$$

M: $j\omega M$ $$V - I_p Z_p = \frac{-j^2\omega^2 M^2 I_p}{Z_s}$$

Substituting -1 for j^2 in the right member,

$$V - I_p Z_p = \frac{\omega^2 M^2 I_p}{Z_s}$$

A: $I_p Z_p$, $$V = I_p Z_p + \frac{(\omega M)^2 I_p}{Z_p}$$

Factoring the right member,

$$V = I_p\left[Z_p + \frac{(\omega M)^2}{Z_s}\right]$$

PROBLEMS 20–11

Solve the following equations:

1. $\sqrt{x} = 5$
2. $\sqrt{R} = 4$
3. $\sqrt{\gamma} = 25$
4. $\sqrt{p} + 3 = 5$
5. $\sqrt{B} - 6 = 24$
6. $\sqrt{L + 5} = 6$
7. $\sqrt{P - 3} = 3$
8. $3\sqrt{\alpha - 6} = 18$
9. $4\sqrt{\lambda + 3} - 2 = 6$
10. $\sqrt{\frac{7K + 4}{2}} = 4$
11. $3\sqrt{\phi + 3} = 2\sqrt{3\phi - 12}$

Given:	Solve for:
12. $E = \sqrt{\frac{\eta\phi}{\omega^2\theta}}$	ϕ
13. $I_x = \varepsilon\sqrt{2L_1L_2}$	L_1
14. $\frac{i_s}{i_n} = \sqrt{\frac{\rho P_s}{e(\Delta f)}}$	P_s
15. $2\pi f = \frac{1}{\sqrt{LC}}$	C
16. $\lambda = \frac{4\pi}{\gamma Q}\sqrt{\frac{KFTS(\Delta f)}{NP_0}}$	$\frac{S}{N}$, P_0
17. $\frac{V}{C} = \sqrt{\frac{1}{\frac{\alpha(v_1 - v_2)^2}{\alpha} + \frac{v_1v_2}{\alpha}}}$	α
18. $\gamma = \sqrt{\frac{1 - \mu_x\eta E}{\omega X}}$	μ_x
19. $Y_n = G\sqrt{\left(\frac{n^2 - 1}{n}\right)^2 Q_2 + 1}$	Q_2
20. $Z_t = R\sqrt{1 + \left(\frac{f}{f_0}\right)^4}$	f_0
21. $G_a = \sqrt{G_1 + \frac{G_1}{R_{eq} + \frac{G_L}{g_m^2}}}$	g_m^2

22. At a resonant frequency of f Hz, the inductive reactance X_L of a circuit of L H is $X_L = \omega L\ \Omega$ and the capacitive reactance X_C of a circuit with a capacitance of C F is $X_C = \frac{1}{\omega C}\Omega$. $\omega = 2\pi f$. At the resonant frequency, with both inductance and capacitance in the circuit, $X_L = X_C$. Solve for the resonant frequency f in terms of π, L, and C.
23. Use the formula for the resonant frequency derived in Prob. 22 to find the value of C in picofarads in a case when $f = 1.4$ MHz and $L = 51.7\ \mu$H.
24. Use the formula that was derived in Prob. 22 to find the value of f when $C = 47$ nF and $L = 15$ nH.
25. $f = \dfrac{1}{\sqrt{2\pi \dfrac{LC_aC_b}{C_a + C_b}}}$. Solve for C_a.
26. In a conductor through which current I flows, energy E_m existing in the magnetic field about the line is $\frac{LI^2}{2}$ J, where L is the inductance of the line per unit length. An equal energy E_c exists in the electrostatic field of the line, equal to $\frac{CV^2}{2}$ J, where C is the capacitance of the line per unit length. If the surge impedance Z_o of the line is $\frac{V}{I}\ \Omega$, show that $Z_o = \sqrt{\frac{L}{C}}$.
27. Given $\Delta = \frac{4}{\pi}\sqrt{1 + \left(\frac{\pi\tau\omega}{4}\right)^2}$, show that

$$\omega = \pm\frac{1}{\pi\tau}\sqrt{(\pi\Delta + 4)(\pi\Delta - 4)}$$

28. Given $\sqrt{\dfrac{1}{\tau_1\tau_2} - \dfrac{1}{4\tau_2{}^2}} = 786$ and $\dfrac{1}{2\tau_2} = 78.6$, solve for τ_1.
29. Show that $KV_p^{\frac{3}{2}} = KV_p\sqrt{V_p}$.
30. A West Coast semiconductor products manufacturer, in a design for a 100-W 10-MHz power amplifier, equates the actual output circuit to its equivalent:

$$\frac{R_L\left(\dfrac{1}{j\omega C_7}\right)}{R_L + \dfrac{1}{j\omega C_7}} = R'_L + \frac{1}{j\omega C'_7}$$

(*a*) Show that

$$C_7 = \frac{1}{\omega R_L}\sqrt{\frac{R_L}{R'_L} - 1}$$

and $$C'_7 = C_7\left[1 + \left(\frac{1}{\omega C_7 R_L}\right)^2\right]$$

(*b*) If $R_L = 50\ \Omega$, $R'_L = 12.5\ \Omega$, and $\omega = 2\pi \times 10^7$, show that $C_7 = 551$ pF and $C'_7 = 735$ pF.

SELF TEST

Perform the indicated operations:

1. $a^p \cdot a^x$
2. $x^{y+z} \div x^{y-z}$
3. $(\theta^2\phi^3)^3$
4. $\left(\dfrac{R^3}{Z^{-2}}\right)^3$
5. $\sqrt{m^4n^2p}$
6. $5y^3\sqrt{361x^5y^3z^2}$
7. $\sqrt{R^2 - \left(\dfrac{R}{2\pi fL}\right)^2}$
8. $\sqrt{80a} + \sqrt{\dfrac{a}{5}} + \sqrt{5a}$
9. $\dfrac{2 + \sqrt{5}}{3 - \sqrt{5}}$
10. Express, using the operator j:

$$\sqrt{\frac{-36}{49}}$$

Perform the indicated operations:

11. $(3 + j7) + (2 + j5)$
12. $(8 + j15) - (4 - j7)$
13. $(2 + j5)(6 - j3)$
14. $\dfrac{2 + j7}{3 - j2}$
15. If $R + jX = 47 - j39$, what is the value of X?
16. Solve for P: $\sqrt{\dfrac{2P + 4}{5}} = 6$

CHAPTER 21

QUADRATIC EQUATIONS

In preceding chapters the study of equations has been limited mainly to equations which contain the unknown quantity in the first degree. This chapter is concerned with equations of the second degree, which are called quadratic equations.

21–1 DEFINITIONS

In common with polynomials (Sec. 11–2), the degree of an equation is defined as the degree of the term of highest degree in it. Thus, if an equation contains the square of the unknown quantity and no higher degree, it is an equation of the second degree, or a *quadratic equation*.

A quadratic equation that contains terms of the second degree only of the unknown is called a *pure quadratic equation*. For example,

$$x^2 = 25 \qquad R^2 - 49 = 0 \qquad 3x^2 = 12 \qquad ax^2 + c = 0$$

are pure quadratic equations.

A quadratic equation that contains terms of *both* the first and the second degree of the unknown is called an *affected* or a *complete quadratic equation;* $x^2 + 3x + 2 = 0$, $3x^2 + 11x = -2$, $ax^2 + bx + c = 0$, etc., are affected, or complete, quadratic equations.

When a quadratic equation is solved, values of the unknown that will satisfy the conditions of the equation are found.

A value of the unknown that will satisfy the equation is called a *solution* or a *root* of the equation.

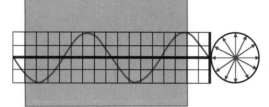

21–2 SOLUTION OF PURE QUADRATIC EQUATIONS

As stated in Sec. 10–5, every number has two square roots that are equal in magnitude but opposite in sign. Hence, all quadratic equations have two roots. In pure quadratic equations, the absolute values of the roots are equal but of opposite sign.

EXAMPLE 1 Solve the equation $x^2 - 16 = 0$.

SOLUTION

Given	$x^2 - 16 = 0$
A: 16,	$x^2 = 16$
$\surd$ (see note below),	$x = \pm 4$

CHECK

Substituting in the equation either +4 or −4 for the value of x, because either squared results in +16, we have

$$(\pm 4)^2 - 16 = 0$$

or

$$16 - 16 = 0$$

NOTE *Hereafter, the radical sign will mean "take the square root of both members of the preceding or designated equation."*

EXAMPLE 2 Solve the equation $5R^2 - 89 = 91$.

SOLUTION

Given	$5R^2 - 89 = 91$
A: 89,	$5R^2 = 180$
D: 5,	$R^2 = 36$
$\surd$,	$R = \pm 6$

CHECK

$$5(\pm 6)^2 - 89 = 91$$
$$5 \times 36 - 89 = 91$$
$$180 - 89 = 91$$
$$91 = 91$$

EXAMPLE 3 Solve the equation

$$\frac{I + 4}{I - 4} + \frac{I - 4}{I + 4} = \frac{10}{3}$$

SOLUTION

Given $$\frac{I + 4}{I - 4} + \frac{I - 4}{I + 4} = \frac{10}{3}$$

Clearing fractions,

$$3(I + 4)(I + 4) + 3(I - 4)(I - 4) = 10(I - 4)(I + 4)$$

Expanding,

$$3I^2 + 24I + 48 + 3I^2 - 24I + 48 = 10I^2 - 160$$

Collecting terms,

$$-4I^2 = -256$$

D: -4, $\qquad I^2 = 64$

$\surd$, $\qquad I = \pm 8$

CHECK

By the usual method.

PROBLEMS 21–1

Solve the following:

1. $x^2 - 36 = 0$
2. $l^2 - 36 = 0$
3. $i^2 + 25 = 183$
4. $\theta^2 - 0.25 = 0$
5. $2V^2 - 66 = 0$
6. $\phi^2 - 0.0004 = 0.0012$
7. $\lambda^2 - \dfrac{9}{121} = 0$
8. $49I^2 - 144 = 0$
9. $5\mu^2 = 3\frac{1}{5}$
10. $5x^2 - 0.0308 = 0.0817$
11. $2(y + 1) - y(y - 3) - 5y = -23$
12. $\dfrac{28}{R^2 - 9} = \dfrac{R + 3}{R - 3} - 1 + \dfrac{R - 3}{R + 3}$
13. $\dfrac{4\theta^2 - 9\theta - 90}{3\theta} = -\dfrac{18 - 3\theta}{6}$
14. $6a(4a - 3) + 3(6a - 16) = 0$
15. $X_C = \dfrac{24 - X_C + (X_C - 1)^3}{2 + X_C^2} - 2$

21–3 COMPLETE QUADRATIC EQUATIONS—SOLUTION BY FACTORING

As an example, let it be assumed that all that is known about two expressions x and y is that $xy = 0$. We know that it is impossible to find the value of either unless the value of the other is known. However, we do know that, if $xy = 0$, *either* $x = 0$ *or* $y = 0$; for the product of two numbers can be zero if, and only if, one of the numbers is zero.

EXAMPLE 4 Solve the equation $x(5x - 2) = 0$.

SOLUTION

Here we have the product of two numbers x and $(5x - 2)$, equal to zero; and in order for the equation to be satisfied, one of the numbers must be equal to zero. Therefore, $x = 0$, or $5x - 2 = 0$. Solving the latter equation, we have $x = \frac{2}{5}$. Hence,

$$x = 0 \qquad \text{or} \qquad x = \frac{2}{5}$$

CHECK

If $x = 0$,

$$x(5x - 2) = 0(5 \cdot 0 - 2) = 0(-2) = 0$$

If $x = \frac{2}{5}$,

$$x(5x - 2) = \frac{2}{5}\left(5 \cdot \frac{2}{5} - 2\right) = \frac{2}{5}(2 - 2) = 0$$

It is evident that the roots of a complete quadratic may be of unequal absolute value and may or may not have the same signs.

It is incorrect to say $x = 0$ *and* $x = \frac{2}{5}$, for actually x cannot be equal to both 0 and $\frac{2}{5}$ at the same time. This will be more apparent in the following examples.

EXAMPLE 5 Solve the equation $(x - 5)(x + 3) = 0$.

SOLUTION

Again we have the product of two numbers, $(x - 5)$ and $(x + 3)$, equal to zero. Hence, either

$$\begin{aligned} x - 5 &= 0 \qquad \text{or} \qquad & x + 3 &= 0 \\ \therefore x &= 5 \qquad \text{or} \qquad & x &= -3 \end{aligned}$$

CHECK

If $x = 5$,

$$\begin{aligned} (x - 5)(x + 3) &= (5 - 5)(5 + 3) \\ &= 0(8) = 0 \end{aligned}$$

If $x = -3$,

$$\begin{aligned} (x - 5)(x + 3) &= (-3 - 5)(-3 + 3) \\ &= (-8)0 = 0 \end{aligned}$$

EXAMPLE 6 Solve the equation $x^2 - x - 6 = 0$.

SOLUTION

Given $\qquad x^2 - x - 6 = 0$

Factoring, $\qquad (x - 3)(x + 2) = 0$

Then, if $x - 3 = 0$, $x = 3$
Also, if $x + 2 = 0$, $x = -2$
$\therefore x = 3$ or -2

CHECK

If $x = 3$,

$$x^2 - x - 6 = 3^2 - 3 - 6 = 9 - 3 - 6 = 0$$

If $x = -2$,

$$x^2 - x - 6 = (-2)^2 - (-2) - 6 = 4 + 2 - 6 = 0$$

EXAMPLE 7 Solve the equation $(V - 3)(V + 2) = 14$.

SOLUTION

Given $(V - 3)(V + 2) = 14$
Expanding, $V^2 - V - 6 = 14$
S: 14, $V^2 - V - 20 = 0$
Factoring, $(V - 5)(V + 4) = 0$
Then, if $V - 5 = 0$, $V = 5$
Also, if $V + 4 = 0$, $V = -4$
$\therefore V = 5$ or -4

CHECK

If $V = 5$, $(V - 3)(V + 2) = (5 - 3)(5 + 2) = (2)(7) = 14$

If $V = -4$, $(V - 3)(V + 2) = (-4 - 3)(-4 + 2) = (-7)(-2) = 14$

PROBLEMS 21–2

Solve by factoring:

1. $y^2 + 5y + 6 = 0$
2. $v^2 - 6v - 7 = 0$
3. $R^2 + 15 = 8R$
4. $x^2 = 5x - 6$
5. $\theta^2 = 10 - 3\theta$
6. $\psi^2 = 17\psi - 60$
7. $X^2 - 20X + 51 = 0$
8. $55 + 6\varepsilon - \varepsilon^2 = 0$
9. $\dfrac{2Q - 13}{Q - 5} = \dfrac{7Q - 5}{5Q - 7}$
10. $\dfrac{8}{\kappa} + \kappa + 2 = \dfrac{2}{\kappa} - 3$
11. $\psi + 35 + \dfrac{18}{\psi} = 18 - \dfrac{24}{\psi}$
12. $\dfrac{160}{I^2} = \dfrac{26}{I} - 1$
13. $\dfrac{4\theta - 12}{2} = \dfrac{35 + 6\theta - \theta^2}{2\theta}$
14. $\dfrac{2F - 6}{17 - F} = 1 - \dfrac{2}{F - 2}$
15. $\dfrac{4}{2i + 2} + \dfrac{i}{3i + 7} - \dfrac{11}{4i + 4} = 0$

21–4 SOLUTION BY COMPLETING THE SQUARE

Some quadratic equations are not readily solved by factoring, but frequently such quadratic equations are readily solved by another method known as *completing the square*.

In Problems 10–5, missing terms were supplied in order to form a perfect trinomial square. This is the basis for the method of completing the square. For example, in order to make a perfect square of the expression $x^2 + 10x$, 25 must be added as a term to obtain $x^2 + 10x + 25$, which is the square of the quantity $x + 5$.

EXAMPLE 8 Solve the equation $x^2 - 10x - 20 = 0$.

SOLUTION

Inspection shows that the given equation cannot be factored with integral numbers. Therefore, the solution will be accomplished by the method of completing the square.

Given $\qquad x^2 - 10x - 20 = 0$

A: 20, $\qquad x^2 - 10x = 20$

Squaring one-half the coefficient of x and adding to both members,

$$x^2 - 10x + 25 = 20 + 25$$

Collecting terms, $\qquad x^2 - 10x + 25 = 45$

Factoring, $\qquad (x - 5)^2 = 45$

$\sqrt{\ }$, $\qquad x - 5 = \pm 6.71$

A: 5, $\qquad x = 5 \pm 6.71$

or $\qquad x = 11.71$ or -1.71

The above answers are correct to three significant figures. The values of x are more precisely stated by maintaining the radical sign in the final roots. That is, if

$$(x - 5)^2 = 45$$

$\sqrt{\ }$, $\qquad x - 5 = \pm\sqrt{45}$

or $\qquad x - 5 = \pm 3\sqrt{5}$

A: 5, $\qquad x = 5 \pm 3\sqrt{5}$

That is, $\qquad x = 5 + 3\sqrt{5} \quad$ or $\quad 5 - 3\sqrt{5}$

EXAMPLE 9 Solve the equation $3x^2 - x - 1 = 0$.

SOLUTION

Given $\qquad 3x^2 - x - 1 = 0$

D: 3 (because the coefficient of x^2 must be 1),

$$x^2 - \tfrac{1}{3}x - \tfrac{1}{3} = 0$$

Transposing the constant term,

$$x^2 - \tfrac{1}{3}x = \tfrac{1}{3}$$

Squaring one-half the coefficient of x and adding to both members,

$$x^2 - \tfrac{1}{3}x + \tfrac{1}{36} = \tfrac{1}{3} + \tfrac{1}{36}$$

Collecting terms, $$x^2 - \tfrac{1}{3}x + \tfrac{1}{36} = \tfrac{13}{36}$$

Factoring,

$$(x - \tfrac{1}{6})^2 = \tfrac{13}{36}$$

$\sqrt{\ }$, $$x - \frac{1}{6} = \pm\frac{\sqrt{13}}{6}$$

$$\therefore x = \frac{1 + \sqrt{13}}{6} \quad \text{or} \quad \frac{1 - \sqrt{13}}{6}$$

To summarize the method, we have the following:

RULES

To solve by completing the square:

1. If the coefficient of the square of the unknown is not 1, divide both members of the equation by the coefficient.
2. Transpose the constant terms (those not containing the unknown) to the right member.
3. Find one-half the coefficient of the unknown of the first degree, square the result, and add this square to both members of the equation. This makes the left member a perfect trinomial square.
4. Take the square root of both members of the equation and write the $\pm$ sign before the square root of the right member.
5. Solve the resulting simple equation.

PROBLEMS 21–3

Solve by completing the square:

1. $x^2 - 9x + 20 = 0$
2. $x^2 - 2x - 48 = 0$
3. $\omega^2 - 18\omega + 72 = 0$
4. $\Omega^2 + 5\Omega + 6 = 0$
5. $x^2 - 22x + 105 = 0$
6. $63 - a^2 = 2a$
7. $\eta^2 - 6\eta + 5 = 0$
8. $e^2 - 6 = e$
9. $M^2 = 22M + 48$
10. $m^2 + 5m + 6 = 0$
11. $v^2 + 5v - 66 = 0$
12. $17I - 42 = I^2 + 2I - 16$
13. $G + \dfrac{91}{G} = 20$
14. $1 + \dfrac{12}{f} + \dfrac{35}{f^2} = 0$
15. $\dfrac{7(R - 4)}{R - 3} - (R - 2) = \dfrac{R - 4}{2}$
16. $\dfrac{Z - 1}{Z + 1} = \dfrac{Z - 2}{Z + 2} - 6$

21–5 STANDARD FORM

Any quadratic equation can be written in the general form

$$ax^2 + bx + c = 0$$

This is called the *standard form* of the quadratic equation. When it is written in this way, a represents the coefficient of the term containing x^2, b represents the coefficient of the term containing x, and c represents the constant term. Note that all terms of the equation, when written in standard form, are in the left member of the equation.

EXAMPLE 10 Given $2x^2 + 5x - 3 = 0$. In this equation, $a = 2$, $b = 5$, and $c = -3$.

EXAMPLE 11 Given $R^2 - 5R - 6 = 0$. In this equation, $a = 1$, $b = -5$, and $c = -6$.

EXAMPLE 12 Given $9E^2 - 25 = 0$. In this equation, $a = 9$, $b = 0$, and $c = -25$.

21–6 THE QUADRATIC FORMULA

Because the standard form

$$ax^2 + bx + c = 0$$

represents *any* quadratic equation, it follows therefore that the roots of $ax^2 + bx + c = 0$ represent the roots of *any* quadratic equation. Therefore, if the standard quadratic equation can be solved for the unknown, the values, or roots, thereby obtained will serve as a formula for finding the roots of *any* quadratic equation.

This formula is derived by solving the standard form by the method of completing the square as follows:

Given $\qquad ax^2 + bx + c = 0$

Divide by a (Rule 1): $\qquad x^2 + \dfrac{bx}{a} + \dfrac{c}{a} = 0$

Transpose the constant term (Rule 2):

$$x^2 + \frac{bx}{a} = -\frac{c}{a}$$

Add the square of one-half the coefficient of x to both members (Rule 3):

$$x^2 + \frac{bx}{a} + \frac{b^2}{4a^2} = \frac{b^2}{4a^2} - \frac{c}{a}$$

Factor the left member, and add terms in the right member:

$$\left(x + \frac{b}{2a}\right)^2 = \frac{b^2 - 4ac}{4a^2}$$

Take the square root of both members:

$$x + \frac{b}{2a} = \pm\frac{\sqrt{b^2 - 4ac}}{2a}$$

Subtract $\frac{b}{2a}$: $$x = -\frac{b}{2a} \pm \frac{\sqrt{b^2 - 4ac}}{2a}$$

Collect terms of the right member:

$$x = \frac{-b \pm \sqrt{b^2 - 4ac}}{2a}$$

This equation is known as the quadratic formula.

Instead of attempting to solve a quadratic equation by factoring or by completing the square, we now make use of the quadratic formula. Upon becoming proficient in the use of the formula, you will find this method a convenience.

EXAMPLE 13 Solve the equation $5x^2 + 2x - 3 = 0$.

SOLUTION

Comparing this equation with the standard form

$$ax^2 + bx + c = 0$$

we have $a = 5$, $b = 2$, and $c = -3$. Substituting in the quadratic formula,

$$x = \frac{-b \pm \sqrt{b^2 - 4ac}}{2a}$$
$$= \frac{-2 \pm \sqrt{2^2 - 4 \cdot 5 \cdot (-3)}}{2 \cdot 5}$$

Hence, $$x = \frac{-2 \pm \sqrt{64}}{10}$$
$$= \frac{-2 \pm 8}{10}$$
$$= \frac{-2 + 8}{10} \quad \text{or} \quad \frac{-2 - 8}{10}$$
$$\therefore x = \tfrac{3}{5} \text{ or } -1$$

CHECK

Substitute the values of x in the given equation.

NOTE *It must be remembered that the expression* $\sqrt{b^2 - 4ac}$ *is the square root of the* quantity $(b^2 - 4ac)$ taken as a whole.

EXAMPLE 14 Solve the equation $\frac{3}{5 - R} = 2R$.

SOLUTION

Clearing the fractions results in $2R^2 - 10R + 3 = 0$. Comparing this equation with the standard form

$$ax^2 + bx + c = 0$$

we have $a = 2$, $b = -10$, and $c = 3$. Substituting in the quadratic formula,

$$x = \frac{-b \pm \sqrt{b^2 - 4ac}}{2a}$$

$$R = \frac{-(-10) \pm \sqrt{(-10)^2 - 4 \cdot 2 \cdot 3}}{2 \cdot 2}$$

Hence,

$$R = \frac{10 \pm \sqrt{76}}{4}$$

Factoring the radicand,

$$R = \frac{10 \pm 2\sqrt{19}}{4}$$

Dividing both terms of the fraction by 2,

$$R = \frac{5 \pm \sqrt{19}}{2}$$

$$= \frac{5 + \sqrt{19}}{2} \quad \text{or} \quad \frac{5 - \sqrt{19}}{2}$$

$$\therefore R = 4.68 \quad \text{or} \quad 0.320$$

These final answers are correct to three significant figures. Check the solution by the usual method.

21–7 TESTING SOLUTIONS

Now that we can obtain solutions to quadratic equations by means of the quadratic formula, there will be two answers that are possible so long as $b^2 - 4ac \neq 0$. One of these answers we may call α:

$$\alpha = \frac{-b + \sqrt{b^2 - 4ac}}{2a}$$

and the other we may call β:

$$\beta = \frac{-b - \sqrt{b^2 - 4ac}}{2a}$$

By suitable combinations of α and β, we can achieve two useful relationships, the proof of which we leave to you as an exercise:

$$\alpha + \beta = \frac{-b}{a} \tag{1}$$

$$\alpha \cdot \beta = \frac{c}{a} \tag{2}$$

Whenever you obtain answers to quadratic equations by means of the formula (or any other means), you may quickly test them for accuracy. The sum of the two answers must equal $-\frac{b}{a}$, and the product of the two must equal $\frac{c}{a}$.

EXAMPLE 15 Solve the equation $6x^2 - 2x - 4 = 0$, and test the answers.

SOLUTION

Using the quadratic formula, $x = 1$ or $x = -\frac{2}{3}$. Applying the tests:

$$\alpha + \beta = 1 - \frac{2}{3} = +\frac{1}{3}$$

$$-\frac{b}{a} = -\frac{-2}{6} = +\frac{1}{3}$$

and

$$\alpha \cdot \beta = (1)\left(-\frac{2}{3}\right) = -\frac{2}{3}$$

$$\frac{c}{a} = \frac{-4}{6} = -\frac{2}{3}$$

The tests show that the solutions obtained are correct. You should make a habit of applying the tests to every solution to quadratic equations that you obtain.

PROBLEMS 21–4

Solve the following equations by using the quadratic formula, and apply the tests of Eqs. (1) and (2):

1. $v^2 = 14 + 5v$
2. $x^2 + 2x = 15$
3. $2R + 15 = R^2$
4. $w^2 - 6w - 7 = 0$
5. $2L + 1 = 24L^2$
6. $3I^2 - 7I + 2 = 0$
7. $2x^2 + 6x = 40$
8. $5(R + 2) = 2R(R - 1)$
9. $6 - \frac{1}{x} = \frac{1}{x^2}$
10. $\frac{2}{I_1} + \frac{3}{I_1} = \frac{1}{I_1^2} - 14$
11. $d - 7 - \frac{11}{d - 2} = \frac{5(d - 6)}{d - 2}$
12. $\frac{2}{\lambda + 3} = \frac{3}{\lambda - 2} - 1$
13. $\frac{7}{\beta - 3} - \frac{1}{2} = \frac{\beta - 2}{\beta - 4}$
14. $\frac{36}{(I + 3)^2} - \frac{I + 2}{I + 3} = 1$
15. $7i + 5 = \frac{21i^3 - 16}{3i^2 - 4}$
16. $4 - E - \frac{1}{2E} = -\frac{E^2 + 25}{7E}$

21–8 THE GRAPH OF A QUADRATIC EQUATION—THE PARABOLA

In Chap. 16 we spent some time on the drawing of graphs, especially graphs of unity-power (first-degree) equations, or linear graphs. Graphs of quadratic equations also may be drawn, and in this section we will investigate the common methods of producing such graphs and also a method of predicting the shape of a graph just from the equation itself in the same way that we learned to use the standard form $y = mx + b$ to predict the slope and y intercept of linear graphs.

All the quadratic equations we have studied so far have contained only one unknown, but that is because we looked at special cases. In the algebraic solution of quadratics, the standard form $ax^2 + bx + c = 0$ is sufficient, because we want to know the values of x which will satisfy this standard form equation. However, to draw a graph requires two variables, an independent one x and a dependent one y, so we rewrite the standard equation:

$$y = ax^2 + bx + c$$

Then, by plotting values of y for given values of x, we can draw the complete graph. Note that the algebraic solutions so far in this chapter have simply let $y = 0$, that is, the algebraic solutions have given us the x intercepts for the equation of the general form

$$y = ax^2 + bx + c$$

EXAMPLE 16 Graph the equation $x^2 - 10x + 16 = 0$.

SOLUTION

Set the equation equal to y:

$$y = x^2 - 10x + 16$$

Make a table of the values of y corresponding to assigned values of x. (See Eq. Fig. 21–1.)

EQ. FIG. 21–1

If $x =$	0	1	2	3	4	5	6	7	8	9	10
Then $x^2 =$	0	1	4	9	16	25	36	49	64	81	100
$10x =$	0	10	20	30	40	50	60	70	80	90	100
$x^2 - 10x =$	0	−9	−16	−21	−24	−25	−24	−21	−16	−9	0
$\therefore y = x^2 - 10x + 16 =$	16	7	0	−5	−8	−9	−8	−5	0	7	16

Plotting the corresponding values of x and y as pairs of coordinates and drawing a smooth curve through the points results in the graph shown in Fig. 21–1.

From Fig. 21–1 on page 370 it is apparent that the graph has two x intercepts at $x = 2$ and $x = 8$. That is, when $y = 0$, the graph crosses the x axis at $x = 2$ and $x = 8$. This is to be expected; for when $y = 0$, the given equation

$$x^2 - 10x + 16 = 0$$

can be solved algebraically to obtain $x = 2$ or 8. Hence, it is evident that the points at which the graph crosses the x axis denote the values of x when $y = 0$, which are the roots of the equation.

Another interesting fact regarding this graph is that the curve goes through a *minimum value*. Suppose it is desired to solve for the coordinates of the point of minimum value. First, if the equation is changed to standard form, we obtain $a = 1$, $b = -10$, and $c = 16$. If the value of $\frac{-b}{2a}$ is computed, the result is

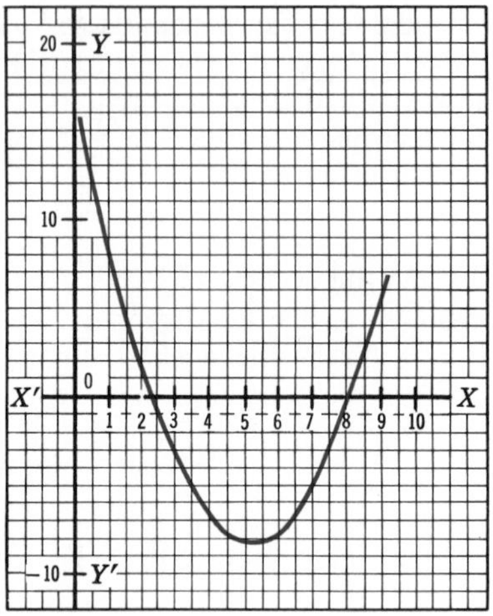

FIG. 21–1 Graph of the equation $y = x^2 - 10x + 16$.

the x value, or abscissa, of the minimum point on the curve. That is,

$$x = -\frac{b}{2a} = -\frac{-10}{2 \times 1} = \frac{10}{2} = 5$$

Substituting this value of x in the original equation,

$$y = x^2 - 10x + 16$$
$$y = 5^2 - 10 \times 5 + 16 = -9$$

Thus, the point $(5, -9)$ is where the curve passes through a minimum value. That is, the dependent variable y is a minimum and equal to -9 when x, the independent variable, is equal to 5.

A third point of interest is that the parabola, as the graph of the quadratic is called, is symmetrical about its turning point, which lies midway between the two intercepts. Indeed, this can be seen from a revision of the quadratic formula:

$$x = \frac{-b \pm \sqrt{b^2 - 4ac}}{2a}$$

which may appropriately be written as

$$x = \frac{-b}{2a} \pm \frac{\sqrt{b^2 - 4ac}}{2a}$$

from which we can see that, with the turning point at $\frac{-b}{2a}$, the values of the x intercepts, or roots, of the graph will be offset from the x value of the turning point by amounts equal to $\pm \frac{\sqrt{b^2 - 4ac}}{2a}$.

Look now at some of the main possibilities concerning the appearance of parabolas:

1. They may open upward or downward (Fig. 21–2).

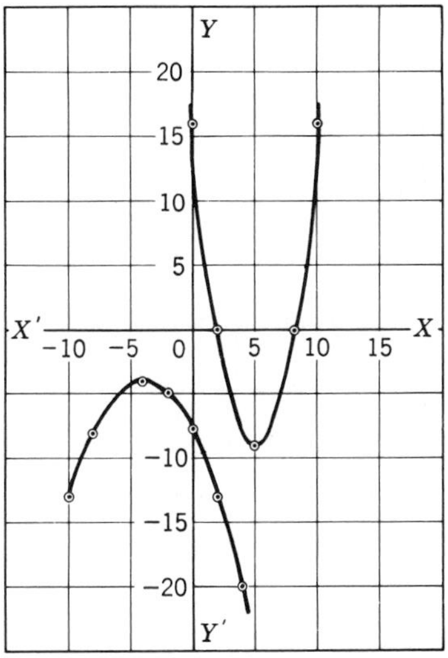

FIG. 21–2 Quadratic graph may open upward or downward.

2. They may be symmetrical about the y axis or about some line parallel to the y axis (Fig. 21–3).

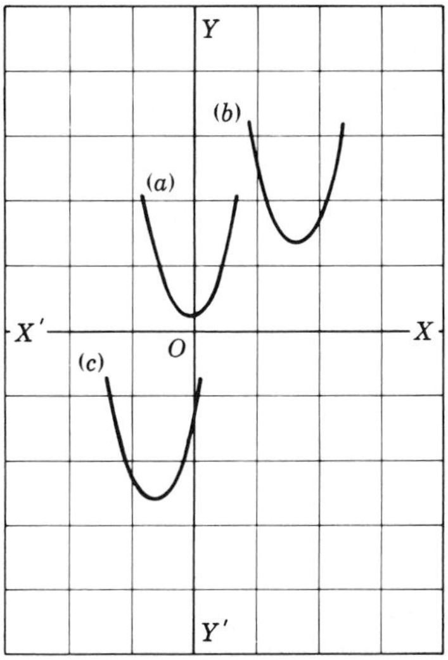

FIG. 21–3 Quadratic graphs may be symmetrical about the y axis or about a line parallel with the y axis.

3. They may (*a*) cut the *x* axis in two places, (*b*) touch the *x* axis (cut it in one place), or (*c*) not touch the *x* axis at all (Fig. 21–4).

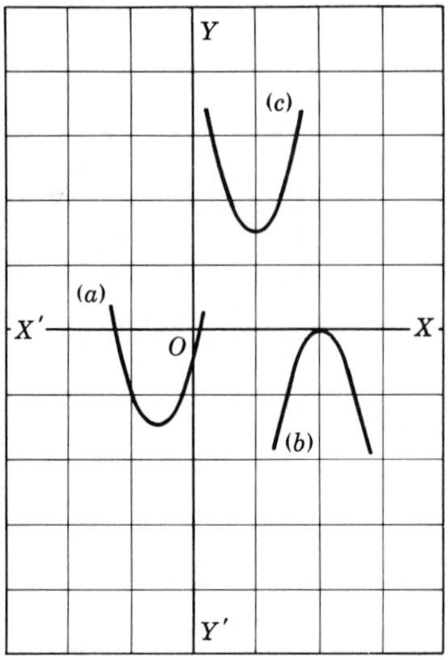

FIG. 21–4 Quadratic graphs may cut the x axis in two places, in one place, or not at all.

It is possible to decide many of these possibilities from the values of a particular quadratic equation. This general equation $y = ax^2 + bx + c$ offers many possibilities and a restriction:

1. *a*, the coefficient of the square term, may be any number, positive or negative, but *not* zero. (Why?)
2. *b*, the coefficient of the unity-power term, may be any number, positive or negative, *or* zero.
3. *c*, the constant term, may be any number, positive or negative, *or* zero.

Now, what is the effect of these algebraic possibilities on the graph? You should, at this point, arm yourself with graph paper and confirm the following statements:

RULE

The effect of *a* on the graph of the quadratic equation:

The value of *a* in the quadratic equation governs the steepness of the parabola. When *a* is large, the parabola is very steep, approaching a needlelike shape. When *a* is small, the parabola is shallow, approaching a dished shape.

Let $b = c = 0$ and plot the comparison graph $y = x^2$, in which the value of *a* is 1. Then plot various graphs of $y = ax^2$, letting *a* equal, in succession, 2, 5, 10, $\frac{1}{2}$, and $\frac{1}{4}$. If all these are plotted on the same graph sheet, with different colors or dashed lines, etc., then the effect of the value of a will be impressed on your mind forever.

RULE

The effect of a on the appearance of the parabola:

The algebraic sign of a will determine the opening of the parabola. $+a$ causes the curve to open upward, and the turning point is the *minimum* value. $-a$ causes the parabola to open downward, and the turning point is the *maximum* value.

You have already plotted a number of graphs with $+a$. Now plot a few graphs of $y = -ax^2$, letting $a = 2, 5, 10, \frac{1}{2}$, and $\frac{1}{4}$.

RULE

The effect of c on the appearance of the parabola:

The constant c in the quadratic equation determines the y intercept, and therefore the amount of vertical shift of the parabola. When c is positive, the curve is raised to cut the y axis above the x axis. When c is negative, the curve cuts the y axis below the x axis.

Let $a = 1$ and $b = 0$ and vary the value of c in the equation $y = x^2 + c$. Draw the curves when $c = +5$ and -5, and compare with the standard parabola $y = x^2$.

RULE

The effect of b on the appearance of the parabola:

The factor b in the quadratic equation determines the rotational shift of the turning point of the graph. When b is positive, the turning point shifts in a positive (ccw) direction about its ''original'' position, and when b is negative, the turning point shifts in a negative (cw) direction about its original position.

Let $a = 1$ and $c = 0$, and vary the value of b in the equation $y = x^2 + bx$. Draw the curves when $b = +2, +5, +10, -2, -5$, and -10. Next, repeat these curves with $a = -1$, and then draw curves for $y = -x^2 - bx$.

EXAMPLE 17 Plot the curve $y = 27 - 3x - 4x^2$.

SOLUTION

Predict, first of all, what effect the various coefficients will have on the graph:

1. The value of a is 4, so that the curve will be reasonably steep.
2. The algebraic sign of a is minus, so that the curve will open downward.
3. The constant term is $+27$, so that the curve cuts the y axis at $+27$, well above the x axis. Since the curve opens downward and the y intercept is above the x axis, the curve will cut the x axis in two places. The special equation $27 - 3x - 4x^2 = 0$ will have two definite solutions.

4. The value of b is -3, so that the turning point will be shifted from the "ideal" value of $x = 0$, $y = 27$ in the clockwise direction. The turning point will then be at a value of y greater than 27 and at some value of x to the left, or minus, side of the y axis.

With these predictions, together with a sketch of the probable appearance of the curve (Fig. 21–5), you may assign values to x and calculate the corresponding values of y:

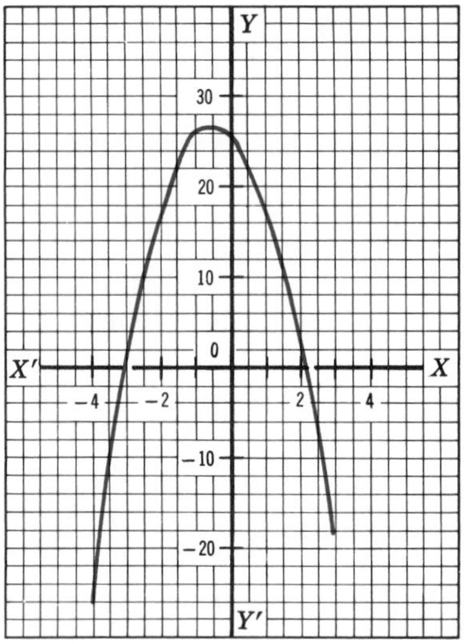

FIG. 21–5 Graph of the equation $y = 27 - 3x - 4x^2$.

If $x =$	-4	-3	-2	-1	0	1	2	3
Then $3x =$	-12	-9	-6	-3	0	3	6	9
$27 - 3x =$	39	36	33	30	27	24	21	18
$x^2 =$	16	9	4	1	0	1	4	9
$4x^2 =$	64	36	16	4	0	4	16	36
$\therefore y = 27 - 3x - 4x^2 =$	-25	0	17	26	27	20	5	-18

Plotting the corresponding values of x and y as pairs of coordinates and drawing a smooth curve through them results in the graph shown in Fig. 21–5.

From the graph of the equation $y = 27 - 3x - 4x^2$, Fig. 21–5, it is observed:

1. The roots (solution) of the equation are denoted by the x intercepts. These are $x = -3$ and $x = 2.25$. They can be checked algebraically to obtain

$$27 - 3x - 4x^2 = 0$$

Factoring,

$$(3 + x)(9 - 4x) = 0$$

$$x = -3 \text{ or } 2.25$$

2. The parabola opens *downward* because the coefficient of x^2 is negative ($a = -4$).
3. Because the parabola opens downward, the graph goes through a *maximum* value. The point of maximum value is found in the same manner as the minimum point of Example 16. That is,

$$x = \frac{-b}{2a} = \frac{-(-3)}{-2(-4)} = -\frac{3}{8}$$

Substituting $-\frac{3}{8}$ for x in the original equation,

$$y = 27 - 3(-\tfrac{3}{8}) - 4(-\tfrac{3}{8})^2 = 27.6$$

Thus, the dependent variable y is a maximum and equal to 27.6 when x, the independent variable, is equal to $-\dfrac{3}{8}$.

EXAMPLE 18 Graph the equations

$$y = x^2 - 8x + 12 \qquad (a)$$
$$y = x^2 - 8x + 16 \qquad (b)$$
$$y = x^2 - 8x + 20 \qquad (c)$$

SOLUTION

1. Based on an analysis of the values of a, b, and c, predict the probable appearance of each curve.
2. As before, and for each equation, make up a table of values of y corresponding to chosen values of x. Using these x and y values as pairs of coordinates, plot the graphs of the equations. These graphs are shown in Fig. 21–6.

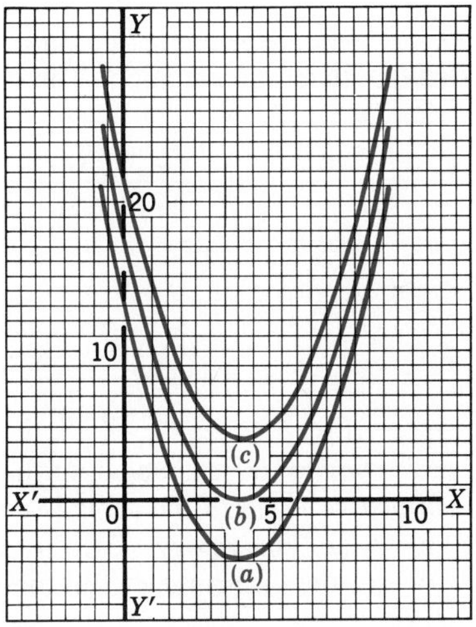

FIG. 21–6 Graphs of the equations of Example 18.

The coefficients of the equations are the same except for the values of the constant term c.

From the graph of the equations of Example 18, it is observed that:

1. The curve of (a) intercepts the x axis at $x = 2$ and $x = 6$, and the roots of the equation are thus denoted as $x = 2$ or 6.
2. The curve of (b) just touches the x axis at $x = 4$. Solving (b) algebraically shows that the roots are *equal,* both roots being 4.
3. The curve of (c) does not intersect or touch the x axis. Solving (c) algebraically results in the imaginary roots $x = 4 \pm j2$.
4. All curves pass through minimum values at points having equal x values. This is as expected, for the x value of a maximum or a minimum is given by $x = \frac{-b}{2a}$, and these values are equal in each of the given equations.
5. Checking the y values of the minima, it is seen that they must be affected by the constant terms, for, as previously mentioned, the other coefficients of the equations are the same.

21–9 GRAPHICAL SOLUTIONS

From the foregoing, it must now be obvious that quadratic equations can be solved graphically by letting the equation $ax^2 + bx + c = 0$ take the more general form $y = ax^2 + bx + c$. Then the two x intercepts of the graph will give the roots of the original equation. It is for this reason that you will often hear the solutions to a quadratic equation referred to as the *zeros* of the equation—they occur when $y = 0$.

PROBLEMS 21–5

Select problems from Problems 21–1, 21–2, 21–3, and 21–4 and solve them graphically to confirm the algebraic solutions. Predict what the graphs will look like before plotting calculated values of x and y.

21–10 THE DISCRIMINANT

The quantity $b^2 - 4ac$ under the radical in the quadratic formula is called the *discriminant* of the quadratic equation. The two roots of the equation are

$$x = \frac{-b + \sqrt{b^2 - 4ac}}{2a}$$

and

$$x = \frac{-b - \sqrt{b^2 - 4ac}}{2a}$$

Now, if $b^2 - 4ac = 0$, it is apparent that the two roots are equal. Also, if $b^2 - 4ac$ is *positive,* each of the roots is a *real* number. But if $b^2 - 4ac$ is *negative,* the roots are *imaginary.* Therefore, there is a direct relationship between the value of the discriminant and the roots, and hence the graph, of a quadratic equation.

For example, the discriminants of the equations of Example 18 in the preceding section are

$$b^2 - 4ac = (-8)^2 - 4 \cdot 1 \cdot 12 = 16$$
$$b^2 - 4ac = (-8)^2 - 4 \cdot 1 \cdot 16 = 0$$
$$b^2 - 4ac = (-8)^2 - 4 \cdot 1 \cdot 20 = -16$$

Upon checking these values with the curves of Fig. 21–6 and also checking the values of the discriminants found in the preceding exercises with their respective curves, it is evident that the roots of a quadratic equation are:

1. Real and unequal if and only if $b^2 - 4ac$ is positive.
2. Real and equal if and only if $b^2 - 4ac = 0$.
3. Imaginary and unequal if and only if $b^2 - 4ac$ is negative.
4. Rational if and only if $b^2 - 4ac$ is a perfect square.

21–11 MAXIMA AND MINIMA

As earlier stated, in the general quadratic equation $ax^2 + bx + c = 0$ the relation $x = \dfrac{-b}{2a}$ gives the value of the independent variable x at which the dependent variable y will be maximum or minimum. Then by substituting this value of x, the independent variable, in the equation, the corresponding value of y can be obtained. Also, it has been shown that the function will be maximum if a, the coefficient of x^2, is negative because the curve opens downward. Similarly, if the coefficient of x^2 is positive, the curve will pass through a minimum because the curve opens upward.

This knowledge facilitates the solution of many problems that heretofore would have involved considerable labor.

EXAMPLE 19 A source of emf V, with an internal resistance r, is connected to a load of variable resistance R. What will be the value of R with respect to r when maximum power is being delivered to the load?

SOLUTION

The circuit can be represented as shown in Fig. 21–7. By Ohm's law, the current flowing through the circuit is

$$I = \frac{V}{r + R} \qquad (a)$$

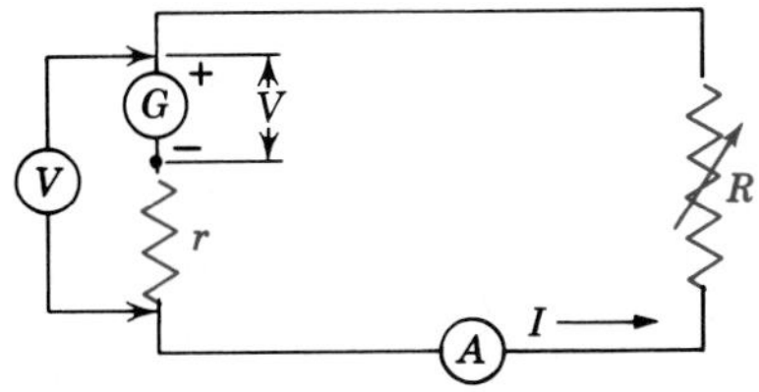

FIG. 21–7 Circuit of Example 19.

The power delivered to the external circuit is

$$P = V_t I = I^2 R \tag{b}$$

where V_t is the terminal voltage of the source and

$$V_t = V - Ir \tag{c}$$

Now the terminal voltage V_t will decrease as the current I increases. Therefore, the power P supplied to the load is a function of the two variables V_t and I. Substituting Eq. (c) in Eq. (b),

$$P = (V - Ir)I = VI - I^2 r$$

that is,

$$P = -rI^2 + VI \tag{d}$$

Equation (d) is a quadratic in I, where

$$a = -r \quad \text{and} \quad b = V$$

Then, since, for maximum conditions, $I = \dfrac{-b}{2a}$,

$$I = \frac{-b}{2a} = \frac{-V}{2(-r)} = \frac{V}{2r} \tag{e}$$

which is the value of the current through the circuit when maximum power is being delivered to the load. Substituting Eq. (e) in Eq. (a),

$$\frac{V}{2r} = \frac{V}{r + R} \tag{f}$$

Solving Eq. (f) for R,

$$R = r \tag{g}$$

Equation (g) shows that maximum power will be delivered to any load when the resistance of that load is equal to the internal resistance of the source of emf. This is one of the important concepts in electronics engineering. For example, we are concerned with obtaining maximum power output from several types of power amplifier. We obtain it when the amplifier load resistance matches the output resistance of the associated components. Also, maximum power is delivered to an antenna circuit when the impedance of the antenna is made to match that of the transmission line that feeds the antenna.

In Fig. 21–8, power delivered to the load is plotted against values of the load resistance R_L when a storage battery with an emf V of 6.6 V and an internal resistance $r = 0.075\ \Omega$ is used. See the circuit in Fig. 21–9.

It is apparent that, when the battery or any other source of emf is delivering maximum power, half the power is lost within the battery. Under these conditions, therefore, the efficiency is 50%.

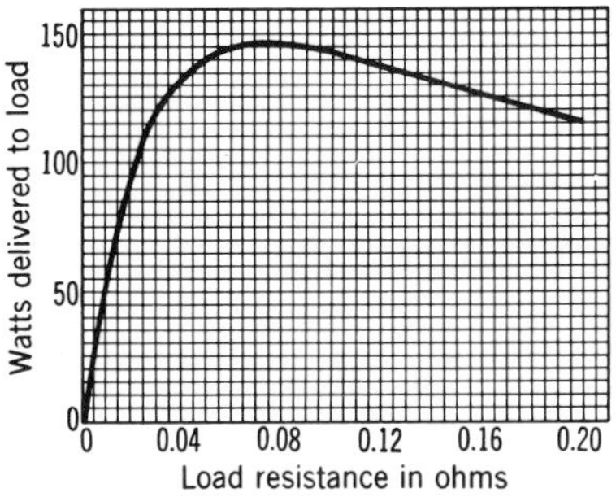

FIG. 21–8 Power delivered to load plotted against load resistance.

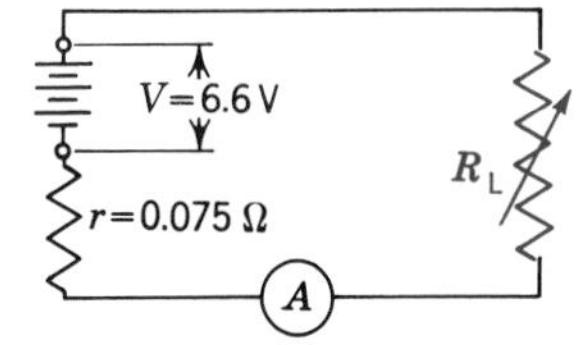

FIG. 21–9 Load resistance R_L is varied to obtain power values plotted in Fig. 21–8.

PROBLEMS 21–6

1. Graph the following equations all on the same sheet with the same axes:
   (*a*) $y = x^2 - 6x - 16$ (*b*) $y = x^2 - 6x - 7$
   (*c*) $y = x^2 - 6x$ (*d*) $y = x^2 - 6x + 5$
   (*e*) $y = x^2 - 6x + 9$ (*f*) $y = x^2 - 6x + 12$
   (*g*) $y = x^2 - 6x + 15$
   Does changing the constant term change only the vertical positions of the graphs and the solutions of the equations? Do all graphs pass through minimum values at the same value of x?
2. Solve the equations of Prob. 1 algebraically. Do these solutions check the graphs of the equations? Test your solutions by means of the quadratic tests.
3. Compute the discriminant for each equation of Prob. 1. Do you see any connection between the value and the graph?
4. Compute the minimum value of the dependent variable y for each equation of Prob. 1. Does the value check with the graph?
5. What do you see from the graphs of Prob. 1 when x is equal to zero?

21–12 SUMMARY

Several methods are available for solving quadratic equations. All quadratic equations can be solved by factoring, by completing the square, by use of the quadratic formula, or by graphical methods. However, some of these methods involve unnecessary work for certain forms or types of quadratic equations; therefore, one tries to choose the most convenient method for a particular equation. For example, a pure quadratic equation is readily solved merely by reducing the equation to its simplest form and extracting the square root of both members of the equation in order to obtain the two roots, which are equal in absolute value but of opposite sign (Sec. 21–2).

In practical problems involving complete quadratic equations the numerical coefficients are such that you will seldom be able to solve the equation readily by factoring. Also, solution by completing the square sometimes can become a chore. Probably the most widely used method is solution by use of the quadratic formula, which, if you forget it, can be found in most handbooks and put to use whenever needed.

Solution by graphical methods allows you to visualize the variation of quantities and serves to check computations. In any event, through solving many problems, you will develop your own methods of attack.

In solving problems involving quadratic equations, care must be used because two answers (roots) are obtained. In all cases both roots will satisfy the mathematics of the equation, but in some cases only one root will satisfy the conditions of the problem. Therefore, we reject the obviously impossible or the impractical answer and retain the one that is consistent with the physical conditions of the problem.

EXAMPLE 20 The square of a certain number plus four times the number is 12. Find the number.

SOLUTION

Let	$x =$	the number
Then	$x^2 =$	the square of the number
and	$4x =$	four times the number
From the problem,	$x^2 + 4x =$	12
S: 12,	$x^2 + 4x - 12 =$	0
Factoring,	$(x + 6)(x - 2) =$	0
Then	$x =$	-6 or 2

Both roots satisfy the equation and the condition of the problem; therefore, both answers are correct.

EXAMPLE 21 Find the dimensions of a right triangle if the hypotenuse of the triangle is 40 m and the base exceeds the altitude by 8 m.

SOLUTION

In any right triangle, Fig. 21–10, $c^2 = a^2 + b^2$.

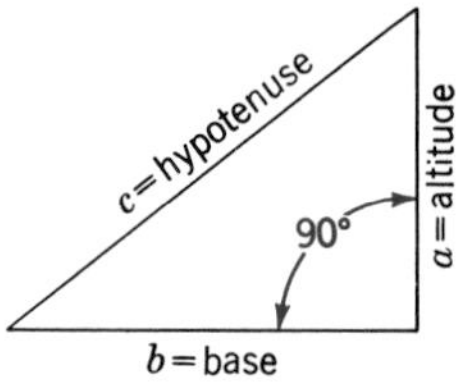

FIG. 21–10 In any right triangle, $c^2 = a^2 + b^2$.

Since	$c = 40$
and	$a = b - 8$
then	$1600 = (b - 8)^2 + b^2$

Are both roots of this equation consistent with the physical conditions of the problem?

EXAMPLE 22 A storage battery has an emf of 6.3 V and an internal resistance of 0.015 Ω. The battery is used to drive a dynamotor requiring 300 W. What current then will the battery deliver to the dynamotor, and what will be the voltage reading across the battery terminals while this current is supplied?

SOLUTION

The circuit is represented in Fig. 21–11.

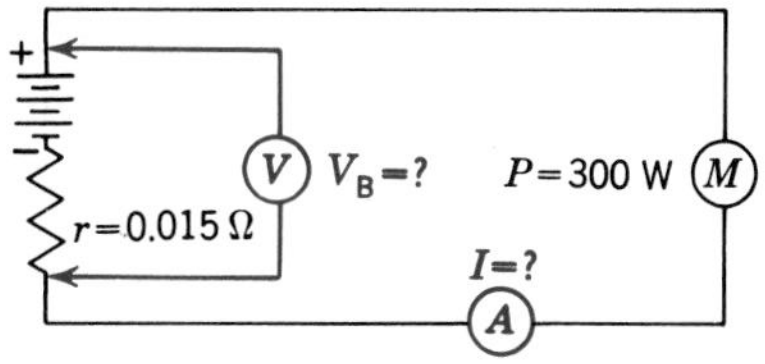

FIG. 21–11 Circuit of Example 22.

Let P = power consumed by dynamotor
= 300 W

V_B = voltage across battery terminals when dynamotor is delivering 300 W

Since $$I = \frac{P}{V_B}$$

then $$I = \frac{300}{V_B}$$

Now $$V_B = 6.3 - rI$$

Substituting for r, $$V_B = 6.3 - 0.015I$$

Substituting for I, $$V_B = 6.3 - 0.015 \times \frac{300}{V_B}$$

Multiplying, $$V_B = 6.3 - \frac{4.5}{V_B}$$

Clearing fractions, $$V_B^2 = 6.3V_B - 4.5$$

Transposing, $$V_B^2 - 6.3V_B + 4.5 = 0$$

This equation is a quadratic in V_B; hence, $a = 1$, $b = -6.3$, and $c = 4.5$. Substituting these values in the quadratic formula,

$$V_B = \frac{-(-6.3) \pm \sqrt{(-6.3)^2 - 4 \cdot 1 \cdot 4.5}}{2 \cdot 1}$$

or $$V_B = \frac{6.3 \pm \sqrt{21.7}}{2}$$

$$\therefore V_B = 5.48 \text{ V or } 0.82 \text{ V}$$

$$I = \frac{300}{V_B} = \frac{300}{5.48} = 54.7 \text{ A}$$

Why was 5.48 V chosen instead of 0.82 V in the above solution?

PROBLEMS 21–7

1. Compute the discriminant, and tell what it shows, in each of these equations:
   (a) $x^2 - 13x + 40 = 0$
   (b) $4x^2 + 16x + 16 = 0$
   (c) $15x^2 + 13x + 7 = 0$

2. Find two positive consecutive even numbers whose product is 168.
3. Find two positive consecutive odd numbers whose product is 255.
4. Can the sides of a right triangle ever be consecutive integers? If so, find the integers.
5. Find the dimensions of a rectangular parking lot whose area is 37 800 m^2 and whose perimeter is 780 m.
6. Separate 182 into two parts such that one part is the square of the other.
7. One number is 20 less than another, and the difference of their squares is 6800. What are the numbers?
8. $F = \dfrac{Wv^2}{32r}$
   (*a*) Solve for v.
   (*b*) If W is doubled and r is halved, what happens to F?
   (*c*) What is W if $F = 15$, $r = 1.75$, and $v = 22$?
9. Given $P = \dfrac{kV^2}{nR}$. Solve for V. If k and n are doubled and P and R are held constant, what happens to V?
10. $R_t = \dfrac{r}{\left(\dfrac{d_o}{d_i}\right)^2} - 1$. Solve for $\dfrac{d_o}{d_i}$.
11. $P = \dfrac{R(r^2 + x^2)}{r(Rr + Xx)}$. Solve for r and x.
12. The following relations exist in the Wien bridge:

$$\omega^2 = \frac{1}{R_1R_2c_1c_1} \quad \text{and} \quad \frac{c_1}{c_2} = \frac{R_b - R_2}{R_aR_1}$$

   Solve for c_1 and c_2 in terms of resistance components and ω.
13. Kinetic energy (KE) is equal to one-half the product of mass m in kilograms and the square of velocity v in meters per second; that is, $\text{KE} = \frac{1}{2}mv^2$ joules. Find the value of v when KE = 2.8 MJ and $m = 8$ kg.
14. A ball rolls down a slope and travels a distance $d = 3t + \frac{1}{4}t^2$ meters in t s. Solve for t.
15. The distance through which an object will fall in t s is $s = \frac{1}{2}gt^2$ meters, where $g = 9.81$ m/s^2. The velocity v attained after t s is $v = gt$ m/s. Solve for the velocity in terms of g and s.
16. If an object is thrown straight up with a velocity of v m/s, its height t s later is given by $h = vt - 4.9t^2$ meters. If a rocket were fired upward with a velocity of 1176 m/s, neglecting air resistance:
   (*a*) At what time would its height be 15 km on the way up?
   (*b*) At what time would its height be 15 km on the way down?
   (*c*) At what time would it attain its maximum height?
   (*d*) What maximum height would it attain?

   Attempt these solutions both graphically and algebraically.

17. Use the formula for height in Prob. 16 to derive a formula for maximum height attained for any initial velocity *v*.

18. In an ac series circuit containing resistance *R* in ohms and inductance *L* in henrys, the current *I* may be computed from the formula

$$I = \frac{V}{\sqrt{R^2 + \omega^2 L^2}} \text{ A}$$

where *V* is the emf in volts applied across the circuit. Find the value of *R* to three significant figures if $V = 140$ V, $I = 1.7$ A, $\omega = 2\pi f$, $f = 100$ Hz, and $L = 0.125$ H.

19. In an ac circuit containing R Ω resistance and X_C Ω reactance, the impedance is

$$Z = \sqrt{R^2 + X_C^2} \quad \Omega$$

Find the value of *R* if $Z = 1$ kΩ and $X_C = 600$ Ω.

20. The susceptance of an ac circuit that contains R Ω resistance and X Ω reactance is

$$B = \frac{X}{R^2 + X^2} \quad \text{siemens (S)}$$

Find the value of *R* to three significant figures when $B = 0.008$ S and $X = 100$ Ω.

21. The equivalent noise resistance R_N of a bipolar NPN transistor depends upon collector current I_C, emitter current I_E, leakage current I_{CO}, and base-emitter internal junction resistance r_e expressed in terms of the mutual transconductance $g_m \cong \frac{1}{r_e}$. The empirical formula relating these parameters is

$$R_N = \frac{2.5 I_C^2}{g_m I_E^2} + \frac{20 I_C I_{CO}}{g_m^2 I_E}$$

Show that the ratio of collector current to emitter current is

$$\frac{I_C}{I_E} = \frac{-20 I_{CO}}{5 g_m} + \frac{1}{5 g_m}\sqrt{400 I_{CO}^2 + 10 R_N g_m^3}$$

22. Did you prove that $\alpha + \beta = \frac{-b}{a}$?

23. Did you prove that $\alpha \cdot \beta = \frac{c}{a}$?

24. Find the two combinations of resistance of R_2 and R_3 that will satisfy the circuit conditions of Fig. 21–12 on page 384.

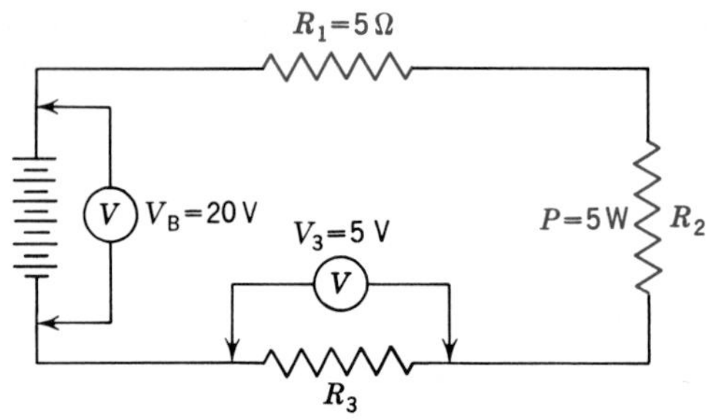

FIG. 21–12 Circuit of Prob. 24.

25. The circuit conditions as shown in Fig. 21–13 existed when the generator G was supplying current to the circuit. When the generator was disconnected, an ohmmeter connected between points A and B read 60 Ω.
    (*a*) What was the circuit current?
    (*b*) What was the generator voltage?
    (*c*) What is the value of each resistor?

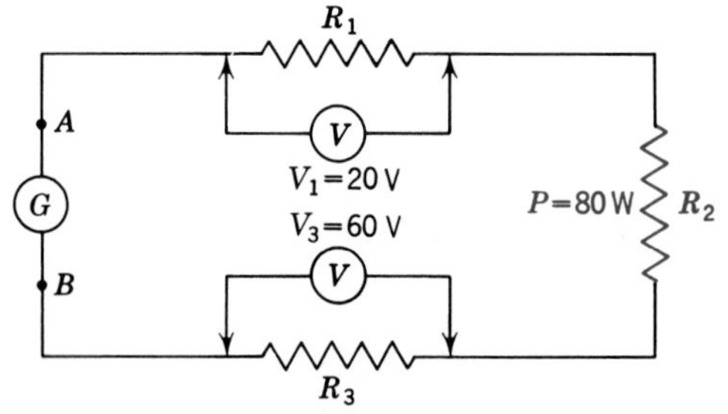

FIG. 21–13 Circuit of Prob. 25.

26. In the circuit of Fig. 21–14, the resistor ABC represents a potentiometer with total resistance (A to C) of 25 000 Ω. R_1 = 5000 Ω, across which is 60 V.
    (*a*) What is the resistance from A to B?
    (*b*) How much current flows from B to C?

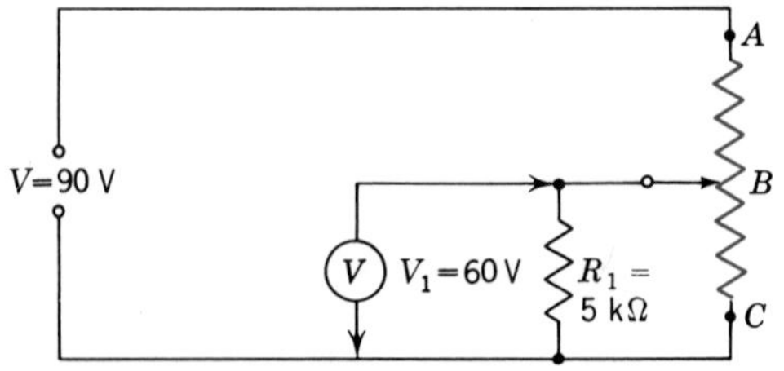

FIG. 21–14 Circuit of Prob. 26.

27. What are the meter readings in the circuit of Fig. 21–15?

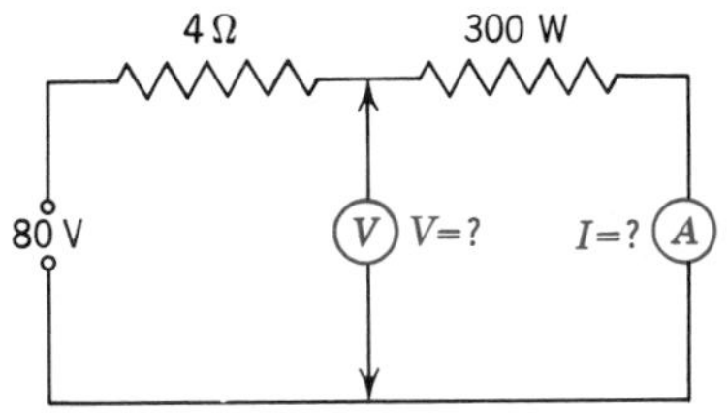

FIG. 21–15 Circuit of Prob. 27.

28. When two capacitors C_1 and C_2 are connected in series, the total capacitance C_t of the combination is always less than either of the two capacitors. That is,

$$C_t = \frac{C_1 C_2}{C_1 + C_2}$$

Suppose we have a tuning capacitor that varies from 200 to 300 pF; that is, it has a *change* in capacitance of 100 pF. What value of fixed capacitor should be connected in series with the tuning capacitor to limit the total *change* of circuit capacitance to 50 pF?

SELF TEST

1. Solve: $\theta^2 - 49 = 0$
2. Solve by factoring: $Q^2 + 5Q - 24 = 0$
3. Solve by completing the square: $x^2 - 6x - 12 = 0$
4. Solve using the quadratic formula:

$$\frac{1}{p-4} - 1 = \frac{p-8}{p-2}$$

5. Plot the graph of $y = x^2 + 2x - 3$ and identify (*a*) the *y*-intercept, (*b*) the *x*-intercepts, and (*c*) the *x* and *y* values of the turning point.
6. What is the value of the discriminant in the solution to the following equation?

$$-5x^2 + 14x + 2 = 0$$

7. One number is 11 greater than another, and the sum of their squares is 821. What are the numbers?
8. $G = \dfrac{R}{R^2 + X^2}$ S. Find R when $G = 0.0025$ S and $X = 40\ \Omega$.

9. Find the two combinations of resistances R_2 and R_3 that will satisfy the conditions of the circuit of Fig. 21–16.

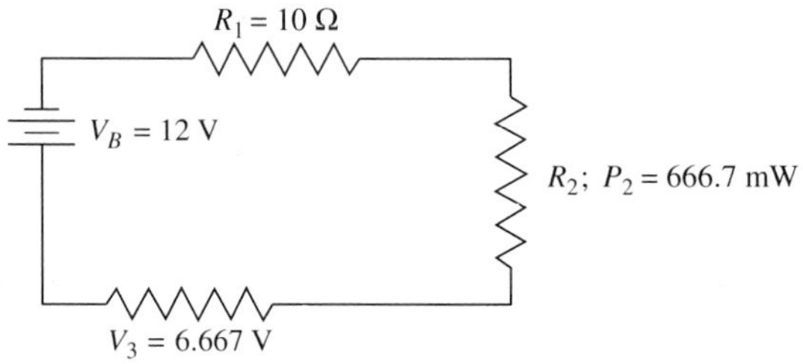

FIG. 21–16 Circuit for self test Prob. 9.

10. What conditions of meter readings satisfy the conditions of the circuit of Fig. 21–17?

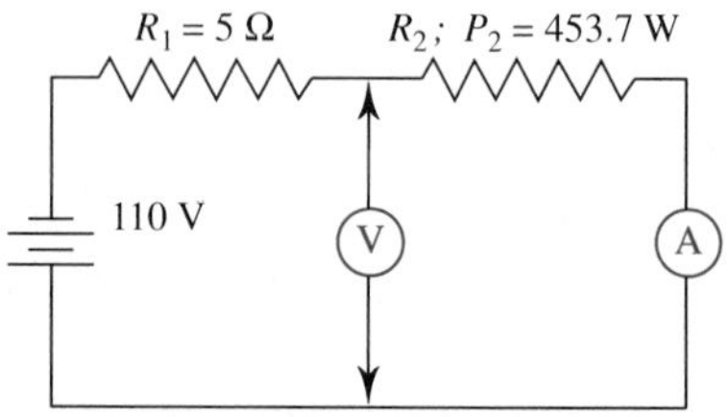

FIG. 21–17 Circuit for self test Prob. 10.

CHAPTER 22

NETWORK SIMPLIFICATION

An understanding of Kirchhoff's laws, plus the ability to apply the laws in analyzing circuit conditions, will give you a better insight into the behavior of circuits. Furthermore, you will be able to solve circuit problems that, with only a knowledge of Ohm's law, would be very difficult in some cases and impossible in others.

22–1 DIRECTION OF CURRENT

As stated in Sec. 8–1, the most generally accepted concept of an electric current is that the current consists of a motion of electrons from a negative toward a more positive point in a circuit. That is, a positively charged body is taken to be one that is deficient in electrons, whereas a negatively charged body carries an excess of electrons. When the two bodies are joined by a conductor, electrons flow from the negatively charged body to the positively charged one. Hence, if two such points in a circuit are *maintained* at a difference of potential, a *continuous* flow of electrons, or current, will take place from negative to positive. Therefore, in the consideration of Kirchhoff's laws, current will be thought of as flowing from the negative terminal of a source of emf, through the external circuit, and back to the positive terminal of the source.

Thus, in Fig. 22–1, the current leaves the negative terminal of the battery, through R_1 and R_2, and returns to the positive terminal of the battery. Note that point b is positive with respect to point a and that point d is positive with respect to point c.

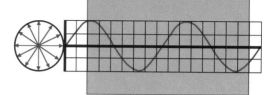

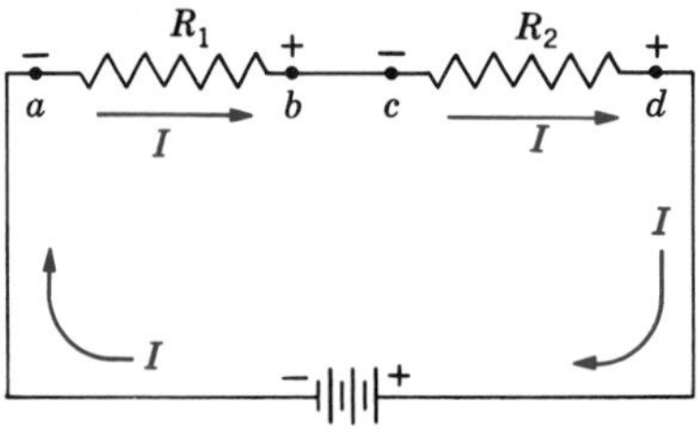

FIG. 22–1 Current I flowing from − to + through the connected circuit.

22–2 STATEMENT OF KIRCHHOFF'S LAWS

In 1847, G. R. Kirchhoff extended Ohm's law by two important statements which have become known as Kirchhoff's laws. These laws can be stated as follows:

1. The algebraic sum of the currents at any junction of conductors is zero.

That is, at any point in a circuit, there is as much current flowing away from the point as there is flowing toward it.

2. The algebraic sum of the applied emf's and voltage drops around any closed circuit is zero.

> applied emf = voltage *rise*
> resistors give voltage *drop*

That is, in any closed circuit, the applied emf is equal to the voltage drops around the circuit.

These laws are straightforward and need no proof here; for the first is self-evident from the study of parallel circuits, and the second was stated in different words in Sec. 8–8. When properly applied, they enable us to set up equations for any circuit and solve for the unknown circuit components, voltages, or currents as required.

22–3 APPLICATION OF SECOND LAW TO SERIES CIRCUITS

The second law is considered first because of its applications to problems with which you are already familiar.

Figure 22–2 represents a 20-V generator connected to three series resistors. The validity of Kirchhoff's second law was shown in Sec. 8–8; that is, in any closed circuit the applied emf will be equal to the sum of the voltage drops around the circuit. Thus, neglecting the internal resistance of the generator and the resistance of the connecting wires in Fig. 22–2,

$$V = IR_1 + IR_2 + IR_3 \qquad (1)$$

or

$$20 = 2I + 3I + 5I$$

Hence,

$$I = 2\text{ A}$$

Equation (1) is satisfactory for a circuit containing one source of emf. By considering the circuit from a different viewpoint, however, the voltage rela-

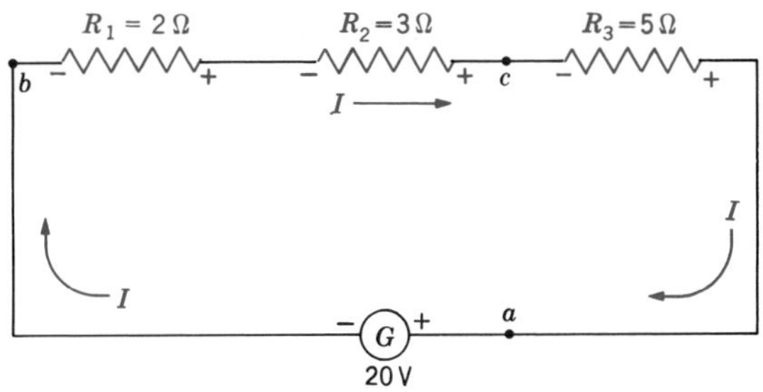

FIG. 22–2 The sum of the voltage drops across the resistors is equal to the applied emf.

tions around the circuit become more understandable. For example, by starting at any point in the circuit, such as point *a,* we proceed completely around the circuit in the direction of current flow, remembering that, when current passes through a resistance, there is a voltage drop that represents a loss and therefore is subtractive. Also, in going around the circuit, sources of emf represent a gain in voltage if they tend to aid current flow and therefore are additive. By this method, according to the second law, the algebraic sum of all emf's and voltage drops around the circuit is zero.

For example, in starting at point *a* in Fig. 22–2 and proceeding around the circuit in the direction of current flow, the first thing encountered is the positive terminal of a source of emf of 20 V. Because this causes current to flow in the direction we are going, it is written +20. This is easily remembered, for the positive terminal was the first one encountered; therefore, write it plus. Next comes R_1, which is responsible for a drop in voltage due to the current I passing through it. Hence, this voltage drop is written $-IR_1$ or $-2I$, for R_1 is known to be 2 Ω. R_2 and R_3 are treated in a similar manner because both represent voltage *drops*. This completes the trip around the circuit, and by equating the algebraic sum of the emf and voltage drops to zero,

$$20 - 2I - 3I - 5I = 0 \qquad (2)$$

or

$$I = 2\ \text{A}$$

Note that Eq. (2) is simply a different form of Eq. (1). If the polarities of the sources of emf are marked, they will serve as an aid in remembering whether to add or subtract. In going around the circuit, if the first terminal of a source of emf is positive, the emf is added; if negative, the emf is subtracted.

The point at which to start around the circuit is purely a matter of choice, for the algebraic sum of all voltages around the circuit is equal to zero. For example, starting at point *b,*

$$-2I - 3I - 5I + 20 = 0$$
$$I = 2\ \text{A}$$

Starting at point *c,*

$$-5I + 20 - 2I - 3I = 0$$
$$I = 2\ \text{A}$$

EXAMPLE 1 Find the amount of current in the circuit that is represented in Fig. 22–3 if the internal resistance of battery V_1 is 0.3 Ω, that of V_2 is 0.2 Ω, and that of V_3 is 0.5 Ω.

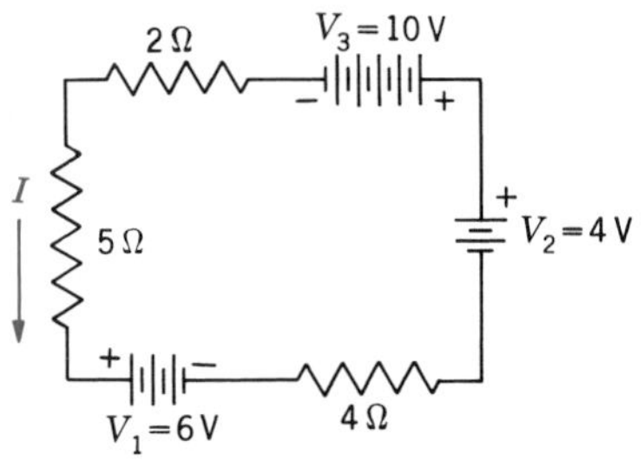

FIG. 22–3 Circuit of Example 1.

SOLUTION

Figure 22–4 is a diagram of the circuit in which the internal resistances are represented in color as an aid in setting up the circuit equation. Beginning at point *a* and going around the circuit in the direction of current flow,

$$6 - 0.3I - 4I - 0.2I - 4 + 10 - 0.5I - 2I - 5I = 0$$

Hence,

$$I = 1 \text{ A}$$

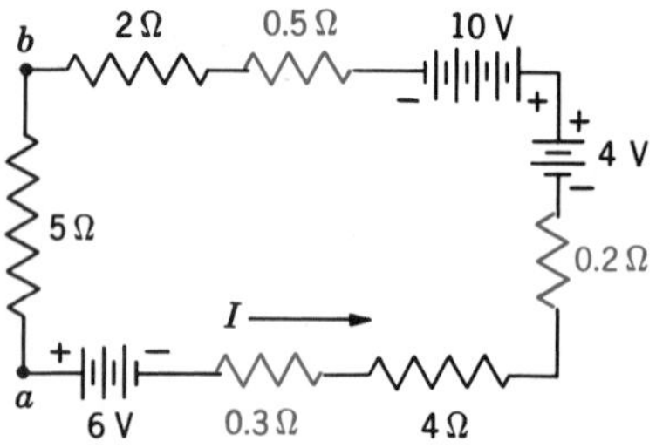

FIG. 22–4 Circuit of Example 1 illustrating internal resistances of the batteries.

In more complicated circuits the direction of the current is often in doubt. However, this need cause no confusion, for the direction of current flow can be *assumed* and the circuit equation can then be written in the usual manner. If the current results in a negative value when the equation is solved, the negative sign denotes that the assumed direction was wrong. As an example, let it be assumed that the current in the circuit of Fig. 22–4 flows in the direction from *a* to *b*. Then, starting at point *a* and going around the circuit in the assumed direction,

$$-5I - 2I - 0.5I - 10 + 4 - 0.2I - 4I - 0.3I - 6 = 0$$

$$\therefore I = -1 \text{ A}$$

As stated above, the minus sign shows that the assumed direction of the current was wrong; therefore, the current flows in the direction from *b* to *a*.

PROBLEMS 22–1

1. Three resistors, $R_1 = 27\ \text{k}\Omega$, $R_2 = 68\ \text{k}\Omega$, and $R_3 = 47\ \text{k}\Omega$, are connected in parallel across a 120-V power supply whose internal resistance is 1.2 kΩ. How much current is drawn from the source?
2. The resistors in Prob. 1 are replaced by new values $R_1 = 2.7\ \text{k}\Omega$, $R_2 = 1.8\ \text{k}\Omega$, and $R_3 = 4.7\ \text{k}\Omega$. How much current will be drawn from the source?
3. Three resistors, $R_1 = 33\ \text{k}\Omega$, $R_2 = 28\ \text{k}\Omega$, and $R_3 = 12\ \text{k}\Omega$, are connected in series across a signal generator whose internal resistance is 5 Ω. If 20 mA flows through the circuit, what is the terminal voltage of the generator?
4. What is the value of R_4 in Fig. 22–5?
5. A motor that draws 20 A at 440 V is connected to a generator through two No. 6 copper feeders each of which is 500 m long. What is the generator terminal voltage?
6. A motor that draws 20 A at 660 V is connected to a generator through two No. 6 copper feeders each of which is 400 m long. What is the generator terminal voltage?
7. (*a*) How much current flows in the circuit of Fig. 22–6?
   (*b*) What is the terminal voltage of the 12-V battery?
8. (*a*) How much current flows in the circuit of Fig. 22–7?
   (*b*) What is the terminal voltage of the generator?

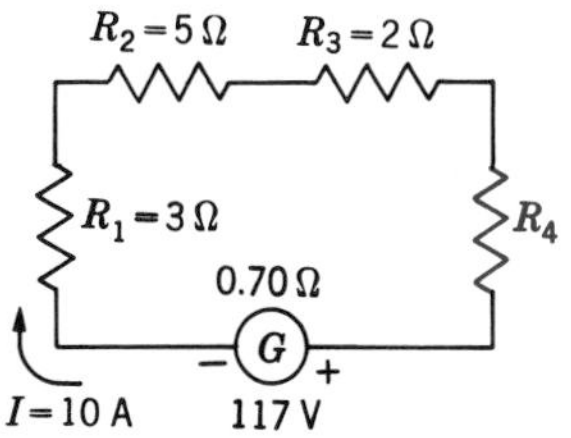

FIG. 22–5 Circuit of Prob. 4.

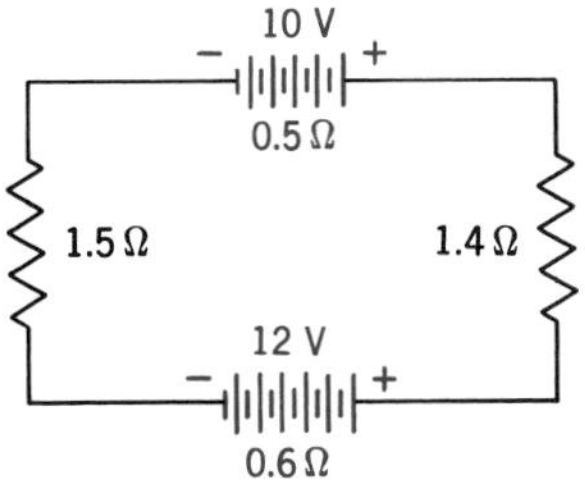

FIG. 22–6 Circuit of Prob. 7.

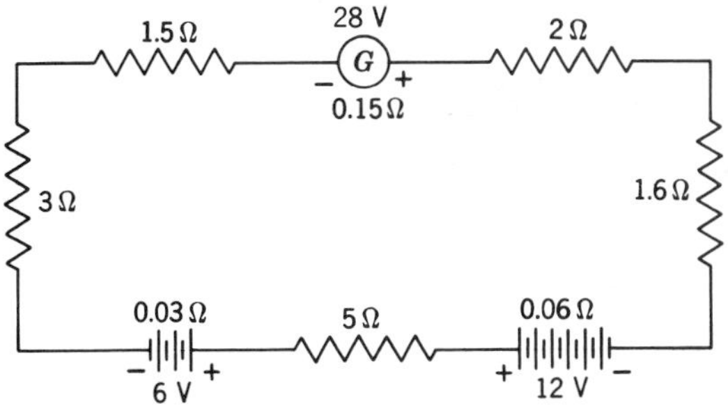

FIG. 22–7 Circuit of Prob. 8.

9. A current of 5 A flows through the circuit of Fig. 22–8. What is the value of *R*?

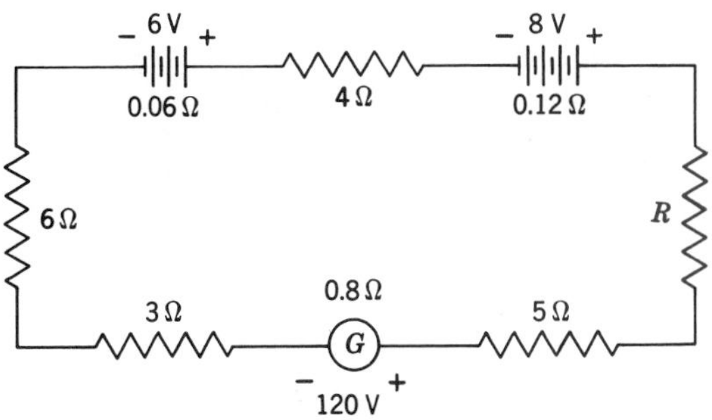

FIG. 22–8 Circuit of Prob. 9.

10. How much current flows in the circuit of Fig. 22–9?

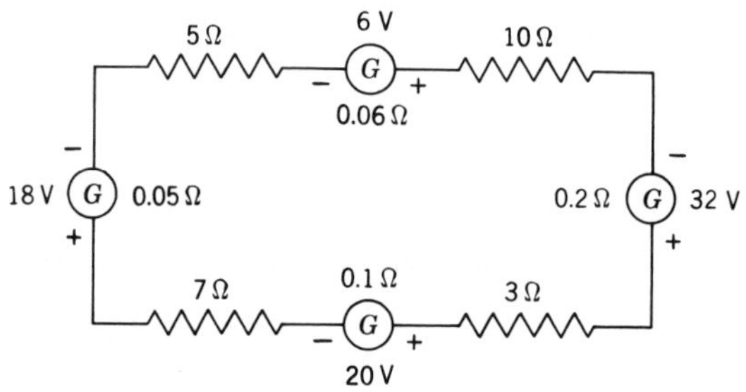

FIG. 22–9 Circuit of Prob. 10.

22–4 SIMPLE APPLICATIONS OF BOTH LAWS

Although the circuits of the following examples can be solved by Ohm's law, they are included here because you are familiar with such circuits. You will have no trouble in solving circuits that appear to be complicated if you understand the applications of Kirchhoff's laws to simple circuits, for all circuits are combinations of the fundamental series and parallel circuits.

EXAMPLE 2 A generator supplies 7 A to two resistances of 40 and 30 Ω connected in parallel. Neglecting the internal resistance of the generator and the resistance of the connecting wires, find the generator voltage and the current through each resistance.

SOLUTION

Figure 22–10 is a diagram of the circuit. From our knowledge of parallel circuits, it is evident that the line current I divides at junction c into the branch currents I_1 and I_2. Similarly, I_1 and I_2 combine at junction f to form the line current I. Therefore,

$$I = I_1 + I_2$$

which is the same as

$$I - I_1 - I_2 = 0 \tag{3}$$

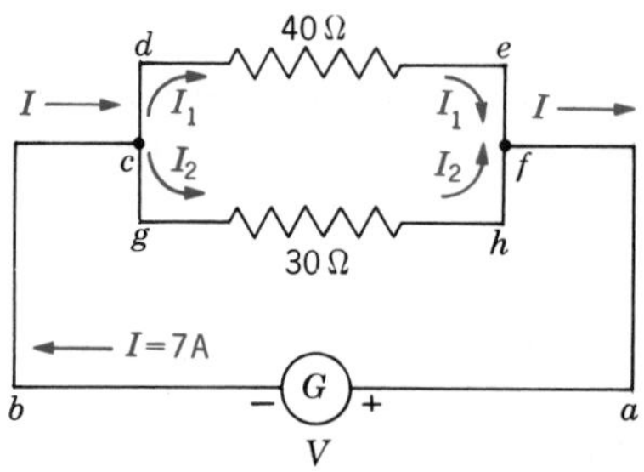

FIG. 22–10 Circuit of Example 2.

These are algebraic expressions for Kirchhoff's first law. When they are used in conjunction with the second law, they facilitate the solution of circuits.

If we start at point a and go around the circuit in the direction of current flow, the equation for the voltages around path $abcdefa$ is

$$V - 40I_1 = 0$$

$$I_1 = \frac{V}{40} \tag{4}$$

The equation for the voltages around path $abcghfa$ is

$$V - 30I_2 = 0$$

$$I_2 = \frac{V}{30} \tag{5}$$

Substituting the known values in Eq. (3),

$$7 - \frac{V}{40} - \frac{V}{30} = 0$$

$$V = 120 \text{ V}$$

$I_1 = 3$ A and $I_2 = 4$ A are found from Eqs. (4) and (5), respectively.

EXAMPLE 3 Two 6-V batteries, each with an internal resistance of 0.05 Ω, are connected in parallel to a load resistance of 9.0 Ω. How much current flows through the load resistance?

SOLUTION

Figure 22–11 is a diagram of the circuit. In the circuit, two identical sources of emf are connected in parallel to supply the line current I to the load resistance. Again,

$$I = I_1 + I_2$$

or

$$I - I_1 - I_2 = 0$$

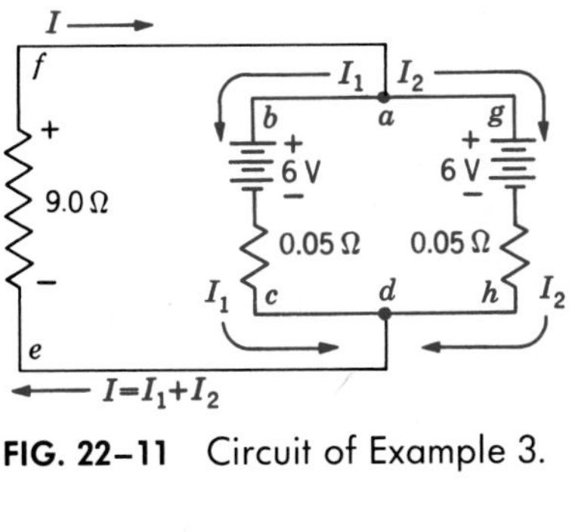

FIG. 22–11 Circuit of Example 3.

Starting at junction *a*, the equation for the voltages around path *abcdefa* is

$$6 - 0.05I_1 - 9I = 0$$

Solving for I_1,

$$I_1 = 120 - 180I \qquad (6)$$

Starting at junction *a*, the equation for the voltages around path *aghdefa* is

$$6 - 0.05I_2 - 9I = 0$$

Solving for I_2,

$$I_2 = 120 - 180I \qquad (7)$$

As would be expected, I_1 and I_2 are equal. Substituting the values of I_1 and I_2 in Eq. (3),

$$I - (120 - 180I) - (120 - 180I) = 0$$

Hence,

$$I = 0.6648 \text{ A}$$

The foregoing solution assumes three unknowns I, I_1, and I_2. However, in writing the equations for the voltages around any path, only two unknowns can be used, for $I = I_1 + I_2$. Thus, around path *abcdefa*,

$$6 - 0.05I_1 - 9(I_1 + I_2) = 0$$

Collecting terms,

$$9.05I_1 + 9I_2 = 6 \qquad (8)$$

Voltages around path *aghdefa*,

$$6 - 0.05I_2 - 9(I_1 + I_2) = 0$$

Collecting terms,

$$9I_1 + 9.05I_2 = 6 \qquad (9)$$

Since Eqs. (8) and (9) are simultaneous equations, they can be solved for I_1 and I_2. Hence,

$$I_1 = 0.3324 \text{ A}$$
$$I_2 = 0.3324 \text{ A}$$

and

$$I = I_1 + I_2 = 0.6648 \text{ A}$$

PROBLEMS 22–2

1. A power supply supplies a total of 1.222 A to two resistors of 120 and 360 Ω connected in parallel. What is the terminal voltage of the power supply?
2. A battery supplies 10.4 A to three resistors of 1.5, 1.8, and 2.2 Ω connected in parallel. What is the terminal voltage of the battery?
3. A generator with an internal resistance of 1 Ω supplies 22.8 A to three resistors of 4.7, 3.3, and 1.8 Ω connected in parallel. What is the generator terminal voltage?
4. A battery supplies 2 A to four resistors that are 150, 50, 100, and 200 Ω connected in parallel. What is the voltage across the resistors?
5. (*a*) What is the value of the current in the circuit of Fig. 22–12?
   (*b*) How much power is expended in each of the batteries?

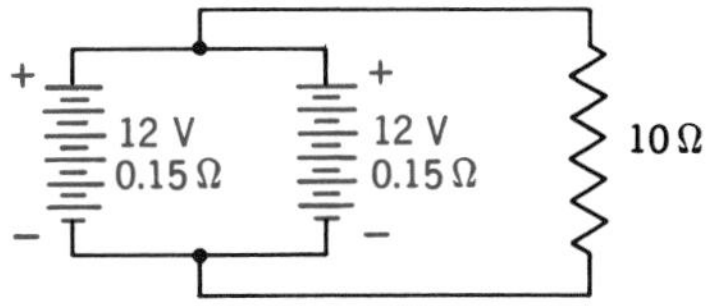

FIG. 22–12 Circuit of Probs. 5 and 6.

6. How much power would be expended in each battery in the circuit of Fig. 22–12 if the load resistance were changed from 10 to 0.5 Ω?
7. (*a*) What is the generator current in the circuit of Fig. 22–13?
   (*b*) In what direction does the current flow?

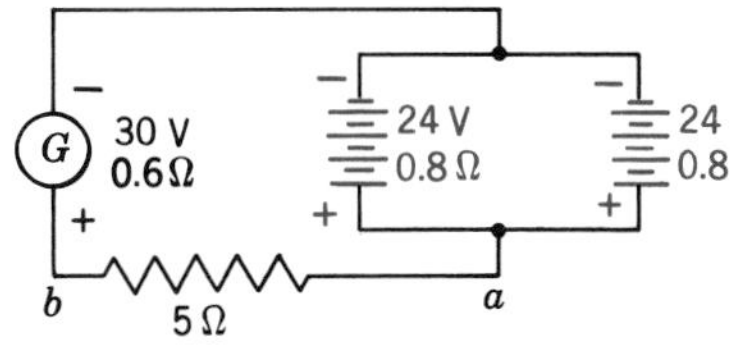

FIG. 22–13 Circuit of Probs. 7 and 8.

8. (*a*) What is the value of the generator current in the circuit shown in Fig. 22–13 if the generator emf voltage is decreased to 12 V?
   (*b*) In what direction does the current flow?

22–5 FURTHER APPLICATIONS OF KIRCHHOFF'S LAWS

In preceding examples and problems, if two sources of emf have been connected to the same circuit, the values of emf and internal resistance have been equal. However, there are many types of circuits that contain more than one source of power, each with a different emf and different internal resistance.

EXAMPLE 4 Figure 22–14 represents two batteries connected in parallel and supplying current to a resistance of 2 Ω. One battery has an emf of 6 V and an internal resistance of 0.15 Ω, and the other battery has an emf of 5 V and an internal resistance of 0.05 Ω. Determine the current through the batteries and the current in the external circuit. Neglect the resistance of the connecting wires.

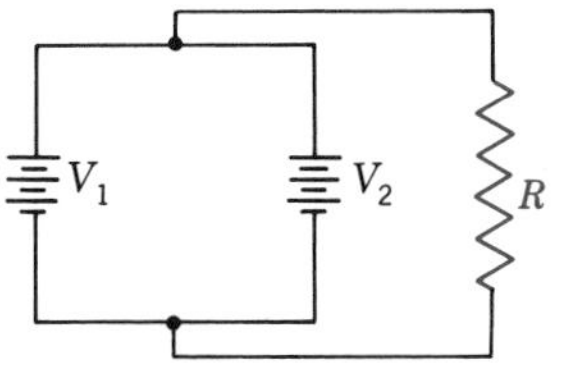

FIG. 22–14 Circuit of Example 4.

SOLUTION

Draw a diagram of the circuit representing the internal resistance of the batteries, and label the circuit with all the known values as shown in Fig. 22–15. Label the unknown currents, and mark the direction in which each current is assumed to flow. There are three currents of unknown value in the circuit: I_1, I_2, and the current I which flows through the external circuit. However, because $I = I_1 + I_2$, the unknown currents can be reduced to two unknowns by considering a current of $I_1 + I_2$ A flowing through the external circuit.

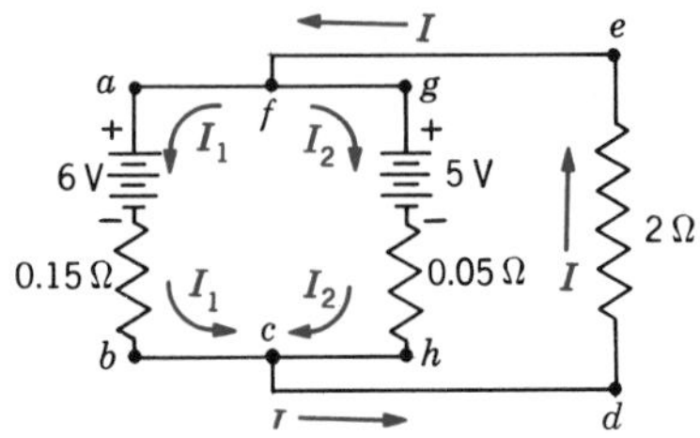

FIG. 22–15 Circuit of Example 4 labeled with known values.

For the path *abcdefa*, $6 - 0.15I_1 - 2(I_1 + I_2) = 0$

Collecting terms, $2.15I_1 + 2I_2 = 6$ (10)

For the path *ghcdefg*, $5 - 0.05I_2 - 2(I_1 + I_2) = 0$

Collecting terms, $2I_1 + 2.05I_2 = 5$ (11)

Equations (10) and (11) are simultaneous equations that, when solved, result in

$$I_1 = 5.64 \text{ A}$$

and $$I_2 = -3.07 \text{ A}$$

The negative sign of the current I_2 denotes that this current is flowing in a direction opposite to that assumed. The value of the line current is

$$I = I_1 + I_2 = 5.64 + (-3.07)$$
$$= 2.57\text{A}$$

Now check this solution by changing the direction of I_2 in Fig. 22–15 and rewriting the voltage equations accordingly, while remembering that now, at junction *f*, for example, $I + I_2 - I_1 = 0$. This will demonstrate that it is immaterial which way the arrows point, for the signs preceding the current values, when found, determine whether or not the assumed directions are correct. As previously mentioned, however, it must be remembered that going through a resistance in a direction opposite to the current arrow represents a voltage (rise) which must be added, whereas going through a resistance in the direction of the current arrow represents a voltage (drop) which must be subtracted.

EXAMPLE 5 Figure 22–16 represents a network containing three unequal sources of emf. Find the current flowing in each branch.

SOLUTION

Assume directions for I_1, I_2, and I_3, and label as shown in Fig. 22–16.

Although three unknown currents are involved, they can be reduced to two unknowns by expressing one current in terms of the other two. This is accomplished by applying Kirchhoff's first law to some junction such as *c*. By

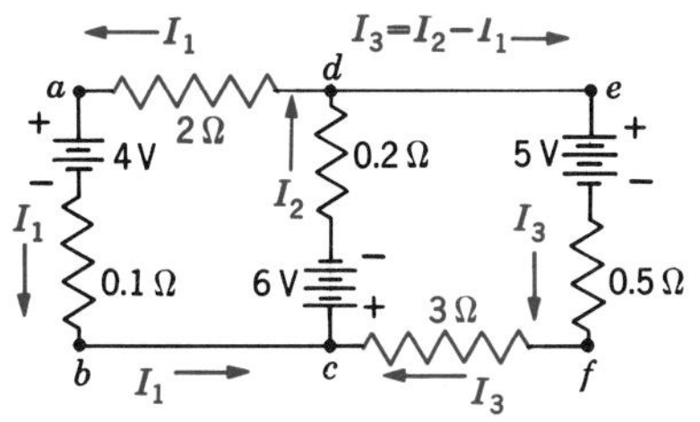

FIG. 22–16 Circuit of Example 5.

considering current flow toward a junction as positive and that flowing away from a junction as negative,

$$I_1 + I_3 - I_2 = 0$$
$$I_3 = I_2 - I_1 \tag{12}$$

Since there are now only two unknown currents I_1 and I_2, Kirchhoff's second law may be applied to any two different closed loops in the network.

For path *abcda,*

$$4 - 0.1I_1 + 6 - 0.2I_2 - 2I_1 = 0$$

Collecting terms, $$2.1I_1 + 0.2I_2 = 10 \tag{13}$$

For path *efcde,*

$$5 - 0.5(I_2 - I_1) - 3(I_2 - I_1) + 6 - 0.2I_2 = 0$$

Collecting terms, $$3.5I_1 - 3.7I_2 = -11 \tag{14}$$

Equations (13) and (14) are simultaneous equations that, when solved, result in

$$I_1 = 4.109 \text{ A}$$

and $$I_2 = 6.860 \text{ A}$$

Substituting in Eq. (12),

$$I_3 = 6.860 - 4.109 = 2.751 \text{ A}$$

The assumed directions of current flow are correct because all values are positive.

The solution can be checked by applying Kirchhoff's second law to a path not previously used. When the current values are substituted in the equation for this path, an identity should result. Thus, for path *adefcba,*

$$2I_1 + 5 - 0.5(I_2 - I_1) - 3(I_2 - I_1) + 0.1I_1 - 4 = 0$$

(continued)

Collecting terms,

$$5.6I_1 - 3.5I_2 = -1 \qquad (15)$$

The substitution of the numerical values of I_1 and I_2 in Eq. (15) verifies the solution within reasonable limits of accuracy.

22–6 OUTLINE FOR SOLVING NETWORKS

In common with all other problems, the solution of a circuit or a network should not be started until the conditions are analyzed and it is clearly understood what is to be found. Then a definite procedure should be adopted and followed until the solution is completed.

To facilitate solutions of networks by means of Kirchhoff's laws, the following procedure is suggested:

1. Draw a large, neat diagram of the network, and arrange the circuits so that they appear in their simplest form.
2. Letter the diagram with all the known values such as sources of emf, currents, and resistances. Carefully mark the polarities of the known emf's.
3. Assign a symbol to each unknown quantity.
4. Indicate with arrows the assumed direction of current in each branch of the network. The number of unknown currents can be reduced by assigning a direction to all but one of the unknown currents at a junction. Then, by Kirchhoff's first law, the remaining current can be expressed in terms of the others.
5. Using Kirchhoff's second law, set up as many equations as there are unknowns to be determined. So that each equation will contain some relation that has not been expressed in another equation, each circuit path followed should cover some part of the circuit not used for other paths.
6. Solve the resulting simultaneous equations for the values of the unknown quantities.
7. Check the values obtained by substituting them in a voltage equation that has been obtained by following a circuit path not previously used.

PROBLEMS 22–3

1. In the circuit of Fig. 22–17, (*a*) what is the current through R_1 and (*b*) how much power is expended in R_3?
2. In the circuit of Fig. 22–17, R_3 becomes short-circuited.
   (*a*) What is the current through the short circuit?
   (*b*) How much power is supplied by generator G_1?

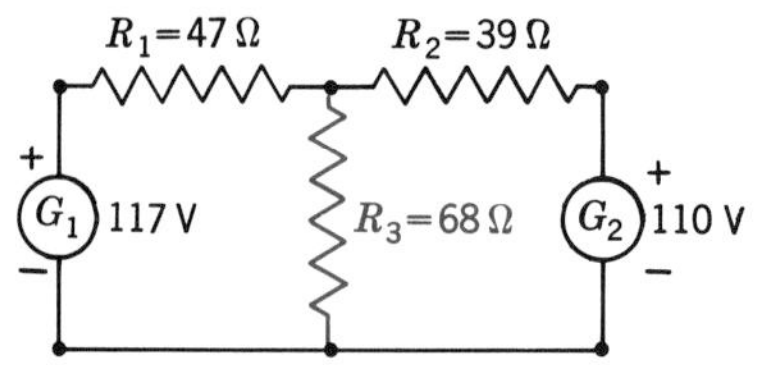

FIG. 22–17 Circuit of Probs. 1 and 2.

3. In the circuit of Fig. 22–18, (*a*) what is the current through R and (*b*) what is the current, and in what direction, through the batteries when R is open-circuited?

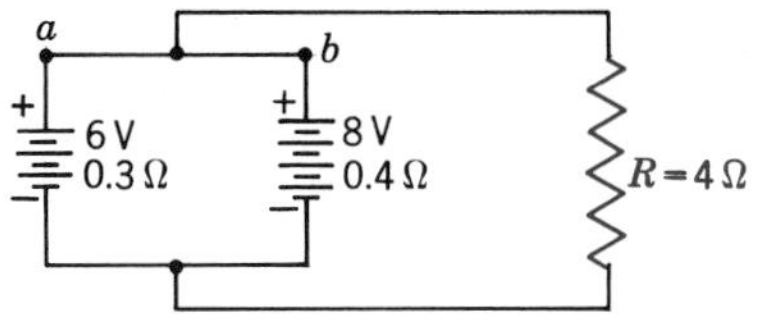

FIG. 22–18 Circuit of Probs. 3 and 4

4. In the original circuit of Fig. 22–18, R is shunted by a resistor of 1 Ω.
   (*a*) How much power is expended in the shunting resistor?
   (*b*) What is the terminal voltage of the 6-V battery?
5. In the circuit of Fig. 22–19, if the internal resistance of the generator is neglected, (*a*) how much power is being supplied by the generator and (*b*) what is the voltage across R?

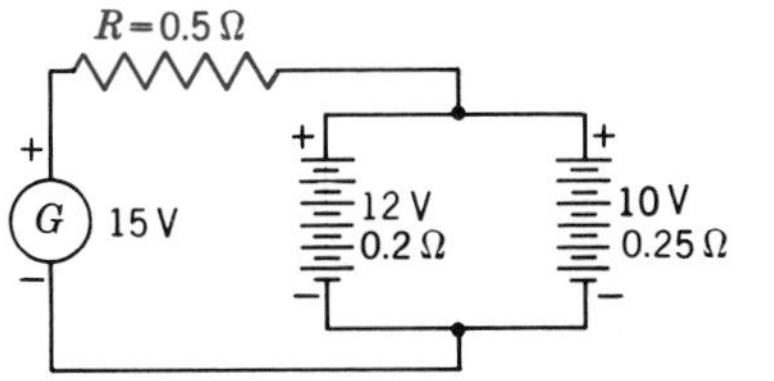

FIG. 22–19 Circuit of Probs. 5 and 6.

6. In the circuit of Fig. 22–19, the generator has an internal resistance of 0.15 Ω. If the connections of the generator are reversed, (*a*) how much power will be dissipated in R and (*b*) what will be the terminal voltage of the 10-V battery?
7. In the circuit of Fig. 22–20, (*a*) how much power is dissipated in R_4 and (*b*) what is the voltage across R_1?
8. If R_1 in the circuit of Fig. 22–20 is short-circuited, (*a*) what is the voltage across R_4 and (*b*) how much power is dissipated in the battery?

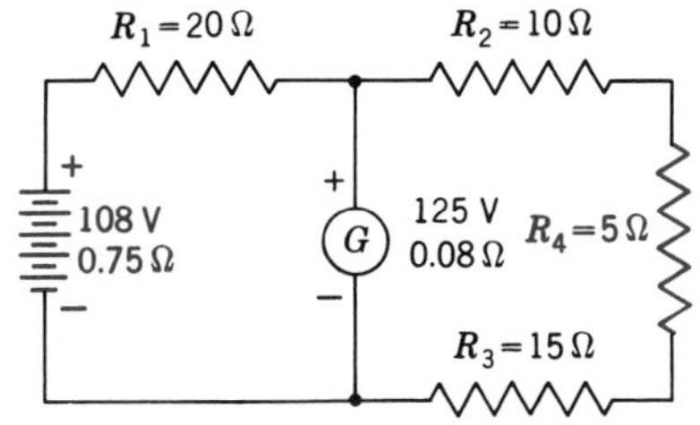

FIG. 22–20 Circuit of Probs. 7 and 8.

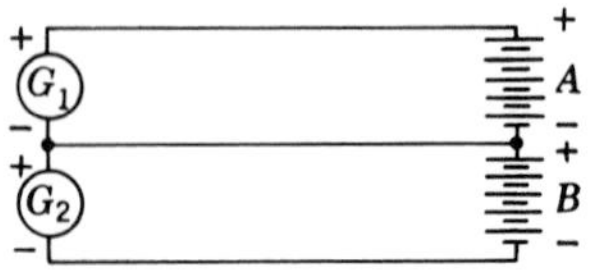

FIG. 22–21 Circuit of Prob. 9.

9. In the circuit of Fig. 22–21, battery A has an emf of 110 V and an internal resistance of 1.2 Ω. Battery B has an emf of 102 V and an internal resistance of 1 Ω. Each generator has an emf of 120 V and an internal resistance of 0.08 Ω. The resistance of each feeder is 0.05 Ω.
   (*a*) What is the current through battery A?
   (*b*) How much power is expended in battery B?

10. In the circuit of Fig. 22–22, (*a*) how much power is expended in R_5 and (*b*) how much power is expended in generator G_2?

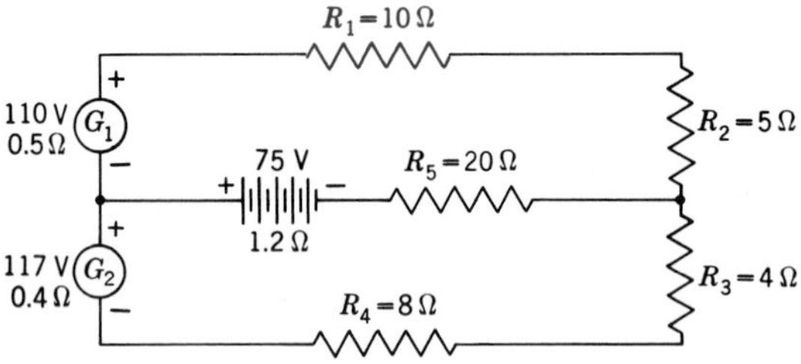

FIG. 22–22 Circuit of Probs. 10 and 11.

11. If the connections of the battery in the circuit of Fig. 22–22 are reversed, (*a*) what is the voltage across R_5 and (*b*) how much power is expended in the entire circuit?

12. Figure 22–23 represents a bank of batteries supplying power to loads R_a and R_b, with R_1, R_2, and R_3 representing the lumped line resistance. R_b is disconnected, and R_a draws 50 A. Neglecting the internal resistance of the generator and batteries, (*a*) what is the voltage across R_2 and (*b*) what is the current in the batteries and in what direction?

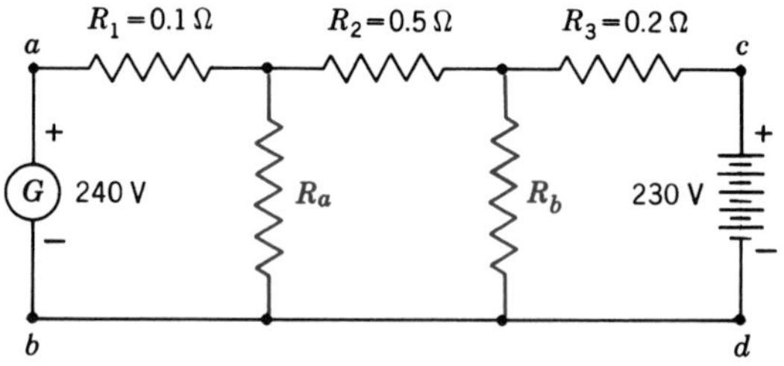

FIG. 22–23 Circuit of Probs. 12, 13, and 14.

13. R_b is connected in the circuit of Fig. 22–23 and draws 75 A. If R_a draws 50 A, (*a*) what is the voltage across R_b and (*b*) how much power is expended in R_2?

14. In the circuit that is shown in Fig. 22–23 the loads are adjusted until R_a draws 150 A and R_b draws 25 A. How much power is lost in R_2?

22–7 EQUIVALENT STAR AND DELTA CIRCUITS

EXAMPLE 6 Determine the currents through the branches of the network shown in Fig. 22–24 and find the equivalent resistance between points a and c.

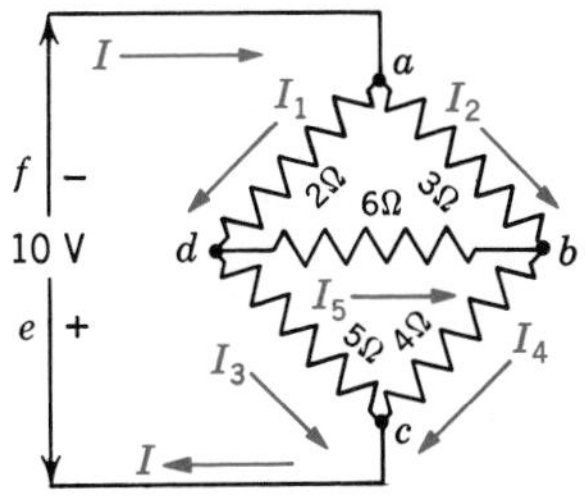

FIG. 22–24 Circuit of Example 6.

SOLUTION

Assume directions for all the currents, and label them on the figure. By Kirchhoff's second law:

Path *efabce*, $$3I - 3I_1 + 4I - 4I_2 = 10 \quad (16)$$

Path *efadce*, $$2I_1 + 5I_2 = 10 \quad (17)$$

Path *adba*, $$2I_1 + 6I_1 - 6I_2 - 3I + 3I_1 = 0 \quad (18)$$

Collecting like terms,

Equation (16) becomes $$7I - 3I_1 - 4I_2 = 10 \quad (19)$$

Equation (17) becomes $$2I_1 + 5I_2 = 10 \quad (20)$$

Equation (18) becomes $$-3I + 11I_1 - 6I_2 = 0 \quad (21)$$

Equations (19), (20), and (21) permit us to write

$$I = \frac{\begin{vmatrix} 10 & -3 & -4 \\ 10 & 2 & 5 \\ 0 & 11 & -6 \end{vmatrix}}{\begin{vmatrix} 7 & -3 & -4 \\ 0 & 2 & 5 \\ -3 & 11 & -6 \end{vmatrix}} = 2.879 \text{ A}$$

The equivalent resistance between points a and c is

$$\frac{V}{I} = \frac{10}{2.879} = 3.47\ \Omega$$

By expressing the branch currents in terms of other currents and labeling the circuit accordingly, this problem can be solved with a smaller number of equations. This is left as an exercise for you.

You will note, from the solution of Example 5, that the solution by Kirchhoff's laws of networks containing such configurations can become complicated. There are many cases, however, in which such networks can be replaced with more convenient equivalent circuits.

The three resistors R_1, R_2, and R_3 in Fig. 22–25*a* are said to be connected in *delta* (Greek letter Δ). R_a, R_b, and R_c in Fig. 22–25*b* are connected in *star*, or Y.

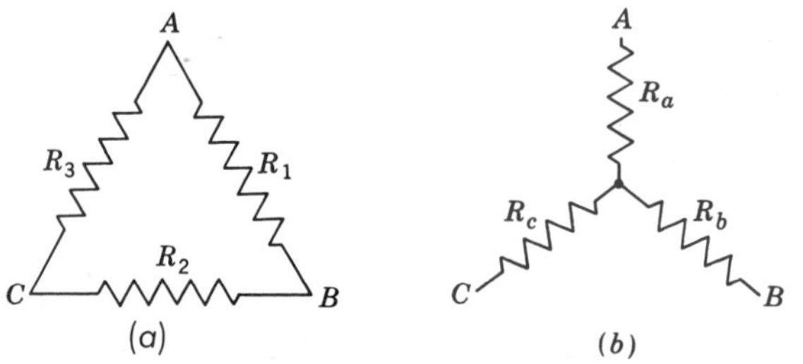

FIG. 22–25 (a) Resistors connected in delta.
(b) Resistors connected in star or Y.

If these two circuits are to be made equivalent, then the resistance between terminals A and B, B and C, and A and C must be the same in each circuit. Hence, in Fig. 22–25a the resistance from A to B is

$$R_{AB} = \frac{R_1(R_2 + R_3)}{R_1 + R_2 + R_3} \tag{22}$$

In Fig. 22–25b the resistance from A to B is

$$R_{AB} = R_a + R_b \tag{23}$$

Equating Eqs. (22) and (23),

$$R_a + R_b = \frac{R_1R_2 + R_1R_3}{R_1 + R_2 + R_3} \tag{24}$$

Similarly,

$$R_b + R_c = \frac{R_1R_2 + R_2R_3}{R_1 + R_2 + R_3} \tag{25}$$

and

$$R_a + R_c = \frac{R_1R_3 + R_2R_3}{R_1 + R_2 + R_3} \tag{26}$$

Equations (24), (25), and (26) are simultaneous and, when solved, result in

$$R_a = \frac{R_1R_3}{R_1 + R_2 + R_3} = \frac{R_1R_3}{\Sigma R_\Delta} \tag{27}$$

$$R_b = \frac{R_1R_2}{R_1 + R_2 + R_3} = \frac{R_1R_2}{\Sigma R_\Delta} \tag{28}$$

and

$$R_c = \frac{R_2R_3}{R_1 + R_2 + R_3} = \frac{R_2R_3}{\Sigma R_\Delta} \tag{29}$$

Since Σ (Greek letter sigma) is used to denote "the summation of,"

$$\Sigma R_\Delta = R_1 + R_2 + R_3$$

EXAMPLE 7 In Fig. 22–25*a*, $R_1 = 2\ \Omega$, $R_2 = 3\ \Omega$, and $R_3 = 5\ \Omega$. What are the values of the resistances in the equivalent Y circuit of Fig. 22–25*b*?

SOLUTION

$$\Sigma R_\Delta = 2 + 3 + 5 = 10\ \Omega$$

Substituting in Eq. (27), $$R_a = \frac{2 \times 5}{10} = 1\ \Omega$$

Substituting in Eq. (28), $$R_b = \frac{2 \times 3}{10} = 0.6\ \Omega$$

Substituting in Eq. (29), $$R_c = \frac{3 \times 5}{10} = 1.5\ \Omega$$

EXAMPLE 8 Determine the equivalent resistance between points *a* and *c* in the circuit of Fig. 22–26*a*.

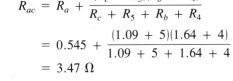

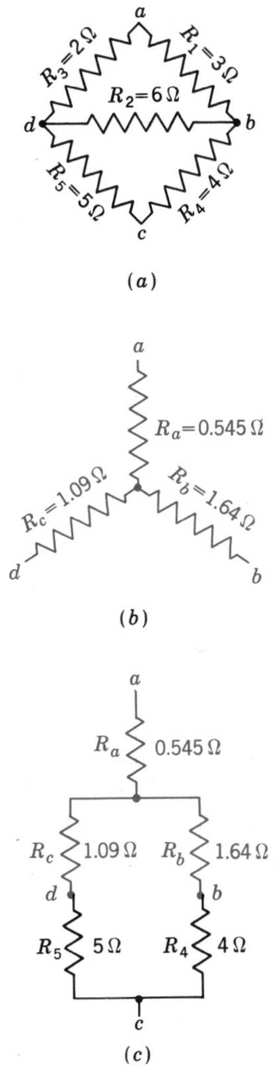

FIG. 22–26 Circuits of Example 8.

SOLUTION

Convert one of the delta circuits of Fig. 22–26*a* to its equivalent Y circuit. Thus, for the delta *abd*,

$$\Sigma R_\Delta = 3 + 6 + 2 = 11\ \Omega$$

The equivalent Y resistances, which are shown in Fig. 22–26*b*, are

$$R_a = \frac{3 \times 2}{11} = 0.545\ \Omega$$

$$R_b = \frac{3 \times 6}{11} = 1.64\ \Omega$$

and $$R_c = \frac{2 \times 6}{11} = 1.09\ \Omega$$

The equivalent Y circuit is connected to the remainder of the network as shown in Fig. 22–26*c* and is solved as an ordinary series-parallel combination. Thus,

$$R_{ac} = R_a + \frac{(R_c + R_5)(R_b + R_4)}{R_c + R_5 + R_b + R_4}$$

$$= 0.545 + \frac{(1.09 + 5)(1.64 + 4)}{1.09 + 5 + 1.64 + 4}$$

$$= 3.47\ \Omega$$

Note that the values of Fig. 22–26 are the same as those of Fig. 22–24.

The equations for converting a Y circuit to its equivalent delta circuit are obtained by solving Eqs. (27), (28), and (29) simultaneously. This results in

$$R_1 = \frac{\Sigma R_Y}{R_c} \qquad (30)$$

$$R_2 = \frac{\Sigma R_Y}{R_a} \qquad (31)$$

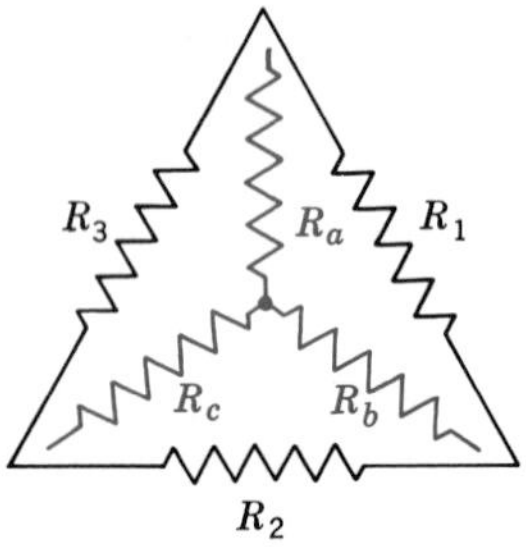

FIG. 22–27 Resistance equivalents.

$$R_3 = \frac{\Sigma R_Y}{R_b} \tag{32}$$

where $\Sigma R_Y = R_aR_b + R_bR_c + R_aR_c$

A convenient method for remembering the Δ to Y and Y to Δ conversions is illustrated in Fig. 22–27.

In converting from Δ to Y, each equivalent Y resistance is equal to the product of the two *adjacent* Δ resistances divided by the sum of the Δ resistances. For example, R_1 and R_3 are adjacent to R_a; therefore,

$$R_a = \frac{R_1R_3}{\Sigma R_\Delta}$$

In converting from Y to Δ, each equivalent Δ resistance is found by dividing ΣR_Y by the *opposite* Y resistance. For example, R_1 is opposite R_c; therefore,

$$R_1 = \frac{\Sigma R_Y}{R_c}$$

NOTE *A comparison of the Y network of Fig. 22–25*b *with the network formed by* R_1, R_2, *and* R_3 *of Fig. 13–9 will show the common interchangeability of the names T and Y and π (pi) and Δ (delta) in electronics circuitry.*

PROBLEMS 22–4

1. In the Δ circuit of Fig. 22–25*a*, $R_1 = 8\ \Omega$, $R_2 = 12\ \Omega$, and $R_3 = 15\ \Omega$. Determine the resistances of the equivalent Y circuit.
2. In the Δ circuit of Fig. 22–25*a*, $R_1 = 150\ \Omega$, $R_2 = 220\ \Omega$, and $R_3 = 330\ \Omega$. Determine the resistances of the equivalent Y circuit.
3. In the π circuit of Fig. 22–25*a*, $R_1 = R_2 = R_3 = 200\ \Omega$. Determine the resistances of the equivalent T circuit.
4. In the Y circuit of Fig. 22–25*b*, $R_a = 10\ \Omega$, $R_b = 15\ \Omega$, and $R_c = 47\ \Omega$. Determine the resistances of the equivalent Δ circuit.
5. In the T circuit of Fig. 22–25*b*, $R_a = 4.8\ \Omega$, $R_b = 4\ \Omega$, and $r_c = 6\ \Omega$. Determine the resistances of the equivalent π circuit.
6. In the T circuit of Fig. 22–25*b*, $R_a = R_b = R_c = 1.5\ \text{k}\Omega$. Determine the resistances of the equivalent π circuit.

In Probs. 7 to 17, solve the circuits by both the Δ to Y conversion and Kirchhoff's laws:

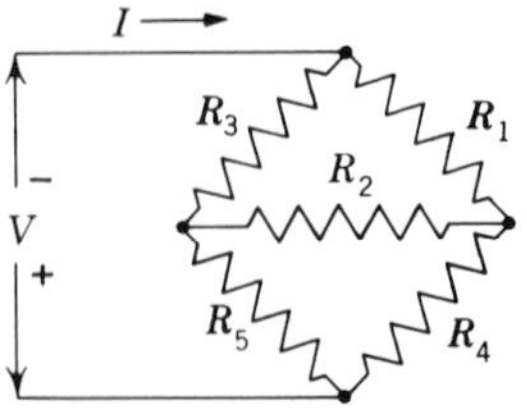

FIG. 22–28 Circuit of Probs. 7 to 12.

7. In the circuit of Fig. 22–28, $R_1 = 20\ \Omega$, $R_2 = 50\ \Omega$, $R_3 = 80\ \Omega$, $R_4 = 10\ \Omega$, $R_5 = 15\ \Omega$, and $V = 10$ V. What is the value of I?
8. What is the current through R_5 of Prob. 7?
9. What is the current through R_2 of Prob. 7?
10. In the circuit of Fig. 22–28, $R_1 = 25\ \Omega$, $R_2 = 10\ \Omega$, $R_3 = 15\ \Omega$, $R_4 = 50\ \Omega$, $R_5 = 30\ \Omega$, and $V = 50$ V. What is the value of I?

11. What is the current through R_2 of Prob. 10?
12. In the circuit of Fig. 22–28, $R_1 = ?$, $R_2 = 10\ \Omega$, $R_3 = 15\ \Omega$, $R_4 = 12\ \Omega$, $R_5 = 8\ \Omega$, $V = 32$ V, and $I = 2.39$ A. What is the resistance of R_1?
13. Determine the value of the current I in Fig. 22–29 if $V = 100$ V.
14. What is the current through R_4 of Prob. 13?
15. What is the current through R_5 of Prob. 13?
16. What is the current through the load resistance R_L in Fig. 22–30?
17. How much current does the signal generator G supply to the circuit of Fig. 22–31?

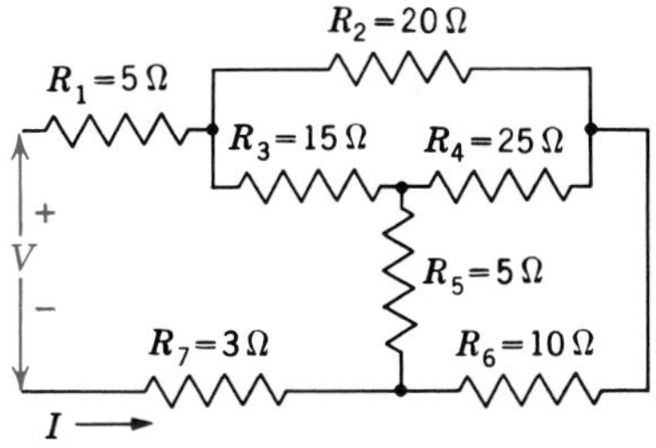

FIG. 22–29 Circuit of Probs. 13, 14, and 15.

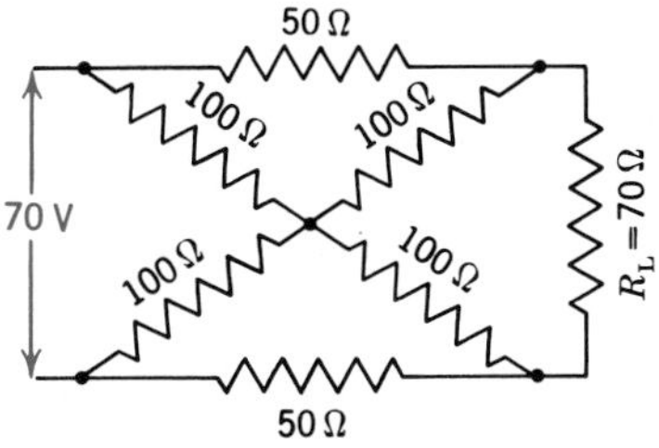

FIG. 22–30 Circuit of Prob. 16.

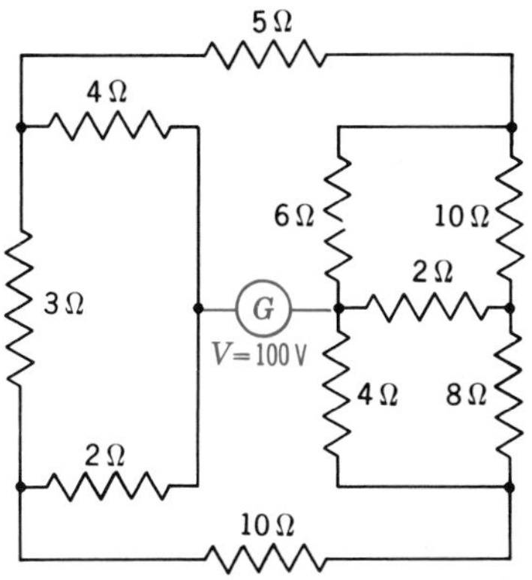

FIG. 22–31 Circuit of Prob. 17.

22–8 THEVENIN AND NORTON EQUIVALENTS

Often a knowledge of the actual components inside a power supply circuit is immaterial so long as we can measure the open-circuit output voltage and the short-circuit output current. From these easy measurements, we can picture a model of the circuit which will behave in exactly the same way as the original so far as any external connected circuit is concerned. Figure 22–32 illustrates this idea.

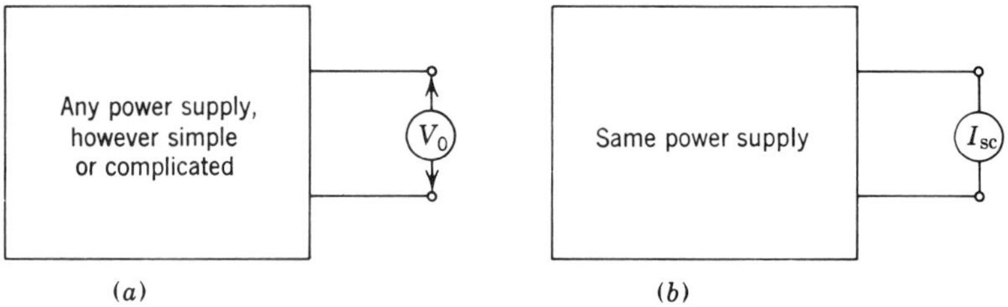

FIG. 22–32 (a) Any power supply will deliver a particular open-circuit (no load) emf V_0, which may be measured by an infinite-resistance voltmeter.
(b) Any power supply will deliver a short-circuit (maximum load) current I_{SC}, which may be measured by a zero-resistance ammeter.

A voltmeter with extremely high resistance can make a reasonable measurement of the open-circuit emf which the power supply generates. And an ammeter with very low resistance can make a reasonable measurement of the short-circuit current which the power supply can deliver. Two such models of power supplies are available to us:

Thevenin's theorem suggests that the power supply of Fig. 22–32 can be pictured as consisting of a simple equivalent source of constant emf V_{Th} in series with an equivalent resistance R_{Th}. Figure 22–33 shows the Thevenin equivalent of the circuit of Fig. 22–32. Obviously,

$$V_{Th} = V_0 \tag{33}$$

$$R_{Th} = \frac{V_0}{I_{sc}} \tag{34}$$

$$I_L = \frac{V_{Th}}{R_{Th} + R_L} \tag{35}$$

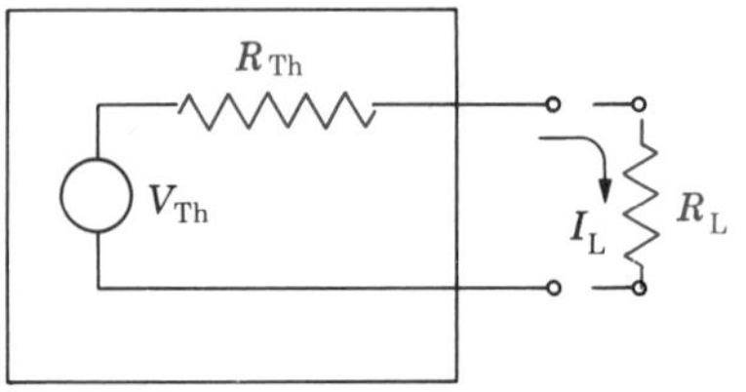

FIG. 22–33 Thevenin's equivalent of power supply of Fig. 22–32; that is, a source of constant emf V_{Th} in series with internal resistance R_{Th}.

Norton's theorem suggests that the power supply of Fig. 22–32 can be pictured as consisting of a simple equivalent source of constant current I_N in parallel with an equivalent resistance R_N. Figure 22–34 shows the Norton equivalent of the circuit of Fig. 22–32. You can see that

$$R_N = R_{Th} \tag{36}$$

$$I_N = \frac{V_{Th}}{R_{Th}} \tag{37}$$

$$I_L = \frac{R_N}{R_N + R_L} \cdot I_N \tag{38}$$

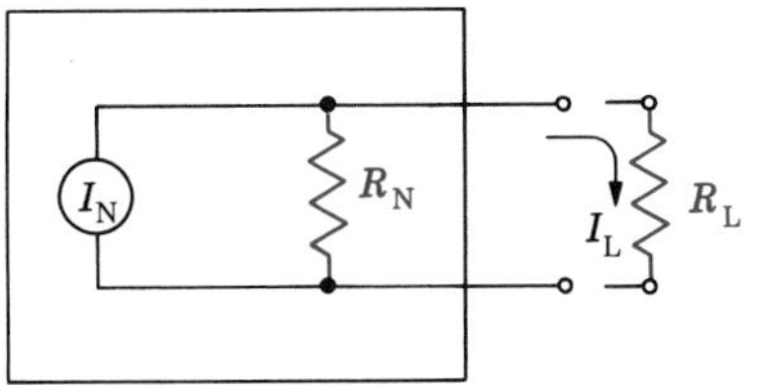

FIG. 22–34 Norton's equivalent of power supply of Fig. 22–32; that is, a source of constant current I_N in parallel with internal resistance R_N.

You should apply your knowledge of parallel resistances in order to prove Eq. (38).

The solution to network problems may sometimes be simplified by applying one or the other of these two theorems.

EXAMPLE 9 Use Thevenin's theorem to solve the current flow through the load resistor R in the circuit of Fig. 22–14 on page 395. $V_1 = 6$ V with an internal resistance of 0.15 Ω and $V_2 = 5$ V with an internal resistance of 0.05 Ω; and $R = 2$ Ω.

SOLUTION

Redraw the circuit to show R as the load to be connected and the rest of the circuit as a power supply (Fig. 22–35 on page 408).

Determine the open-circuit voltage which would appear across the terminals *ab* of Fig. 22–35. A high resistance voltmeter would measure

$$V_{Th} = V_0 = V_2 + I_c r_2$$

The circulating current I_c is found by applying Ohm's law to the internal circuit:

$$I_c = \frac{6 - 5}{0.15 + 0.05} = \frac{1}{0.20} = 5 \text{ A}$$

and

$$V_{Th} = 5 + 5(0.05) = 5 + 0.25$$
$$= 5.25 \text{ V}$$

Determine the circuit resistance which would be seen by an ohmmeter connected to terminals *ab* with the sources of emf shorted and represented by their

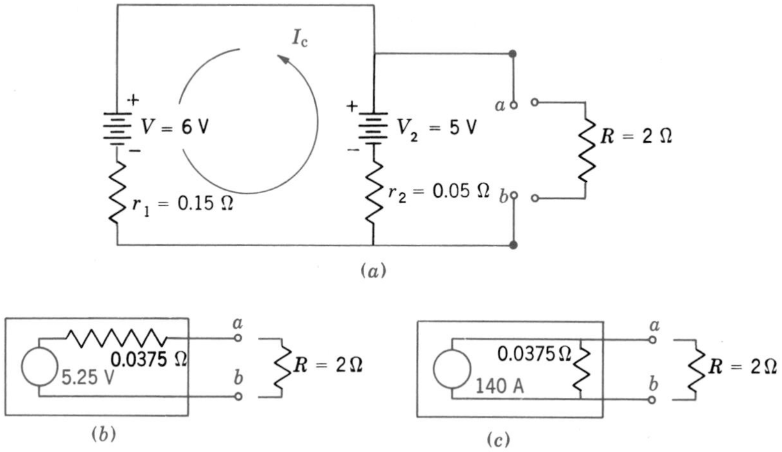

FIG. 22–35 (a) Redrawn from Fig. 22–14 for Thevenin's solution.
(b) Redrawn from Fig. 22–14 for Thevenin's equivalent circuit.
(c) Redrawn from Fig. 22–14 for Norton's equivalent circuit.

internal resistances. Under such circumstances, an ohmmeter would see r_1 and r_2 in parallel:

$$R_{\text{Th}} = \frac{0.15 \times 0.05}{0.15 + 0.05} = 0.0375\ \Omega$$

Thus, the Thevenin equivalent circuit (Fig. 22–35*b*) is a constant source of 5.25 V in series with 0.0375 Ω. Then the current through the 2-Ω "load" is

$$I_{\text{R}} = \frac{5.25}{2.0375} = 2.58\ \text{A}$$

(Compare with Example 4, Sec. 22–5.)

EXAMPLE 10 Solve Example 9 by using Norton's theorem.

SOLUTION

As before, determine the equivalent internal resistance of the "power supply" as seen by a connected load:

$$R_{\text{N}} = R_{\text{Th}} = 0.0375\ \Omega$$

Then determine the current which the "power supply" would drive through a short circuit across terminals *ab*. This may be done by using a Thevenin open-circuit approach and finding

$$I_{\text{N}} = \frac{V_{\text{Th}}}{R_{\text{Th}}} = \frac{5.25}{0.0375} = 140\ \text{A}$$

from which

$$I_R = \frac{0.0375}{2.0375} \times 140 = 2.58 \text{ A}$$

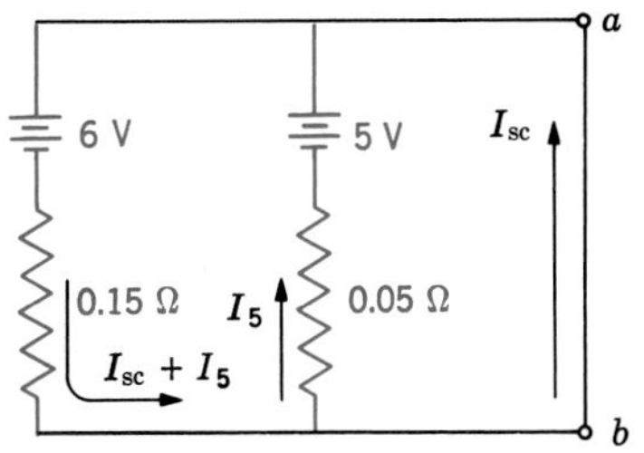

FIG. 22–36 Determination of $I_N = I_{SC}$ for Fig. 22–35.

Alternatively, determine from first principles what the short-circuit current through *ab* would be. Figure 22–36 shows this approach. Using Kirchhoff's laws,

$$0.15(I_{sc} + I_5) + 0.05I_5 = 1$$
$$0.015(I_{sc} + I_5) = 6$$

from which $I_{sc} = 140$ A

and $I_R = 2.58$ A

22–9 OUTLINE FOR THEVENIN AND NORTON SOLUTIONS

The following systematic procedure will simplify the utilization of these two circuit simplification theorems:

1. Determine the *leg* of a circuit through which the current flow is to be determined and redraw the circuit, omitting that part.
2. Consider the balance of the circuit to be a power supply whose terminals are eventually to deliver current to the part omitted. Often it is helpful to letter all connecting points in the original circuit to make sure that the equivalent has been drawn correctly.
3. Determine the voltage which would be indicated by a voltmeter connected across the open-circuit terminals of the "power supply." This is V_{Th}.
4. Short-circuit all the internal sources of emf, leaving them represented by their internal resistances, and determine the resistance which would be indicated by an ohmmeter connected across the open-circuit terminals of the power supply. This is $R_{Th} = R_N$.
5. Determine $I_N = \dfrac{V_{Th}}{R_{Th}}$, or
6. Determine the value of I_N as the current which the power supply would drive through an ammeter connected across its terminals.
7. Use Eq. (35) or (38) to determine the current flow through the reconnected "load."

EXAMPLE 11 Determine the current I_5 through the 6-Ω resistor of Fig. 22–24 on page 401.

SOLUTION

Redraw the circuit. Omit the 6-Ω bridging resistor and let the balance of the circuit be a power supply which will later serve the 6-Ω load (Fig. 22–37).

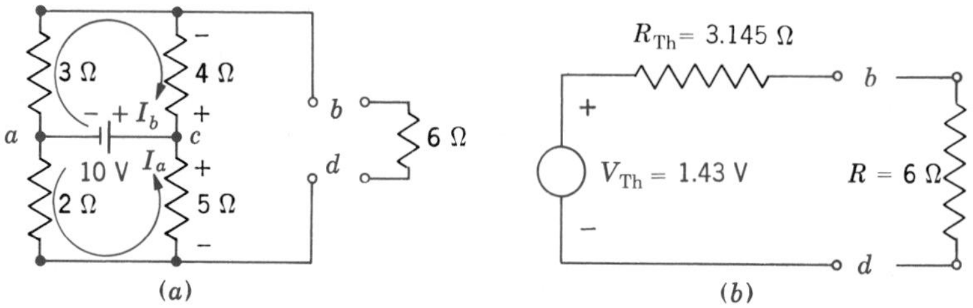

FIG. 22–37 (a) Redrawn from Fig. 22–24 for Thevenin's solution of current I_5 through resistor across points *bd*.
(b) Thevenin's equivalent circuit for (a).

When terminals *bd* are open-circuited, the 10-V source will drive currents I_a and I_b through the power supply internal circuitry, thereby producing voltage drops across the 4- and 5-Ω resistors with the polarities indicated:

$$I_a = \frac{10}{7} = 1.43 \text{ A}$$
$$I_b = \frac{10}{7} = 1.43 \text{ A}$$
$$V_4 = 1.43 \times 4 = 5.72 \text{ V}$$
$$V_5 = 1.43 \times 5 = 7.15 \text{ V}$$

A voltmeter across terminals *bd* will measure

$$V_{Th} = 7.15 - 5.72 = 1.43 \text{ V}$$

When the 10-V internal source is shorted, its internal resistance being zero, an ohmmeter across terminals *bd* will measure

$$R_{Th} = \frac{3 \times 4}{3 + 4} + \frac{2 \times 5}{2 + 5}$$
$$= 1.715 + 1.43$$
$$= 3.145 \ \Omega$$

Then the Thevenin equivalent to the power supply is a constant 1.43 V in series with 3.145 Ω.

$$I_5 = \frac{1.43}{6 + 3.145} = 156 \text{ mA}$$

(Compare with Example 6, Sec. 22–7.)

PROBLEMS 22–5

1. A power supply delivers an open-circuit emf of 100 V. An ammeter that is connected across its terminals measures a short-circuit current of 180 A.
   (a) What is the Thevenin circuit equivalent to the power supply so far as any connected load is concerned?
   (b) What is the Norton equivalent to the power supply?
2. A power supply delivers an open-circuit emf of 7.2 V. An ammeter that is connected across its terminals measures a short-circuit current of 120 mA.
   (a) What is the Thevenin equivalent circuit to the power supply, so far as any connected load is concerned?
   (b) What is the Norton equivalent to the power supply?
3. What is the Thevenin equivalent circuit of the ''power supply'' portion of the circuit of Fig. 22–28 for the solution of the current through R_2 if $R_1 = 4\ \Omega$, $R_2 = 10\ \Omega$, $R_3 = 3\ \Omega$, $R_4 = 2\ \Omega$, $R_5 = 5\Omega$, and $V = 12$ V? What is the current flow through R_2?
4. What is the Thevenin equivalent circuit of the power supply portion of the circuit of Fig. 22–28 for the solution of the current through R_2 if $R_1 = 25\ \Omega$, $R_2 = 10\ \Omega$, $R_3 = 15\ \Omega$, $R_4 = 50\ \Omega$, $R_5 = 30\ \Omega$, and $V = 50$ V? What is the current through R_3?
5. What is the Norton equivalent circuit of the power supply portion of the circuit of Prob. 3 for the solution of the current through R_5? What is the current through R_5?
6. What is the Thevenin equivalent circuit of the power supply portion of the circuit of Prob. 4 for the solution of the current through R_1? What is the current through R_1?

SELF TEST

1. Three resistors, $R_1 = 10$ kΩ, $R_2 = 12$ kΩ, and $R_3 = 27$ kΩ, are connected in series across a signal source whose internal resistance is 49 kΩ. If the source delivers a loaded terminal voltage of 120 mV, how much current flows through the load?
2. Two loads with $R_1 = 27$ kΩ, and $R_2 = 40$ kΩ, are connected in parallel across the output terminals of a generator and that has an internal resistance of 2.7 kΩ that delivers an open-circuit emf of 445 V. What is the terminal voltage of the generator?
3. In the circuit of Fig. 22–38 (next page), how much power is expended in R_5?
4. Determine the equivalent Y circuit of the Δ circuit of Fig. 22–39.
5. Determine the value of the current flowing through R_4 in the Wheatstone bridge circuit of Fig. 22–40.

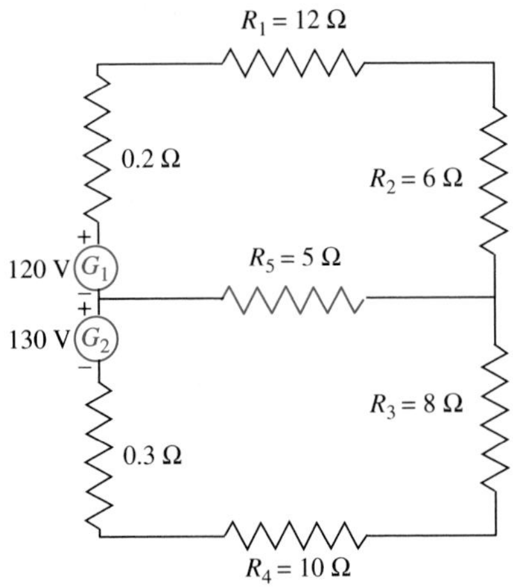

FIG. 22–38 Circuit for self test Prob. 3.

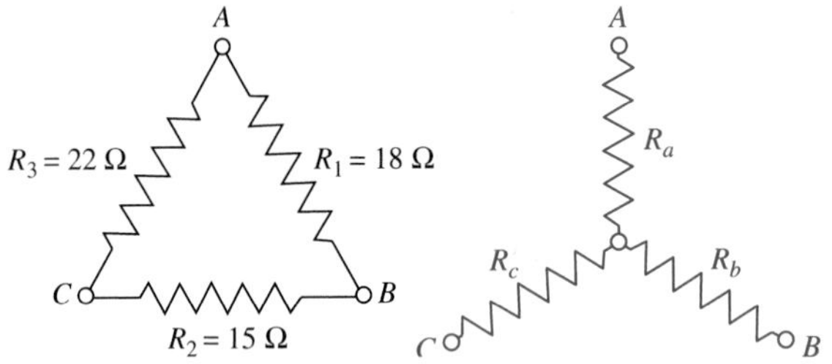

FIG. 22–39 Circuit for self test Prob. 4.

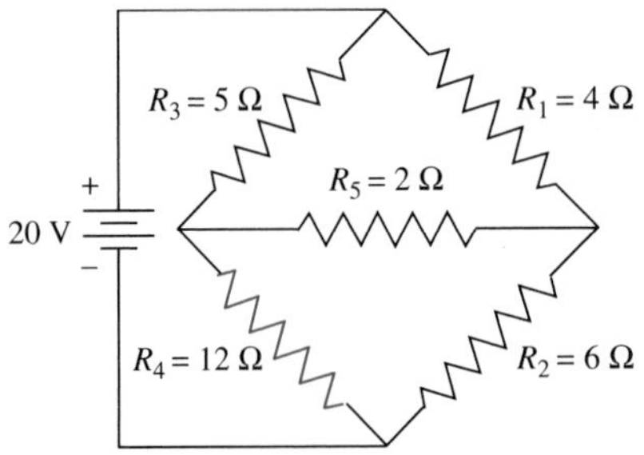

FIG. 22–40 Wheatstone bridge circuit for self test Prob. 5.

6. Given the original circuit of Prob. 5, what is the Thevenin equivalent power supply to determine the current flow through R_1?
7. Given the original circuit of Prob. 5, what is the Norton equivalent power supply to determine the current flow through R_5?

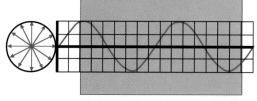

CHAPTER 23

ANGLES

This chapter deals with the study of angles as an introduction to the branch of mathematics called *trigonometry.* The word "trigonometry" is derived from two Greek words meaning "measurement" or "solution" of triangles.

Trigonometry is both algebraic and geometric in nature. It is not confined to the solution of triangles but forms a basis for more advanced subjects in mathematics. A knowledge of the subject paves the way for a clear understanding of ac and related circuits.

23–1 ANGLES

In trigonometry, we are concerned primarily with the many relations that exist among the sides and angles of triangles. In order to understand the meaning and measurement of angles, it is essential that you thoroughly understand these relationships.

An angle is formed whenever two straight lines meet at a point.

In Fig. 23–1*a,* lines *OA* and *OX* meet at the point *O* to form the angle *AOX*. Too, in Fig. 23–1*b,* the angle *BOX* is formed by lines *OB* and *OX* meeting at the point *O*. This point is called the *vertex* of the angle, and the two lines are called the *sides* of the angle. The size, or magnitude, of an angle is a measure of the difference in directions of the sides. Thus, in Fig. 23–1, angle *BOX* is a larger angle than *AOX*. The lengths of the sides of an angle have no bearing on the size of the angle.

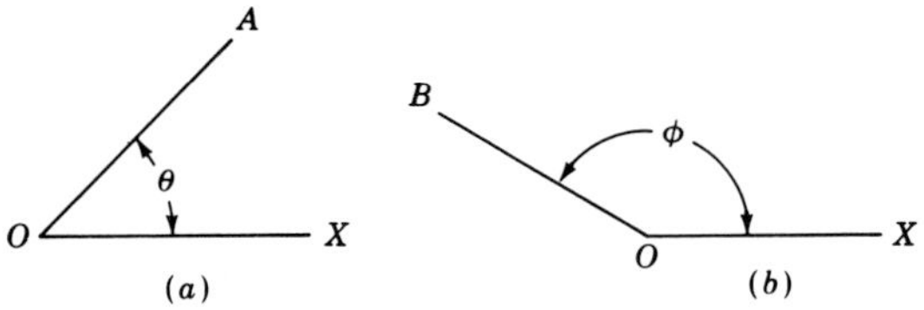

FIG. 23–1 Formation of angles.

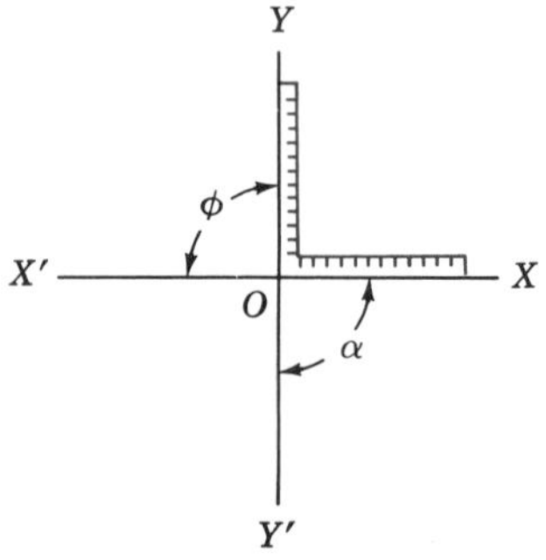

FIG. 23–2 Right angles.

In geometry it is customary to denote an angle by the symbol ∠. If this notation is used, "angle *AOX*" would be written ∠*AOX*. An angle is also denoted by the letter at the vertex or by a supplementary letter placed inside the angle. Thus, angle *AOX* is correctly denoted by ∠*AOX*, ∠*O*, or ∠θ. Also, *BOX* could be written ∠*BOX*, ∠*O*, or ∠ϕ.

If equal angles are formed when one straight line intersects another, the angles are called *right angles*. In Fig. 23–2, angles *XOY*, ϕ, *X′OY′*, and α are all right angles.

An *acute angle* is an angle that is less than a right angle. For example, in Fig. 23–3*a*, ∠α is an acute angle.

An *obtuse angle* is an angle that is greater than a right angle. For example, in Fig. 23–3*b*, ∠β is an obtuse angle.

Two angles whose sum is one right angle are called *complementary angles*. Either one is said to be the *complement* of the other.

For example, in Fig. 23–3*c*, angles ϕ and θ are complementary angles; ϕ is the complement of θ, and θ is the complement of ϕ.

Two angles whose sum is two right angles (a straight line) are called *supplementary angles*. Either one is said to be the *supplement* of the other. Thus, in Fig. 23–3*d*, angles *b* and *a* are supplementary angles; *b* is the supplement of *a*, and *a* is the supplement of *b*.

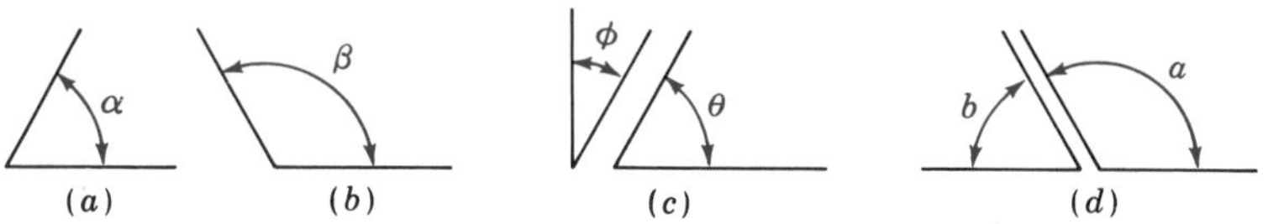

FIG. 23–3 (a) Acute angle; (b) obtuse angle; (c) complementary angles; (d) supplementary angles.

23–2 GENERATION OF ANGLES

In the study of trigonometry, it becomes necessary to extend our concept of angles beyond the geometric definitions stated in Sec. 23–1. An angle should be thought of as being generated by a line (line segment or half ray) that starts in a certain initial position and rotates about a point called the *vertex* of the angle until it stops at its final position. The original position of the rotating line is called the *initial side* of the angle, and the final position is called the *terminal side* of the angle.

An angle is said to be in *standard position* when its vertex is at the origin of a system of rectangular coordinates and its initial side extends in the positive direction along the *x* axis. Thus, in Fig. 23–4, the angle θ is in standard position. The vertex is at the origin, and the initial side is on the positive *x* axis. The angle has been generated by the line *OP* revolving, or sweeping, from *OX* to its final position.

An angle is called a *positive angle* if it is generated by a line revolving counterclockwise. If the generating line revolves clockwise, the angle is called a *negative angle*. In Fig. 23–5, all angles are in standard position. ∠θ is a positive angle that was generated by the line *OM* revolving counterclockwise

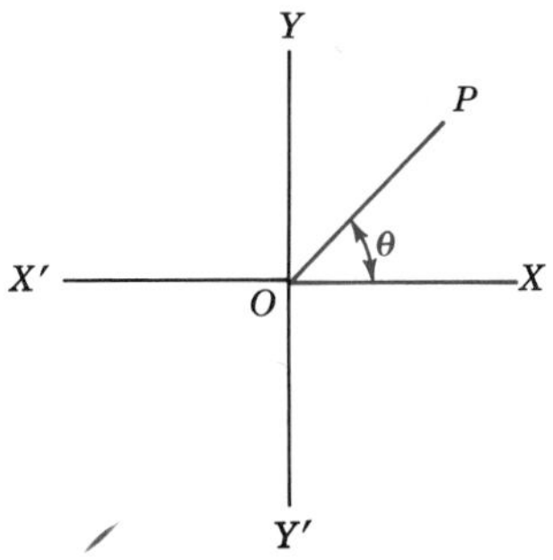

FIG. 23–4 Angle θ in standard position.

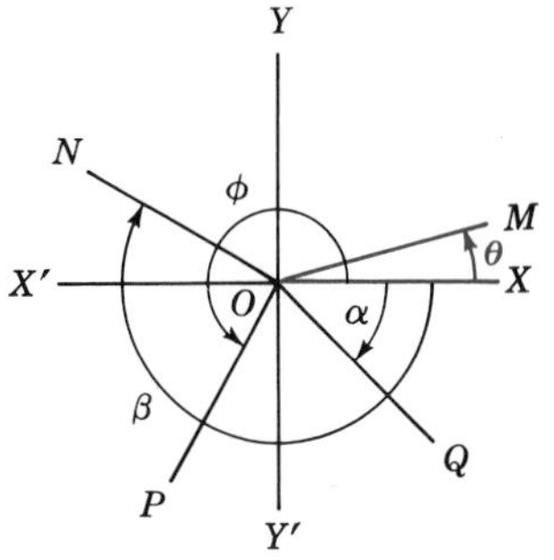

FIG. 23–5 Generation of angles.

from *OX*. ϕ is also a positive angle whose terminal side is *OP*. α is a negative angle that was generated by the line *OQ* revolving in a clockwise direction from the initial side *OX*. β is also a negative angle whose terminal side is *ON*.

If the terminal side of an angle that is in standard position lies in the first quadrant, then that angle is said to be *an angle in the first quadrant,* etc. Thus, θ in Fig. 23–4 and θ in Fig. 23–5 are in the first quadrant.

Similarly, in Fig. 23–5, β is in the second quadrant, ϕ is in the third quadrant, and α is in the fourth quadrant.

23–3 THE SEXAGESIMAL SYSTEM

There are several systems of angular measurement. The three most commonly used are the right angle, the circular (or natural) system, and the sexagesimal system. The right angle is almost always used as a unit of angular measure in plane geometry and is constantly used by builders, surveyors, etc. However, for the purposes of trigonometry, it is an inconvenient unit because of its large size.

The unit most commonly used in trigonometry is the *degree,* which is one-ninetieth of a right angle. The degree is defined as the angle formed by one three hundred sixtieth part of a revolution of the angle-generating line. The degree is divided into 60 equal parts called *minutes,* and the minute into 60 equal parts called *seconds*. The word ''sexagesimal'' is derived from a Latin word pertaining to the number 60.

Instead of dividing the degrees into minutes and seconds, we shall divide them decimally for convenience. For example, instead of expressing an angle of 43 degrees 36 minutes as 43°36′, we write 43.6°.

The actual measurement of an angle consists in finding how many degrees and a decimal part of a degree there are in the angle. This can be accomplished with a fair degree of accuracy by means of a *protractor,* which is an instrument for measuring or constructing angles.

To measure an angle *XOP,* as in Fig. 23–6, p. 416, place the center of the protractor indicated by *O* at the vertex of the angle with, say, the line *OX* coinciding with one edge of the protractor as shown in Fig. 23–7 on p. 416. The magnitude of the angle, which is 60°, is indicated where the line *OP* crosses the graduated scale.

To construct an angle, say 30° from a given line *OX,* place the center of the protractor on the vertex *O*. Pivot the protractor about this point until *OX* is on

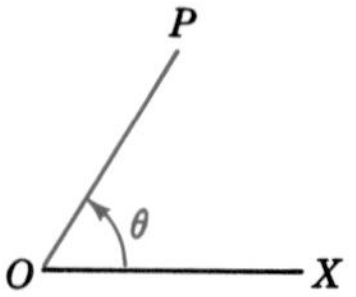

FIG. 23–6 Angle to be measured.

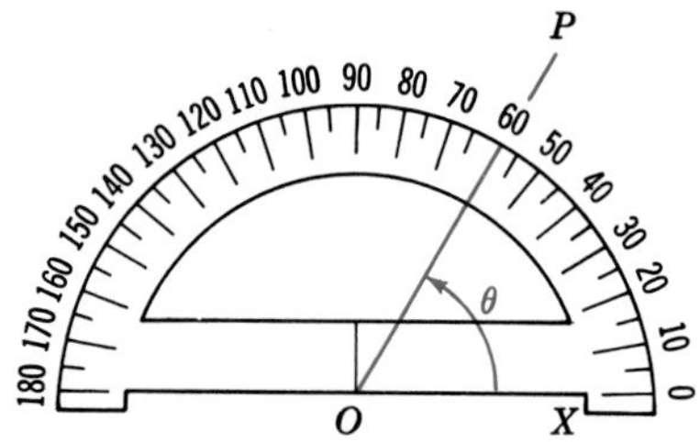

FIG. 23–7 Using protractor to measure angle *XOP* of Fig. 23–6.

a line with the 0° mark on the scale. In this position, 30° on the scale now marks the terminal side *OP* as shown in Figs. 23–8 and 23–9.

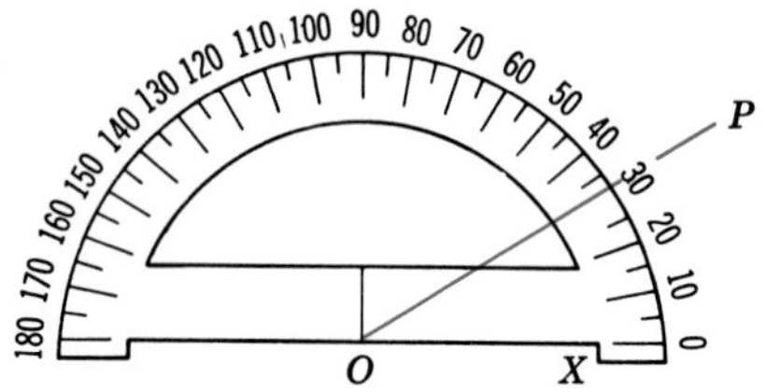

FIG. 23–8 Using protractor to construct angle.

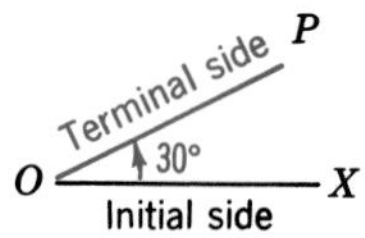

FIG. 23–9 30° angle constructed by protractor in Fig. 23–8.

23–4 ANGLES OF ANY MAGNITUDE

In the study of trigonometry, it will be necessary to extend our concept of angles in order to include angles greater than 360°, either positive or negative. Thinking of an angle being generated, as was explained in Sec. 23–2, permits consideration of angles of any size; for the generating line can rotate from its initial position in a positive or negative direction to produce any size angle, even one greater than 360°. Figure 23–10 illustrates how an angle of +750° is generated. However, for the purpose of ordinary computation, we consider such an angle to be in the same quadrant as its terminal side, with a magnitude equal to the remainder after the largest multiple of 360° it will contain has been subtracted from it. Thus, in Fig. 23–10, the angle is in the first quadrant and, geometrically, is equal to $750° - 720° = 30°$.

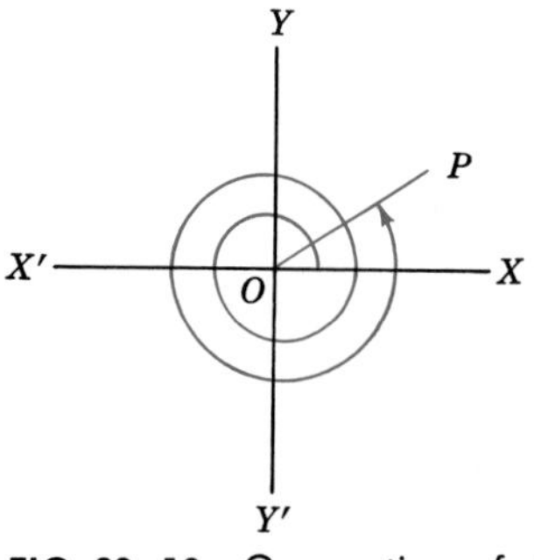

FIG. 23–10 Generation of 750° angle.

PROBLEMS 23–1

1. What is the complement of (*a*) 57°, (*b*) 18°, (*c*) 46°, (*d*) 152°, (*e*) 240°, (*f*) −25°?
2. What is the supplement of (*a*) 65°, (*b*) 160°, (*c*) 235°, (*d*) 270°, (*e*) 340°, (*f*) −130°?
3. Construct two complementary angles each in standard position on the same pair of axes.
4. Construct two supplementary angles each in standard position on the same pair of axes.
5. By using a protractor, construct the following angles and place them in standard position on rectangular coordinates. Indicate by arrows the direction and amount of rotation necessary to generate these angles: (*a*) 45°, (*b*) 160°, (*c*) 220°, (*d*) 315°, (*e*) 405°, (*f*) −60°, (*g*) −315°, (*h*) −300°, (*i*) −390°, (*j*) −850°.
6. Through how many degrees does the minute hand of a clock turn in (*a*) 20 min, (*b*) 40 min?
7. Through how many right angles does the minute hand of a clock turn from 8:30 A.M. to 5:30 P.M. of the same day?
8. Through how many degrees per minute do (*a*) the second hand, (*b*) the minute hand, (*c*) the hour hand of a clock rotate?
9. A motor armature has a speed of 1800 rev/min. What is the angular velocity (speed) in degrees per second?
10. The shaft of the motor armature in Prob. 9 is directly connected to a pulley 300 mm in diameter. What is the pulley rim speed in meters per second?

23–5 THE CIRCULAR, OR NATURAL, SYSTEM

The circular, or natural, system of angular measurement is sometimes called *radian measure* or *π measure*. The unit of measure is the *radian*. [In this book the abbreviation for *radian* is "rad" when used with units (0.55 rad/s); but an angle of 0.55 radian is written symbolically with a Roman superscript "r" (0.55^r) in order to parallel the use of the degree symbol (288°).]

A radian is an angle that, when placed with its vertex at the center of a circle, intercepts an arc equal in length to the radius of the circle. Thus, in Fig. 23–11, if the length of the arc *AP* equals the radius of the circle, then angle *AOP* is equal to one radian. Figure 23–12 shows a circle divided into radians.

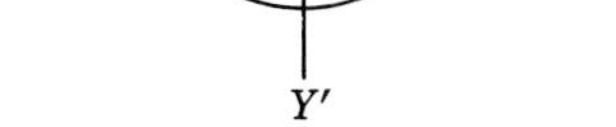

FIG. 23–11
Angle *AOP* = 1^r.

Does your calculator have a π key? Does it let you calculate decimal degrees as well as minutes and seconds? Does it provide conversions between systems?

The circular system of measure is used extensively in electrical and electronics formulas and is almost universally used in the higher branches of mathematics.

From geometry, it is known that the circumference of a circle is given by the relation

$$C = 2\pi r \tag{1}$$

where r is the radius of the circle. Dividing both sides of Eq. (1) by r, we have

$$\frac{C}{r} = 2\pi \qquad (2)$$

> Refer to your computer user's manual. If your computer is capable of working directly in degrees, it will probably respond to the command DEG.
>
> Your computer probably works directly in radians (most computers do). If you are given an angle in degrees, you must convert it to radians in order to provide your computer with acceptable input. See the rules in green in the text.

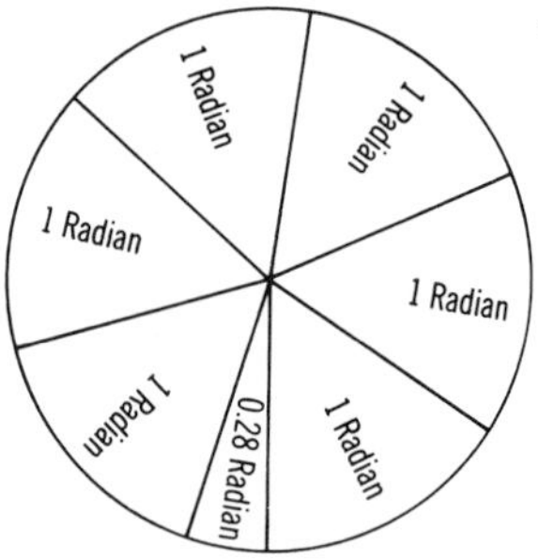

FIG. 23–12 Circle divided into $2\pi^r$.

Now Eq. (2) says simply that the ratio of the circumference to the radius is 2π; that is, the circumference is 2π times longer than the radius. Therefore, a circle must contain 2π radians ($2\pi^r$). Also, since the circumference subtends 360°, it follows that

$$2\pi^r = 360°$$

$$\pi^r = 180°$$

or

$$1^r = \frac{180°}{\pi} = 57.2959° \cong 57.3° \qquad (3)$$

From Eq. (3), the following is evident:

- To reduce radians to degrees, multiply the number of radians by $\frac{180°}{\pi^r}$ ($\cong$57.3).
- To reduce degrees to radians, multiply the number of degrees by $\frac{\pi^r}{180°}$ ($\cong$0.017 45).

> If there is no π key on your computer keyboard, use 4*ARCTAN(1). This will give you a numerical value (3.141 592 654) that is correct to as many decimal places as your computer can work. (Arctan is explained in Sec. 25–2.)
>
> For $\frac{\pi}{180}$, use
>
> (4*ARCTAN(1))/180
>
> For $\frac{180}{\pi}$, use
>
> 180/(4*ARCTAN(1))

EXAMPLE 1 Reduce 1.7^r to degrees.

SOLUTION

$$1^r = 57.3°$$

Hence,

$$1.7^r = 1.7 \times 57.3 = 97.4°$$

EXAMPLE 2 Convert 15.6° to radians.

SOLUTION

$$1^\circ = 0.017\,45^r$$

Hence,

$$15.6^\circ = 15.6 \times 0.017\,45 = 0.272^r$$

PROBLEMS 23–2

1. Express the following angles in radians, first in terms of π and second as decimals: (*a*) 30°, (*b*) 60°, (*c*) 120°, (*d*) 210°, (*e*) 225°, (*f*) 12°.
2. Express the following angles in degrees: (*a*) 1^r, (*b*) 0.5^r, (*c*) $\frac{1^r}{\pi}$, (*d*) $\frac{\pi^r}{5}$, (*e*) $\frac{2\pi^r}{3}$, (*f*) $0.785\,40^r$.
3. Through how many radians does the second hand of a clock turn between 7:25 A.M. and 10:40 A.M. of the same day?
4. Through how many radians does the hour hand of a clock turn in 40 min?
5. Through how many radians does the minute hand of a clock turn in 2 h 45 min?
6. What is the angular velocity in radians per second of (*a*) the second hand, (*b*) the minute hand, (*c*) the hour hand of a clock?
7. The speed of a rotating switch is 600 rev/min. What is its angular velocity in radians per second?
8. A radar antenna rotates at 6 rev/min. What is its angular velocity in radians per second?
9. A radar antenna has an angular velocity of π rad/s. What is its speed of rotation in revolutions per minute?
10. What is the approximate angular velocity of the earth in radians per minute (rad/min)?

23–6 GONS AND GRADS

In some parts of Europe, as a stage in furthering decimalization, the right angle is divided into one hundred equal parts known as *grads* (from the German) or, internationally, as *gons* (from the Greek). Each gon (grad) may be subdivided into 100 centigons (centigrads). Sometimes angles measured in this system will be written 20^g, sometimes 20 gon or 20 grad (no *s* for plural). The manner of notation will be a matter for future international agreement. Some calculators now available to electronics technicians offer alternative angle calculations in gons (grads).

$$1 \text{ right angle} = 90^\circ = \frac{\pi^r}{2} = 100^g \tag{4}$$

From Eq. (4), the following is evident:

- To reduce gons to degrees, multiply the number of gons by $\frac{90^\circ}{100^g}$ (= 0.9).

- To reduce degrees to gons, multiply the number of degrees by $\frac{100^g}{90^\circ}$ $(= 1.111)$.
- To reduce gons to radians, multiply the number of gons by $\frac{\pi/2^r}{100^g}$ $(\cong 0.0157)$.
- To reduce radians to gons, multiply the number of radians by $\frac{100^g}{\pi/2^r}$ $(\cong 63.7)$.

PROBLEMS 23–3

1. Express the following angles in gons: (*a*) 30°, (*b*) 45°, (*c*) 60°, (*d*) 135°, (*e*) 225°, (*f*) 330°.
2. Express the following angles in degrees: (*a*) 30^g, (*b*) 50^g, (*c*) 60^g, (*d*) 120^g, (*e*) 150^g, (*f*) 300^g.
3. Express the following angles in radians: (*a*) 33.3^g, (*b*) 50^g, (*c*) 66.7^g, (*d*) 133^g, (*e*) 250^g, (*f*) 300^g.
4. Express the following angles in gons: (*a*) $\frac{\pi^r}{4}$, (*b*) $\frac{\pi^r}{6}$, (*c*) $\frac{2\pi^r}{5}$, (*d*) $\frac{3\pi^r}{2}$, (*e*) $\frac{5\pi^r}{6}$, (*f*) 1.5708^r.

If you have an electronic calculator which gives angles in gons (grads), you may convert angles less than 90° as follows:

- Enter degrees, 45°; call for sin and read 0.707 11; convert to gons and call for arcsin to read 50^g.
- Enter gons, 50^g; call for sin and read 0.707 11; convert to rad and call for arcsin to read $0.785\ 40^r$.

23–7 SIMILAR TRIANGLES

Two triangles are said to be *similar* when their corresponding angles are equal. That is, similar triangles are identical in shape but may not be the same in size. The important characteristic of similar triangles is that a direct proportionality exists between corresponding sides. The three triangles of Fig. 23–13 have

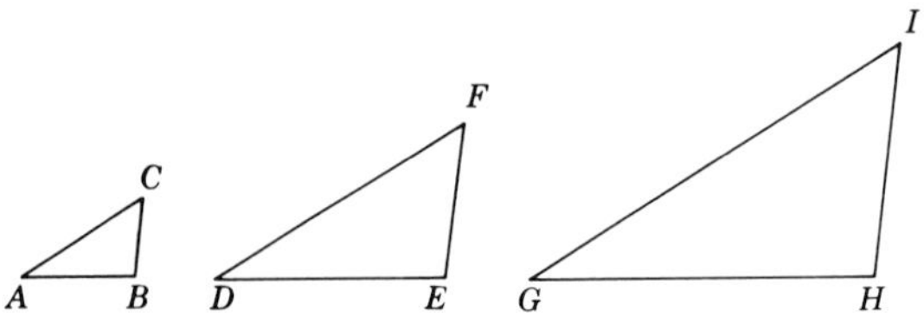

FIG. 23–13 Similar triangles.

been so constructed that their corresponding angles are equal. Therefore, the three triangles are similar, and their corresponding sides are proportional. This leads to the proportions

$$\frac{AB}{AC} = \frac{DE}{DF} = \frac{GH}{GI}$$
$$\frac{BC}{AB} = \frac{EF}{DE} = \frac{HI}{GH} \quad \text{etc.}$$

As an example, if $AB = 0.5$ cm, $DE = 1$ cm, and $GH = 1.5$ cm, then DF is twice as long as AC, and GI is three times as long as AC. Similarly, HI is three times as long as BC, and EF is twice as long as BC.

The properties of similar triangles are used extensively in measuring distance, such as the distances across bodies of water or other obstructions and the heights of various objects. In addition, the relationship between similar triangles forms the very basis of trigonometry.

Since the sum of the three angles of any triangle is 180°, it follows that if two angles of a triangle are equal to two angles of another triangle, the third angle of one must also be equal to the third angle of the other. Therefore, two triangles are similar if two angles of one are equal to two angles of the other.

If the numerical values of the necessary parts of a triangle are known, the triangle can be drawn to scale by using compasses, protractor, and ruler. The completed figure can then be measured with protractor and ruler to obtain the numerical values of the unknown parts. This is conveniently accomplished on squared paper.

PROBLEMS 23–4

1. The sides of a triangular plot are 60, 80, and 100 cm. The shortest side of a scale triangle is 3 cm. How long are the other two sides of the smaller triangle?
2. Two triangles are similar. The sides of the first are 18, 30, and 36 in. The longest side of the second is 20 mm. How long are the other two sides of the second triangle?

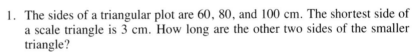

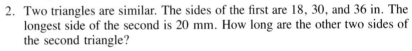

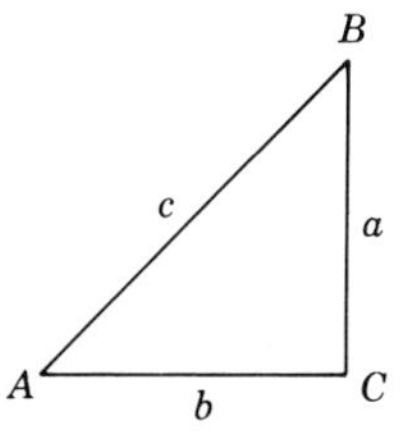

FIG. 23–14 Triangle for Probs. 3 to 10.

NOTE *In the following problems the sides and angles of all triangles will be as represented in Fig. 23–14. That is, the angles will be represented by the capital letters* A, B, *and* C *and the sides opposite the angles will be correspondingly lettered* a, b, *and* c.

Solve the following triangles by graphical methods:

3. $b = 4$, $A = 36.9°$, $C = 90°$
4. $a = 15$, $b = 20$, $c = 25$
5. $b = 12$, $A = 60°$, $C = 80°$
6. $b = 8$, $c = 5.50$, $A = 80°$
7. $a = 12$, $B = 120°$, $C = 20°$
8. $a = 4.95$, $c = 7$, $B = 45°$
9. $a = 15.4$, $b = 20$, $C = 29.3°$
10. $a = 35$, $c = 35$, $A = 60°$

23–8 THE RIGHT TRIANGLE

If one of the angles of a triangle is a right angle, the triangle is called a *right triangle*. Then, since the sum of the angles of any triangle is 180°, a right triangle contains one right angle and two acute angles. Also, the sum of the acute angles must be 90°. This relation enables us to find one acute angle when the other is given. For example, in the right triangle shown in Fig. 23–15, if $\theta = 30°$, then $\phi = 60°$.

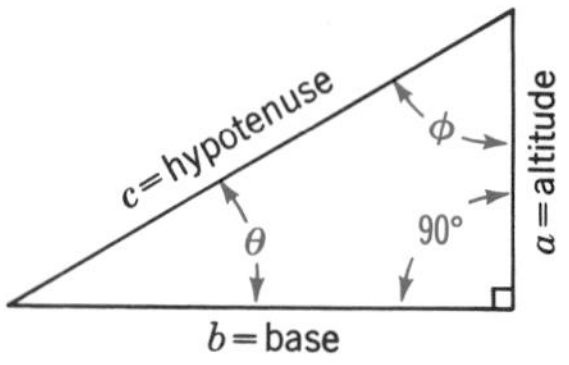

FIG. 23–15 Right triangle.

Since all right angles are equal, if an acute angle of one right triangle is equal to an acute angle of another right triangle, the two triangles are similar.

The side of a right triangle opposite the right angle is called the *hypotenuse*. Thus, in Fig. 23–15, the side c is the hypotenuse. When a right triangle is in standard position as in Fig. 23–15, the side a is called the *altitude* and the side b is called the *base*.

Another very important property of a right triangle is that the square of the hypotenuse is equal to the sum of the squares of the other two sides. That is,

$$c^2 = a^2 + b^2$$

This relationship provides a means of computing any one of the three sides if two sides are given.

EXAMPLE 3 A chimney is 40 m high. What is the length of its shadow at a time when a vertical post 2 m high casts a shadow that is 2.1 m long?

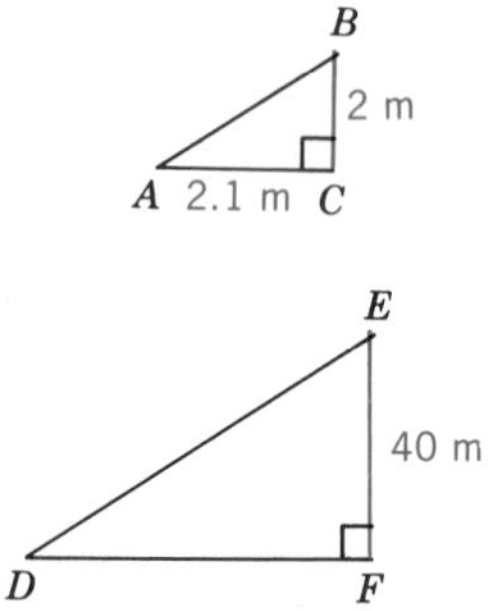

FIG. 23–16 Similar right triangles of Example 3.

SOLUTION

BC in Fig. 23–16 represents the post, and EF represents the stack. Because the rays of the sun strike both chimney and post at the same angle, right triangles ABC and DEF are similar. Then, since

$$\frac{DF}{AC} = \frac{EF}{BC}$$

by substituting,

$$\frac{DF}{2.1} = \frac{40}{2}$$

or

$$DF = 42 \text{ m}$$

EXAMPLE 4 What is the length of a in the triangle of Fig. 23–17?

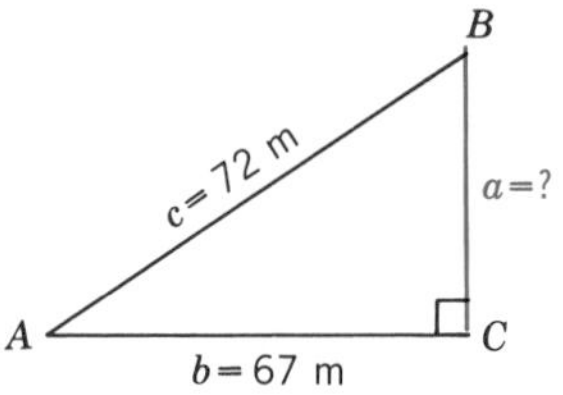

FIG. 23–17 Right triangle of Example 4.

SOLUTION

Given $\quad c^2 = a^2 + b^2$

Transposing, $\quad a^2 = c^2 - b^2$

$\sqrt{\ }$, $\quad a = \sqrt{c^2 - b^2}$

Substituting, $\quad a = \sqrt{72^2 - 67^2} = \sqrt{695}$

$a = 26.4$ m

PROBLEMS 23–5

In the following right triangles, solve for the indicated elements:

1. $a = 48$, $b = 12$, $A = 76°$. Find c and B.
2. $a = 36$, $b = 48$, $A = 53.1°$. Find c and B.
3. $b = 110$, $c = 117$, $B = 70°$. Find a and A.

4. An instrument plane flies north at the rate of 650 kn, and a hurricane hunter flies east at 1100 kn. If both planes start from the same place at the same time, how far apart will they be in 2 h?

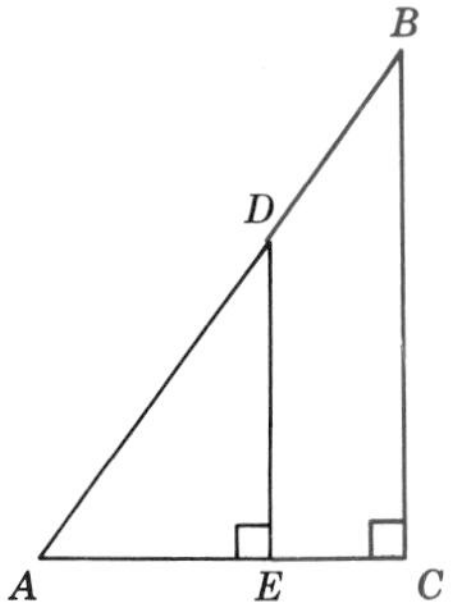

FIG. 23–18 Similar right triangles of Probs. 5, 6, and 7.

5. In Fig. 23–18, if $AC = 20$ m, $BC = 48$ m, and $AE = 15$ m, find the length of DE.
6. In Fig. 23–18, if $AD = 30$ cm, $DB = 20$ cm, and $BC = 40$ cm, what is the length of DE?
7. In Fig. 23–18, $AE = 32$ m, $EC = 18$ m, and $AB = 130$ m. What is the length of DE?
8. The top of an antenna tower is 40 m above the ground. The tower is to be guyed at a point 6 m below its top to a point on the ground 18 m from the base of the tower. What is the length of the guy?
9. A transmitter antenna tower casts a shadow 192 m long at a time when a meterstick held upright with one end touching the ground casts a shadow 1.6 m long. What is the height of the tower?
10. The tower in Prob. 9 is to be guyed from its top with a 230-m guy wire. How far out from the base of the tower may the guy be anchored?

SELF TEST

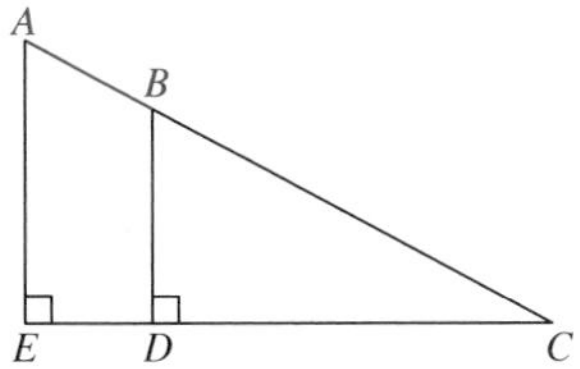

FIG. 23–19 Similar triangles of self test Prob. 4.

1. Through how many degrees does a timing mark on a motor armature turn in 2 μs if its rotational speed is 3600 rev/min?
2. What is the angular velocity in radians per second of the armature of Prob. 1?
3. An angle is measured as 120^g. What is this measurement in degrees?
4. In Fig. 23–19, $BD = 6$ mm, $DC = 8$ mm, and $DE = 4$ mm. How long is AB?
5. A telephone pole stands 15 m above the ground. Its guy wire, fastened 2.5 m below its top, is anchored 8 m horizontally from the base of the pole. Allowing 1.5 m for connections, how long a cable should be cut for the guy?

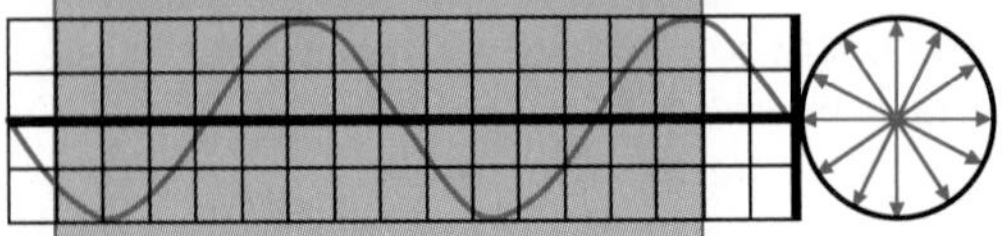

CHAPTER 24

TRIGONOMETRIC FUNCTIONS

What is the numerical value identified as sine 30°? Anyone with a calculator in *degree mode* can key 30. *Sin* may be identified as a direct key function, or as a "second function", or by color as a *g* or *h* or some other function. Call for the sin and read 0.500. But what is the relationship between 30° and the number 0.500? If you are to be able to instruct your calculator correctly and to understand its readouts perfectly, it is not enough to just punch keys on a calculator. Make sure you understand clearly the basic material covered in this chapter.

24–1 TRIGONOMETRIC FUNCTIONS ARE RATIOS

In Sec. 23–7 we saw that triangles may be similar regardless of their respective sizes. For example, in Fig. 24–1, the two triangles *ABC* and *DEF* are similar, and

$$\frac{AB}{AC} = \frac{DE}{DF} \qquad \frac{BC}{AC} = \frac{EF}{DF} \qquad \text{etc.}$$

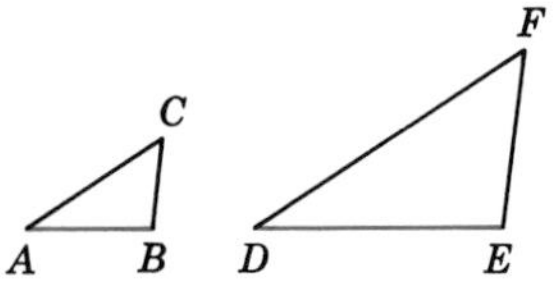

FIG. 24–1 Similar triangles.

Even if one of the pair of similar triangles is tilted (Fig. 24–2), the ratios still hold, since the triangles themselves have not changed in any of their dimensions. We may, however, have to look a little harder to see that this is so.

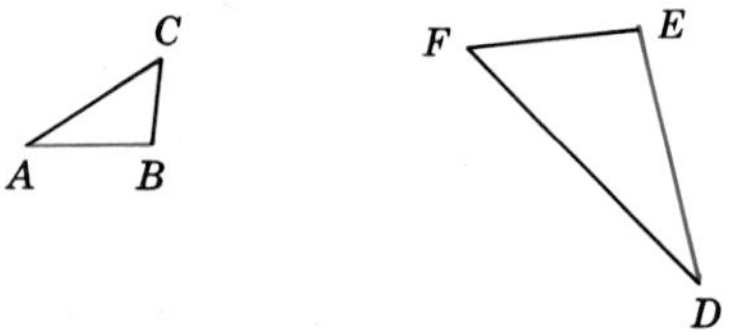

FIG. 24–2 Similar triangles of Fig. 24–1, except triangle *DEF* has been rotated.

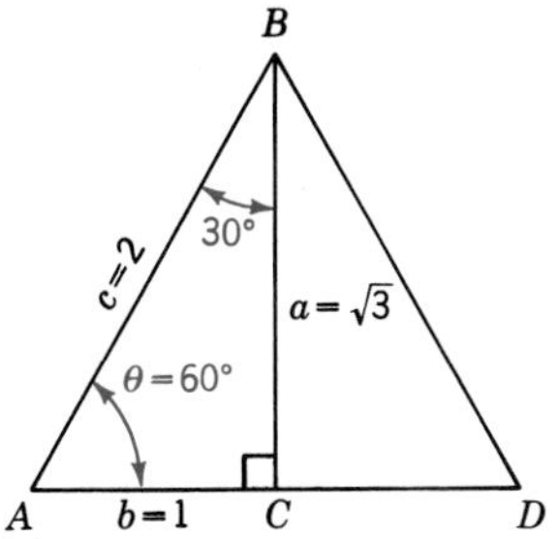

FIG. 24–3 Equilateral triangle divided into two equal 30°–60°–90° right triangles.

Consider the 30°–60°–90° triangle developed by bisecting an equilateral triangle (Fig. 24–3). First of all, you should confirm that, if the hypotenuse is 2 units long, then the base *AC* will be 1 unit long and the altitude *CB* will be $\sqrt{3}$ units long. Then consider the truth of the following statement:

In the 30°–60°–90° triangle, regardless of its size, the ratio of the base to the hypotenuse will always be 0.5000.

You should draw several 30°–60°–90° triangles of different sizes and prove to your complete satisfaction that this statement *must* always be true.

If the triangle were now rotated so that the side *CB* were the base and *AC* the altitude, the above statement would have to be adjusted. Therefore, we should rename the parts of the triangle so that there can be no possibility of misunderstanding a statement about it. The most convenient way to refer to a side of a triangle is to relate the side to the angles in the triangle. For instance, the hypotenuse is always the longest side, it is always opposite the right angle, and it is always adjacent to (forms) each of the other two angles. We can always refer to it as simply the hypotenuse without introducing any possibility of being misunderstood.

In the 30°–60°–90° triangle with which we are dealing, the side *AC* is always the side *opposite* the 30° angle, and it is always the side *adjacent* to the 60° angle, regardless of the letter designation given it or the orientation of the triangle.

Similarly, the side *CB* is always opposite the 60° angle, and it is always adjacent to the 30° angle, regardless of the symbols used to identify the side or how the triangle is tilted. These side-angle relationships are illustrated in Fig. 24–4. They must be memorized, because they will be used continuously henceforth.

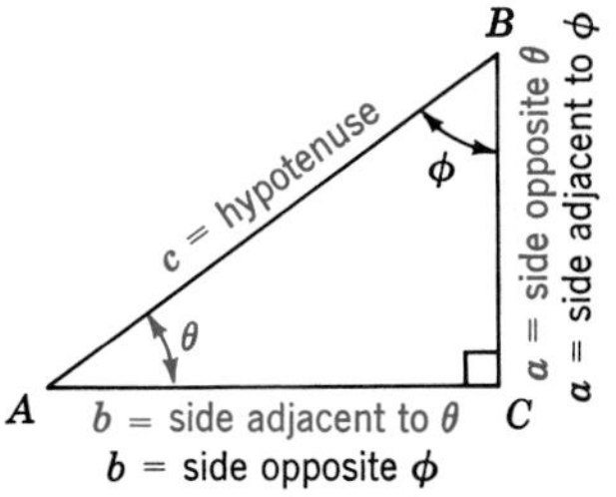

FIG. 24–4 Side-angle relationships in the standard right triangle.

For the rest of this chapter and the next, we shall be dealing only with right triangles. The hypotenuse is always the longest side and is opposite the right angle. The other two sides will be designated according to their relationships to the acute angles.

You should immediately confirm, while using sketches as required, the truth of the following statements relating to the sides of the 30°–60°–90° triangle, first as they apply to the 30° angle and then as they apply to the 60° angle:

1. In the 30°–60°–90° triangle, regardless of its size or orientation, the ratio of the side opposite the 30° angle to the hypotenuse will always be 0.5000.
2. In the 30°–60°–90° triangle, regardless of its size or orientation, the ratio of the side adjacent to the 30° angle to the hypotenuse will always be 0.866.
3. In the 30°–60°–90° triangle, regardless of its size or orientation, the ratio of the side opposite the 30° angle to the side adjacent to the 30° angle will always be 0.577.
4. In the 30°–60°–90° triangle, regardless of its size or orientation, the ratio of the side opposite the 60° angle to the hypotenuse will always be 0.866.
5. In the 30°–60°–90° triangle, regardless of its size or orientation, the ratio of the side adjacent to the 60° angle to the hypotenuse will always be 0.5000.
6. In the 30°–60°–90° triangle, regardless of its size or orientation, the ratio of the side opposite the 60° angle to the side adjacent to the 60° angle will always be 1.732.

It is left as an exercise for you to develop the three similar statements for the 45°–45°–90° triangle. (Why only three statements?)

Now, student, stop and look at these statements. See what they really mean. Make sure that their message is plain. When you fully understand the import of the relationships between sides of triangles, you will have trigonometry in the palm of your hand forever. We do not say that all of trigonometry is simple. But to grasp quickly the fact that the trigonometric functions are merely ratios of sides of triangles is to resolve most of the difficulties which stand in the way of students who have never properly understood how simple the functions of trigonometry really are.

The word "trigonometry" just means "measurement of triangles," and one of the most useful tools in the measurement of triangles is the ratios of sides.

"In the triangle, regardless of its size or orientation" means that, so long as the angles made by the sides are specified, the triangle itself may be formed by:

1. Three lines on a piece of paper
2. A ladder, the ground, and the wall of a house
3. An antenna mast, its shadow on the ground, and the line of sight from the end of the shadow to the top of the mast
4. The lines of sight between two surveyors and a distant landmark
5. A mast, a guy wire, and the ground between the foot of the mast and the guy anchor
6. Any other system which uses three straight lines to form three enclosed angles

The entire statement, "In the . . . triangle . . . will always be . . ." is quite a mouthful, far too lengthy for convenience, and it is often abbreviated. For instance, statement 1 on p. 425 becomes

$$\frac{\text{opp } 30°}{\text{hyp}} = 0.500$$

or

$$\frac{\text{opp}}{\text{hyp}} 30° = 0.500$$

and all the other parts of the statement are understood to apply. Statement 2 on p. 425 becomes

$$\frac{\text{adj}}{\text{hyp}} 30° = 0.866$$

and statement 3 on that page becomes

$$\frac{\text{opp}}{\text{adj}} 30° = 0.577$$

Using statements on p. 425, you should now write similar abbreviations for 4, 5, and 6 and check your work for the 45°–45°–90° triangle to show your own statements 7, 8, and 9 may be written

$$\frac{\text{opp}}{\text{hyp}} 45° = 0.7071$$

$$\frac{\text{adj}}{\text{hyp}} 45° = 0.7071$$

$$\frac{\text{opp}}{\text{adj}} 45° = 1.000$$

EXAMPLE 1 A triangular piece of farm land is to be used as an "antenna farm." It is in the shape of a 30°–60°–90° triangle, the shortest side of which is 600 m long (Fig. 24–5). What are the dimensions of the other two sides?

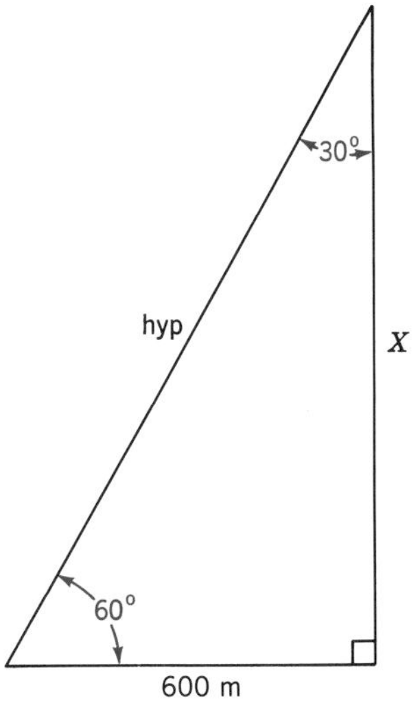

FIG. 24–5 Triangle of Example 1.

SOLUTION

By using the ratios which have been discovered above and drawing a sketch of the triangle to show the relationships between the sides and angles, we find that the 600-m side must be adjacent to the 60° angle. Then we have

$$\frac{600}{\text{hyp}} 60° = 0.500$$

from which $\text{hyp} = \frac{600}{0.5} = 1200 \text{ m}$

and $\frac{600}{\text{adj}} 30° = 0.577$

from which $\text{adj } 30° = \frac{600}{0.577} = 1040 \text{ m}$

Even these abbreviations are more than we require for everyday use, and we now introduce the proper *trigonometric names* for the different ratios (*functions*). θ is the "general angle," just as x is the "general number."

1. The ratio $\frac{\text{opp}}{\text{hyp}}$ θ is properly called sine θ, abbreviated to sin θ.
2. The ratio $\frac{\text{adj}}{\text{hyp}}$ θ is properly called cosine θ, abbreviated to cos θ.
3. The ratio $\frac{\text{opp}}{\text{adj}}$ θ is properly called tangent θ, abbreviated to tan θ.

It must be clearly understood that the names sine, cosine, and tangent are meaningless in themselves; you must relate them to angles of triangles. To say simply "cosine" means nothing. But "cos 60°" means, very specifically, the ratio of the side adjacent to the 60° angle of a 30°–60°–90° triangle to the hypotenuse of the same triangle.

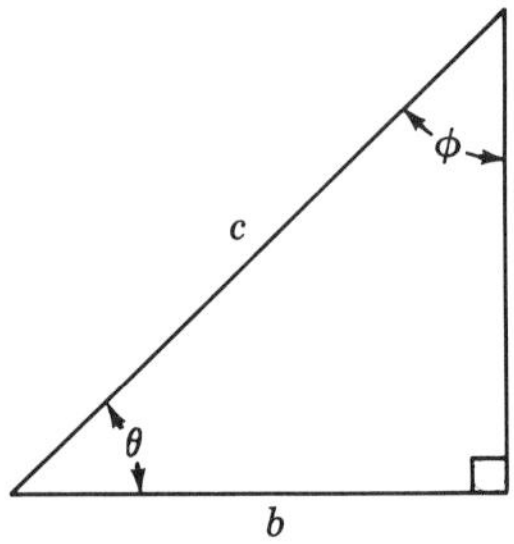

FIG. 24–6 Standard right triangle, as used in electronics problems.

In the general triangle, Fig. 24–6, it will be seen that there exist *six* possible trigonometric functions. Three of them we have already discovered, and the others are reciprocals of those three.

$$\frac{\text{opp}}{\text{hyp}}\theta = \sin\theta = \frac{a}{c}$$

$$\frac{\text{adj}}{\text{hyp}}\theta = \cos\theta = \frac{b}{c}$$

$$\frac{\text{opp}}{\text{adj}}\theta = \tan\theta = \frac{a}{b}$$

$$\frac{\text{hyp}}{\text{opp}}\theta = \text{cosecant}\ \theta = \csc\theta = \frac{c}{a}$$

$$\frac{\text{hyp}}{\text{adj}}\theta = \text{secant}\ \theta = \sec\theta = \frac{c}{b}$$

$$\frac{\text{adj}}{\text{opp}}\theta = \text{cotangent}\ \theta = \cot\theta = \frac{b}{a}$$

The cosecant, secant, and cotangent should always be thought of as the reciprocals of the sine, cosine, and tangent, respectively. This is shown easily by considering the reciprocal of sin θ:

$$\frac{1}{\sin\theta} = \frac{1}{\frac{a}{c}} = \csc\theta$$

You should confirm the other two reciprocal functions.

These definitions should be memorized so thoroughly that you can tell instantly any ratio of either acute angle of a right triangle, regardless of its position.

The sine, cosine, and tangent are the ratios most frequently used in practical work. If they are carefully learned, the others are easily remembered because they are reciprocals. In fact almost any mathematical aid will only provide sin, cos, and tan. You can always adjust any right triangle so that you can solve it using only these three functions. If you ever need one of the reciprocal functions, just use your $\frac{1}{x}$ key.

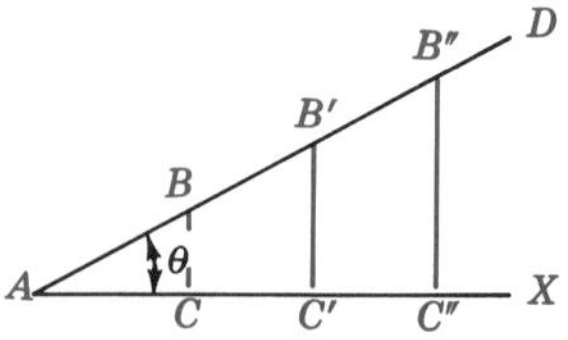

FIG. 24–7 The values of the functions depend only on the size of the angle.

The fact that the numerical value of any one of the trigonometric functions (ratios) depends only upon the magnitude of the angle θ is of fundamental importance. This is established by considering Fig. 24–7. There, the angle θ is generated by the line *AD* revolving about the point *A*. From the points *B*, *B′*, and *B″*, perpendiculars are let fall to the initial line, or adjacent side, *AX*. These form similar triangles *ABC*, *AB′C′*, and *AB″C″* because all are right triangles that have a common acute angle θ (Sec. 23–8). Hence,

$$\frac{BC}{AB} = \frac{B'C'}{AB'} = \frac{B''C''}{AB''}$$

Each of these ratios defines the sine of θ. Similarly, it can be shown that this property is true for each of the other functions. Therefore, the size of the right triangle is immaterial, for only the *relative* lengths of the sides are of importance.

Each one of the six ratios will change in value whenever the angle changes in magnitude. Thus, it is evident that the ratios are really functions of the angle under consideration. If the angle is considered to be the independent variable, then the six functions (ratios) and the relative lengths of the sides of the triangles are dependent variables.

EXAMPLE 2 Calculate the functions of the angle θ in the right triangle shown in Fig. 24–6 if $a = 6$ mm and $c = 10$ mm.

SOLUTION

Since $c^2 = a^2 + b^2$,

then
$$b = \sqrt{c^2 - a^2} = \sqrt{100 - 36}$$
$$= \sqrt{64} = 8 \text{ mm}$$

Applying the definitions of the six functions,

$$\sin\theta = \frac{6}{10} = \frac{3}{5} \quad \cos\theta = \frac{8}{10} = \frac{4}{5}$$
$$\tan\theta = \frac{6}{8} = \frac{3}{4} \quad \cot\theta = \frac{8}{6} = \frac{4}{3}$$
$$\sec\theta = \frac{10}{8} = \frac{5}{4} \quad \csc\theta = \frac{10}{6} = \frac{5}{3}$$

What would be the values of the above functions if $a = 6$ m, $b = 8$ m, and $c = 10$ m?

24–2 FUNCTIONS OF COMPLEMENTARY ANGLES

By applying the definitions of the six functions to the angle ϕ given in Fig. 24–8 and noting the positions of the adjacent and opposite sides for this angle, we obtain

$$\sin\phi = \frac{\text{opp}}{\text{hyp}} = \frac{b}{c} \qquad \csc\phi = \frac{\text{hyp}}{\text{opp}} = \frac{c}{b}$$
$$\cos\phi = \frac{\text{adj}}{\text{hyp}} = \frac{a}{c} \qquad \sec\phi = \frac{\text{hyp}}{\text{adj}} = \frac{c}{a}$$
$$\tan\phi = \frac{\text{opp}}{\text{adj}} = \frac{b}{a} \qquad \cot\phi = \frac{\text{adj}}{\text{opp}} = \frac{a}{b}$$

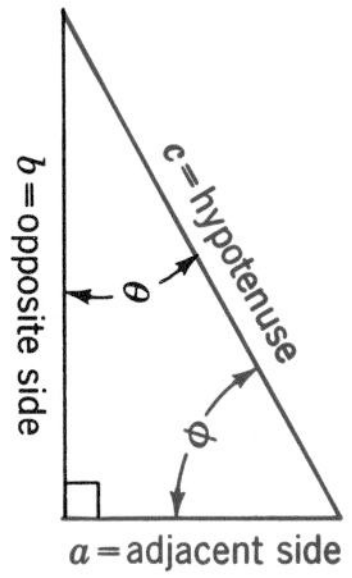

FIG. 24–8 Right triangle for determining functions of angle ϕ.

Upon comparing these with the original definitions given for the triangle of Fig. 24–2, we find the following relations:

$$\sin\phi = \cos\theta \qquad \cos\phi = \sin\theta$$
$$\tan\phi = \cot\theta \qquad \cot\phi = \tan\theta$$
$$\sec\phi = \csc\theta \qquad \csc\phi = \sec\theta$$

Since $\phi = 90° - \theta$, the above relations can be written

$$\begin{aligned} \sin(90° - \theta) &= \cos\theta & \cos(90° - \theta) &= \sin\theta \\ \tan(90° - \theta) &= \cot\theta & \cot(90° - \theta) &= \tan\theta \\ \sec(90° - \theta) &= \csc\theta & \csc(90° - \theta) &= \sec\theta \end{aligned}$$

The above can be stated in words as follows: A function of an acute angle is equal to the cofunction of its complementary angle. This enables us to find the function of every acute angle greater than 45° if we know the functions of all angles less than 45°. For example, $\sin 56° = \cos 34°$, $\tan 63° = \cot 27°$, $\cos 70° = \sin 20°$, etc.

24–3 CONSTRUCTION OF AN ANGLE WHEN ONE FUNCTION IS GIVEN

When the trigonometric function of an acute angle is given, the angle can be constructed geometrically by using the definition for the given function. Also, the magnitude of the resulting angle can be measured by the use of a protractor.

EXAMPLE 3 Construct the acute angle whose tangent is $\frac{9}{10}$.

SOLUTION

Erect perpendicular lines *AC* and *BC*, preferably on cross-sectional paper. Measure off 10 units along *AC* and 9 units along *BC*. Join *A* and *B* and thus form the right triangle *ABC*. Since $\tan A = \frac{9}{10}$, *A* is an angle of approximately 42°. See the construction in Fig. 24–9.

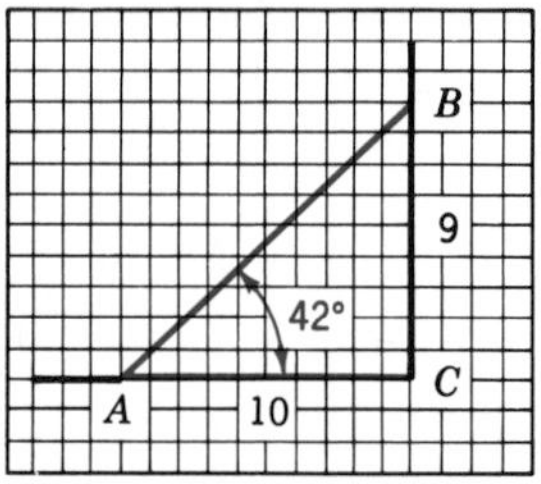

FIG. 24–9 Construction of acute angle whose tangent is $\frac{9}{10}$.

EXAMPLE 4 Find by construction the acute angle whose cosine is $\frac{3}{4}$.

SOLUTION

Erect perpendicular lines *AC* and *BC*. Measure off three units along *AC*. (Let three divisions of the cross-sectional paper be equal to one unit for greater accuracy.) With *A* as a center and with a radius of 4 units, draw an arc to intersect the perpendicular at B. Connect A and B. $\cos A = \frac{3}{4}$; therefore A is the required angle. Measuring A with a protractor shows it to be an angle of approximately 41.4°. The construction is shown in Fig. 24–10.

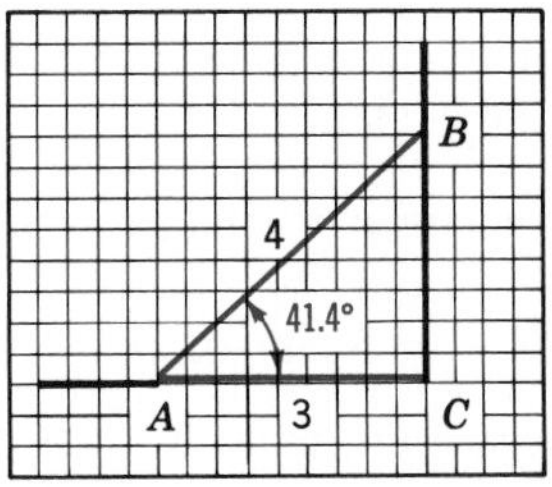

FIG. 24–10 Construction of acute angle whose cosine is $\frac{3}{4}$.

PROBLEMS 24–1

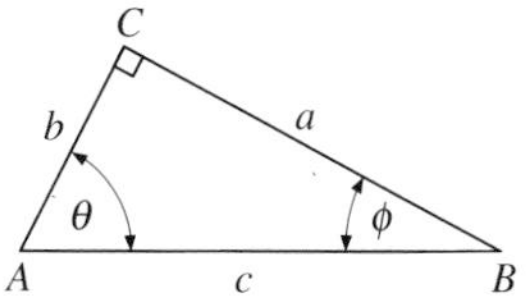

FIG. 24–11 Right triangle of Prob. 1.

1. In Fig. 24–11, what are the values of the trigonometric functions for the angles θ and ϕ in terms of ratios of the sides, a, b, and c?
2. As shown in Fig. 24–12, (*a*) $\sin \alpha = ?$ (*b*) $\sin \beta = ?$ (*c*) $\cot \beta = ?$ (*d*) $\sec \alpha = ?$ (*e*) $\tan \alpha = ?$
3. In Fig. 24–12, (*a*) $\frac{OP}{OR} = \tan ?$ (*b*) $\frac{PR}{PO} = \sec ?$ (*c*) $\frac{OR}{PR} = \cos ?$ (*d*) $\frac{OP}{RP} = \sin ?$ (*e*) $\frac{PR}{RO} = \csc ?$
4. The three sides of a right triangle are 5, 12, and 13. Let α be the acute angle opposite the side 5 and let β be the other acute angle. Write the six functions of α and β.
5. In Fig. 24–13, if $X = R$, find the six functions of θ.
6. In Fig. 24–13, if $R = \frac{1}{2}Z$, find the sine, cosine, and tangent of θ.

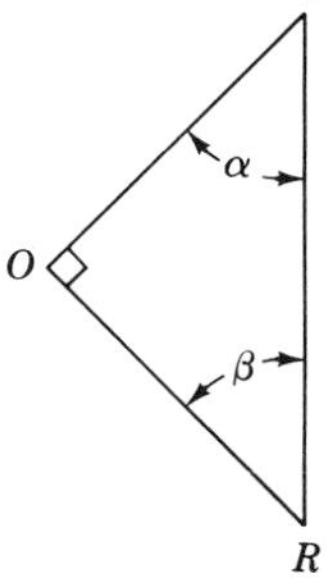

FIG. 24–12 Right triangle of Probs. 2 and 3.

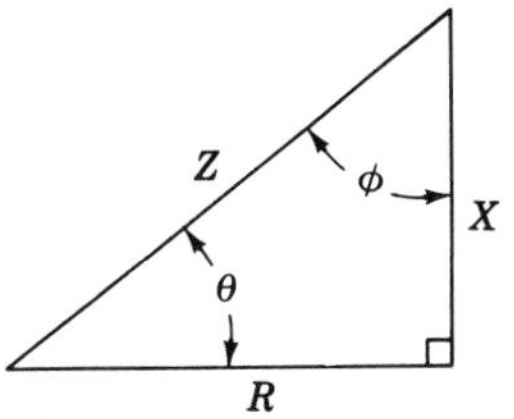

FIG. 24–13 Right triangle of Probs. 5, 6, and 7.

7. In Fig. 24–13, if $X = 2R$, find the sine, cosine, and tangent of θ.
8. (*a*) $\sin\theta = \frac{2}{3}$, $\csc\theta = ?$ (*b*) $\sec\alpha = 2$, $\cos\alpha = ?$
   (*c*) $\cot\beta = \frac{7}{8}$, $\tan\beta = ?$ (*d*) $\cos\phi = \frac{5}{16}$, $\sec\phi = ?$
   (*e*) $\tan\phi = 12$, $\cot\phi = ?$ (*f*) $\csc\alpha = 4$, $\sin\alpha = ?$
9. The three sides of a right triangle are 5, 12, and 13. Write the six functions of the largest acute angle.
10. Write the other functions of an acute angle whose cosine is $\frac{4}{5}$.
11. In a right triangle, $c = 5$ cm and $\cos A = \frac{4}{5}$. Construct the triangle, and write the functions of the angle B.
12. State which of these is greater if $\theta \neq 0°$ and is less than 90°: (*a*) $\sin\theta$ or $\tan\theta$, (*b*) $\cos\theta$ or $\cot\theta$, (*c*) $\sec\theta$ or $\tan\theta$, (*d*) $\csc\theta$ or $\cot\theta$.

24–4 FUNCTIONS OF ANY ANGLE

The notion of trigonometric functions has been introduced from the point of view of right triangles because this allows for an easy introduction which most students can follow with assurance. However, the total concept applies to far more than just right triangles and to far more than angles between 0° and 90°. In Chap. 27 we shall investigate a few interesting and useful relationships in nonright triangles. For the moment, we will concentrate on the trigonometric functions of any angle.

In Chap. 23 we found the concepts of angles were extended to include angles in any quadrant and both positive and negative angles. In Fig. 24–14 the line $\boldsymbol{r}$ revolves about the origin of the rectangular coordinate system in a counterclockwise (positive) direction. This line, which generates the angle θ, is known as the *radius vector*. The initial side of θ is the positive x axis, and the terminal side is the radius vector. If a perpendicular is let fall from any point P along the radius vector, in any of the quadrants, a right triangle $xy\boldsymbol{r}$ will be

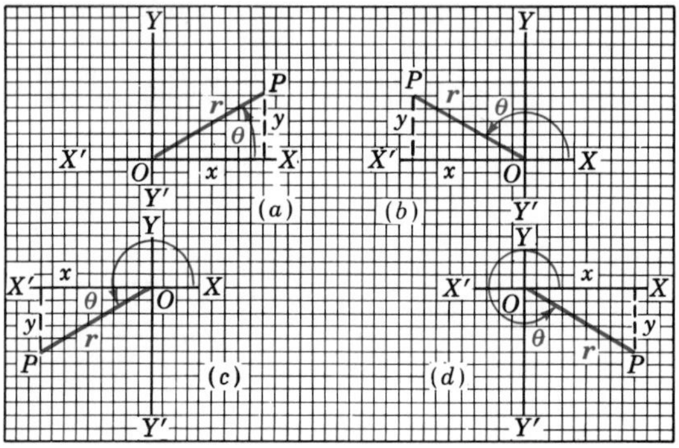

FIG. 24–14 Radius vector $\boldsymbol{r}$ generating angles.

formed with $\boldsymbol{r}$ as a hypotenuse of constant unit length and with x and y having lengths equal to the respective coordinates of P.

We then define the trigonometric functions of θ as follows:

$$\sin\theta = \frac{y}{\boldsymbol{r}} = \frac{\text{ordinate}}{\text{radius}} \qquad \csc\theta = \frac{\boldsymbol{r}}{y} = \frac{\text{radius}}{\text{ordinate}}$$

$$\cos\theta = \frac{x}{\boldsymbol{r}} = \frac{\text{abscissa}}{\text{radius}} \qquad \sec\theta = \frac{\boldsymbol{r}}{x} = \frac{\text{radius}}{\text{abscissa}}$$

$$\tan\theta = \frac{y}{x} = \frac{\text{ordinate}}{\text{abscissa}} \qquad \cot\theta = \frac{x}{y} = \frac{\text{abscissa}}{\text{ordinate}}$$

Since the values of the six trigonometric functions are entirely independent of the position of the point P along the radius vector, it follows that they depend only upon the position of the radius vector, or the size of the angle. Therefore, for every angle there is one, and only one, value of each function.

24–5 SIGNS OF THE FUNCTIONS

The signs of the functions of angles in various quadrants are very important. If you remember the signs of the abscissas (x values) and the ordinates (y values) in the four quadrants, you will encounter no trouble.

For angles in the first quadrant, as shown in Fig. 24–14*a*, the x and y values are positive. Since the length of the radius vector $\boldsymbol{r}$ is always considered positive, it is evident that all functions of angles in the first quadrant are positive. For angles in the second quadrant, as in Fig. 24–14*b*, the x values are negative values and the y values are positive. Therefore, the sine and its reciprocal are positive and the other four functions are negative. Similarly, the signs of all the functions can be checked from their definitions as given in the preceding section. You should verify each part of Table 24–1.

TABLE 24–1

Quadrant	sin θ	cos θ	tan θ	cot θ	sec θ	csc θ
I	+	+	+	+	+	+
II	+	−	−	−	−	+
III	−	−	+	+	−	−
IV	−	+	−	−	+	−

If the proper signs for the sine and cosine are fixed in mind, the other signs will be remembered because of an important relationship.

$$\frac{\sin\theta}{\cos\theta} = \frac{\dfrac{y}{\boldsymbol{r}}}{\dfrac{x}{\boldsymbol{r}}} = \frac{y}{x}$$

Since

$$\tan\theta = \frac{y}{x}$$

then

$$\frac{\sin\theta}{\cos\theta} = \tan\theta$$

If the sine and the cosine have like signs, the tangent is positive, and if they have unlike signs, the tangent is negative. Because the signs of the sine, cosine, and tangent always agree with signs of the respective reciprocals, the cosecant, secant, and cotangent, the signs for the latter are obtainable from the signs of the sine and cosine as outlined above. Figure 24–15 will serve as an aid in remembering the signs.

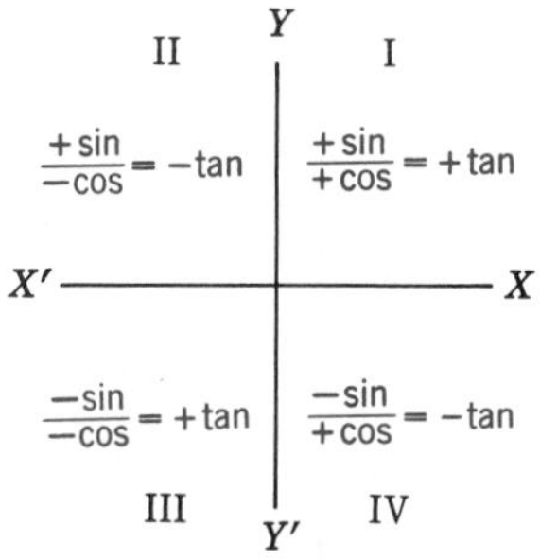

FIG. 24–15 Signs of functions in quadrants.

PROBLEMS 24–2

In what quadrant or quadrants is θ for each of the following conditions?

1. sin θ is positive.
2. cos θ is positive.
3. sin θ is negative.
4. tan θ is negative.
5. cos θ is negative.
6. sin θ is positive, cos θ negative.
7. tan θ and sin θ both positive.
8. cot θ negative, cos θ negative.
9. tan θ negative, cos θ positive.
10. All functions of θ are positive.
11. tan θ = 6
12. cos θ = $-\frac{3}{4}$
13. Is there an angle whose cosine is negative and whose secant is positive?
14. When tan θ = $\frac{3}{4}$, find the value of

$$\frac{\sin\theta - \csc\theta}{\cot\theta - \sec\theta}$$

Give the signs of the sine, cosine, and tangent of each of the following angles:

15. 49°
16. 210°
17. 120°
18. 350°
19. −135°
20. $\frac{\pi^r}{3}$
21. $\frac{-2\pi^r}{3}$
22. −72°
23. 800°

Find the value of the radius vector ***r*** for each of the following positions of *P*, and then find the trigonometric functions of the angle θ (∠*XOP*). Keep answers in fractional form.

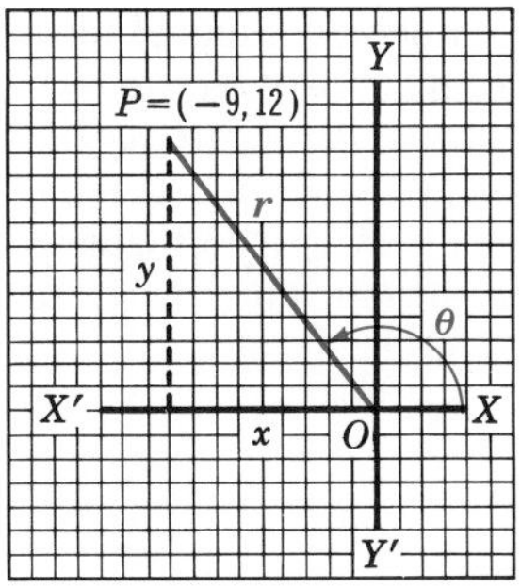

FIG. 24–16 Diagram of Prob. 24.

24. (−9, 12)

Solution: Draw the radius vector $\boldsymbol{r}$ from O to $P = (-9, 12)$ as shown in Fig. 24–16. Hence, θ is an angle in the second quadrant with a side adjacent that has an x value of −9 and a side opposite that has a y value of 12.

Then

$$\boldsymbol{r} = \sqrt{x^2 + y^2} = \sqrt{(-9)^2 + (12)^2} = 15$$

Hence, by definition,

$$\sin\theta = \frac{y}{\boldsymbol{r}} = \frac{12}{15} = \frac{4}{5} \qquad \csc\theta = \frac{\boldsymbol{r}}{y} = \frac{15}{12} = \frac{5}{4}$$

$$\cos\theta = \frac{x}{\boldsymbol{r}} = \frac{-9}{15} = -\frac{3}{5} \qquad \sec\theta = \frac{\boldsymbol{r}}{x} = \frac{15}{-9} = -\frac{5}{3}$$

$$\tan\theta = \frac{y}{x} = \frac{12}{-9} = -\frac{4}{3} \qquad \cot\theta = \frac{x}{y} = \frac{-9}{12} = -\frac{3}{4}$$

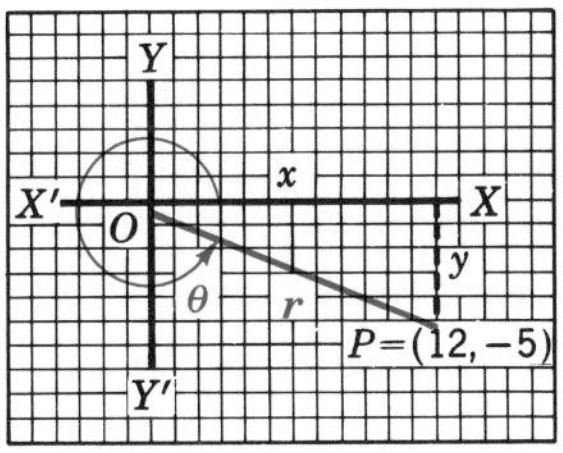

FIG. 24–17 Diagram of Prob. 25.

25. (12, −5)

Solution: Draw the radius vector $\boldsymbol{r}$ from O to P as in Fig. 24–17. θ is an angle in the fourth quadrant with a side adjacent that has an x value of 12 and a side opposite that has a y value of −5.
Then

$$\boldsymbol{r} = \sqrt{x^2 + y^2} = \sqrt{12^2 + (-5)^2} = 13$$

Hence, by definition,

$$\sin\theta = \frac{y}{\boldsymbol{r}} = -\frac{5}{13} \qquad \csc\theta = \frac{\boldsymbol{r}}{y} = -\frac{13}{5}$$

$$\cos\theta = \frac{x}{\boldsymbol{r}} = \frac{12}{13} \qquad \sec\theta = \frac{\boldsymbol{r}}{x} = \frac{13}{12}$$

$$\tan\theta = \frac{y}{x} = -\frac{5}{12} \qquad \cot\theta = \frac{x}{y} = -\frac{12}{5}$$

26. (3, 4)
27. (12, 5)
28. (−3, 4)
29. (−4, −5)
30. (3, 3)
31. (4, −3)
32. (−8, 6)
33. (−5, −3)
34. (8, 8)

24–6 COMPUTATION OF THE FUNCTIONS

In Sec. 24–1 we developed the functions of 30°, 45°, and 60° by merely using simple notions about right triangles. These angles are very important and will be used often, so they and their trigonometric functions are worthy of the time you spent in this development. At the same time, their use will make it easy for some students to quickly relearn trigonometry a few years hence if their work has been such that they haven't required it immediately. In Chap. 25 we will extend our notions of trigonometric functions and investigate the limits of most hand-held calculators when dealing with angles greater than 90°.

24–7 FUNCTIONS OF 0°

For an angle of 0°, both the initial and terminal sides are on OX. At any distance a from O, choose the point P as shown in Fig. 24–18. Then the coordinates of P are $(a, 0)$. That is, the x value is equal to a units, and the y value is zero. Since the radius vector $\boldsymbol{r}$ is equal to a, by definition,

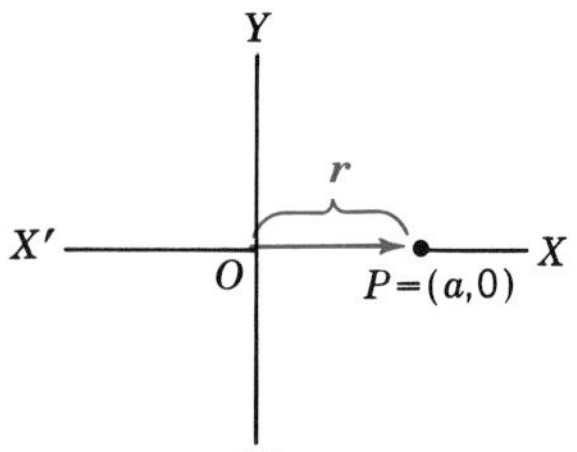

FIG. 24–18 $\theta = 0°, x = a$, and $y = 0$.

$$\sin 0° = \frac{y}{\boldsymbol{r}} = \frac{0}{\boldsymbol{r}} = 0 \qquad \csc 0° = \frac{\boldsymbol{r}}{y} = \frac{a}{0} = \infty$$

$$\cos 0° = \frac{x}{\boldsymbol{r}} = \frac{a}{a} = 1 \qquad \sec 0° = \frac{\boldsymbol{r}}{x} = \frac{a}{a} = 1$$

$$\tan 0° = \frac{y}{x} = \frac{0}{a} = 0 \qquad \cot 0° = \frac{x}{y} = \frac{a}{0} = \infty$$

By $\frac{a}{0} = \infty$ is meant the value of $\frac{a}{y}$ as y approaches zero without limit. Thus, as y gets nearer and nearer to zero, $\frac{a}{y}$ gets larger and larger. Therefore, $\frac{a}{y}$ is said to *approach* infinity as y approaches zero. However, $\frac{a}{0}$ does not actually result in a quotient of infinity, for division by zero is meaningless.

Determining the functions of 90°, 180°, and 270° is accomplished by the same method as that used for 0°. This is left as an exercise for you.

24–8 THE RANGES OF THE FUNCTIONS

As the radius vector $\boldsymbol{r}$ starts from OX and revolves about the origin in a positive (counterclockwise) direction, the angle θ is generated and varies in magnitude continuously from 0° to 360° through the four quadrants. Figure 24–19 illustrates the manner in which the sine, cosine, and tangent vary as the angle θ changes in value.

Quadrant I. As θ increases from 0° to 90°,

- x is positive and decreases from $\boldsymbol{r}$ to 0.
- y is positive and increases from 0 to $\boldsymbol{r}$.

Therefore,

$\sin\theta = \frac{y}{\boldsymbol{r}}$ is *positive* and increases from 0 to 1.

$\cos\theta = \frac{x}{\boldsymbol{r}}$ is *positive* and decreases from 1 to 0.

$\tan\theta = \frac{y}{x}$ is *positive* and increases from 0 to ∞.

Quadrant II. As θ increases from 90° to 180°,

- x is negative and increases from 0 to $-\boldsymbol{r}$.
- y is positive and decreases from $\boldsymbol{r}$ to 0.

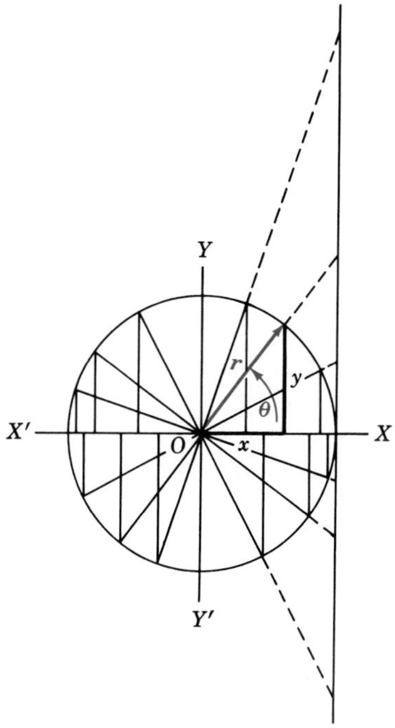

FIG. 24–19 Lengths of lines showing the ranges of sin θ, cos θ, and tan θ.

Therefore,

$\sin\theta = \dfrac{y}{\boldsymbol{r}}$ is *positive* and decreases from 1 to 0.

$\cos\theta = \dfrac{x}{\boldsymbol{r}}$ is *negative* and increases from 0 to -1.

$\tan\theta = \dfrac{y}{x}$ is *negative* and decreases from $-\infty$ to 0.

Quadrant III. As θ increases from 180° to 270°,

- x is negative and decreases from $-\boldsymbol{r}$ to 0.
- y is negative and increases from 0 to $-\boldsymbol{r}$.

Therefore,

$\sin\theta = \dfrac{y}{\boldsymbol{r}}$ is *negative* and increases from 0 to -1.

$\cos\theta = \dfrac{x}{\boldsymbol{r}}$ is *negative* and decreases from -1 to 0.

$\tan\theta = \dfrac{y}{x}$ is *positive* and increases from 0 to ∞.

Quadrant IV. As θ increases from 270° to 360°,

- x is positive and increases from 0 to $\boldsymbol{r}$.
- y is negative and decreases from $-\boldsymbol{r}$ to 0.

Therefore,

$\sin\theta = \frac{y}{\boldsymbol{r}}$ is *negative* and decreases from -1 to 0.

$\cos\theta = \frac{x}{\boldsymbol{r}}$ is *positive* and increases from 0 to 1.

$\tan\theta = \frac{y}{x}$ is *negative* and decreases from $-\infty$ to 0.

Students often become confused in comparing the variations of the functions, when represented as lines, with the actual numerical values of the functions. For example, in quadrant II as the angle θ increases from 90 to 180°, we say that $\cos\theta$ increases from 0 to $-\boldsymbol{r}$. Actually, the abscissa representing the cosine is getting *longer;* confusion results from not remembering that a negative number is always greater than zero in the defined negative direction. The *lengths* of the lines representing the functions, when compared with the radius vector, indicate only the *magnitude* of the function. The positions of the lines, with respect to the x or y axis, specify the signs of the functions.

24–9 LINE REPRESENTATION OF THE FUNCTIONS

By representing the functions as lengths of lines, we are able to obtain a mental picture of the manner in which the functions vary as the radius vector $\boldsymbol{r}$ revolves and generates angles. Since we are primarily concerned with the sine, cosine, and tangent, only these functions will be represented graphically.

In Fig. 24–20 the radius vector $\boldsymbol{r}$, with a length of one unit, is revolving about the origin and generating the angle θ. Then, in each of the four quadrants,

$$\sin\theta = \frac{BC}{\boldsymbol{r}} = \frac{BC}{1} = BC$$

and

$$\cos\theta = \frac{OC}{\boldsymbol{r}} = \frac{OC}{1} = OC$$

It is evident *that the sine of an angle can be represented by the ordinate (y value) of any point where the end of the radius vector coincides with the circumference of the circle*. Hence, the length BC represents $\sin\theta$ in all quadrants, as shown in Fig. 24–20. Note that the ordinate gives both the sign and the magnitude of the sine in any quadrant. Thus, in quadrants I and II, $\sin\theta = +0.6$; in quadrants III and IV, $\sin\theta = -0.6$. That is, when the radius vector is above the x axis, the ordinate and therefore the sine are positive. When the radius vector is below the x axis, the ordinate and therefore the sine are negative.

Similarly, *the cosine of an angle can be represented by the abscissa (x value) of any point where the end of the radius vector coincides with the circumference of the circle*. Hence, the length OC represents $\cos\theta$ in all quadrants, as shown in Fig. 24–20. The abscissa gives both the sign and the magnitude of the cosine in any quadrant. Thus, in quadrants I and IV,

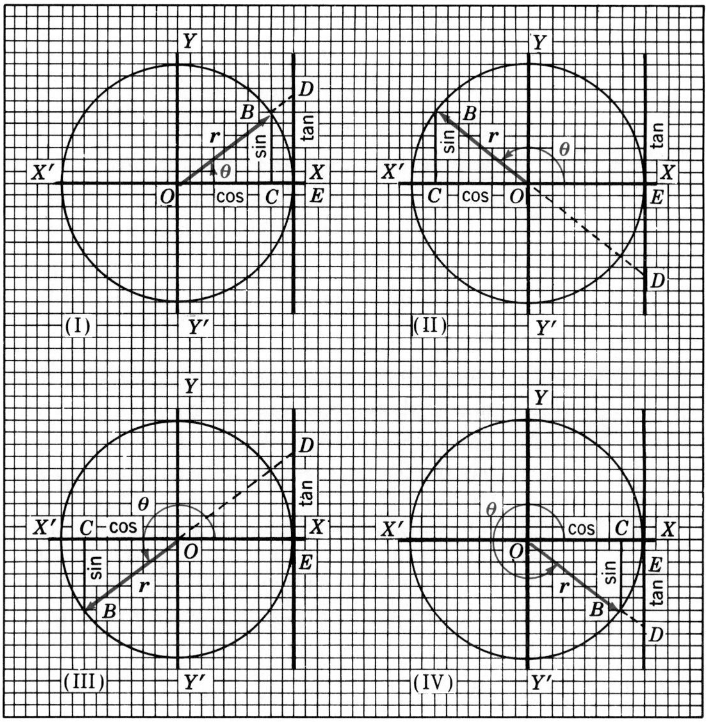

FIG. 24–20 Line representation of functions.

$\cos \theta = +0.8$; in quadrants II and III, $\cos \theta = -0.8$. That is, when the radius vector is to the right of the y axis, the abscissa and therefore the cosine are positive. When the radius vector is to the left of the y axis, the abscissa and therefore the cosine are negative.

In Fig. 24–20, the radius vector has been extended to intersect the line *DE*, which has been drawn tangent to the circle at the positive x axis. Since by construction, *DE* is perpendicular to *OX*, *OBC* and *ODE* are similar right triangles, for they have a common acute angle *BOC*. From the similar triangles,

$$\frac{BC}{OC} = \frac{DE}{OE}$$

Then, in each of the four quadrants,

$$\tan \theta = \frac{BC}{OC} = \frac{DE}{OE} = \frac{DE}{1} = DE$$

From the above, it is evident that *the tangent of an angle can be represented by the ordinate (y value) of any point where the extended radius vector intersects the tangent line*. The ordinate gives both the sign and the magnitude of the tangent in any quadrant. Thus, in quadrants I and III, $\tan \theta = +0.75$; in quadrants II and IV, $\tan \theta = -0.75$.

PROBLEMS 24–3

1. What is the least value sin θ may have?
2. What is the least value cos θ may have?
3. What is the greatest value csc θ may have in the first quadrant?
4. What is the greatest value sec θ may have in the fourth quadrant?
5. Can the secant and cosecant have values between −1 and +1?
6. What is the greatest value sin θ may have in the (*a*) first quadrant, (*b*) second quadrant, (*c*) third quadrant, and (*d*) fourth quadrant?
7. What is the greatest value cos θ may have in the (*a*) first quadrant, (*b*) second quadrant, (*c*) third quadrant, and (*d*) fourth quadrant?

SELF TEST

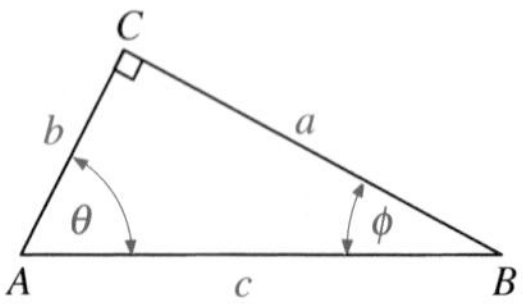

FIG. 24–21 Triangle for self test Probs. 1, 2, and 3.

1. In Fig. 24–21, what is the sine of angle ϕ?
2. In Fig. 24–21, what is the cosine of the angle at *A*?
3. In Fig. 24–21, which angle has a tangent $\frac{a}{b}$?
4. In which quadrant is an angle α, if sin α and tan α are both positive?
5. What is the sign of sin 310°?
6. What is the sign of cos 122°?
7. What is the sign of tan 168°?
8. If $\sin \lambda = \frac{2}{3}$, what is cos λ?
9. If $\cos \alpha = \frac{7}{16}$, what is tan α?
10. What is the maximum negative value that cos θ may have?

CHAPTER 25

TRIGONOMETRIC VALUES

There was a time when it was necessary for people to possess printed *tables* of trigonometric functions. Inside the front cover of this book is a three-place table that will prove useful in emergencies. Notice that this table covers angles between 0° and 90°. You must apply your knowledge of the ranges of functions and their algebraic signs in order to use it correctly.

25–1 GIVEN AN ANGLE—TO FIND THE DESIRED FUNCTION

How to use the calculator for natural functions is best illustrated by examples. *Tables* are available in a number of printed forms, in addition to being programmed into a variety of ''scientific'' calculators. You should check with your instructors to determine whether or not printed tables are available for your use in the event of any emergency. It is left as an exercise for you to become acquainted with printed tables. Trigonometric values in this book are taken from a scientific hand-held calculator rounded off to five decimal places. You should carefully check the instruction manual for your own calculator as you follow the examples below. At the same time, refer to the convenient three-place tables inside the front covers, and compare their values with those on your calculator.

EXAMPLE 1 Find the sine of 36.7°.

SOLUTION

Key 36.7
Key SIN
Read 0.597 63

$$\sin 36.7° = 0.597\ 63$$

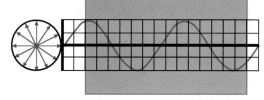

EXAMPLE 2 Find the cosine of 7.9°.

SOLUTION

Key 7.9
Key COS
Read 0.990 51

$$\cos 7.9° = 0.990\ 51$$

EXAMPLE 3 Find the tangent of 79.1°.

SOLUTION

Key 79.1
Key TAN
Read 5.192 93

$$\tan 79.1° = 5.192\ 93$$

EXAMPLE 4 Find the sine of 26.42°.

SOLUTION

Key 26.42
Key SIN
Read 0.444 95

$$\sin 26.42° = 0.444\ 95$$

EXAMPLE 5 Find the cosine of 53.77°.

SOLUTION

$$\cos 53.77° = 0.591\ 03$$

EXAMPLE 6 Find the tangent of 48.13°.

SOLUTION

$$\tan 48.13° = 1.115\ 69$$

PROBLEMS 25–1

1. Find the sine, cosine, and tangent of (*a*) 3.5°, (*b*) 4.8°, (*c*) 24°, (*d*) 52°, (*e*) 77°.
2. Find the sine, cosine, and tangent of (*a*) 12°, (*b*) 88.7°, (*c*) 70.2°, (*d*) 0.8°, (*e*) 20.1°.
3. Find the sine, cosine, and tangent of (*a*) 1.94°, (*b*) 57.36°, (*c*) 38.91°, (*d*) 40.28°, (*e*) 55.37°.
4. Find the sine, cosine, and tangent of (*a*) 7.39°, (*b*) 12.18°, (*c*) 32.65°, (*d*) 41.55°, (*e*) 3.17°.

25–2 INVERSE TRIGONOMETRIC FUNCTIONS

Frequently some form of notation is needed in order to express an angle in terms of one of its functions. For example, in Sec. 24–3 Example 3 dealt with an angle whose tangent was $\frac{9}{10}$. Similarly, in Example 4 of the same section, we considered an angle whose cosine was $\frac{3}{4}$.

> If your computer is capable of directly delivering arcsin θ, its command is probably ASN.
> Most computers are programmed to deliver only arctangent, and usually use ATAN or ATN as commands for arctangent.

If $\sin \theta = x$, then θ is an angle whose sine is x. It has been agreed to express such a relation by the notation

$$\theta = \sin^{-1} x \qquad \text{or} \qquad \theta = \arcsin x$$

Both are read "θ is equal to the angle whose sine is x" or "the inverse sine of x." For example, the tangent of 32.7° is 0.759 04. Stated as an inverse function, this would be written

$$37.2° = \arctan 0.759\ 04$$

Similarly, in the case of a right triangle labeled as in Fig. 24–8, we should write $\theta = \arctan \frac{a}{b}$, $\theta = \arccos \frac{b}{c}$, etc. In this book, we shall not use the notation "$\theta = \sin^{-1} x$" (although it appears on some calculators), for we prefer not to use an exponent when no exponent is intended. Although this form of notation is used in a number of texts, you will find that nearly all recent mathematics and engineering texts are using the "$\theta = \arcsin x$" form of notation. Because more advanced mathematics employs trigonometric functions affected by exponents, it is evident that confusion would eventually result from utilizing the other notation for specifying the inverse functions.

25–3 GIVEN A FUNCTION—TO FIND THE CORRESPONDING ANGLE

As in Sec. 25–1, the use of the calculator is best illustrated by examples. The results in degrees are rounded off to two decimal places.

EXAMPLE 7 Find the angle whose sine is 0.235 14.

SOLUTION

Key 0.235 14
Key INV SIN
Read 13.60

$$\arcsin 0.235\ 14 = 13.6°$$

EXAMPLE 8 Find θ if cos θ = 0.033 16.

SOLUTION

Key 0.033 16
Key INV COS
Read 88.09 97

arccos 0.033 16 = 88.10°

EXAMPLE 9 Find θ if θ = arctan 1.142 29.

SOLUTION

Key 1.142 29
Key INV TAN
Read 48.79

arctan = 1.142 29 = 48.79°

EXAMPLE 10 Find θ if θ = arcsin 0.445 26.

SOLUTION

arcsin 0.445 26 = 26.44°

EXAMPLE 11 Find θ if cos θ = 0.373 15.

SOLUTION

θ = 68.09°

EXAMPLE 12 Find θ if θ = arctan 0.591 87.

SOLUTION

θ = 30.62°

25–4 ACCURACY

In our considerations of ac circuits, we shall confine our accuracy of component values to three significant figures and angles to the nearest tenth of a degree. This, except for isolated cases, will more than meet all practical requirements.

Inside the front cover of this book is a three-place table of sines, cosines, and tangents for each degree from 0° to 90°. With the confidence gained from working with the components that form all but the most precise circuits, you will find that this table will serve many of your needs.

You should study the tables at this point, and compare them with your calculator to satisfy yourself that, for angles up to about 6°, the values of sin θ and tan θ are within 0.55% of each other and, at 10°, the difference is only 1.52%.

Compare your own calculator with other brands. You will find many different readings in the seventh and eighth decimal places, but identical values to five places.

PROBLEMS 25–2

1. Find the angles having the following values as sines:
(*a*) 0.453 99, (*b*) 0.116 67, (*c*) 0.878 82, (*d*) 0.644 12, (*e*) 0.037 34.
2. Find the angles having the following values as cosines:
(*a*) 0.965 93, (*b*) 0.190 81, (*c*) 0.998 72, (*d*) 0.866 90, (*e*) 0.343 17.
3. Find the angles whose tangents are (*a*) 0.048 85, (*b*) 0.225 36,
(*c*) 0.568 08, (*d*) 2.525 71, (*e*) 7.806 22.
4. Find θ if:
(*a*) θ = arctan 1.356 37
(*b*) θ = arccos 0.486 34
(*c*) θ = arcsin 0.273 96
(*d*) θ = arccos 0.048 85
(*e*) θ = arcsin 0.518 03
5. Find θ if:
(*a*) θ = arccos 0.973 74
(*b*) θ = arctan 0.009 25
(*c*) θ = arcsin 0.963 06
(*d*) θ = arctan 0.893 15
(*e*) θ = arcsin 0.732 66

25–5 FUNCTIONS OF ANGLES GREATER THAN 90°

You will note that the trigonometric functions inside the front cover have been tabulated only for angles of 0° to 90°. The signs and magnitudes for angles in all quadrants were considered in the preceding chapter, and it is evident that any table of functions must be combined with a method of expressing any angle in terms of an angle of the first quadrant in order to make use of the table of functions. Similarly, all calculators provide the ability to deal with angles greater than 90°.

25–6 TO FIND THE FUNCTIONS OF AN ANGLE IN THE SECOND QUADRANT

In Fig. 25–1, let θ represent any angle in the second quadrant. From any point P on the radius vector $\boldsymbol{r}$, draw the perpendicular y to the horizontal axis. The acute angle that $\boldsymbol{r}$ makes with the horizontal axis is designated by ϕ. Consequently, because $\theta + \phi = 180°$, θ and ϕ are supplementary angles. Hence,

$$\phi = 180° - \theta$$

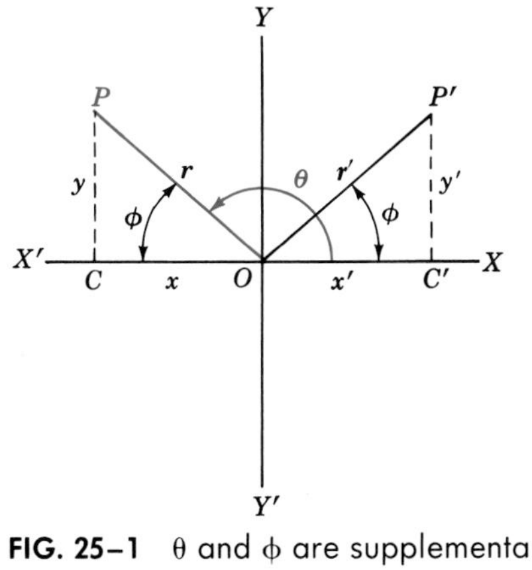

FIG. 25–1 θ and ϕ are supplementary angles; $\theta + \phi = 180°$.

Now construct the angle XOP' in the first quadrant equal to ϕ, make $\boldsymbol{r}'$ equal to $\boldsymbol{r}$, and draw y' perpendicular to OX. Since the right triangles OPC and $OP'C'$ are equal, $x = -x'$ and $y = y'$. Then

$$\sin(180° - \theta) = \frac{y}{r} = \frac{y'}{r'} = \sin\phi$$

$$\cos(180° - \theta) = \frac{x}{r} = \frac{-x'}{r'} = -\cos\phi$$

$$\tan(180° - \theta) = \frac{y}{x} = \frac{y'}{-x'} = -\tan\phi$$

These relationships show that, in all respects, the function of an angle has the same absolute value as the same function of its supplement. That is, if two angles are supplementary, their sines are equal in all respects and their cosines and tangents are equal in magnitude but opposite in sign.

EXAMPLE 13

$$\begin{aligned}\sin 140° &= \sin(180° - 140°)\\ &= \sin 40° = 0.642\ 79\end{aligned}$$

$$\begin{aligned}\cos 100° &= -\cos(180° - 100°)\\ &= -\cos 80° = -0.173\ 65\end{aligned}$$

$$\begin{aligned}\tan 175° &= -\tan(180° - 175°)\\ &= -\tan 5° = -0.087\ 49\end{aligned}$$

Note that your calculator gives the correct algebraic sign as well as the numerical value.

25–7 TO FIND THE FUNCTIONS OF AN ANGLE IN THE THIRD QUADRANT

In the triangles shown in Fig. 25–2, let θ represent any angle in the third quadrant and let ϕ be the acute angle that the radius vector $\boldsymbol{r}$ makes with the horizontal axis.

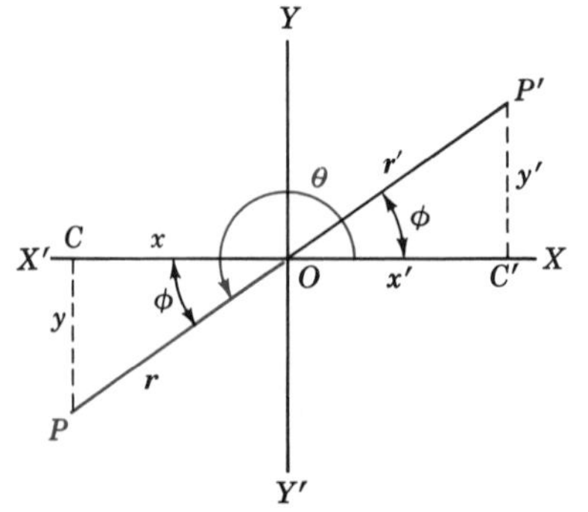

FIG. 25–2 θ is in the third quadrant; $\phi = \theta - 180°$.

Then

$$\phi = \theta - 180°$$

Now construct the angle XOP' in the first quadrant equal to ϕ, make $\mathbf{r}'$ equal to $\mathbf{r}$, and draw y and y' perpendicular to the horizontal axis. Since the right triangles OPC and $OP'C'$ are equal, $x = -x'$ and $y = -y'$. Then

$$\sin(\theta - 180°) = \frac{y}{\mathbf{r}} = \frac{-y'}{\mathbf{r}'} = -\sin\phi$$

$$\cos(\theta - 180°) = \frac{x}{\mathbf{r}} = \frac{-x'}{\mathbf{r}'} = -\cos\phi$$

$$\tan(\theta - 180°) = \frac{y}{x} = \frac{-y'}{-x'} = \tan\phi$$

These relationships show that the function of an angle in the third quadrant has the same absolute value as the same function of the acute angle between the radius vector and the horizontal axis. The signs of the functions are the same as for any angle in the third quadrant, as discussed in Sec. 24–5.

EXAMPLE 14

$$\begin{aligned}
\sin 200° &= -\sin(200° - 180°) \\
&= -\sin 20° = -0.342\ 02 \\
\cos 260° &= -\cos(260° - 180°) \\
&= -\cos 80° = -0.173\ 65 \\
\tan 234° &= \tan(234° - 180°) \\
&= \tan 54° = 1.376\ 38
\end{aligned}$$

25–8 TO FIND THE FUNCTIONS OF AN ANGLE IN THE FOURTH QUADRANT

In Fig. 25–3, let θ represent any angle in the fourth quadrant and let ϕ be the acute angle that the radius vector $\mathbf{r}$ makes with the horizontal axis. Then

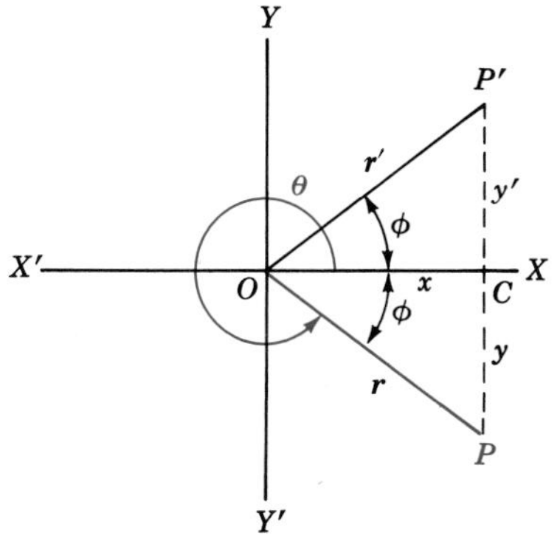

FIG. 25–3 θ is in the fourth quadrant; $\phi = 360° - \theta$.

Now construct the angle XOP' in the first quadrant equal to ϕ, make $\mathbf{r}'$ equal to $\mathbf{r}$, and draw y and y' perpendicular to the horizontal axis. Since the right triangles OPC and $OP'C$ are equal, $y = -y'$. Then

$$\sin(360° - \theta) = \frac{y}{r} = \frac{-y'}{r'} = -\sin\phi$$

$$\cos(360° - \theta) = \frac{x}{r} = \frac{x}{r'} = \cos\phi$$

$$\tan(360° - \theta) = \frac{y}{x} = \frac{-y'}{x} = -\tan\phi$$

These relationships show that the functions of an angle in the fourth quadrant have the same absolute value as the same functions of an acute angle in the first quadrant equal to $360° - \theta$. The signs of the functions, however, are those for an angle in the fourth quadrant, as discussed in Sec. 24–5.

EXAMPLE 15

$$\begin{aligned}\sin 300° &= -\sin(360° - 300°)\\ &= -\sin 60° = -0.866\ 03\\ \cos 285° &= \cos(360° - 285°)\\ &= \cos 75° = 0.258\ 82\\ \tan 316° &= -\tan(360° - 316°)\\ &= -\tan 44° = -0.965\ 69\end{aligned}$$

25–9 TO FIND THE FUNCTIONS OF AN ANGLE GREATER THAN 360°

Any angle θ greater than 360° has the same trigonometric functions as θ minus an integral multiple of 360°. That is, a function of an angle larger than 360° is found by dividing the angle by 360° and finding the required function of the remainder. Thus θ in Fig. 25–4 is a positive angle of 955°. To find any function of 955°, divide 955° by 360°, which gives 2 with a remainder of 235°. Hence,

$$\begin{aligned}\sin 955° &= \sin 235° = -\sin(235° - 180°)\\ &= -\sin 55° = -0.819\ 15\\ \cos 955° &= \cos 235° = -\cos(235° - 180°)\\ &= -\cos 55° = -0.573\ 58\\ \tan 955° &= \tan 235° = \tan(235° - 180°)\\ &= \tan 55° = 1.428\ 15\end{aligned}$$

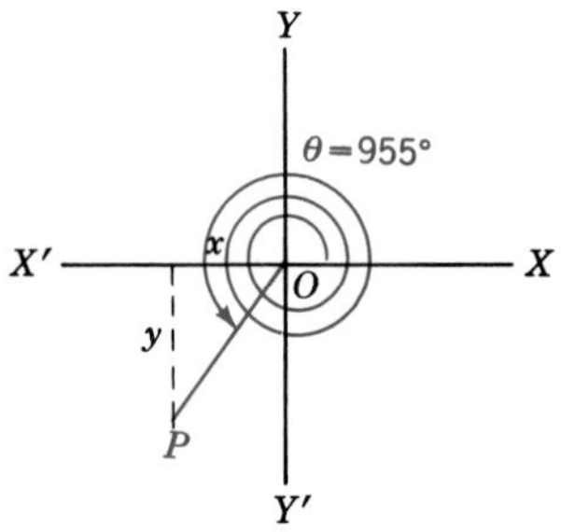

FIG. 25–4 $\theta = 955°$.

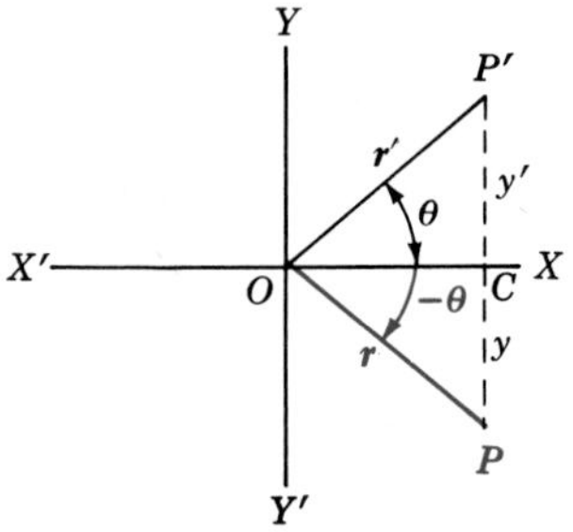

FIG. 25–5 $-\theta$ is generated by clockwise rotation.

When using a calculator it is not necessary to go through the above divisions and subtractions. Simply enter 955 and enter the required functions. The calculator will give the correct sign and result.

25–10 TO FIND THE FUNCTIONS OF A NEGATIVE ANGLE

In Fig. 25–5, let $-\theta$ represent a negative angle in the fourth quadrant made by the radius vector $\boldsymbol{r}$ and the horizontal axis. Construct the angle θ in the first quadrant equal to $-\theta$, make $\boldsymbol{r}'$ equal to $\boldsymbol{r}$, and draw y and y' perpendicular to the horizontal axis. Since the right triangles OPC and $OP'C$ are equal, $y = -y'$. Then

$$\sin(-\theta) = \frac{y}{\boldsymbol{r}} = \frac{-y'}{\boldsymbol{r}'} = -\sin\theta$$

$$\cos(-\theta) = \frac{x}{\boldsymbol{r}} = \frac{x}{\boldsymbol{r}'} = \cos\theta$$

$$\tan(-\theta) = \frac{y}{x} = \frac{-y'}{x'} = -\tan\theta$$

These relationships are true for any values of $-\theta$ regardless of the quadrant or the magnitude of the angle.

EXAMPLE 16

$$\begin{aligned}
\sin(-65°) &= -\sin 65° &&= -0.906\ 31\\
\cos(-150°) &= -\cos 150° &&= -\cos(180° - 150°)\\
&= -\cos 30° &&= -0.866\ 03\\
\tan(-287°) &= -\tan 287° &&= -\tan(360° - 287°)\\
&= -(-\tan 73°) &&= 3.270\ 85
\end{aligned}$$

Test your calculator, as in Sec. 25–10:
Key 65
Key +/−
Key SIN
Read −0.906 31

25–11 TO REDUCE THE FUNCTIONS OF ANY ANGLE TO THE FUNCTIONS OF AN ACUTE ANGLE

It has been shown in the preceding sections that all angles can be reduced to terms of $(180° - \theta)$, $(\theta - 180°)$, $(360° - \theta)$, or θ. These results can be summarized as follows:

RULE

To find any function of any angle θ, take the same function of the acute angle formed by the terminal side (radius vector) and the *horizontal* axis and prefix the proper algebraic sign for that quadrant.

When finding the functions of angles, you should make a sketch showing the approximate location of the angle. This procedure will clarify the trigonometric relationships, and many errors will be avoided by using it.

EXAMPLE 17 Find the functions of 143°.

SOLUTION

Construct the angle 143°, and mark the signs of the radius vector, abscissa, and ordinate, as shown in Fig. 25–6. (The radius vector is always positive.) Since 180° − 143° = 37°, the acute angle for the functions is 37°. Hence,

$$\begin{aligned} \sin 143^\circ &= \sin 37^\circ &&= 0.601\ 82 \\ \cos 143^\circ &= -\cos 37^\circ &&= -0.798\ 64 \\ \tan 143^\circ &= -\tan 37^\circ &&= -0.753\ 55 \end{aligned}$$

In this and the following examples, confirm that your calculator gives both the values and the signs of the functions.

EXAMPLE 18 Find the functions of 245°.

SOLUTION

Construct the angle 245° as shown in Fig. 25–7.
Since 245° − 180° = 65°, the acute angle for the functions is 65°. Hence,

$$\begin{aligned} \sin 245^\circ &= -\sin 65^\circ &&= -0.906\ 31 \\ \cos 245^\circ &= -\cos 65^\circ &&= -0.422\ 62 \\ \tan 245^\circ &= \tan 65^\circ &&= 2.144\ 51 \end{aligned}$$

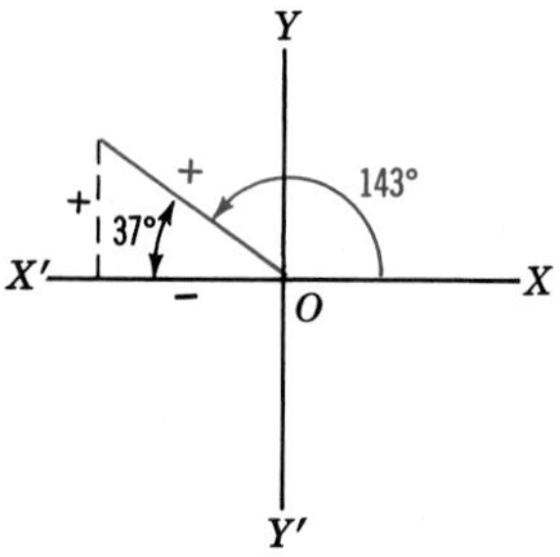

FIG. 25–6
180° − 143° = 37°.

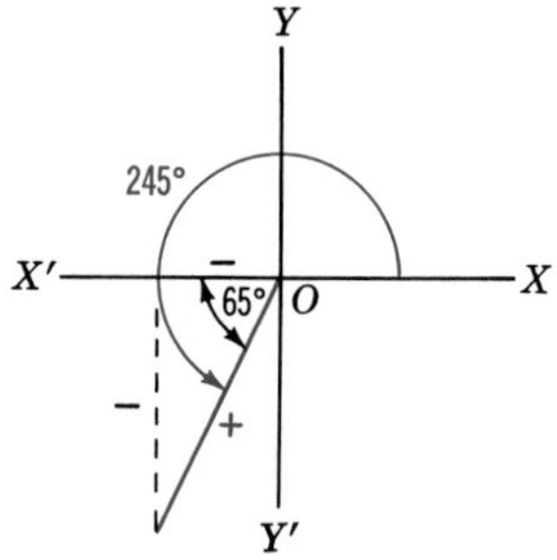

FIG. 25–7
245° − 180° = 65°

EXAMPLE 19 Find the functions of 312°.

SOLUTION

Construct the angle 312° as shown in Fig. 25–8.
Since 360° − 312° = 48°, the acute angle for the functions is 48°. Hence,

$$\begin{aligned} \sin 312^\circ &= -\sin 48^\circ &&= -0.743\ 14 \\ \cos 312^\circ &= \cos 48^\circ &&= 0.669\ 13 \\ \tan 312^\circ &= -\tan 48^\circ &&= -1.110\ 61 \end{aligned}$$

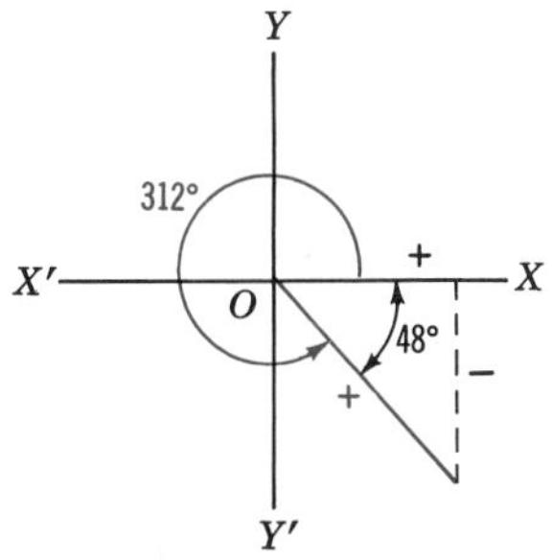

FIG. 25–8
360° – 312° = 48°

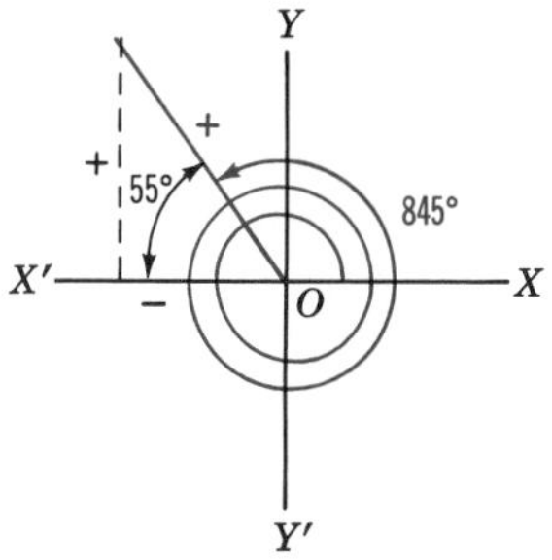

FIG. 25–9 Functions of 845° are the same as those of 125°.

EXAMPLE 20 Find the functions of 845°.

SOLUTION

$845° \div 360° = 2 + 125°$. Therefore, the functions of 125° will be identical with those of 845°. The construction is shown in Fig. 25–9.

Since $180° - 125° = 55°$, the acute angle for the functions is 55°. Hence,

$$\begin{aligned} \sin 845° &= \sin 55° &&= 0.819\ 15 \\ \cos 845° &= -\cos 55° &&= -0.573\ 58 \\ \tan 845° &= -\tan 55° &&= -1.428\ 15 \end{aligned}$$

EXAMPLE 21 Find the functions of −511°.

SOLUTION

$-511° \div 360° = -(1 + 151°)$. Therefore, the functions of −151° will be identical with those of −511°. The construction is shown in Fig. 25–10. Since $180° - 151° = 29°$, the acute angle for the functions is 29°. Hence,

$$\begin{aligned} \sin(-151°) &= -\sin 29° &&= -0.484\ 81 \\ \cos(-151°) &= -\cos 29° &&= -0.874\ 62 \\ \tan(-151°) &= \tan 29° &&= 0.554\ 31 \end{aligned}$$

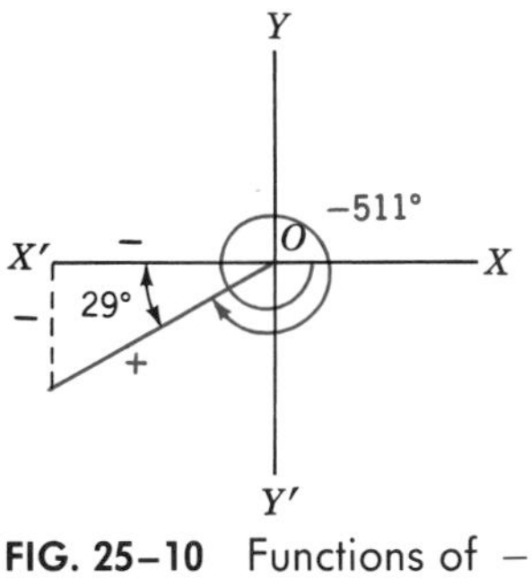

FIG. 25–10 Functions of −511° are the same as those of −151°.

25–12 ANGLES CORRESPONDING TO INVERSE FUNCTIONS

Now that we are able to express all angles as acute angles in order to use the table of functions from 0° to 90°, it has probably occurred to you that an important distinction exists between the direct trigonometric functions and the inverse trigonometric functions. Each trigonometric function of any given angle has only one value, whereas a given function corresponds to an infinite number of angles. For example, an angle of 30° has only one sine value, which is 0.500 00, but an angle whose sine is 0.500 00 (arcsin 0.500 00) may be taken as 30°, 150°, 390°, 510°, etc.

To avoid confusion, it has been agreed that the values of arcsin θ and arctan θ which lie between +90° and −90°, in the first and fourth quadrants, are to be known as the *principal values* of arcsin θ and arctan θ. The principal value is often denoted by using a capital letter, as Arcsin θ. Thus,

	Arcsin 0.575 01	=	35.1°	
			Key	0.575 01
			Key	INV SIN
			Read	35.1
and	Arcsin (−0.998 03)	=	−86.4°	
			Key	0.998 03
			Key	+/−
			Key	INV SIN
			Read	−86.4
Also	Arctan 1.482 56	=	56°	
and	Arctan (−0.069 93)	=	−4°	

The principal values of arccos θ are taken as the values between 0° and 180° and are denoted by Arccos θ. Thus,

Arccos 0.173 65 = 80°

and Arccos (−0.981 63) = 169°

Take special note of the principal angles because functions of angles greater than 90° may lead to errors of interpretation.

EXAMPLE 22

Key 120
Key SIN
Read 0.866 03
Key INV SIN
Read 60

The calculator does not "remember" that it was originally working from a second quadrant angle.

This is another good reason for sketching angles when solving problems—keep track of the quadrants so that the calculator readouts do not lead to errors.

EXAMPLE 23

Key 240
Key COS
Read −0.500 00
Key INV COS
Read 120

EXAMPLE 24

Key 300
Key COS
Read 0.500 00
Key INV COS
Read 60

EXAMPLE 25

Key 240
Key TAN
Read 1.732 05
Key INV TAN
Read 60

PROBLEMS 25–3

1. Find the sine, cosine, and tangent of (*a*) 107°, (*b*) 160°, (*c*) 130.1°, (*d*) 147.5°, (*e*) 176.2°.
2. Find the sine, cosine, and tangent of (*a*) 183°, (*b*) 235°, (*c*) 217.8°, (*d*) 180.9°, (*e*) 268.1°.
3. Find the sine, cosine, and tangent of (*a*) 280°, (*b*) 318°, (*c*) 349.9°, (*d*) 300.1°, (*e*) 359.5°.
4. Find the sine, cosine, and tangent of (*a*) 461°, (*b*) 510°, (*c*) 480.5°, (*d*) 523.2°, (*e*) 539.3°.
5. Find the sine, cosine, and tangent of (*a*) 905°, (*b*) −17.1°, (*c*) 940.7°, (*d*) −362.6°, (*e*) 1260.2°.
6. Find θ if:
   (*a*) $\theta = \text{Arccos } 0.969\ 02$
   (*b*) $\theta = \text{Arcsin } 0.582\ 12$
   (*c*) $\theta = \text{Arccos } (-0.455\ 55)$
   (*d*) $\theta = \text{Arctan } (-3.510\ 53)$
   (*e*) $\theta = \text{Arcsin } (-0.377\ 84)$
7. Find ϕ if:
   (*a*) $\phi = \text{Arctan } (-1.076\ 13)$
   (*b*) $\phi = \text{Arccos } (-0.027\ 92)$
   (*c*) $\phi = \text{Arcsin } 0.780\ 43$
   (*d*) $\phi = \text{Arccos } (-0.976\ 30)$
   (*e*) $\phi = \text{Arctan } (-2.732\ 63)$
8. The illumination on a surface that is not perpendicular to the rays of light from a light source is given by the formula

$$E = \frac{F \cos \theta}{d^2} \qquad \text{lux (lx)}$$

where E = illumination at a point on the surface, lx
F = intensity of light output of source, lumens (lm)
d = distance of source of light to surface, m
θ = angle between incident light ray and a line perpendicular to the surface

Solve for F, d, and θ.

9. In the formula of Prob. 8, find the value of d if $F = 1200$ lm, $\theta = 56°$, and $E = 350$ lx.
10. A 100-W lamp has a total light output of 1700 lm. Disregarding reflection, compute the illumination at a point on a surface 3 m from the lamp if the plane of the surface is at an angle of 30° to the incident rays.
11. In the formula of Prob. 8, at what angle of the plane of the surface to the incident ray will the illumination be the greatest?
12. The illumination on a horizontal surface from a source of light at a given vertical distance from the surface is given by the formula

$$E_h = \frac{F}{h^2}\cos^3\theta \text{ lx}$$

where E_h = illumination at a point on horizontal surface, lx
F = intensity of light output from source of light, lm
h = vertical distance from horizontal surface to source of light, m
θ = angle between incident ray and vertical line, as shown in Fig. 25–11

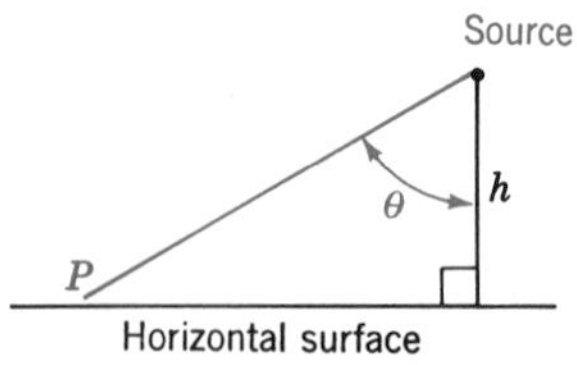

FIG. 25–11 Illumination at P from source.

NOTE *$\cos^3\theta$ means $(\cos\theta)$ raised to the third power.*

Solve for F, h, and θ.

13. Use the formula of Prob. 12 to solve for E_h if $F = 4750$ lm, $h = 2.1$ m, and $\theta = 22°$.
14. Use the formula of Prob. 12 to solve for F if $E_h = 330$ lx, $h = 4.5$ m, and $\theta = 50°$.
15. According to illumination experts, 1000 to 1500 lx of illumination on a printed page should be provided for study purposes. A 60-W, 850-lm lamp is suspended 1.5 m above a reading table. The reflector used projects 60% of the light downward. Does this produce a satisfactory amount of illumination on a book directly below the lamp?
16. To produce 1250 lx on the book in Prob. 15, what lumen-rating lamp should be used?
17. Snell's law states that, when a wave of electromagnetic energy passes from one dielectric material to another, the ratio of the sines of the angles of incidence θ_1 and refraction θ_2 is inversely proportional to the square root of the ratio of the dielectric relative permittivities (Fig. 25–12). That is,

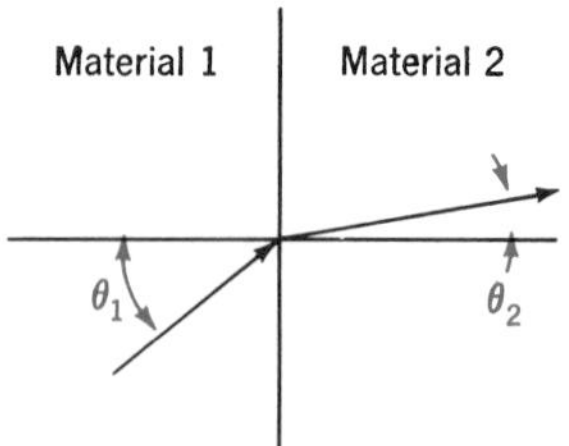

FIG. 25–12 Diagram for Prob. 17.

$$\frac{\sin\theta_1}{\sin\theta_2} = \sqrt{\frac{\varepsilon_2}{\varepsilon_1}}$$

If the angle of incidence $\theta_1 = 70°$, material 1 is lucite, $\varepsilon_1 = 2.6$, and material 2 is mica, $\varepsilon_2 = 5.4$, what is the angle of refraction θ_2?

SELF TEST

Your instructor will determine whether tables or calculators may be used. Trigonometric functions should be given to five significant figures and angles to two decimal places.

1. What is the numerical value of sin 37.2°?
2. What is the numerical value of cos 184.4°?
3. What is the principal angle whose tangent is $-1.158\ 51$?
4. What is θ if arcsin = 0.863 22?
5. What is cos 115.55°?
6. What is ϕ if $\tan \phi = -1.493\ 01$?
7. What is θ if $\sin \theta = 0.959\ 57$?
8. What is cos 532.88°?
9. What is Arccos $-0.743\ 73$?
10. What is arctan 0.334 01?

CHAPTER 26

SOLUTION OF RIGHT TRIANGLES

One of the most important applications of trigonometry is the solution of triangles, both right and oblique. This chapter is concerned with the former. The right triangle is probably the most universally used geometric figure; with the aid of trigonometry, it is applied to numerous problems in measurement that otherwise might be impossible to solve.

A large percentage of the problems relating to the analysis of ac circuits and networks involves the solution of the right triangle in one form or another.

26–1 FACTS CONCERNING RIGHT TRIANGLES

Before we proceed with the actual solutions of right triangles, we will review the following useful facts regarding the properties of the right triangle:

1. The square of the hypotenuse is equal to the sum of the squares of the other two sides ($c^2 = a^2 + b^2$).
2. The acute angles are complements of each other; that is, the sum of the two acute angles is 90° ($A + B = 90°$).
3. The hypotenuse is greater than either of the other two sides and is less than their sum.
4. The greater angle is opposite the greater side, and the greater side is opposite the greater angle.

These facts will often be a material aid in checking computations made by trigonometric methods.

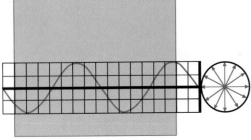

26–2 PROCEDURE FOR SOLUTION OF RIGHT TRIANGLES

Every triangle has three sides and three angles, and these are called the six *elements* of the triangle. To *solve* a triangle is to find the values of the unknown elements.

A triangle can be solved by two methods:

1. By constructing the triangle accurately from known elements with scale, protractor, and compasses. The unknown elements can then be measured with the scale and the protractor.
2. By computing the unknown elements from those that are known.

The first method has been used to some extent in preceding chapters. However, as previously discussed, the graphical method is cumbersome and has a limited degree of accuracy.

Trigonometry, combined with simple algebraic processes, furnishes a powerful tool for solving triangles by the second method previously listed. Moreover, the degree of accuracy is limited only by the number of significant figures to which the elements have been measured and the number of significant figures available in the table of functions or the calculator used for the solution.

As pointed out in earlier chapters, every type of problem should be approached and solved in a planned and systematic manner. Only in this way are the habits of clear and ordered thinking developed, the principles of the problem understood, and the possibility of errors reduced to a minimum. With the foregoing in mind, we list the following suggestions for solving right triangles as a guide:

1. Make a reasonable sketch of the triangle, and mark the known (given) elements. This shows the relation of the elements, helps you choose the functions needed, and will serve as a check for the solution. List what is to be found.
2. To find an unknown element, select a formula that contains two known elements and the required unknown element. Substitute the known elements in the formula, and solve for the unknown.
3. As a rough check on the solution, compare the results with the drawing. To check the values accurately, note whether they satisfy relationships different from those already employed for the solution of the values being checked. A convenient check for the sides of a right triangle is the relation

$$a^2 = c^2 - b^2 = (c + b)(c - b)$$

4. In the computations, round off the numbers representing the lengths of sides to three significant figures and all angles to the nearest tenth of a degree. This means that the values of the functions employed in computations are to be used to only three significant figures. As previously stated, such accuracy is sufficient for ordinary *practical circuit* computations.

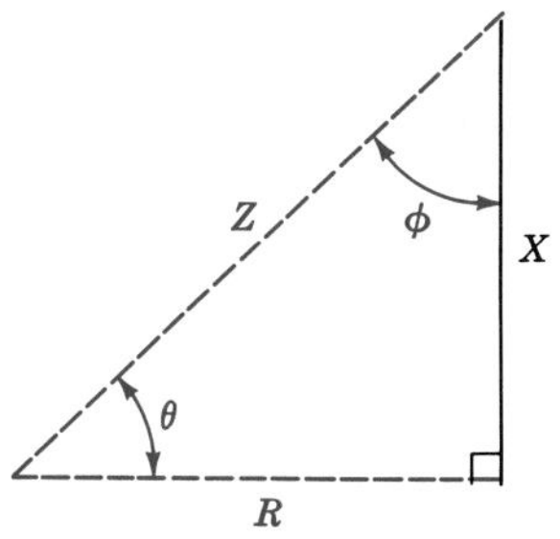

FIG. 26–1 Lettering of "standard" electrical right triangle.

Heretofore, the right triangles used in figures for illustrative examples have been lettered in the conventional manner, as shown in Figs. 24–4, 24–11, etc. At this point the notation for the various elements will be changed to that of Fig. 26–1. In no way does this change of lettering have any effect on the fundamental relations existing among the elements of a right triangle, nor are any new ideas involved in connection with the trigonometric functions. Because certain ac problems will employ this form of notation, this is a convenient place to introduce it in order that you may become accustomed to solving right triangles lettered in this manner.

The following sections illustrate all the possible conditions encountered in the solution of right triangles.

26–3 GIVEN AN ACUTE ANGLE AND A SIDE NOT THE HYPOTENUSE

EXAMPLE 1 Given $R = 30.0$ and $\theta = 25.0°$. Solve for Z, X, and ϕ.

SOLUTION

The construction is shown in Fig. 26–2.

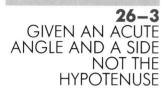

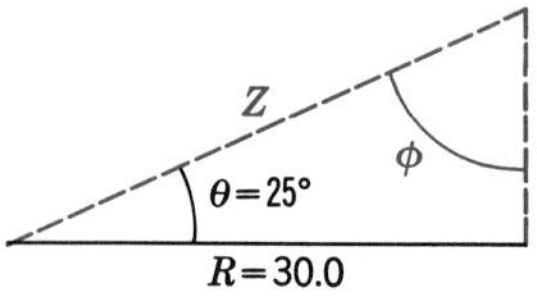

FIG. 26–2 Construction for solution of Example 1.

$$\phi = 90° - \theta = 90° - 25° = 65°$$

An equation containing the two known elements and one unknown is

$$\tan\theta = \frac{X}{R}$$

Solving for X,

$$X = R\tan\theta$$

Substituting the values of R and $\tan\theta$,

$$X = 30 \times 0.466 = 14.0$$

Also, since

$$\sin\theta = \frac{X}{Z}$$

Solving for Z,

$$Z = \frac{X}{\sin\theta}$$

Substituting the values of X and $\sin\theta$,

$$Z = \frac{14.0}{0.423} = 33.1$$

This solution can be checked by using some relation other than the relations already used. Thus, substituting values in

$$X^2 = (Z + R)(Z - R)$$

results in

$$14.0^2 = (33.1 + 30.0)(33.1 - 30.0)$$
$$196 = 63.1 \times 3.10 = 196$$

Since all results were rounded off to three significant figures, the check shows the solution to be correct for this degree of accuracy.

The value of Z can be checked by employing a function not used in the solution. Thus, since

$$R = Z \cos \theta$$

by substituting the values,

$$30 = 33.1 \times 0.906$$

Still another check could be made by use of an inverse function employing two of the elements found in the solution. For example,

$$\begin{aligned} \phi &= \arccos \frac{X}{Z} = \arccos \frac{14.0}{33.1} \\ &= \arccos 0.423 \\ &= 65° \end{aligned}$$

EXAMPLE 2 Given $X = 106$ and $\theta = 36.4°$. Solve for Z, R, and ϕ.

SOLUTION

The construction is shown in Fig. 26–3.

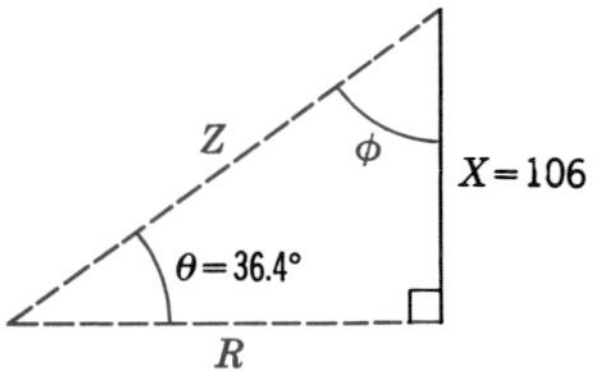

FIG. 26–3 Triangle of Example 2.

$$\phi = 90° - \theta = 90° - 36.4° = 53.6°$$

An equation containing two known elements and one unknown is

$$\sin \theta = \frac{X}{Z}$$

Solving for Z,

$$Z = \frac{X}{\sin \theta}$$

Substituting the values of X and $\sin \theta$,

$$Z = \frac{106}{0.593} = 179$$

Also, since

$$\cos \theta = \frac{R}{Z}$$

solving for R,

$$R = Z \cos \theta$$

Substituting the values of Z and $\cos \theta$,

$$R = 179 \times 0.805 = 144$$

Check the solution by one of the methods previously explained.

EXAMPLE 3 Given $R = 8.35$ and $\phi = 62.7°$. Find Z, X, and θ.

SOLUTION

The construction is shown in Fig. 26–4.

$$\theta = 90° - \phi = 90° - 62.7° = 27.3°$$

When θ is found, the methods to be used in the solution of this example become identical with those of Example 1. Hence,

$$X = R \tan \theta = 8.35 \tan 27.3°$$
$$= 8.35 \times 0.516 = 4.31$$
$$Z = \frac{X}{\sin \theta} = \frac{4.31}{\sin 27.3°}$$
$$= \frac{4.31}{0.459} = 9.39$$

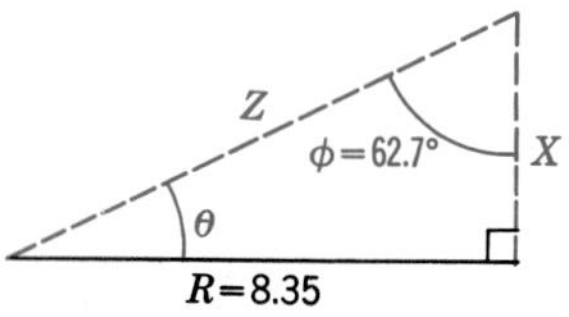

FIG. 26–4 Triangle of Example 3.

Check the solution by the method considered most convenient.

EXAMPLE 4 Given $X = 1290$ and $\phi = 41.9°$. Find Z, R, and θ.

SOLUTION

The construction is shown in Fig. 26–5.

$$\theta = 90° - \phi = 90° - 41.9° = 48.1°$$

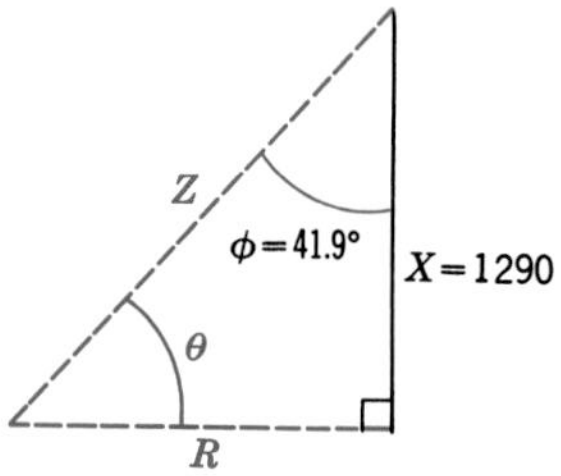

FIG. 26–5 $X = 1290$, $\phi = 41.9°$.

When θ is found, the methods to be used in the solution of this example become identical with those of Example 2. Hence,

$$Z = \frac{X}{\sin \theta} = \frac{1290}{\sin 48.1°} = \frac{1290}{0.744} = 1730$$
$$R = Z \cos \theta = 1730 \cos 48.1°$$
$$= 1730 \times 0.688 = 1190$$

Check the solution by the method considered most convenient.

With the exception of finding the unknown acute angle, which involves subtraction, any of the foregoing examples and the following problems may be solved using previously described calculator keystrokes.

PROBLEMS 26–1

Solve the following right triangles for the unknown elements. Check each by making a construction and by substituting into a formula not used in the solution:

1. R = 40.0, $\theta = 22.6°$
2. $X = 7.25$, $\phi = 68.5°$
3. $X = 296$, $\phi = 35°$
4. $R = 6.4$, $\phi = 44°$
5. $R = 118.5$, $\theta = 73.8°$
6. $X = 1530$, $\theta = 73.5°$
7. $R = 2.66 \times 10^3$, $\theta = 34.6°$
8. $R = 222$, $\phi = 26.3°$
9. $X = 420$, $\theta = 40°$
10. $R = 0.230$, $\theta = 77°$

11. $R = 0.697$, $\theta = 28.5°$
12. $X = 0.0929$, $\theta = 6.4°$
13. $R = 0.850$, $\theta = 48.5°$
14. $R = \frac{2}{3}$, $\theta = 51.9°$
15. $X = \frac{3}{8}$, $\theta = 82.4°$
16. $R = \frac{1}{\sqrt{2}}$, $\theta = 45°$

26–4 GIVEN AN ACUTE ANGLE AND THE HYPOTENUSE

EXAMPLE 5 Given $Z = 45.3$ and $\theta = 20.3°$. Find R, X, and ϕ.

SOLUTION

The construction is shown in Fig. 26–6.

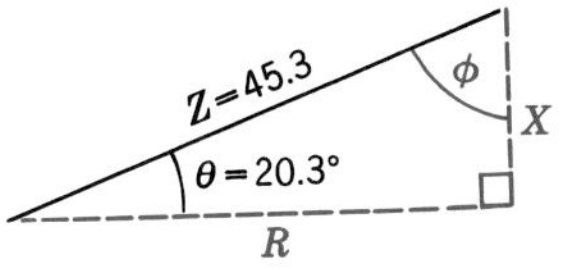

FIG. 26–6 $Z = 45.3$, $\theta = 20.3°$.

$$\phi = 90° - \theta = 90° - 20.3° = 69.7°$$

An equation containing two known elements and one unknown is

$$\cos\theta = \frac{R}{Z}$$

Solving for R,
$$R = Z\cos\theta$$

Substituting the values of Z and $\cos\theta$,

$$R = 45.3 \times 0.938 = 42.5$$

Another convenient equation is

$$\sin\theta = \frac{X}{Z}$$

Solving for X,
$$X = Z\sin\theta$$

Substituting the values of Z and $\sin\theta$,

$$X = 45.3 \times 0.347 = 15.7$$

The solution can be checked by any of the usual methods.

EXAMPLE 6 Given $Z = 265$ and $\phi = 22.4°$. Find R, X, and θ.

SOLUTION

The construction is shown in Fig. 26–7.

$$\theta = 90° - \phi = 90° - 22.4° = 67.6°$$

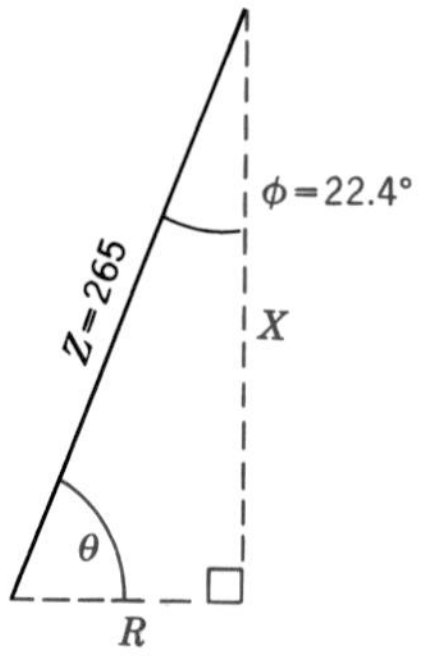

FIG. 26–7 $Z = 265,\ \phi = 22.4°$.

When θ is found, this triangle is solved by the methods used in Example 1. Hence,

$$R = Z \cos\theta = 265 \cos 67.6° = 265 \times 0.381 = 101$$
$$X = Z \sin\theta = 265 \sin 67.6° = 265 \times 0.924 = 245$$

Check the solution by one of the several methods.

PROBLEMS 26–2

Solve the following right triangles for the unknown elements. Check each by construction and by substituting into an equation not used in the solution.

1. $Z = 12.5,\ \phi = 62°$
2. $Z = 625,\ \theta = 34.2°$
3. $Z = 47.6,\ \theta = 69.1°$
4. $Z = 126,\ \phi = 69.7°$
5. $Z = 2.7 \times 10^3,\ \phi = 72.5°$
6. $Z = 40,\ \theta = 51.4°$
7. $Z = 0.948,\ \phi = 79.6°$
8. $Z = 610,\ \phi = 79.7°$
9. $Z = 16.4,\ \theta = 32.7°$
10. $Z = 0.342,\ \phi = 73.2°$

26–5 GIVEN THE HYPOTENUSE AND ONE OTHER SIDE

EXAMPLE 7 Given $Z = 38.3$ and $R = 23.1$. Find X, θ, and φ.

SOLUTION

The construction is shown in Fig. 26–8.

An equation containing two known elements and one unknown is

$$\cos\theta = \frac{R}{Z}$$

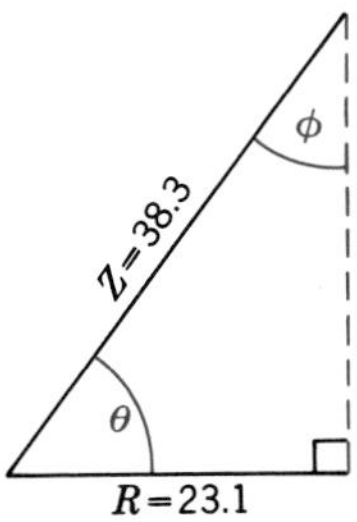

FIG. 26–8 Triangle of Example 7.

Substituting the values of R and Z,

$$\cos\theta = \frac{23.1}{38.3} = 0.603$$
$$\therefore \theta = 52.9^\circ$$
$$\phi = 90^\circ - \theta$$
$$= 90^\circ - 52.9^\circ$$
$$= 37.1^\circ$$

Then, since $\quad \sin\theta = \dfrac{X}{Z}$

Solving for X, $\quad X = Z\sin\theta$

Substituting the values of Z and $\sin\theta$,

$$X = 38.3 \times 0.798 = 30.6$$

EXAMPLE 8 Given $Z = 10.7$ and $X = 8.10$. Find R, θ, and ϕ.

SOLUTION

The construction is shown in Fig. 26–9.

An equation containing two known elements and one unknown is

$$\sin\theta = \frac{X}{Z}$$

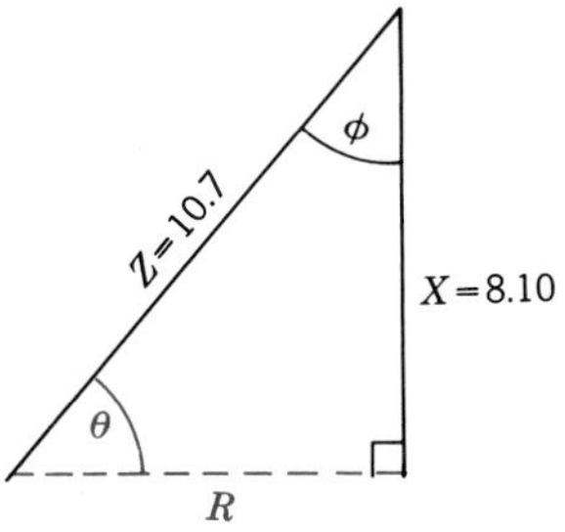

FIG. 26–9 Triangle of Example 8.

Substituting the values of X and Z,

$$\sin\theta = \frac{8.10}{10.7} = 0.757$$
$$\therefore \theta = 49.2^\circ$$
$$\phi = 90^\circ - \theta$$
$$= 90^\circ - 49.2^\circ$$
$$= 40.8^\circ$$

Then, since $\quad \cos\theta = \dfrac{R}{Z}$

Solving for R, $\quad R = Z\cos\theta$

Substituting the values of Z and $\cos\theta$,

$$R = 10.7 \times 0.653 = 6.99$$

PROBLEMS 26–3

Solve the following right triangles and check each graphically and algebraically as in the preceding problems:

1. $Z = 185$, $X = 140$
2. $Z = 1840$, $R = 860$
3. $Z = 36.9$, $R = 14$
4. $Z = 3100$, $R = 3060$
5. $Z = 1.273$, $R = 0.695$
6. $Z = 407$, $X = 57.0$
7. $Z = 1.4 \times 10^3$, $X = 4.8 \times 10^2$
8. $Z = 39.7$, $R = 11.4$
9. $Z = 426$, $R = 208$
10. $Z = 0.342$, $R = 0.327$

26–6 GIVEN TWO SIDES NOT THE HYPOTENUSE

EXAMPLE 9 Given $R = 76.0$ and $X = 37.4$. Find Z, θ, and ϕ.

SOLUTION

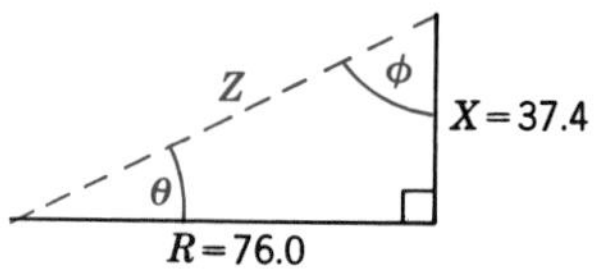

FIG. 26–10 Triangle of Example 9.

The construction is shown in Fig. 26–10.

An equation containing two known elements and one unknown is

$$\tan\theta = \frac{X}{R}$$

Substituting the values of X and R,

$$\tan\theta = \frac{37.4}{76.0} = 0.492$$
$$\therefore \theta = 26.2°$$
$$\phi = 90° - \theta = 90° - 26.2° = 63.8°$$

$Z = 84.7$ can be found by one of the methods explained in the preceding sections.

PROBLEMS 26–4

Solve the following right triangles and check as in the preceding problems:

1. $R = 20.7$, $X = 12.4$
2. $R = 14.2$, $X = 16.6$
3. $X = 6.08$, $R = 8.26$
4. $R = 85.2$, $X = 222$
5. $X = 20.3$, $R = 430$
6. $X = 50.6$, $R = 10.3$
7. $R = 12.8$, $X = 44.3$
8. $R = \frac{\sqrt{3}}{2}$, $X = \frac{1}{2}$
9. $X = 0.290$, $R = 0.280$
10. $X = 2.06$, $R = 4.19$

26–7 TERMS RELATING TO MISCELLANEOUS TRIGONOMETRIC PROBLEMS

If an object is higher than an observer's eye, the *angle of elevation* of the object is the angle between the horizontal and the line of sight to the object. This is illustrated in Fig. 26–11.

If an object is lower than an observer's eye, the *angle of depression* of the object is the angle between the horizontal and the line of sight to the object. This is illustrated in Fig. 26–12.

The *horizontal distance* between two points is the distance from one of the two points to a vertical line that is drawn through the other. Thus, in Fig. 26–13, the line AC is a vertical line through the point A and CB is a horizontal line through the point B. Then the horizontal distance from A to B is the distance between C and B.

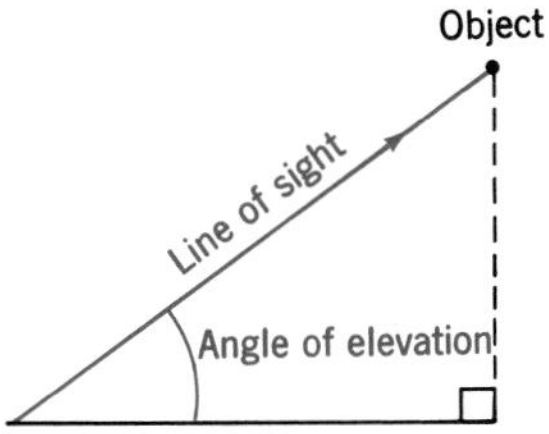

FIG. 26–11 Angle of elevation.

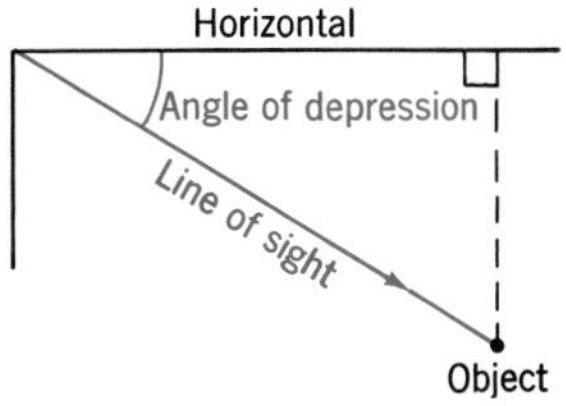

FIG. 26–12 Angle of depression.

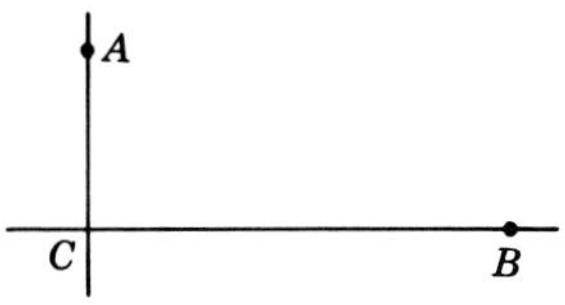

FIG. 26–13 Vertical and horizontal distances.

The *vertical distance* between two points is the distance from one of the two points to the horizontal line drawn through the other. Thus, the vertical distance from A to B, in Fig. 26–13, is the distance between A and C.

Calculations of distance in the vertical plane are made by means of right triangles having horizontal and vertical sides. The horizontal side is usually called the *run*, and the vertical side is called the *rise* or *fall*, as the case may be.

The *slope* or *grade* of a line is the rise or fall divided by the run. Thus, if a road rises 5 m in a run of 100 m, the grade of the road is

$$5 \div 100 = 0.05 = 5\%$$

PROBLEMS 26–5

1. What is the angle of inclination of a stairway with the floor if the steps have a tread of 19 cm and a rise of 30 cm?
2. What angle does an A-frame rafter make with the horizontal if it has a rise of 3.68 m in a run of 1.36 m?
3. A transmission line rises 3.8 m in a run of 35 m. What is the angle of elevation of the line with the horizontal?
4. A radio tower casts a shadow 174 m long, and at the same time the angle of elevation of the sun is 41.7°. What is the height of the tower?
5. An antenna mast 130 m tall casts a shadow 53 m long. What is the angle of elevation of the sun?
6. At a horizontal distance of 85 m from the foot of a radio tower, the angle of elevation of the top is found to be 31°. How high is the tower?
7. A telephone pole 12.2 m high is to be guyed from its middle, and the guy is to make an angle of 60° with the ground. Allowing 1 m extra for splicing, how long must the guy wire be?
8. An extension ladder 15 m long rests against a vertical wall with its foot 4 m from the wall. (Do not use Pythagoras' theorem to solve.)
   (*a*) What angle does the ladder make with the ground?
   (*b*) How far up the wall does the ladder reach?
9. A ladder 15 m long can be so placed that it will reach a point on a wall 12 m above the ground. By tipping it back without moving its foot, it can be made to reach a point on another wall 10 m above the ground. What is the horizontal distance between the walls?

10. From the top of a cliff 58 m high, the angle of depression of a boat is 28.6°. How far out is the boat?
11. In order to find the width *BC* of a river, a distance *AB* was laid off along the bank, the point *B* being directly opposite a tree *C* on the opposite side, as shown in Fig. 26–14. If the angle *BAC* was observed to be 52.3° and *AB* was measured at 60 m, find the width of the river.
12. In order to measure the distance *AC* across a swamp, a surveyor lays off a line *AB* such that the angle *BAC* = 90°, as shown in Fig. 26–15. At point *B*, 240 m from *A*, the surveyor observes that angle *ABC* = 59.1°. Find the distance *AC*.

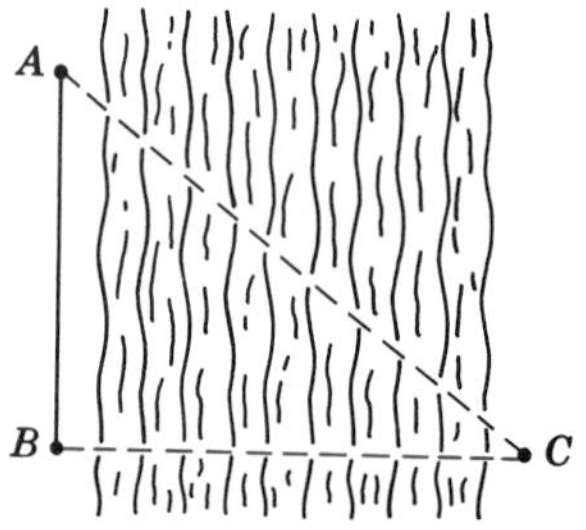

FIG. 26–14 Measuring across a river.

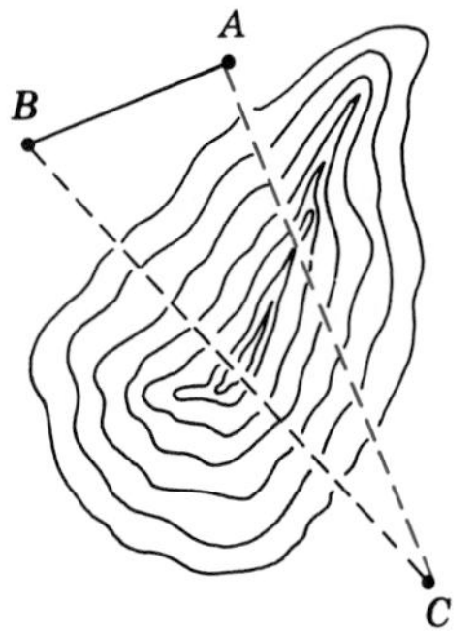

FIG. 26–15 Measuring across a pond or swamp.

26–8 THE AREA OF TRIANGLES

A convenient use of trigonometry is the calculation of the area of a triangle. In Fig. 26–16, the area of the triangle *ABC*, from previous knowledge, is known to be

$$A = \frac{1}{2}ab$$

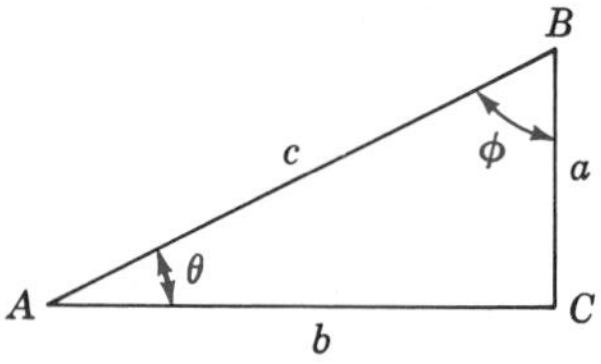

FIG. 26–16 Area of right triangle.

But $b = c \sin \phi$ and $a = c \sin \theta$, from which we can write

$$A = \frac{1}{2}ac \sin \phi$$

or

$$A = \frac{1}{2}bc \sin \theta$$

Either of these expressions may be stated:

The area of a triangle is one-half the product of any two sides times the sine of the angle between them.

You should prove that the formula holds for the more general case of the triangle of Fig. 26–17.

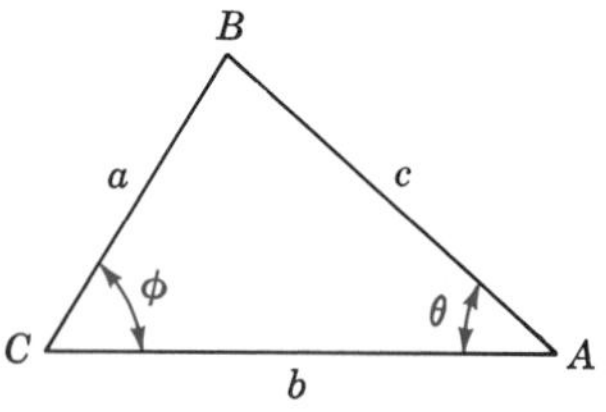

FIG. 26–17 Area of an oblique triangle.

▶ **HINT** Draw an altitude perpendicular to the base.

PROBLEMS 26–6

1. In the right triangle of Fig. 26–16, $a = 25$ m and $c = 60$ m.
   (*a*) What is the angle ϕ?
   (*b*) What is the area of the triangle by the sine formula?
   (*c*) What is the length of *b?*
   (*d*) What is the area by the formula $A = \frac{1}{2}$(base)(altitude)?
2. In the triangle of Fig. 26–17, $a = 77$ mm, $b = 96.4$ mm, $c = 72$ mm, $\phi = 47.5°$, and $\theta = 52°$.
   (*a*) What is the angle opposite side *b?*
   (*b*) What is the area of the triangle by the sine formula?
   (*c*) What is the length of altitude *h?*
   (*d*) What is the area by the formula $A = \frac{1}{2}$(base)(altitude)?
3. In the triangle of Fig. 26–17, $a = 85$ mm, $c = 64.2$ mm, and angle $CBA = 110°$. What is the area of the triangle?

SELF TEST

Refer to the triangle of Fig. 26–1:

1. Given $R = 12.2$ and $\theta = 34.6°$, evaluate *X, Z,* and ϕ.
2. Given $Z = 0.825$ and $\phi = 67.7°$, evaluate *X, R,* and θ.
3. Given $Z = 648$ and $X = 265$, evaluate *R*, θ, and ϕ.
4. Given $R = 144$ and $X = 216$, evaluate *Z*, θ, and ϕ.
5. From a point 10 m out from the base of a tower, the angle of elevation to the top of the tower is 56°. What is the height of the tower?
6. In the triangle of Fig. 26–17, side $a = 244$ m, side $c = 283$ m, and the angle at $B = 97°$. What is the area of the triangle?

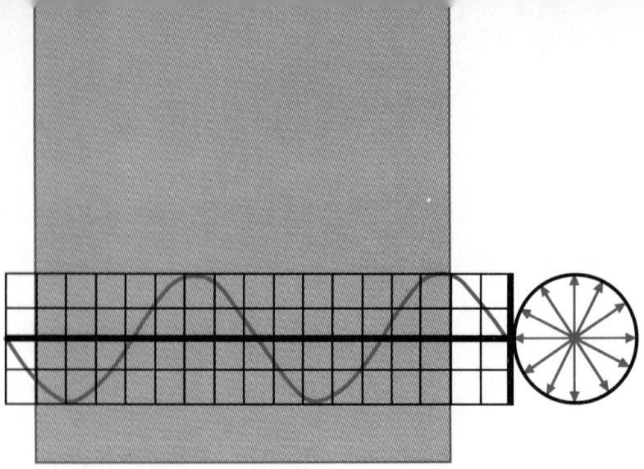

CHAPTER 27

TRIGONOMETRIC IDENTITIES AND EQUATIONS

So far, our studies in trigonometry have been confined to the solution of *right triangles,* but there are times when other types of problems must be considered. In this chapter, we shall develop some useful relationships between the trigonometric functions and also solve oblique triangles.

27–1 SIMPLE IDENTITIES

FIG. 27–1 "Standard" right triangle.

Consider the right triangle *ABC* (Fig. 27–1). From our studies in trigonometry we know that:

$$\sin\theta = \frac{X}{Z}$$

and

$$\cos\theta = \frac{R}{Z}$$

The ratio of these two functions is

$$\frac{\sin\theta}{\cos\theta} = \frac{\frac{X}{Z}}{\frac{R}{Z}} = \frac{X}{R} = \tan\theta \qquad (1)$$

This interesting and useful relationship is the simplest of a group of trigonometric interrelationships called *identities*. We shall develop a few of the simpler identities and then tabulate them for convenience.

27–2 THE PYTHAGOREAN IDENTITIES

In the triangle of Fig. 27–1, we can readily see that

$$X^2 + R^2 = Z^2$$

the statement of Pythagoras' theorem. Dividing the entire equation by Z^2:

$$\frac{X^2}{Z^2} + \frac{R^2}{Z^2} = \frac{Z^2}{Z^2}$$

from which we can see that

$$(\sin \theta)^2 + (\cos \theta)^2 = 1$$

which is usually written (as in 12 of Problems 25–3)

$$\sin^2 \theta + \cos^2 \theta = 1 \qquad (2)$$

This is the first of the interrelationships known as the *Pythagorean identities* because they are derived from Pythagoras' theorem. You should now repeat the process twice, dividing first by X^2 and then by R^2 to develop the other two Pythagorean identities:

$$1 + \cot^2 \theta = \csc^2 \theta \qquad (3)$$

$$\tan^2 \theta + 1 = \sec^2 \theta \qquad (4)$$

These relationships will prove quite useful in the advanced study of electronics because many of the mathematical descriptions of electrical and electronics phenomena are described by rather complicated combinations of trigonometric functions, which may often be simplified by the use of identities. Here we shall confine ourselves to achieving some practice in manipulation of identities.

No set rule may be established about simplifying or proving identities. Usually, one side of the identity is manipulated until it is shown to be equal to the other side. Sometimes, each side is developed into the same equivalent in order to arrive at an obvious equality.

EXAMPLE 1 Show that $\dfrac{\tan^2 \theta}{\sec^2 \theta} + \dfrac{\cot^2 \theta}{\csc^2 \theta} = 1$.

SOLUTION

(*a*) One possible method of solution uses the fundamental relationships between the trigonometric functions:

$$\frac{\tan^2 \theta}{\sec^2 \theta} + \frac{\cot^2 \theta}{\csc^2 \theta} = \frac{\left(\dfrac{\sin \theta}{\cos \theta}\right)^2}{\left(\dfrac{1}{\cos \theta}\right)^2} + \frac{\left(\dfrac{1}{\tan \theta}\right)^2}{\left(\dfrac{1}{\sin \theta}\right)^2}$$

$$= \sin^2 \theta + \sin^2 \theta\left(\frac{\cos^2 \theta}{\sin^2 \theta}\right)$$

$$= 1$$

(*b*) An alternative solution is to start with the Pythagorean identities, which suggests itself from the square relationships in the problem:

$$\begin{aligned}\frac{\tan^2\theta}{\sec^2\theta}+\frac{\cot^2\theta}{\csc^2\theta} &= \frac{\sec^2\theta-1}{\sec^2\theta}+\frac{\csc^2\theta-1}{\csc^2\theta}\\ &= 1-\frac{1}{\sec^2\theta}+1-\frac{1}{\csc^2\theta}\\ &= 2-(\cos^2\theta+\sin^2\theta)\\ &= 2-1=1\end{aligned}$$

PROBLEMS 27–1

Prove that the following equations are identities:

1. $\cos\theta\tan\theta=\sin\theta$
2. $(\sec\phi+\tan\phi)(\sec\phi-\tan\phi)=1$
3. $\cos^2\lambda-\sin^2\lambda=1-2\sin^2\lambda$
4. $\sin^4\alpha-\cos^4\alpha=\sin^2\alpha-\cos^2\alpha$
5. $\dfrac{1+\cos\beta}{\sin\beta}+\dfrac{\sin\beta}{1+\cos\beta}=2\csc\beta$
6. $\dfrac{\cos^2\phi}{1-\sin\phi}=1+\sin\phi$
7. $(1+\tan^2\beta)\cos^2\beta=1$
8. $\tan\theta+\cot\theta=\sec\theta\csc\theta$
9. $(\sin\theta+\cos\theta)^2+(\sin\theta-\cos\theta)^2=2$
10. $1-2\sin^2\omega=2\cos^2\omega-1$
11. $\tan^2\psi-\sin^2\psi=\tan^2\psi\sin^2\psi$
12. $\dfrac{1-2\cos^2\alpha}{\sin\alpha\cos\alpha}=\dfrac{\sin^2\alpha-\cos^2\alpha}{\sin\alpha\cos\alpha}$
13. $\dfrac{1-2\sin^2\alpha}{1+\tan^2\alpha}=2\cos^4\alpha-\cos^2\alpha$
14. $\sec\phi-\cos\phi=\sqrt{(\tan\phi+\sin\phi)(\tan\phi-\sin\phi)}$
15. $\cot\theta\cos\theta=\csc\theta-\sin\theta$
16. $\dfrac{\sin\theta+\tan\theta}{1+\tan^2\theta}=(\sin\theta+\tan\theta)\cos^2\theta$
17. $\tan\lambda+\cot\lambda=\dfrac{\csc^2\lambda+\sec^2\lambda}{\csc\lambda\sec\lambda}$
18. $(\tan\alpha-\sin\alpha)^2+(1-\cos\alpha)^2=(1-\sec\alpha)^2$
19. $\dfrac{1-\sin\omega}{1+\sin\omega}=(\sec\omega-\tan\omega)^2$
20. $\dfrac{\tan\alpha+\tan\beta}{\cot\alpha+\cot\beta}=\tan\alpha\tan\beta$

27–3 LAW OF SINES

Consider the triangle *ABC* (Fig. 27–2). This is not a right triangle, and therefore we have no relationships which we can use to solve it, that is, to relate the various sides and angles in order to find the unknown dimensions in an actual numerical problem. But if we were to develop within it our own right triangles, we might derive some useful relationships.

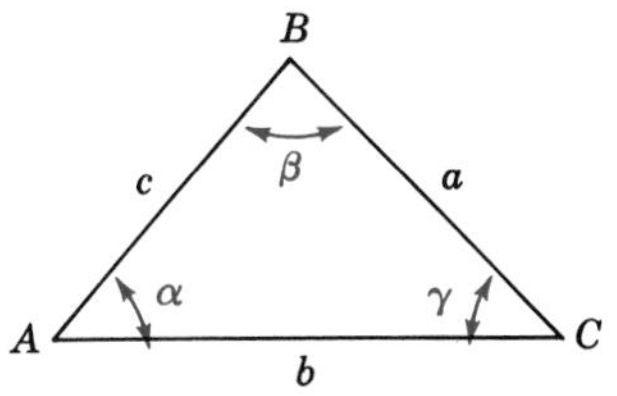

FIG. 27–2 Nonright triangle cannot be solved by simple trigonometric relationships.

First of all, we redraw the triangle, Fig. 27–3, and from the vertex *B* we drop the altitude *h* perpendicular to the base *b*. This yields two right triangles, from which we develop the relationships

$$h = c \sin \alpha \qquad \text{and} \qquad h = a \sin \gamma$$

Then, equating things equal to the same thing (Axiom 5, Sec. 5–2),

$$c \sin \alpha = a \sin \gamma$$

We rewrite this equation in the simple easy-to-remember form

$$\frac{a}{\sin \alpha} = \frac{c}{\sin \gamma} \tag{5}$$

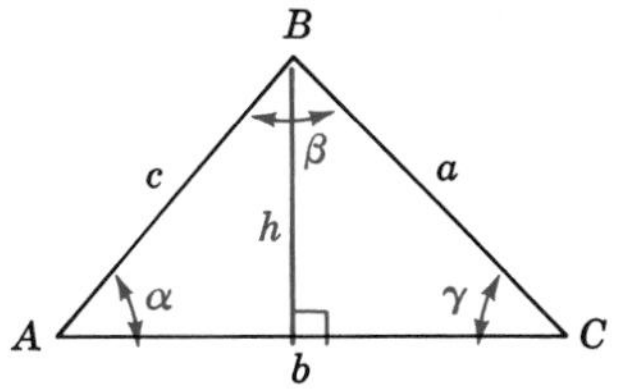

FIG. 27–3 Redrawn from Fig. 27–2 with altitude *h* perpendicular to base *b*.

You should immediately prove the more general statement:

$$\frac{a}{\sin \alpha} = \frac{b}{\sin \beta} = \frac{c}{\sin \gamma} \tag{6}$$

EXAMPLE 2 Given the triangle *MPL*, Fig. 27–4, find the values of *m* and *p*.

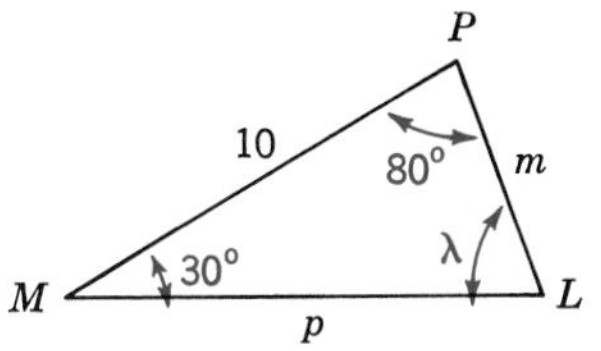

FIG. 27–4 Triangle of Example 2.

SOLUTION

First of all, solve for

$$\lambda = 180° - (80° + 30°) = 70°$$

Then, using the law of sines,

$$\frac{10}{\sin 70°} = \frac{p}{\sin 80°} = \frac{m}{\sin 30°}$$

so that

$$p = \frac{10 \sin 80°}{\sin 70°} = 10.5$$

and

$$m = \frac{10 \sin 30°}{\sin 70°} = 5.32$$

NOTE *To be able to use the law of sines, it is necessary for us to know certain specific data: two sides and the angle opposite one of them or two angles and the side opposite one of them.*

PROBLEMS 27–2

Referring to Fig. 27–2, solve the following triangles:

1. $a = 10.5$, $\alpha = 42°$, $\beta = 68°$
2. $a = 32$, $\beta = 77°$, $\gamma = 84°$
3. $b = 38.8$, $\alpha = 42°$, $\beta = 64°$
4. $c = 760$, $\alpha = 68°$, $\beta = 42°$

5. $b = 76$, $\alpha = 20°$, $\beta = 52°$
6. $b = 3.26$, $\alpha = 25°$, $\beta = 41°$
7. $c = 0.456$, $\beta = 36°$, $\gamma = 102°$
8. $a = 600$, $\beta = 17.6°$, $\gamma = 105.9°$
9. $b = 58$, $\alpha = 9.2°$, $\gamma = 115.3°$
10. $c = 635$, $\alpha = 15.5°$, $\beta = 26°$
11. Two observers who are 1200 m apart on a horizontal plane observe a radiosonde balloon in the same vertical plane as themselves and between themselves. The angles of elevation are 51.7° and 68.2°. Find the height of the balloon.
12. A 50-m antenna mast stands on the edge of the roof of the studio building. From a point on the ground at some distance from the base of the building, the angles of elevation of the top and bottom of the mast are respectively 76.5° and 54.5°. How high is the building?

27–4 LAW OF COSINES

Sometimes we are not given data suitable for solving a triangle by means of the law of sines, but another useful relationship can be readily developed. Using the triangle *ABC* of Fig. 27–2, copied as Fig. 27–5 and adjusted with an altitude h perpendicular to the base and rising to the vertex, so that the base is divided into parts x and y,

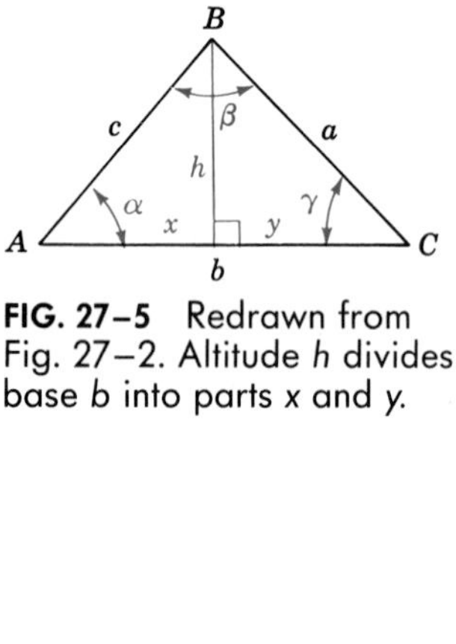

FIG. 27–5 Redrawn from Fig. 27–2. Altitude h divides base b into parts x and y.

$$h^2 = c^2 - x^2 = a^2 - y^2$$

from which

$$\begin{aligned} a^2 &= c^2 - x^2 + y^2 \\ &= c^2 - x^2 + (b - x)^2 \\ &= c^2 - x^2 + b^2 - 2bx + x^2 \\ &= b^2 + c^2 - 2bx \end{aligned}$$

but

$$x = c \cos \alpha$$

and

$$a^2 = b^2 + c^2 - 2bc \cos \alpha \qquad (7)$$

See how straightforward this statement may be: "In any triangle, the square of any one side is equal to the sum of the squares of the other two sides minus twice their product times the cosine of the angle between them." You should prove that this statement holds true for right triangles, to become Pythagoras' theorem.

Like the law of sines, the law of cosines has a rhythm that makes it easy to memorize one part and simply rotate the other parts into duplicate statements. However, besides merely memorizing the result, you should prove that all parts of the full statement of the law of cosines are true:

$$a^2 = b^2 + c^2 - 2bc \cos \alpha$$

$$b^2 = a^2 + c^2 - 2ac \cos \beta \qquad (8)$$

$$c^2 = a^2 + b^2 - 2ab \cos \gamma \qquad (9)$$

The careful use of these three equations, together with what we have learned about the *signs* of the cosine, will enable us to prepare any triangle so that we may complete its solution by means of the law of sines.

EXAMPLE 3 Acute triangle. Solve the triangle of Fig. 27–6.

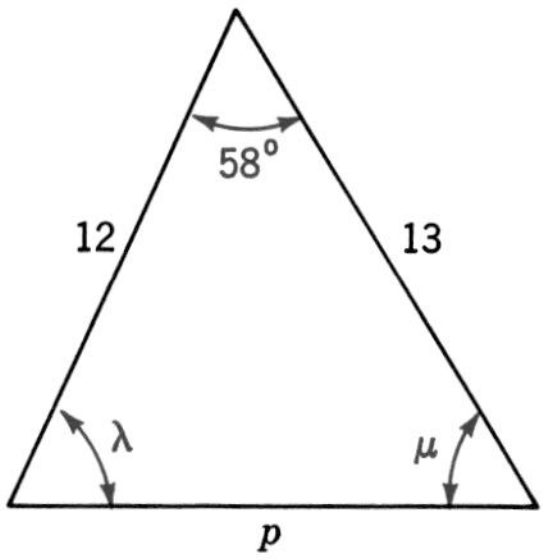

FIG. 27–6 Triangle of Example 3.

SOLUTION

Using the law of cosines:

$$\begin{aligned} p^2 &= 12^2 + 13^2 - 2 \times 12 \times 13 \times \cos 58^\circ \\ &= 144 + 169 - 312 \cos 58^\circ \\ &= 147.7 \\ p &= 12.15 \end{aligned}$$

Now, having at least two sides and the angle opposite one of them, we may, if we wish, complete the solution by means of the law of sines instead of repeating the cosine solution.

$$\frac{12.2}{\sin 58^\circ} = \frac{12}{\sin \mu}$$

from which $\mu = 56.9^\circ$

Similarly $\lambda = \arcsin \dfrac{13 \sin 58^\circ}{12.15} = 65.1^\circ$

TEST

$$58^\circ + 56.9^\circ + 65.1^\circ = 180.0^\circ$$

EXAMPLE 4 Oblique triangle. Solve the triangle of Fig. 27–7.

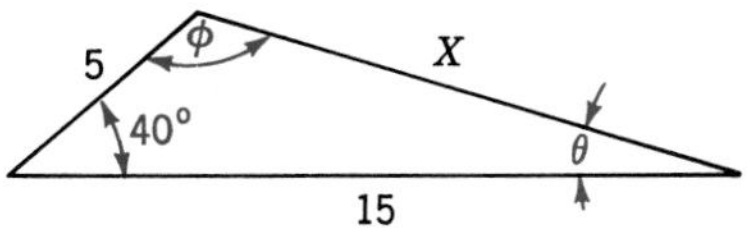

FIG. 27–7 Triangle of Example 4.

SOLUTION

Since the information given is not sufficient to use the law of sines, check to see if the law of cosines may be applied. Knowing two sides and the angle between them is sufficient:

$$\begin{aligned} X^2 &= 5^2 + 15^2 - 2 \times 5 \times 15 \times \cos 40^\circ \\ &= 135.1 \\ X &= 11.6 \end{aligned}$$

Then, using the law of sines,

$$\theta = \arcsin \frac{5 \sin 40^\circ}{11.6} = 16.1^\circ$$

and $$\phi = \arcsin \frac{15 \sin 40°}{11.6} = 56.1°$$

TEST

$$40° + 16.1° + 56.1° = 112.2° \qquad \text{Oh.}$$

From Fig. 27–7, the side of length 15, being the longest side, *must* be opposite the largest angle, which we have calculated as 56.1°. Since this must be the largest angle, since it *could* be obtuse (greater than 90°, an angle in the second quadrant), and since all that our calculations guarantee is that $\phi = \arcsin 0.831$, then perhaps ϕ is $180° - 56.1° = 123.9°$.
Testing this possibility,

$$40° + 16.1° + 123.9° = 180°$$

We have arrived at the correct solution.
Be sure to test your solutions.

EXAMPLE 5 The three sides of a triangle are given, and it is required to solve the angles. (Note that if just three angles are given, there is an infinite number of solutions.) Solve the triangle of Fig. 27–8.

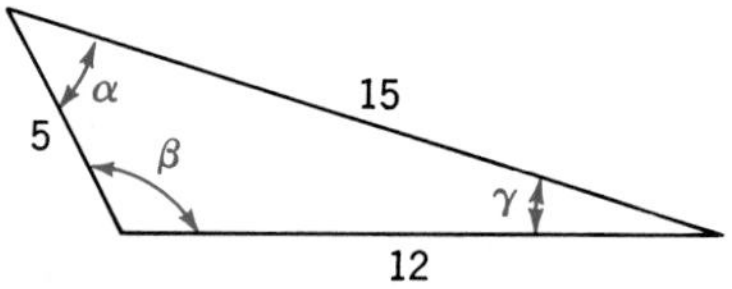

FIG. 27–8 Triangle of Example 5.

SOLUTION
If any angle in the triangle can be obtuse, it will be angle β. (Why?) We will defer solving for β for now. Consider the angle α. It is related, by the law of cosines, as follows:

$$12^2 = 5^2 + 15^2 - 2 \times 5 \times 15 \times \cos \alpha$$

from which $$\alpha = \arccos \frac{5^2 + 15^2 - 12^2}{2 \times 5 \times 15} = 45.1°$$

You should confirm that

$$\gamma = \arccos \frac{12^2 + 15^2 - 5^2}{2 \times 12 \times 15} = 17.2°$$

Then $$\beta = 180° - (45.1° + 17.2°) = 117.7°$$

Alternatively, starting the solution for β,

$$15^2 = 5^2 + 12^2 - 2 \times 5 \times 12 \times \cos \beta$$

from which

$$\beta = \arccos(-0.466)$$

This negative cosine indicates immediately that β must be an angle between 90° and 180°, and we find it to be

$$180° - 62.3° = 117.7°$$

PROBLEMS 27–3

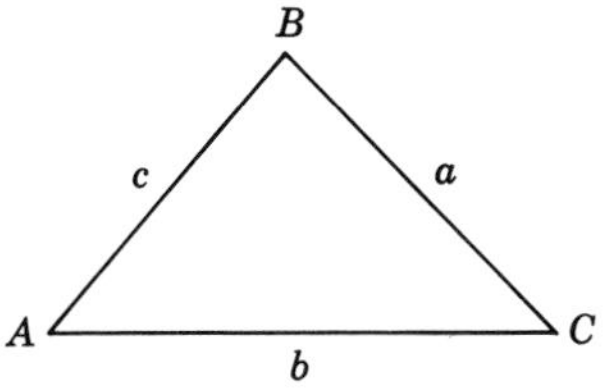

Referring to the figure to the left, solve the following triangles:

1. $b = 12.8$, $c = 18.5$, $\alpha = 68°$
2. $a = 474$, $b = 791$, $\gamma = 69°$
3. $a = 0.26$, $b = 0.825$, $\gamma = 144°$
4. $a = 2.6$, $c = 8.45$, $\beta = 48.8°$
5. $a = 1450$, $b = 3000$, $\gamma = 130.5°$
6. $b = 0.0945$, $c = 0.0980$, $\alpha = 5°$
7. $a = 4$, $b = 6$, $c = 9$
8. $a = 2000$, $b = 4000$, $c = 6000$
9. $a = 950$, $b = 2200$, $c = 3100$
10. $a = 25$, $b = 30$, $c = 50$
11. The diagonals of a parallelogram are 130 mm and 180 mm, and they intersect at an angle of 38°. What are the sides of the parallelogram?
12. Using the data of Prob. 11, but *not* your results, what is the area of the parallelogram? (After obtaining a solution, check it by means of a different computational method.)

27–5 THE SUM IDENTITIES

Often in the solution of antenna and modulation problems we come upon various combinations such as $\sin(\theta + \phi)$ and $\cos(\theta - \phi)$. It is often convenient to resolve these forms into the products of simple trigonometric functions.

Consider triangle *PQR*, Fig. 27–9, with the altitude *h* dividing the angle *RPQ* into the angles α and β. Since the area of the whole triangle must be equal to the sum of the areas of the two component triangles,

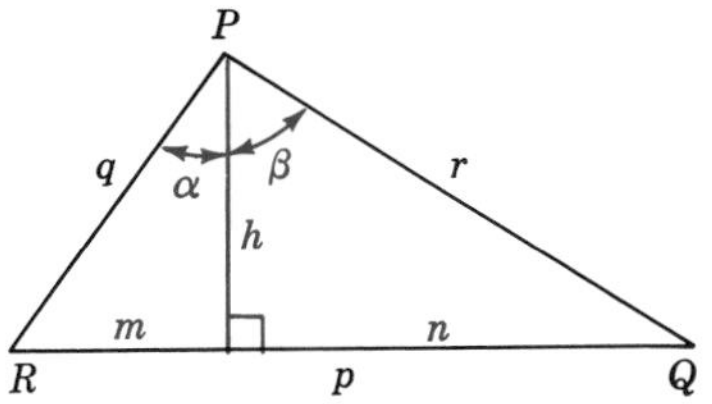

FIG. 27–9 Triangle adjusted for development of the sum identities.

$$\frac{1}{2}qr \sin(\alpha + \beta) = \frac{1}{2}qh \sin\alpha + \frac{1}{2}rh \sin\beta$$

from which $$\sin(\alpha + \beta) = \frac{h}{r}\sin\alpha + \frac{h}{q}\sin\beta$$

which yields $$\sin(\alpha + \beta) = \sin\alpha\cos\beta + \cos\alpha\sin\beta \qquad (10)$$

Again, using the same triangle, Fig. 27–9, and the law of cosines,

$$(m + n)^2 = q^2 + r^2 - 2qr\cos(\alpha + \beta)$$

from which $$\cos(\alpha + \beta) = \frac{q^2 + r^2 - m^2 - n^2 - 2mn}{2qr}$$

$$= \frac{q^2 - m^2}{2qr} + \frac{r^2 - n^2}{2qr} - \frac{2mn}{2qr}$$

$$= \frac{2h^2}{2qr} - \frac{2mn}{2qr}$$

$$= \frac{h}{q}\cdot\frac{h}{r} - \frac{m}{q}\cdot\frac{n}{r}$$

which converts to $$\cos(\alpha + \beta) = \cos\alpha\cos\beta - \sin\alpha\sin\beta \qquad (11)$$

27–6 THE DIFFERENCE IDENTITIES

Sometimes, instead of functions of the sum of two angles, it is necessary to deal with the differences of two angles: In triangle PQR, Fig. 27–10, the line q

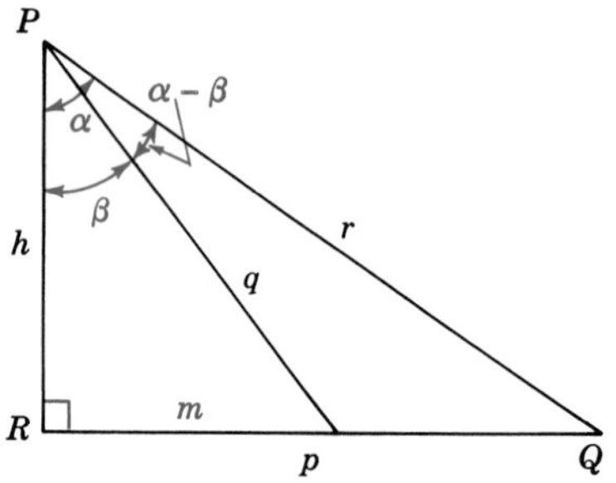

FIG. 27–10 Triangle adjusted for the development of the difference identities.

divides the vertex into two angles, β and α − β. As in the sum identity, the area of the whole triangle is equal to the sum of the parts:

$$\frac{1}{2}hr \sin\alpha = \frac{1}{2}hq \sin\beta + \frac{1}{2}qr \sin(\alpha - \beta)$$

from which $$\sin(\alpha - \beta) = \frac{hr\sin\alpha - hq\sin\beta}{qr}$$

$$= \frac{h}{q}\sin\alpha - \frac{h}{r}\sin\beta$$

which yields $$\sin(\alpha - \beta) = \sin\alpha\cos\beta - \cos\alpha\sin\beta \tag{12}$$

And, as before, using the law of cosines:

$$(p - m)^2 = q^2 + r^2 - 2qr\cos(\alpha - \beta)$$

from which

$$\begin{aligned}\cos(\alpha - \beta) &= \frac{q^2 + r^2 - p^2 - m^2 + 2mp}{2qr}\\ &= \frac{q^2 - m^2}{2qr} + \frac{r^2 - p^2}{2qr} + \frac{2mp}{2qr}\\ &= \frac{h}{q}\cdot\frac{h}{r} + \frac{m}{q}\cdot\frac{p}{r}\end{aligned}$$

which yields $$\cos(\alpha - \beta) = \cos\alpha\cos\beta + \sin\alpha\sin\beta \tag{13}$$

EXAMPLE 6 Simplify the expression

$$\sin(\theta + 45°) + \cos(\theta + 45°)$$

SOLUTION

Using Eqs. (10) and (11) and substituting the equivalent product expressions:

$$\begin{aligned}&\sin(\theta + 45°) + \cos(\theta + 45°)\\ &\quad = \sin\theta\cos 45° + \cos\theta\sin 45° + \cos\theta\cos 45° - \sin\theta\sin 45°\\ &\quad = 0.7071\sin\theta + 0.7071\cos\theta + 0.7071\cos\theta - 0.7071\sin\theta\\ &\quad = 1.4142\cos\theta\end{aligned}$$

TABLE 27–1 TRIGONOMETRIC IDENTITIES AND USEFUL RELATIONSHIPS

$$\tan\theta = \frac{\sin\theta}{\cos\theta} \qquad \cot\theta = \frac{\cos\theta}{\sin\theta}$$

$$\begin{aligned}\sin^2\theta + \cos^2\theta &= 1\\ 1 + \tan^2\theta &= \sec^2\theta\\ 1 + \cot^2\theta &= \csc^2\theta\end{aligned}$$

$$\frac{a}{\sin\alpha} = \frac{b}{\sin\beta} = \frac{c}{\sin\gamma}$$

$$a^2 = b^2 + c^2 - 2bc\cos\alpha$$

$$\begin{aligned}\sin(\theta + \phi) &= \sin\theta\cos\phi + \cos\theta\sin\phi\\ \cos(\theta + \phi) &= \cos\theta\cos\phi - \sin\theta\sin\phi\\ \sin(\theta - \phi) &= \sin\theta\cos\phi - \cos\theta\sin\phi\\ \cos(\theta - \phi) &= \cos\theta\cos\phi + \sin\theta\sin\phi\end{aligned}$$

PROBLEMS 27–4

Using the sum and difference relationships, simplify:

1. $\sin(\theta + 30°) + \cos(\theta + 30°)$
2. $\sin(45° - \theta) - \cos(45° + \theta)$
3. $\sin(\theta - 60°) + \cos(\theta + 60°)$
4. $\sin(\theta - 30°) - \cos(\theta - 45°)$

Given $\sin\theta = \frac{3}{5}$ and $\sin\phi = \frac{5}{12}$, evaluate:

5. $\cos(\theta + \phi)$
6. $\sin(\theta - \phi) - \cos(\theta - \phi)$
7. Use Eq. (10) to show that $\sin 2\theta = 2\sin\theta\cos\theta$.
8. Use Eq. (11) to show that $\cos 2\theta = \cos^2\theta - \sin^2\theta$.
9. When a VHF direction-finding array is fed in modulation phase quadrature, the two fields about the antennas are

$$V_1 = K\cos\theta\cos pt\cos\omega t$$
$$V_2 = K\sin\theta\sin pt\cos\omega t$$

Show that the total field $V_t = V_1 + V_2 = K\cos\omega t\cos(pt - \theta)$.

10. Use Eqs. (11) and (13) to show that

$$\tfrac{1}{2}\cos(\omega t - pt) - \tfrac{1}{2}\cos(\omega t + pt) = \sin pt\sin\omega t$$

It is based on this relationship that an amplitude-modulated carrier wave is shown to consist of a fundamental and two sidebands. The equation of the modulated carrier wave is

$$v = V\sin\omega t + mV\sin\omega t\sin pt$$

where m is the depth of modulation, and your work in this problem shows the correctness of the substitution:

$$v = V\sin\omega t + \tfrac{1}{2}mV\cos(\omega t - pt) - \tfrac{1}{2}mV\cos(\omega t + pt)$$

where $V\sin\omega t$ represents the original carrier and the other two parts represent the difference and sum sideband frequencies whose amplitudes are each one-half that of the carrier voltage when $m = 1$.

Many computers provide three direct trigonometric functions, sin, cos, and tan, but only one inverse function, arctan. Apply the lessons of this chapter to prove the following relationships:

11. $\tan\theta = \sqrt{\dfrac{1 - \cos^2\theta}{\cos^2\theta}}$

12. $\tan\theta = \dfrac{1}{\dfrac{\sqrt{1-\sin^2\theta}}{\sqrt{\sin^2\theta}}}$

13. $\tan\theta = \sqrt{\sec^2\theta - 1}$

14. $\tan\theta = \dfrac{1}{\sqrt{\csc^2\theta - 1}}$

15. Show that $\arccos 0.5 = 90° - \arctan\left(\dfrac{0.5}{\sqrt{1-0.5^2}}\right)^{\circ}$

16. Show that $\arcsin 0.8660 = \arctan\left(\dfrac{0.8660}{\sqrt{1-0.8660^2}}\right)^{\circ}$

SELF TEST

1. Reduce the following expression to a single simple function:

$$\sqrt{\frac{1-\cos^2\theta}{\cos^2\theta}}$$

2. In the triangle of Fig. 27–11, evaluate λ, P, and Q.
3. In the triangle of Fig. 27–12, evaluate X, α, and β.
4. Use the sum and difference relationships to simplify:

$$Q = \sin(\theta + 40°) - \cos(\theta - 22°)$$

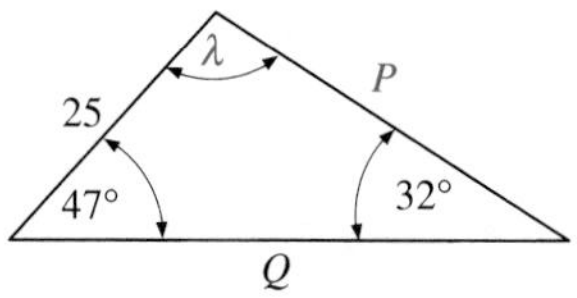

FIG. 27–11 Triangle of self test Prob. 2.

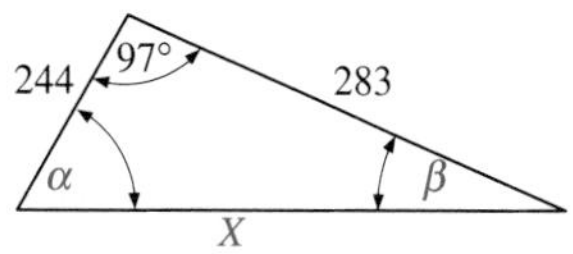

FIG. 27–12 Triangle of self test Prob. 3.

CHAPTER 28

ELEMENTARY PLANE VECTORS

Many physical quantities can be expressed by specifying a certain number of units. For example, the volume of a tank may be expressed as so many cubic meters, the temperature of a room as a certain number of degrees, and the speed of a moving object as a number of linear units per unit of time such as kilometers per hour or meters per second. Such quantities are *scalar quantities,* and the numbers that represent them are called *scalars*. A scalar quantity is one that has only magnitude; that is, it is a quantity fully described by a number, but it does not involve any concept of direction.

28–1 VECTORS

Many other types of physical quantities need to be expressed more definitely than is possible by specifying magnitude alone. For example, the velocity of a moving object has a direction as well as a magnitude. Also, a force due to a push or a pull is not completely described unless the direction as well as the magnitude of the force is given. In addition, electric circuit analysis is built up around the idea of expressing the directions and magnitudes of voltages and currents. Quantities which have both magnitude and direction are called *vector quantities*. A vector quantity is conveniently represented by a directional straight-line segment called a magnitude and whose head points in the direction of the vector quantity.

EXAMPLE 1 If a vessel steams northeast at a speed of 15 kn, its speed can be represented by a line whose length represents 15 kn, to some convenient scale, as shown in Fig. 28–1. The direction of the line represents the direction in which the vessel is traveling. Thus the line ***OA*** is a vector that completely describes the velocity of the vessel.

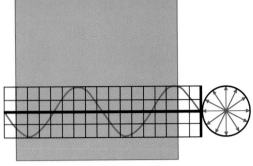

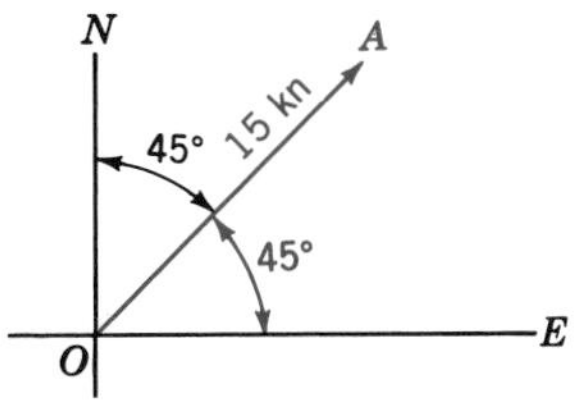

FIG. 28–1 Vector *OA* of Example 1.

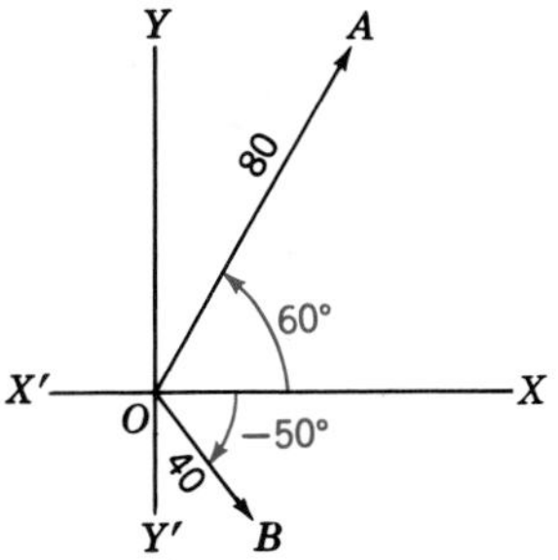

FIG. 28–2 Vector diagram of Example 2.

EXAMPLE 2 In Fig. 28–2, the vector ***OA*** represents a force of 80 N pulling on a body at *O* in a direction of 60°. The vector ***OB*** represents a force of 40 N acting on the same body in a direction of 310° or $-50°$.

Two vectors are equal if they have the same magnitude and direction. Thus, in Fig. 28–3, vectors ***A, B,*** and ***C*** are equal.

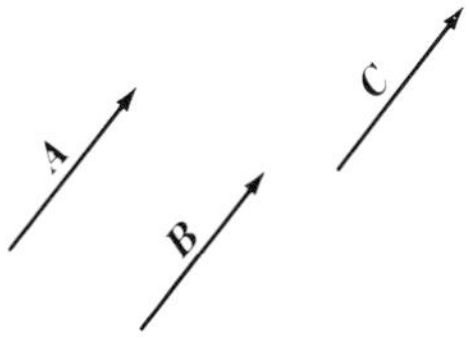

FIG. 28–3 Vectors ***A, B,*** and ***C*** are equal.

28–2 NOTATION

As you progress in the study of vectors, you will find that vectors and scalars satisfy different algebraic laws. For example, a scalar when reduced to its simplest terms is simply a number and as such obeys all the laws of ordinary algebraic operations. Since a vector involves direction, in addition to magnitude, it does not obey the usual algebraic laws and therefore has an analysis peculiar to itself.

From the foregoing, it is apparent that it is desirable to have a notation that indicates clearly which quantities are scalars and which are vectors. Several methods of notation are used, but you will find little cause for confusion; for most authors specify and explain their particular system of notation.

A vector can be denoted by two letters, the first indicating the origin, or initial point, and the second indicating the head, or terminal point. This form of notation was used in Examples 1 and 2 of the preceding section. Sometimes a small arrow is placed over these letters to emphasize that the quantity considered is a vector. Thus $\overrightarrow{OA}$ could be used to represent the vector from *O* to *A* as in Fig. 28–2. In most texts, vectors are indicated by boldface type; thus, ***A*** denotes the vector *A*. Other common forms of specifying a vector quantity, as, for example, the vector *A*, are $\bar{A}$, $\dot{A}$, $\underset{\cdot}{A}$, and $\underset{\sim}{A}$.

28–3 ADDITION OF VECTORS

Scalar quantities are added algebraically.

Thus 20 cents + 8 cents = 28 cents
and 16 insulators − 7 insulators = 9 insulators

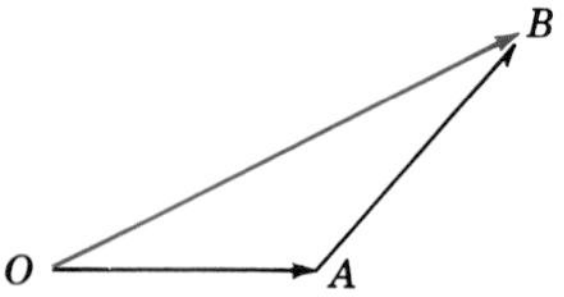

FIG. 28–4 Vector ***OB*** is the vector sum of ***OA*** and ***AB.***

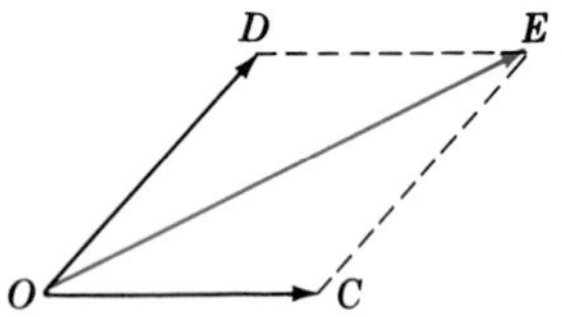

FIG. 28–5 Resultant vector ***OE*** is the vector sum of ***OC*** and ***OD.***

Since vector quantities involve direction as well as magnitude, they cannot be added algebraically unless their directions are parallel. Figure 28–4 illustrates vectors ***OA*** and ***AB.*** Vector ***OA*** can be considered as a motion from *O* to *A*, and vector ***AB*** as a motion from *A* to *B*. Then the sum of the vectors represents the sum of the motions from *O* to *A* and from *A* to *B*, which is the motion from *O* to *B*. This sum is the vector ***OB;*** that is, the vector sum of ***OA*** and ***AB*** is ***OB.*** Therefore, the sum of two vectors is the vector joining the initial point of the first to the terminal point of the second if the initial point of the second vector is joined to the terminal point of the first vector as shown in Fig. 28–4.

In Fig. 28–5, vectors ***OC*** and ***OD*** are equal to vectors ***OA*** and ***AB,*** respectively, of Fig. 28–4. In Fig. 28–5, however, the vectors start from the same origin. That their sum can be represented by the diagonal of a parallelogram of which the vectors are adjacent sides is evident by comparing Figs. 28–4 and 28–5. This is known as the *parallelogram law* for the composition of forces, and it holds for the composition or addition of all vector quantities.

The addition of vectors that are not at right angles to each other will be considered in Sec. 28–7. At this time, it is sufficient to know that two forces acting simultaneously on a point, or an object, can be replaced by a single force called the *resultant*. That is, the resultant force will produce the same effect on the object as the joint action of the two forces. Thus, in Fig. 28–4 the vector ***OB*** is the resultant of vectors ***OA*** and ***AB.*** Similarly, in Fig. 28–5, the vector ***OE*** is the resultant of the vectors ***OC*** and ***OD.*** Note that ***OB*** = ***OE.***

EXAMPLE 3

Three forces, **A, B,** and **C,** are acting on point *O*, as is shown in Fig. 28–6. Force *A* exerts 150 N at an angle of 60°; *B* exerts 100 N at an angle of 135°; and *C* exerts 150 N at an angle of 260°. What is the resultant force on point *O*?

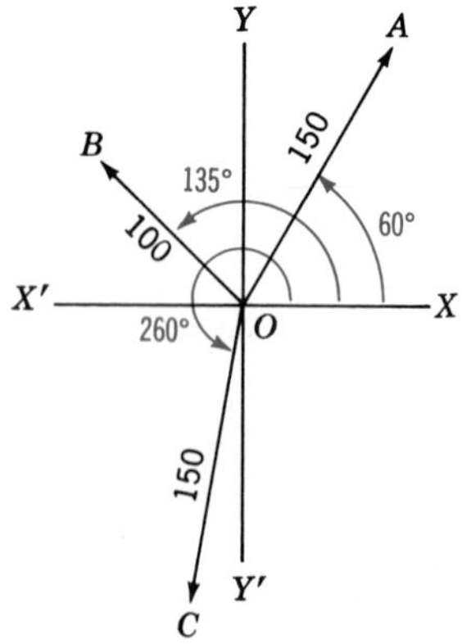

FIG. 28–6 Vector diagram of Example 3.

SOLUTION

The resultant of vectors ***A, B,*** and ***C*** can be found graphically by either of two methods.

(*a*) First draw the vectors to scale. Find the resultant of any two vectors, such as ***OA*** and ***OC,*** by constructing a parallelogram with ***OA*** and ***OC*** as adjacent sides. Then the resultant of ***OA*** and ***OC*** will be the diagonal ***OD*** of the parallelogram *OADC* as was illustrated in Fig. 28–7. In effect, there are now two forces, ***OB*** and ***OD***, acting on point *O*. The resultant of these two forces is found as before by constructing a parallelogram with ***OB*** and ***OD*** as adjacent sides. The resultant force on point *O* is then the diagonal ***OE*** of the parallelogram *OBED*. By measurement with scale and protractor, *OE* is found to be 57 N acting at an angle of 112°.

(*b*) Draw the vectors to scale as shown in Fig. 28–8, joining the initial point of *C* to the terminal point of *B*. The vector drawn from the point *O* to the terminal point of *C* is the resultant force, and measurements show it to be the same as that found by the method that is illustrated in Fig. 28–7.

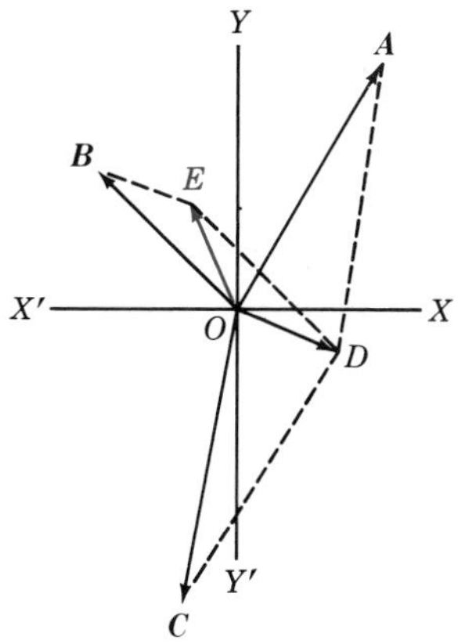

FIG. 28–7 *OE* is the vector sum of vectors *A*, *B*, and *C*.

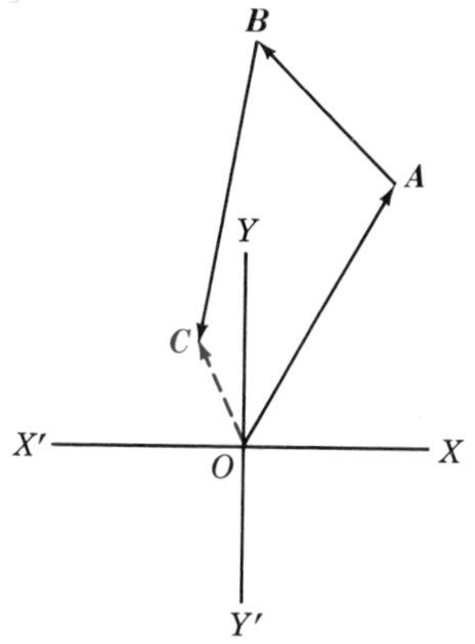

FIG. 28–8 *OC* is the vector sum of vectors *A*, *B*, and *C*.

A figure such as *OABCO*, in Fig. 28–8, is called a *polygon of forces*. The vectors can be joined in any order as long as the initial point of one vector joins the terminal point of another vector and the vectors are drawn with the proper magnitude and direction. The length and direction of the line that is necessary to close the polygon, that is, the line from the original initial point to the terminal point of the last vector drawn, constitute a vector that represents the magnitude and the direction of the resultant.

PROBLEMS 28–1

1 to 4. Find the magnitude and direction, with respect to the positive x axis, of the resultants of the vectors shown in Figs. 28–9 to 28–12.

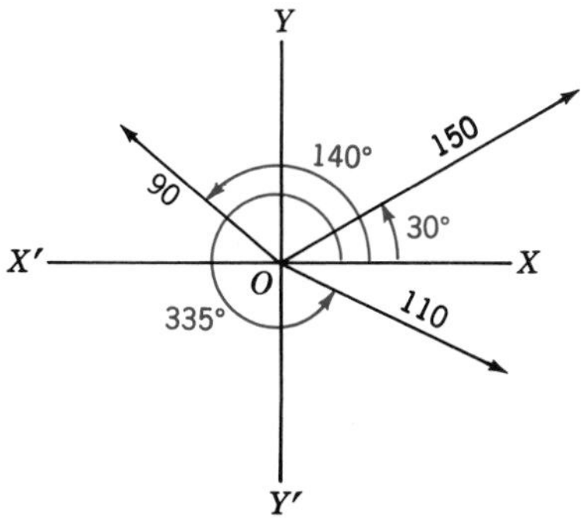

FIG. 28–9 Vector diagram of Prob. 1.

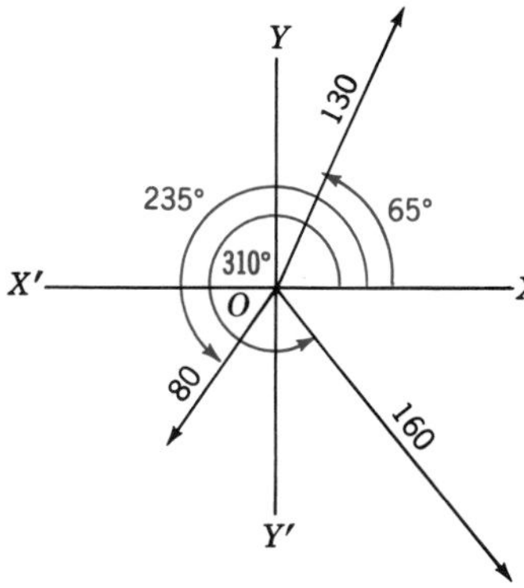

FIG. 28–10 Vector diagram of Prob. 2.

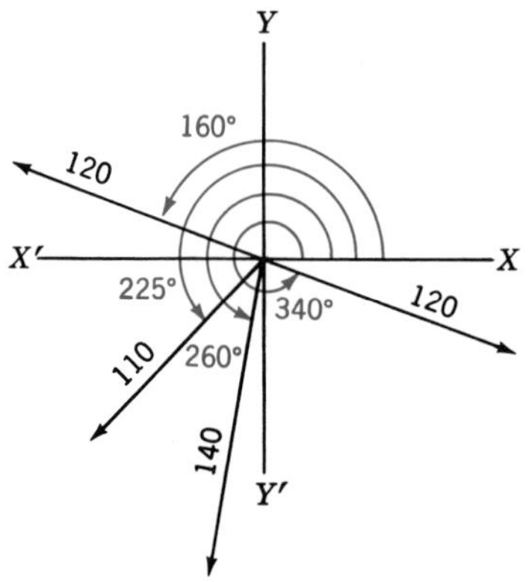

FIG. 28–11 Vector diagram of Prob. 3.

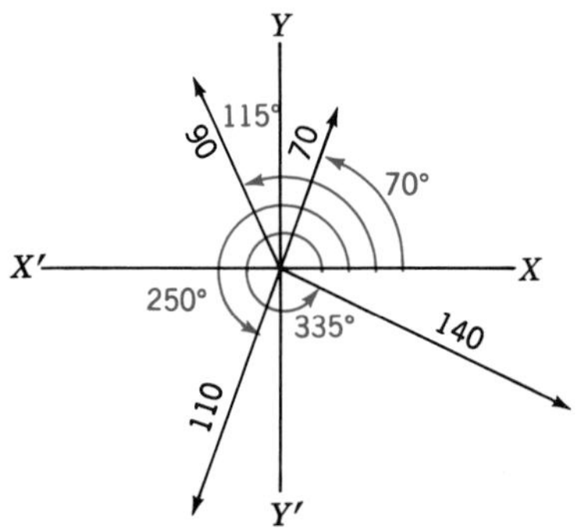

FIG. 28–12 Vector diagram of Prob. 4.

28–4 COMPONENTS OF A VECTOR

From what has been considered regarding combining or adding vectors, it follows that a vector can be resolved into components along any two specified directions. For example, in Fig. 28–4, the vectors ***OA*** and ***AB*** are components of the vector ***OB***. If the directions of the components are so chosen that they are at right angles to each other, the components are called *rectangular components*.

By placing the initial point of a vector at the origin of the x and y axes, the rectangular components are readily obtained either graphically or mathematically.

EXAMPLE 4 A vector with a magnitude of 10 makes an angle of 53.1° with the horizontal. What are the vertical and horizontal components?

SOLUTION

The vector is illustrated in Fig. 28–13 as the directed line segment ***OA***. Its length drawn to scale represents the magnitude of 10, and it makes an angle of 53.1° with the x axis.

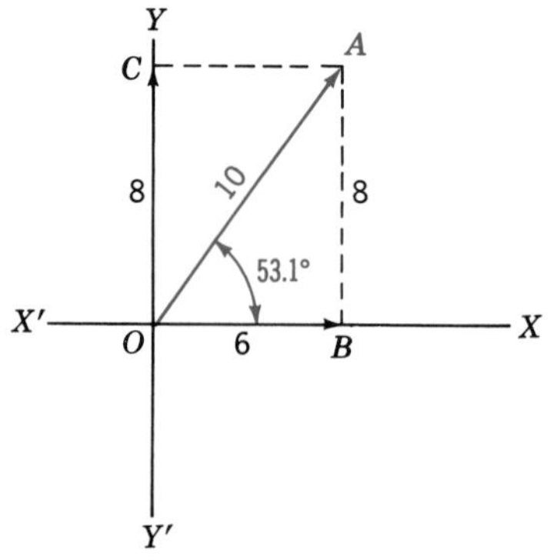

FIG. 28–13 Vertical and horizontal components of vector.

The *horizontal component* of $\boldsymbol{OA}$ is the horizontal distance (see Sec. 26–7) from O to A and is found graphically by projecting the vector $\boldsymbol{OA}$ upon the x axis. Thus the vector $\boldsymbol{OB}$ is the horizontal component of $\boldsymbol{OA}$.

The *vertical component* of $\boldsymbol{OA}$ is the vertical distance from O to A and is found graphically by projecting the vector $\boldsymbol{OA}$ upon the y axis. Similarly, the vector $\boldsymbol{OC}$ is the vertical component of $\boldsymbol{OA}$. Finding the horizontal and vertical components of $\boldsymbol{OA}$ by mathematical methods is simply a problem in solving a right triangle as was outlined in Sec. 26–4. Hence,

$$\boldsymbol{OB} = 10 \cos 53.1° = 6$$

and

$$\boldsymbol{OC} = \boldsymbol{BA} = 10 \sin 53.1° = 8$$

CHECK

$$\theta = \arctan \frac{8}{6} = \arctan 1.33 = 53.1°$$

$$10^2 = 6^2 + 8^2 = 36 + 64 = 100$$

The foregoing can be summarized as follows:

RULE

1. The horizontal component of a vector is the projection of the vector upon a horizontal line, and it equals the magnitude of the vector multiplied by the cosine of the angle made by the vector with the horizontal.
2. The vertical component of a vector is the projection of the vector upon a vertical line, and it equals the magnitude of the vector multiplied by the sine of the angle made by the vector with the horizontal.

EXAMPLE 5 An airplane is flying on a course of 40° at a speed of 400 km/h. How many kilometers per hour is the plane advancing in a due eastward direction? In a direction due north?

SOLUTION

Draw the vector diagram as shown in Fig. 28–14. (Courses are measured from the north.) The vector $\boldsymbol{OB}$, which is the horizontal component of $\boldsymbol{OA}$, represents the velocity of the airplane in an eastward direction. The vector $\boldsymbol{OC}$, which is the vertical component of $\boldsymbol{OA}$, represents the velocity of the airplane in a northward direction.

Again, the process of finding the magnitude of $\boldsymbol{OB}$ and $\boldsymbol{OC}$ resolves into a problem of solving the right triangle OBA. Hence,

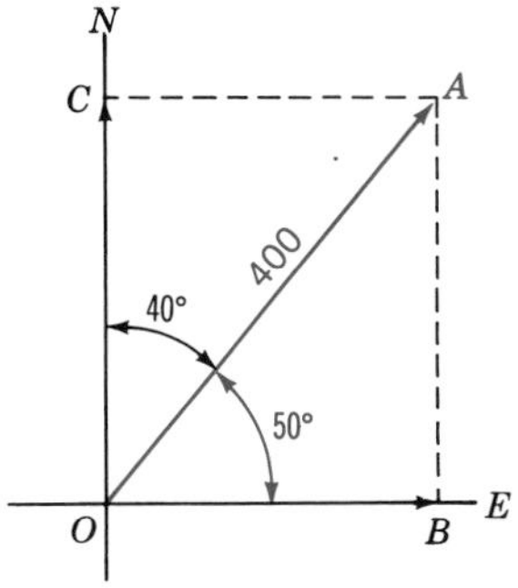

FIG. 28–14 Vector diagram of Example 5.

$$\boldsymbol{OB} = 400 \cos 50° = 257 \text{ km/h eastward}$$

$$\boldsymbol{OC} = \boldsymbol{BA} = 400 \sin 50° = 306.5 \text{ km/h northward}$$

If the vector diagram has been drawn to scale, an approximate check can be made by measuring the lengths of $\boldsymbol{OB}$ and $\boldsymbol{OC}$. Such a check will disclose any large errors in the mathematical solution.

EXAMPLE 6 A radius vector of unit length is rotating about a point with a velocity of $2\pi^r$/s. What are its horizontal and vertical components (*a*) at the end of 0.15 s, (*b*) at the end of 0.35 s, (*c*) at the end of 0.75 s?

SOLUTION

(*a*) In 15 s the rotating vector will generate $2\pi \times 0.15 = 0.942^r$, or $0.942 \times 57.3° = 54°$ as shown in Fig. 28–15.

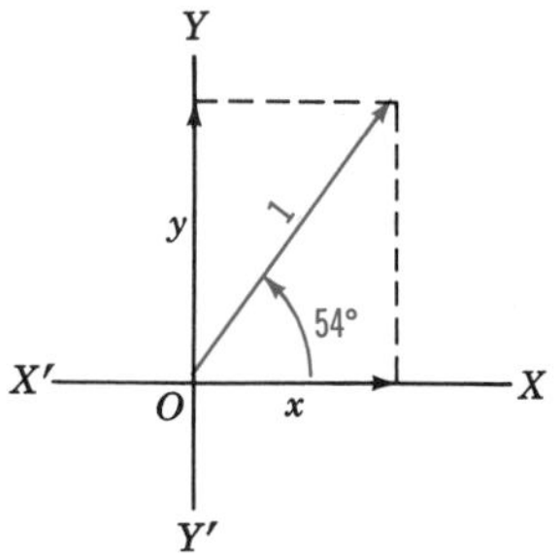

FIG. 28–15 When $t = 0.15$ s, angle $\theta = 54°$.

The horizontal component, measured along the *x* axis, is

$$x = 1 \cos 54° = 0.588$$

The vertical component, measured along the *y* axis, is

$$y = 1 \sin 54° = 0.809$$

Check the solution by measurement or any other method considered convenient.

(*b*) At the end of 0.35 s the rotating vector will have generated an angle of $2\pi \times 0.35 = 2.20^r$, or $2.20 \times 57.3° = 126°$ as shown in Fig. 28–16. The horizontal component, measured along the *x* axis, is

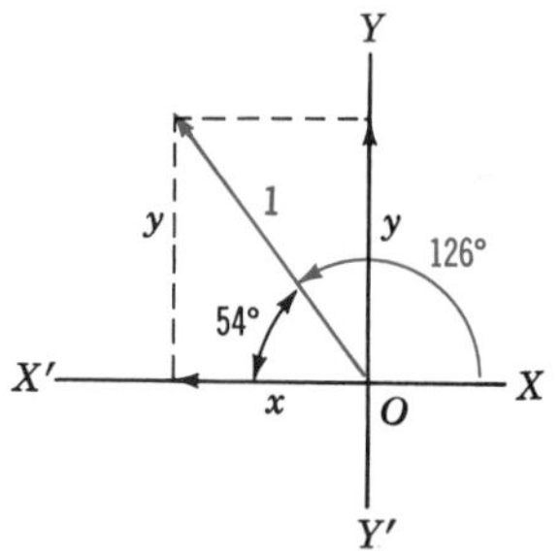

FIG. 28–16 When $t = 0.35$ s, angle $\theta = 126°$.

$$x = 1 \cos 126° = 1(-\cos 54°)$$
$$= -0.588 \qquad \text{(Sec. 25–11)}$$

The vertical component, measured along the *y* axis, is

$$y = 1 \sin 126° = 1 \sin 54° = 0.809 \qquad \text{(Sec. 25–11)}$$

Check by some convenient method.

(*c*) At the end of 0.75 s the rotating vector will have generated an angle of $2\pi \times 0.75 = 4.71^r$, or $4.71 \times 57.3° = 270°$ as shown in Fig. 28–17.

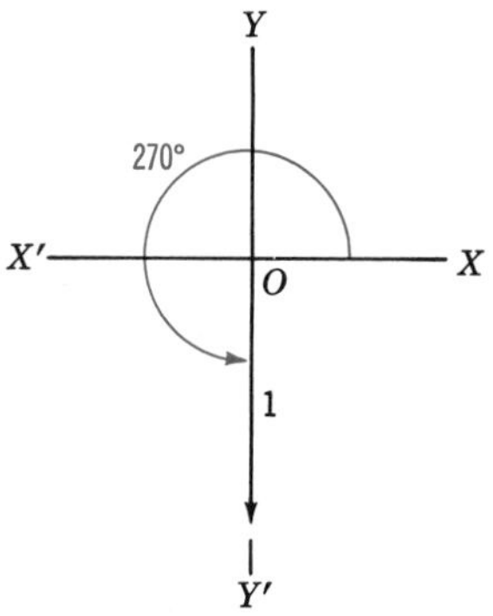

FIG. 28–17 When $t = 0.75$ s, angle $\theta = 270°$.

The horizontal component is

$$x = 1 \cos 270° = 0$$

The vertical component is

$$y = 1 \sin 270° = -1$$

PROBLEMS 28–2

Find the horizontal and vertical components, denoted by x and y, respectively, of the following vectors. Check the mathematical solution of each by drawing a vector diagram to scale.

1. 45 at 52° (This is commonly written $45\underline{/\ 52°}$)
2. $68\underline{/\ 34.6°}$
3. $1.36\underline{/\ 71.8°}$
4. $1800\underline{/\ 120°}$
5. $51.3\underline{/\ 180°}$
6. $0.987\underline{/\ 295.5°}$
7. $210.6\underline{/\ 247.5°}$
8. $27.8275\underline{/\ 90°}$
9. $30.8\underline{/\ 157.3°}$
10. $1600\underline{/\ 270°}$
11. The resultant of two forces acting at right angles is a force of 1120 N which makes an angle of 26.7° with one of the forces. Find the component forces.
12. A test missile was fired at an angle of 82° from the horizontal. At a particular instant its velocity was 1950 km/h. Find its horizontal velocity at that instant in meters per second.
13. A jet fighter leaves its base and flies 1800 km southeast. How far east does it go?
14. Resolve a force of 250 N into two rectangular components, one of which is 155 N.
15. The resultant of two forces acting at right angles is 288 N. One of the forces is 122 N. What is the other?

28–5 PHASORS

Early in this chapter we discovered the difference between scalar quantities, which involve magnitude only, and vectors, which involve both magnitude and direction. When electrical units are shown on paper, with the length of the line indicating the magnitude and the direction of the line indicating the phase relationship, they may be thought of as *vectors*. However, when an emf is impressed across a circuit, its *polarity* is not *direction* in the sense of vector definition. The paper representation as vectors serves a valuable purpose in our circuit calculations, but the electrical quantities are not true vectors. Since the angular separation of electrical units always represents *time* revealed as a *phase* relationship, scientists and engineers prefer to use the term *phasors* when discussing electrical "vectors."

On paper (in a "uniplanar" representation) there is no difference between phasors and vectors. The operations of conversion between rectangular and polar forms are the same. The summation of perpendicular components is the same. But since our purpose is to study the mathematics of electronics in an electronics environment and our communication is with electronics and scientific people, we will use the expressions *phasor* and *phasor summation* throughout the remaining chapters of this book.

28–6 PHASOR SUMMATION OF RECTANGULAR COMPONENTS

If two forces that are at right angles to each other are acting on a body, their resultant can be found by the usual methods of phasor summation as outlined in Sec. 28–3. However, the resultant can be obtained by geometric or trigonometric methods; for the problem is that of solving for the hypotenuse of a right triangle when the other two sides are given, as outlined in Sec. 26–6.

EXAMPLE 7 Two phasors are acting at a point. One with a magnitude of 6 is directed along the horizontal to the right of the point, and the other with a magnitude of 8 is directed vertically above the point. Find their resultant.

SOLUTION 1

In Fig. 28–18 the horizontal phasor, with a magnitude of 6, is shown as ***OB.*** The vertical phasor, with a magnitude of 8, is shown as ***OC***. The resultant of these two phasors can be obtained graphically by completing the parallelogram of forces *OCAB*, as was outlined in Sec. 28–3. Thus, the magnitude of the resultant will be represented by the length of ***OA*** in Fig. 28–18. The angle, or direction of the resultant, can be measured with the protractor.

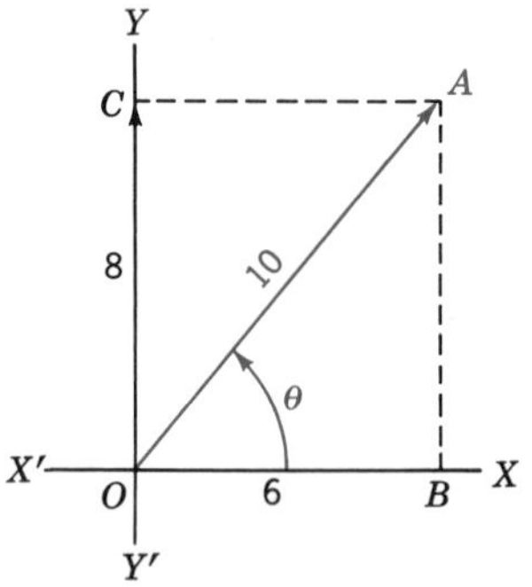

FIG. 28–18 Addition of rectangular components.

Graphical methods have a limited degree of accuracy, as pointed out in earlier sections. They should be used as an approximate check for more precise mathematical methods.

SOLUTION 2

Since ***BA*** = ***OC*** in Fig. 28–18, *OBA* is a right triangle the hypotenuse of which is the resultant ***OA***. Therefore, the magnitude of the resultant is

$$\begin{aligned} \boldsymbol{OA} &= \sqrt{\boldsymbol{OB}^2 + \boldsymbol{BA}^2} = \sqrt{6^2 + 8^2} \\ &= 10 \end{aligned}$$

The angle, or direction of the resultant, is

$$\begin{aligned} \theta &= \arctan \frac{\boldsymbol{BA}}{\boldsymbol{OB}} = \arctan \frac{8}{6} = \arctan 1.33 \\ &= 53.1° \end{aligned}$$

Although the method of Solution 2 is accurate and mathematically correct, there are several operations involved. For example, in finding the magnitude, 6 and 8 must be squared, these squares must be added, and then the square root of this sum must be extracted. This involves four operations.

SOLUTION 3

Since *OBA* is a right triangle for which ***OB*** and ***BA*** are given, the hypotenuse (resultant) can be computed as explained in Sec. 26–6. Hence,

$$\tan\theta = \frac{BA}{OB} = 1.33$$

$$\therefore \theta = 53.1^\circ$$

Then $$OA = \frac{OB}{\cos 53.1^\circ} = \frac{6}{0.6} = 10$$

or $$OA = \frac{BA}{\sin 53.1^\circ} = \frac{8}{0.8} = 10$$

The method of Solution 3 is to be preferred, owing to the minimum number of operations involved.

It should be noted that Example 4 of Sec. 28–4 involves the same quantities as those that were used in the example of this section and that Figs. 28–13 and 28–18 are alike. In the earlier example a vector that is resolved into its rectangular components is given. In the example of this section, the same components are given as vectors which are added vectorially to obtain the vector of the first example. From this it is apparent that resolving a vector into its rectangular components and adding vectors that are separated by 90° are inverse operations. Basically, either problem resolves itself into the solution of a right triangle.

PROBLEMS 28–3

Find the resultants of the following sets of phasors:

1. $41.6\underline{/0^\circ}$ and $188\underline{/90^\circ}$
2. $12.4\underline{/0^\circ}$ and $1.97\underline{/90^\circ}$
3. $4.27\underline{/90^\circ}$ and $5.36\underline{/0^\circ}$
4. $45.4\underline{/0^\circ}$ and $153\underline{/90^\circ}$
5. $295\underline{/0^\circ}$ and $110\underline{/90^\circ}$
6. $459\underline{/0^\circ}$ and $405\underline{/0^\circ}$
7. $202\underline{/0^\circ}$ and $95\underline{/180^\circ}$
8. $5.27\underline{/180^\circ}$ and $6.0\underline{/90^\circ}$
9. $295\underline{/270^\circ}$ and $170\underline{/180^\circ}$
10. $323\underline{/270^\circ}$ and $323\underline{/0^\circ}$
11. $5.15\underline{/180^\circ}$ and $9.26\underline{/270^\circ}$
12. $84.2\underline{/0^\circ}$, $34.4\underline{/90^\circ}$, and $37\underline{/90^\circ}$
13. $28.2\underline{/270^\circ}$, $37\underline{/90^\circ}$, $21.4\underline{/0^\circ}$, and $47\underline{/180^\circ}$
14. $167\underline{/270^\circ}$, $252\underline{/0^\circ}$, $143.8\underline{/180^\circ}$, and $81.3\underline{/90^\circ}$
15. $12.1\underline{/0^\circ}$, $72.3\underline{/270^\circ}$, $51.9\underline{/90^\circ}$, $2.7\underline{/270^\circ}$, $8.6\underline{/90^\circ}$, and $31.6\underline{/180^\circ}$
16. Check your calculated answers graphically.

28–7 PHASOR SUMMATION OF NONRECTANGULAR COMPONENTS

Often we are called upon to resolve into a resultant a set of phasors which are not themselves perpendicular (Fig. 28–19). The best analytical method of arriving at a solution is to apply the methods already developed in this chapter.

The first step is to find the perpendicular components of each of the phasors to be added and determine their magnitudes and directions. These are shown in Fig. 28–19 as h_A and v_A, the components of phasor A, and h_B and v_B, the components of phasor B.

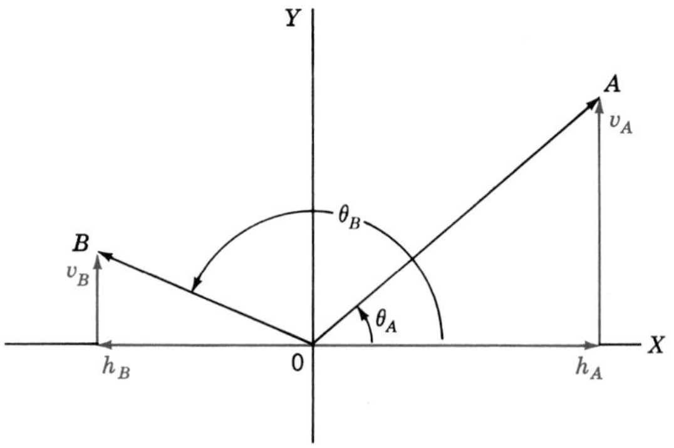

FIG. 28–19 Summation of nonrectangular phasors by resolution into rectangular components.

Second, these components are added algebraically. The horizontal components are added to obtain the resultant horizontal phasor, and then the vertical components are added to obtain the resultant vertical phasor:

$$h_R = h_A + h_B$$

and

$$v_R = v_A + v_B$$

taking into consideration the signs as well as the magnitudes of the components.

Finally, the resultant is the phasor summation of the new perpendicular components:

$$R = \sqrt{h_R^2 + v_R^2}$$

$$\theta_R = \arctan \frac{v_R}{h_R}$$

$$R = \frac{h_R}{\cos \theta} = \frac{v_R}{\sin \theta}$$

EXAMPLE 8 Find the resultant of two phasors $500\underline{/\ 36.9°}$ and $142\underline{/\ 135°}$.

SOLUTION

Sketch the two phasors in the standard position (Fig. 28–20), and then resolve each phasor into its perpendicular components:

$$h_{500} = 500 \cos 36.9° = 400$$

$$h_{142} = 142 \cos 135° = -142 \cos 45° = -100$$

$$v_{500} = 500 \sin 36.9° = 300$$

$$v_{142} = 142 \sin 135° = 142 \sin 45° = 100$$

Add these components algebraically to obtain the new horizontal and vertical resultants:

$$h_R = +400 - 100 = 300$$

$$v_R = +300 + 100 = 400 \qquad \text{(Fig. 28–21)}$$

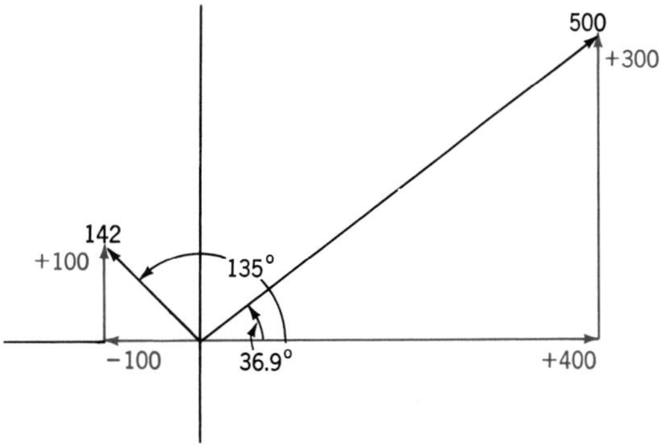

FIG. 28–20 Nonrectangular phasor summation of Example 8.

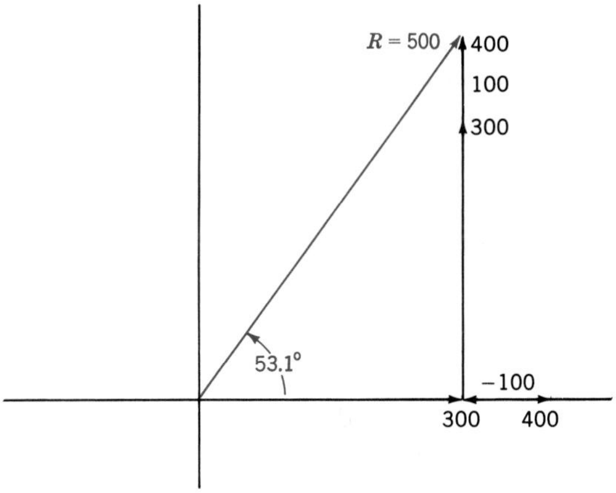

FIG. 28–21 Perpendicular components of Fig. 28–20 resolved into resultant *R*.

The angle θ_R, which R makes with the x axis, is

$$\theta = \arctan \frac{400}{300} = 53.1°$$

and the resultant R of the two resultant perpendicular components is

$$R = \frac{300}{\cos 53.1°}$$

or

$$R = \frac{400}{\sin 53.1°} = 500$$

This process of analysis of phasors into their components and synthesis of resultant components into a final phasor resultant may be applied to any number of phasors.

PROBLEMS 28–4

Find the resultants of the following sets of phasors. Check your solutions graphically:

1. $198\angle 51.3^\circ$ and $115\angle 38.1^\circ$
2. $647\angle 51.5^\circ$ and $215\angle 135^\circ$
3. $64.4\angle 35.1^\circ$ and $22.2\angle 310^\circ$
4. $7.65\angle 17.8^\circ$ and $4.34\angle 137.5^\circ$
5. $9.1\angle 27.3^\circ$, $39.8\angle 78.4^\circ$, and $58.4\angle 235.6^\circ$

Your calculator may be programmed to automatically convert polar forms to rectangular and vice versa. Look for a key labelled P→R or some simple variation. It is time to consult your instruction manual to determine the steps necessary for the conversion by internal operations. Attempt each of the conversions in this chapter using the P→R. Once you have become proficient, you may want to use the key regularly in the following chapters.

In addition, check your manual for the possibility that your calculator has been programmed to automatically add or subtract numbers in complex form. Your calculator may require an input in only polar or only rectangular form, but it may be programmed to accept the two parts of a complex number and add (or subtract) them to the two parts of another complex number. Then the result can be easily converted to the equivalent form.

SELF TEST

1. Find the magnitude and direction with respect to the positive X axis, of the resultant of the vectors in Fig. 28–22.

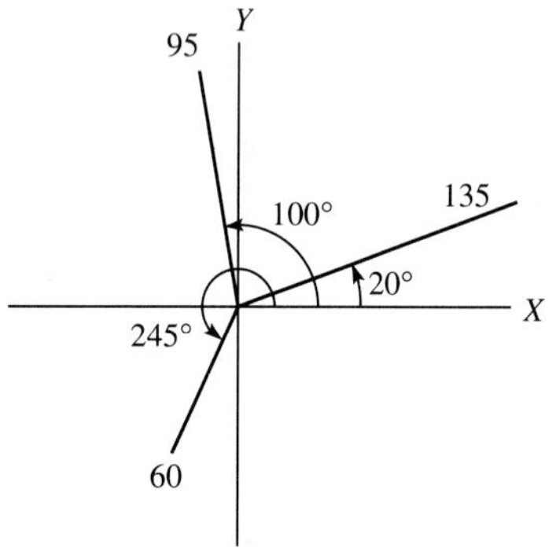

FIG. 28–22 Vectors of self test Prob. 1.

2. The resultant of two forces acting at right angles is 2.40 kN. One of the forces is 800 N. What is the other?
3. What is the resultant of $120\angle 90^\circ$ and $210\angle 180^\circ$?
4. What is the resultant of $84\angle 30.6^\circ$ and $125\angle 47.2^\circ$?

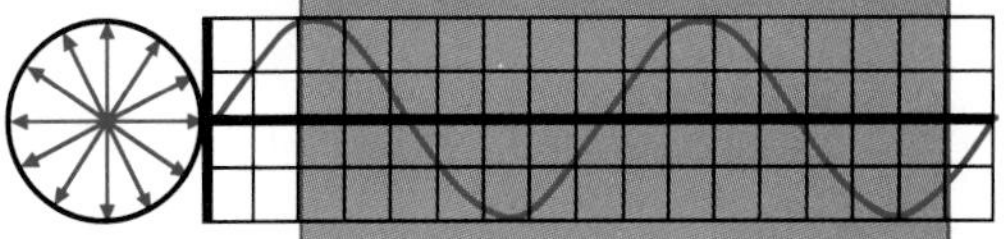

CHAPTER 29

PERIODIC FUNCTIONS

In Sec. 24–9, it was shown that the trigonometric functions could be represented by the ratios of lengths of certain lines to the unit radius vector. Also, in Sec. 24–8, the variation of the functions was represented by lines.

The complete variation of the functions is more clearly illustrated and better understood by plotting the continuous values of the functions on rectangular coordinates.

29–1 THE GRAPH OF THE SINE CURVE $y = \sin x$

The equation $y = \sin x$ can be plotted just as the graphs of algebraic equations are plotted, that is, by assigning values to the angle x (the independent variable), computing the corresponding value of y (the dependent variable), plotting the points whose coordinates are thus obtained, and drawing a smooth curve through the points. This is the same procedure as used for plotting linear equations in Chap. 16 and for plotting quadratic equations for Chap. 21.

The first questions that come to mind in preparing to graph this equation are, "What values will be assigned to x? Will they be in radians or degrees?" Either might be used, but it is more reasonable to use radians. In Sec. 23–5, it was shown that an angle measured in radians can be represented by the arc intercepted by the angle on the circumference of a circle of unit radius. Since, as previously mentioned, the functions of an angle can be represented by suitable lengths of lines, it follows that if an angle is expressed in radian measure, both the angle and its functions can be expressed in terms of a common unit of length. Therefore, we shall select a suitable unit of length and plot both x and y values in terms of that unit. Then, to graph the equation $y = \sin x$, the procedure is as follows:

1. Assign values to x.
2. From the calculator or a table, determine the corresponding values of y (Table 29–1).

TABLE 29–1

x, degrees	x, radians (π measure)	x, radians (unit measure)	y ($\sin x$)	Point
0	0	0	0	$P_0 = (0, 0)$
30	$\frac{\pi}{6}$	0.52	0.50	$P_1 = (0.52, 0.50)$
60	$\frac{\pi}{3}$	1.05	0.87	$P_2 = (1.05, 0.87)$
90	$\frac{\pi}{2}$	1.57	1.00	$P_3 = (1.57, 1.00)$
120	$\frac{2\pi}{3}$	2.09	0.87	$P_4 = (2.09, 0.87)$
150	$\frac{5\pi}{6}$	2.62	0.50	$P_5 = (2.62, 0.50)$
180	π	3.14	0	$P_6 = (3.14, 0)$

3. Take each pair of values of x and y as coordinates of a point, and plot the point.
4. Draw a smooth curve through the points.

It is not necessary to tabulate values of $\sin x$ between π and 2π radians (180 to 360°), for these values are negative but equal in magnitude to the sines of the angles between 0 and π radians (0 to 180°). To plot the curve, the angle and the function should have the same unit or scale; that is, one *unit* on the y axis should be the same length as that representing 1 *radian* on the x axis. When the curve is so plotted, it is called a *proper sine curve,* as shown in Fig. 29–1. This wave-shaped curve is called the *sine curve* or *sinusoid*.

If additional values of x are chosen, both positive and negative, the curve continues indefinitely in both directions while repeating in value. Note that, as x increases from 0 to $\frac{\pi}{2}$ $\left(\text{or } \frac{1}{2}\pi\right)$, $\sin x$ increases from 0 to 1; as x increases from $\frac{1}{2}\pi$ to π, $\sin x$ decreases from 1 to 0; as x increases from π to $\frac{3\pi}{2}$, $\sin x$ increases from 0 to -1; and as x increases from $\frac{3\pi}{2}$ to 2π, $\sin x$ decreases from -1 to 0. Thus the curve repeats itself for every multiple of 2π radians.

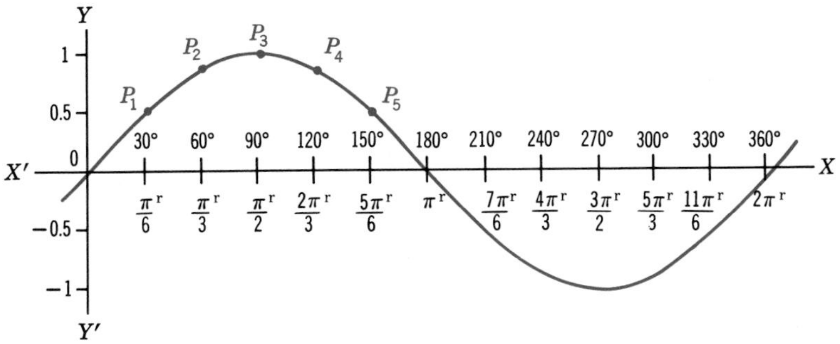

FIG. 29–1 Graph of the equation $y = \sin x$.

29–2 THE GRAPH OF THE COSINE CURVE $y = \cos x$

By following the procedure for plotting the sine curve, you can easily verify that the graph of $y = \cos x$ appears as shown in Fig. 29–2.

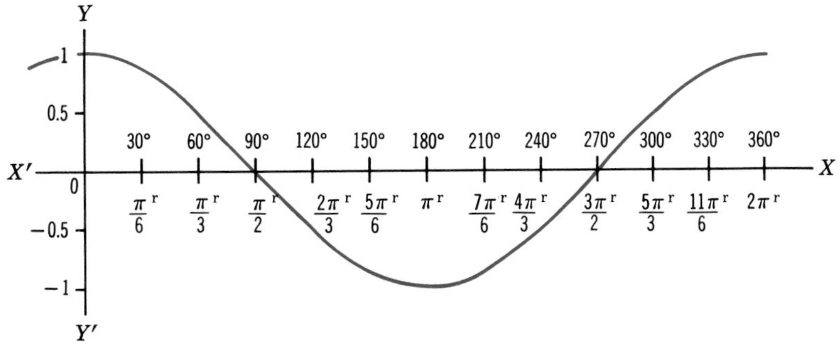

FIG. 29–2 Graph of the equation $y = \cos x$.

Note that, as x increases from 0 to $\frac{1}{2}\pi$, $\cos x$ decreases from 1 to 0; as x increases from $\frac{1}{2}\pi$ to π, $\cos x$ increases from 0 to -1; as x increases from π to $\frac{3\pi}{2}$, $\cos x$ decreases from -1 to 0; and as x increases from $\frac{3\pi}{2}$ to 2π, $\cos x$ increases from 0 to 1. If additional values of x are chosen, both positive and negative, the curve will repeat itself indefinitely in both directions. The cosine curve is identical in shape with the sine curve except that there is a difference of 90° between corresponding points on the two curves. Another similarity between the curves is that both curves repeat their values for every multiple of 2π radians ($2\pi^r$).

29-3 THE GRAPH OF THE TANGENT CURVE $y = \tan x$

The graph of the equation $y = \tan x$, shown in Fig. 29-3, has characteristics different from those of the sine or cosine curve. The curve slopes upward and to the right. At points where x is an odd multiple of $\frac{1}{2}\pi$, the curve is discontinuous. This is to be expected from the discussion of the tangent function in Sec. 24-8.

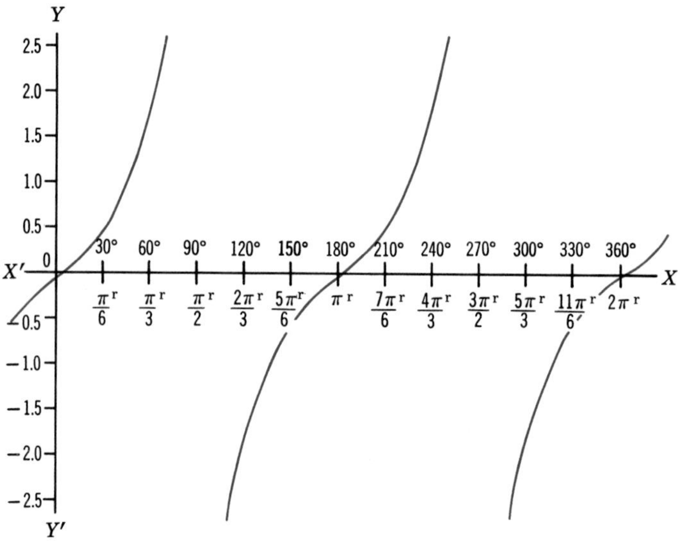

FIG. 29-3 Graph of the equation $y = \tan x$.

The tangent curve repeats itself at intervals of π radians (π^r), and it is thus seen to be a series of separate curves, or branches, rather than a continuous curve.

PROBLEMS 29-1

1. Plot the equation $y = \sin x$ from -2π to $2\pi^r$.
2. Plot the equation $y = \cos x$ from -2π to $2\pi^r$.
3. Plot the equation $y = \cot x$ from 0 to $2\pi^r$.
4. Plot the equation $y = \sec x$ from 0 to $2\pi^r$.
5. Plot the equation $y = \csc x$ from 0 to $2\pi^r$.
6. Plot the equations $y = \sin^2 x$ and $y = \cos^2 x$ on the same coordinates and to the same scale. In computing points, remember that when a negative number is squared, the result is positive. Add the respective ordinates of the curves for several different values of angle, and plot the results. What conclusion do you draw from these results?

29–4 PERIODICITY

From the graphs plotted in the preceding figures and from earlier considerations of the trigonometric functions, it is evident that each trigonometric function repeats itself exactly in the same order and at regular intervals. A function that repeats itself periodically is called a *periodic function*. From the definition, it is apparent that the trigonometric functions are periodic functions.

Owing to the fact that many natural phenomena are periodic in character, the sine and cosine curves lend themselves ideally to graphical representation and mathematical analysis of these recurrent motions. For example, the rise and fall of tides, motions of certain machines, the vibrations of a pendulum, the rhythm of our bodily life, sound waves, and water waves are familiar happenings that can be represented and analyzed by the use of sine and cosine curves. An alternating current follows these variations, as will be shown in Chap. 30, and it is because of this fact that you must have a good grounding in trigonometry. It is essential that you understand the mathematical expressions for various periodic functions and especially their applications to ac circuits.

The tangent, cotangent, secant, and cosecant curves are not used to represent recurrent happenings, for although these curves are periodic, they are discontinuous for certain values of angles.

29–5 ANGULAR MOTION

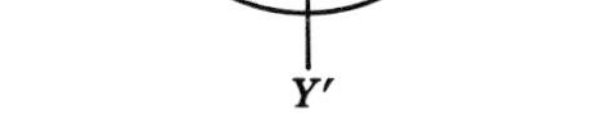

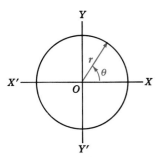

FIG. 29–4 Radius vector ***r*** generates angle θ.

The *linear velocity* of a point or object moving in a particular direction is the rate at which distance is traveled by the point or object. The unit of velocity is the distance traveled in unit time when the motion of the point or object is uniform, such as kilometers per hour, meters per second, or centimeters per second.

The same concept is used to measure and define *angular velocity.*

In Fig. 29–4 the radius vector $\boldsymbol{r}$ is turning about the origin in a counterclockwise direction to generate the angle θ. The *angular velocity* of such a rotating line is the rate at which an angle is generated by rotation. When the rotation is uniform, the unit of angular velocity is the angle generated per unit of time. Thus, angular velocity is measured in degrees per second or radians per second, the latter being the more widely used.

Angular velocity may be expressed as revolutions per minute or revolutions per second. For example, if f is the number of revolutions per second of the vector of Fig. 29–4, then $2\pi f$ is the number of radians generated per second. The angular velocity in radians per second is denoted by ω (Greek letter omega). Thus, if the radius vector is rotating f revolutions per second,

$$\omega = 2\pi f \qquad \text{rad/s}$$

If the armature of a generator is rotating at 1800 rev/min, which is 30 rev/s, it has an angular velocity of

$$\omega = 2\pi f = 2\pi \times 30 = 188.4^{r}/\text{s}$$

where we have again used ^{r} as the symbol for radians.

The total angle θ generated by a rotating line in t s at an angular velocity of ω^r/s is

$$\theta = \omega t \qquad \text{rad}$$

Thus the angle generated by the armature in 0.01 s is

$$\theta = \omega t = 188.4 \times 0.01 = 1.884^r$$

or

$$\theta = 1.884 \times \frac{180}{\pi} = 108°$$

EXAMPLE 1 A flywheel has a velocity of 300 rev/min. (*a*) What is its angular velocity? (*b*) What angle will be generated in 0.2 s? (*c*) How much time is required for the wheel to generate 628^r?

SOLUTION

(*a*)
$$f = \frac{300 \text{ rev/min}}{60} = 5 \text{ rev/s}$$

Then
$$\omega = 2\pi f = 2\pi \times 5 = 10\pi \text{ or } 31.4^r\text{/s}$$

(*b*)
$$\theta = \omega t = 10\pi \times 0.2 = 2\pi^r$$
$$\theta = 360°$$

(*c*) Since
$$\theta = \omega t$$

then
$$t = \frac{\theta}{\omega} = \frac{628}{10\pi} = 20 \text{ s}$$

PROBLEMS 29–2

1. What is the angular velocity, in terms of π^r/s, of (*a*) the hour hand of a clock, (*b*) the minute hand of a clock, and (*c*) the second hand of a clock?
2. Express the angular velocity of 3600 rev/min in (*a*) radians per second and (*b*) degrees per second.
3. If a satellite circles the earth in 78 min, what is its average angular velocity in (*a*) degrees per minute and (*b*) radians per second?
4. A revolution counter on an armature shaft recorded 1200 rev in 30 s. What was the value of the shaft's angular velocity in (*a*) radians per minute and (*b*) degrees per minute?
5. The radius vector $\boldsymbol{r}$ of Fig. 29–4 is rotating at the rate of 1200 rev/s. What is the value of θ in radians at the end of:
(*a*) 0.01 s, (*b*) 0.001 s, and (*c*) 0.5 ms?
6. If the radius vector $\boldsymbol{r}$ of Fig. 29–4 is rotating at the rate of 1 rev/s, what is the value of $\sin \omega t$ at the end of:
(*a*) 0.001 s, (*b*) 0.1 s, (*c*) 0.5 s, and (*d*) 0.95 s?

29–6 PROJECTION OF A POINT HAVING UNIFORM CIRCULAR MOTION

In Fig. 29–5 the radius vector $\boldsymbol{r}$ rotates about a point in a counterclockwise direction with a uniform angular velocity of 1 rev/s. Then every point on the radius vector, such as the end point P, rotates with uniform angular velocity. If the radius vector starts from 0°, at the end of $\frac{1}{12}$ s it will have rotated 30° or 0.5236^r, to P_1; at the end of $\frac{1}{6}$ s it will have rotated to P_2 and generated an angle of 60°, or 1.047^r, etc.

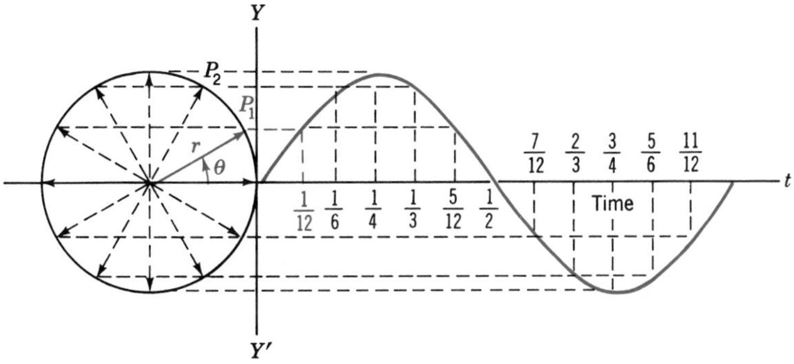

FIG. 29–5 Radius vector generating sine curve.

The projection of the end point of the radius vector, that is, its ordinate value at any time, can be plotted as a curve. The plotting is accomplished by extending the horizontal diameter of the circle to the right for use as an x axis along which time is to be plotted. Choose a convenient length along the x axis and divide it into as many intervals as there are angle values to be plotted. In Fig. 29–5, projections have been made every 30°, starting from 0°. Therefore, the x axis is divided into 12 divisions; and since one complete revolution takes place in 1 s, each division on the time axis will represent $\frac{1}{12}$ s, or 30° rotation.

Through the points of division on the time axis (x axis), construct vertical lines, and through the corresponding points (made by the end point of the radius vector at that particular time) draw lines parallel to the time axis. Draw a smooth curve through the points of intersection. Thus the resulting sine curve traces the ordinate of the end point of the radius vector for any time t, and from it we would obtain the sine value for any angle generated by the radius vector.

As the vector continues to rotate, successive revolutions will generate repeating, or periodic, curves.

Since the y value of the curve is proportional to the sine of the generated angle and the length of the radius vector, we have

$$y = \boldsymbol{r} \sin \theta$$

Then, since the radius vector rotates through $2\pi^r$ in 1 s, the y value at any time t is

$$y = \boldsymbol{r} \sin 2\pi t$$

or

$$y = \boldsymbol{r} \sin 6.28t$$

which is the equation of the sine curve of Fig. 29–5.

From the foregoing considerations, it is apparent that if a straight line of length $\boldsymbol{r}$ rotates about a point with a uniform angular velocity of ω^r per unit time, starting from a horizontal position when the time $t = 0$, the projection y of the end point upon a vertical straight line will have a motion that can be represented by the relationship

$$y = \boldsymbol{r} \sin \omega t \qquad (1)$$

This equation is of fundamental importance in describing the motion of any object or quantity that varies periodically, or with *simple harmonic motion*. Thus the value of an alternating emf at any instant can be completely described in terms of such an equation; that is, if the motion or variation can be represented by a sine curve, it is said to be *sinusoidal* or to vary *sinusoidally.*

EXAMPLE 2 A crank 150 mm long, starting from 0°, turns in a counterclockwise direction at the rate of 1 rev in 10 s.

(*a*) What is the equation for the projection of the crank handle upon a vertical line at any instant? That is, what is the vertical distance from the crankshaft at any time?

(*b*) What is the vertical distance from the handle to the shaft at the end of 3 s?

(*c*) At the end of 8 s?

SOLUTION

(*a*) The general equation for the projection of the end point on a vertical line is

$$y = \boldsymbol{r} \sin \omega t \tag{1}$$

where $\boldsymbol{r}$ = length of rotating object
ω = angular velocity, rad/s
t = time at any instant, s

Then, since the crank makes 1 rev, or $2\pi^{r}$, in 10 s, the angular velocity is

$$\omega = \frac{2\pi}{10} = \frac{\pi}{5}, \text{ or } 0.628^{r}/\text{s}$$

Substituting the values of r and ω in Eq. (1),

$$y = 150 \sin 0.628t \text{ mm}$$

(*b*) At the end of 3 s, the crank will have turned through

$$0.628 \times 3 = 1.88^{r}$$

which is $1.88 \times \frac{180}{\pi} = 108°$. Substituting this value for $0.628t$ in Eq. (1) results in

$$y = 150 \sin 108° = 150 \times 0.951 = 142.6 \text{ mm}$$

which is the vertical distance of the handle from the shaft at the end of 3 s.

(*c*) At the end of 8 s the crank will have turned through

$$0.628 \times 8 = 5.02^{r}$$

Which is $5.02 \times \frac{180}{\pi} = 288°$. Substituting this value for $0.628t$ in the above equation results in

$$y = 150 \sin 288° = 150 \times (-0.951) = -142.6 \text{ mm}$$

which is the vertical distance of the handle from the shaft at the end of 8 s. The negative sign denotes that the handle is *below* the shaft; that is, the distance is measured downward, whereas the distance in (*b*) above was taken as positive, or *above* the shaft.

If it is desired to express the projection of the end point of the radius vector upon the horizontal, the relation is

$$y = \mathbf{r} \cos \omega t \tag{2}$$

which, when plotted, results in a cosine curve. Thus, in the foregoing example, the horizontal distance (Sec. 26–7) between the handle and shaft at the end of 8 s will be

$$\begin{aligned} y = 150 \cos 288° &= 150 \times 0.309 \\ &= 46.4 \text{ mm} \end{aligned}$$

29–7 AMPLITUDE

The graphs of Figs. 29–1, 29–2, and 29–5 have an equal amplitude of 1, that is, an equal vertical displacement from the horizontal axis. The value of the radius vector $\boldsymbol{r}$ determines the amplitude of a general curve, and for this reason the factor $\boldsymbol{r}$ in the general equation

$$y = \boldsymbol{r} \sin \omega t$$

is called the *amplitude factor*. Thus the amplitude of a periodic curve is taken as the maximum displacement, or value, of the curve. It is apparent that, if the length of the radius vector which generates a sine wave is varied, the amplitude of the sine wave will be varied accordingly. This is illustrated in Fig. 29–6.

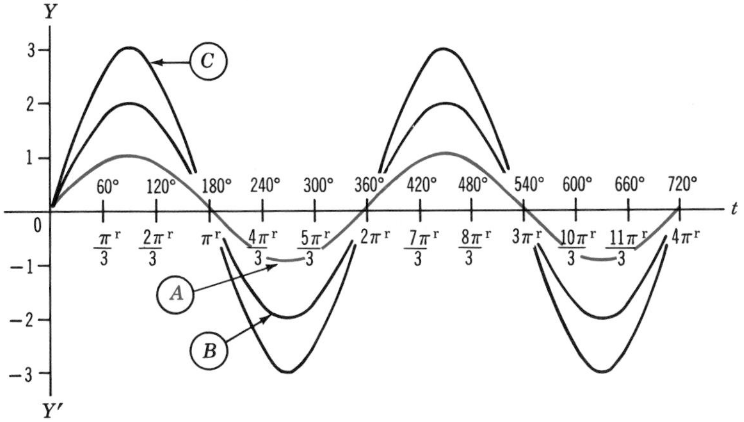

FIG. 29–6 A: $y = \sin \theta$, B: $y = 2 \sin \theta$, C: $y = 3 \sin \theta$.

29–8 FREQUENCY

When the radius vector makes one complete revolution, regardless of its starting point, it has generated one complete sine wave; hence, we say the sine wave has gone through one complete *cycle*. Thus the number of cycles occurring in a periodic curve in a unit of time is called the *frequency* of the curve. For example, if the radius vector rotated 5 rev/s, the curve describing its motion would go through 5 cycles in 1 s of time. The frequency f in hertz is obtained by dividing the angular velocity ω by 360° when the latter is measured in degrees or by 2π when measured in radians. That is,

$$f = \frac{\omega}{2\pi} \qquad \text{Hz} \tag{3}$$

Curves for different frequencies are shown in Fig. 29–7.

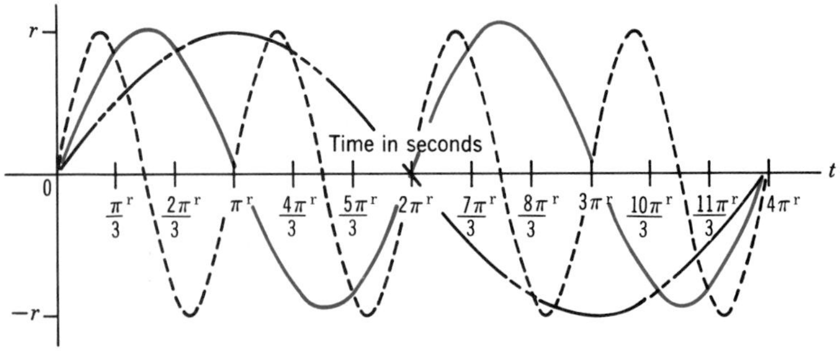

FIG. 29–7 $y = \boldsymbol{r} \sin t$ ____, $y = \boldsymbol{r} \sin 2t$ _ _ _, $y = \boldsymbol{r} \sin \frac{1}{2}t$ ___ _ _ ___.

In the equation $y = \boldsymbol{r} \sin \frac{1}{2}t$, since $\omega t = \frac{1}{2}t$, the angular velocity ω is 0.5 rad/s. That is, at the end of 2π, or 6.28 s, the curve has gone through one-half cycle, or 3.14^r of an angle, as shown in Fig. 29–7.

In the equation $y = \boldsymbol{r} \sin t$, since $\omega t = t$, the angular velocity ω is 1 rad/s. Thus at the end of 2π s the curve has gone through one complete cycle, or $2\pi^r$ of angle.

Similarly, in the equation $y = \boldsymbol{r} \sin 2t$, the angular velocity ω is 2 rad/s. Then at the end of 2π s the curve has completed two cycles, or $4\pi^r$ of angle.

29–9 PERIOD

The time T required for a periodic function, or curve, to complete one cycle is called the *period*. Hence, if the frequency f is given by

$$f = \frac{\omega}{2\pi} \qquad \text{Hz}$$

it follows that

$$T = \frac{2\pi}{\omega} = f^{-1} \qquad \text{s} \tag{4}$$

For example, if a curve repeats itself 60 times in 1 s, it has a frequency of 60 Hz and a period of

$$T = \frac{1}{60} = 0.0167 \text{ s}$$

Similarly, in Fig. 29–7, the curve represented by $y = \boldsymbol{r} \sin \frac{1}{2}t$ has a frequency of

$$\frac{\omega}{2\pi} = \frac{0.5}{2\pi} = 0.0796 \text{ Hz}$$

and a period of 12.6 s. The curve of $y = \boldsymbol{r} \sin t$ has a frequency of

$$\frac{\omega}{2\pi} = \frac{1}{2\pi} = 0.159 \text{ Hz}$$

and a period of 6.28 s. The curve of $y = \boldsymbol{r} \sin 2t$ has a frequency of 0.318 Hz and a period of 3.14 s.

29–10 PHASE

In Fig. 29–8, two radius vectors are rotating about a point with equal angular velocities of ω and separated by the constant angle θ. That is, if $\boldsymbol{r}$ starts from the horizontal axis, then $\boldsymbol{r}_1$ starts ahead of $\boldsymbol{r}$ by the angle θ and maintains this angular difference.

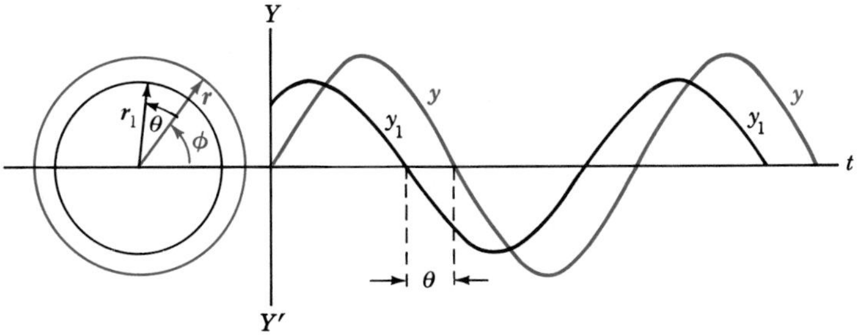

FIG. 29–8 $y = \boldsymbol{r} \sin \omega t$, $y_1 = \boldsymbol{r}_1 \sin (\omega t + \theta)$.

When $t = 0$, $\boldsymbol{r}$ starts from the horizontal axis to generate the curve $y = \boldsymbol{r} \sin \omega t$. At the same time, $\boldsymbol{r}_1$ is ahead of $\boldsymbol{r}$ by an angle θ; hence, $\boldsymbol{r}_1$ generates the curve

$$y_1 = \boldsymbol{r}_1 \sin (\omega t + \theta)$$

It will be noted that this *displaces* the y_1 curve along the horizontal by an angle θ as shown in the figure.

The angular difference θ between the two curves is called the *phase angle*, and since y_1 is *ahead* of y, we say that y_1 leads y. Thus, in the equation $y_1 = \boldsymbol{r}_1 \sin (\omega t + \theta)$, θ is called the *angle of lead*. In Fig. 29–8, y_1 leads y by 30°; therefore, the equation for y_1 becomes

$$y_1 = \boldsymbol{r}_1 \sin (\omega t + 30°)$$

In Fig. 29–9, the radius vectors $\boldsymbol{r}$ and $\boldsymbol{r}_1$ are rotating about a point with equal angular velocities of ω, except that now $\boldsymbol{r}_1$ is *behind* $\boldsymbol{r}$ by a constant

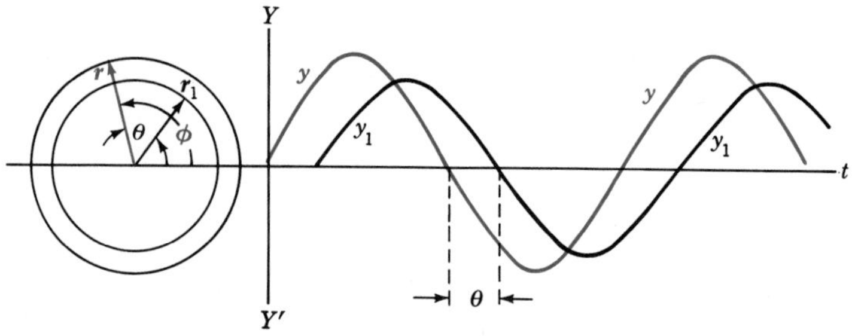

FIG. 29–9 $y = \boldsymbol{r} \sin \omega t$, $y_1 = \boldsymbol{r}_1 \sin (\omega t - \theta)$.

angle θ. The phase angle between the two curves is θ, but in this case, y_1 lags y. Hence the equation for the curve generated by $\boldsymbol{r}_1$ is

$$y_1 = \boldsymbol{r}_1 \sin (\omega t - \theta)$$

In Fig. 29–9, the radius vectors $\boldsymbol{r}$ and $\boldsymbol{r}_1$ are rotating about a point with equal angular velocities of ω, except that now $\boldsymbol{r}_1$ is *behind* $\boldsymbol{r}$ by a constant angle θ. The phase angle between the two curves is θ, but in this case, y_1 lags y. Hence the equation for the curve generated by $\boldsymbol{r}_1$ is y_1

$$y_1 = \boldsymbol{r}_1 \sin (\omega t - \theta)$$

29–11 SUMMARY

The general equation

$$y = \boldsymbol{r} \sin (\omega t \pm \theta) \tag{5}$$

describes a periodic event, and its graph results in a periodic curve. By choosing the proper values for the three arbitrary constants $\boldsymbol{r}$, ω, and θ, you can describe or plot any periodic sequence of events because a change in any one of them will change the curve accordingly. Hence,

If $\boldsymbol{r}$ is changed, the *amplitude* of the curve will be changed proportionally. For this reason, $\boldsymbol{r}$ is called the *amplitude factor*.
If ω is changed, the *frequency*, or period, of the curve will be changed. Thus, ω is called the *frequency factor*.
If θ is changed, the curve is moved along the time axis with no other change. Thus, if θ is made larger, the curve is displaced to the left and results in a leading phase angle. If θ is made smaller, the curve is moved to the right and results in a lagging phase angle. Hence the angle θ in the general equation is called the *phase angle* or *the angle of lead or lag*.

EXAMPLE 3 Discuss the equation

$$y = 147 \sin (377t + 30°)$$

SOLUTION

Given $y = 147 \sin (377t + 30°)$.

Comparing the given equation with the general equation, it is seen that $\boldsymbol{r} = 147$, $\omega = 377$ rad/s, and $\theta = 30°$. Therefore, the curve represented by this equation is a sine curve with an amplitude of 147. The angular velocity is 377; hence, the frequency is

$$f = \frac{\omega}{2\pi} = \frac{377}{2\pi} = 60 \text{ Hz}$$

and the period is

$$T = f^{-1} = \frac{1}{60} = 0.0167 \text{ s}$$

The curve has been displaced to the left 30°; that is, it leads the curve $y = \boldsymbol{r} \sin 377t$ by a phase angle of 30°. Therefore, when $t = 0$, the curve begins at an angle of 30° with a value of

$$\begin{aligned} y &= 147 \sin (\omega t + 30°) \\ &= 147 \sin (0° + 30°) \\ &= 147 \times 0.5 = 73.5 \end{aligned}$$

PROBLEMS 29–3

In the following equations of periodic curves, specify (*a*) amplitude, (*b*) angular velocity, (*c*) frequency, (*d*) period, and (*e*) angle of lead or lag with respect to a curve of the same frequency but having no displacement angle.

1. $y = 250 \sin (2\pi t + 20°)$
2. $y = 170 \sin (377t - 15°)$
3. $i = 2.55 \sin (628t + 10°)$
4. $i = I_{max} \sin (2513t - 22°)$
5. $v = 184 \sin (157t - 22°)$
6. $i_c = I_{c_{max}} \sin (1000\pi t + 37°)$

For problems 7–14, plot the curves that represent the following motions:

7. $y = \sin 2\pi t$
8. $y = 10 \sin 10t$
9. $v = 141 \sin 120t$
10. $i = 0.5 \sin (120t + 30°)$
11. $i = 1.3 \sin (120t - 20°)$
12. $y = 16 \sin (377t + 10°)$
13. A parabolic radar antenna 120 cm across is pivoted at its center to produce an "effective horizontal length" of 60 cm. It rotates in a horizontal plane at 20 rev/s in a counterclockwise direction, starting from east.
    (*a*) Plot the curve that shows the projection of the antenna on a north-south centerline at any time.
    (*b*) Write the equation for the curve.
    (*c*) What is the distance of the end of the antenna from the east-west line at the end of 0.08 s?
    (*d*) What is the distance of the end of the antenna from the north-south line at the end of 0.1 s?
    (*e*) Through how many radians will the antenna turn in 0.25 s?

14. A radar scope scanning line rotates on the face of the oscilloscope just as a spoke on a wheel rotates with the wheel. If a scan line that is 175 mm long rotates in a positive direction at 12 sweeps/s, starting from a position 40° below the horizontal:
    (*a*) Plot the curve that shows the projection of the line upon a vertical reference line at any time.
    (*b*) Write the equation of the curve.
    (*c*) Find the vertical projection of the line at the end of 0.0375 s.
    (*d*) Find the horizontal projection of the line at the end of 0.833 s.
    (*e*) Through how many radians will the line sweep in 2.5 s?

SELF TEST

An almanac shows the time of moonrise at a particular latitude on two successive days to be 5:36 P.M. and 6:17 P.M. What is the apparent angular velocity of the moon in terms of π r/s?

2. If $y = 277 \sin (120\pi t + 18°)$, what is:
    (*a*) the amplitude?
    (*b*) the angular velocity?
    (*c*) the frequency?
    (*d*) the period?
    (*e*) the angle of lead or lag?
    (*f*) the value of y when $t = 6$ ms?

CHAPTER 30

ALTERNATING CURRENTS—FUNDAMENTAL IDEAS

Thus far we have considered direct voltages and direct currents, that is, voltages that do not change in polarity and currents that do not change in their directions of flow.

In this chapter, you will begin the study of mathematics as applied to alternating currents. An *alternating current* is one that alternates, or changes its direction, periodically.

The fact that over 90% of the electric energy produced is generated in the form of alternating current makes this subject very important, for the operation of all radio and communication circuits is based on ac phenomena. The first requisite in the study of electronics engineering is a solid foundation in the principles of alternating currents.

30–1 GENERATION OF AN ALTERNATING ELECTROMOTIVE FORCE

A coil of wire that has its ends connected to slip rings and is rotating in a counterclockwise direction in a uniform magnetic field is illustrated in Fig. 30–1. That an alternating emf will be generated in the coil is apparent from a consideration of generated currents. For example, when the side of the coil *ab* moves from its present position away from the S pole, the emf generated in it will be directed from *b* to *a;* that is, *a* will be positive with respect to *b*. At the same time, the side of the coil *cd* is moving away from the N pole, thus cutting magnetic lines of force with a motion opposite that of *ab*. Then the emf generated in *cd* will be directed from *c* to *d* and will add to the emf from *b* to *a* to send a current I_1 through the resistance R.

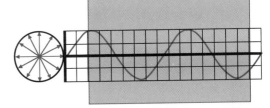

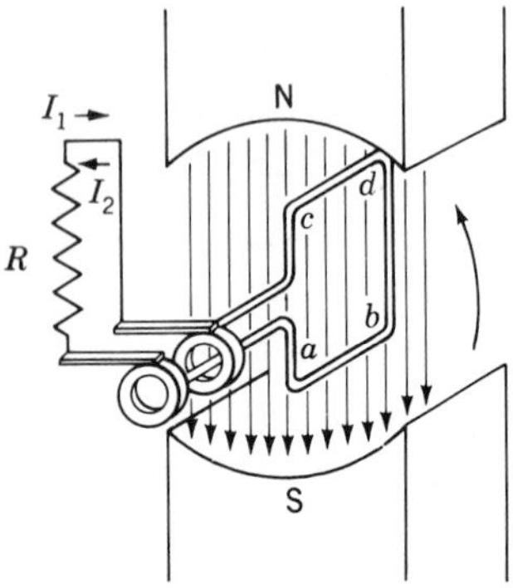

FIG. 30–1 Representation of elementary alternator.

When the coil has rotated 90° from the position shown in Fig. 30–1, the plane of the coil is perpendicular to the magnetic field, and at this instant the sides of the coil are moving parallel to the magnetic field, thus cutting no lines of force. There is no emf generated at this instant.

As the side of the coil *ab* begins to move up toward the N pole, the emf generated in it will now be directed from *a* to *b*. Similarly, because the side of the coil *cd* is now moving down toward the S pole, the emf in *cd* will be directed from *d* to *c*. This reversal of the direction of generated emf is due to a change of direction of motion with respect to the direction of the lines of force. Therefore, the flow of current I_2 through R will be in the direction indicated by the arrow.

When the coil rotates so that the plane of the coil is again perpendicular to the lines of force (270° from the position shown in Fig. 30–1), no emf will be generated at that instant. Rotation beyond this position, however, causes the generation of an emf such that current flows in the original direction I_1. Such an emf, which periodically reverses its direction, is known as an *alternating electromotive force,* and the resulting current is known as an *alternating current*.

In some engineering textbooks the generation of an emf is explained as due to the change of magnetic flux through the rotating coil. In the final analysis, the results are the same. Here we are interested mainly in the behavior of the circuits connected to sources of alternating currents.

30–2 VARIATION OF AN ALTERNATING ELECTROMOTIVE FORCE

The first questions that come to mind are, "In what manner does an alternating emf vary? How can we represent that variation graphically?"

Figure 30–2 illustrates a cross section of the elementary alternator of Fig. 30–1. The circles represent either side of the rotating coil at successive instants during the rotation.

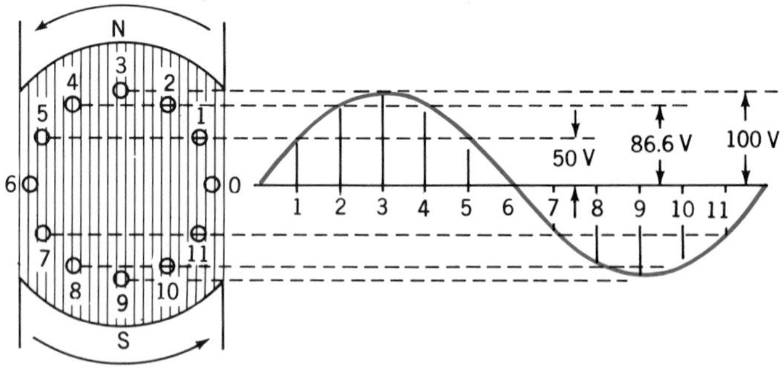

FIG. 30–2 Generation of voltage sine wave.

When a conductor passes through a magnetic field, there must be a component of its velocity at right angles to the lines of force in order to generate an emf. For example, a conductor must actually *cut* lines in order to develop an emf the amount of which will be proportional to the number of lines cut and the rate of cutting.

From studies of rotation and a consideration of Fig. 30–2, it is evident that the component of horizontal velocity of the rotating conductor is proportional to the sine of the angle of rotation. Because the horizontal velocity is perpendicular to the magnetic field, it is this component that develops an emf. For example, at position 0, where the angle of rotation is zero, the conductor is moving parallel to the field; hence, no voltage is generated. As the conductor rotates toward 90°, the component of horizontal velocity becomes greater, thus generating a higher voltage. Therefore, the sine curve of Fig. 30–2 is a graphical representation of the induced emf in a conductor rotating in a uniform magnetic field. The voltage starts from zero, increases in a positive direction to a maximum value (100 V in the figure) at 90°, decreases to zero at 180°, increases in the opposite or negative direction until it attains maximum negative value at 270°, and finally decreases to zero value again at 360°. It follows, then, that the induced emf can be completely described by the relation

$$v = \boldsymbol{V}_{\max} \sin \theta \qquad \text{V} \tag{1}$$

where v = instantaneous value of emf at any angle θ, V
$\boldsymbol{V}_{\max}$ = maximum value of emf, V
θ = angular position of coil

30–3 VECTOR REPRESENTATION

Since the sine wave of emf is a periodic function, a simpler method of representing the relation of the emf induced in a coil to the angle of rotation is available. The rotating conductor can be replaced by a rotating radius vector whose length represents the magnitude of the maximum generated voltage $\boldsymbol{V}_{\max}$. Then the instantaneous value for any position of the conductor can be represented by the vertical component of the vector (Sec. 28–4).

In Fig. 30–3, which is the vector diagram for the conductor at position 0 in Fig. 30–2, the sector $\boldsymbol{V}_{\max}$ is at 0° position and therefore has no vertical component. Thus the value of the emf in this position is zero. Or, since

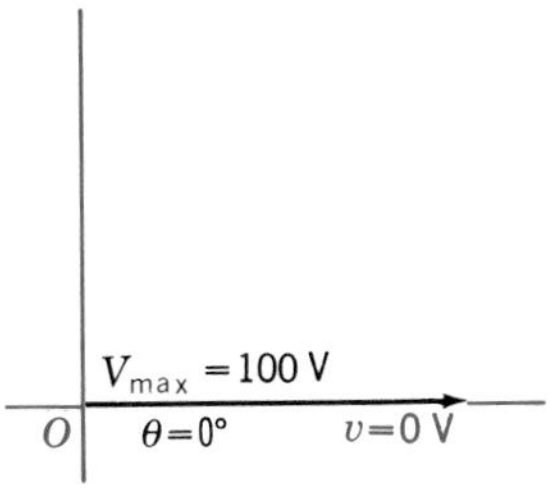

FIG. 30–3 $v = \sin 0° = 0$ V.

$$v = \boldsymbol{V}_{\max} \sin \theta$$

by substituting the values of $\boldsymbol{V}_{\max}$ and θ,

$$v = 100 \sin 0° = 0$$

In Fig. 30–4, which is the vector diagram for the conductor at position 2 in Fig. 30–2, the coil has moved 60° from the zero position. The vector $\boldsymbol{V}_{\max}$ is therefore at an angle of 60° from the reference axis, and the instantaneous value of the induced emf is represented by the vertical component of $\boldsymbol{V}_{\max}$. Then, since

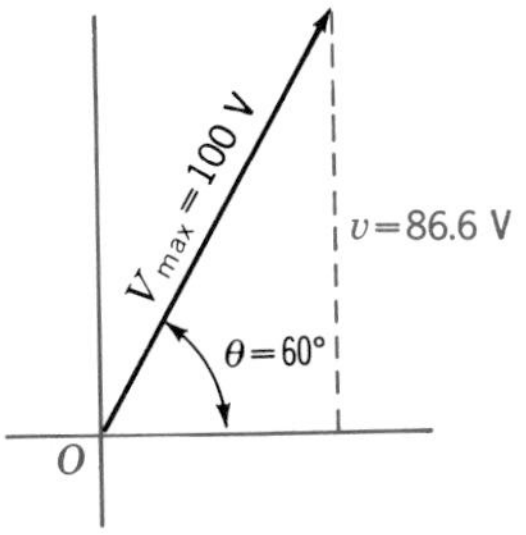

FIG. 30–4
$v = 100 \sin 60°$
$= 86.6$ V.

$$v = \boldsymbol{V}_{\max} \sin \theta$$

by substituting the values of $\boldsymbol{V}_{\max}$ and θ,

$$v = 100 \sin 60° = 86.6 \text{ V}$$

EXAMPLE 1 What is the instantaneous value of an alternating emf that has reached 58° of its cycle? The maximum value is 500 V.

SOLUTION

Draw the vector diagram to scale as shown in Fig. 30–5. The instantaneous value is the vertical component of the vector $\boldsymbol{V}_{max}$. Then, since

$$v = \boldsymbol{V}_{max} \sin \theta$$

by substituting the values of $\boldsymbol{V}_{max}$ and θ,

$$v = 500 \sin 58° = 424 \text{ V}$$

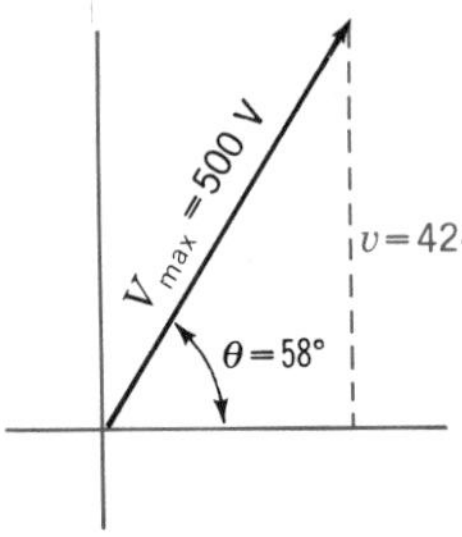

FIG. 30–5 $v = 500 \sin 58° = 424$ V.

EXAMPLE 2 What is the instantaneous value of an alternating emf when the emf has reached 216° of its cycle? The maximum value is 163 V.

SOLUTION

Draw the vector diagram to scale as shown in Fig. 30–6. The instantaneous value is the vertical component of the vector $\boldsymbol{V}_{max}$. Then, since

$$v = \boldsymbol{V}_{max} \sin \theta$$

by substituting the values of $\boldsymbol{V}_{max}$ and θ,

$$\begin{aligned} v = 163 \sin 216° &= 163[-\sin (216° - 180°)] \\ &= 163(-\sin 36°) = -95.8 \text{ V} \end{aligned}$$

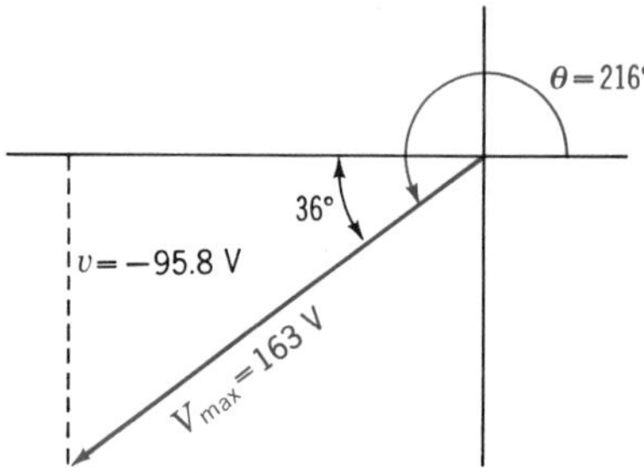

FIG. 30–6 $v = 163 \sin (-36°) = -95.8$ V.

A vector diagram drawn to scale should be made for every ac problem. It will give you a better insight into the functioning of alternating currents and at the same time serve as a good check on the mathematical solution.

Since the current in a circuit is proportional to the applied voltage, it follows that an alternating emf which varies periodically will produce a current of similar variation. Hence, the instantaneous current of a sine wave of alternating current is given by

$$i = \boldsymbol{I}_{max} \sin \theta \qquad \text{A} \qquad (2)$$

where i = instantaneous value of current, A
I_{max} = maximum value of current, A
θ = angular position of coil

PROBLEMS 30–1

1. An alternating current has a maximum value of 170 mA. What are the instantaneous values of this current at the following points in its cycle: (*a*) 10°, (*b*) 60°, (*c*) 135°, (*d*) 225°, (*e*) 320°?
2. The instantaneous value of an alternating emf at 24° is 18.4 V. What is its maximum value?
3. The instantaneous current at 196° is −121 mA. What is the maximum value?
4. An alternating current has a maximum value of 750 mA. What are the instantaneous values of the current at the following points in its cycle: (*a*) 26°, (*b*) 341°, (*c*) 210°, (*d*) 297°, (*e*) 162°?
5. The instantaneous value of an alternating emf is 110 V at 71°. What will the value be at 232°?
6. The instantaneous value of an alternating emf at 289° is −22 V. What will the value be at 142°?
7. The instantaneous value of voltage at 59° is 3 V. What is the maximum value?
8. An alternating current has a maximum value of 365 mA. At what angles will it be 80% of its positive maximum value?
9. At what angles are the instantaneous values of an alternating current equal to 50% of the maximum negative value?
10. What is the instantaneous value of an alternating emf 92° after its maximum positive value of 156 V?

30–4 CYCLES, FREQUENCY, AND POLES

Each revolution of the coil in Fig. 30–1 results in one complete *cycle* which consists of one positive and one negative loop of the sine wave (Sec. 29–8). The number of cycles generated in 1 s is called the *frequency* of the alternating emf, and the *period* is the time required to complete one cycle. One half cycle is called an *alternation*. Thus, by a 60-Hz alternating current is meant that the current passes through 60 cycles per second, which results in a period of 0.0167 s. Also, a 60-Hz current completes 120 alternations per second.

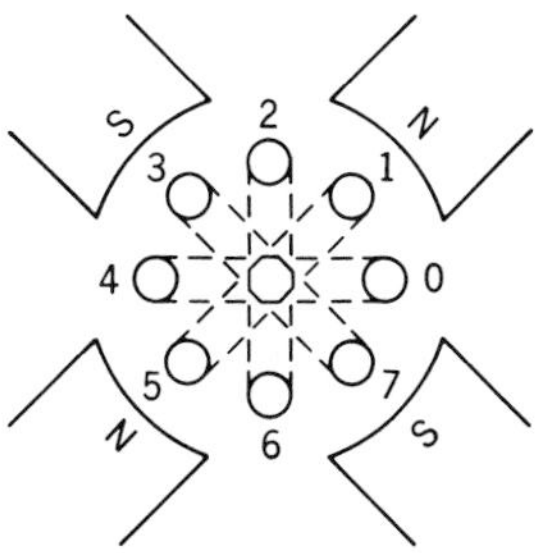

FIG. 30–7 Elementary four-pole alternator.

Figure 30–7 represents a coil rotating in a four-pole machine. When one side of the coil has rotated from position 0 to position 4, it has passed under the influence of an N and an S pole, thus generating one complete sine wave, or electrical cycle. This corresponds to 2π electrical radians, or 360 electrical degrees, although the coil has rotated only 180 space degrees. Therefore, in one complete revolution the coil will generate two complete cycles, or 720 electrical degrees, so that for every *space degree* there result two *electrical time degrees*.

In any alternator the armature, or field, must move an angular distance equal to the angle formed by two consecutive like poles in order to complete one cycle. It is evident, then, that a two-pole machine must rotate at twice the speed of a four-pole machine to produce the same frequency. Therefore, to find the frequency of an alternator in hertz (cycles per second), *the number of pairs of poles is multiplied by the speed of the armature in revolutions per second.* That is,

$$f = \frac{PS}{60} \qquad \text{Hz} \qquad (3)$$

where f = frequency, Hz
P = number of pairs of poles
S = rotational speed of armature, or field, rev/min

EXAMPLE 3 What is the frequency of an alternator that has four poles and rotates at a speed of 1800 rev/min?

SOLUTION

$$f = \frac{2 \times 1800}{60} = 60 \text{ Hz}$$

30–5 EQUATIONS OF VOLTAGES AND CURRENTS

Since each cycle consists of 360 electrical degrees, or 2π electrical radians, the variation of an alternating emf can be expressed in terms of time. Thus, a frequency of f Hz results in $2\pi f^{\text{r}}/\text{s}$, which is denoted by ω (Sec. 29–5). Hence, the instantaneous emf at any time t is given by the relation

$$v = V_{max} \sin \omega t \qquad \text{V} \qquad (4)$$

The instantaneous current is

$$i = I_{max} \sin \omega t \qquad \text{A} \qquad (5)$$

You should review Secs. 29–6 to 29–10 to ensure a complete understanding of the relations between the general equation for a periodic function and Eqs. (4) and (5). Thus, V_{max} and I_{max} are the amplitude factors of their respective equations, and ω is the frequency factor.

EXAMPLE 4 Write the equation of a 60-Hz alternating voltage that has a maximum value of 156 V.

SOLUTION
The angular velocity ω is 2π times the frequency, or

$$2\pi \times 60 = 377^{\text{r}}/\text{s}$$

Substituting 156 V for V_{max} and 377 for ω in Eq. (4),

$$v = 156 \sin 377t \qquad \text{V}$$

EXAMPLE 5 Write the equation of a radio-frequency (RF) current of 700 kHz that has a maximum value of 21.2 A.

SOLUTION

$I_{max} = 21.2$ A and $f = 700$ kHz $= 7 \times 10^5$ Hz. Then

$$\omega = 2\pi f = 2\pi \times 7 \times 10^5$$
$$= 4.4 \times 10^6$$

Substituting these values in Eq. (5),

$$i = 21.2 \sin (4.4 \times 10^6)t \qquad \text{A}$$

EXAMPLE 6 If the time $t = 0$ when the voltage of Example 4 is zero and increasing in a positive direction, what is the instantaneous value of the voltage at the end of 0.002 s?

SOLUTION

Substituting 0.002 for t in the equation for the voltage,

$$v = 156 \sin (377 \times 0.002)$$
$$= 156 \sin 0.754^r \qquad \text{V}$$

where 0.754 is the time angle in *radians*. Then, since $1^r \cong 57.3°$,

$$v = 156 \sin (0.754 \times 57.3°)$$
$$= 156 \sin 43.2°$$

Hence, $$v = 107 \text{ V}$$

PROBLEMS 30–2

1. An alternator with 32 poles has a speed of 1800 rev/min and develops a maximum emf of 311 V.
   (*a*) What is the frequency of the alternating emf?
   (*b*) What is the period of the alternating emf?
   (*c*) Write the equation for the instantaneous emf at any time t.
2. An alternator with 12 poles has a speed of 1200 rev/min and develops a maximum voltage of 170 V.
   (*a*) What is the frequency of the alternating emf?
   (*b*) Write the equation for the instantaneous value of the emf at any time t.
3. A 400-Hz generator which develops a maximum emf of 250 V has a speed of 1200 rev/min.
   (*a*) How many poles has it?
   (*b*) Write the equation of the voltage.
   (*c*) What is the value of the voltage when the time $t = 2$ ms?

4. An 800-Hz alternator generates a maximum of 163 V at the rate of 4000 rev/min.
   (*a*) How many poles has it?
   (*b*) Write the equation for the voltage.
   (*c*) What is the value of the emf when time $t = 500\ \mu s$?
5. At what speed must a 16-pole 400-Hz alternator be driven in order to develop its rated frequency?
6. The equation for a certain alternating current is

$$i = 84.6 \sin 377t \qquad \text{mA}$$

   What is the frequency of the current?
7. The equation for an alternating emf is

$$v = 0.05 \sin (3.14 \times 10^6)t \qquad \text{V}$$

   What is the frequency of the emf?
8. A 400-mV peak signal has a frequency of 4.3 MHz. It is applied across a 180-kΩ resistor.
   (*a*) At what time will its instantaneous value be 346 mV?
   (*b*) What is the maximum instantaneous current that will flow in the 180-kΩ resistor?
9. A 500-MHz current has a maximum instantaneous value of 30 μA. Write the equation describing the current.
10. A broadcasting station operating at 1430 kHz develops a maximum potential of 0.362 mV across a listener's antenna. Write the equation for this emf.

30–6 AVERAGE VALUE OF CURRENT OR VOLTAGE

Since an alternating current or voltage is of sine-wave form, it follows that the average current or voltage of one cycle is zero owing to the reversal of direction each half-cycle. The term *average value* is usually understood to mean the average value of one alternation without regard to positive or negative values. The average value of a sine wave, such as that shown in Fig. 30–2, can be computed to a fair degree of accuracy by taking the average of many instantaneous values between two consecutive zero points of the curve, the values chosen being separated by equal values of angle. Thus, the average value is equal to the average height of any voltage or current loop. The exact average value is $2 \div \pi \cong 0.637$ times the maximum value. Thus, if I_{av} and V_{av} denote the average values of alternating current and voltage, respectively, we obtain

$$I_{av} = \frac{2}{\pi} I_{max} \cong 0.637 I_{max} \qquad \text{A} \tag{6}$$

and

$$V_{av} = \frac{2}{\pi} V_{max} \cong 0.637 V_{max} \qquad \text{V} \tag{7}$$

EXAMPLE 7 The maximum value of an alternating voltage is 622 V. What is the average value?

SOLUTION

$$V_{av} = 0.637V_{max} = 0.637 \times 622$$
$$= 396 \text{ V}$$

30–7 EFFECTIVE VALUE OF CURRENT OR VOLTAGE

If a direct current of I A is caused to flow through a resistance of $R\ \Omega$, the resulting energy converted into heat equals I^2R W. We should not expect an alternating current with a maximum value of 1 A to produce as much heat as a direct current of 1 A, for the former does not maintain a constant value. Thus, the above ac ampere is not as effective as the dc ampere. The *effective value* of an alternating current is rated in terms of direct current; that is, an alternating current has an effective value of 1 A if, when it flows through a given resistance, it produces heat at the same rate as one dc ampere would.

The effective value of a sine wave of current can be computed to a fair degree of accuracy by taking equally spaced instantaneous values and extracting the square root of their average, or mean, squared values. For this reason, the effective value is often called the *root-mean-square* (rms) value. The exact effective value of an alternating current or voltage is $1/\sqrt{2} \cong 0.707$ times the maximum value. Thus, if I and V denote the effective values of current and voltage, respectively, we obtain

$$I = \frac{I_{max}}{\sqrt{2}} \cong 0.707I_{max} \qquad \text{A} \qquad (8)$$

and

$$V = \frac{V_{max}}{\sqrt{2}} = 0.707V_{max} \qquad \text{V} \qquad (9)$$

It should be noted that all meters, unless marked to the contrary, read effective values of current and voltage.

EXAMPLE 8 The maximum value of an alternating voltage is 311 V. What is the effective value?

SOLUTION

$$V = 0.707V_{max} = 0.707 \times 311$$
$$= 220 \text{ V}$$

EXAMPLE 9 An ac ammeter reads 15 A. What is the maximum value of the current?

SOLUTION 1

Since $$I = 0.707I_{max}$$

then $$I_{max} = \frac{I}{0.707}$$

Substituting 15 A for I, $$I_{max} = \frac{15}{0.707} = 21.2 \text{ A}$$

SOLUTION 2

Since $$I = \frac{I_{max}}{\sqrt{2}}$$

then $$I_{max} = I\sqrt{2} = 1.41I$$

Substituting for I, $$I_{max} = 1.41 \times 15 = 21.2 \text{ A}$$

Hence the maximum value of an alternating current or voltage is equal to 1.41 times the effective value.

PROBLEMS 30–3

1. What is the average value of an alternating emf whose maximum value is 12.5 V?
2. What is the maximum value of an alternating current whose average value is 120 mA?
3. The average value of an alternating emf is 10.5 V. What is the maximum value?
4. The maximum value of an alternating current is 173 μA. What is the average value?
5. What is the average value of a sine wave of voltage whose effective value is 77 V?
6. An rms voltmeter indicates 117 V of alternating emf. What is the maximum value of the emf?
7. What is the effective value of an alternating current which has a maximum value of 15 A?
8. What is the effective value of an alternating emf which has an average value of 125 V?
9. The average value of a sine wave of current is 140 mA. What is the rms value?
10. An rms ammeter indicates an alternating current reading of 33.8 A. What is the average value of the current?

30–8 PHASE RELATIONS—PHASE ANGLES

Nearly all ac circuits contain elements, or components, that cause the voltage and current to pass through their corresponding zero values at different times. The effects of such conditions are given detailed consideration in the next chapter.

If an alternating voltage and the resulting alternating current of the same frequency pass through corresponding zero values at the same instant, they are said to be *in phase*.

If the current passes through a zero value before the corresponding zero value of the voltage, the current and voltage are *out of phase* and the current is said to *lead* the voltage.

Figure 30–8 illustrates a phasor diagram and the corresponding sine waves for a current of i A leading a voltage of v V by a *phase angle* of θ

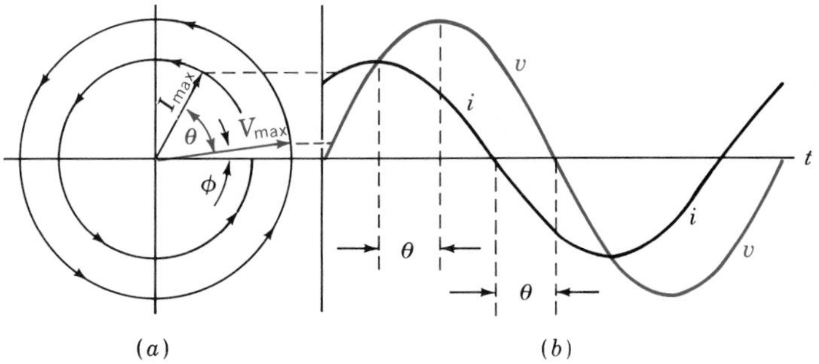

FIG. 30–8 Current i leads voltage v by phase angle θ.

(see Sec. 29–10). Hence, if the voltage is taken as reference, the general equation of the voltage is

$$v = V_{max} \sin \omega t \qquad \text{V} \tag{10}$$

and the current is given by

$$i = I_{max} \sin (\omega t + \theta) \qquad \text{A} \tag{11}$$

The instantaneous values of the voltage and current for any angle φ of the voltage are

$$v = V_{max} \sin \phi \qquad \text{V} \tag{12}$$

and

$$i = I_{max} \sin (\phi + \theta) \qquad \text{A} \tag{13}$$

EXAMPLE 10 In Figure 30–8, the maximum values of the voltage and the current are 156 V and 113 A, respectively. The frequency is 60 Hz, and the current leads the voltage by 40°.

(*a*) Write the equation for the voltage at any time *t*.
(*b*) Write the equation for the current at any time *t*.
(*c*) What is the instantaneous value of the current when the voltage has reached 10° of its cycle?

SOLUTION
Given

$$\begin{aligned} \text{Maximum voltage} &= V_{max} = 156 \text{ V} \\ \text{Maximum current} &= I_{max} = 113 \text{ A} \\ \text{Frequency} &= f = 60 \text{ Hz} \\ \text{Phase angle} &= \theta = 40^\circ \text{ lead} \\ \text{Voltage angle} &= \phi = 10^\circ \end{aligned}$$

Draw a vector diagram as shown in Fig. 30–8*a*. (The circles are not necessary; they simply denote rotation of the vectors.)
(*a*) Substituting given values in Eq. (10),

$$v = 156 \sin 2\pi \times 60t$$

or

$$v = 156 \sin 377t \text{ V}$$

(*b*) Substituting given values in Eq. (11),

$$i = 113 \sin (377t + 40°) \text{ A}$$

NOTE *The quantity 377*t *is in* radians.

(*c*) Substituting given values in Eq. (13),

$$i = 113 \sin (10° + 40°)$$

or

$$i = 113 \sin 50° = 86.6 \text{ A}$$

Figure 30–9 illustrates a vector diagram and the corresponding sine waves for a current of i A lagging a voltage of v V by a *phase angle* of θ. Therefore,

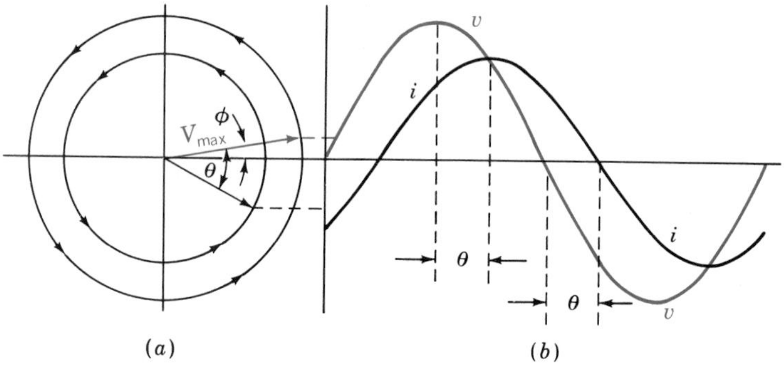

FIG. 30–9 Current i lags voltage v by phase angle θ

if the voltage is taken as reference, the general equation of the voltage will be as given by Eq. (10) and the current will be

$$i = I_{max} \sin (\omega t - \theta) \qquad \text{A}$$

The instantaneous value of the current for any angle ϕ of the voltage is

$$i = I_{max} \sin (\phi - \theta) \qquad \text{A}$$

EXAMPLE 11 In Fig. 30–9, the maximum values of the voltage and the current are 170 V and 14.1 A, respectively. The frequency is 800 Hz, and the current lags the voltage by 40°.

(*a*) Write the equation for the voltage at any time t.

(*b*) Write the equation for the current at any time t.

(*c*) What is the instantaneous value of the current when the voltage has reached 10° of its cycle?

SOLUTION

Given

$$\text{Maximum voltage} = V_{max} = 170 \text{ V}$$
$$\text{Maximum current} = I_{max} = 14.1 \text{ A}$$

$$\text{Frequency} = f = 800 \text{ Hz}$$
$$\text{Phase angle} = \theta = 40° \text{ lag}$$
$$\text{Voltage angle} = \phi = 10°$$

Draw a vector diagram as shown in Fig. 30–9*a*.

(*a*) Substituting given values in Eq. (10),

$$v = 170 \sin 2\pi \times 800t$$

or

$$v = 170 \sin 5030t \text{ V}$$

(*b*) Substituting given values in Eq. (14),

$$i = 14.1 \sin (5030t - 40°) \text{ A}$$

(*c*) Substituting given values in Eq. (15),

$$i = 14.1 \sin (10° - 40°)$$

or

$$i = 14.1 \sin (-30°) = -7.05 \text{ A}$$

EXAMPLE 12 In a certain ac circuit a current of 14 A lags a voltage of 220 V by an angle of 60°. What is the instantaneous value of the voltage when the current has completed 245° of its cycle?

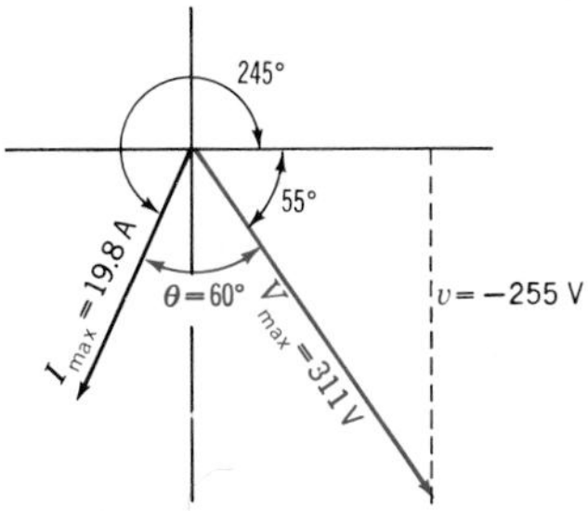

FIG. 30–10 Phasor diagram of Example 12.

NOTE *Unless otherwise specified, all voltages and currents are to be considered* effective *values.*

SOLUTION

Draw the vector diagram as shown in Fig. 30–10.

$$V_{max} = \sqrt{2}V = \sqrt{2} \times 220 = 311 \text{ V}$$
$$\phi = 245° + \theta = 245° + 60°$$
$$= 305° = -55°$$

Then, substituting the values of V_{max} and θ in Eq. (12),

$$v = 311 \sin (-55°) = -255 \text{ V}$$

PROBLEMS 30–4

1. A 120-Hz alternator generates a maximum emf of 156 V and delivers a maximum current of 15 A. The current leads the voltage by an angle of 21°.
   (*a*) Write the equation for the current at any time *t*.
   (*b*) What is the instantaneous value of the current when the emf has completed 60° of its cycle?
2. A 25-Hz alternator generates 6.6 kV at 700 A. The current leads the voltage by an angle of 22°.
   (*a*) Write the equation for the current at any time *t*.
   (*b*) How much of the voltage cycle will have been completed the first time that the instantaneous current rises to 465 A?

3. In the alternator of Prob. 1, what will be the instantaneous value of the current when the voltage has completed 200° of its cycle?
4. In the alternator of Prob. 2, what will be the instantaneous value of the current when the voltage has completed 350° of its cycle?
5. A 400-Hz alternator generates 600 V with a current of 40 A. The phase angle is 20° lagging.
   (*a*) Write the equation for the current at any time t.
   (*b*) What is the instantaneous value of the current when the voltage has completed 192° of its cycle?
6. In the alternator of Prob. 5, what is the instantaneous value of the current when the voltage has completed 17° of its cycle?
7. A 60-Hz alternator generates a maximum of 170 V and delivers a maximum current of 42.4 A. If the instantaneous value of the current is 22.5 A when the instantaneous value of the emf is 112 V, what is the phase angle between the current and the emf?
8. In Prob. 7, what will be the instantaneous value of the emf when the instantaneous value of the current is −39.3 A for the first time?
9. A 400-Hz alternator develops 30 A at 230 V. If the instantaneous value of the emf is −85.8 V when the instantaneous value of the current is 23.5 A, what is the phase angle between current and emf?
10. (*a*) Write the equation for the current in Prob. 9.
    (*b*) In Prob. 9, what will be the instantaneous value of the current when the emf has reached its maximum value negatively?

SELF TEST

1. The maximum value of an alternating voltage is 170. What is its instantaneous value when it has completed 65° of its cycle?
2. The equation of an alternating current is

$$i = 220 \sin (120\pi \times 10^9)t \qquad \mu\text{A}$$

   (*a*) What is its frequency?
   (*b*) What will be its instantaneous amplitude when it has completed 27.5° of its cycle?
3. What is the average value of the current of Prob. 2?
4. What is the effective value of the current of Prob. 2?
5. A 440-Hz alternator delivers 600 V and a current of 62 A with a lagging phase angle of 18°.
   (*a*) Write the equation of the voltage.
   (*b*) Write the equation of the current.
   (*c*) What is the instantaneous value of the current when $t = 0.72$ ms?

CHAPTER 31

PHASOR ALGEBRA

In Sec. 1–2 we commented briefly on the use of mathematics as a tool in electronics. One of the most valuable of all the mathematical tools, certainly the most valuable in the solution of ac circuits, is the operator j together with complex numbers. The complex number operations to be developed in this chapter are so important in electronics that some of the more advanced electronic calculators have special function keys to simplify even further the mathematical processes involved.

31–1 PHASOR DEFINITIONS

REAL AND IMAGINARY NUMBERS Complex numbers were introduced in Secs. 20–14 to 20–18, and, in keeping with traditional methods of notation, Secs. 20–13 and 20–14 used the expressions ''real number'' and ''imaginary number.'' In the complex number $a + jb$, theoretical mathematicians refer to a as the *real part,* and to b as the *imaginary part,* of the complex number.

PHASE Since our development of the operator j did not depend upon any imaginary features, we now abandon the traditional definitions in favor of the more realistic expressions of electrical engineering. We will refer to a as the *in-phase* portion of the complex number and to b as the *out-of-phase* portion. These expressions are more in keeping with the correct understanding of the phase relationships of ac circuits which will be developed further in Chap. 32.

RECTANGULAR FORM The form $a + jb$ is referred to as the *rectangular form* of the complex number. Later, in circuit analyses, we shall use the form $R + jX$.

POLAR FORM When the rectangular components a and b of a complex number are resolved into a single magnitude $\boldsymbol{r}$ rotated through an angle θ from

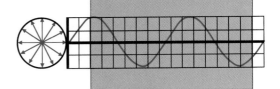

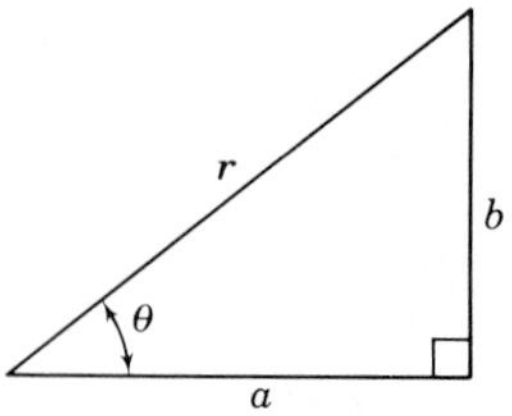

FIG. 31–1 Interrelations of rectangular and phasor forms of complex numbers:

$a = \boldsymbol{r} \cos \theta$ and $b = \boldsymbol{r} \sin \theta$

$\theta = \arctan \frac{b}{a}$

and $\boldsymbol{r} = \frac{a}{\cos \theta} = \frac{b}{\sin \theta}$

a reference axis, the resultant form $\boldsymbol{r}/\underline{\theta}$ is referred to as the *polar form*. Later, in circuit analyses, we shall use the form $\boldsymbol{Z}/\underline{\theta}$.

Any complex number is fully described by either its rectangular form or its polar form.

You must be wholly satisfied with the following relationships, which are described in Fig. 31–1, before continuing the study of this chapter.

If $a + jb = \boldsymbol{r}/\underline{\theta}$

then $a = \boldsymbol{r} \cos \theta$

and $b = \boldsymbol{r} \sin \theta$

and $\theta = \arctan \frac{b}{a}$

and $\boldsymbol{r} = \frac{a}{\cos \theta} = \frac{b}{\sin \theta}$

In Chap. 28 we developed the idea of *vectors* and introduced the *phasor* as the two-dimensional electrical refinement of a vector to describe the polar or phasor relationships. In Sec. 28–6 we developed *phasor summation,* and you can see, by comparing that section with Sec. 20–15, that a phasor relationship may be simply described by a complex number. *Phasor algebra* is merely the systematic analytical form of right triangle calculations developed in Chaps. 26 and 28.

31–2 ADDITION AND SUBTRACTION OF PHASORS IN RECTANGULAR FORM

As stated in Sec. 20–15, complex numbers, or phasors in rectangular form, can be added or subtracted by treating them as ordinary binomials.

EXAMPLE 1 Add 4.60 + j2.82 and 2.11 − j8.10.

SOLUTION

$$\begin{array}{r} 4.60 + j2.82 \\ 2.11 - j8.10 \\ \hline 6.71 - j5.28 \end{array}$$

Express the sum in polar form,

$$6.71 - j5.28 = 8.54/\underline{-38.2^\circ}$$

EXAMPLE 2 Subtract 3.7 + j4.62 from 14.6 − j8.84.

SOLUTION

$$\begin{array}{r} 14.6 - j8.84 \\ 3.7 + j4.62 \\ \hline 10.9 - j13.46 \end{array}$$

Express the result in polar form,

$$10.9 - j13.46 = 17.3\underline{/-51^\circ}$$

PROBLEMS 31–1

Perform the indicated operations and express the answers in both rectangular and polar forms. Check your results by graphical methods:

1. $(2.8 - j8.6) + (11.1 + j18.3)$
2. $(18.4 + j25) + (81.2 - j110)$
3. $(400 + j298) + (700 + j102)$
4. $(16.95 - j17.8) + (-11.33 - j22.2)$
5. $(137 + j844) + (-221 - j215)$
6. $(-488 - j603) + (172 + j168)$
7. $(36.1 - j52.3) - (14.8 - j9.9)$
8. $(8.37 - j3.4) - (-6.53 + j10.2)$
9. $(1100 - j200) - (-1400 - j600)$
10. $(75.3 - j38.7) - (137.4 + j47.1)$
11. $(5.6 + j32.6) - (3.4 + j22.6)$
12. $(-16.5 - j13.7) - (-16.5 + j86.3)$

31–3 MULTIPLICATION OF PHASORS IN RECTANGULAR FORM

Multiplication of complex numbers was explained in Sec. 20–16, where it was shown that phasors expressed in terms of their rectangular components are multiplied by treating them as ordinary binomials.

EXAMPLE 3 Multiply 8 + j5 by 10 + j9.

SOLUTION

$$\begin{array}{l} 8 + j5 \\ \underline{10 + j9 \qquad\qquad} \\ 80 + j50 \\ \underline{\quad + j72 + j^2 45} \\ 80 + j122 + j^2 45 \end{array}$$

Since $j^2 = -1$, the product is

$$80 + j122 + (-1)45 = 80 + j122 - 45 = 35 + j122$$

Expressing the product in polar form,

$$35 + j122 = 127\underline{/74^\circ}$$

EXAMPLE 4 Multiply 80 + j39 by 35 − j50.

SOLUTION

$$\begin{array}{r} 80 + j39 \\ 35 - j50 \\ \hline 2800 + j1365 \qquad\qquad \\ - j4000 - j^2 1950 \\ \hline 2800 - j2635 - j^2 1950 \end{array}$$

Since $j^2 = -1$, the product is

$$\begin{aligned} 2800 - j2635 - (-1)1950 &= 2800 - j2635 + 1950 \\ &= 4750 - j2635 \end{aligned}$$

Expressing the product in polar form,

$$4750 - j2635 = 5430\underline{/ -29^\circ}$$

31–4 DIVISION OF PHASORS IN RECTANGULAR FORM

As explained in Sec. 20–17, division of complex numbers, or phasors in rectangular form, is accomplished by rationalizing the denominator in order to obtain an in-phase number for a divisor. Multiplying a complex number by its conjugate always results in a product that is a simple number not affected by the operator j.

EXAMPLE 5 Find the quotient of $\dfrac{50 + j35}{8 + j5}$.

SOLUTION

Multiply both dividend and divisor (numerator and denominator) by the conjugate of the divisor, which is 8 − j5. Thus,

$$\begin{aligned} \frac{50 + j35}{8 + j5} \cdot \frac{8 - j5}{8 - j5} &= \frac{400 + j30 - j^2 175}{64 - j^2 25} \\ &= \frac{575 + j30}{89} \end{aligned}$$

That is,

$$\begin{aligned} \frac{575 + j30}{89} &= \frac{575}{89} + j\frac{30}{89} \\ &= 6.46 + j0.337 \end{aligned}$$

Express the quotient in polar form,

$$6.46 + j0.337 \cong 6.46\underline{/ 3.0^\circ}$$

EXAMPLE 6 Simplify $\dfrac{10}{3 + j4}$.

SOLUTION

Multiply both numerator and denominator by the conjugate of the denominator, which is 3 − j4. Thus,

$$\frac{10}{3 + j4} \cdot \frac{3 - j4}{3 - j4} = \frac{10(3 - j4)}{9 - j^2 16}$$
$$= \frac{30 - j40}{25} = 1.2 - j1.6$$

Express the quotient in polar form,

$$1.2 - j1.6 = 2.0\underline{/ -53.1^\circ}$$

PROBLEMS 31–2

Perform the indicated operations and express the answers in both rectangular and polar form:

1. $(2 + j4)(5 - j6)$
2. $(12 + j14)(22 + j17)$
3. $(4.6 + j8.2)(2.7 - j4.2)$
4. $(470 - j35.0)(330 + j0.621)$
5. $(8.1 - j3.5) \div (4.6 - j9.7)$
6. $(2.7 - j9)(12 - j8)$
7. $(23 + j27) \div (14.3 + j19.3)$
8. $(7 - j5) \div (10 - j14)$
9. $(20 - j16)(3 + j5)$
10. $1 \div (12 - j9)$

31–5 ADDITION AND SUBTRACTION OF POLAR PHASORS

As explained in preceding sections, phasors expressed in polar form can be added or subtracted by graphical methods only if their directions are parallel.

In order to add or subtract them algebraically, phasors must be expressed in terms of their rectangular components.

EXAMPLE 7 Add $5.40\underline{/ 31.5^\circ}$ and $8.37\underline{/ -75.4^\circ}$.

SOLUTION

Converting the phasors into their rectangular components,

$$5.40\underline{/ 31.5^\circ} = 5.40(\cos 31.5^\circ + j \sin 31.5^\circ) = 4.60 + j2.82$$
$$8.37\underline{/ -75.4^\circ} = 8.37(\cos 75.4^\circ - j \sin 75.4^\circ) = \underline{2.11 - j8.10}$$

Adding, $\text{Sum} = 6.71 - j5.28$

Expressing the sum in polar form,

$$6.71 - j5.28 = 8.54\underline{/ -38.2^\circ}$$

Note that the phasors of this example are the same as those of Example 1 of Sec. 31–2.

EXAMPLE 8 Subtract $5.92\underline{/ 51.3^\circ}$ from $17.1\underline{/ -31.2^\circ}$.

SOLUTION

Converting the phasors into their rectangular components,

$$17.1\underline{/ -31.2^\circ} = 17.1(\cos 31.2^\circ - j \sin 31.2^\circ) = 14.6 - j8.86$$
$$5.92\underline{/ 51.3^\circ} = 5.92(\cos 51.3^\circ + j \sin 51.3^\circ) = \underline{3.7 + j4.62}$$

Subtracting, $\text{Result} = 10.9 - j13.48$

Expressing the result in polar form,

$$10.9 - j13.48 = 17.3\underline{/-51^\circ}$$

Note that the phasors of this example are the same as those of Example 2 of Sec. 31–2.

PROBLEMS 31–3

Perform the indicated operations and express the results in both polar and rectangular form. Check results graphically.

1. $12\underline{/-24.8^\circ} + 32.4\underline{/57.9^\circ}$
2. $31\underline{/53.7^\circ} + 137\underline{/-53.7^\circ}$
3. $400\underline{/53.1^\circ} + 820\underline{/4.95^\circ}$
4. $24.6\underline{/-46.4^\circ} + 24.9\underline{/242.8^\circ}$
5. $933\underline{/82.9^\circ} + 590\underline{/198.3^\circ}$
6. $777\underline{/-129^\circ} + 241\underline{/44.3^\circ}$
7. $40.4\underline{/-55.5^\circ} - 18.6\underline{/-21.8^\circ}$
8. $9\underline{/-22.2^\circ} - 12.1\underline{/57.4^\circ}$
9. $1110\underline{/10.3^\circ} - 1510\underline{/203.4^\circ}$
10. $85\underline{/-27.15^\circ} - 145\underline{/18.91^\circ}$
11. $1000\underline{/-53.1^\circ} - 1500\underline{/-53.1^\circ}$
12. $10.64\underline{/-53^\circ} - 22.35\underline{/62.5^\circ}$

31–6 MULTIPLICATION OF POLAR PHASORS

In Example 3 of Sec. 31–3, it was shown that

$$(8 + j5)(10 + j9) = 127\underline{/74^\circ}$$

Now $\quad 8 + j5 = 9.44\underline{/32^\circ}$

and $\quad 10 + j9 = 13.45\underline{/42^\circ}$

Multiplying the magnitudes and adding the angles,

$$(9.44 \times 13.45)\underline{/32^\circ + 42^\circ} = 127\underline{/74^\circ}$$

which is the same product as that obtained by multiplying the phasors when expressed in terms of their rectangular components.

Similarly, in Example 4 of Sec. 31–3, it was shown that

$$(80 + j39)(35 - j50) = 5430\underline{/-29^\circ}$$

Now $\quad 80 + j39 = 89.0\underline{/26^\circ}$

and $\quad 35 - j50 = 61.0\underline{/-55^\circ}$

Multiplying the magnitudes and adding the angles,

$$(89 \times 61.0)\underline{/\ 26^\circ + (-55^\circ)} = 5430\underline{/\ -29^\circ}$$

which is the same product as that obtained by multiplying the phasors when the phasors are expressed in terms of their rectangular components.

From the foregoing, it is evident that the product of two polar phasors is found by multiplying the magnitudes and adding the angles of the phasors algebraically.

31–7 DIVISION OF POLAR PHASORS

In Example 5 of Sec. 31–4, it was shown that

$$\frac{50 + j35}{8 + j5} = 6.46\underline{/\ 3.0^\circ}$$

Now $$50 + j35 = 61.0\underline{/\ 35^\circ}$$

and $$8 + j5 = 9.44\underline{/\ 32^\circ}$$

Dividing the magnitudes and subtracting the angle of the divisor from the angle of the dividend,

$$\frac{61.0\underline{/\ 35^\circ}}{9.44\underline{/\ 32^\circ}} = \frac{61.0}{9.44}\underline{/\ 35^\circ - 32^\circ} = 6.46\underline{/\ 3.0^\circ}$$

which is the same quotient as that obtained by dividing the phasors when expressed in terms of their rectangular components.

Similarly, in Example 6 of Sec. 31–4, it was shown that

$$\frac{10}{3 + j4} = 2.0\underline{/\ -53.1^\circ}$$

Since 10 is a positive number, it is plotted on the 0° axis (Sec. 3–5) and expressed as

$$10\underline{/\ 0^\circ}$$

Now $$3 + j4 = 5\underline{/\ 53.1^\circ}$$

Dividing the magnitudes and subtracting the angle of the divisor from the angle of the dividend,

$$\frac{10\underline{/\ 0^\circ}}{5\underline{/\ 53.1^\circ}} = \frac{10}{5}\underline{/\ 0^\circ - 53.1^\circ}$$
$$= 2.0\underline{/\ -53.1^\circ}$$

which is the same quotient as that obtained by dividing the phasors when expressed in terms of their rectangular components.

From the foregoing, it is evident that the quotient of two polar phasors is found by dividing the magnitudes of the phasors and subtracting the angle of the divisor from the angle of the dividend.

31–8 EXPONENTIAL FORM

In the preceding two sections it has been demonstrated that angles are added when phasors are multiplied and they are subtracted when one phasor is divided by another. These operations can be further justified from a consideration of the sine and cosine when expanded in series form.

By Maclaurin's theorem, a treatment of which is beyond the scope of this book, cos θ and sin θ can be expanded into series form as follows:

$$\cos\theta = 1 - \frac{\theta^2}{2!} + \frac{\theta^4}{4!} - \frac{\theta^6}{6!} + \cdots \tag{1}$$

$$\sin\theta = \theta - \frac{\theta^3}{3!} + \frac{\theta^5}{5!} - \frac{\theta^7}{7!} + \cdots \tag{2}$$

The symbol $n!$ denotes the product of 1, 2, 3, 4, . . ., n and is read "factorial n." Thus, 5! (factorial five) is $1 \times 2 \times 3 \times 4 \times 5$. Similarly, it can be shown that

$$e^{j\theta} = 1 + j\theta - \frac{\theta^2}{2!} - j\frac{\theta^3}{3!} + \frac{\theta^4}{4!} + j\frac{\theta^5}{5!} - \frac{\theta^6}{6!} - j\frac{\theta^7}{7!} + \cdots \tag{3}$$

where e (formerly ε) is the base of the natural system of logarithms ≅ 2.718. By collecting and factoring j terms, Eq. (3) can be written

$$e^{j\theta} = \left(1 - \frac{\theta^2}{2!} + \frac{\theta^4}{4!} - \frac{\theta^6}{6!} + \cdots\right) + j\left(\theta - \frac{\theta^3}{3!} + \frac{\theta^5}{5!} - \frac{\theta^7}{7!} + \cdots\right) \tag{4}$$

Note that the first term of the right member of Eq. (4) is cos θ as given in Eq. (1) and that the second term in the right member of Eq. (4) is j sin θ. Therefore,

$$e^{j\theta} = \cos\theta + j\sin\theta \tag{5}$$

This expression, cos θ + j sin θ, is often referred to as cis θ, and some texts will actually refer to the *cis function*. You should bear in mind that *cis* is simply an abbreviation for cos + j sin.

Since a phasor, such as $Z\underline{/\theta}$, can be expressed in terms of its rectangular components by the relation

$$Z\underline{/\theta} = Z(\cos\theta + j\sin\theta) \tag{6}$$

it follows from Eqs. (5) and (6) that

$$Z\underline{/\theta} = Ze^{j\theta} \tag{7}$$

Similarly, it can be shown that

$$Z\underline{/-\theta} = Ze^{-j\theta} \tag{8}$$

Equations (7) and (8) show that the angles of phasors can be treated as exponents.

Two vectors $Z_1\underline{/\ \theta}$ and $Z_2\underline{/\ \phi}$ are multiplied by multiplying the magnitudes and adding the angles of the phasors algebraically. That is,

$$(Z_1\underline{/\ \theta})(Z_2\underline{/\ \phi}) = Z_1Z_2\underline{/\ \theta + \phi}$$

Also,

$$\frac{Z_1\underline{/\ \theta}}{Z_2\underline{/\ \phi}} = \frac{Z_1}{Z_2}\underline{/\ \theta - \phi}$$

and

$$\frac{Z_a\underline{/\ \theta}}{Z_b\underline{/\ -\phi}} = \frac{Z_a}{Z_b}\underline{/\ \theta + \phi}$$

EXAMPLE 9 Multiply $Z_1 = 8.4\underline{/\ 15^\circ}$ by $Z_2 = 10.5\underline{/\ 20^\circ}$.

SOLUTION

$$\begin{aligned} Z_1Z_2 &= 8.4 \times 10.5\underline{/\ 15^\circ + 20^\circ} \\ &= 88.2\underline{/\ 35^\circ} \end{aligned}$$

EXAMPLE 10 Multiply $Z_a = 164\underline{/\ -39^\circ}$ by $Z_b = 2.2\underline{/\ -26^\circ}$.

SOLUTION

$$\begin{aligned} Z_aZ_b &= 164 \times 2.2\underline{/\ -39^\circ + (-26^\circ)} \\ &= 361\underline{/\ -65^\circ} \end{aligned}$$

EXAMPLE 11 Divide $Z_1 = 54.2\underline{/\ 47^\circ}$ by $Z_2 = 18\underline{/\ 16^\circ}$.

SOLUTION

$$\frac{Z_1}{Z_2} = \frac{54.2}{18}\underline{/\ 47^\circ - 16^\circ} = 3.01\underline{/\ 31^\circ}$$

EXAMPLE 12 Divide $Z_a = 886\underline{/\ 18^\circ}$ by $Z_b = 31.2\underline{/\ -50^\circ}$.

SOLUTION

$$\begin{aligned} \frac{Z_a}{Z_b} &= \frac{886}{31.2}\underline{/\ 18^\circ - (-50^\circ)} \\ &= 28.4\underline{/\ 68^\circ} \end{aligned}$$

31–9 POWERS AND ROOTS OF POLAR PHASORS

In addition to following the laws of exponents for multiplication and division, phasor angles can be used as any other exponents are used when powers or roots of phasors are desired. For example, to square a phasor, the magnitude is squared and the angle is multiplied by 2. Similarly, the root of a phasor is found by extracting the root of the magnitude and dividing the angle by the index of the root.

EXAMPLE 13 Find the square of $Z_1 = 14\underline{/18°}$.

SOLUTION

$$Z_1^2 = (14\underline{/18°})^2 = 14^2\underline{/18° \times 2}$$
$$= 196\underline{/36°}$$

EXAMPLE 14 Find the square root of $Z_a = 625\underline{/60°}$.

SOLUTION

$$\sqrt{Z_a} = \sqrt{625\underline{/60°}}$$
$$= \sqrt{625}\underline{/60° \div 2}$$
$$= \pm 25\underline{/30°}$$

Our treatment of this subject at this time is necessarily limited to the features which are of immediate use to us in our present studies. You will find in advanced work in mathematics that DeMoivre's theorem proves that there are as many answers to a root problem as there are roots to be taken: the third root of a phasor has three answers, each of the same magnitude but at a different angle. For our immediate purposes, however, Examples 13 and 14 show the basic operations.

PROBLEMS 31–4

Perform the indicated operations and express the results in both polar and rectangular form:

1. $10\underline{/45°} \times 2.8\underline{/-60°}$
2. $21.4\underline{/52.6°} \times 25.5\underline{/25.6°}$
3. $(10.7\underline{/42.2°})(5.1\underline{/75.3°})$
4. $(183.3\underline{/-11°})(3.26\underline{/11°})$
5. $(8.24\underline{/-34°})(9.07\underline{/-52.6°})$
6. $(9.5\underline{/-71.6°})(8.26\underline{/-7.6°})$
7. $10\underline{/53.2°} \div 5\underline{/36.8°}$
8. $92.3\underline{/-12.5°} \div 81\underline{/-64.6°}$
9. $4.24\underline{/-64.5°} \div 16.8\underline{/47°}$
10. $5\underline{/20°} \div 30\underline{/-51.9°}$
11. $\dfrac{66.8\underline{/13°}}{4.73\underline{/24°}}$
12. $\dfrac{1.87\underline{/-180°}}{3.54\underline{/-180°}}$

Perform the indicated operations:

13. $(12\underline{/30°})^2$
14. $\sqrt{1024\underline{/-17°}}$
15. $\sqrt{2.89\underline{/44°}}$
16. $(0.31\underline{/-60°})^2$
17. $(4\underline{/90°})^3$
18. $\sqrt[3]{1331\underline{/17.40°}}$
19. $\sqrt[3]{27\underline{/33°}}$
20. $(2\underline{/-16°})^5$
21. Given that $\sqrt{a + jb} = x + jy$, show that

$$y = \sqrt{\tfrac{1}{2}(\sqrt{a^2 + b^2} - a)} \qquad \text{and} \qquad x = \frac{b}{2y}$$

At the end of Chap. 28, we drew your attention to the polar-rectangular conversion key available on some calculators. Now it is time for you to check on some other calculator features which may be available to you. If your calculator offers Σ+ and Σ− features, it may also offer an opportunity to add or subtract rectangular components which have been computed and are stored in scratchpad memory. Check your instruction manual for this added service. You may wish to practice it on the examples and problems of this chapter and the following chapters.

SELF TEST

Perform the indicated operations. Show all your work for every step.

1. $(112.6 + j84.7) + (45.9 + j28.6)$
2. $(-12.6 + j14.4) - (6.8 - j5.2)$
3. $(3 + j6)(10 - j5)$
4. Perform all calculations, and show your answer in rectangular form:

$$\frac{2 + j6}{1 - j4}$$

5. Answer in both polar and rectangular form:

$$12\angle 41.8^\circ + 16\angle -35^\circ$$

6. Answer in both polar and rectangular form:

$$\frac{31\angle 27.5^\circ \times 12.4\angle 13.8^\circ}{31\angle 27.5^\circ + 12.4\angle 13.8^\circ}$$

7. Answer in both polar and rectangular form:

$$\sqrt{512\angle 40^\circ}$$

8. Answer in both polar and rectangular form:

$$(3 + j4)^2$$

CHAPTER 32

ALTERNATING CURRENTS—SERIES CIRCUITS

Because of the phenomena that occur in them, ac circuits make a very interesting subject for study. In several ways, they are unlike circuits that carry direct currents. The product of the voltage and current is seldom equal to the reading of a wattmeter connected in the circuit; the current may lag or lead the voltage; and the potential difference across an inductance or capacitance may be several times the supply voltage. This chapter deals with the computation of such effects in series circuits.

32–1 DEFINITIONS

In Chap. 9 we investigated *resistance* and defined it as the amount of opposition to current within a conductor. It may be helpful to think of it as the electrical phenomenon which always tends to oppose the flow of electric current and which always converts some of the energy of the current electricity into heat energy. This heat energy is dissipated, usually by radiation, and is *lost* so far as the circuit is concerned. In some cases, of course, the purpose of the circuit is to provide a conversion of electric energy into heat energy. This heat energy is then radiated away from the circuit, and it represents lost energy so far as the circuit is concerned.

In this chapter, we will also investigate relationships which are involved when alternating current flows under the influence of alternating emf's because, when inductance and/or capacitance is involved in the circuit, we must abandon Ohm's law as a specific method of computation.

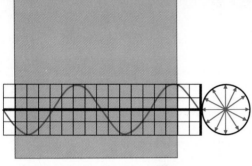

Inductance is the electrical phenomenon which always tends to oppose a change in electric current and which always converts some of the energy of current electricity into stored electromagnetic energy. This electromagnetic energy is stored by the inductance when the current increases, and is released into the circuit when the current decreases. It is found that the current flow through an inductance lags the applied emf by 90 electrical degrees.

Capacitance is the electrical phenomenon which always tends to oppose a change in voltage and which converts some of the energy of current electricity into stored electrostatic energy. This electrostatic energy is stored by the capacitance as an electric charge on the plates of a capacitor when the applied emf is increasing, and it is released into the circuit when the applied emf is decreasing. It is found that the voltage across a capacitor lags the current flow "through" the capacitor by 90 electrical degrees.

It is the 90° phase angles between voltage and current in *ac* circuits containing inductance and capacitance, together with their associated resistances, that really bring the trigonometric functions into play. You should make a special effort to resolve any difficulties which may still exist in your ability to solve right triangles by trigonometry (Chap. 26) and the j operator (Chaps. 20 and 31). Also, you should be confident in the use of the trigonometric operations on your calculator.

32–2 THE RESISTIVE CIRCUIT

Figure 32–1 represents a 60-Hz alternator supplying 220 V to two resistances connected in series.

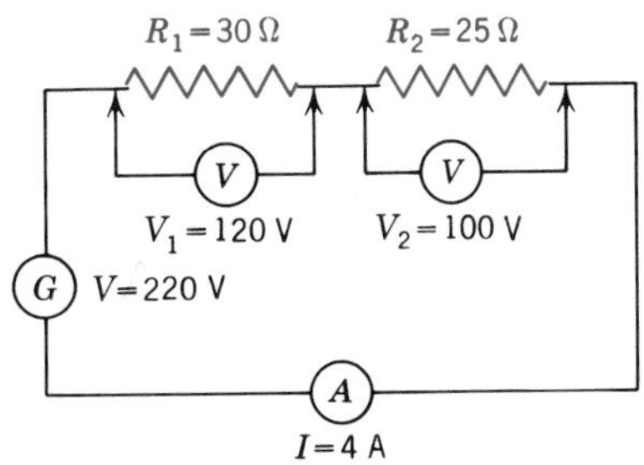

FIG. 32–1 Alternator supplying resistive circuit.

This circuit contains resistance only; therefore, Ohm's law applies in every respect. The internal resistance of the alternator and the resistance of the connecting wires being neglected, the current through the circuit is given by the familiar relation

$$I = \frac{V}{R_t} = \frac{V}{R_1 + R_2} = \frac{220}{30 + 25} = \frac{220}{55}$$
$$= 4\ \text{A}$$

Again, as with direct currents, the voltage drops, or potential differences, across the resistances are

$$\begin{aligned} V_1 &= IR_1 = 4 \times 30 = 120\ \text{V} \\ V_2 &= IR_2 = 4 \times 25 = \underline{100\ \text{V}} \\ &\text{Applied voltage} = 220\ \text{V} \end{aligned}$$

In an ac circuit containing only resistance, the voltage and current are in phase; that is, the voltage and current pass through corresponding parts of their cycles at the same instant.

From the above it follows that if

$$v = V_{\max} \sin \omega t = 311 \sin 377t \text{ V}$$

is the equation for the alternator voltage of Fig. 32–1, then the current through the circuit is

$$i = I_{\max} \sin (\omega t + \theta) = I_{\max} \sin (\omega t + 0°)$$
$$= 5.66 \sin 377t \text{ A}$$

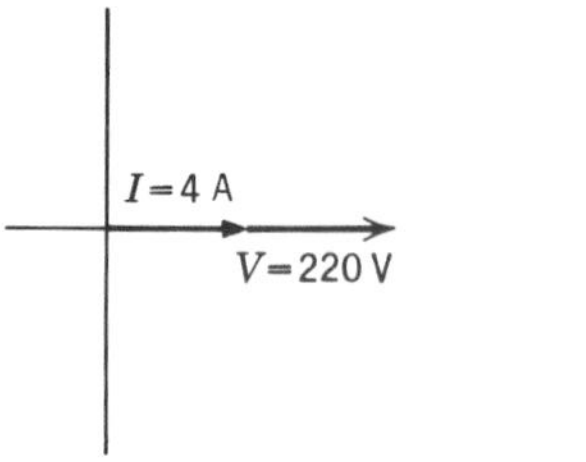

FIG. 32–2 Phasor diagram for circuit of Fig. 32–1.

Figure 32–2 is the phasor diagram for the circuit of Fig. 32–1. It will be noted that the voltage phasor and the current phasor coincide. This is what the equations for the voltage and current would make you expect, for they differ only in amplitude factors. The frequency factors are equal, and the phase angle is 0° (Secs. 29–7 to 29–10).

It is evident that Ohm's law says nothing about maximum, average, or effective values of current and voltage. Any of these values can be used; that is, maximum voltage can be used to find maximum current, average voltage can be used to find average current, etc. Naturally, maximum voltage is not used to find effective current unless the proper conversion constant is introduced into the equation. As previously stated, all voltage and current values here are to be considered as effective values unless otherwise specified (Sec. 30–8).

32–3 POWER IN THE RESISTIVE CIRCUIT

In dc circuits the power is equal to the product of the voltage and the current (Sec. 8–5). This is true of ac circuits for *instantaneous values* of voltage and current. That is, the *instantaneous power* is

$$p = vi \qquad \text{VA} \tag{1}$$

and is measured in *voltamperes* or *kilovoltamperes,* abbreviated VA and kVA (or sometimes V·A and kV·A), respectively.*

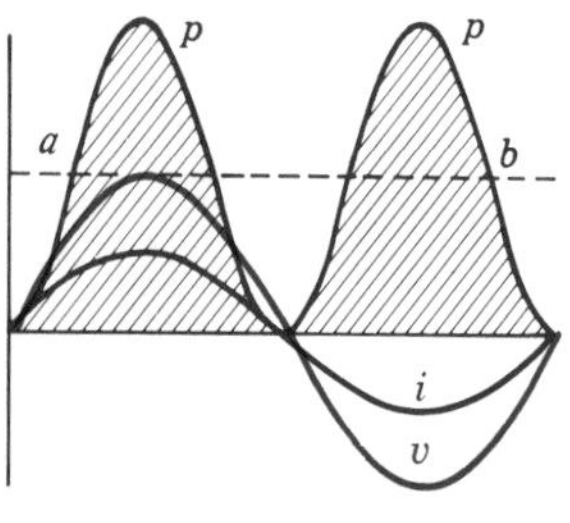

FIG. 32–3 Instantaneous power curves for circuit containing only resistance.

When a sine wave of voltage is impressed across a resistance, the relations among voltage, current, and power are as shown in Fig. 32–3. The voltage existing across the resistance is in phase with the current flowing through the resistance. The power delivered to the resistance at any instant is represented by the height of the power curve, which is the product of the instantaneous values of voltage and current at that instant. The shaded area under the power curve represents the total power delivered to the circuit during one complete cycle of voltage. It will be noted that the power curve is of sine-wave voltage. Also, the

*The dot is preferred in general physics relationships, but it is customarily omitted in electricity and electronics use.

power curve lies entirely above the x axis; there are no negative values of power.

The maximum height of the power curve is the product of the maximum values of voltage and current. Stated as an equation

$$\boldsymbol{P}_{\max} = \boldsymbol{V}_{\max}\boldsymbol{I}_{\max} \tag{2}$$

The average power delivered to a resistance load is represented by the height of the line *ab* in Fig. 32–3, which is half the maximum height of the power curve, or its average height. Then, since

$$\text{Average power} = P = \tfrac{1}{2}\boldsymbol{P}_{\max}$$

by dividing both members of Eq. (2) by 2 we obtain

$$\tfrac{1}{2}\boldsymbol{P}_{\max} = \tfrac{1}{2}\boldsymbol{V}_{\max}\boldsymbol{I}_{\max}$$

Substituting for the value of $\frac{1}{2}\boldsymbol{P}_{\max}$ and factoring the denominator of the right member,

$$P = \frac{\boldsymbol{V}_{\max}\boldsymbol{I}_{\max}}{\sqrt{2}\sqrt{2}}$$

Substituting for the values in the right member (Sec. 30–7),

$$P = VI \qquad \text{W} \tag{3}$$

Hence, the alternating power consumed by a resistance load is equal to the product of the effective values of voltage and current. As in dc circuits, alternating power is measured in watts and kilowatts.

EXAMPLE 1 What is the power expended in the resistances of Fig. 32–1?

SOLUTION

$$\begin{aligned}
\text{Voltage across } R_1 &= V_1 = 120 \text{ V}\\
\text{Voltage across } R_2 &= V_2 = 100 \text{ V}\\
\text{Current through circuit} &= I \;\; = 4\text{A}\\
\text{Power expended in } R_1 &= P_1 = V_1I = 120 \times 4 = 480 \text{ W}\\
\text{Power expended in } R_2 &= P_2 = V_2I = 100 \times 4 = \underline{400 \text{ W}}\\
&\qquad\qquad\qquad\qquad\quad \text{Total} = 880 \text{ W}
\end{aligned}$$

Also, the total power is

$$P_t = VI = 220 \times 4 = 880 \text{ W}$$

Because $P = VI$, the usual Ohm's law relations hold for resistances in ac circuits. Hence,

$$P = I^2R \qquad \text{W} \tag{4}$$

and

$$P = \frac{V^2}{R} \qquad \text{W} \tag{5}$$

Thus, the power consumed by R_1 of Fig. 32–1 can be computed by using Eq. (4) or (5). Hence,

$$P_1 = I^2R_1 = 4^2 \times 30 = 480 \text{ W}$$

or

$$P_1 = \frac{V_1^2}{R_1} = \frac{120^2}{30} = 480 \text{ W}$$

PROBLEMS 32–1

1. A 60-Hz alternator supplies 110 V across a combination of three series resistors of 180, 47, and 33 Ω.
   (*a*) What is the current in the circuit?
   (*b*) Write the equation for the alternator voltage at any time *t*.
   (*c*) Write the equation for the circuit current at any time *t*.
   (*d*) What is the voltage measured across the 47-Ω resistor?
   (*e*) How much power is dissipated by the 33-Ω resistor?
   (*f*) What is the instantaneous value of the current when the instantaneous emf is 26 V?
2. Given the circuit of Fig. 32–4:
   (*a*) Write the equation for the emf of the alternator at any time *t*.
   (*b*) Write the equation for the total current of the circuit.
   (*c*) What is the voltage across R_3?
   (*d*) How much power is dissipated in R_2?
   (*e*) What is the current through R_1?
   (*f*) What is the instantaneous value of the total current when the instantaneous alternator emf is 36.5 V?

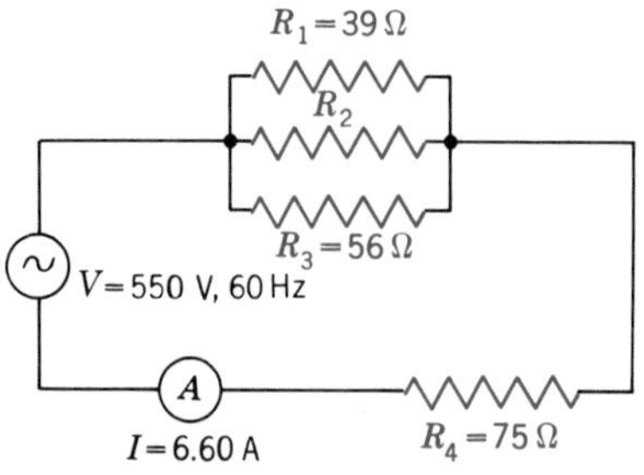

FIG. 32–4 Circuit of Probs. 2 and 3.

3. In Fig. 32–4, what is the instantaneous value of the voltage across R_2 when the instantaneous current through R_4 is 2.75 A?

4. A 10-kHz signal generator is connected to a 600-Ω resistive load. A milliwattmeter indicates that the resistor is dissipating 800 mW. What is the maximum instantaneous voltage developed at the generator terminals?
5. What is the equation of the current in Prob. 4?

32–4 THE INDUCTIVE CIRCUIT

A circuit, or an inductance coil, has the property of inductance when there is set up in it an emf due to a *change* of current through it. Thus, a circuit has an inductance of 1 H when a change of current of 1 A/s induces an emf of 1 V. Expressed as an equation,

$$V_{av} = L\frac{I}{t} \qquad \text{V} \tag{6}$$

where V_{av} is the average voltage induced in a circuit of L H by a *change* of current of I A in t s.

An alternating current of $\boldsymbol{I}_{max}$ A makes *four changes* during each cycle. These are

1. From zero to maximum positive value
2. From maximum positive value to zero
3. From zero to maximum negative value
4. From maximum negative value to zero

The time required for one complete cycle of alternating currents is $T = f^{-1}$ s (Sec. 29–9), and each of the above changes occurs in one-quarter of the time required for the completion of one cycle. Then the time for each change is $(4f)^{-1}$ s. Substituting this value of t, and $\boldsymbol{I}_{max}$ for I, in Eq. (6), we have

$$V_{av} = L\frac{\boldsymbol{I}_{max}}{(4f)^{-1}} = 4fL\boldsymbol{I}_{max} \tag{7}$$

Equation (7) is cumbersome if used in its present form, for it contains an average-voltage term and a maximum-current term. The equation can be expressed in terms of the relation between average and maximum values as given in Sec. 30–6:

$$V_{av} = \frac{2}{\pi}\boldsymbol{V}_{max}$$

Substituting in Eq. (7) for this value of V_{av}, we have

$$\frac{2}{\pi}\boldsymbol{V}_{max} = 4\pi fL\boldsymbol{I}_{max}$$

which becomes

$$\boldsymbol{V}_{max} = 2\pi fL\boldsymbol{I}_{max} \tag{8}$$

Because both voltage and current in Eq. (8) are now in terms of maximum values, effective values can be used. Thus,

$$V = 2\pi fLI \qquad V \tag{9}$$

The factors $2\pi fL$ in Eqs. (8) and (9) represent a reaction due to the frequency of the alternating current and the amount of inductance contained in the circuit. Hence, the alternating voltage of V required to cause a current of I A with a frequency of f Hz to flow through an inductance of L H is given by Eq. (9). That is, the voltage must overcome the reaction $2\pi fL$, which is called the *inductive reactance*. From Eq. (9) the inductive reactance, which is denoted by X_L and expressed in ohms, is given by

$$\frac{V}{I} = 2\pi fL$$

or

$$X_L = 2\pi fL = \omega L \qquad \Omega \tag{10}$$

where f = frequency, Hz
L = inductance, H

Note the similarity of the relations between voltage and current for inductive reactance and resistance. Both inductive reactance and resistance offer an opposition to a flow of alternating current; both are expressed in ohms; and both are equal to the voltage divided by the current. Here the similarity ends; there is no inductive reactance to steady-state direct currents because there is no *change* in current, and, as explained later, inductive reactances consume no alternating power.

Figure 32–5 represents a 60-Hz alternator delivering 220 V to a coil having an inductance of 0.165 H. The opposition, or inductive reactance, to the flow of current is

$$\begin{aligned} X_L &= 2\pi fL \\ &= 2\pi \times 60 \times 0.165 \\ &= 62.2\ \Omega \end{aligned}$$

It is impossible to construct an inductance containing no resistance; but in order to simplify basic considerations, we shall consider the coil of Fig. 32–5 as being an inductance with negligible resistance. (The effects of inductance

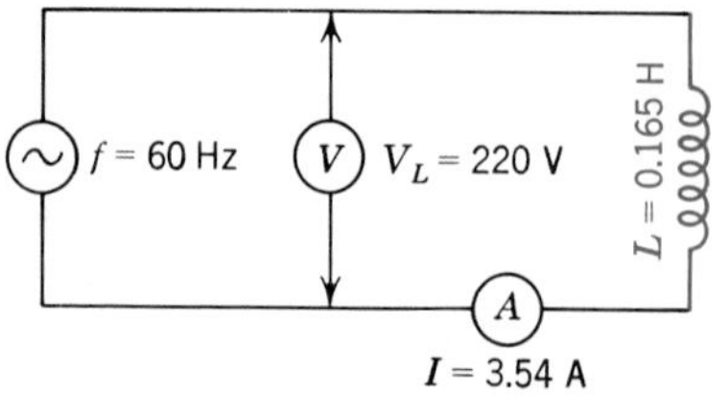

FIG. 32–5 V_L = 220 V, L = 0.165 H.

and resistance acting together are discussed in Sec. 32–8.) The current in the circuit due to the action of voltage and inductive reactance is

$$I = \frac{V_L}{X_L} = \frac{220}{62.2} = 3.54 \text{ A}$$

EXAMPLE 2 What is the inductive reactance of an inductance of 17 μH at a frequency of 2500 kHz?

SOLUTION

$$f = 2500 \text{ kHz} = 2.5 \times 10^6 \text{ Hz}$$
$$L = 17 \ \mu\text{H} = 1.7 \times 10^{-5} \text{ H}$$
$$X_L = 2\pi f L = 2\pi \times 2.5 \times 10^6 \times 1.7 \times 10^{-5}$$
$$= 2\pi \times 1.7 \times 2.5 \times 10 = 267 \ \Omega$$

EXAMPLE 3 An inductor is connected to 115 V 60 Hz. An ammeter connected in series with the coil reads 0.714 A. On the assumption that the coil contains negligible resistance, what is its inductance?

SOLUTION

$$V_L = 115 \text{ V}$$
$$f = 60 \text{ Hz}$$
$$I = 0.714 \text{ A}$$
$$X_L = \frac{V_L}{I} = \frac{115}{0.714} = 161 \ \Omega$$

Since $$X_L = 2\pi f L$$

then $$L = \frac{X_L}{2\pi f} = \frac{161}{2\pi \times 60} = 0.427 \text{ H}$$

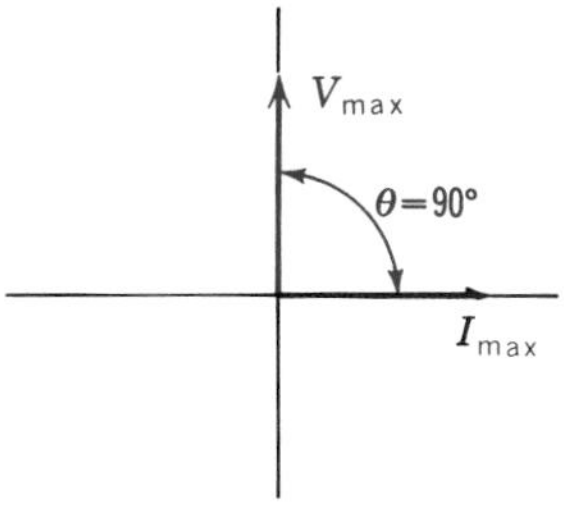

FIG. 32–6 Current lags voltage by 90°.

In a circuit containing inductance, a change of current induces an emf of such polarity that it always opposes the change of current. Because an alternating current is constantly changing, in an inductive circuit there is always present a reaction that opposes this change. The net effect, in a *purely inductive circuit,* is to cause the *current to lag the voltage by* 90°. This is illustrated by the phasor diagram of Fig. 32–6, which shows the voltage of the circuit of Fig. 32–5 to be at maximum positive value when the current is passing through zero.

The instantaneous voltage across the inductance is given by

$$v = \boldsymbol{V}_{\text{max}} \sin \omega t \qquad \text{V}$$

or

$$v = 311 \sin 377t \qquad \text{V}$$

Since the current lags the voltage by a phase angle θ of 90°, the equation for the current through the inductance is

$$i = \boldsymbol{I}_{\text{max}} \sin (\omega t - \theta) \qquad \text{A} \qquad (11)$$

or

$$i = 5 \sin (377t - 90°) \qquad \text{A} \qquad (12)$$

If the voltage has completed $\phi°$ of its cycle, the instantaneous current is

$$i = 5 \sin (\phi - 90°) \qquad \text{A} \qquad (13)$$

EXAMPLE 4 What is the instantaneous value of the current in Fig. 32–5 when the voltage has completed 120° of its cycle?

SOLUTION

Draw a phasor diagram of the current and voltage relations as shown in Fig. 32–7. The instantaneous value of the current is found from Eq. (13) and is

$$\begin{aligned} i &= I_{max} \sin (\phi - 90°) \\ &= 5 \sin (120° - 90°) \\ &= 5 \sin 30° \\ &= 2.5 \text{ A} \end{aligned}$$

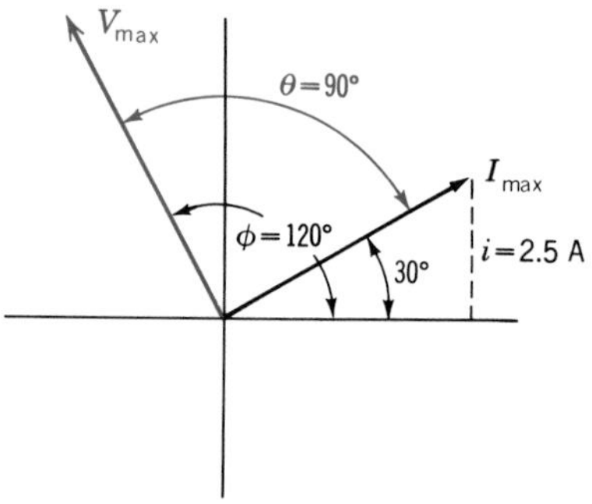

FIG. 32–7 Phasor diagram of Example 4.

PROBLEMS 32–2

1. What is the reactance of a 12-mH coil at 400 Hz?
2. What is the reactance of a 12-mH coil at 1 kHz?
3. What inductance value will produce a reactance of 1590 Ω when the frequency is 1 MHz?
4. What is the inductance of a coil that exhibits a reactance of 754 Ω at a frequency of 400 Hz?
5. A tuning coil in a radio transmitter has an inductance of 210 μH. What is its reactance at a frequency of 2.6 MHz?
6. At what frequency will a television set coil with an inductance of 3.25 μH offer a reactance of 3740 Ω?
7. Assuming negligible resistance, what would be the current flow through an inductance of 0.067 H at a voltage of 100 V, 800 Hz?
8. What would be the equation of the current in Prob. 7?
9. A 12-μH coil with a negligible resistance drops 9 V when a current of 120 μA is passed through it. What frequency will the applied signal voltage have?

10. An emf described by the equation $v = 311 \sin 314t$ V is applied to an inductor of 1.65 H. What is the equation of the current flow, assuming negligible resistance?
11. What is the instantaneous value of the current in Prob. 10 when the emf has completed 40° of its cycle?
12. What is the instantaneous value of the applied voltage in Prob. 10 when the current has completed 210° of its cycle?
13. What happens to the inductive reactance of a circuit when the inductance is fixed but the frequency of the applied emf is (*a*) doubled, (*b*) tripled, (*c*) halved?
14. What happens to the inductive reactance of a circuit when the frequency of the applied emf is held constant and the inductance is varied?

32–5 THE CAPACITIVE CIRCUIT

A capacitance is formed between two conductors when there is an insulating material between them. A circuit, or a capacitor, is said to have a capacitance of one farad when a *change* of one volt per second produces a current of one ampere. Expressed as an equation,

$$I_{av} = C\frac{V}{t} \qquad \text{A} \tag{14}$$

where I_{av} is the average current in amperes that is caused to flow through a capacitance of C by a *change* of V V in t s.

In all probability the above definition does not clearly indicate to you *how much* electricity, or charge, a given capacitor will contain. Perhaps a more understandable definition is that a circuit, or a capacitor, has a capacitance of one farad when a difference of potential of one volt will produce on it one coulomb of charge. Expressed as an equation,

$$Q = CV \qquad \text{C} \tag{15}$$

where Q is the charge in coulombs placed on a capacitor of C F by a difference of potential of V V across the capacitor.

It was shown in Sec. 32–4 that the time t required for one change of an alternating emf was $(4f)^{-1}$ s. Thus, if an alternating emf of $\boldsymbol{V}_{max}$ V at a frequency of f Hz is impressed across a capacitor of C F, by substituting the above value of t and $\boldsymbol{V}_{max}$ for V in Eq. (14),

$$I_{av} = C\frac{\boldsymbol{V}_{max}}{(4f)^{-1}} = 4fC\boldsymbol{V}_{max} \qquad \text{A} \tag{16}$$

Again, as in Eq. (7), the above equation contains an average term and a maximum term. As given in Sec. 30–6,

$$I_{av} = \frac{2}{\pi}\boldsymbol{I}_{max} \qquad \text{A}$$

Substituting in Eq. (16) for this value of I_{av}, we have

$$\frac{2}{\pi}\boldsymbol{I}_{\max} = 4fC\boldsymbol{V}_{\max}$$

which becomes

$$\boldsymbol{I}_{\max} = 2\pi fC\boldsymbol{V}_{\max} \qquad \text{A} \tag{17}$$

Because both voltage and current in Eq. (17) are now in terms of maximum values, effective values can be used. Thus,

$$I = 2\pi fCV \qquad \text{A} \tag{18}$$

The factors $2\pi fC$ represent a reaction due to the frequency of the alternating emf and the amount of capacitance; hence, it is evident that the amount of current in a purely capacitive circuit depends upon these factors. As in the case of resistive circuits and inductive circuits, the opposition to the flow of current is obtained by dividing the voltage by the current. Then, from Eq. (18),

$$\frac{V}{I} = \frac{1}{2\pi fC} \qquad \Omega \tag{19}$$

The right member of Eq. (19), which represents the opposition to a flow of alternating current in a purely capacitive circuit, is called the *capacitive reactance*. It is denoted by X_C and expressed in ohms. Thus,

$$X_C = \frac{1}{2\pi fC} = \frac{1}{\omega C} \qquad \Omega \tag{20}$$

where f = frequency, Hz
C = capacitance, F

Figure 32–8 represents a 60-Hz alternator delivering 220 V to a capacitor having a capacitance of 14.5 μF. The opposition, or capacitive reactance, to the flow of current is

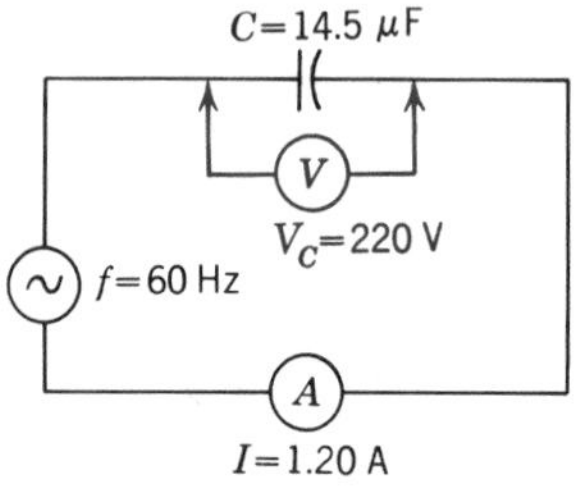

FIG. 32–8
V_C=220 V, C = 14.5 μF.

$$X_C = \frac{1}{2\pi fC} = \frac{1}{2\pi \times 60 \times 14.5 \times 10^{-6}}$$
$$= \frac{10^4}{2\pi \times 6 \times 1.45} = 183\ \Omega$$

Neglecting the resistance of the connecting leads and the extremely small losses at low frequencies in a well-constructed capacitor, the current in the circuit due to the action of the voltage and capacitive reactance is

$$I = \frac{V_C}{X_C} = \frac{220}{183} = 1.20\ \text{A}$$

EXAMPLE 5 What is the capacitive reactance of a 350-pF capacitor at a frequency of 1200 kHz?

SOLUTION

$$f = 1200 \text{ kHz} = 1.2 \times 10^6 \text{ Hz}$$
$$C = 350 \text{ pF} = 3.5 \times 10^{-10} \text{ F}$$
$$X_C = \frac{1}{2\pi fC}$$
$$= \frac{1}{2\pi \times 1.2 \times 10^6 \times 3.5 \times 10^{-10}}$$
$$= \frac{10^4}{2\pi \times 1.2 \times 3.5} = 379\ \Omega$$

EXAMPLE 6 A capacitor is connected across 110 V 60 Hz. A milliammeter connected in series with the capacitor reads 350 mA. What is the capacitance of the capacitor?

SOLUTION

$$V_C = 110 \text{ V}$$
$$f = 60 \text{ Hz}$$
$$I = 350 \text{ mA} = 0.350 \text{ A}$$
$$X_C = \frac{V_C}{I} = \frac{110}{0.35} = 314\ \Omega$$

since $$X_C = \frac{1}{2\pi fC}$$

often $$C = \frac{1}{2\pi f X_C} = \frac{1}{2\pi \times 60 \times 314}$$
$$= \frac{10^{-3}}{2\pi \times 6 \times 3.14}$$
$$= 8.44 \times 10^{-6} \text{ F} = 8.44\ \mu\text{F}$$

Because there is an alternating current (when we would expect a capacitor to produce an open circuit), we customarily say that there is a current *through* a capacitive circuit when an alternating voltage is impressed across it. Furthermore, the greatest amount of current will flow when the voltage is changing most rapidly, and this occurs when the voltage passes through zero value. This property, in conjunction with the effects of the counter emf, *causes the current to lead the voltage by* 90° *in a purely capacitive circuit.* This is illustrated by the vector diagram of Fig. 32–9, which shows the current through the circuit of Fig. 32–8 to be at maximum positive value when the voltage is passing through zero.

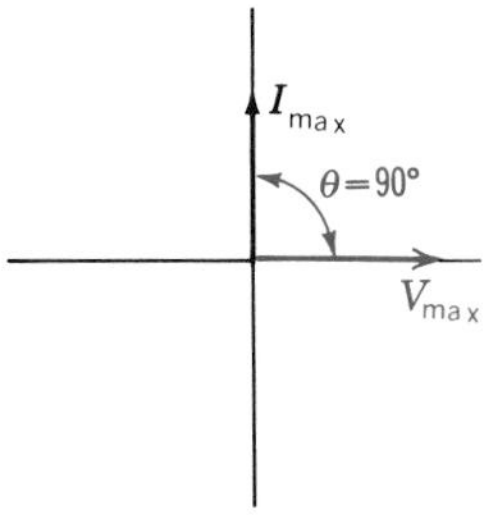

FIG. 32–9 Current leads voltage by 90°.

The instantaneous voltage across the capacitor is given by

$$v = V_{max} \sin \omega t \qquad \text{V} \tag{21}$$

or

$$v = 311 \sin 377t \text{ V} \tag{22}$$

Therefore, the equation for the current is

$$i = I_{max} \sin (377t + \theta) \quad \text{A} \tag{23}$$

or

$$i = 1.70 \sin (377t + 90°) \text{ A} \tag{24}$$

If the voltage has completed $\phi°$ of its cycle, the instantaneous current is

$$i = I_{max} \sin (\phi + 90°) \quad \text{A} \tag{25}$$

EXAMPLE 7 What is the instantaneous value of the current in the circuit shown in Fig. 32–8 when the voltage has completed 35° of its cycle?

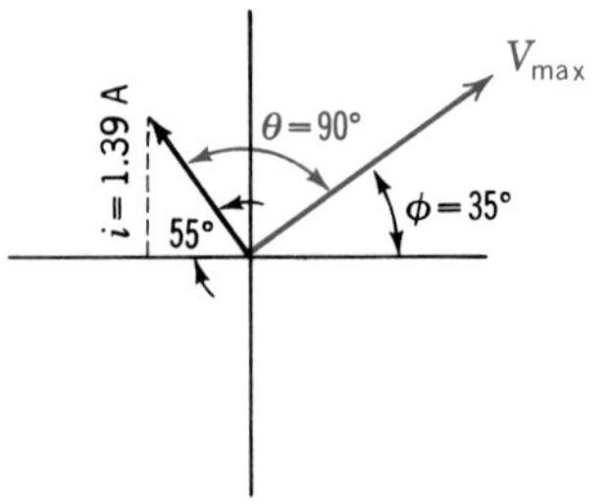

FIG. 32–10 Phasor diagram for Example 7.

SOLUTION

Draw a phasor diagram of the current and voltage relations as shown in Fig. 32–10. The instantaneous value of the current is found from Eq. (25) and is

$$\begin{aligned} i &= I_{max} \sin (\phi + 90°) \\ &= 1.70 \sin (35° + 90°) \\ &= 1.70 \sin 125° = 1.39 \text{ A} \end{aligned}$$

32–6 CAPACITORS IN SERIES

Figure 32–11 represents two capacitors C_1 and C_2 connected in series with a voltage V across the combination. Because the capacitors are in series, the same quantity of electricity must be sent into each of them. Then, if V_1 and V_2 represent the potential differences across C_1 and C_2, respectively, Q represents the quantity of electricity in each capacitor and C_t is the capacitance of the combination. Hence,

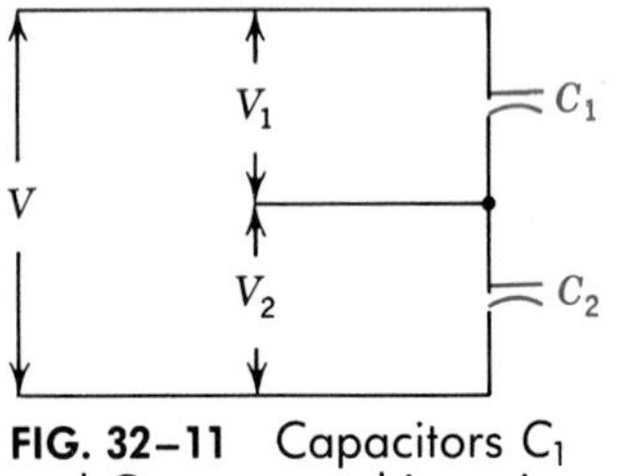

FIG. 32–11 Capacitors C_1 and C_2 connected in series.

$$V = \frac{Q}{C_t}$$

$$V_1 = \frac{Q}{C_1}$$

and

$$V_2 = \frac{Q}{C_2}$$

Since

$$V = V_1 + V_2 \tag{26}$$

by substituting the values for all voltages into Eq. (26),

$$\frac{Q}{C_t} = \frac{Q}{C_1} + \frac{Q}{C_2}$$

or

$$\frac{1}{C_t} = \frac{1}{C_1} + \frac{1}{C_2} \tag{27}$$

Equation (27) resolves into

$$C_t = \frac{C_1 C_2}{C_1 + C_2} \tag{28}$$

The above illustrates the fact that capacitors in series combine like resistances in parallel; that is, the reciprocal of the combined capacitance of capacitors in series is equal to the sum of the reciprocals of the capacitances of the individual capacitors.

EXAMPLE 8 What is the capacitance of a 6-μF capacitor in series with a capacitor of 4 μF?

SOLUTION

$$C_t = \frac{6 \times 4}{6 + 4} = 2.4\ \mu\text{F}$$

PROBLEMS 32–3

1. What is the capacitive reactance of a 50-μF capacitor at a frequency of 200 Hz?
2. What is the capacitive reactance of a 50-μF capacitor at a frequency of 1 kHz?
3. What is the capacitive reactance of a 22-μF capacitor at a frequency of 100 kHz?
4. What is the reactance of a 50-pF capacitor at a frequency of 12 GHz?
5. The filter capacitance in a radio receiver is 0.0016 μF. What is its reactance at a frequency of 720 kHz?
6. What is the reactance of the capacitor of Prob. 5 if the frequency is increased to 1320 kHz?
7. How much current will flow in a capacitor of 2.5 pF when 125 V at 1.5 kHz is impressed across the capacitor, neglecting resistance?
8. What will be the current in the capacitor of Prob. 7 if the frequency is increased to 12 kHz?
9. When a 100-V 1-kHz emf is impressed across a capacitor, the current flow is 3.25 A. What is the capacitance?
10. There is a current of 452 mA through a 5-μF capacitor when the frequency of the applied emf is 60 Hz. What is the voltage?
11. What is the equation for the current in Prob. 10?
12. What is the instantaneous value of the current in Prob. 10 when the emf has completed 230° of its cycle?
13. What is the resulting capacitance when a 220-pF capacitor is connected in series with a 500-pF capacitor?
14. Two capacitors, 20 and 200 pF, are connected in series. What is the resultant capacitance?
15. If an emf of 80 V at 15 kHz is impressed across the series circuit of Prob. 14, what will be the resultant current flow, neglecting resistance?
16. What happens to the capacitive reactance of a circuit when the capacitance is fixed but when the frequency of the applied emf is (*a*) doubled, (*b*) tripled, (*c*) halved?
17. What happens to the capacitive reactance of a circuit when the frequency of the applied emf is held constant and the capacitance is varied?

18. Neglecting the resistance of the connecting wires in Fig. 32–12:
    (*a*) Write the equation for the emf of the alternator.
    (*b*) Write the equation for the circuit current.
    (*c*) What is the voltage across C_1?
    (*d*) What is the voltage across C_2?

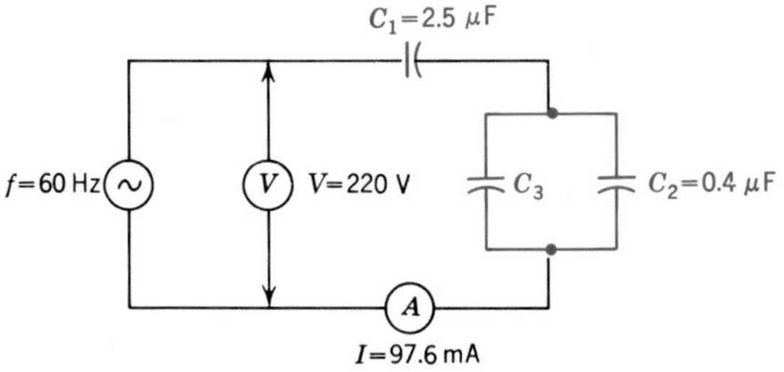

FIG. 32–12 Circuit of Prob. 18.

19. A capacitor in an oscillator circuit is 52 pF. What will be the reactance of the capacitor if the circuit oscillates at 1.32 MHz?
20. The capacitor of Prob. 19 is also to be used with an adjustable inductor which varies the frequency. What will be the reactance of the capacitor as the circuit is tuned to (*a*) 1027 kHz and (*b*) 750 kHz?

32–7 POWER IN CIRCUITS CONTAINING ONLY INDUCTANCE OR CAPACITANCE

Figure 32–13 illustrates the voltage, current, and power relations when a sine wave of emf is impressed across an inductor whose resistance is negligible.

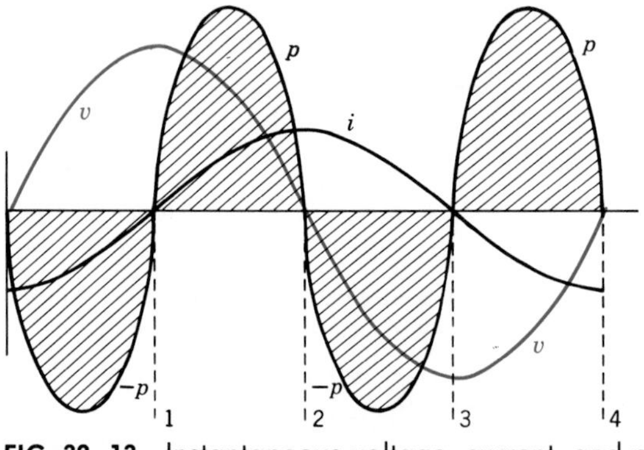

FIG. 32–13 Instantaneous voltage, current, and power in an inductive circuit.

When the current is increasing from zero to maximum positive value, during the time interval from 1 to 2, power is being taken from the source of emf and is being stored in the magnetic field about the coil. As the current through the inductor decreases from maximum positive value to zero, during the time from 2 to 3, the magnetic field is collapsing, thus returning its power to the circuit. Thus during the intervals from 1 to 2 and from 3 to 4, the inductor is taking power from the source that is represented by the *positive* power in the figure. During the intervals from 0 to 1 and 2 to 3, the inductor is returning power to the source that is represented by the *negative* power in the figure. As previously stated, the instantaneous power is equal to the product of the voltage and current; it is positive when the voltage and current are of like sign and negative when they are of unlike sign. Note that between points 3 and 4, although both the voltage and the current are negative, the power is positive.

When an alternating emf is impressed across a capacitor, power is taken from the source and stored in the capacitor as the voltage increases from zero to maximum positive value. As the voltage decreases from maximum positive value to zero, the capacitor discharges and returns power to the source. As in the case of the inductor, half of the power loops are positive and half are negative; therefore, no power is expended in either circuit, for the power alternately flows to and from the source. This power is called *reactive* or *apparent power* and is given by the relation

$$P = VI \qquad \text{VA}$$

32–8 RESISTANCE AND INDUCTANCE IN SERIES

It has been explained that in a circuit containing only resistance the voltage applied across the resistance and the current through the resistance are in phase and that in a circuit containing only reactance the voltage and current are 90° out of phase. However, circuits encountered in practice contain both resistance and reactance. Such a condition is shown in Fig. 32–14, where an alternating emf of 100 V is impressed across a combination of 6 Ω resistance in series with 8 Ω inductive reactance.

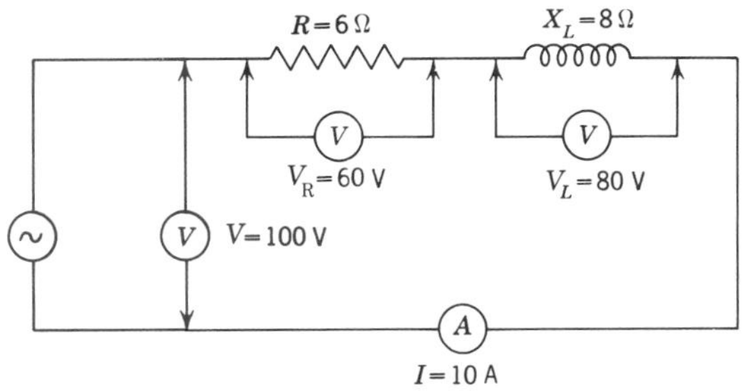

FIG. 32–14 Series circuit containing resistance and inductance.

As with dc circuits, the sum of the voltage drops around the circuit comprising the load must equal the applied emf. In the consideration of resistance and reactance, however, we are dealing with voltages that can no longer be added or subtracted arithmetically. That is because the voltage drop across the resistance is in phase with the current and the voltage drop across the inductive reactance is 90° ahead of the current.

Because the current is the same in all parts of a series circuit, we can use it as a reference and plot the voltage across the resistance and that across the inductive reactance, as shown in Fig. 32–15. The resultant of these two volt-

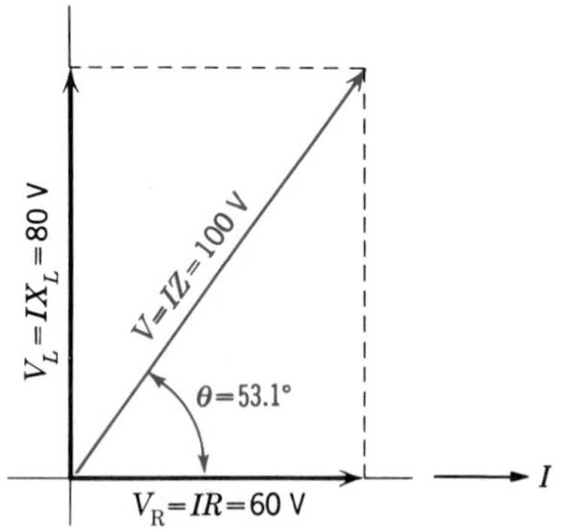

FIG. 32–15 Phasor diagram for circuit of Fig. 32–14.

ages, which can be treated as rectangular components (see Sec. 28–4), must be equal to the applied emf. Hence, if IR and IX_L are the potential differences across the resistance and inductive reactance, respectively,

$$V = \sqrt{(IR)^2 + (IX_L)^2} \quad \text{V} \tag{29}$$

or

$$V = \sqrt{60^2 + 80^2} = 100 \text{ V}$$

The phase angle θ between voltage and current can be found by using any of the trigonometric functions. For example,

$$\tan\theta = \frac{IX_L}{IR} = \frac{80}{60} = 1.33$$

$$\therefore \theta = 53.1°$$

and it is apparent from the phasor diagram that the current through the circuit lags the applied voltage by this amount.

Although the foregoing demonstrates that the *phasor summation* of the voltage across the circuit is equal to the applied emf, no relation between applied voltage and circuit current has been given as yet.

Since $\quad V = \sqrt{(IR)^2 + (IX_L)^2}$

then $\quad V = \sqrt{I^2R^2 + I^2X_L^2}$

Factoring, $\quad V = \sqrt{I^2(R^2 + X_L^2)}$

Hence, $\quad V = I\sqrt{R^2 + X_L^2} \quad \text{V} \quad (30)$

As previously stated, the applied voltage divided by the current results in a quotient that represents the opposition offered to the flow of current. Hence, from Eq. (30),

$$\frac{V}{I} = \sqrt{R^2 + X_L^2} \tag{31}$$

The expression $\sqrt{R^2 + X_L^2}$ is called the *impedance* of the circuit. It is denoted by Z and measured in ohms. Therefore

$$Z = \sqrt{R^2 + X_L^2} \quad \Omega \tag{32}$$

Applying Eq. (32) to the circuit of Fig. 32–14,

$$Z = \sqrt{6^2 + 8^2} = 10\ \Omega$$

and

$$I = \frac{V}{Z} = 10\ \text{A}$$

From Eq. (31), Eq. (32) can be written

$$V = IZ = I\sqrt{R^2 + X_L^2}$$

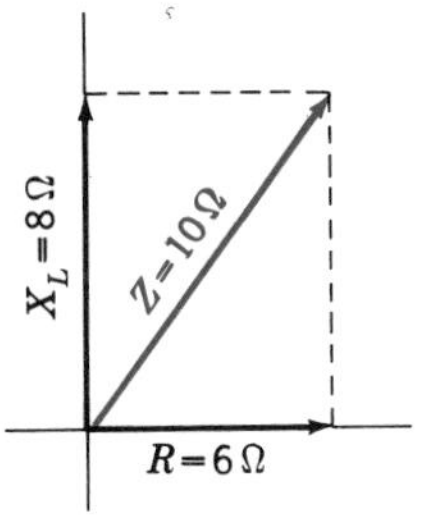

FIG. 32–16 Z can be plotted as phasor sum of R and X_L.

The foregoing illustrates that the factor I is common to all expressions, which is the same as saying that the current is the same in all parts of the circuit. Because this condition exists, it is permissible to plot the resistance and reactance as rectangular components as shown in Fig. 32–16; hence, the impedance of a series circuit is simply the phasor sum of the resistance and reactance. The various methods used in solving for the impedance are the same as those given for phasor summation of rectangular components in Example 4 of Sec. 28–4. Note that the values are identical.

EXAMPLE 9 A circuit consisting of 120 Ω resistance in series with an inductance of 0.35 H is connected across a 440-V 60-Hz alternator. Determine (*a*) the phase angle between voltage and current, (*b*) the impedance of the circuit, and (*c*) the current through the circuit.

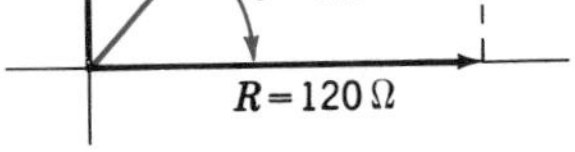

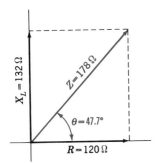

FIG. 32–17 Impedance phasor diagram for circuit of Example 9.

SOLUTION

(*a*) Drawing and labeling the circuit is left to you. The inductive reactance is

$$X_L = 2\pi fL = 2\pi \times 60 \times 0.35$$
$$= 132\ \Omega$$

Draw the phasor impedance diagram as shown in Fig. 32–17. Then, since

$$\tan\theta = \frac{X_L}{R} = \frac{132}{120} = 1.10$$
$$\therefore \theta = 47.7°$$

Note that the phase angle denotes the position of the applied voltage with respect to the current, which is taken as a reference. Thus an inductive series circuit always has a "lagging" phase angle which is a *positive angle* when resistance, reactance, and impedance are plotted vectorially.

(*b*)
$$Z = \frac{R}{\cos\theta} = \frac{120}{\cos 47.7°} = 178\ \Omega$$

or
$$Z = \frac{X_L}{\sin\theta} = \frac{132}{\sin 47.7°} = 178\ \Omega$$

(*c*)
$$I = \frac{V}{Z} = \frac{440}{178} = 2.47\ \text{A}$$

32–9 RESISTANCE AND CAPACITANCE IN SERIES

Figure 32–18 represents a circuit in which an alternating emf of 100 V is applied across a combination of 6 Ω resistance in series with 8 Ω capacitive

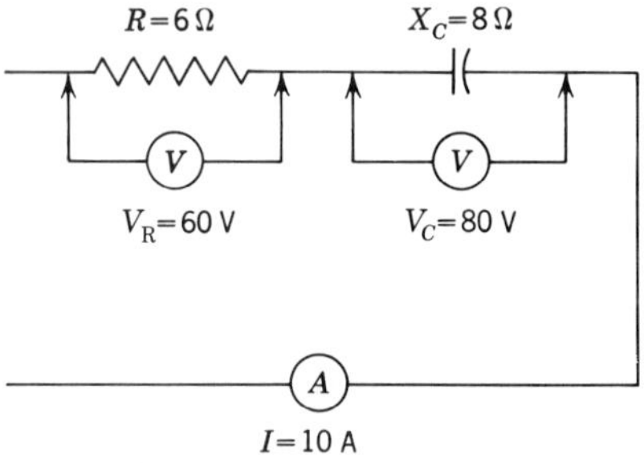

FIG. 32–18 Series circuit consisting of resistance and capacitance.

reactance. Note the similarity between the circuits shown in Figs. 32–14 and 32–18. Both have the same values of resistance and absolute values of reactance. However, in the circuit of Fig. 32–18 the voltage drop across the capacitive reactance is 90° behind the current. Again using the current as a reference, because it is the same in all parts of the circuit, the voltage across the resistance and the voltage across the capacitive reactance are plotted as shown in Fig. 32–19 and treated as rectangular components of the applied emf. The impedance of the circuit is found in the same manner as that of the inductive circuit, that is, by phasor summation of the rectangular components. The phase angle is found by the same method.

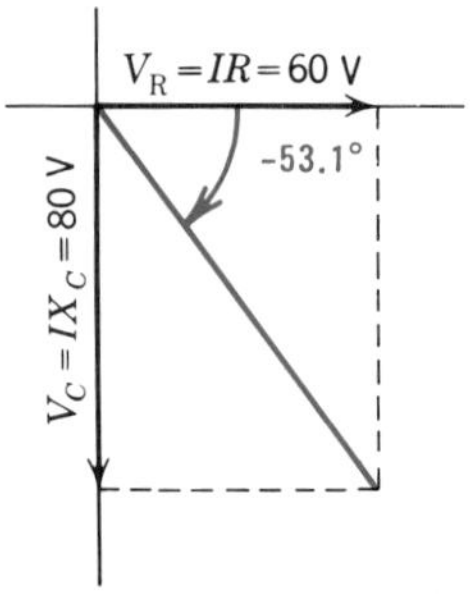

FIG. 32–19 Impedance phasor diagram for circuit of Fig. 32–18.

$$\tan\theta = \frac{X_C}{R} = \frac{8}{6} = 1.33$$
$$\therefore \theta = -53.1°$$

In the capacitive circuit the current leads the voltage, and we prefix the impedance angle with a minus sign because of its position (Sec. 23–2).

EXAMPLE 10 A circuit consisting of 175 Ω resistance in series with a capacitor of 5.0 μF is connected across a source of 150 V 120 Hz. Determine (*a*) the phase angle between voltage and current, (*b*) the impedance of the circuit, and (*c*) the current through the circuit.

SOLUTION

(*a*) Drawing and labeling the circuit is left to you. The capacitive reactance is

$$X_C = \frac{1}{2\pi f C} = \frac{1}{2\pi \times 120 \times 5 \times 10^{-6}} = \frac{10^4}{2\pi \times 1.2 \times 5} = 265\ \Omega$$

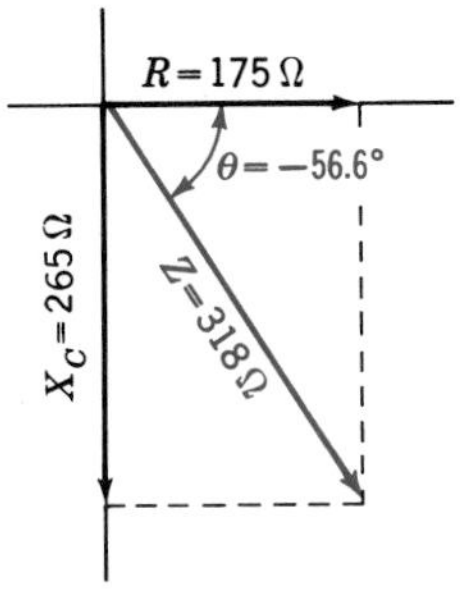

FIG. 32–20 Impedance phasor diagram for Example 10.

Draw the impedance diagram as shown in Fig. 32–20. Then, since

$$\tan\theta = \frac{X_C}{R} = \frac{265}{175} = 1.51$$

$$\therefore \theta = -56.6^\circ$$

Thus the current is leading the voltage by 56.6°, as shown by the impedance phasor diagram.

(*b*)

$$Z = \frac{R}{\cos\theta} = \frac{175}{\cos 56.6^\circ} = 318\ \Omega$$

or

$$Z = \frac{X_C}{\sin\theta} = \frac{265}{\sin 56.6^\circ} = 318\ \Omega$$

(*c*)

$$I = \frac{V}{Z} = \frac{150}{318} = 0.472\ \text{A}$$

PROBLEMS 32–4

1. A series circuit consists of a 1.8-H inductor which has a resistance of 400 Ω. It is supplied with 120 V 60 Hz. Find
   (*a*) The inductive reactance.
   (*b*) The impedance of the coil.
   (*c*) The current flowing through the coil.
   (*d*) The equation of the current.
   (*e*) The voltage across the resistance of the coil.
   (*f*) The voltage across the inductance of the coil.
   (*g*) Why (*e*) + (*f*) does not equal 120 V.
2. A 500-V 8-MHz source is connected to a series circuit consisting of a 3.3-kΩ resistor and a 500-μH inductor of negligible resistance. Find
   (*a*) The inductive reactance of the inductor.
   (*b*) The impedance of the circuit.
   (*c*) The current flowing through the circuit.
   (*d*) The phase angle of the current.
   (*e*) The voltage across the resistor.
   (*f*) The voltage across the inductor.

3. In the circuit of Prob. 2, the applied emf is held constant while the frequency is decreased.
   (*a*) Why will this cause the current to rise?
   (*b*) When the current is twice that found in Prob. 2, find the impedance, the frequency, and the phase angle.
4. A 2.8-H choke has a measured resistance of 1210 Ω at 400 Hz. This choke is connected across 50 V at 120 Hz. Find (*a*) the impedance of the choke and (*b*) the current flow.
5. A 120-V 60-Hz source energizes a series circuit consisting of a 330-Ω resistor and a 22-μF capacitor. Find
   (*a*) The capacitive reactance.
   (*b*) The impedance of the circuit.
   (*c*) The current flow through the circuit.
   (*d*) The voltage across the resistor.
   (*e*) The voltage across the capacitor.
6. If the frequency of the 120-V source in Prob. 5 is doubled, what will be the current flow through the circuit?
7. What will be the impedance of the circuit of Prob. 5 if a 50-μF capacitor is connected in series with the original circuit?
8. What will be the current flow in the circuit of Prob. 5 if a 6.7-kΩ resistor is connected in series with the original circuit?
9. A series circuit consisting of a 1-kΩ resistor and a 200-pF capacitor produces an impedance of $1031\underline{/-14^\circ}$ Ω and a current of $582\underline{/0^\circ}$ mA. Determine:
   (*a*) The applied voltage.
   (*b*) The frequency of the applied voltage.
   (*c*) The voltage across the capacitor.
   (*d*) The voltage across the resistor.
10. When a 50-pF capacitor is added in series to the circuit of Prob. 9, the voltage source now has a lagging phase angle of 51.3°. Find:
    (*a*) The new current value.
    (*b*) The voltage across the resistor.
    (*c*) The voltage across the 50-pF capacitor.
    (*d*) The voltage across the 200-pF capacitor.
    (*e*) The impedance of the new circuit.

32–10 RESISTANCE, INDUCTANCE, AND CAPACITANCE IN SERIES

It has been shown that inductive reactance causes the current to lag the voltage and that capacitance reactance causes the current to lead the voltage; hence, these two reactions are exactly opposite in effect. Figure 32–21 represents a series circuit consisting of resistance, inductance, and capacitance connected across an alternator that supplies 220 V, 60 Hz. Now

$$\omega = 2\pi f = 2\pi \times 60 = 377$$
$$\therefore X_L = \omega L = 377 \times 0.35 = 132\ \Omega$$

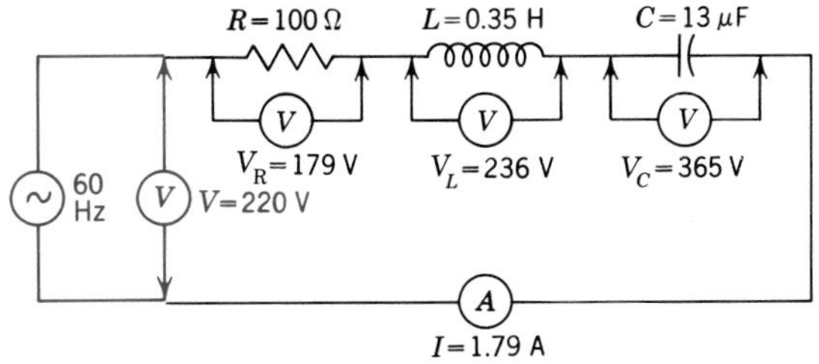

FIG. 32–21 Series circuit consisting of *R*, *L*, and *C*.

and

$$X_C = \frac{1}{\omega C} = \frac{1}{377 \times 13 \times 10^{-6}}$$

$$= \frac{10^3}{3.77 \times 1.3} = 203\ \Omega$$

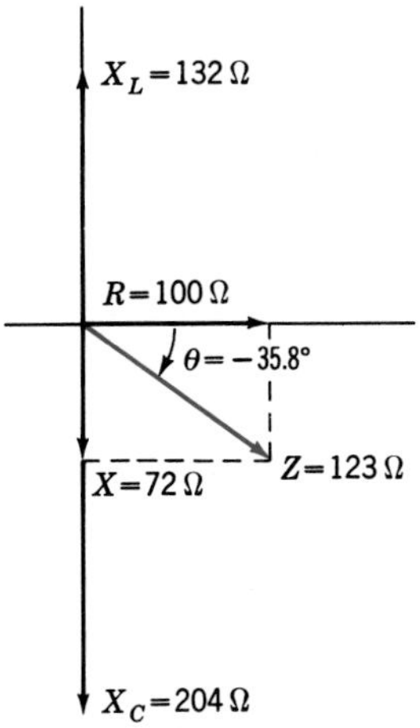

FIG. 32–22 Impedance phasor diagram for circuit of Fig. 32–21.

Figure 32–22 is an impedance phasor diagram of the conditions existing in the circuit. Since X_L and X_C are oppositely directed phasors, it is evident that the resultant reactance will have a magnitude equal to their algebraic sum and will be in the direction of the greater. Therefore, the net reactance of the circuit is a capacitive reactance of 72 Ω as illustrated in Fig. 32–22. Thus the entire circuit could be replaced by an equivalent series circuit consisting of 100 Ω resistance and 72 Ω capacitive reactance, provided that the frequency of the alternator remained constant.

The impedance, current, and potential differences are found by the usual methods.

$$\tan \theta = \frac{X_C}{R} = \frac{72}{100} = 0.72$$

$$\therefore \theta = -35.8°$$

$$Z = \frac{X}{\sin \theta} = \frac{72}{\sin 35.8°} = 123\ \Omega$$

$$I = \frac{V}{Z} = \frac{220}{123} = 1.79\ \text{A}$$

$$V_R = IR = 1.79 \times 100 = 179\ \text{V}$$

$$V_L = IX_L = 1.79 \times 132 = 236\ \text{V}$$

$$V_C = IX_C = 1.79 \times 204 = 365\ \text{V}$$

Note that the potential difference across the reactances is greater than the emf impressed across the entire circuit. This is reasonable, for the applied emf is across the impedance of the circuit, which is a smaller value, in ohms, than the reactances. Because the current is common to all circuit components, it follows that the greatest potential difference will exist across the component offering the greatest opposition.

32–11 POWER IN A SERIES CIRCUIT

It has been shown that, in a circuit consisting of resistance only, no power is returned to the source of emf. Also, it has been shown that a circuit containing reactance alone consumes no power; that is, a reactance alternately receives and returns all power to the source. It is evident, therefore, that in a circuit containing both resistance and reactance there must be some power expended in the resistance and also some returned to the source by the reactance. Figure 32–23 represents the relation among voltage, current, and power in the circuit of Fig. 32–21.

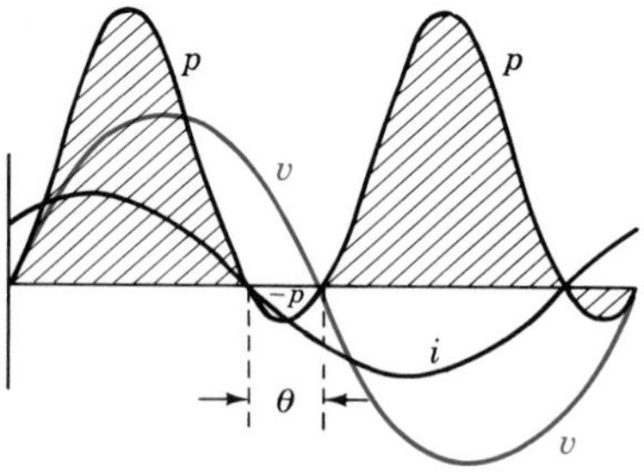

FIG. 32–23 Instantaneous voltage, current, and power relations for circuit shown in Fig. 32–21.

As previously stated, the instantaneous power in the circuit is equal to the product of the applied voltage and the current through the circuit. When the voltage and current are of the same sign, they are acting together and taking power from the source. When their signs are unlike, they are operating in opposite directions and power is returned to the source. The *apparent power* is

$$P_a = VI \qquad \text{VA} \tag{33}$$

and the actual power taken by the circuit, which is called the *true power* or *active power,* is

$$P = I^2R \qquad \text{W} \tag{34}$$

or

$$P = V_RI \qquad \text{W} \tag{35}$$

where V_R is the potential difference across the resistance of the circuit.

The *power factor* (PF) of a circuit is the ratio of the true power to the apparent power. That is,

$$\text{PF} = \frac{P}{P_a} \tag{36}$$

Substituting the value of P from Eq. (34) and that of P_a in Eq. (33),

$$\text{PF} = \frac{I^2R}{VI} = \frac{IR}{V}$$

Then, since

$$V = IZ$$

$$PF = \frac{IR}{IZ}$$

or

$$PF = \frac{R}{Z} \tag{37}$$

Hence, the power factor of a series circuit can be obtained by dividing the resistance of a circuit by its impedance. The power factor is often expressed in terms of the angle of lead or lag. From preceding vector diagrams, it is evident that

$$\frac{R}{Z} = \cos\theta$$

$$\therefore PF = \cos\theta \tag{38}$$

From Eq. (36), $P = (P_a)(PF)$

Substituting for P_a, $P = (VI)(PF)$

Substituting for the PF, $P = VI\cos\theta$ (39)

From the foregoing it is seen that the power expended in a circuit can be obtained by utilizing different relations. For example, in the circuit of Fig. 32–21,

$$P = I^2R = 1.79^2 \times 100 = 320 \text{ W}$$

$$P = V_RI = 179 \times 1.79 = 320 \text{ W}$$

and

$$\begin{aligned} P &= VI\cos\theta \\ &= 220 \times 1.79 \times \cos 35.8^\circ \\ &= 320 \text{ W} \end{aligned}$$

The power factor of a circuit can be expressed as a decimal or as a percent. Thus the power factor of this circuit is

$$\cos\theta = \cos 35.8^\circ = 0.812$$

Expressed as percent,

$$PF = 100 \cos 35.8^\circ = 81.2\%$$

32–12 NOTATION FOR SERIES CIRCUITS

In Sec. 3–5, it was shown that positive and negative "real" numbers could be represented graphically by plotting them along a horizontal line. The positive numbers were plotted to the right of zero, and the negative numbers were plotted to the left. This idea was expanded in Sec. 16–3, where the original horizontal line was made the x axis of a system of rectangular coordinates.

In Sec. 20–12 the system of representation was extended to include what we referred to as "imaginary" numbers by agreeing to plot the numbers along the y axis, the letter j being used as a symbol of 90° operation. Thus, when some number is prefixed with j, it means that the vector which the number represents is to be rotated through an angle of 90°. The rotation is positive, or in a counterclockwise direction, when the sign of j is positive and negative, or in a clockwise direction, when the sign of j is negative. In Sec. 31–1 we saw that "real" and "imaginary" are more properly called "in phase" and "out of phase."

From the foregoing, it is evident that resistance, when plotted on an impedance phasor diagram, is considered as an in-phase number because it is plotted along the x axis. In this instance the term *real* may well define resistance, for only the opposition to the flow of current consumes power.

Since reactances are displaced 90° from resistance in an impedance phasor diagram, it follows that inductive reactance can be prefixed with a plus j and capacitive reactance with a minus j. Thus, an inductive reactance of 75 Ω would be written j75 Ω and plotted on the positive y axis and a capacitive reactance of 86 Ω would be written −j86 Ω and plotted on the negative y axis.

In Sec. 31–1 we saw that a vector can be completely described in terms of either its polar or its rectangular components. For example, the circuit of Fig. 32–14 can be described as consisting of an impedance of 10 Ω at an angle of 53.1°, which would be written

$$Z = 10\underline{/\ +53.1^\circ}\ \Omega$$

where the angle sign may be included for emphasis and the number of degrees denotes the angle that the vector makes with the positive x axis. This is known as *polar form*. Since this impedance is made up of 6 Ω of resistance and 8 Ω of inductive reactance, we can write

$$Z = R + jX_L = 6 + j8\ \Omega$$

This is known as *rectangular form*.

In converting from rectangular to polar form, use the usual methods of solution of right triangles that were developed in Chap. 26 and shown in Fig. 32–24.

The rectangular form is a very convenient method of notation. For example, instead of writing, "A series circuit of 4 Ω resistance and 3 Ω capacitive reactance," we can write, "A series circuit of 4 − j3 Ω." Figure 32–24 shows the various types of series circuits with their proper impedance phasor diagram and corresponding notation.

Note that the sign of the phase angle is the same as that of j in the rectangular form. 4 − j3 converts to a polar form with a negative angle $5\underline{/\ -36.9^\circ}$. It must be understood that neither the rectangular form nor the polar form is a method for solving series circuits. The two forms are simply convenient kinds of notation that completely describe circuit conditions from both electrical and mathematical viewpoints.

Circuit	Impedance phasor	Z Rectangular form	Z Polar form
$R = 10\ \Omega$	$R = 10\ \Omega$	$Z = 10 + j0\ \Omega$	$Z = 10\angle 0°\ \Omega$
$X_L = 7\ \Omega$	$X_L = j7\ \Omega$, θ	$Z = 0 + j7\ \Omega$	$Z = 7\angle 90°\ \Omega$
$X_C = 6\ \Omega$	θ, $X_C = -j6\ \Omega$	$Z = 0 - j6\ \Omega$	$Z = 6\angle -90°\ \Omega$
$R = 4\ \Omega$, $X_L = 3\ \Omega$	$X_L = j3\ \Omega$, Z, θ, $R = 4\ \Omega$	$Z = 4 + j3\ \Omega$	$Z = 5\angle 36.9°\ \Omega$
$R = 6\ \Omega$, $X_C = 8\ \Omega$	$R = 6\ \Omega$, θ, Z, $X_C = -j8\ \Omega$	$Z = 6 - j8\ \Omega$	$Z = 10\angle -53.1°\ \Omega$
$R = 7\ \Omega$, $X_C = 40\ \Omega$, $R = 13\ \Omega$, $X_L = 20\ \Omega$	$R = 20\ \Omega$, θ, Z, $X_C = -j20\ \Omega$	$Z = 20 - j20\ \Omega$	$Z = 28.2\angle -45°\ \Omega$

FIG. 32–24 Phasor notation for series circuits.

EXAMPLE 11 Find the phasor impedance of the following series circuit:

$$250 - j100\ \Omega$$

SOLUTION

Given

$$Z = R - jX = 250 - j100\ \Omega$$

$$\tan\theta = \frac{X}{R} = \frac{100}{250} = 0.400$$

$$\therefore \theta = -21.8°$$

$$Z = \frac{X}{\sin\theta} = \frac{100}{\sin 21.8°} = 269\ \Omega$$

or

$$Z = \frac{R}{\cos\theta} = \frac{250}{\cos 21.8°} = 269\ \Omega$$

Hence,

$$Z = 269\angle -21.8°\ \Omega$$

Converting from polar form, in which the magnitude and angle are given, to rectangular form is simplified by making use of the trigonometric functions.

Since

$$R = Z \cos \theta \qquad \Omega$$

$$X = Z \sin \theta \qquad \Omega$$

and
$$Z = R \pm jX \tag{40}$$

by substitution,
$$Z = Z \cos \theta + jZ \sin \theta \tag{41}$$

Factoring,
$$Z = Z(\cos \theta + j \sin \theta) \qquad \Omega \tag{42}$$

The ± sign is omitted in Eqs. (41) and (42) because, if the proper angles are used (positive or negative), the respective sine values will determine the proper sign of the reactance component.

EXAMPLE 12 A series circuit has an impedance of 269 Ω with a leading power factor of 0.928. What are the reactance and resistance of the circuit?

SOLUTION

Given Z = 269 Ω and PF = 0.928. The power factor, when expressed as a decimal, is equal to the cosine of the phase angle. Hence,

if
$$0.928 = \cos \theta$$
then
$$\theta = -21.9°$$

The angle was given the minus sign because a ''leading power factor'' means the current leads the voltage. Therefore,

$$Z = 269\underline{/-21.9°}\ \Omega$$

Substituting these values in Eq. (41),

$$\begin{aligned} Z &= 269 \cos 21.9° - j269 \sin 21.9° \\ &= 250 - j100\ \Omega \end{aligned}$$

32–13 THE GENERAL SERIES CIRCUIT

In a series circuit consisting of several resistances and reactances, the total resistance of the circuit is the sum of all the series resistances and the total reactance is the algebraic sum of the series reactances. That is, the total resistance is

$$R_t = R_1 + R_2 + R_3 + \cdots$$

and the reactance of the circuit is

$$X = j(\omega L_1 + \omega L_2 + \omega L_3 + \cdots) - j\left(\frac{1}{\omega C_1} + \frac{1}{\omega C_2} + \frac{1}{\omega C_3} + \cdots\right)$$

Hence, the impedance is

$$Z = R_t \pm jX \qquad \Omega$$

As an alternate method, such a circuit can always be reduced to an equivalent series circuit by combining inductances and capacitances before computing reactances. Thus, the total inductance is

$$L_t = L_1 + L_2 + L_3 + \cdots$$

and the capacitance of the circuit is obtained from

$$\frac{1}{C_t} = \frac{1}{C_1} + \frac{1}{C_2} + \frac{1}{C_3} + \cdots$$

However, when voltage drops across individual reactances are desired, it is best to find the equivalent circuit by combining reactances.

EXAMPLE 13 Given the circuit of Fig. 32–25, which is supplied by 220 V 60 Hz. Find the (*a*) equivalent series circuit, (*b*) impedance of the circuit, (*c*) current, (*d*) power factor, (*e*) power expended in the circuit, (*f*) apparent power, (*g*) voltage drop across C_1, and (*h*) power expended in R_2.

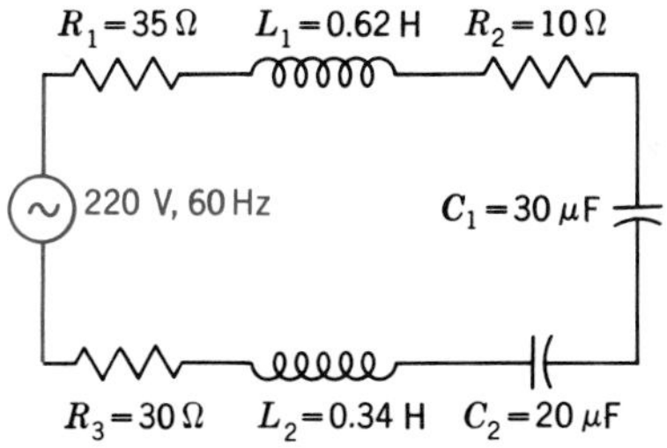

FIG. 32–25 Series circuit of Example 13.

SOLUTION

(*a*)

$$\begin{aligned}
R_t &= R_1 + R_2 + R_3 = 35 + 10 + 30 \\
&= 75\ \Omega \\
\omega &= 2\pi f = 2\pi \times 60 = 377 \\
L_t &= L_1 + L_2 = 0.62 + 0.34 = 0.96\ \text{H} \\
X_L &= \omega L = 377 \times 0.96 = 362\ \Omega \\
X_{C_1} &= \frac{1}{\omega C_1} = \frac{1}{377 \times 30 \times 10^{-6}} \\
&= \frac{10^3}{3.77 \times 3} = 88.4\ \Omega \\
X_{C_2} &= \frac{1}{\omega C_2} = \frac{1}{377 \times 20 \times 10^{-6}} \\
&= \frac{10^3}{3.77 \times 2} = 132.6\ \Omega \\
X_C &= 88.4 + 132.6 = 221\ \Omega \\
X &= X_L - X_C = 362 - 221 = 141\ \Omega
\end{aligned}$$

The equivalent series circuit consists of a resistance of 75 Ω and an inductive reactance of 141 Ω. That is,

$$Z = 75 + j141\ \Omega$$

The impedance phasor diagram for the equivalent circuit is shown in Fig. 32–26.

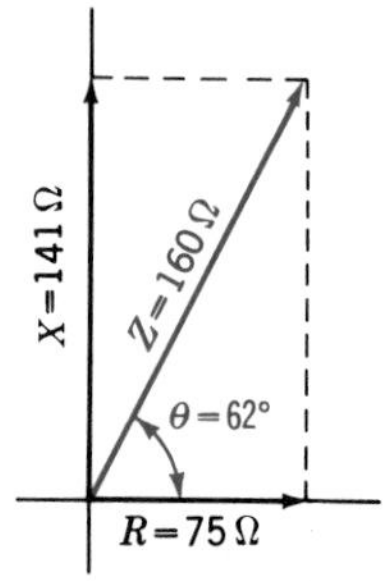

FIG. 32–26 Impedance phasor diagram for circuit of Fig. 32–25.

(*b*) $$\tan\theta = \frac{X}{R_t} = \frac{141}{75} = 1.88$$

$$\therefore \theta = 62°$$

$$Z = \frac{R}{\cos\theta} = \frac{75}{\cos 62°} = 160\ \Omega$$

Hence, $$Z = 160\angle 62°\ \Omega$$

(*c*) $$I = \frac{V}{Z} = \frac{220}{160} = 1.38\text{ A}$$

(*d*) $$\text{PF} = \cos\theta = \cos 62° = 0.470$$

Expressed as a percent, PF = 47.0%

(*e*) $$P = VI\cos\theta = 220 \times 1.38 \times \cos 62° = 143\text{ W}$$

or $$P = I^2R = 1.38^2 \times 75 = 143\text{ W}$$

(*f*) $$P_a = VI = 220 \times 1.38 = 304\text{ VA}$$

(*g*) $$V_{C_1} = IX_{C_1} = 1.38 \times 88.4 = 122\text{ V}$$

(*h*) $$P_{R_2} = I^2R_2 = 1.38^2 \times 10 = 19\text{ W}$$

You will find it convenient to compute the value of the angular velocity $\omega = 2\pi f$ for all ac problems, for this factor is common to all reactance equations.

As with all electric circuit problems, a neat diagram of the circuit, with all known circuit components, voltages, and currents clearly marked, should be made. In addition, a phasor or impedance diagram should be drawn to scale in order to check the mathematical solution.

PROBLEMS 32–5

Given the circuit of Fig. 32–27, with values as listed in Table 32–1. Draw an impedance phasor diagram for each circuit and find (*a*) the impedance of the circuit, (*b*) the current flowing through the circuit, (*c*) the equation of the current, (*d*) the PF of the circuit, and (*e*) the power expended in the circuit.

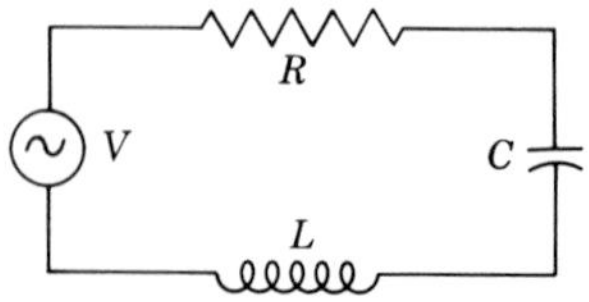

FIG. 32–27 Circuit of Probs. 1 to 10.

11. A choke coil, when connected to a 220-V 60-Hz supply, draws 550 mA. When it is connected to a 220-V dc source, it draws 1.1 A. Determine:
    (*a*) The resistance of the coil.
    (*b*) The reactance of the coil.
    (*c*) The inductance of the coil.
12. Assuming that the resistance of the coil in Prob. 11 is unchanged, how much power would the coil draw when it is connected across 230 V, 400 Hz?

TABLE 32–1 PROBLEMS 1 TO 10

Problems	*V*, V	*f*	*R*	*L*	*C*
1.	120	60 Hz	400 Ω	1.8 H	20 μF
2.	450	1 kHz	67 Ω	5 mH	50 μF
3.	120	100 Hz	1.5 kΩ	2 H	5 μF
4.	850	400 Hz	500 Ω	2.5 H	100 μF
5.	1.5 mV	20 MHz	1 kΩ	52 μH	50 pF
6.	1000	8 GHz	330 Ω	0.08 μH	0.005 pF
7.	117	60 Hz	15 Ω	4.5 mH	2500 μF
8.	2	10 kHz	27 Ω	3.5 μH	1.5 μF
9.	120	3 MHz	90 Ω	10 μH	200 pF
10.	110	60 Hz	50 Ω	300 mH	22.0 μF

13. The following 60-Hz impedances are connected in series:

$$Z_1 = 30 - j40\ \Omega \qquad Z_2 = 5 + j12\ \Omega$$
$$Z_3 = 8 - j6\ \Omega \qquad Z_4 = 4 + j4\ \Omega$$

(*a*) What is the resultant impedance of the circuit?
(*b*) What value of pure reactance must be added in series to make the PF of the circuit 80% leading?

14. The meters represented in Fig. 32–28 are connected such a short distance from an inductive load that line drop from meters to load is negligible. What is the equivalent series circuit of the load?

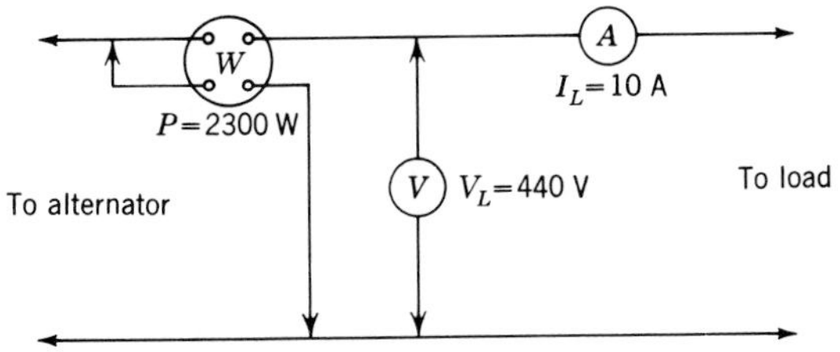

FIG. 32–28 Circuit of Prob. 14.

15. A single-phase induction motor, with 440 V across its input terminals, delivers 10.8 mechanical horsepower at an efficiency of 90% and a PF of 86.6%.
(*a*) What is the line current?
(*b*) How much power is taken by the motor?

16. Given any series circuit, for example, 110 V at 60 Hz applied across 3 + j4 Ω. On the same set of axes and to the same scale, plot instantaneous values of the applied emf *v*, the potential difference across the resistance *R*, and the potential difference across the reactance *X*. What is your conclusion?

32–14 SERIES RESONANCE

It has been shown that the inductive reactance of a circuit varies directly as the frequency and that the capacitive reactance varies inversely as the frequency. That is, the inductive reactance will increase and the capacitive reactance will decrease as the frequency is increased, and vice versa. Then, for any value of inductance and capacitance in a circuit, there is a frequency at which the inductive reactance and the capacitive reactance are equal. This is called the *resonant frequency* of the circuit. Since, in a series circuit,

$$Z = R + j\left(\omega L - \frac{1}{\omega C}\right) \quad \Omega$$

at resonance,

$$\omega L = \frac{1}{\omega C} \tag{43}$$

Hence,

$$Z = R$$

Therefore, at the resonant frequency of a series circuit, the resistance is the only circuit component that limits the flow of current, for the net reactance of the circuit is zero. Thus the current is in phase with the applied voltage, which results in a circuit power factor of 100%.

EXAMPLE 14 There is impressed 10 V at a frequency of 1 MHz across a circuit consisting of a coil of 92.2 μH in series with a capacitance of 275 pF. The effective resistance of the coil at this frequency is 10 Ω, and both the resistance of the connecting wires and the capacitance are negligible.

(*a*) What is the impedance of the circuit?
(*b*) How much current flows through the circuit?
(*c*) What are the voltages across the reactances?

SOLUTION

The resistance of the coil is treated as being in series with its inductive reactance.

(*a*)

$$\omega = 2\pi f = 6.28 \times 10^6$$
$$X_L = \omega L$$
$$= 6.28 \times 10^6 \times 92.2 \times 10^{-6}$$
$$= 6.28 \times 92.2 = 579\ \Omega$$
$$X_C = \frac{1}{\omega C}$$
$$= \frac{1}{6.28 \times 10^6 \times 275 \times 10^{-12}}$$
$$= \frac{10^4}{6.28 \times 2.75} = 579\ \Omega$$

Since

$$X_L = X_C$$

then

$$Z = R = 10\ \Omega$$

(*b*)

$$I = \frac{V}{Z} = \frac{10}{10} = 1\ \text{A}$$

(*c*)

$$V_C = IX_C = 1 \times 579 = 579\ \text{V}$$
$$V_L = IX_L = 1 \times 579 = 579\ \text{V}$$

Note that the voltages across the inductance and capacitance are much greater than the applied voltage.

An inductance has a *quality* or *merit,* denoted by Q, that is defined as the ratio of its inductive reactance to its resistance at a given frequency. Thus,

$$Q = \frac{\omega L}{R} \tag{44}$$

Then, at resonance,

$$V_C = V_L = I\omega L \qquad \text{V}$$

Substituting for I,

$$V_C = V_L = \frac{V\omega L}{R} \qquad \text{V}$$

Substituting for $\frac{\omega L}{R}$,

$$V_C = V_L = VQ \qquad \text{V} \tag{45}$$

Because the average radio circuit has purposely been designed for high Q values, it is seen that very high voltages can be developed in resonant series circuits.

32–15 RESONANT FREQUENCY

The resonant frequency of a circuit can be determined by rewriting Eq. (43). Thus,

$$2\pi fL = \frac{1}{2\pi fC}$$

$$\therefore f = \frac{1}{2\pi\sqrt{LC}} \qquad \text{Hz} \tag{46}$$

where f, L, and C are in the usual units, hertz, henrys, and farads, respectively, when $Q > 10$.

EXAMPLE 15 A series circuit consists of an inductance of 500 μH and a capacitor of 400 pF. What is the resonant frequency of the circuit?

SOLUTION

$$L = 500\ \mu\text{H} = 5 \times 10^{-4}\ \text{H}$$

$$C = 400\ \text{pF} = 4 \times 10^{-10}\ \text{F}$$

$$f = \frac{1}{2\pi\sqrt{LC}}$$

$$= \frac{1}{2\pi\sqrt{5 \times 10^{-4} \times 4 \times 10^{-10}}}$$

$$= \frac{10^7}{2\pi\sqrt{20}}$$

$$= 356\ 000\ \text{Hz}$$

or

$$f = 356\ \text{kHz}$$

From Eq. (46) it is evident that the resonant frequency of a series circuit depends *only* upon the LC product. This means there is an infinite number of combinations of L and C that will resonate to a particular frequency.

EXAMPLE 16 How much capacitance is required to obtain resonance at 1500 kHz with an inductance of 45 μH?

SOLUTION

$$f = 1500 \text{ kHz} = 1.5 \times 10^6 \text{ Hz}$$

$$L = 45 \ \mu\text{H} = 4.5 \times 10^{-5} \text{ H}$$

$$\omega = 2\pi f = 2\pi \times 1.5 \times 10^6 = 9.42 \times 10^6$$

From Eq. (46),

$$C = \frac{1}{(2\pi f)^2 L} = \frac{1}{\omega^2 L}$$

$$\therefore C = \frac{1}{(9.42 \times 10^6)^2 \times 4.5 \times 10^{-5}}$$

$$= 250 \text{ pF}$$

PROBLEMS 32–6

1. 80 V 15 kHz is impressed across a series circuit consisting of a 500-pF capacitor of negligible resistance and a 265-mH coil with effective resistance of 1.10 kΩ.
   (*a*) How much current flows through the circuit?
   (*b*) How much power does the circuit absorb from the source?
   (*c*) What are the voltages across the capacitor and the coil?
2. What is the Q of the coil in Prob. 1?
3. At what frequency would the circuit of Prob. 1 be resonant?
4. What type and value of "pure reactance" must be added to the circuit of Prob. 1 to make the circuit resonant at 10 kHz?
5. A tuning capacitor has a range of capacitance between 20 pF and 350 pF.
   (*a*) What inductance must be connected in series with it to provide a lowest resonant frequency of 550 kHz?
   (*b*) What will then be the highest resonant frequency?
6. What is the equivalent circuit of a series circuit when operating (*a*) at resonant frequency, (*b*) at a frequency less than resonant frequency, and (*c*) at a frequency higher than resonant frequency?
7. An inductor of 0.239 mH with 15 Ω internal resistance is connected in series with a 100-pF capacitor, and supplied with a signal voltage of 0.3 mV at 1 MHz.
   (*a*) What is the circuit impedance?
   (*b*) What is the circuit current?
8. The circuit of Prob. 7 is to be made resonant at 1 MHz. What value of capacitor should be added to achieve this condition?
   (*a*) What will be the Q of the coil?
   (*b*) What voltage appears across the 15-Ω resistance?
   (*c*) What voltage appears across the coil?
   (*d*) What is the phase angle between the generator current and voltage?

SELF TEST

1. Three resistors, $R_1 = 12\ \Omega$, $R_2 = 18\ \Omega$, and R_3, are connected in parallel. This parallel combination is in series with resistor $R_4 = 33\ \Omega$. The total circuit is energized by a power source that delivers 220 V at 440 Hz and a total current of 5.87 A.
   (*a*) Write the equation for the instantaneous voltage of the power source.
   (*b*) Write the equation for the total current.
   (*c*) What is the voltage across R_3?
   (*d*) How much power is dissipated in R_4?
   (*e*) How much current flows through R_1?
   (*f*) What is the instantaneous value of the total current when the instantaneous voltage of the source is 82 V?
2. A 12-μH coil with a negligible dc resistance drops 1.2 V when it conducts an ac current of 22 mA. What is the frequency of the current?
3. Two capacitors, $C_1 = 22\ \mu F$, and $C_2 = 50\ \mu F$, are connected in series. What is their total capacitance?
4. A 12-H coil has a dc resistance of 1.2 kΩ. In a circuit, it exhibits a voltage drop of 32.5 V. If the frequency of the circuit current is 60 Hz, what is:
   (*a*) the inductive reactance?
   (*b*) the impedance of the coil?
   (*c*) the current through the coil?
   (*d*) the voltage across the resistance of the coil?
   (*e*) the voltage across the inductance of the coil?
   (*f*) the algebraic sum of the voltages in parts (*d*) and (*e*)?
   (*g*) the phasor sum of the voltages in parts (*d*) and (*e*)?
5. A series circuit consists of a "pure" inductance of 8 H, a "pure" capacitance of 1 μF, and a total resistance of 680 Ω, supplied with 100 V at 60 Hz. What is:
   (*a*) the inductive reactance?
   (*b*) the capacitive reactance?
   (*c*) the impedance of the circuit?
   (*d*) the current through the circuit?
   (*e*) the voltage across the inductance?
   (*f*) the voltage across the capacitance?
   (*g*) the voltage across the resistance?
   (*h*) the power delivered to the circuit?
6. What is the resonant frequency of the circuit of Prob. 5?

CHAPTER 33

ALTERNATING CURRENTS—PARALLEL CIRCUITS

Parallel circuits are the kind of circuit most commonly encountered. The average distribution circuit has many types of loads all connected in parallel with each other: lighting circuits, motors, transformers for various uses, etc. The same is true of electronic circuits, which range from the most simple parallel circuits to complex networks.

This chapter deals with the solution of parallel circuits. Such a solution may reduce a parallel circuit to an equivalent series circuit that, when connected to the same source of emf as the given parallel circuit, would result in the same line current and phase angle; that is, the alternator would "see" the same load.

33–1 RESISTANCES IN PARALLEL

It was explained in Secs. 32–1 and 32–2 that, in an ac circuit containing resistance only, the voltage, current, and power relations are the same as in dc circuits. However, in order to build a foundation from which all parallel circuits can be analyzed, the case of paralleled resistances must be considered from a phasor viewpoint.

Figure 33–1 represents a 60-Hz 220-V alternator connected to three resistances in parallel.

Neglecting the internal resistance of the alternator and the resistance of the connecting wires, the emf of the alternator is impressed across each of the three resistances. If I_1, I_2, and I_3 represent the currents flowing through R_1, R_2, and R_3, respectively, then by Ohm's law,

$$I_1 = 2.5 \text{ A}$$
$$I_2 = 0.5 \text{ A}$$
$$I_3 = 2.0 \text{ A}$$

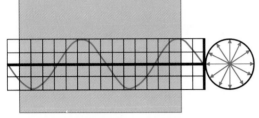

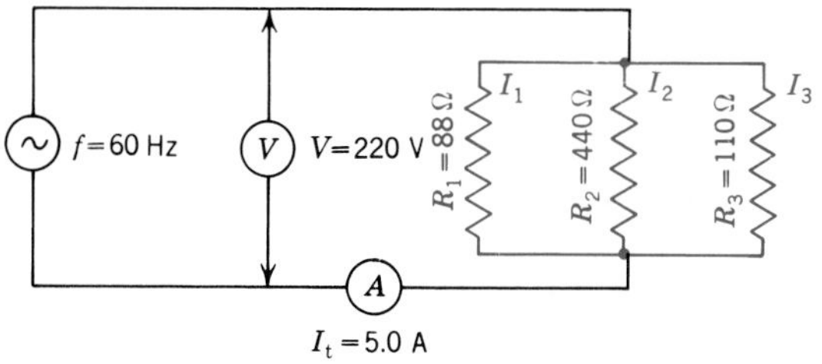

FIG. 33–1 Alternator connected to three resistors in parallel.

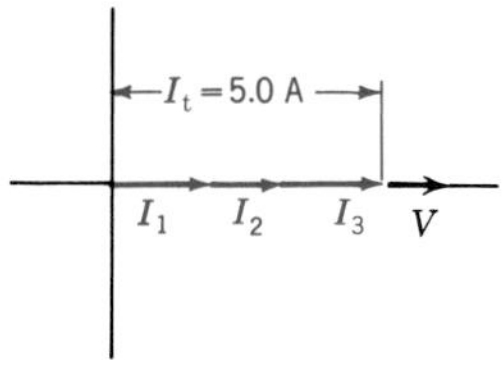

FIG. 33–2 Phasor diagram for the circuit of Fig. 33–1.

Since all currents are in phase, the total current flowing in the line, or external circuit, will be equal to the sum of the branch currents or 5.0 A. The phasor diagram for the three currents is illustrated in Fig. 33–2. All currents are plotted in phase with the applied emf, which is used as a reference phasor because the voltage is common to all resistances. Then, using rectangular phasor notation,

$$I_1 = 2.5 + j0 \text{ A}$$
$$I_2 = 0.5 + j0 \text{ A}$$
$$I_3 = 2.0 + j0 \text{ A}$$
$$I_t = 5.0 + j0 \text{ A} = 5.0\angle 0^\circ \text{ A}$$

As with all other circuits, the equivalent series impedance, which in this case is a pure resistance, is found by dividing the voltage across the circuit by the total current. That is,

$$Z = \frac{V}{I_t} = \frac{220}{5} = 44\ \Omega = 44\angle 0^\circ\ \Omega$$

33–2 CAPACITORS IN PARALLEL

Figure 33–3 represents two capacitors C_1 and C_2 connected in parallel across a voltage V. The quantity of charge in capacitor C_1 will be

$$Q_1 = C_1 V \tag{1}$$

and that in capacitor C_2 will be

$$Q_2 = C_2 V \tag{2}$$

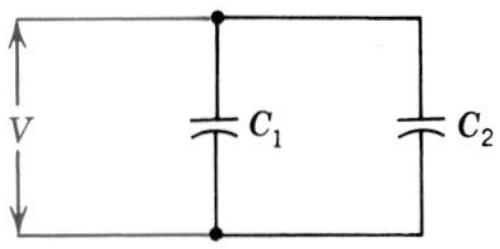

FIG. 33–3 Capacitors C_1 and C_2 connected in parallel.

Since the total quantity in both capacitors is $Q_1 + Q_2$, then

$$Q_1 + Q_2 = C_p V \tag{3}$$

where C_p is the total capacitance of the parallel combination.
Then adding Eqs. (1) and (2),

$$Q_1 + Q_2 = C_1 V + C_2 V$$

or

$$Q_1 + Q_2 = (C_1 + C_2)V$$

Substituting the value of $Q_1 + Q_2$ from Eq. (3),

$$C_pV = (C_1 + C_2)V$$

which results in

$$C_p = C_1 + C_2 \tag{4}$$

From the foregoing, it is apparent that capacitors in parallel combine like resistances in series; that is, the capacitance of paralleled capacitors is equal to the sum of the individual capacitances.

EXAMPLE 1 What is the capacitance of a 6-μF capacitor in parallel with a 4-μF capacitor?

SOLUTION

$$C_p = 6 + 4 = 10\ \mu\text{F}$$

33–3 INDUCTANCE AND CAPACITANCE IN PARALLEL

When a purely inductive reactance and a capacitive reactance are connected in parallel, as shown in Fig. 33–4, the currents flowing through these reactances differ in phase by 180°.

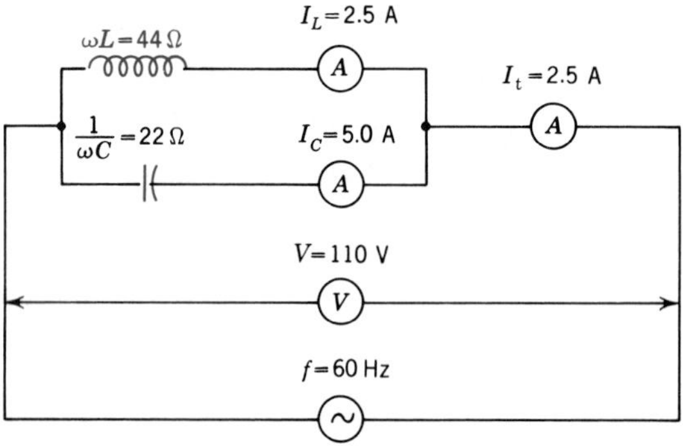

FIG. 33–4 X_L and X_C connected in parallel.

The current flowing through the inductor is

$$I_L = \frac{V}{X_L} = \frac{V}{\omega L} = \frac{110}{44} = 2.5\ \text{A}$$

and that through the capacitor is

$$I_C = \frac{V}{X_C} = \omega CV = \frac{110}{22} = 5.0\ \text{A}$$

In series circuits, the current was used as the reference phasor because the current is the same in all parts of the circuit. In parallel circuits there are

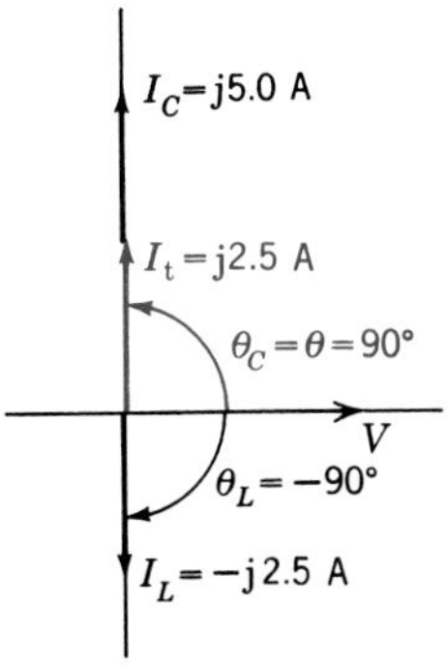

FIG. 33–5 Phasor diagram for the circuit of Fig. 33–4.

different values of currents in various parts of a circuit; therefore, the current cannot be used as the reference phasor.

Since the same voltage exists across two or more parallel branches, the applied voltage can be used as the reference phasor as illustrated in Fig. 33–5.

Note that the current I_L through the inductor is plotted as *lagging* the alternator voltage by 90° and the current I_C through the capacitor is *leading* the voltage by 90°. The total line current I_t, which is the phasor sum of the branch currents, is leading the applied voltage by 90°. That is, using rectangular phasor notation,

$$I_L = 0 - j2.5 \text{ A}$$
$$I_C = 0 + j5.0 \text{ A}$$
$$I_t = 0 + j2.5 \text{ A} = 2.5\angle 90° \text{ A}$$

Since the line current leads the alternator voltage by 90°, the equivalent series circuit consists of a capacitive reactance of

$$\frac{V}{I_t} = \frac{110}{2.5} = 44\ \Omega$$

That is, the parallel circuit could be replaced by a 60.3-μF capacitor which would result in a current of 2.5 A leading the voltage by 90°; in other words, the alternator would not sense the difference.

Note the difference between reactances in series and reactances in parallel. In a series circuit the *greatest* reactance of the circuit results in the equivalent series circuit containing the same kind of reactance. For this reason, it is said that reactances, or voltages across reactances, are the controlling factors of series circuits. In a parallel circuit the *least* reactance of the circuit, which passes the greatest current, results in the equivalent series circuit containing the same kind of reactance. For this reason, it is said that currents are the controlling factors of parallel circuits.

33–4 ASSUMED VOLTAGES

The solutions of the great majority of parallel circuits are facilitated by assuming a voltage to exist across a parallel combination. The currents through each branch, due to the assumed voltage, are then added vectorially to obtain the total current. The assumed voltage is then divided by the total current, the quotient being the joint impedance of the parallel branches.

To avoid small decimal quantities, the assumed voltage should be greater than the largest impedance of any parallel branch.

EXAMPLE 2 Given the circuit of Fig. 33–6. What are the impedance and the power factor of the circuit at a frequency of 2.5 MHz?

SOLUTION

C_1 and C_2 are in parallel; hence, the total capacitance is

$$C_p = C_1 + C_2 = 200 + 125 = 325 \text{ pF}$$

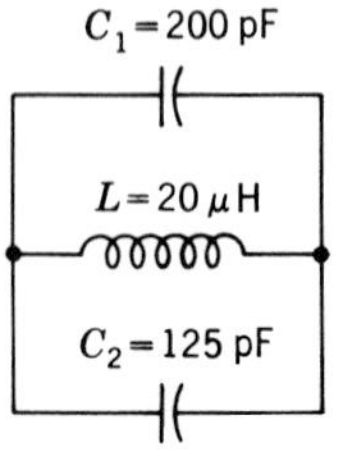

FIG. 33–6 Circuit of Example 2.

This simplifies the circuit to a capacitor C of 325 pF in parallel with an inductance L of 20 μH.

$$\omega = 2\pi f = 2\pi \times 2.5 \times 10^6 = 1.57 \times 10^7$$

$$X_L = \omega L = 1.57 \times 10^7 \times 2 \times 10^{-5} = 314\ \Omega$$

$$X_C = \frac{1}{\omega C} = \frac{1}{1.57 \times 10^7 \times 325 \times 10^{-12}}$$

$$= \frac{10^3}{1.57 \times 3.25} = 196\ \Omega$$

Assume 1000 V across the parallel branch. Then the current through the capacitors is

$$I_C = \frac{V_a}{X_C} = \frac{1000}{196} = 5.10\ \text{A}$$

and the current through the inductance is

$$I_L = \frac{V_a}{X_L} = \frac{1000}{314} = 3.18\ \text{A}$$

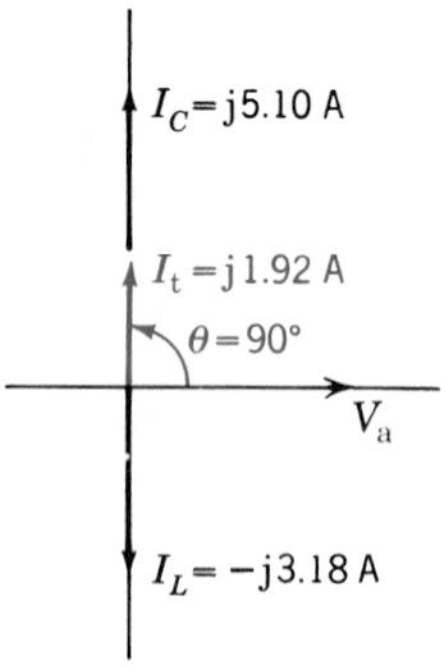

FIG. 33–7 Phasor diagram for circuit of Example 2.

Since I_C leads the assumed voltage by 90° and I_L lags the assumed voltage by 90°, they are plotted with the assumed voltage as reference phasor as shown in Fig. 33–7. Then the total current I_t that would flow because of assumed voltage would be the phasor summation of I_C and I_L. Performing phasor summation:

$$I_C = 0 + j5.10\ \text{A}$$
$$I_L = 0 - j3.18\ \text{A}$$
$$I_t = 0 + j1.92\ \text{A} = 1.92\angle 90^\circ\ \text{A}$$

Again, since the total current leads the voltage by 90°, the equivalent series circuit consists of a capacitor whose capacitive reactance is

$$\frac{V_a}{I_t} = \frac{1000}{1.92} = 521\ \Omega$$

Since $\theta = 90^\circ$, $\quad \text{PF} = \cos\theta = 0$

You should solve the circuit of Fig. 33–6 with different values of assumed voltages.

33–5 RESISTANCE AND INDUCTANCE IN PARALLEL

When a resistance and an inductive reactance are connected in parallel, as represented in Fig. 33–8, the currents differ in phase by 90°. The current flowing through the resistance is

$$I_R = \frac{V}{R} = \frac{120}{20} = 6.0\ \text{A}$$

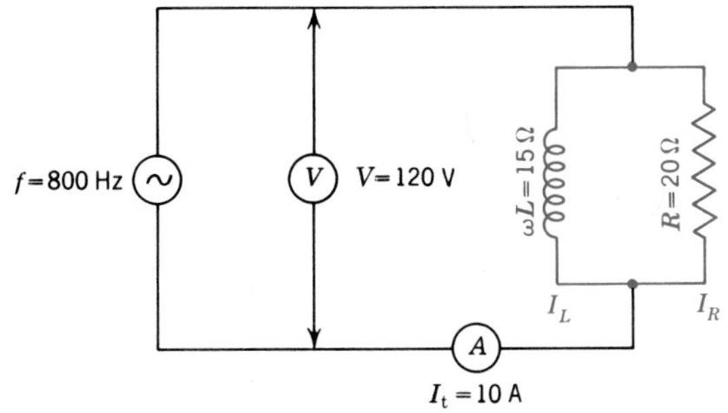

FIG. 33–8 *R* and X_L in parallel.

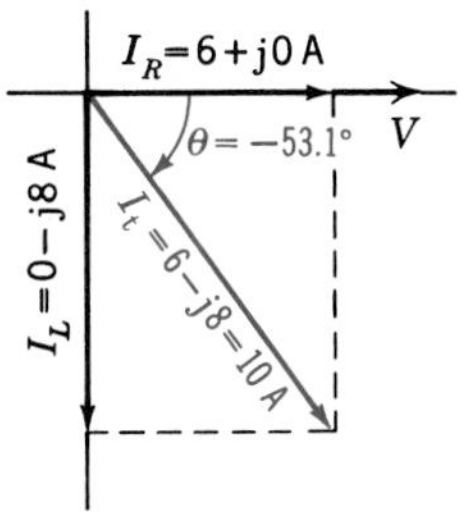

FIG. 33–9 Phasor diagram for circuit of Fig. 33–8.

and that through the inductance is

$$I_L = \frac{V}{\omega L} = \frac{120}{15} = 8.0 \text{ A}$$

Since the current through the resistance is in phase with the applied voltage and the current through the inductance lags the applied voltage by 90°, I_R and I_L are plotted with the applied emf as reference phasor as shown in Fig. 33–9. Then the total current I_t, or line current, is the phasor sum of I_R and I_L. Performing phasor summation,

$$\begin{aligned} I_R &= 6.0 + j0 \text{ A} \\ I_L &= 0 - j8.0 \text{ A} \\ \hline I_t &= 6.0 - j8.0 \text{ A} \end{aligned}$$

Hence, the total current, which consists of an in-phase component of 6.0 A and a 90° lagging component of 8.0 A, is expressed in terms of its rectangular components. The magnitude and phase angle are then found by the usual trigonometric methods. Thus,

$$I_t = 10\angle -53.1^\circ \text{ A}$$

The power factor of the circuit is

$$\begin{aligned} \text{PF} &= \cos\theta = \cos(-53.1^\circ) \\ &= 0.60 \text{ lagging} \end{aligned}$$

The true power expended in the circuit is

$$\begin{aligned} P &= VI\cos\theta = 120 \times 10 \times 0.60 \\ &= 720 \text{ W} \end{aligned}$$

or

$$P = I_R{}^2R = 6^2 \times 20 = 720 \text{ W}$$

The equivalent impedance, or total impedance, of the circuit is

$$Z_t = \frac{V}{I_t} = \frac{120}{10} = 12\ \Omega$$

Since the entire circuit has a lagging PF of 0.60, it follows that the equivalent series circuit consists of a resistance and an inductive reactance in series, the phasor sum of which is 12 Ω at a phase angle θ such that cos θ = 0.60. Therefore, θ = 53.1°, and

$$\begin{aligned} Z_t &= 12\angle 53.1^\circ\ \Omega \\ &= 12\,(\cos 53.1^\circ + j \sin 53.1^\circ) \\ &= 7.2 + j9.6\ \Omega \end{aligned}$$

From the above, it is evident that the parallel circuit of Fig. 33–8 could be replaced by a series circuit of 7.2 Ω resistance and 9.6 Ω inductive reactance and that the alternator would be working under exactly the same load conditions as before.

In order to justify such solutions, solve for the equivalent impedance of the circuit of Fig. 33–8 by using an assumed voltage and then using the *actual* voltage to obtain the power.

33–6 RESISTANCE AND CAPACITANCE IN PARALLEL

When resistance and capacitive reactance are connected in parallel, as represented in Fig. 33–10, the current through the resistance is in phase with the voltage across the parallel combination, and the current through the capacitive reactance leads this voltage by 90°.

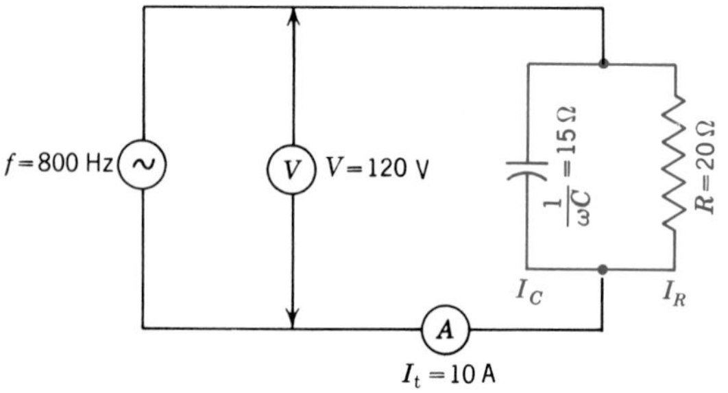

FIG. 33–10 *R* and X_C in parallel.

The circuit of Fig. 33–10 is similar to that of Fig. 33–8 except that Fig. 33–10 contains a capacitive reactance of 15 Ω in place of the inductive reactance of 15 Ω. The phasor diagram of currents is illustrated in Fig. 33–11, and it is evident that the total current is

$$I_t = 6.0 + j8.0\ \text{A} = 10\angle 53.1^\circ\ \text{A}$$

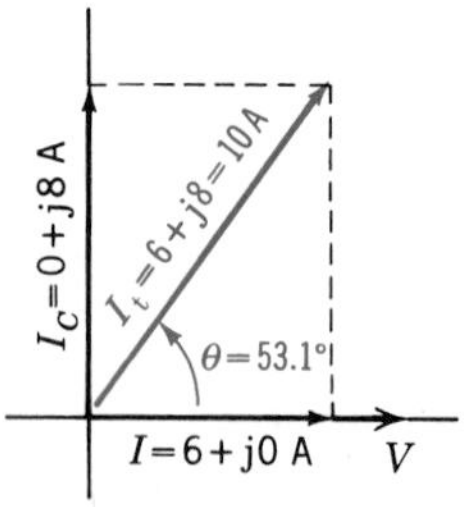

FIG. 33–11 Phasor diagram for circuit of Fig. 33–10.

The power factor of the circuit is

$$\begin{aligned} \text{PF} &= \cos\theta = \cos 53.1^\circ \\ &= 0.60 \text{ leading} \end{aligned}$$

Similarly, the total impedance of the circuit is 12 Ω; and since the circuit has a leading PF of 0.60, it follows that the equivalent series circuit consists of resistance and capacitive reactance in series the phasor sum of which is 12 Ω at a phase angle θ such that cos θ = 0.60. Therefore,

$$\theta = -53.1°$$

and

$$Z_t = 12\angle -53.1° \ \Omega$$
$$= 7.2 - j9.6 \ \Omega$$

If the parallel circuit of Fig. 33–10 were replaced by a series circuit of 7.2 Ω resistance and 9.6 Ω capacitive reactance, the alternator would be working under exactly the same load conditions as before.

33–7 RESISTANCE, INDUCTANCE, AND CAPACITANCE IN PARALLEL

When resistance, inductive reactance, and capacitive reactance are connected in parallel, as represented in Fig. 33–12, the line current is the phasor sum of the several currents.

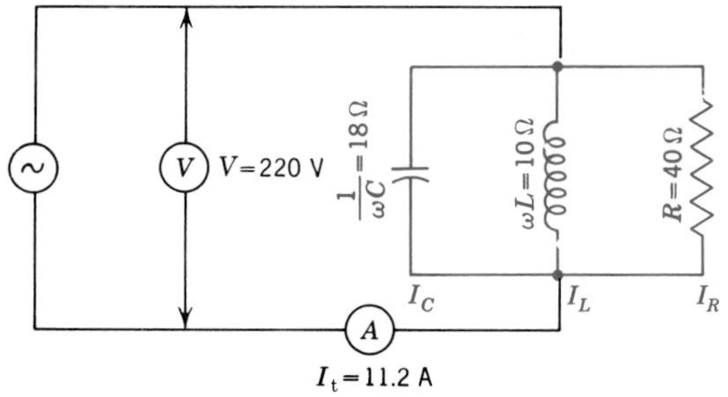

FIG. 33–12 *L, R,* and *C* in parallel.

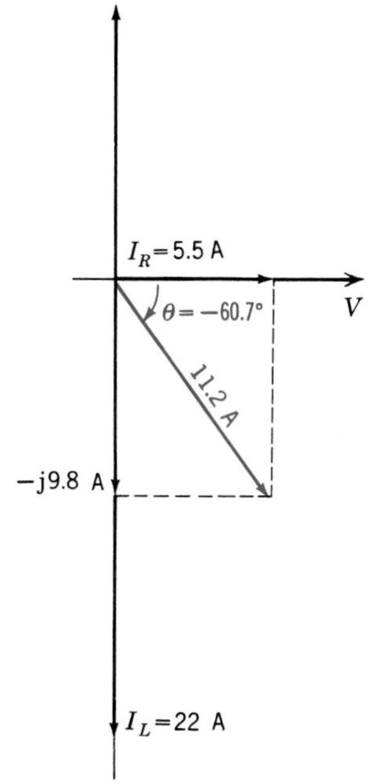

FIG. 33–13 Phasor diagram for circuit of Fig. 33–12.

The currents through the branches are

$$I_R = \frac{220}{40} = 5.5 \text{ A}$$
$$I_L = \frac{220}{10} = 22 \text{ A}$$
$$I_C = \frac{220}{18} = 12.2 \text{ A}$$

Performing phasor summation of these currents as shown in Fig. 33–13,

$$I_R = 5.5 + j0 \quad \text{A}$$
$$I_L = 0 - j22 \quad \text{A}$$
$$I_C = 0 + j12.2 \text{ A}$$
$$I_t = 5.5 - j9.8 \text{ A} = 11.2\angle -60.7° \text{ A}$$
$$\text{PF} = \cos(-60.7°) = 0.489 \text{ lagging}$$

The total impedance is

$$Z_t = \frac{V}{I_t} = \frac{220}{11.2} = 19.6\ \Omega$$

Since the circuit has a lagging PF of 0.489, the equivalent series circuit consists of a resistance and an inductive reactance. The phasor sum of these must be 19.6 Ω at a phase angle θ such that cos θ = 0.489. Therefore, θ = 60.7° and

$$Z_t = 19.6\underline{/\ 60.7^\circ}\ \Omega = 9.59 + j17.1\ \Omega$$

which are the values which constitute the equivalent series circuit.

EXAMPLE 3 Given the circuit represented in Fig. 33–14. Solve for the equivalent series circuit at a frequency of 5 MHz.

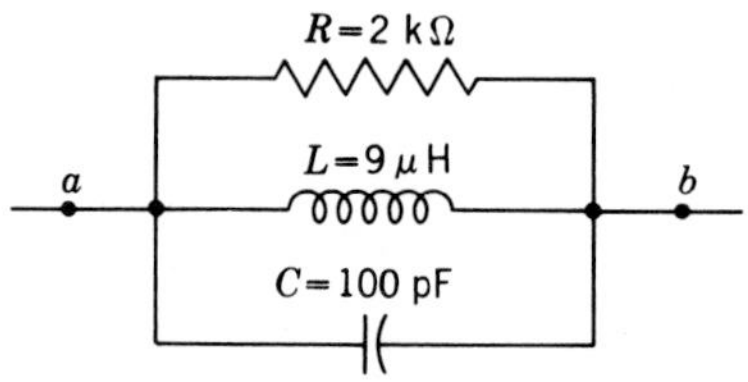

FIG. 33–14 Circuit of Example 3.

SOLUTION

$$\begin{aligned} f &= 5\ \text{MHz} = 5 \times 10^6\ \text{Hz} \\ L &= 9\ \mu\text{H} = 9 \times 10^{-6}\ \text{H} \\ C &= 100\ \text{pF} = 10^{-10}\ \text{F} \\ \omega &= 2\pi f = 2\pi \times 5 \times 10^6 = 3.14 \times 10^7 \\ X_L &= \omega L = 3.14 \times 10^7 \times 9 \times 10^{-6} \\ &= 283\ \Omega \\ X_C &= \frac{1}{\omega C} = \frac{1}{3.14 \times 10^7 \times 10^{-10}} = \frac{10^3}{3.14} \\ &= 318\ \Omega \end{aligned}$$

Assume $V_a = 1000$ V applied between a and b.

$$\begin{aligned} I_R &= \frac{V_a}{R} = \frac{1000}{2000} = 0.50\ \text{A} \\ I_L &= \frac{V_a}{X_L} = \frac{1000}{283} = 3.54\ \text{A} \\ I_C &= \frac{V_a}{X_C} = \frac{1000}{318} = 3.14\ \text{A} \end{aligned}$$

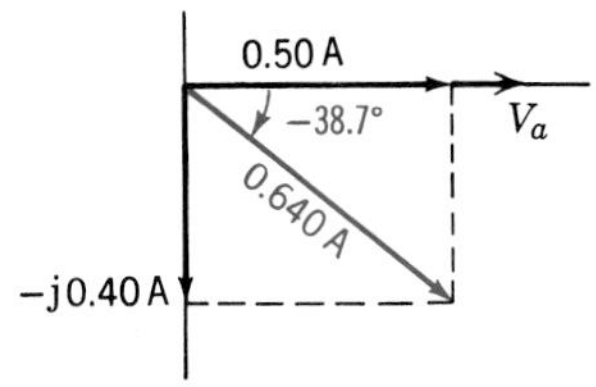

FIG. 33–15 Phasor diagram for circuit of Fig. 33–14.

The total current I_t is the phasor sum of the three branch currents as represented in the phasor diagram of Fig. 33–15. Adding vectorially,

$$\begin{aligned} I_R &= 0.50 + j0 \\ I_L &= 0 \quad - j3.54\text{ A} \\ \underline{I_C} &= \underline{0 \quad + j3.14\text{ A}} \\ I_t &= 0.50 - j0.40\text{ A} = 0.640\,\underline{/-38.7°}\text{ A} \\ \text{PF} &= \cos(-38.7°) = 0.78\text{ lagging} \end{aligned}$$

The total impedance Z_t, which is the impedance between points a and b, is

$$Z_t = Z_{ab} = \frac{V_a}{I_t} = \frac{1000}{0.64} = 1560\ \Omega$$

Since the current is lagging the voltage, the equivalent series circuit consists of a resistance and an inductive reactance. The phasor sum of these is 1560 Ω at a phase angle θ such that cos θ = 0.78. Therefore, θ = 38.7° and

$$Z_t = 1560\underline{/38.7°}\ \Omega = 1220 + j976\ \Omega$$

That is, the equivalent series circuit is a resistance of $R = 1220\ \Omega$ and an inductive reactance of $\omega L = 976\ \Omega$. Since

$$\omega L = 976\ \Omega$$

then

$$L = \frac{976}{\omega} = \frac{976}{3.14 \times 10^7} = 31.1\ \mu\text{H}$$

which results in the equivalent circuit as represented in Fig. 33–16 with the impedance phasor diagram of Fig. 33–17.

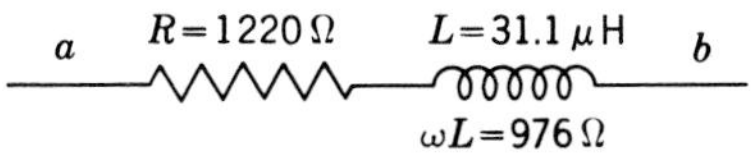

FIG. 33–16 Equivalent series circuit of Example 3.

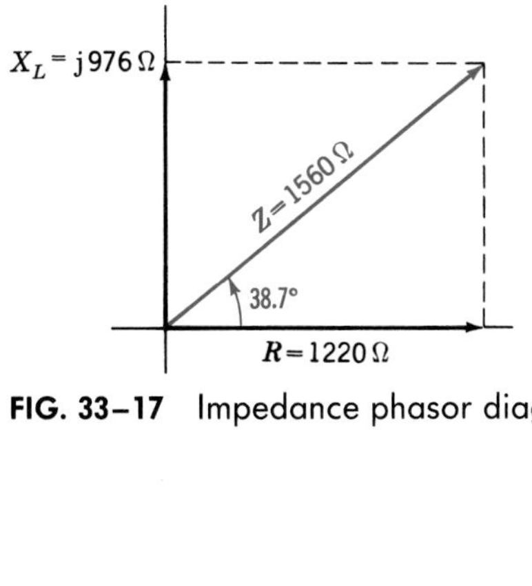

FIG. 33–17 Impedance phasor diagram for equivalent series circuit.

PROBLEMS 33–1

1. What is the resulting capacitance when a 650-pF capacitor is connected in parallel with a 200-pF capacitor?
2. Two capacitors, 50 and 500 pF, are connected in parallel. A current of 200 mA, 2.7 GHz, flows through the 500-pF capacitor. How much current flows through the 50-pF capacitor?
3. Neglecting the resistance of the connecting wires in Fig. 32–12,
   (*a*) Write the equation for the emf of the alternator.
   (*b*) Write the equation for the circuit current.
   (*c*) What is the voltage across C_1?
   (*d*) What is the capacitance of C_3?
   (*e*) How much current flows through C_2?
4. In Fig. 33–18, $R = 200\ \Omega$, $L = 2$ H, $C = 5\ \mu$F, $V = 220$ V, and $f = 60$ Hz.
   (*a*) What is the ammeter reading?
   (*b*) How much power is expended in the circuit?
   (*c*) What is the equivalent series circuit?
   (*d*) What is the power factor?
   (*e*) What is the equation of the current flowing through the ammeter?

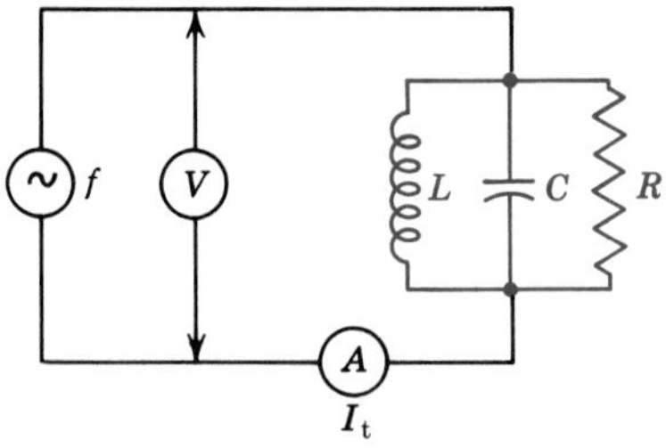

FIG. 33–18 Circuit of Probs. 4 to 6.

5. Using the other values of Prob. 4, what must be the value of the inductance in the circuit in order to obtain a PF of (*a*) 0.8 lagging and (*b*) 1.0?
6. In Fig. 33–18, $R = 500\ \Omega$, $L = 6$ nH, $C = 0.02$ pF, $V = 1$ kV, and $f = 8$ GHz.
   (*a*) What is the reading of the ammeter?
   (*b*) What parallel capacitance must be added to the circuit in order to achieve unity PF?

33–8 PHASOR IMPEDANCES IN PARALLEL

It was shown in Sec. 13–2 that the reciprocal of the equivalent resistance R_p of several resistances in parallel is expressed by the relation

$$\frac{1}{R_p} = \frac{1}{R_1} + \frac{1}{R_2} + \frac{1}{R_3} + \frac{1}{R_4} + \cdots$$

and that when two resistances R_1 and R_2 are connected in parallel, the equivalent resistance is

$$R_p = \frac{R_1 R_2}{R_1 + R_2}$$

An analogous condition exists when two or more impedances are connected in parallel. By following the line of reasoning used for resistances in parallel, the reciprocal of the equivalent impedance of several impedances in parallel is found to be

$$\frac{1}{Z_p} = \frac{1}{Z_1} + \frac{1}{Z_2} + \frac{1}{Z_3} + \frac{1}{Z_4} + \ldots \qquad (5)$$

Similarly, the equivalent impedance Z_p of two impedances Z_1 and Z_2 connected in parallel is

$$Z_p = \frac{Z_1 Z_2}{Z_1 + Z_2} \qquad (6)$$

Note that the impedances of Eqs. (5) and (6) are in polar form.

EXAMPLE 4 Find the equivalent impedance of the circuit of Fig. 33–19.

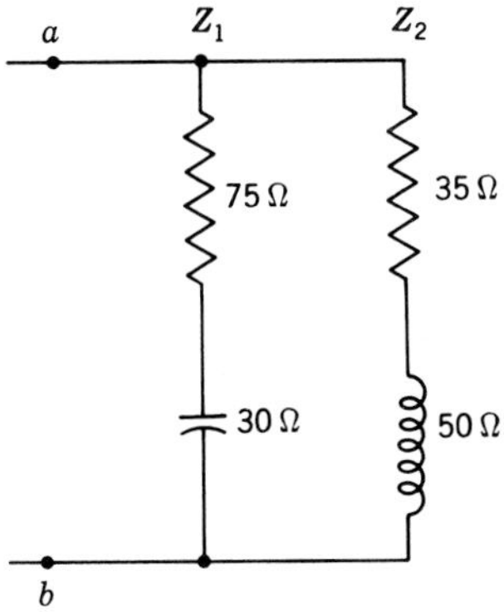

FIG. 33–19 Circuit of Example 4.

SOLUTION

First express the given impedances in both rectangular and polar forms.

$$Z_1 = 75 - j30 = 80.8\angle -21.8^\circ\ \Omega$$
$$Z_2 = 35 + j50 = 61.0\angle 55^\circ\ \Omega$$

As pointed out in Sec. 31–5, phasors in polar form cannot be added algebraically; they must be added in terms of their rectangular components. Therefore, when the given impedance values are substituted in Eq. (6), the impedances in the denominator must be in rectangular form so that the indicated addition can be carried out. Substituting,

$$Z_p = \frac{(80.8\angle -21.8^\circ)(61.0\angle 55^\circ)}{(75 - j30) + (35 + j50)} = \frac{4930\angle 33.2^\circ}{110 + j20}$$

Because the denominator is in rectangular form and the numerator is in polar form, the denominator must be converted to polar form so the indicated division can be completed. Thus, performing phasor summation of the terms of the denominator,

$$Z_p = \frac{4930\angle 33.2^\circ}{112\angle 10.3^\circ} = \frac{4930}{112}\angle 33.2^\circ - 10.3^\circ = 44\angle 22.9^\circ\ \Omega$$

EXAMPLE 5 Find the equivalent impedance of the circuit of Fig. 33–20.

SOLUTION

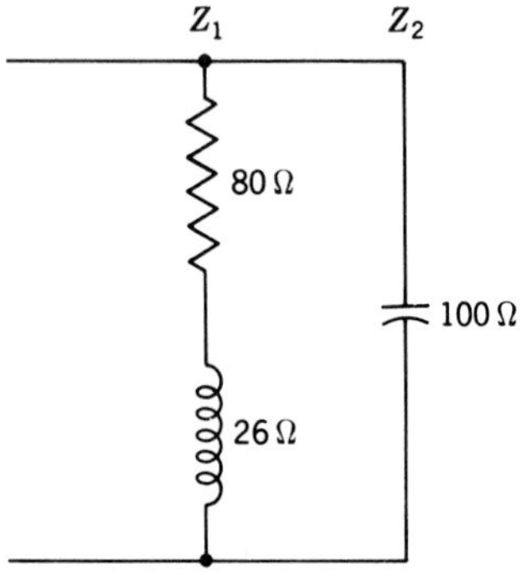

FIG. 33–20 Circuit of Example 5.

Expressing the impedance in rectangular and polar form,

$$Z_1 = 80 + j26 = 84.1\underline{/\ 18°}\ \Omega$$
$$Z_2 = 0 - j100 = 100\ \underline{/\ -90°}\ \Omega$$

Substituting these values in Eq. (6),

$$Z_p = \frac{(84.1\underline{/\ 18°})(100\underline{/\ -90°})}{(80 + j26) + (0 - j100)} = \frac{8410\underline{/\ -72°}}{80 - j74}$$

Performing the phasor summation in the denominator,

$$Z_p = \frac{8410\underline{/\ -72°}}{109\underline{/\ -42.8°}}$$

$$\therefore Z_p = 77.2\underline{/\ -29.2°}\ \Omega$$

The equivalent series circuit is found by the usual method of converting from polar form to rectangular form, namely,

$$\begin{aligned} 77.2\underline{/\ -29.2°} &= 77.2(\cos 29.2° - j \sin 29.2°) \\ &= 77.2 \cos 29.2° - j77.2 \sin 29.2° \\ &= 67.4 - j37.7\ \Omega \end{aligned}$$

33–9 SERIES-PARALLEL CIRCUITS

An equation for the equivalent impedance of a series-parallel circuit is obtained in the same manner as the equation for the equivalent resistance of a combination of resistances in series and parallel as was outlined in Sec. 13–3. For example, in the circuit represented in Fig. 33–21, the total impedance is

$$Z_t = Z_s + \frac{Z_1 Z_2}{Z_1 + Z_2} \tag{7}$$

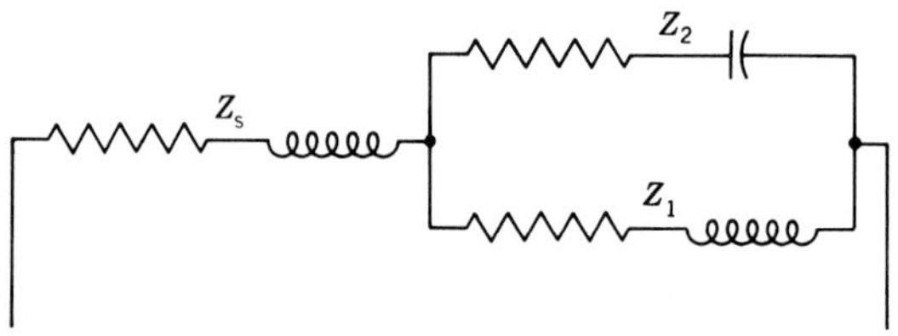

FIG. 33–21 Series-parallel circuit of Example 6.

EXAMPLE 6 In the circuit which is shown in Fig. 33–21, $Z_s = 12.4 + j25.6\ \Omega$, $Z_1 = 45 + j12.9\ \Omega$, and $Z_2 = 35 - j75\ \Omega$. Determine the equivalent impedance of the circuit.

SOLUTION

Since Z_1 and Z_2 must be multiplied, it is necessary to express them in polar form.

$$Z_1 = 45 + j12.9 = 46.8\angle 16^\circ\ \Omega$$

and

$$Z_2 = 35 - j75 = 82.8\angle -65^\circ\ \Omega$$

Substituting the values in Eq. (7),

$$Z_t = (12.4 + j25.6) + \frac{(46.8\angle 16^\circ)(82.8\angle -65^\circ)}{(45 + j12.9) + (35 - j75)}\ \Omega$$

The solution is completed in the usual manner and results in

$$Z_t = 53.2\angle 20^\circ\ \Omega$$

From the foregoing examples, it is evident that an equation for the impedance of a network is expressed exactly as in direct-current problems, impedances in polar form being substituted for the resistances.

PROBLEMS 33–2

1. What is the equivalent impedance if $Z_1 = 240\angle 12.5^\circ\ \Omega$ and $Z_2 = 85\angle 48^\circ\ \Omega$ are connected in parallel?
2. What is the equivalent impedance if $Z_1 = 148.5\angle 42.2^\circ\ \Omega$ and $Z_2 = 145\angle -12.7^\circ\ \Omega$ are connected in parallel?
3. What is the equivalent impedance when $Z_1 = 50.6 - j45.2\ \Omega$ and $Z_2 = 40 + j40\ \Omega$ are connected in parallel?
4. What is the equivalent impedance when $Z_1 = 276 - j180\ \Omega$ and $Z_2 = 117 - j18.6\ \Omega$ are connected in parallel?
5. What is the equivalent impedance when $Z_z = 180\angle -12^\circ\ \Omega$ and $Z_L = 0 + j110\ \Omega$ are connected in parallel?
6. What is the equivalent impedance when two impedances $Z_1 = 355\angle 12^\circ\ \Omega$ and $Z_2 = 0 - j100\ \Omega$ are connected in parallel?
7. What is the equivalent impedance when two impedances $Z_z = 251\angle -3^\circ\ \Omega$ and $Z_L = 0 + j70\ \Omega$ are connected in parallel?
8. The joint impedance of two parallel impedances is $53.5\angle -42.4^\circ\ \Omega$. One of the impedances is $168\angle 27^\circ\ \Omega$. What is the other?
9. What impedance must be connected in parallel with $64.9 + j45.4\ \Omega$ to produce $43.7 + j155.5\ \Omega$?
10. In Fig. 33–22, $Z_s = 9.4 + j6.6\ \Omega$, $Z_1 = 78.5 - j35\ \Omega$, and $Z_2 = 33.6 + j48\ \Omega$. What is the single equivalent impedance Z_t?

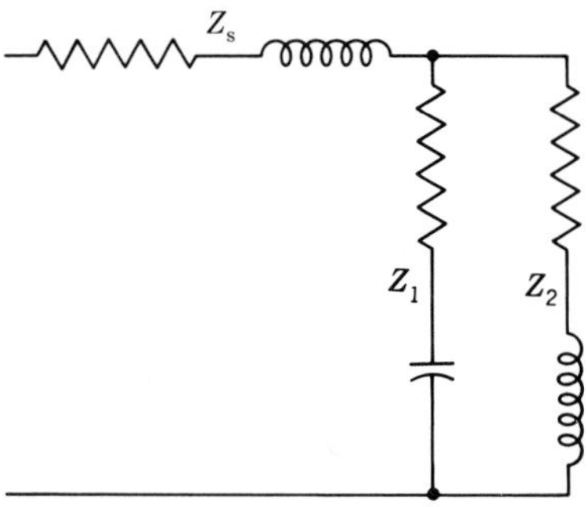

FIG. 33–22 Circuit for Probs. 10, 11, and 12.

11. In Fig. 33–22, $Z_s = 111.5\angle 21°\Omega$, $Z_1 = 27.7 - j50\ \Omega$, and $Z_2 = 150 + j76.2\ \Omega$. What is Z_t?
12. In Fig. 33–22, $Z_s = 5 + j3.9\ \Omega$, $Z_1 = 57.2\angle -61°\ \Omega$, and $Z_2 = 168\angle 27°\ \Omega$. What is Z_t?
13. The primary current I_p of a coupled circuit is expressed by the equation

$$I_p = \frac{V}{Z_p + \dfrac{(\omega M)^2}{Z_s}} \qquad \text{A}$$

Compute the value of I_p when $V = 110\angle 0°$ V, $Z_p = 1.2 + j4\ \Omega$, $Z_s = 1.8 + j5\ \Omega$, and $\omega M(= 2\pi f \times \text{mutual inductance}) = 15$.

14. The secondary current I_s of a coupled circuit is expressed by the equation

$$I_s = \frac{-j\omega MV}{Z_pZ_s + (\omega M)^2} \qquad \text{A}$$

Compute the value of I_s if $\omega M = 15$, $V = 20$ V, $Z_p = 6 + j8\ \Omega$, and $Z_s = 20 + j12\ \Omega$.

33–10 PARALLEL RESONANCE

Communication circuits and electronic networks contain resonant parallel circuits. Figure 33–23 represents a typical parallel circuit consisting of an inductor and capacitor in parallel. The resistance of the capacitor, which is very small, can be neglected, and the resistance R represents the effective resistance of the inductor.

At low frequencies the inductive reactance is a low value whereas the capacitive reactance is high. Hence, a large current flows through the inductive branch and a small current flows through the capacitive branch. The phasor sum of these currents causes a large lagging line current which, in effect, results in an equivalent series circuit of low impedance consisting of resistance and

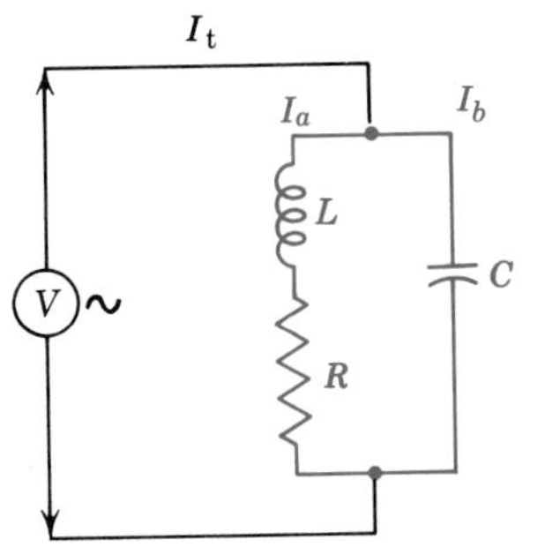

FIG. 33–23 Parallel LC circuit. R represents effective resistance of L.

inductive reactance. At high frequencies the inductive reactance is large and the capacitive reactance is small. This results in a large leading line current with an attendant equivalent series circuit of low impedance consisting of resistance and capacitive reactance.

There is one frequency, between those mentioned previously, at which the lagging component of current through the inductive branch is equal to the leading current through the capacitive branch. This condition results in a small line current that is in phase with the voltage across the parallel circuit and therefore an impedance that is equivalent to a very high resistance.

The resonant frequency of a parallel circuit is often a source of confusion to the student studying parallel resonance for the first time. The reason is that different definitions for the resonant frequency are encountered in various texts. Thus, the resonant frequency of a parallel circuit can be defined by any one of the following as:

1. The frequency at which the parallel circuit acts as a pure resistance.
2. The frequency at which the line current becomes minimum.
3. The frequency at which the inductive reactance equals the capacitive reactance. This is the same definition as that for the resonant frequency of a series circuit. That is,

$$\omega L = \frac{1}{\omega C}$$

or

$$f_r = \frac{1}{2\pi\sqrt{LC}} \tag{8}$$

A little consideration of these definitions will convince you that, in high-Q circuits, the three resonant frequencies differ by an amount so small as to be negligible.

In the circuit of Fig. 33–23,

$$I_b = \frac{V}{\dfrac{1}{\omega C}} = \omega CV$$

Also,

$$I_a = \frac{V}{R + j\omega L}$$

Rationalizing (Sec. 20–17),

$$I_a = \frac{V}{R + j\omega L} \cdot \frac{R - j\omega L}{R - j\omega L} = \frac{V(R - j\omega L)}{R^2 + (\omega L)^2}$$

$$= \frac{VR}{R^2 + (\omega L)^2} - j\frac{\omega LV}{R^2 + (\omega L)^2}$$

In order to satisfy the first definition for resonant frequency, the line current must be in phase with the applied voltage; that is, the out-of-phase, or quadrature, component of the current through the inductive branch must be equal to the current through the capacitive branch. Thus,

$$\frac{\omega LV}{R^2 + (\omega L)^2} = \omega CV$$

D: ωV,

$$\frac{L}{R^2 + (\omega L)^2} = C$$

M: $[R^2 + (\omega L)^2]$,

$$L = [R^2 + (\omega L)^2]C \tag{9}$$

or

$$\frac{L}{C} - R^2 = (\omega L)^2$$

Hence,

$$\omega = \frac{\sqrt{\frac{L}{C} - R^2}}{L}$$

$$= \sqrt{\frac{1}{LC} - \frac{R^2}{L^2}}$$

Substituting $2\pi f$ for ω,

$$2\pi f = \sqrt{\frac{1}{LC} - \frac{R^2}{L^2}}$$

Thus, the resonant frequency is

$$f = \frac{1}{2\pi}\sqrt{\frac{1}{LC} - \frac{R^2}{L^2}} \tag{10}$$

If the Q of the inductance is at all large, then $\omega L \gg R$, which, for all practical purposes, makes the term $\frac{R^2}{L^2}$ in Eq. (10) of such low value that it can be neglected, and Eq. (10) is thus reduced to Eq. (8).

Work out several examples with different circuit values, and compare the resonant frequencies obtained from the formulas. In this connection, it is left to you as an exercise to show that in a parallel-resonant circuit, as represented in Fig. 33–23, the line current and applied voltage will be in phase (unity power factor) when

$$R^2 = X_L(X_C - X_L) \tag{11}$$

33–11 IMPEDANCE OF PARALLEL-RESONANT CIRCUITS

When a parallel circuit is operating at the frequency at which it acts as a pure resistance, it has unity PF, and the line current I_t (Fig. 33–23) consists of the in-phase component of I_a. That is,

$$I_t = \frac{VR}{R^2 + (\omega L)^2} \quad \text{A} \tag{12}$$

Then, since

$$Z_t = \frac{V}{I_t} \quad \Omega$$

substituting in Eq. (12) for I_t,

$$\frac{V}{Z_t} = \frac{VR}{R^2 + (\omega L)^2}$$

Hence, $$Z_t = \frac{R^2 + (\omega L)^2}{R} \quad \Omega \tag{13}$$

From Eq. (9), $$R^2 + (\omega L)^2 = \frac{L}{C}$$

Substituting this value in Eq. (13),

$$Z_t = \frac{L}{CR} \quad \Omega \tag{14}$$

EXAMPLE 7 In the circuit of Fig. 33–23, let $L = 203\ \mu\text{H}$, $C = 500\ \text{pF}$, and $R = 6.7\ \Omega$.
(*a*) What is the resonant frequency of the circuit?
(*b*) What is the impedance of the circuit at resonance?

SOLUTION

(*a*)
$$\begin{aligned} f &= \frac{1}{2\pi\sqrt{LC}} \\ &= \frac{1}{2\pi\sqrt{2.03 \times 10^{-4} \times 5 \times 10^{-10}}} \\ &= 500\ \text{kHz} \end{aligned}$$

(*b*)
$$\begin{aligned} Z_t &= \frac{L}{CR} \\ &= \frac{203 \times 10^{-6}}{500 \times 10^{-12} \times 6.7} \\ &= \frac{203}{5 \times 6.7} \times 10^4 \\ &= 60.6\ \text{k}\Omega \end{aligned}$$

If the value of C is unknown, Eq. (14) can be used in different form. Thus, by multiplying both numerator and denominator by ω,

$$Z_t = \frac{\omega L}{\omega CR} = \frac{1}{\omega C}\frac{\omega L}{R}$$

Since at resonance,

$$\omega L = \frac{1}{\omega C}$$

then $$Z_t = \frac{(\omega L)^2}{R} \quad \Omega \tag{15}$$

Moreover, since $$Q = \frac{\omega L}{R}$$

substituting in Eq. (15),

$$Z_t = \omega L Q \quad \Omega \tag{16}$$

EXAMPLE 8 In the circuit of Fig. 33–23, let $L = 70.4\ \mu\text{H}$ and $R = 5.31\ \Omega$. If the resonant frequency of the circuit is 1.2 MHz, determine (*a*) the impedance of the circuit at resonance and (*b*) the capacitance of the capacitor.

SOLUTION

$$f = 1.2\text{ MHz} = 1.2 \times 10^6\text{ Hz}$$

$$\omega = 2\pi f = 2\pi \times 1.2 \times 10^6 = 7.54 \times 10^6$$

(*a*)

$$Z_t = \frac{(\omega L)^2}{R} = \frac{(7.54 \times 10^6 \times 70.4 \times 10^{-6})^2}{5.31}$$

$$= 53.1\text{ k}\Omega$$

(*b*) Since, at resonance, $\omega L = \dfrac{1}{\omega C}$ and $\omega L = 531\ \Omega$.

then

$$\frac{1}{\omega C} = 531\ \Omega$$

Hence,

$$C = \frac{1}{531\omega} = 250\text{ pF}$$

What is the Q of this circuit?

PROBLEMS 33–3

1. An inductor of 12 μH and a capacitor of 60 pF are connected in parallel as shown in Fig. 33–23. If the effective resistance of the coil is 28 Ω, find:
   (*a*) The resonant frequency of the circuit according to definition 1 (Sec. 33–10).
   (*b*) The resonant frequency according to definition 3.
   (*c*) The Q of the coil by using the frequency of part (*b*).
2. Repeat Prob. 1 for an effective resistance of the coil of 44 Ω.
3. An inductor of 10 mH with a Q of 800 is connected in parallel with a 200-pF capacitor.
   (*a*) What is the resonant frequency of the circuit?
   (*b*) What is the impedance of the circuit at resonance?
   (*c*) What is the effective resistance of the inductor?
4. If the circuit of Prob. 3 is energized with 600 V at the resonant frequency, how much power will it absorb?
5. A coil with a Q of 64.5 is connected in parallel with a capacitor, and the circuit is found to resonate at 310 kHz. The impedance at resonance is measured at 87 kΩ. What is the value of the capacitor?
6. An inductor is connected in parallel with a 254-pF capacitor, and the circuit is found to resonate at 999 kHz. A circuit magnification meter indicates that the Q of the inductor is 90.
   (*a*) What is the value of the inductance?
   (*b*) What is the effective resistance of the inductor?
   (*c*) What is the impedance of the circuit at resonance?
7. If the circuit of Prob. 6 is connected to 20 V at the resonant frequency, how much power will it absorb?

8. If the circuit of Prob. 6 is connected to a 20-V source at 499 kHz, (*a*) how much power will it absorb and (*b*) what will be the PF of the circuit?
9. If the circuit of Prob. 6 is connected to a 20-V source at 1499 kHz, what will be the PF?
10. An inductor with a measured Q of 100 resonates with a capacitor at 7.496 MHz with an impedance of 65.9 kΩ. What is the value of the inductance?
11. What is the capacitance of the test capacitor in Prob. 10?
12. 18.9 mA is the total current drain when a capacitor is in resonance with an inductor at 1.5 MHz and the parallel circuit is energized with a 1-kV source. The Q of the inductor is measured at 99.7. What is the value of the capacitor?
13. A 1-MHz 400-V source ''sees'' $Z_t = 38.8 + j50.7$ kΩ when connected across a parallel resistive inductive circuit. One branch of the parallel combination draws a current $i_1 = 1.15\underline{/\ -57.2°}$ m*A*. Find:
    (*a*) The resistor values.
    (*b*) The inductor values.
    (*c*) The total current and phase angle.
    (*d*) The current through the other branch, i_2.
14. The circuit of Prob. 13 is to be replaced with its series equivalent circuit, and then made resonant at 1 MHz. Find:
    (*a*) the circuit current at 1 MHz.
    (*b*) The series resistor and inductor.
    (*c*) The capacitor value for resonance.
    (*d*) The voltage across the resistor at 1 MHz.
15. Use the data from Prob. 13 and the relationship

$$\frac{1}{Z_t} = \frac{1}{Z_1} + \frac{1}{Z_2} \qquad \Omega$$

to evaluate Z_2.

33–12 EQUIVALENT Y AND Δ CIRCUITS

When networks contain complex impedances, the equations for converting from a Δ network to an equivalent Y network, or vice versa, are derived by methods identical with those of Sec. 22–7. Thus, in Fig. 33–24, each equivalent Y impedance is equal to the product of the two *adjacent* Δ impedances divided by the summation of the Δ impedances, or

$$Z_a = \frac{Z_1Z_3}{\Sigma Z_\Delta} \quad (17) \qquad Z_b = \frac{Z_1Z_2}{\Sigma Z_\Delta} \quad (18) \qquad \text{and} \quad Z_c = \frac{Z_2Z_3}{\Sigma Z_\Delta} \quad (19)$$

where

$$\Sigma Z_\Delta = Z_1 + Z_2 + Z_3$$

and all impedances are expressed in polar form.

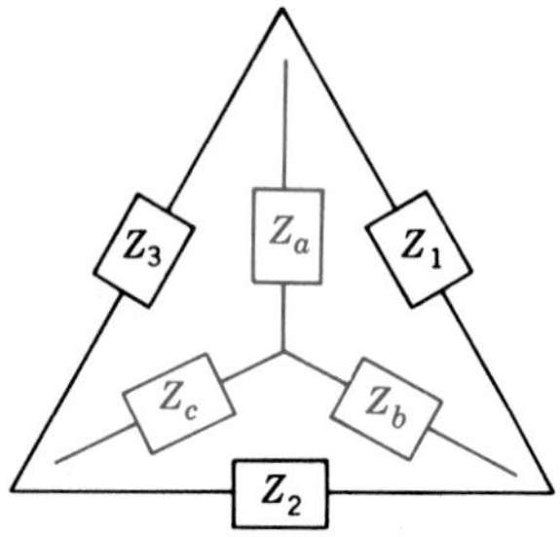

FIG. 33–24 Equivalent Y and Δ impedances.

Similarly, each equivalent Δ impedance is equal to the summation of the Y impedances divided by the *opposite* Y impedance. Thus,

$$Z_1 = \frac{\Sigma Z_Y}{Z_c} \tag{20}$$

$$Z_2 = \frac{\Sigma Z_Y}{Z_a} \tag{21}$$

and

$$Z_3 = \frac{\Sigma Z_Y}{Z_b} \tag{22}$$

where

$$\Sigma Z_Y = Z_aZ_b + Z_bZ_c + Z_aZ_c$$

and all impedances are expressed in polar form.

EXAMPLE 9 In Fig. 33–24,

$$Z_1 = 7.07 + j7.07\ \Omega$$
$$Z_2 = 4 + j3\ \Omega$$

and

$$Z_3 = 6 - j8\ \Omega$$

What are the values of the equivalent Y circuit?

SOLUTION

Express all impedances in both rectangular and polar forms.

$$\begin{aligned} Z_1 &= 7.07 + j7.07 = 10\underline{/\ 45^\circ}\ \Omega \\ Z_2 &= 4 + j3 = 5\underline{/\ 36.9^\circ}\ \Omega \\ Z_3 &= 6 - j8 = 10\underline{/\ -53.1^\circ}\ \Omega \\ \Sigma Z_\Delta &= (7.07 + j7.07) + (4 + j3) + (6 - j8) \\ &= 17.2\underline{/\ 6.91^\circ}\ \Omega \end{aligned}$$

Substituting in Eq. (17),
$$Z_a = \frac{(10\underline{/\ 45^\circ})(10\underline{/\ -53.1^\circ})}{17.2\underline{/\ 6.91^\circ}} = 5.62 - j1.51\ \Omega$$

Substituting in Eq. (18),
$$Z_b = \frac{(10\underline{/\ 45^\circ})(5\underline{/\ 36.9^\circ})}{17.2\underline{/\ 6.91^\circ}} = 0.752 + j2.81\ \Omega$$

Substituting in Eq. (19),
$$Z_c = \frac{(5\underline{/\ 36.9^\circ})(10\underline{/\ -53.1^\circ})}{17.2\underline{/\ 6.91^\circ}} = 2.67 - j1.14\ \Omega$$

The solution can be checked by converting the above Y-network equivalents back to the original Δ by using Eqs. (20), (21), and (22).

EXAMPLE 10 Determine the equivalent impedance between points *a* and *c* shown in Fig. 33–25.

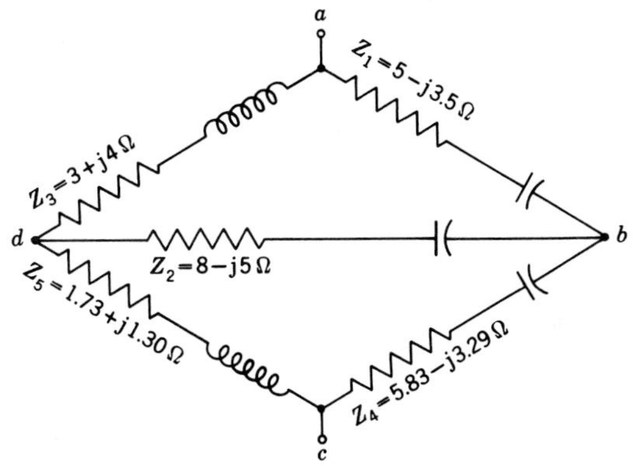

FIG. 33–25 Circuit of Example 10.

SOLUTION

Convert one of the Δ circuits of Fig. 33–25 to its equivalent Y circuit. Thus, for the delta *abd,*

$$Z_1 = 5 - j3.5 = 6.1\underline{/-35^\circ}\ \Omega$$
$$Z_2 = 8 - j5 = 9.44\underline{/-32^\circ}\Omega$$
$$Z_3 = 3 + j4 = 5\underline{/53.1^\circ}\ \Omega$$
$$\Sigma Z_\Delta = (5 - j3.5) + (8 - j5) + (3 + j4)$$
$$= 16.6\underline{/-15.7^\circ}\ \Omega$$

Substituting in Eq. (17),
$$Z_a = \frac{(6.1\underline{/-35^\circ})(5\underline{/53.1^\circ})}{16.6\underline{/-15.7^\circ}}$$
$$= 1.84\underline{/33.8^\circ}$$
$$= 1.53 + j1.02\ \Omega$$

Substituting in Eq. (18),
$$Z_b = \frac{(6.1\underline{/-35^\circ})(9.44\underline{/-32^\circ}}{16.6\underline{/-15.7^\circ}}$$
$$= 3.47\underline{/-51.3^\circ}$$
$$= 2.17 - j2.71\ \Omega$$

Substituting in Eq. (19),
$$Z_c = \frac{(9.44\underline{/-32^\circ})(5\underline{/53.1^\circ})}{16.6\underline{/-15.7^\circ}}$$
$$= 2.84\underline{/36.8^\circ}$$
$$= 2.27 + j1.70\ \Omega$$

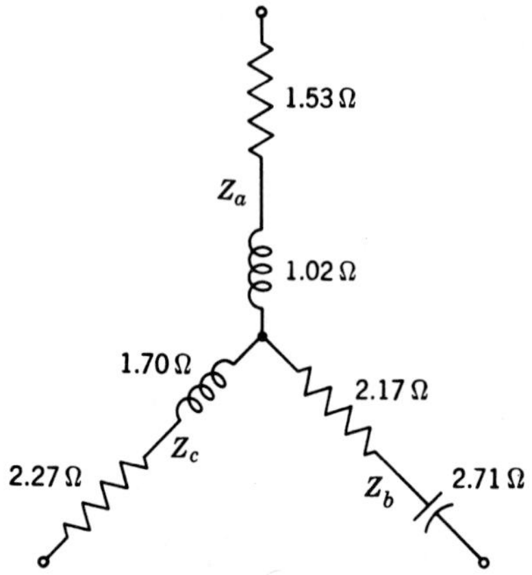

FIG. 33–26 Equivalent Y impedances for circuit of Fig. 33–25.

The equivalent Y impedances are shown in Fig. 33–26.

The equivalent Y impedances are connected to the remainder of the circuit as shown in Fig. 33–27 and solved as an ordinary series-parallel circuit. See the following equations for Fig. 33–27.

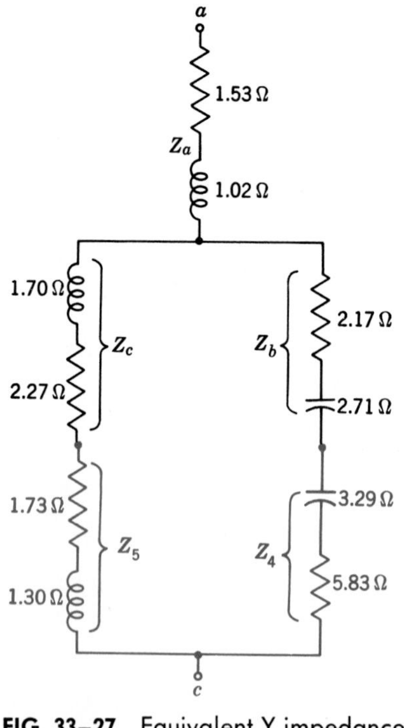

FIG. 33–27 Equivalent Y impedances connected to remainder of circuit of Fig. 33–25.

$$Z_{ac} = Z_a + \frac{(Z_c + Z_5)(Z_b + Z_4)}{Z_c + Z_5 + Z_b + Z_4}$$

$$= 1.53 + j1.02 + \frac{[(2.27 + j1.70) + (1.73 + j1.30)][(2.17 - j2.71) + (5.83 - j3.29)]}{(2.27 + j1.70) + (1.73 + j1.30) + (2.17 - j2.71) + (5.83 - j3.29)}$$

$$= 5.45 + j2.0\ \Omega$$

As we saw in Sec. 22–7, the Δ network is more generally referred to in electronics as a π network and the Y or star network is often known as the T network. In the problems which follow, the two sets of expressions are used interchangeably.

PROBLEMS 33–4

1. In the circuit of Fig. 33–24, $Z_1 = 30 + j40\ \Omega$, $Z_2 = 20 + j40\ \Omega$, $Z_3 = 60 - j15\ \Omega$. Find the impedances of the equivalent Y circuit.
2. In the circuit of Fig. 33–24, $Z_1 = 3 + j4\ \Omega$, $Z_2 = 12 + j15\ \Omega$, $Z_3 = 8 - j6\ \Omega$. Find the equivalent Y-circuit values.
3. In the circuit of Fig. 33–24,

$$Z_a = 51.6\angle 72.45^\circ\ \Omega$$
$$Z_b = 48.7\angle -42.6^\circ\ \Omega$$
$$Z_c = 61.9\angle -32.8^\circ\ \Omega$$

Find the impedances of the equivalent π circuit.

4. In the circuit of Fig. 33–24, $Z_a = 50.9\angle 86.8^\circ\ \Omega$, $Z_b = 62.7\angle -20.2^\circ\ \Omega$, and $Z_c = 44.5\angle 8.8^\circ\ \Omega$. Find the equivalent Δ-circuit values.
5. In the circuit of Fig. 33–28, $Z_1 = 64\angle 18.5^\circ\ \Omega$, $Z_2 = 74\angle -54.7^\circ\ \Omega$, $Z_3 = 40\angle 45^\circ\ \Omega$, $Z_4 = 55\angle -68.8^\circ\ \Omega$, and $Z_5 = 90\angle 53.1^\circ\ \Omega$. Find Z_{ab}.
6. In Prob. 5, if $V = 100\angle 0^\circ$ V, find the current through impedance Z_4.
7. In the circuit of Fig. 33–28, $Z_1 = 102 + j190\ \Omega$, $Z_2 = 134 - j33\ \Omega$, $Z_3 = 380 - j210\ \Omega$, $Z_4 = 30 - j40\ \Omega$, and $Z_5 = 80 - j60\ \Omega$. What is the equivalent impedance Z_{ab}?
8. In Prob. 7, if $V = 440$ V, what is the current through Z_5?
9. In Prob. 7, if $V = 200$ V, what is the power expended in Z_4?
10. In Prob. 7, if $V = 200$ V, what is the current through Z_2?
11. In Fig. 33–28, $Z_1 = 90 - j120\ \Omega$, $Z_2 = 115 - j18\ \Omega$, $Z_3 = 168 - j58\ \Omega$, $Z_4 = 50 + j0\ \Omega$, and $Z_5 = 0 + j25\ \Omega$. Determine the equivalent impedance Z_{ab}.
12. In Prob. 11, if $V = 100$ V, what is the current through Z_5?
13. In Prob. 11, if $V = 100$ V, how much power is expended in Z_1?
14. In Fig. 33–29, $Z_1 = 3 + j4\ \Omega$, $Z_2 = 37\angle 77.5^\circ\ \Omega$, $Z_3 = 40\angle -80^\circ\ \Omega$, $Z_4 = 64 - j50\ \Omega$, $Z_5 = 15 + j85\ \Omega$, $Z_6 = 40 - j36\ \Omega$, $Z_7 = 10\angle -53.1^\circ\ \Omega$, and $V = 120$ V. What is the current through Z_7?

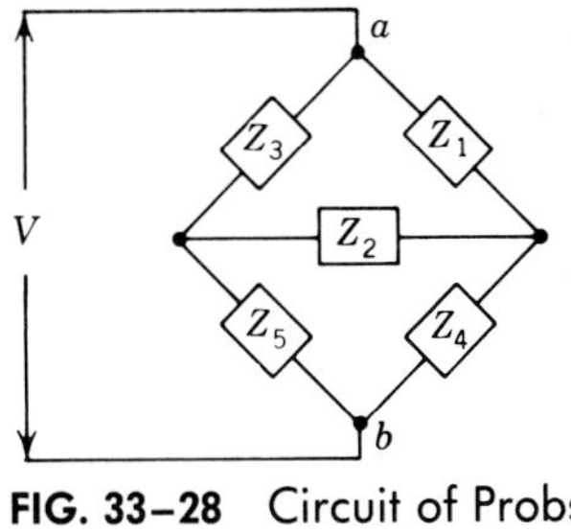

FIG. 33–28 Circuit of Probs. 5 to 13.

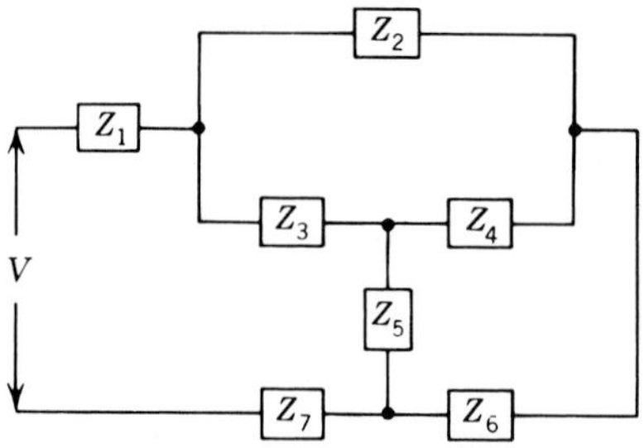

FIG. 33–29 Circuit of Prob. 14.

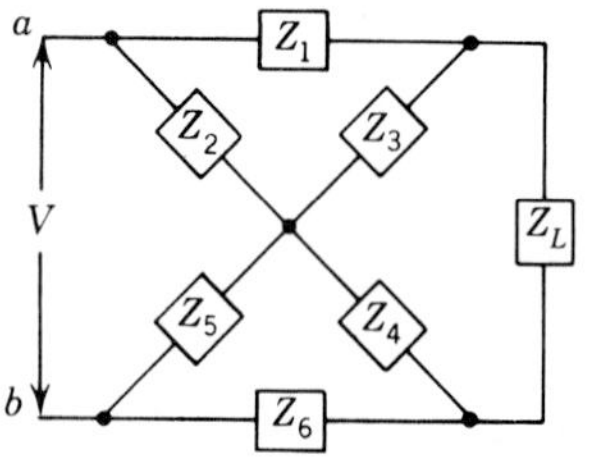

FIG. 33–30 Circuit of Probs. 15 to 18.

15. In Fig. 33–30, $Z_1 = 254\angle 88.6°\ \Omega$, $Z_2 = 306\angle 86.1°\Omega$, $Z_3 = 437\angle -73.6°\ \Omega$, $Z_4 = 177\angle -87°\ \Omega$, $Z_5 = 288\angle 87.5°\ \Omega$, $Z_6 = 250\angle 89.1°\ \Omega$, and $Z_L = 680\angle 0°\ \Omega$. Determine the equivalent impedance Z_{ab}.
16. In Prob. 15, if $V = 475$ V, what is the current through the load impedance Z_L?
17. In Fig. 33–30, $Z_1 = 63 + j5\ \Omega$, $Z_2 = 12 + j60\ \Omega$, $Z_3 = 20 + j90\ \Omega$, $Z_4 = 18 + j86\ \Omega$, $Z_5 = 8 + j52\ \Omega$, $Z_6 = 47 + j2\ \Omega$, $Z_L = 600 + j0\ \Omega$. Determine the equivalent impedance Z_{ab}.
18. In Prob. 17, if $V = 135$ V, how much power is dissipated in the load impedance Z_L?

SELF TEST

1. What is the resultant capacitance when a 50-μF capacitor is connected in parallel with a 22-μF capacitor?
2. Impedance $Z_1 = 25\angle 14.6°\ \Omega$ is connected in parallel with $Z_2 = 60\angle -85°\ \Omega$. What is the resultant impedance?
3. What is the impedance of the circuit shown in Fig. 33–31?

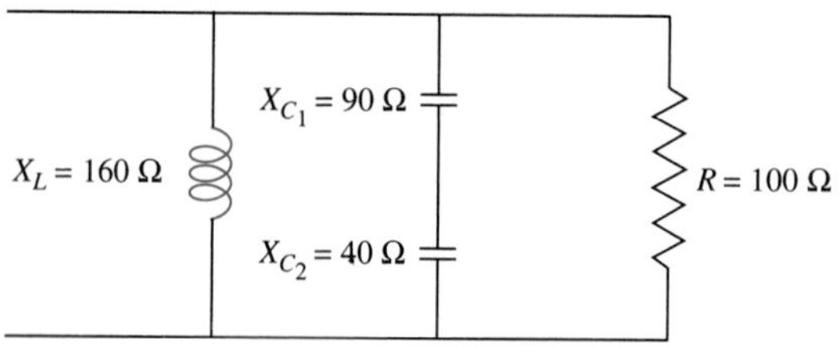

FIG. 33–31 Circuit of self test Prob. 3.

4. An 8-mH coil with a dc resistance of 1.1 kΩ is connected in parallel with a 20-pF capacitor.
   (*a*) What is the resonant frequency of the circuit?
   (*b*) What is the Q of the coil at resonance?
5. Given the delta circuit of Fig. 33–32, in which $Z_1 = 60\angle 22°\ \Omega$, $Z_2 = 80\angle 65°\ \Omega$, and $Z_3 = 100\angle -50°\ \Omega$, determine the impedance Z_a of the equivalent Y network.

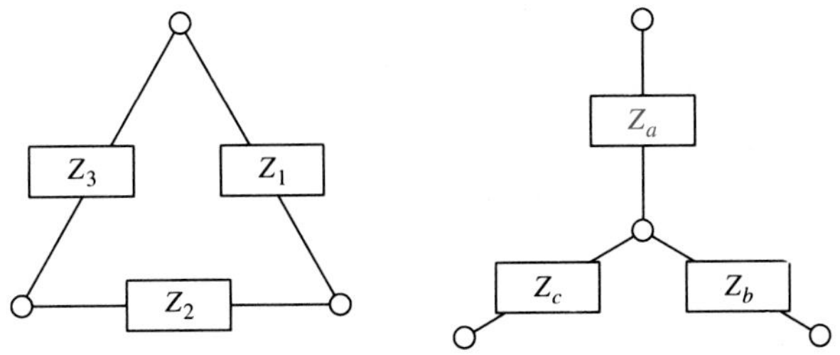

FIG. 33–32 Delta and equivalent Y circuit of self test Prob. 5.

CHAPTER 34

LOGARITHMS

In the days before calculators, it was often good practice to use logarithms to simplify calculations. Credit for the invention of logarithms is chiefly due to John Napier, whose tables appeared in 1614. This was an extremely important event in the development of mathematics, for by the use of logarithms:

1. Multiplication is reduced to addition.
2. Division is reduced to subtraction.
3. Raising to a power is reduced to one multiplication.
4. Extracting a root is reduced to one division.

Now that we have calculators to do all the donkey work of long multiplication and division, these applications of logarithms have become less important. A few examples have been included at the end of this chapter for your general interest. However, even though we now have little need in electronics to perform calculations *by means of* logarithms, we have considerable need for performing calculations *involving* logarithms.

34–1 DEFINITION

The *logarithm* of a quantity is the exponent of the power to which a given number, called the *base,* must be raised in order to equal the quantity.

EXAMPLE 1 Since $10^3 = 1000$, then $3 =$ logarithm of 1000 to the base 10.

EXAMPLE 2 Since $2^3 = 8$, then $3 =$ logarithm of 8 to the base 2.

EXAMPLE 3 Since $a^x = b$, then $x =$ logarithm of b to the base a.

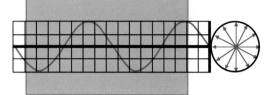

34–2 NOTATION

If

$$b^x = N \tag{1}$$

then x is the logarithm of N to the base b. It may be helpful to mentally translate this expression to "x is the power to which b must be raised to obtain N." This statement is abbreviated by writing

$$x = \log_b N \tag{2}$$

It is evident that Eqs. (1) and (2) mean the same thing and are simply different methods of expressing the same relation among b, x, and N. Eq. (1) is called the *exponential form,* and Eq. (2) is called the *logarithmic form.*

As an aid in remembering that a *logarithm is an exponent,* Eq. (1) can be written in the form

$$(\text{Base})^{\log} = \text{number}$$

The following example illustrates relations between exponential and logarithmic forms.

EXAMPLE 4

Exponential Notation	Logarithmic Notation
$2^4 = 16$	$4 = \log_2 16$
$3^5 = 243$	$5 = \log_3 243$
$25^{0.5} = 5$	$0.5 = \log_{25} 5$
$10^2 = 100$	$2 = \log_{10} 100$
$10^4 = 10\ 000$	$4 = \log_{10} 10\ 000$
$a^b = c$	$b = \log_a c$
$\varepsilon^x = y$	$x = \log_\varepsilon y$

From the foregoing examples, it is apparent that any positive number, other than 1, can be selected as a base for a system of logarithms. Because 1 raised to any power is 1, it cannot be used as a base.

Based on the definitions in Eqs. (1) and (2), you should satisfy yourself with the correctness of the following statement:

$$\log_a a^b = b$$

PROBLEMS 34–1

Express the following equations in logarithmic form:

1. $10^2 = 100$
2. $10^3 = 1000$
3. $2^3 = 8$
4. $3^4 = 81$
5. $4^{0.5} = 2$
6. $\varepsilon^1 = \varepsilon$
7. $a^1 = a$
8. $10^1 = 10$
9. $a^0 = 1$
10. $1 = 10^0$

Express the following equations in exponential form:

11. $3 = \log_{10} 1000$
12. $5 = \log_{10} 100\,000$
13. $2 = \log_5 25$
14. $2 = \log_7 49$
15. $0 = \log_6 1$
16. $0 = \log_a 1$
17. $5 = \log_4 1024$
18. $0.5 = \log_{25} 5$
19. $t = \log_s r$
20. $2x = \log_3 M$

Find the value of x:

21. $4^x = 256$
22. $2^x = 64$
23. $10^x = 10\,000$
24. $x = \log_2 32$
25. $2^x = 128$
26. $3^x = 81$

27. Show that $\log_{10} 100 = \log_{10} 100\,000 - \log_{10} 1000$.
28. Show that $\log_7 1 = \log_p 1$
29. What are the logarithms to the base 2 of 2, 4, 8, 16, 32, 64, 128, 256, and 512?
30. What are the logarithms to the base 3 of 3, 9, 27, 81, 243, 729, and 2187?

34–3 LOGARITHM OF A PRODUCT

The logarithm of a product is equal to the sum of the logarithms of the factors.

Consider the two factors M and N, and let x and y be their respective logarithms to the base a; then,

$$x = \log_a M \tag{3}$$

and

$$y = \log_a N \tag{4}$$

Writing Eq. (3) in exponential form,

$$a^x = M \tag{5}$$

Writing Eq. (4) in exponential form,

$$a^y = N \tag{6}$$

Then

$$M \cdot N = a^x \cdot a^y = a^{x+y}$$

$$\therefore \log_a (M \cdot N) = x + y = \log_a M + \log_a N$$

EXAMPLE 5

$$2 = \log_{10} 100 \quad \text{or} \quad 10^2 = 100$$

$$4 = \log_{10} 10\,000 \quad \text{or} \quad 10^4 = 10\,000$$

Then

$$100 \times 10\,000 = 10^2 \cdot 10^4$$

$$= 10^{2+4} = 10^6$$

$$\therefore \log_{10} (100 \times 10\,000) = 2 + 4$$

$$= \log_{10} 100 + \log_{10} 10\,000$$

The above proposition is also true for the product of more than two factors. Thus, by successive applications of the proof, it can be shown that

$$\log_a (A \cdot B \cdot C \cdot D) = \log_a A + \log_a B + \log_a C + \log_a D$$

34–4 LOGARITHM OF A QUOTIENT

The logarithm of the quotient of two numbers is equal to the logarithm of the dividend minus the logarithm of the divisor.

As in Sec. 34–3, let

$$x = \log_a M \qquad (3)$$

and

$$y = \log_a N \qquad (4)$$

Writing Eq. (3) in exponential form,

$$a^x = M \qquad (5)$$

Writing Eq. (4) in exponential form,

$$a^y = N \qquad (6)$$

Dividing Eq. (5) by Eq. (6),

$$\frac{a^x}{a^y} = \frac{M}{N}$$

That is,

$$a^{x-y} = \frac{M}{N} \qquad (7)$$

Writing Eq. (7) in logarithmic form,

$$x - y = \log_a \frac{M}{N} \qquad (8)$$

Substituting in Eq. (8) for the values of x and y,

$$\log_a M - \log_a N = \log_a \frac{M}{N}$$

EXAMPLE 6

$$2 = \log_{10} 100 \quad \text{or} \quad 10^2 = 100$$
$$4 = \log_{10} 10\,000 \quad \text{or} \quad 10^4 = 10\,000$$

Then

$$\frac{10\,000}{100} = \frac{10^4}{10^2} = 10^{4-2} = 10^2$$

$$\therefore \log_{10} \frac{10\,000}{100} = 4 - 2 = \log_{10} 10\,000 - \log_{10} 100$$

34–5 LOGARITHM OF A POWER

The logarithm of a power of a number equals the logarithm of the number multiplied by the exponent of the power.

Again, let $$x = \log_a M \tag{3}$$

Then $$M = a^x \tag{9}$$

Raising both sides of Eq. (9) to the nth power,

$$M^n = a^{nx} \tag{10}$$

Writing Eq. (10) in logarithmic form,

$$\log_a M^n = nx \tag{11}$$

Substituting in Eq. (11) for the value of x,

$$\log_a M^n = n \log_a M$$

EXAMPLE 7

$$2 = \log_{10} 100 \qquad \text{or} \qquad 100 = 10^2$$

Since $$(10^2)^2 = 10^{2\cdot 2} = 10^4 = 10\ 000$$
then $$\log_{10} 10\ 000 = 4$$
$$\therefore \log_{10} 100^2 = 2 \log_{10} 100 = 2 \cdot 2 = 4$$

34–6 LOGARITHM OF A ROOT

The logarithm of a root of a number is equal to the logarithm of the number divided by the index of the root.

Again, let $$x = \log_a M \tag{3}$$

Then $$M = a^x \tag{9}$$

Extracting the nth root of both sides of Eq. (9),

$$M^{1/n} = a^{x/n} \tag{12}$$

Writing Eq. (12) in logarithmic form,

$$\log_a M^{1/n} = \frac{x}{n} \tag{13}$$

Substituting in Eq. (13) for the value of x,

$$\log_a M^{1/n} = \frac{\log_a M}{n}$$

EXAMPLE 8

$$4 = \log_{10} 10\,000 \quad \text{or} \quad 10\,000 = 10^4$$

Since $$\sqrt{10\,000} = \sqrt{10^4} = 10^{4/2} = 10^2 = 100$$

then $$\log_{10} \sqrt{10\,000} = \frac{\log_{10} 10\,000}{2} = \frac{4}{2} = 2$$

34–7 SUMMARY

It is evident that if the logarithms of numbers instead of the numbers themselves are used for computations, then *multiplication, division, raising to powers,* and *extracting roots* are replaced by *addition, subtraction, multiplication,* and *division,* respectively. Because you are familiar with the laws of exponents, especially as applied to the powers of 10, the foregoing operations with logarithms involve no new ideas. The sole idea behind logarithms is that every positive number can be expressed as a power of some base. That is,

$$\text{Any positive number} = (\text{base})^{\log}$$

34–8 THE COMMON SYSTEM OF LOGARITHMS

Since 10 is the base of our number systems, both integral and decimal, the base 10 has been chosen for a system of logarithms. This system is called the *common system* or *Briggs's system*. The natural system, of which the base to five decimal places is 2.718 28, is discussed below.

Hereafter, when no other base is stated, the base will be 10. For example, $\log_{10} 625$ will be written log 625, the base 10 being understood.

34–9 THE NATURAL SYSTEM OF LOGARITHMS

In the number system there exist certain special numbers whose value is not absolutely determined, but which are themselves extremely valuable to us. You are already familiar with π, which has a value which is approximately $\frac{22}{7}$.

Another useful number is e (formerly ε), which has a value of approximately 2.718 28. This unusual number turns out to be extremely valuable when used as a base for logarithms. Because it can be shown to be related to *natural* events, like the decay of charge on a capacitor which is discharged through a resistor or the decay of current when the magnetic field about an inductance collapses, it is called the base of the *natural logarithms*. Tables of natural logarithms, or logarithms to the base e, are to be found in many published books of tables. A good "scientific" calculator will deliver powers of e and logarithms to the base e (ln x). In Sec. 34–13 we will see how to change logarithms to the base 10 into logarithms to the base e or to other bases.

The notation for logarithms to the base e is shown variously as log_e, $\log_\varepsilon$, or ln (pronounced "lon").

34–10 DEVELOPING A TABLE OF LOGARITHMS

Table 34–1 illustrates the connection between the power of 10 and the logarithms of certain numbers.

TABLE 34–1

Exponential Form	Logarithmic Form
$10^4 = 10\ 000$	$\log 10\ 000 = 4$
$10^3 = 1000$	$\log 1000 = 3$
$10^2 = 100$	$\log 100 = 2$
$10^1 = 10$	$\log 10 = 1$
$10^0 = 1$	$\log 1 = 0$
$10^{-1} = 0.1$	$\log 0.1 = -1$
$10^{-2} = 0.01$	$\log 0.01 = -2$
$10^{-3} = 0.001$	$\log 0.001 = -3$
$10^{-4} = 0.0001$	$\log 0.0001 = -4$

Inspection of Table 34–1 shows that only whole-number powers of ten have integers for logarithms. Also, it is evident that the logarithm of any number between 10 and 100, for example, is between 1 and 2; that is, it is 1 plus a decimal. Similarly, the logarithm of any number between 100 and 1000 is between 2 and 3, and so on. Therefore, to represent all numbers, it is necessary for us to develop the fractional powers which represent numbers between 1 and 10. Then, by using powers of 10 to convert any number to a number between 1 and 10 times the appropriate power of 10 (Chap. 6), we may use our new fractional powers of 10 instead of just integral powers of 10 to find the logarithm of any number.

In Sec. 20–4 we saw that $a^{1/2} = \sqrt{a}$. Accordingly, we can see that

$$10^{0.5} = 10^{1/2} = \sqrt{10} = 3.162\ 277\ 66$$

which gives us the first intermediate step in our table of logarithms between 1 and 10:

$$\log_{10} 3.162\ 277\ 66 = 0.5$$

Similarly,

$$10^{0.25} = (10^{0.5})^{0.5} = \sqrt{3.162\ 277\ 66}$$
$$= 1.778\ 279$$

or

$$\log_{10} 1.778\ 279 = 0.25$$

By repeating the square root operation time after time, we can obtain

$$\log_{10} 1.334 = 0.125 \quad \text{etc.}$$

Then, by applying the laws of exponents developed in Sec. 4–3 and summarized in Sec. 20–1, we can determine that

$$3.162\,277\,66 \times 1.778\,279 = 10^{0.5} \times 10^{0.25}$$
$$= 10^{0.75} = 5.623\,41$$

or

$$\log 5.623\,41 = 0.75$$

Repeated applications of this method give us such additional logarithms as

$$\log 4.2173 = 0.625$$

and

$$\log 2.37 = 0.375$$

You should use the values now developed to prove that
$10^{0.75} \times 10^{0.25} = 10$, as a check on our method.

These various values can be plotted on a graph, as in Fig. 34–1, and the more convenient logarithms can be picked off the curve, or other more sophisticated methods of higher mathematics may be applied to yield Table 34–2 of logarithms of numbers between 1 and 10:

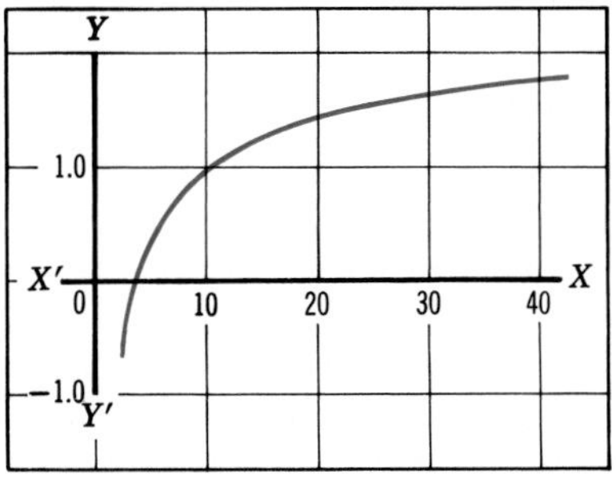

Fig. 34–1 Graph of the equation $y = \log_{10} x$.

TABLE 34–2 LOGARITHMS OF NUMBERS BETWEEN 1 AND 10

Number	Logarithm	Number	Logarithm
1	0.000 00	6	0.778 15
2	0.301 03	7	0.845 10
3	0.477 12	8	0.903 09
4	0.602 06	9	0.954 24
5	0.698 97	10	1.000 00

Since we convert every number to its equivalent number between 1 and 10 times the appropriate power of 10, every logarithm we will ever look up will be a decimal fraction. Because of this universality of decimals as logarithms, almost every table of logarithms published omits the decimal point: log 2 will appear as simply 301 03 instead of 0.301 03.

From the foregoing discussion it will be evident that every logarithm has two parts: a decimal part which we read from the table of logarithms and an integer which we must provide each time from our knowledge of powers of 10.

EXAMPLE 9 Determine the logarithm of 200.

SOLUTION

First, rewrite the number in standard form:

$$200 = 2.00 \times 10^2$$

Since log 2.00 = 0.301 03, this number could be written

$$2.00 \times 10^2 = 10^{0.301\,03} \times 10^2 = 10^{2.301\,03}$$

This power to which 10 is raised to be equal to 200 is the logarithm of 200. In other words,

$$\log 200 = 2.301\ 03$$

This logarithm is made up of two parts: the decimal part from the table and the integer part which we developed from the "power of 10."

Similarly, log 2000 = 3.301 03.

In the same manner, referring to Table 34–1, it follows that the logarithm of a number between 0.1 and 0.01 will be −2 followed by a decimal number from the table and the logarithm of a number between 0.001 and 0.0001 will be −4 followed by a decimal. When logarithms had to be taken from tables, it was necessary to distinguish between the *positive fraction* (the *mantissa*) read from the table and the *negative integer* (the *characteristic*) representing the power of ten. With calculators, this distinction is no longer necessary.

EXAMPLE 10 What is the logarithm of 0.002?

SOLUTION

$$\begin{aligned} 0.002 &= 2 \times 10^{-3} \\ &= 10^{0.30103} \times 10^{-3} \\ &= 10^{-2.69897} \end{aligned}$$

Your calculator will confirm that log 0.002 = −2.69897. You will often see texts and notes that retain the distinction between the positive value 0.30103 and the negative value −3. Using this notation log 0.002 might be written $\bar{3}.30103$, with $\bar{3}$ being pronounced "bar three", or it might be written 7.30103 − 10; the positive 7 coupled with the negative 10 has a net value of −3.

This combination of negative characteristic and positive mantissa was necessary for taking antilogarithms from tables. With your calculator, you can work directly with negative logarithms such as −2.69897 without having to distinguish between the −3 and the +0.30103. (Be thankful!)

You will find a table of three-place common logarithms inside the front cover of this book that will help you to make reasonable approximations when your calculator is not at hand.

34–11 LOGARITHMS FROM A CALCULATOR

Use the instruction manual that came with your calculator to identify the method of determining logarithms. You should have a choice of *common logs,* that is, logarithms to the base 10, probably identified as LOG or log, and *natural logs,* that is, logarithms to the base e (or ε, see Sec. 34–9), probably identified as LN or ln. These might be direct keys or second (maybe yellow) functions. In addition, the same keys, or the second or third function of those keys, will read antilogarithms, probably identified as 10^x and e^x.

Sort out the appropriate keys for your particular calculator, and achieve 100% in Problems 34–2.

PROBLEMS 34–2

Determine the common logarithms of the following numbers:
(Read your calculator to five decimal places)

1. 7.27
2. 72.7
3. 727
4. 0.000 727
5. 95.816
6. 1002
7. 10.02
8. 0.000 100 2
9. 100 200 0
10. 3.3×10^4

34–12 TO FIND THE NUMBER CORRESPONDING TO A GIVEN LOGARITHM

Use your calculator instruction manual to determine the second (or third, or yellow or blue) function of your log key. We read the second function as 10^x. Enter 0.301 03, control function as required, and call for 10^x. Read 2.00. When you have the necessary steps clearly in mind, check through Example 11.

EXAMPLE 11 What is the number that corresponds to the common logarithm 2.698 97?

SOLUTION

(*a*) The characteristic 2 represents 10^2.
By calculator, $10^{0.698\ 97} = 5.00$
Thus, antilog 2.698 97 $= 5 \times 10^2 = 500.00$

(*b*) From the three-place table of logarithms, 0.699 is the log of 5. Couple this antilog of the mantissa with the antilog of the characteristic 2 to yield:

antilog 2.699 $= 5 \times 10^2$

EXAMPLE 12 Find the number whose logarithm is 3.630 43.

SOLUTION

(*a*) 3.630 43 = 0.630 43 × 10^3
Enter 0.630 43 in your calculator,
Key 10^x and read 4.270
Use your knowledge of characteristics to determine:
antilog 3.630 43 = 4.270 × 10^3

(*b*) Enter 3.630 43 in your calculator
Key 10^x and read 4270

A change in the characteristic changes only the position of the decimal point (the power of 10). Note how an electronic calculator in the "scientific" mode reads out the antilogarithm as a number between 1 and 10 multiplied by the appropriate power of 10.

PROBLEMS 34–3

Use your calculator to determine the numbers corresponding to the following common logarithms; adjust to engineering notation.

1. 1.968 70
2. 5.968 70
3. −2.031 302 2
4. −4.031 302 2
5. 6.879 10
6. 0.879 10
7. 3.879 10 − 10
8. $\bar{4}.879\ 10$
9. 814.134 34 × 10^{-3}
10. −2.185 865 6

34–13 CHANGE OF BASE

In Problems 34–1 we found logarithms of numbers to many bases besides 10, and it is often convenient for us to be able to find the logarithms of numbers to certain bases other than 10 without developing a set of tables for other bases. An interesting development shows us how this may be achieved.

$$N = a^x \tag{14}$$

which we may rewrite

$$x = \log_a N \tag{15}$$

Taking logarithms of both sides of Eq. (14) to the base b:

$$\log_b N = \log_b a^x \tag{16}$$

Substituting Eq. (11) into Eq. (16),

$$\log_b N = x \log_b a \tag{17}$$

Substituting Eq. (15) into Eq. (17),

$$\log_b N = \log_a N \cdot \log_b a \tag{18}$$

Since it can be shown that

$$\log_b a = \frac{1}{\log_a b} \tag{19}$$

Equation (18) may be written in the form

$$\log_b N = \frac{\log_a N}{\log_a b} \tag{20}$$

If, then, we have a table of logarithms to the base 10 and find it necessary to produce the logarithm of any number to any other base *b*, we simply divide the logarithm to the base 10 of the given number by the logarithm to the base 10 of the other base number *b:*

$$\log_b N = \frac{\log_{10} N}{\log_{10} b} \tag{21}$$

Refer to your computer user's manual. If your computer is capable of working in common logarithms, it will use one or more of these commands:

LOG 10
CLOG
CLG
LGT

Most computers automatically work in natural logarithms $\log_\varepsilon$, $\log_e$, lon, or ln using such commands as

LN
LOG
LOGE

To achieve common logs ($\log_{10}$), you will have to use a conversion factor from Eqs. (21) and (23):

$$\log_{10} N = \frac{\log_e N}{\log_e 10} = \frac{\log_e N}{2.302\ 58} = 0.4343 \log_e N$$

Your computer may respond to LOG(N)/LOG(10)
LOG(2)/LOG(10) which should read out
as 0.30103

We are primarily concerned with the natural system of logarithms, which has for its base the number $e = 2.718\ 28 \ldots$ (Sec. 34–9). Many relationships in electronics as well as other branches of science involve logarithms to this base.

Although we will be developing and using logarithms to the base *e* in Sec. 34–14, you will often find that only tables of logarithms to the base 10 are immediately available. Using the relationship expressed in formulas (18) and (20), you will be able to perform the necessary operations.

$$\log_e N = 2.302\ 59 \log_{10} N \tag{22}$$

$$\log_{10} N = 0.434\ 29 \log_e N \tag{23}$$

EXAMPLE 13

$$\begin{aligned}\log_e 1000 &= 2.302\ 59 \log_{10} 1000\\ &= 2.302\ 59 \times 3\\ &= 6.907\ 77\end{aligned}$$

EXAMPLE 14

$$\begin{aligned}\log_{10} 100 &= 0.434\ 29 \log_e 100\\ &= 0.434\ 29 \times 4.6052\\ &= 2.0000\end{aligned}$$

EXAMPLE 15 Given $x = \log_e 48$. Solve for x.

SOLUTION

$$\begin{aligned}\log_e 48 &= 2.302\ 59 \log_{10} 48\\ &= 2.302\ 59 \times 1.681\ 24\\ x &= 3.8712\end{aligned}$$

34–14 NATURAL LOGARITHMS

Because so many calculations in electronics do involve logarithms to the base e, you will use your calculator keys LN or lnx often. If you do not have these keys, keep Eqs. (22) and (23) close at hand. A few special notes will simplify your use of these natural logarithms:

1. The laws of logarithms [Eqs. (6), (8), (11), and (13)] apply to any logarithmic system, regardless of the base used. Therefore, natural logarithms may be used instead of common logarithms, if you prefer, for any problem involving multiplication, division, raising to powers, or extracting roots.
2. For convenience, we often replace the notation $\log_\varepsilon$ with the special symbol ln, pronounced lon.
3. Since $\ln e = 1$, the characteristics in natural logarithms do not represent powers of 10. (They represent powers of e.) Accordingly, tables, as well as your calculator, give the *entire* natural logarithm of a number, and not just its mantissa.

> To display the numerical value of e to as many places as your calculator or computer can deliver, call up the value of e^1.
>
> EXP(1) will read
> 2.718 281 8 . . .

EXAMPLE 16

$$\begin{aligned}\ln 2.70 &= 0.993\ 25\\ \ln 2.72 &= 1.000\ 63\\ \ln 5.05 &= 1.619\ 39\\ \ln 7.38 &= 1.998\ 77\\ \ln 7.39 &= 2.000\ 13\end{aligned}$$

4. Because the characteristic represents a power of e, it is necessary to build up natural logarithms of very large and very small numbers, using Eq. (6) for the purpose.

EXAMPLE 17 Using your calculator, find the natural logarithm of 127.4.

SOLUTION

Key 127.4
Key LN or lnx
Read 4.847 33

EXAMPLE 18 Find ln 0.001 274.

SOLUTION

Key .001 274
Key LN or lnx
Read −6.665 59

34–15 GRAPH OF $y = \log_{10} x$

The graph of $y = \log_{10} x$ is shown in Fig. 34–1. A study of the graph shows the following:

1. A negative number has no real logarithm.
2. The logarithm of a positive number less than 1 (a decimal between 0 and 1) is negative.
3. The logarithm of 1 is zero.
4. The logarithm of a positive number greater than 1 is positive.
5. As the number approaches zero, its logarithm decreases without limit.
6. As the number increases indefinitely, its logarithm increases without limit.

Is the method of interpolation that treats a short distance on the logarithmic curve as a straight line sufficiently accurate for computation?

34–16 LOGARITHMIC EQUATIONS

An equation in which there appears the logarithm of some expression involving the unknown quantity is called a *logarithmic equation*.

Logarithmic equations have wide application in electric circuit analysis. In addition, the communications engineer uses them in computations involving decibels and transmission line characteristics.

EXAMPLE 19 Solve the equation $4 \log x + 3.796\ 00 = 4.699\ 09 + \log x$.

SOLUTION

Given	$4 \log x + 3.796\ 00 = 4.699\ 09 + \log x$
Transposing,	$4 \log x - \log x = 4.699\ 09 - 3.796\ 00$
Collecting terms,	$3 \log x = 0.903\ 09$
D: 3,	$\log x = 0.301\ 03$

Using a calculator,
Key 0.301 03
Key INV LOG *or* 10^x
Read *2.00*

In solving logarithmic equations, the logarithm of the unknown, as $\log x$ in Example 19, is considered as any other literal *coefficient*. That is, in general, the rules for solving ordinary algebraic equations apply in logarithmic equations.

A common error made by students in solving logarithmic equations is confusing coefficients of logarithms with coefficients of the unknown. For example,

$$3 \log x \neq \log 3x$$

because the left member denotes the product of 3 times the logarithm of x, whereas the right member denotes the logarithm of the quantity 3 times x, that is, $\log (3x)$.

EXAMPLE 20 Given $500 = 276 \log \dfrac{d}{0.05}$. Solve for d.

SOLUTION 1

Given $\quad 500 = 276 \log \dfrac{d}{0.05}$

Then $\quad 500 = 276 (\log d - \log 0.05)$

D: 276, $\quad 1.81 = \log d - \log 0.05$

Transposing, $\quad \log d = 1.81 + \log 0.05$

Substituting 8.698 97 − 10 for log 0.05,

$$\log d = 1.81 + 8.698\,97 - 10$$

Collecting terms, $\quad \log d = 0.508\,97$

Using a calculator, Key 10^x, read $\quad d = 3.23$

SOLUTION 2

Given $\quad 500 = 276 \log \dfrac{d}{0.05}$

D: 276, $\quad 1.81 = \log \dfrac{d}{0.05}$

Taking antilogs of both members, $\quad 64.57 = \dfrac{d}{0.05}$

Solving for d, $\quad d = 3.23$

34–17 EXPONENTIAL EQUATIONS

An equation in which the unknown appears in an exponent is called an *exponential equation*. In the equation

$$x^3 = 125$$

it is necessary to find some value of x that, when cubed, will equal 125. In this equation *the exponent is a constant*.

In the *exponential equation*

$$5^x = 125$$

the situation is different. The *unknown appears as an exponent,* and it is now necessary to find to what power 5 must be raised to obtain 125.

Some exponential equations can be solved by inspection. For example, the value of x in the foregoing equation is 3. In general, taking the logarithms of both sides of an exponential equation will result in a logarithmic equation that can be solved by the usual methods.

EXAMPLE 21 Given $4^x = 256$. Solve for x.

SOLUTION

Given

$$4^x = 256$$

Taking the logarithms of both members,

$$\log 4^x = \log 256$$

or

$$x \log 4 = \log 256$$

D: log 4,

$$x = \frac{\log 256}{\log 4}$$

From the log tables,

$$x = \frac{2.408}{0.602} = 4$$

Using a calculator,

Key 256
Key LOG
Read 2.40824

Key 4
Key LOG
Read 0.602 06
Key ÷

$$x = \frac{2.408\ 24}{0.602\ 06} = 4$$

EXAMPLE 22 Given $x^2 = 124$, solve for x.

SOLUTION

$$x^2 = 124$$
$$2\log x = \log 124$$
$$\log x = \frac{\log 124}{2}$$
$$= \frac{2.093\ 42}{2}$$
$$= 1.046\ 71$$

taking antilogs:

$$x = 10^{1.046\ 71} = 11.135\ 53$$

(Confirmation: $\sqrt{124} = 11.135\ 53$)

CHECK

$$4^4 = 256$$

EXAMPLE 23 Given $5^{x-3} = 52$. Solve for x.

SOLUTION

Given $$5^{x-3} = 52$$

Taking the logarithms of both members,

$$\log 5^{x-3} = \log 52$$

or $$(x - 3) \log 5 = \log 52$$

D: log 5, $$x - 3 = \frac{\log 52}{\log 5}$$

Using a calculator, $$x - 3 = \frac{1.716\ 00}{0.698\ 97}$$

A: 3, $$x = \frac{1.716\ 00}{0.698\ 97} + 3$$

or $$x = 5.455$$

How would you check this solution?

PROBLEMS 34–4

Solve the following equations:

1. $x = \log_e 599$
2. $x = \log_e 4.38$
3. $\log x + 2 \log x = 6$
4. $\log x + \log 6x = 8.5$

(▶ **HINT** $\log 6x = \log 6 + \log x$)

5. $\log 2x + 2 \log x = 6$
6. $\log \frac{P}{3} = 0.573$
7. $\log \frac{P_1}{15} = 2.123\ 20$
8. $\log \frac{12}{V} = 3$
9. $\log x^2 - \log x = 4.542$
10. $x^4 = 462$
11. $4^x = 512$
12. $5^x = 37.3$
13. $2^m = 0.941\ 66$
14. $3^{q-3} = 14$
15. $4^{3x} = 18.6$
16. $m^{2.5} = 80$
17. $y^{1.6} = 47$
18. $\log \frac{y^2}{4} = 2.2$
19. $\log \frac{15}{x^3} = 0.888$
20. $10 \log \frac{P}{1.5} = 32$
21. $\log x^5 - \log x^2 = 1.766$
22. $\log x^{5.44} - \log x^{2.78} = 1.786$
23. $\log x^3 - \log x = 2.442$
24. $x = \log_6 1296$
25. $x = \log_3 2187$
26. If $L_1 = \sqrt[3]{L_2^2}$, solve for L_2.

27. If $20 \log \dfrac{2Z_1}{2Z_1 - Z_a} = 20 \log \dfrac{-Z_b}{-Z_b + \dfrac{Z_1}{2}}$,

solve for Z_1 in terms of Z_a and Z_b.

28. If $V_g = \dfrac{2.3T}{11\ 600} \log \dfrac{I_0}{I_g}$, solve for I_0.

29. If $i = \dfrac{V}{L} te^{S_c t}$, solve for S_c.

> Some computer dialects use EXP to indicate powers of e, such as
>
> $\dfrac{V}{R}(1 - \text{EXP}(Rt/L))$
>
> $= \dfrac{V}{R}(1 - e^{R_t/L})$

30. If $i_c = \dfrac{V}{R} e^{-(t/RC)}$, solve for (*a*) *V*, (*b*) *C*, (*c*) *t*.

31. If $I_k = AT^2e^{-(B/T)}$, solve for (*a*) *A*, (*b*) *B*.

32. If $i_L = \dfrac{V}{R}(1 - e^{-Rt/L})$, solve for (*a*) *V*, (*b*) *L*, (*c*) *t*.

33. If $q = CV(1 - e^{-t/RC})$, solve for (*a*) *V*, (*b*) *R*, (*c*) *t*.

34. If $I_p + I_g = K\left(V + \dfrac{V_p}{\mu}\right)^{\frac{3}{2}}$, solve for (*a*) *V*, (*b*) V_p, (*c*) μ.

35. In an inductive circuit, the equation for the growth of current is given by

$$i = \frac{V}{R}(1 - e^{-Rt/L})\ A \qquad (24)$$

where i = current, A
t = any elapsed time after switch is closed, s
V = constant impressed voltage, V
L = inductance of the circuit, H
R = circuit resistance, Ω
e = base of natural system of logarithms.

A circuit of 0.75 H inductance and 15 Ω resistance is connected across a 12-V battery. What is the value of the current at the end of 0.06 s after the circuit is closed?

Solution:

The circuit is shown in Fig. 34–2.

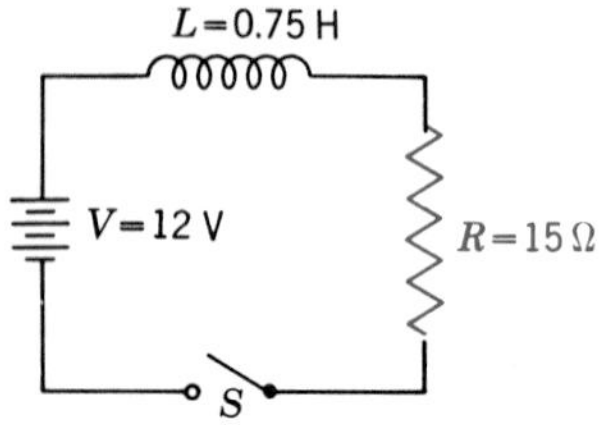

Fig. 34–2 Circuit of Probs. 35 to 38.

Given $$i = \frac{V}{R}(1 - e^{-(Rt/L)})$$

Substituting the known values,

$$i = \frac{12}{15}(1 - e^{-(15 \times 0.06/0.75)})$$

$$i = 0.8(1 - e^{-1.2})$$

Multiplying, $$i = 0.8 - 0.8e^{-1.2}$$

or $$i = 0.8 - \frac{0.8}{e^{1.2}} \qquad (25)$$

Now evaluate $e^{1.2}$,

$$\log_{10} e^{1.2} = 1.2 \log_{10} e = 1.2 \times 0.434\ 29$$
$$= 0.521\ 15$$

Taking antilogs, $e^{1.2} = 3.32$

Substituting the value of $e^{1.2}$ in Eq. (26),

$$i = 0.8 - \frac{0.8}{3.32} = 0.559 \text{ A}$$

The growth of the current in the circuit of Fig. 34–2 is shown graphically in Fig. 34–3.

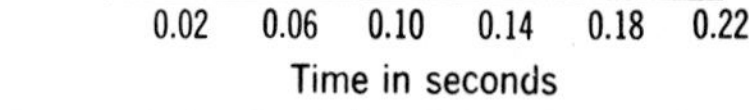

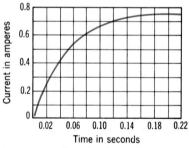

Fig. 34–3 Graph of current in *RL* circuit of Prob. 35.

36. The inductance of the circuit in Fig. 34–2 is halved and the resistance is thus reduced to 0.71 times its original value. If other circuit values remain the same, what will be the value of the current 0.08 s after the switch is closed?
37. Using the circuit values for the circuit of Fig. 34–2, what will be the value of the current (*a*) 0.005 s after the switch is closed and (*b*) 0.5 s after the switch is closed?
38. In the circuit of Fig. 34–2, after the switch is closed, how long will it take the current to reach 50% of its maximum value?

39. If $\frac{L}{R}$ is substituted for t in the equation

$$i = \frac{V}{R}\left(1 - e^{-(Rt/L)}\right)$$

show that the value of the current will be 63.2% of its steady-state value. The numerical value of L/R in seconds is known as the *time constant* of the inductive circuit. It is useful in determining the rapidity with which current rises or falls in one inductive circuit in comparison with others.

40. A 220-V generator shunt field has an inductance of 12 H and a resistance of 80 Ω. How long after the line voltage is applied does it take for the current to reach 75% of its maximum value?

41. A relay of 1.2 H inductance and 500 Ω resistance is to be used for keying a radio transmitter. The relay is to be operated from a 110-V line, and 0.175 A is required to close the contacts. How many words per minute will the relay carry if each word is considered as five letters of five impulses per letter? The time of opening of the contacts is the same as the time required to close them.

▶ **HINT** $0.175 = \frac{110}{500}\left(1 - e^{-(500/1.2)}\right)$. t is the time required to close the relay.

42. How many words per minute would the relay of Prob. 41 carry if 50 Ω resistance were connected in series with it? The line voltage remains at 110 V.

43. In a capacitive circuit the equation for the current is given by

$$i = \frac{V}{R} e^{-(t/RC)} \quad A \qquad (26)$$

where i = current, A
t = any elapsed time after switch is closed, s
V = impressed voltage, V
C = capacitance of the circuit, F
R = circuit resistance, Ω
e = base of natural system of logarithms

A capacitance of 500 μF in series with 1 kΩ is connected across a 50-V generator.
(*a*) What is the value of the current at the instant the switch is closed?

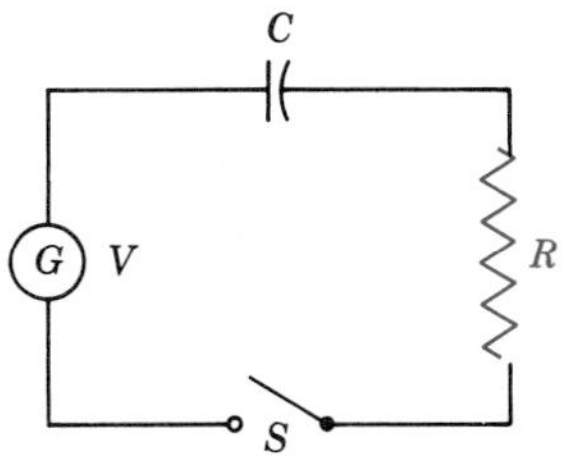

Fig. 34–4 Circuit of Probs. 43 and 44.

▶ **HINT** $t = 0$.

(*b*) What is the value of the current 0.02 s after the switch is closed? The circuit is shown in Fig. 34–4.

44. In the circuit of Fig. 34–4, how long after the switch is closed will the current have decayed to 30% of its initial value if V = 110 V, R = 500 Ω, C = 20 μF, and $i = \frac{0.3\ V}{R}$? t = ?

Solution: $i = \dfrac{0.3V}{R} = \dfrac{0.3 \times 110}{500} = 0.066 \quad \text{A}$

Substituting in Eq. (26), $0.066 = \dfrac{110}{500} e^{-(t/500 \times 20 \times 10^{-6})}$

Simplifying, $0.066 = 0.22e^{-(t/10^{-2})}$

or $0.066 = 0.22e^{-100t}$

D: 0.22, $0.3 = e^{-100t}$

By the law of exponents, $0.3 = \dfrac{1}{e^{100t}}$

M: e^{100t}, $0.3e^{100t} = 1$

D: 0.3, $e^{100t} = 3.33$

Taking logarithms, $\log_{10} e^{100t} = \log_{10} 3.33$

That is, $100t \log_{10} e = \log_{10} 3.33$

Then $100t \times 0.4343 = 0.5224$

or $43.43t = 0.5224$

$\therefore t = 0.012 \text{ s}$

The decay of the current in the circuit of Fig. 34–4 is shown graphically in Fig. 34–5.

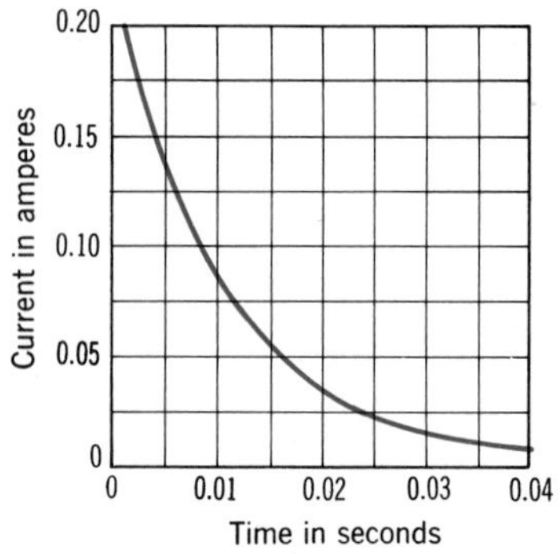

Fig. 34–5 Graph of current in *RC* circuit of Prob. 44.

45. A 20-μF capacitor in series with a resistance of 680 Ω is connected across a 110-V source.
    (*a*) What is the initial value of the current?
    (*b*) How long after the switch is closed will the current have decayed to 36.8% of its initial value?
    (*c*) Is the time obtained in (*b*) equal to *CR* s? The product of *CR*, in seconds, is the time constant of a capacitive circuit.

46. The quantity of charge on a capacitor is given by

$$q = CV(1 - e^{-(t/CR)}) \quad C \tag{27}$$

where q is the quantity of electricity in coulombs.

(*a*) Calculate the charge q in coulombs on a capacitor of 50 μF in series with a resistance of 3.3 kΩ, 0.008 s after being connected across a 70-V source.

(*b*) What is the voltage across the capacitor at the end of 0.02 s?

47. A key-click filter consisting of a 2-μF capacitor in series with a resistance is connected across the keying contacts of a transmitter. If the average time of impulse is 0.004 s, calculate the value of the series resistance required in order that the capacitor can discharge 90% in this time.

▶ **HINT** Under steady-state conditions, $q = CV$. Then

$$0.9CV = CV(1 - e^{-(t/RC)})$$

48. The emission current in amperes of a heated filament is given by

$$I = AT^2 e^{-(B/T)} \quad A \tag{28}$$

For a tungsten filament, $A = 60$ and $B = 52\,400$. Find the current of such a filament at a temperature $T = 2500$ K.

49. An important triode formula is

$$I_p + I_g = K\left(V_g + \frac{V_p}{\mu}\right)^{\frac{3}{2}} \quad A \tag{29}$$

where I_p = plate current, A
I_g = grid current, A
V_g = grid voltage, V
V_p = plate voltage, V
μ = amplification factor

Calculate $I_p + I_g$ if $K = 0.0005$, $V_g = 6$ V, $V_p = 270$ V, and $\mu = 15$.

50. The diameter of No. 0000 wires is 11.68 mm, and that of No. 36 is 0.127 mm. There are 38 wire sizes that are between No. 0000 and No. 36; therefore, the ratio between cross-sectional areas of successive sizes is the thirty-ninth root of the ratio of the area of No. 0000 wire to that of No. 36 wire, or $\sqrt[39]{\frac{11.68^2}{0.127^2}}$. Compute the value of this ratio. Because this ratio is nearly equal to $\sqrt[3]{2}$, we can use the approximation that the cross-sectional area of a wire doubles for every decrease of three sizes, as explained in Sec. 9–4. Calculate the percent error introduced by using $\sqrt[3]{2}$.

51. In the "tempered" musical scale there are 12 notes from the first note of an octave to the first note of the next higher octave, which is double the frequency of the previous first note. Find the 12th root of 2 to find:
   (*a*) The multiplier between frequencies of adjacent notes.
   (*b*) The frequency of B♭ if the next lower note, A, has a frequency of 440 Hz.

34–18 CALCULATIONS BY MEANS OF LOGARITHMS

When calculations are performed *by means of logarithms,* it is best to set out a table, with calculating indications based on our recall of the four rules:

1. For multiplication, add logs.
2. For division, subtract logs.
3. For raising to a power, multiply the log.
4. For extracting a root, divide the log.

EXAMPLE 24 Evaluate: $0.5\left(\dfrac{21.866 \times 2.47^3}{\sqrt[4]{188}}\right)$

SOLUTION

Plan a table, with a sequence that will simplify the work, showing the operation to be performed for each line:

log 2.47	= 0.			
	3	×		
log 2.47^3	=			
log 21.866	= 1.	+		
log numerator	=	=		
log 188	= 2.			
	4	÷		
log denom	=	=	____	−
log bracket	=	=		
log 0.5	=	=	−0.	+
log answer	=	=		
answer	= antilog	=	____	

Only after the skeleton is complete, and the various arithmetic operations have been checked, do we insert the logarithms, evaluate, and take antilogs:

log 2.47	= 0.392 70		
	3 ×		
	1.178 09		
log 21.866	= 1.339 77 +		
log num	= 2.517 86	= 2.517 86	
log 188	= 2.274 16		

	4 ÷		
log denom	= 0.568 54	= 0.568 54	−
log bracket	=	1.949 32	
log 0.5	=	−0. 301 03	+
log answer		= 1.648 29	
answer	= antilog 1.648 29	= 44.492 90	

Of course, you will want to check your result with your calculator. For your interest, attempt exercises in various problem sets. Bear in mind that there are no logarithms of negative numbers. If a negative number is involved, determine the sign of the answer, and take logs of all positive numbers.

PROBLEMS 34–5

Use logarithms to evaluate the following:

1. 2.79×684
2. $(-9.5)(26)$
3. $(14.83)(-2.222)(0.1123)$
4. $(-4627)(9126)(-7336)$
5. $948 \div 237$
6. $(-2325) \div (4.023)$
7. $(0.000\ 517\ 9) \div (-3.648)$
8. $(0.002\ 69) \div (-\ 0.001\ 08)$
9. $\dfrac{6.28 \times 0.000\ 159 \times 326}{0.003\ 68 \times 436 \times 0.0278}$
10. $\dfrac{(1.12)(1.23)(3.21 \times 10^{-7})}{1.3776 \times 10^{2} \times 4.17}$
11. 12^3
12. 0.0563^5
13. $\sqrt[3]{815}$
14. $\sqrt[4]{0.009\ 55}$
15. $0.75\left(\dfrac{27.69 \times 3.74^3}{5\sqrt[4]{277}}\right)$

SELF TEST A

Test A is to be performed without the use of tables, calculators, or computers.

1. Express in logarithmic form: $2^5 = 32$
2. Express in exponential form: $\log_{16}4096 = 3$
3. Solve for x: $3^x = 81$
4. Solve for x: $x = \log_2 16$
5. Write in expanded logarithmic form:

$$\left[\frac{(2x^2)(3y^4)}{5z}\right]^2$$

6. If log 3200 = 3.505 15, what is the characteristic?
7. If log 175 = 2.243 04, what is the mantissa?
8. What is the characteristic of log 0.000 71?
9. If log 6 = 0.778 15 and log 16 = 1.204 12, evaluate 2 log 6 + 0.5 log 16
10. What is $\log 6^3$?
11. If $10^{0.54407} = 3.500$, what is antilog 3.544 07?
12. What is $\log_7 343$?
13. log 24.5 = 1.389 17. Evaluate $\log(\sqrt{24.5})^3$

SELF TEST B

Your instructor will determine whether Test B may be performed using tables or calculators.

1. Evaluate x: $4.5^x = 126$
2. Evaluate x: $x = \log_7 444\,67$
3. Evaluate x: $x = \log 1.6^3$
4. Evaluate x: $x = \sqrt[5]{1800}$
5. Evaluate P: $P = \ln 177$
6. Use logarithms to evaluate x:

$$x = \frac{298 \times 0.004\,16}{0.573}$$

7. If $\log_{10} 167 = 2.222\,72$, what is $\log_8 167$?
8. Solve for x: $\log x + \log x^4 = 2.005$
9. Evaluate Y: $Y^{1.6} = 28.164\,68$
10. If $i = \dfrac{V}{R}e^{-t/RC}$, what is V when $i = 4.4$ mA, $R = 25$ kΩ, $C = 50$ μF, and $t = 220$ μs?

CHAPTER 35

APPLICATIONS OF LOGARITHMS

We have seen that logarithms can be extremely useful in the performance of arithmetic operations. Multiplication, division, raising to powers, and extracting roots are important applications of logarithms which will be explored further in this chapter.

Similarly, proficiency in the use of logarithmic equations is an essential part of the electronics technician's mathematical toolbox. The broad application of these equations to computers, power measurement, amplification, attenuators, and transmission lines all testify to the equations' importance.

In this chapter, we will see how logarithmic calculations are applied to the fields mentioned above and we will investigate briefly two extremely important applications of logarithms to our everyday work in electronics—preferred values and decibels.

35–1 PREFERRED VALUES

In the determination of the values of resistors, capacitors, and inductors which may be required in a circuit, such as those calculated in Sec. 15–2, we often find that the values available off the shelf are not identical with our calculated values. We may desire to have a 620-Ω resistor, and the lab assistant says, "Use a 560- or a 680-Ω. Either will be close enough." How can the lab assistant say so, unhesitatingly? How does the lab assistant know? In other words, how do we arrive at *preferred values*?

Under the prompting of the industry as a whole, the Electrical Industries Association has established lists of suggested figures for the guidance of manufacturers and technicians. Several series of values are normally listed, depending on the quality of service required. Most commonly used are the *R*6 and *R*12 series, which list the 6 and 12 values that cover all the requirements for 20% and 10% tolerances, respectively. Becoming more and more called upon is the $R24$ series, which gives values for 5% tolerance. Naturally, the price of the more exact values is considerably higher than the price of the other values, and the $R6$ and $R12$ values meet the demands of ordinary service quite satisfactorily.

Each of the series is developed from a logarithmic progression based on an appropriate root of 10. To develop the $R6$ series, we take the $\frac{1}{6}$, $\frac{2}{6}$, $\frac{3}{6}$, $\frac{4}{6}$, $\frac{5}{6}$, and $\frac{6}{6}$ roots of 10, in order. Table 35–1 shows the development of the $R6$ series of preferred values.

TABLE 35–1 *R6* SERIES OF PREFERRED VALUES

x	$10^{\frac{x}{6}}$	Preferred Value	Difference	Percent Difference	Max % Error
0	1.000	1.0	0.5	50	±20
1	1.468	1.5	0.7	46	18.9
2	2.155	2.2	1.1	50	20
3	3.162	3.3	1.4	42.5	17.5
4	4.642	4.7	2.1	44.6	18.3
5	6.813	6.8	3.2	47	19.1
6	10.	10.	5.0	50	20
		15			

You should confirm, by using logarithms, that $10^{\frac{4}{6}} \cong 4.642$. Now, the calculated values may be rounded off to easy-to-remember two-significant-figure numbers in order to arrive at the preferred values. Naturally, all these values may be multiplied by any power of 10, so that memorizing six numbers is all that is needed to cover the entire range of 20% values. The maximum error of ±20% has been arrived at by choosing desired values midway between the two preferred values and determining the percentage error. If we required a 4-kΩ resistor, choosing either 3.3-kΩ or 4.7-kΩ would not introduce more than a 20% error. Obviously, then, any value closer to a preferred value than one midway between the two must be closer than 20% tolerance. The advantages to manufacturers, sales agencies, and technicians will be obvious at once.

When greater accuracy (less tolerance) is required, we may use the $R12$ series for ±10% or even the $R24$ series for ±5% values. Naturally, the 5% shows all the values in the 10% and 20% series plus intermediate values to round out the series.

PROBLEMS 35–1

1. Using successive twelfth roots of 10, by logarithms, list the preferred values and the maximum percentage errors for the $R12$ series of preferred values.
2. Using successive twenty-fourth roots of 10, by logarithms, list the preferred values and the maximum percentage errors for the $R24$ series of preferred values.

3. The standard published values of capacitors made by a prominent manufacturer follow the $R10$ series. By using successive tenth roots of 10, determine the nominal value of electrolytic capacitors available from this manufacturer between 100 and 1000 pF. What will be the probable published tolerance?
4. The permeability ratings of a popular line of potentiometer cores follow the $R5$ series. By using successive fifth roots of 10, develop the nominal values between 1 and 100 mH. What will be the probable published tolerance?
5. To what R series do the following resistors belong: (*a*) 51 Ω ±5%, (*b*) 680 Ω ± 10%, (*c*) 1.1 kΩ ±5%, (*d*) 470 Ω ±20%?
6. Which, if any, of the resistors in Prob. 5 belong to more than one R series of resistances?

35–2 POWER RATIOS—THE DECIBEL

The Weber-Fechner law states that "the minimum change in stimulus necessary to produce a perceptible change in response is proportional to the stimulus already existing." With respect to our sense of hearing, this means that the ear considers as equal changes of sound intensity those changes which are in the same *ratio*.

The above is more easily understood from a consideration of sound intensities. Any volume of sound must be changed approximately 25% before the ear notes a change in volume. If the volume is increased by this amount, in order for the ear to detect another increase in volume, the new value must be increased by an additional 25%. For example, the output of an amplifier delivering 16 W would have to be increased to a new output of 20 W in order for the ear to discern the increase in volume. Then, in order for the ear to detect an additional increase in volume, the output would have to be increased 25% of 20 W to a new output of 25 W.

From the foregoing it is apparent that a *change* of volume, for example, from 10 to 20 mW (a 10-mW change), would seem the same as the *change* from 100 to 200 mW (a 100-mW change) because $\frac{20}{10} = \frac{200}{100}$. Since these changes in hearing response are equally spaced on a logarithmic scale, it follows that the ear responds logarithmically to variations in sound intensity. Therefore, any unit used for expressing power gains or losses in communication circuits must, in order to be practical, vary logarithmically.

One of the earliest of such units was the international transmission unit, the *bel* (B), so called to honor the inventor of the telephone, Alexander Graham Bell. The definition of the bel is

$$\text{bel} = \log_{10} \frac{P_2}{P_1}$$

where P_1 is the initial, or reference, power and P_2 is the final, or referred, power.

In normal practice, the number of bels is quite small and is invariably a decimal number; and so a derived unit, the *decibel,* is used as the practical indicator of power ratio. The abbreviation for decibel is dB. A difference of 1 dB between two sound intensities is just discernible to the ear. Since deci means one-tenth, a decibel is one-tenth the size of a bel, and

$$\text{Number of decibels} = \text{dB} = 10 \log \frac{P_2}{P_1} \qquad (1)$$

You should refer to the second paragraph of this section and prove that the difference between two discernible sound intensities is 0.969 dB.

EXAMPLE 1 A power of 10 mW is required to drive an AF amplifier. The output of the amplifier is 120 mW. What is the gain, expressed in decibels?

SOLUTION

$P_1 = 10$ mW, and $P_2 = 120$ mW. dB = ? Substituting in Eq. (1),

$$\text{dB} = 10 \log \frac{120}{10} = 10 \log 12$$
$$= 10.8 \text{ dB gain}$$

EXAMPLE 2 A network has a loss of 16 dB. What power ratio corresponds to this loss?

SOLUTION

Given $\qquad \text{dB} = 10 \log \frac{P_2}{P_1} \qquad (1)$

Substituting 16 for db, $\qquad 16 = 10 \log \frac{P_2}{P_1}$

D: 10, $\qquad 1.6 = \log \frac{P_2}{P_1}$

Taking antilogs of both members, $\qquad 39.8 = \frac{P_2}{P_1}$

Thus, a loss of 16 dB corresponds to a power ratio of 39.8:1.

Because dB is 10 times the log of the power ratio, it is evident that power ratios of 10 = 10 db, 100 = 20 dB, 1000 = 30 dB, etc. Therefore, it could have been determined by inspection that the 16-dB loss in the preceding example represented a power ratio somewhere between 10 and 100. This is evident by the figure 1 of 16 dB. The second digit 6 of 16 dB is ten times the logarithm of 3.98; hence, 16 dB represents a power ratio of 39.8.

A loss in decibels is customarily denoted by the minus sign. Thus, a loss of 16 dB is written −16 dB.

Expressing the gain or loss of various circuits or apparatus in decibels obviates the necessity of computing gains or losses by multiplication and division. Because the decibel is a logarithmic unit, the total gain of a circuit is found by adding the individual decibel gains and losses of the various circuit components.

EXAMPLE 3 A dynamic microphone with an output of −85 dB is connected to a preamplifier with a gain of 60 dB. The output of the preamplifier is connected through an attenuation pad with a loss of 10 dB to a final amplifier with a gain of 90 dB. What is the total gain?

SOLUTION

In this example, all decibel values have been taken from a common reference level. Because the microphone is 85 dB below reference level, the preamplifier brings the level up to $-85 + 60 = -25$ dB. The attenuation pad then reduces the level to $-25 - 10 = -35$ dB, and the final amplifier causes a net gain of $-35 + 90 = 55$ dB. Hence, it is apparent that the overall gain in any system is simply the algebraic sum of the decibel gains or losses of the associated circuit components. Thus, $-85 + 60 - 10 + 90 = 55$ dB gain.

35–3 POWER REFERENCE LEVELS

It is essential that you remember that the decibel is not an absolute quantity; it merely represents a change in power relative to the level at some different time or place. It is meaningless to say that a given amplifier has an output of so many dB unless that output is referred to a specific power level. If we know what the output power is, then the *ratio* of that output power to the specific input power may be expressed in dB.

Several reference levels (''zero-reference'' or ''zero-dB'') have been developed within the industry. Some of these have already been dropped generally; some are used in isolated communities or within individual companies; others are in general use throughout the entire electronics industry. Some of the more common levels are discussed below.

dBm The most common reference level used in the telephone industry is one milliwatt. And since many radio and television programs are carried between studio and transmitter by telephone systems, we should be able to understand telephone transmission engineers when they talk about relative powers. The rather widespread use of the expression ''decibels above or below one milliwatt'' is usually abbreviated ±dBm. Signal power in communications systems is almost always being amplified (multiplication) or attenuated (division). It is far more convenient to add or subtract dB than to calculate the power in milliwatts or watts by long processes of multiplication or division. Thus, when a telephone engineer speaks of a power level of 25 dBm, the listeners can readily understand that, if $P_1 = 1$ mW, P_2 is 25 dB higher.

EXAMPLE 4 What is the output power represented by a level of 25 dBm?

SOLUTION

dBm means ''decibels referred to a reference power level of 1 mW''; that is, $P_1 = 1$ mW. Then, an amplification of 25 dB means:

$$25 = 10 \log_{10} \frac{P_2}{1\text{ mW}}$$

$$\log P_2 = 2.5$$

$$P_2 = 316.23\text{ mW}$$

Because circuits do not amplify or attenuate all frequencies by the same amount, the industry often reserves the term dBm for an input signal of a single-frequency (pure) sine wave (often 400 Hz or 1 kHz). However, dBm is often applied to more complex waveforms because of the convenience of calculations.

6 mW Several radio receiver and audio amplifier manufacturers use 0.006 W (6 mW) as their reference, or zero-dB, level.

EXAMPLE 5 How much power is represented by a gain of 23 dB if zero level is 6 mW?

SOLUTION 1

Substituting 23 for dB and 6 for P_1 in Eq. (1),

$$23 = 10 \log \frac{P_2}{6}$$

D: 10,

$$2.3 = \log \frac{P_2}{6}$$

Taking antilogs of both members,

$$199.5 = \frac{P_2}{6}$$

$$\therefore P_2 = 1197 \text{ mW}$$

CHECK

$$23 = 10 \log \frac{1197}{6}$$

$$23 = 10 \log 199.5$$

$$23 = 10 \times 2.3$$

SOLUTION 2

$$2.3 = \log \frac{P_2}{6}$$

or

$$2.3 = \log P_2 - \log 6$$

Transposing,

$$\log P_2 = 2.3 + \log 6$$

Substituting the value of log 6,

$$\log P_2 = 2.3 + 0.778$$

$$\log P_2 = 3.078$$

Taking antilogs,

$$P_2 = 1197 \text{ mW}$$

EXAMPLE 6 How much power is represented by −64 dB if zero level is 6 mW?

SOLUTION 1

Substituting −64 for dB and 6 for P_1 in Eq. (1),

$$-64 = 10 \log \frac{P_2}{6}$$

D: 10,

$$-6.4 = \log \frac{P_2}{6}$$

The left member of the above equation is a logarithm with a negative mantissa because the entire number 6.4 is negative. Hence, to express this logarithm with a positive mantissa, the equation is written

$$3.6 - 10 = \log \frac{P_2}{6}$$

Taking antilogs of both members,

$$3.98 \times 10^{-7} = \frac{P_2}{6}$$

$$\therefore P_2 = 2.39 \times 10^{-6} \text{ mW}$$

CHECK

$$\begin{aligned} -64 &= 10 \log \frac{2.39 \times 10^{-6}}{6} \\ &= 10 \log 3.98 \times 10^{-7} \\ -64 &= 10(3.6 - 10) \\ -64 &= -64 \end{aligned}$$

SOLUTION 2

$$-6.4 = \log \frac{P_2}{6}$$

Then $\quad -6.4 = \log P_2 - \log 6$

Transposing, $\quad \log P_2 = \log 6 - 6.4$

Substituting the value of log 6, $\quad \log P_2 = 0.778 - 6.4$

$$= (10.78 - 10) - 6.4$$

$$\therefore P_2 = 2.39 - 10^{-6} \text{ mW}$$

If the larger power is always placed in the numerator of the power ratio, the quotient will always be greater than 1; therefore, the characteristic of the logarithm of the ratio will always be zero or a positive value. In this manner the use of a negative characteristic is avoided. As an illustration, from Example 6,

$$-6.4 = \log \frac{P_2}{6}$$

which is the same as $\quad 6.4 = \log 6 - \log P_2$

Hence, $\quad 6.4 = \log \frac{6}{P_2}$

It is always apparent whether there is a gain or a loss in decibels; therefore, the proper sign can be affixed after working the problem.

VU The volume unit, abbreviated VU, is used in broadcasting, and it is based on the amplitude of the program frequencies throughout the system. The standard volume indicator (VU meter) is calibrated in decibels with zero level corresponding to 1 mW of power in a 600-Ω line under steady-state conditions, usually at a frequency between 35 Hz and 10 kHz. Owing to the ballistic characteristics of the instrument, the scale markings are referred to as volume units and correspond to dBm only in the case of steady-state sine-wave signals.

dBRN AND dBA The signal-to-noise ratio is very important in most electronic amplifiers and communications circuits. When engineers establish a reference noise level, then the signal power may be expressed as being so many dB above this arbitrary reference level. The expression "decibels referred to an arbitrary reference noise level" is abbreviated dBRN. Often this reference noise level is set at -90 dBm. You should confirm that this represents 1 pW of power.

Then, when an original established reference noise level is adjusted to some new level, as it sometimes is in the telephone industry, the abbreviation dBA indicates "decibels referred to some adjusted reference noise level."

dBRAP A sound may be heard by "the average human ear" (whatever that is) if it has a power of 10^{-16} W or more. This minimum power represents the threshold of hearing, and it is called reference acoustical power. Any noise or signal of any kind must be above the minimum power to be heard, and it may then be compared with the minimum power. Thus, dBRAP means a power ratio in dB when $P_1 = 10^{-16}$ W. Sound engineers often call the number of dBRAP by the name *phons*.

OTHER SPECIALIZED TERMS Other reference levels used in more specialized fields are:

- dBW dB referred to 1 W as zero-dB reference level.
- dBk dB referred to 1 kW as reference level.
- dBV dB referred to 1 V as zero reference signal level.

These, and many other zero reference levels, need introduce no great problem to you. It is only necessary to remember that dB represents a power ratio which must be referred to some original or arbitrary reference level.

35–4 CURRENT AND VOLTAGE RATIOS

Fundamentally, the decibel is a measure of the ratio of two powers. However, voltage ratios and current ratios can be utilized for computing the decibel gain or loss provided that the input and output impedances are taken into account.

In the following derivations, P_1 and P_2 will represent the power input and power output, respectively, and R_1 and R_2 will represent the input and output impedances, respectively. Then

$$P_1 = \frac{V_1^2}{R_1} \qquad \text{and} \qquad P_2 = \frac{V_2^2}{R_2}$$

Since

$$\text{dB} = 10 \log \frac{P_2}{P_1}$$

substituting for P_1 and P_2,

$$\text{dB} = 10 \log \frac{\dfrac{V_2^2}{R_2}}{\dfrac{V_1^2}{R_1}}$$

$$\therefore \text{dB} = 10 \log \frac{V_2^2 R_1}{V_1^2 R_2}$$

$$= 10 \log \left(\frac{V_2}{V_1}\right)^2 \frac{R_1}{R_2}$$

$$= 10 \log \left(\frac{V_2}{V_1}\right)^2 + 10 \log \frac{R_1}{R_2}$$

$$= 20 \log \frac{V_2}{V_1} + 10 \log \frac{R_1}{R_2} \tag{2}$$

$$= 20 \log \frac{V_2\sqrt{R_1}}{V_1\sqrt{R_2}} \tag{3}$$

Similarly, $P_1 = I_1^2 R_1$ and $P_2 = I_2^2 R_2$

Then, since $\text{dB} = 10 \log \dfrac{P_2}{P_1}$

by substituting for P_1 and P_2,

$$\text{dB} = 10 \log \frac{I_2^2 R_2}{I_1^2 R_1}$$

$$= 20 \log \frac{I_2}{I_1} + 10 \log \frac{R_2}{R_1} \tag{4}$$

$$= 20 \log \frac{I_2\sqrt{R_2}}{I_1\sqrt{R_1}} \tag{5}$$

If, in both the above cases, the impedances R_1 and R_2 are *equal,* they will cancel and the following formulas will result:

$$\text{Number of dB} = 20 \log \frac{V_2}{V_1} \tag{6}$$

and

$$\text{Number of dB} = 20 \log \frac{I_2}{I_1} \tag{7}$$

It is evident that voltage or current ratios can be translated into decibels *only* when the impedances across which the voltages exist or into which the currents flow are taken into account.

EXAMPLE 7 An amplifier has an input resistance of 200 Ω and an output resistance of 6400 Ω. When 0.5 V is applied across the input, a voltage of 400 V appears across the output. (*a*) What is the power output of the amplifier? (*b*) What is the gain in decibels?

SOLUTION

(*a*)

$$\text{Power output} = P_o = \frac{V_o^2}{R_o} = \frac{400^2}{6400} = 25 \text{ W}$$

(*b*)

$$\text{Power input} = P_i = \frac{V_i^2}{R_i} = \frac{0.5^2}{200} = 1.25 \times 10^{-3} \text{ W}$$

$$\text{Power gain} = 10 \log \frac{P_o}{P_i} = 10 \log \frac{25}{1.25 \times 10^{-3}} = 43 \text{ dB}$$

Check the solution by substituting the values of the voltages and resistances in Eq. (3).

$$\text{dB} = 20 \log \frac{V_o}{V_i}\sqrt{\frac{R_i}{R_o}} = 20 \log \frac{400}{0.5}\sqrt{\frac{200}{6400}} = 43$$

35–5 THE MERIT, OR GAIN, OF AN ANTENNA

The merit of an antenna, especially one designed for directive transmission or reception, is usually expressed in terms of antenna *gain*. The gain is generally taken as the ratio of the power that must be supplied some standard-comparison antenna to the power that must be supplied the antenna under test in order to produce the same field strengths in the desired direction at the receiving antenna. Similarly, the gain of one antenna over another could be taken as the ratio of the respective radiated fields.

The "effective radiated power" of an antenna is the product of the antenna power and the antenna power gain.

EXAMPLE 8 One kilowatt is supplied to a rhombic antenna, which results in a field strength of 20 μV/m at the receiving station. In order to produce the same field strength at the receiving station, a half-wave antenna, properly oriented and located near the rhombic, must be supplied with 16.6 kW. What is the gain of the rhombic?

SOLUTION

Because the same antenna is used for reception, both transmitting antennas deliver the same power to the receiver. Hence,

$$\text{dB} = 10 \log \frac{P_2}{P_1} = 10 \log \frac{16.6}{1} = 12.2$$

PROBLEMS 35–2

1. How many decibels correspond to a power ratio of (*a*) 15, (*b*) 25, (*c*) 40, (*d*) $\frac{1}{140}$?
2. Referred to equal impedances, how many decibels correspond to a voltage ratio of (*a*) 32, (*b*) 90, (*c*) $\frac{1}{120}$, (*d*) $\frac{1}{150}$?
3. If 0 dB is taken as 6 mW, how much voltage across a 75-Ω load does it represent?
4. If 0 dB is taken as 6 mW, how much voltage across a 600-Ω load does it represent?
5. What is the voltage across a 300-Ω line at zero dBm?
6. What is the voltage across a 600-Ω line at 12 dBm?
7. If a reference level is taken as 10 mW, how much voltage across a 300-Ω load does it represent?
8. If a reference level is taken as 12.5 mW, (*a*) how much voltage across a 73-Ω load does it represent? (*b*) How much current flows through the load?
9. If 0 dB is 6 mW, compute the output power in the appropriate units and determine the voltage across the 300-Ω load for the following output power meter readings: (*a*) 3 dB, (*b*) 10 dB, (*c*) −10 dB, (*d*) −30 dB, (*e*) −80 dB.
10. If 0 dB is 1 mW, compute the power in milliwatts and the voltage across a 600-Ω load for the following output meter readings: (*a*) 5 dB, (*b*) 10 dB, (*c*) 20 dB, (*d*) −10 dB.
11. An amplifier is rated as having a 75-dB gain. What power ratio does this represent?
12. The amplifier of Prob. 11 has equal input and output impedances. What is the ratio of the output current to the input current?

13. An amplifier has a gain of 80 dB. If the input power is 1 mW, what is the output power?
14. If a high-selectivity tuned circuit has a very high Q, spurious signals which are 10% lower or higher in frequency will be attenuated at least 50 dB. What power ratio is represented by this level?
15. The manufacturer of a high-fidelity 100-W power amplifier claims that hum and noise in the amplifier is 90 dB below full power output. How much hum and noise power does this represent?
16. In the amplifier of Prob. 15, what will be the dB level of noise to signal when the amplifier is producing 3 W of output power?
17. A network has a loss of 80 dB. What power ratio corresponds to this loss?
18. If the network in Prob. 17 has equal input and output impedances, what is the ratio of the output voltage to the input voltage?
19. In single-sideband operation, the signals appearing in the unwanted set of sidebands should be attenuated by at least 30 dB. What is the ratio of output powers of the desired signal to the unwanted signal?
20. The noise level of a certain telephone line used for wired music programs is 60 dB down from the program level of 12.5 mW. How much noise power is represented by this level?
21. A certain crystal microphone is rated at −80 dB. There is on hand a final AF amplifier rated at 60 dB. How much gain must be provided by a preamplifier in order to drive the final amplifier to full output if an attenuator pad between the microphone and preamplifier has a loss of 20 dB? (All dB ratings are taken from the same reference.)
22. The output of a 200-Ω dynamic microphone is rated at −81.5 dB from a reference level of 6 mW. This microphone is to be used with an amplifier which is to have a power output of 25 W. What gain must be provided between the microphone and the amplifier output?
23. If the amplifier of Prob. 22 has an output impedance of 2.7 kΩ, what is the overall voltage ratio from microphone output to amplifier output?
24. What is the equivalent power amplification in the amplifier discussed in Prob. 23?
25. It is desired to use the amplifier of Prob. 22 with a phonograph pickup which is rated at −20 dBm. To keep from overloading the amplifier, how much loss must be introduced between pickup and input?
26. An amplifier has a normal output of 30 W. A selector switch is arranged to reduce the output in 5-dB steps. What power outputs correspond to reductions in output of 5, 10, 15, 20, 25, and 30 dB?
27. A two-stage television RF amplifier has a 300-μV input signal into 75 Ω. The second stage has a gain of 50 dB. When matched input-output impedances are used, the voltage output of the second stage must be 4.22 V to allow distribution of the signal.
    Determine:
    (*a*) The input voltage of the second stage.
    (*b*) The dB gain of the first stage.
    (*c*) The overall gain of the two amplifiers when all impedances are 75 Ω.

28. A type 2N45 transistor has the following ratings when used as a class A power amplifier:

■ Collector voltage V,	−20
■ Emitter current, mA	5
■ Input impedance, Ω	10
■ Source impedance, Ω	50
■ Load impedance, Ω	4500
■ Power output, mW	45
■ Power gain, dB	23

What is the power input?

29. An amplifier has an input impedance of 600 Ω and an output impedance of 6000 Ω. The power output is 30 W when 1.9 V is applied across the input.
(*a*) What is the voltage gain of the amplifier?
(*b*) What is the power gain in decibels?
(*c*) What is the power input?

30. A television tuner amplifier has an input impedance of 300 Ω and an output impedance of 3500 Ω; when a 300-mV signal is applied at the input, a 250-V signal appears at the output.
(*a*) What is the power output of the amplifier?
(*b*) What is the power gain in decibels?
(*c*) What is the voltage gain of the amplifier?

31. A dynamic microphone with an output level of −72 dB is connected to a speech amplifier consisting of three voltage amplifier stages. The first voltage amplifier stage has a voltage gain of 100, and the second has a voltage gain of 9. The interstage transformer between the second and third voltage amplifier stages has a step-up ratio of 3:1, and the third stage has a voltage gain of 8. The driver stage and modulator have a gain of 23 dB. If zero power level is 6 mW, what is the output power of the modulator?

▶ **(HINT** Transformers do not introduce *power changes*.)

Microphone −72 dB	▶	Amplifier 100	▶	Amplifier 9	▶	Transformer 3	▶	Amplifier 8	▶	Modulator 23 dB

32. How many decibels gain is necessary to produce a 60-μW signal in 600-Ω telephones if the received signal supplies 9 μV to the 80-Ω line that feeds the receiver?

33. In the receiver of Prob. 32, if the overall gain is increased to 96 dB, what received signal will produce the 60-μW signal in the telephones?

34. The voltage across the 600-Ω telephones is adjusted to 1.73 V. When the AF filter is cut in, the voltage is reduced to 1.44 V. What is the "insertion loss" of the filter?

35. The input power to a 120-km line is 10 mW, and 40 μW is delivered at the end of the line. What is the attenuation in decibels per kilometer?

36. It is desired to raise the power level at the end of the line discussed in Prob. 35 to that of the original output. What is the voltage gain of the required amplifier?
37. In Prob. 35, what is the ratio of input power to output power?
38. One of the original attenuation units was the *neper,* which is given by

$$\text{Number of nepers} = \log_e \frac{I_1}{I_2}$$

Since

$$\text{Number of dB} = 20 \log_{10} \frac{I_1}{I_2}$$

what is the relation between nepers and decibels for equal impedances?

▶ **HINT** $\log_e \frac{I_1}{I_2} = 2.30 \log_{10} \frac{I_1}{I_2}$

39. A television transmitting antenna has a power gain of 8.6 dB. If the power input to the antenna is 15 kW, what is the effective radiated power?
40. When 500 W is supplied to a directive antenna, the result is a field strength of 5 μV/m at a receiving station. In order to produce the same field strength at the same receiving station, the standard-comparison antenna must be supplied with 8 kW. What is the decibel gain of the directive antenna?
41. A rhombic transmitting antenna produces a field strength of 98 μV/m at a receiving test station. The standard-comparison antenna delivers a field strength of 5 μV/m. What is the decibel gain of the rhombic antenna?
42. A broadcasting station is rated at 1 kW. If the received signals vary as the square root of the radiated power, how much gain in decibels would be apparent to a nearby listener if the broadcasting station doubled its power?

35–6 TRANSMISSION LINES

A transmission line is a device that consists of one or more electric conductors and is designed for the purpose of transferring electric energy from one point to another. It has a wide variety of uses: in one form it can carry electric power to a city several kilometers from the power plant; in another form it can be used for carrying network programs from one studio to several broadcast stations; and in still another form it can carry RF energy from a radio transmitter to an antenna or from an antenna to a radio receiver.

The most common types of transmission lines are:

1. The two-wire open-air line as shown in Fig. 35–1*a*. This line consists of two parallel conductors whose spacing is carefully held constant.
2. The concentric-conductor line, as illustrated in Fig. 35–1*b*, which consists of tubular conductors one inside the other.
3. The four-wire open-air line as shown in Fig. 35–1*c*. In this type of line the diagonally opposite wires are connected to each other for effecting an electrical balance.
4. The twisted-pair line, as shown in Fig. 35–1*d*, which may consist of lamp cord, a telephone line, or other insulated conductors.

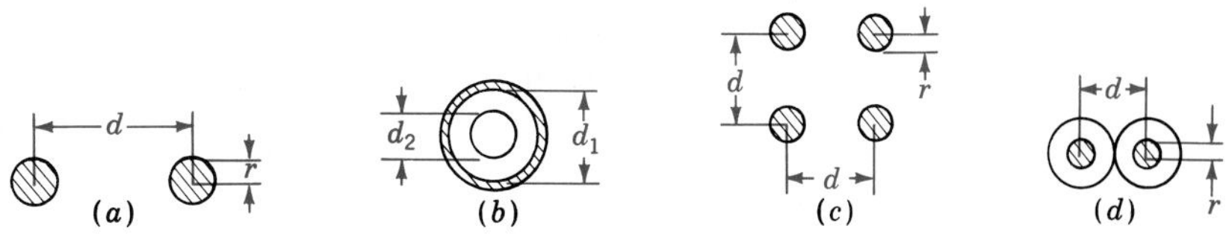

FIG. 35–1 Types of transmission lines.

Any conductor has a definite amount of self-inductance, capacitance, and resistance per unit length. These properties account for the behavior of transmission lines in their various forms and uses.

The derivations of the transmission line equations that follow can be found in advanced engineering texts.

35–7 THE INDUCTANCE OF A LINE

The inductance of a two-wire open-air line is given by the equation

$$L = 0.621\, l\left(0.161 + 1.48 \log_{10} \frac{d}{r}\right) \qquad \text{mH} \qquad (8)$$

where L = inductance of line and return, mH
l = length of line, km
d = distance between conductor centers
r = radius of each wire, same units as d

EXAMPLE 9 What is the inductance of a line 145 km long consisting of No. 0000 copper wire spaced 1.5 m apart?

SOLUTION

Diameter of No. 0000 copper wire = 11.68 mm; thus radius = 5.84 mm.

$$\frac{d}{r} = \frac{1500}{5.84} = 256.85$$

$$\log_{10} 256.85 = 2.409\,68\text{, say } 2.41$$

Then

$$L = 0.621 \times 145(0.161 + 1.48 \times 2.41) \text{ mH}$$
$$= 336 \text{ mH}$$

For radio frequencies, more accurate results are obtained by the approximate relation

$$L \cong 9.21 \times 10^{-9} \log_{10} \frac{d}{r} \qquad \text{H/cm} \qquad (9)$$

where L is the inductance in henrys per centimeter and d and r have the same values as in Eq. (8).

35–8 THE CAPACITANCE OF A LINE

The capacitance of a two-wire open-air line is

$$C = \frac{0.0121\, l}{\log \frac{d}{r}} \qquad \mu\text{F} \qquad (10)$$

where C = capacitance of line, μF
l = length of line, km
d = distance between wire centers
r = radius of wire, in same units as d

EXAMPLE 10 What is the capacitance per kilometer of a line consisting of No. 00 copper wire spaced 1.2 m apart?

SOLUTION

Diameter of No. 00 copper wire = 9.266 mm; thus radius = 4.633 mm.

$$\frac{d}{r} = \frac{1200}{4.633} = 259$$

$$\log_{10} 259 = 2.413\ 30 \qquad \text{say } 2.413$$

Then

$$C = \frac{0.0121}{2.413}\ \mu\text{F/km}$$
$$= 5.01\ \text{nF/km}$$

For radio frequencies, more accurate results are obtained by the equation

$$C \cong \frac{1}{9.21 \times 10^{-9} c^2 \log_{10} \frac{d}{r}} \qquad \text{F/cm} \qquad (11)$$

where C is the capacitance in farads per centimeter, c is the velocity of light (3×10^{10} cm/s), and d and r have the same values as in Eq. (10).

The capacitance of submarine cables and of cables laid in metal sheaths is given by

$$C = \frac{0.0241\, Kl}{\log_{10} \frac{d_1}{d_2}} \qquad \mu\text{F} \qquad (12)$$

where C = capacitance of line, μF
K = relative dielectric constant of insulation
l = length of line, km
d_1 = inside diameter of outer conductor
d_2 = outside diameter of inner conductor

EXAMPLE 11 A No. 14 copper wire is lead-sheathed. The wire is insulated with 3 mm gutta percha ($K = 4.1$). What is the capacitance of 1 km of this cable?

SOLUTION

$$d_2 = \text{diameter of No. 14} = 1.63 \text{ mm}$$

$$d_1 = 1.63 + 2(3) = 7.63 \text{ mm}$$

$$\log \frac{d_1}{d_2} = \log \frac{7.63}{1.63} = \log 4.687 = 0.6703$$

$$l = 1 \text{ km}$$

Then

$$C = \frac{0.0241Kl}{\log \dfrac{d_1}{d_2}} = \frac{0.0241 \times 4.1 \times 1}{0.6703} = 0.147 \ \mu\text{F}$$

PROBLEMS 35–3

1. What is the inductance of a 140-km line consisting of two No. 0 wires spaced 1 m between centers?
2. What is the inductance of a 30-km line consisting of two No. 6 copper wires spaced 60 cm between centers?
3. A 5-km transmission line consists of two No. 2 solid copper wires spaced 40 cm between centers. Determine (*a*) the inductance and (*b*) the capacitance of the line.
4. If the spacing of the line of Prob. 3 were 1 m between centers, what would be (*a*) the inductance and (*b*) the capacitance?
5. A 40-km-long two-wire line is to be constructed of No. 0 solid copper wire. What must be the minimum spacing between centers to keep the capacitance below 0.250 μF?
6. A 22-km two-wire line consisting of No. 00 solid copper wire is spaced 180 mm between wire centers. What is the capacitance of the line in microfarads per kilometer?
7. A lead-sheathed underground cable is to be constructed with solid copper wire covered with 12.5 mm of rubber insulation ($K = 4.3$). If then the maximum capacitance per kilometer must be limited to 0.15 μF, ±10%, what size conductor should be used?

8. A lead-sheathed cable which consists of No. 0 copper wire with 12.5 mm of rubber insulation ($K = 4.3$) is broken. A capacitance bridge measures 0.26 μF between the conductor and the sheath. How far out is the open circuit?
9. What is the capacitance per kilometer of the cable of Prob. 8?
10. The cable of Prob. 7 becomes open-circuited 5 km out. What reading will be given on a capacitance bridge?
11. The value of the current in a line at a point l km from the source of power is given by

$$i = I_0 e^{-\kappa l}$$

where I_0 is the current at the source and κ is the attenuation constant. In a certain line, with $\kappa = 0.02$ dB/km, find the length of line where i is 10% of the original current I_0.
12. If the attenuation of a line is 0.012 dB/km, how far out from the power source will the current have decreased to 70.7% of its original value?
13. A two-wire open-air transmission line is used to couple a receiving antenna to the receiver. The line is 155 m long, and it consists of No. 10 wire spaced 15 cm between centers. Using Eqs. (9) and (11), find:
(*a*) Inductance per centimeter of line
(*b*) Capacitance per centimeter of line
(*c*) Inductance of the entire line
(*d* Capacitance of the entire line
14. A two-wire open-air transmission line is used to couple a radio transmitter to an antenna. The line is 250 m long, and it consists of No. 14 wire spaced 14 cm between centers. Using Eqs. (9) and (11), find the (*a*) inductance of the line and (*b*) capacitance of the line.

35–9 CHARACTERISTIC IMPEDANCES OF RF TRANSMISSION LINES

The most important characteristic of a transmission line is the *characteristic impedance,* denoted by Z_0 and expressed in ohms. This impedance is often called *surge impedance, surge resistance,* or *iterative impedance.*

The value of the characteristic impedance is determined by the construction of the line, that is, by the size of the conductors and their spacing. At radio frequencies, the characteristic impedance can be considered to be a resistance the value of which is given by

$$Z_0 = \sqrt{\frac{L}{C}} \quad \Omega \tag{13}$$

where L and C are the inductance and capacitance, respectively, per unit length of line as given in Eqs. (9) and (11). The unit of length selected for L and C is immaterial as long as the *same* unit is used for both.

Substituting the values of L and C for a two-wire open-air transmission line in Eq. (13) results in

$$Z_0 = 276 \log_{10} \frac{d}{r} \quad \Omega \tag{14}$$

where d is the spacing between wire centers and r is the radius of the conductors *in the same units as d*. Note that the characteristic impedance is *not* a function of the length of the line.

Equation (14) is valid when you can conveniently neglect the capacitance of each line to ground. If you cannot, replace $\frac{d}{r}$ with $\frac{d}{D}$, where D is the diameter of the conductors.

EXAMPLE 12 A transmission line is made of No. 10 wire spaced 30 cm between centers. What is the characteristic impedance of the line?

SOLUTION

$d = 30$ cm $= 300$ mm. Diameter of No. 10 wire $= 2.588$ mm; therefore $r = 1.294$ mm.

$$\begin{aligned} Z &= 276 \log \frac{d}{r} = 276 \log \frac{300}{1.294} \\ &= 276 \log 231.8 = 276 \times 2.365 \\ &= 653\ \Omega \end{aligned}$$

The characteristic impedance of a concentric line is given by

$$Z_0 = 138 \log_{10} \frac{d_1}{d_2} \qquad \Omega \tag{15}$$

where d_1 is the inside diameter of the outer conductor and d_2 is the outside diameter of the inner conductor.

EXAMPLE 13 The outer conductor of a concentric transmission line consists of copper tubing 1.6 mm thick with an outside diameter of 25 mm. The copper tubing which forms the inner conductor is 0.8 mm thick with an outside diameter of 6 mm. What is the characteristic impedance of the line?

SOLUTION

$$\begin{aligned} d_1 &= 25 - (2 \times 1.6) = 21.8 \text{ mm} \\ d_2 &= 6 \text{ mm} \\ Z_0 &= 138 \log \frac{d_1}{d_2} = 138 \log \frac{21.8}{6} \\ &= 138 \log 3.633 \\ &= 138 \times 0.5603 = 77.3\ \Omega \end{aligned}$$

PROBLEMS 35–4

1. What is the characteristic impedance of a two-wire open-air transmission line consisting of No. 12 wire spaced 120 mm between centers?
2. It is desired to use No. 16 wire to provide a transmission line with a characteristic impedance of approximately 500 Ω. What logical spacing between centers should be used?
3. If a 40-mm spacing is used for the line of Prob. 2, what percentage of error is introduced by assuming that the line does have a characteristic impedance of 500 Ω?

4. It is necessary to construct a 600-Ω transmission line to couple a radio transmitter to its antenna, and No. 10 wire is readily available. What should be the spacing between wire centers?
5. The impedance at the center of a half-wave antenna is approximately 74 Ω. For maximum power transfer between transmission line and antenna, the impedance of the line must match that of the antenna. Is it physically possible to construct an *open-wire* line with a characteristic impedance as low as 74 Ω?
6. Plot a graph of the characteristic impedance in ohms against the ratio $\frac{d}{r}$ for two-wire open-air transmission lines. Use values of $\frac{d}{r}$ between 1 and 150.
7. It is desired to construct a 300-Ω two-wire line at a certain television station. In the stock room there is a large number of 37.5-mm spreader insulators. That is, these spreaders will space the *wires* 37.5 mm. What size wire should be used to obtain as nearly as possible the desired impedance if the 37.5-mm spreaders are used?

▶ **HINT** $d = 30 + 2r$.

8. What outside-diameter tubing should be used to construct a quarter-wave matching stub having an impedance of approximately 300 Ω if spreaders 40 mm long are used?
9. The outer conductor of a concentric transmission line is a copper pipe 3.5 mm thick with an outside diameter of 60 mm. The inner conductor is a copper rod 5 mm in diameter. What is the characteristic impedance of the line?
10. The inside diameter of the outer conductor of a coaxial line is 12 mm. The surge impedance is 75 Ω. What is the outside diameter of the inner conductor?
11. Plot a graph of the characteristic impedance in ohms against the ratio $\frac{d_1}{d_2}$ for concentric transmission lines. Use values of $\frac{d_1}{d_2}$ between 2 and 10.
12. A particular grade of twisted-pair transmission line, which has a surge impedance of 72 Ω, has a loss of 0.2 dB/m. For a 30-m length of line, determine (*a*) the total loss in decibels and (*b*) the efficiency of transmission.

▶ **HINT** $\% \text{ efficiency} = \dfrac{\text{power output}}{\text{power input}} \times 100$

13. The twisted-pair line of Prob. 12 is replaced by a coaxial cable that has a loss of 0.01 dB/m. What is the new efficiency of transmission?
14. For a two-wire transmission line, the attenuation in decibels *per meter of wire* is given by the equation

$$\alpha = \frac{0.0157\, R_{ac}}{\log_{10} \dfrac{d}{r}} \qquad \text{dB/m} \qquad (16)$$

where R_{ac} is the ac resistance of one meter of *wire*. One kilowatt of power, at a frequency of 16 MHz, is delivered to a 460-m line consisting of No. 8 wire spaced 30 cm between centers. The RF resistance of No. 8 wire is 49 times the dc resistance. (*a*) What is the line loss in decibels and (*b*) what is the efficiency of transmission?

15. If the spacing of the line in Prob. 14 should be changed to 20 cm between centers, what will be (*a*) the line loss in decibels and (*b*) the efficiency of transmission?

16. For a concentric transmission line, the attenuation in decibels per meter of *line* is expressed by the relation

$$\alpha = \frac{15.09\sqrt{f}\,(d_1 + d_2)10^{-6}}{d_1 d_2 \log_{10}\dfrac{d_1}{d_2}} \qquad \text{dB/m} \qquad (17)$$

where d_1 and d_2 are in centimeters and have the same meaning as in Eq. (15), and f is the frequency in megahertz. A concentric line 400 m long consists of an outer conductor with an inside diameter of 3.2 cm and an inner conductor that is 0.8 cm in diameter. At a frequency of 27.8 MHz, what is (*a*) the line loss in dB and (*b*) the efficiency of transmission?

17. The capacitance of a vertical antenna which is shorter than one-quarter wavelength at its operating frequency can be computed by the equation

$$C_a = \frac{55.77l}{\left(\log_e \dfrac{200l}{d} - 1\right)\left[1 - \left(\dfrac{fl}{75}\right)^2\right]} \qquad \text{pF} \qquad (18)$$

where C_a = capacitance of antenna, pF
l = height of antenna, m
d = diameter of antenna conductor, cm
f = operating frequency, MHz

Determine the capacitance of a vertical antenna that is 85 m high and consists of 1.5-cm wire. The antenna is being operated at a frequency of 214 kHz.

18. The RF resistance of a copper concentric transmission line can be computed by

$$R_{ac} = 8.33\sqrt{f}\left(\frac{1}{d_1} + \frac{1}{d_2}\right) \times 10^{-3} \qquad \Omega/\text{m} \qquad (19)$$

where f = frequency, MHz
d_1 = inside diameter of outer conductor, cm
d_2 = outside diameter of inner conductor, cm

What is the resistance of a concentric line 80 m long operating at 130 MHz if $d_1 = 4.8$ cm and $d_2 = 0.38$ cm?

19. If an antenna is matched to a coaxial transmission line, the percent efficiency is given by

$$\eta = \frac{100R_T}{Z_0 + R_T} \quad \% \qquad (20)$$

where Z_0 = characteristic impedance of the concentric line
R_T = effective resistance of the line due to attenuation, obtainable from the line constants

$$R_T = Z_0\left(e^{\frac{rl}{Z_0}} - 1\right) \quad \Omega \qquad (21)$$

where r = RF resistance per meter of line as found in Eq. (19)
l = length of line, m

Find the efficiency of transmission of a matched concentric transmission line with a characteristic impedance of 300 Ω. The line is 24 m long, and it has an RF resistance of 0.72 Ω/m.

20. What is the efficiency of transmission of a matched concentric transmission line with a characteristic impedance of 90 Ω if the line is 335 m long and has an RF resistance of 0.33 Ω/m?

35–10 GRAPHS INVOLVING LOGARITHMIC VALUES

In your further studies in mathematics for electronics, you will spend considerable time investigating graphic displays involving logarithmic values. In Chap. 16 we used only graph paper with equally spaced lines, so the blank sheets consisted of a multitude of little squares. However, in some cases we mathematically changed the *values* of the equal-sized steps by changing the scales. For instance, we plotted milliamperes against volts, not against millivolts. We did not squeeze our graph into a tiny corner of the sheet in order to plot very small values in amperes; we "stretched" the scale of current by a factor of 1000 to best use the graph sheet.

Sometimes it happens that one set of values to be graphed covers a very large range. For instance, it is quite common to hear of amplifiers capable of handling frequencies between 20 and 20 000 Hz without appreciable distortion. (Some manufacturers will actually specify the capabilities of their audio equipment to 200 000 Hz!) Normally, there is very little, if any, change over large ranges of frequency, and the changes with which we are normally concerned occur at the very low and very high ends of the graph. So rather than plot values on equally spaced graph paper, we compress the available spaces by plotting the frequency logarithmically. Figure 35–2 illustrates a typical frequency response curve for an audio amplifier. From this figure you can see that the "roll-off" for the amplifier starts at about 5 kHz and continues to roll off at approximately 3 dB per octave.

You will encounter various other types of semilog and logarithmic graph papers as you continue your studies. Do not downgrade the value of logarithms. You will find logarithms to be useful in almost any field of electronics application.

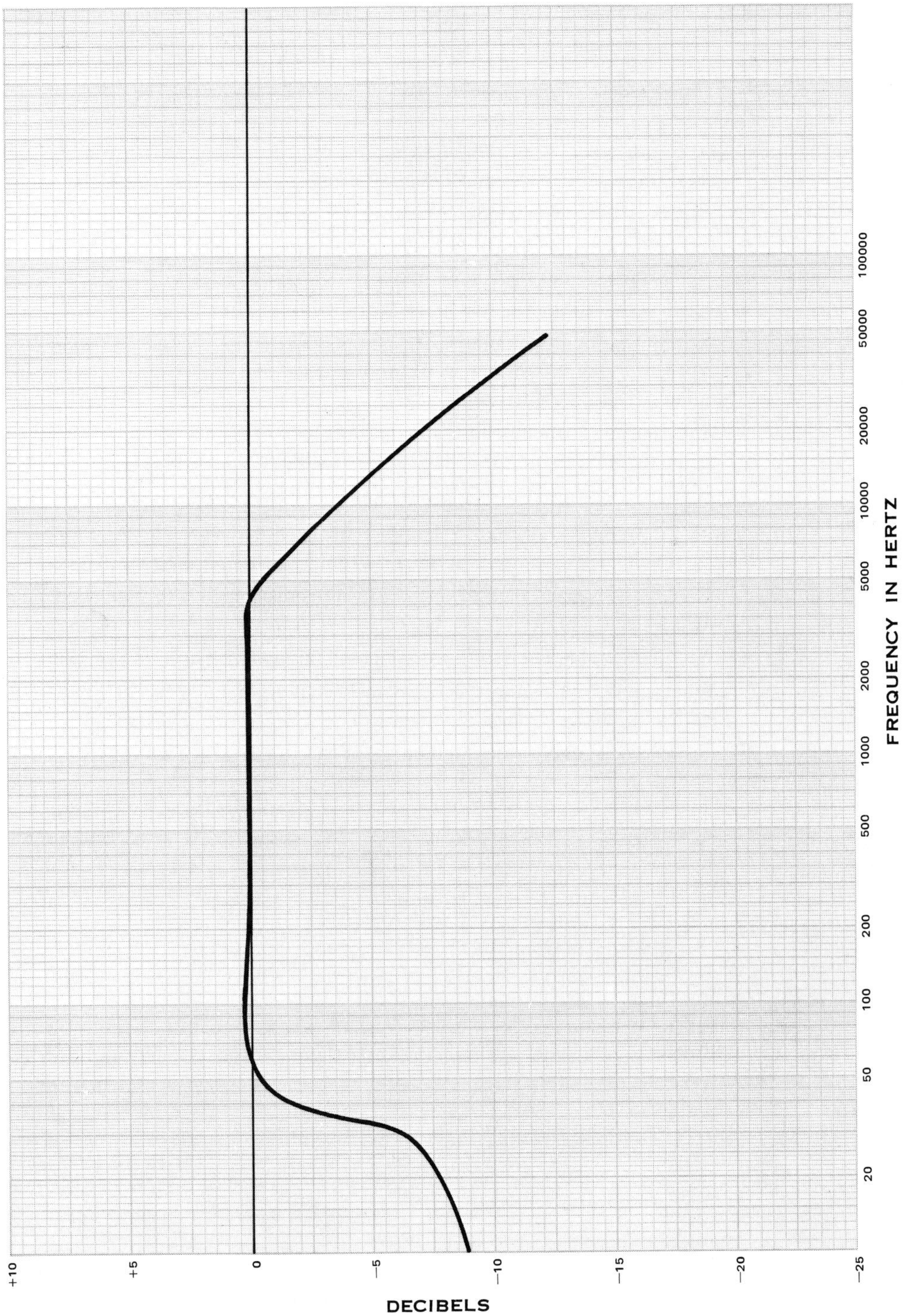

FIG. 35–2 Typical record-playback response curve. Frequency plotted logarithmically; decibels (a logarithmic value) plotted linearly. Commercially published graphs may show horizontal scale (and vertical lines) of 1, 2, 5 values only. (*Courtesy of Keuffel and Esser Company, Morristown, NJ.*)

SELF TEST

1. Determine to three significant figures the successive fourth roots of 10 between 10 and 100, with their actual tolerances.
2. Express in decibels a power ratio of 48:1.
3. If 0 dB is taken as 3 mW, what output power, to two significant figures, is represented as 65 dB?
4. What is the voltage across a 93-Ω transmission line at 10 dBm?
5. An amplifier has an input impedance of 1 kΩ and an output impedance of 24 Ω. If the input voltage is 100 mV and the output voltage is 70 V, what is the dB rating of the amp?
6. A transmitting antenna has a rated power gain of 12 dB. If the power input to the antenna is 50 kW, what is its effective radiated power?
7. If $C = \dfrac{0.0121\, l}{\log \dfrac{d}{r}}$ μF where l = length of line, km,
   d = distance between wire centers
   r = radius of wire, same units as d,

   what is the capacitance of an 800-m transmission line of No. 10 copper wire spaced 30 cm apart?
8. What is the characteristic impedance of the transmission line of Prob. 7?
9. What should be the distance between centers of the wire of Prob. 7 to achieve a characteristic impedance of 300 Ω?
10. For a coaxial cable, the attenuation of the line is given by

$$\alpha = \frac{15.09\sqrt{f}(d_1 + d_2)\, l \times 10^{-6}}{d_1 d_2 \log_{10} \dfrac{d_1}{d_2}} \text{ dB}$$

where f = frequency, MHz
d_1 = inside diameter of outer conductor, cm
d_2 = outside diameter of inner conductor, cm
l = length of line, m

What is the attenuation of a 150-m cable whose inner conductor is No. 20 copper wire inside a 2-mm thick insulation, used to carry a 100-MHz signal?

CHAPTER 36

NUMBER SYSTEMS FOR COMPUTERS

Have you ever wondered about our numbering systems—the seldom-discussed ''philosophy'' of how we count? In this chapter, we shall explore the background of counting systems and apply the knowledge gained to the electronic computing field.

36–1 NUMBERS IN GENERAL

Recall from Sec. 6–14 how we referred to the problem of adding 5×10^3 to 3×10^2:

$$\begin{aligned} 5 \times 10^3 &= 5000 \\ \underline{3 \times 10^2} &= \underline{\ \ 300} \\ 5 \times 10^3 + 3 \times 10^2 &= 5300 = 5.3 \times 10^3 \end{aligned}$$

In other words, a number like 5300 may be thought of as being made up of two separate parts, 5×10^3 and 3×10^2. Similarly, all the numbers in our decimal system may be broken down into different factors multiplied by suitable powers of 10. For example, 5328 may be thought of as—indeed, it really is:

$$\begin{array}{rcl} 5000 & \text{or} & 5 \times 10^3 \\ 300 & & 3 \times 10^2 \\ 20 & & 2 \times 10^1 \\ 8 & & 8 \times 10^0 \end{array}$$

and we could write 5328 in the form

$$5 \times 10^3 + 3 \times 10^2 + 2 \times 10^1 + 8 \times 10^0$$

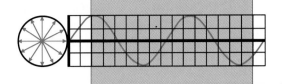

In fact, the very way we place the digits in their appropriate places carries out the sense of powers of 10. In many elementary schools, students learn what the "place names" of the digits are in a long number like the one that follows:

7 654 321
- 1 — units
- 2 — tens
- 3 — hundreds
- 4 — thousands
- 5 — ten thousands
- 6 — hundred thousands
- 7 — millions

and so on, and we would pronounce the whole number by using most of those place names: "seven million, six hundred fifty-four thousand, three hundred twenty-one."

36–2 BINARY NUMBERS

When we talk about decimal numbers, or the decimal number system, we mean we are counting in units of 10. That is, our numbering system has a *radix* of 10.

In the *binary* system, which is used extensively in digital computers, the radix is 2 and every number in the system represents an appropriate factor times the suitable power of 2:

or, in binary

$$0 = 0 \times 2^0 = 0_2$$
$$1 = 1 \times 2^0 = 1_2$$
$$2 = 1 \times 2^1 + 0 \times 2^0 = 10_2$$
$$3 = 1 \times 2^1 + 1 \times 2^0 = 11_2$$
$$4 = 1 \times 2^2 + 0 \times 2^1 + 0 \times 2^0 = 100_2$$

Stop and be sure. 100_2, that is, what at first appears to be one hundred in the binary numbering system, means: from the position of the digits,

$$1 \times 2^2 + 0 \times 2^1 + 0 \times 2^0$$

or

$$4 + 0 + 0 = 4$$

If you are sure, go on. If you are not sure, go back to the introduction and start the chapter again. When you are sure of the notion that a power of 2 must be connected with each digit in the binary number and the particular power depends upon the location of the digit in the number, go on and prove the following extension of the binary table:

$$5 = 101_2 \qquad 8 = 1000_2$$
$$6 = 110_2 \qquad 9 = 1001_2$$
$$7 = 111_2 \qquad 10 = 1010_2$$

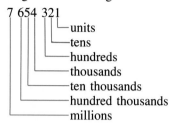

$$
\begin{aligned}
11 &= 1011_2 & 14 &= 1110_2 \\
12 &= 1100_2 & 15 &= 1111_2 \\
13 &= 1101_2 & 16 &= 10000_2
\end{aligned}
$$

EXAMPLE 1 Write the decimal equivalent of the number 10011001_2.

SOLUTION

Taking our cue from the position of the digits in the number and keeping track of the appropriate powers of 2, we convert each digit to its decimal equivalent, evaluating from the right:

$$
\begin{aligned}
1 \times 2^0 &= 1_{10} \\
0 \times 2^1 &= 0_{10} \\
0 \times 2^2 &= 0_{10} \\
1 \times 2^3 &= 8_{10} \\
1 \times 2^4 &= 16_{10} \\
0 \times 2^5 &= 0_{10} \\
0 \times 2^6 &= 0_{10} \\
1 \times 2^7 &= \underline{128_{10}} \\
10011001_2 &= 153_{10}
\end{aligned}
$$

You should use the subscripts to designate the *system* in which you are counting until you are satisfied with your confidence in intersystem conversions.

PROBLEMS 36–1 Write the following binary numbers in decimal form:

1. 000111
2. 001010
3. 11000011
4. 001011
5. 0010001
6. 100111
7. 101111
8. 110001
9. 100011
10. 111101

Now let us consider the reverse operation: converting a decimal number into its binary equivalent. Again we are looking for factors (either 1 or 0) times suitable powers of 2. The number 153, for instance, contains 128, which is 2^7. The remainder, $153 - 128 = 25$, contains 16, which is 2^4. The next remainder, $25 - 16 = 9$, contains 8, which is 2^3, and the last remainder, $9 - 8 = 1$, is 2^0. However, to write the complete binary equivalent, we must show the factors (zero) of 2^6, 2^5, 2^2, and 2^1.

$$153_{10} = 10011001_2$$

Obviously, it would be a tremendous help to know the whole 2^x table, and students anticipating advanced studies in computer designing, programming, or servicing will make these conversions by memorizing the 2^x table, say, to $2^{10} = 1024$. However, a ready mechanical method of arriving at the same binary number, without forgetting the missing powers of 2, is to convert the multiplication process into one of repeated division:

Radix Divisor	Decimal Number to Be Converted	Remainder
2	) 153	1
2	) 76	0
2	) 38	0
2	) 19	1
2	) 9	1
2	) 4	0
2	) 2	→ 0
	1	
	Read Up	

Writing the quotient and the remainders in order "backwards," we arrive at

$$153_{10} = 10011001_2$$

PROBLEMS 36–2

Convert the following decimal numbers into binary form:

1. 7
2. 12
3. 75
4. 23
5. 38
6. 88
7. 97
8. 126
9. 717
10. 361

36–3 OCTAL NUMBERS

Modern computers and calculators work in binary numbers, but it is often convenient for us to reduce the number of digits by working in octal (base 8) or hexadecimal (base 16) systems.

$$
\begin{aligned}
& & & \textit{Or} \\
0_{10} &= 0 \times 8^0 & &= 0_8 \\
1 &= 1 \times 8^0 & &= 1_8 \\
2 &= 2 \times 8^0 & &= 2_8 \\
3 &= 3 \times 8^0 & &= 3_8 \\
4 &= 4 \times 8^0 & &= 4_8 \\
5 &= 5 \times 8^0 & &= 5_8 \\
6 &= 6 \times 8^0 & &= 6_8 \\
7 &= 7 \times 8^0 & &= 7_8 \\
8 &= 1 \times 8^1 + 0 \times 8^0 & &= 10_8
\end{aligned}
$$

$$9 = 1 \times 8^1 + 1 \times 8^0 = 11_8$$
$$10 = 1 \times 8^1 + 2 \times 8^0 = 12_8$$
$$11 = 1 \times 8^1 + 3 \times 8^0 = 13_8$$
$$12 = 1 \times 8^1 + 4 \times 8^0 = 14_8$$
$$13 = 1 \times 8^1 + 5 \times 8^0 = 15_8$$
$$14 = 1 \times 8^1 + 6 \times 8^0 = 16_8$$
$$15 = 1 \times 8^1 + 7 \times 8^0 = 17_8$$
$$16 = 2 \times 8^1 + 0 \times 8^0 = 20_8$$

Just as the binary system uses digits up to, but not including, 2, so the octal system uses only digits below its radix, 8.

Write the following number in decimal form:

$$2731_8$$

As in the binary, we take our cue from the position of the digits in the number and introduce the appropriate powers of 8, reading from the right:

$$1 \times 8^0 = 1$$
$$3 \times 8^1 = 24$$
$$7 \times 8^2 = 448$$
$$2 \times 8^3 = 1024$$
$$2731_8 = 1497_{10}$$

PROBLEMS 36–3

Convert the following octal numbers into their decimal equivalents:

1. 00005
2. 00017
3. 00027
4. 00102
5. 00077
6. 00100
7. 02307
8. 01035
9. 06270
10. 22453

The conversion of decimal numbers to octal equivalents is achieved in the same fashion as in the binary, except that the divisor is the radix 8 instead of 2:

EXAMPLE 2

Radix Divisor	Decimal Number to Be Converted	Remainder
8	) 1497	1
8	) 187	3
8	) 23	7
	2	
	Read up	

$1497_{10} = 2731_8$

PROBLEMS 36–4

Convert the following decimal numbers to their octal equivalents:

1. 29
2. 37
3. 77
4. 127
5. 561
6. 477
7. 916
8. 1062
9. 8329
10. 5000

36–4 SYSTEMS WITH ANY RADIX

Just as we have developed binary numbers with radix 2 or octal numbers with radix 8, so we may develop any number system. Consider, for example, quinary numbers: the digits in a quinary number will consist of appropriate factors times suitable powers of 5. The factors may be 0, 1, 2, 3, and 4, but not 5 or higher.

$$22_5 = 2 \times 5^1 + 2 \times 5^0 = 12_{10}$$

PROBLEMS 36–5

Write the following decimal numbers in the systems of the indicated radices:

Number:	Radix:	Number:	Radix:
1. 25	5	6. 565	3
2. 12	4	7. 1827	8
3. 32	5	8. 1728	7
4. 256	16	9. 2750	8
5. 201	3	10. 672	5

Write the decimal equivalents of the numbers given:

11. 1042_6
12. 163_7
13. 333_4
14. 201_3
15. 003_{12}
16. 0725_9
17. 0106_8
18. 2388_9
19. 1540_6
20. 73006_8

36–5 CONVERSION BETWEEN SYSTEMS

We have already seen how to convert from any numbering system to decimal and from decimal to any other system. Thus, if we should be required to convert a number with any given radix a into a system with some other radix b, we could do so in two steps: (1) convert the given number into its decimal equivalent and (2) convert the decimal equivalent into the new system.

EXAMPLE 3 Convert 5134_6 into its binary equivalent.

SOLUTION

In the first step, convert 5134_6 to 1138_{10}. In the second step, convert 1138_{10} to 10001110010_2.

Actually, most of the conversions which concern us are between the binary and octal and hexadecimal systems.

EXAMPLE 4 Convert 1772_8 into its binary equivalent.

SOLUTION

$$1772_8 = 1018_{10} = 1111111010_2 = 1\ 111\ 111\ 010_2$$

In the various numbering systems, no change of value is introduced if we add zeros to the *left* of a number, so that we may change the appearance of $1\ 111\ 111\ 010_2$ to $001\ 111\ 111\ 010_2$ without introducing any value change but yielding a number which consists of a quantity of groups of three binary digits. (BInary digiTS are often referred to as *bits*.) By good advance planning, (1) the octal numbering system uses ordinary arabic numerals up to 7 and (2) the largest binary number consisting of three digits is 7 ($= 111_2$). If we evaluate each digit in the octal number into its three-bit binary equivalent, we arrive at

$$\begin{array}{cccc} 1 & 7 & 7 & 2_8 \\ 001 & 111 & 111 & 010_2 \end{array}$$

Thus, $1772_8 = 001\ 111\ 111\ 010_2$.

EXAMPLE 5 Convert 5317_8 into its binary equivalent.

SOLUTION

Replace each octal digit in turn with its binary three-bit equivalent:

$$5317_8 = 101\ 011\ 001\ 111_2$$

EXAMPLE 6 Convert 10110011001_2 to its octal equivalent.

SOLUTION

From the right, mark off the given binary number into groups of three bits:

$$010\ 110\ 011\ 001$$

Replace each three-bit group with its regular decimal equivalent to arrive at the octal equivalent of the number:

$$010\ 110\ 011\ 001_2 = 2631_8$$

PROBLEMS 36–6

Convert the following octal numbers to their binary equivalents:

1. 163_8
2. 277_8
3. 456_8
4. 527_8
5. 160_8
6. 645_8
7. 6256_8
8. 5267_8
9. 7777_8
10. 1000_8

Convert the following binary numbers to their octal equivalents:

11. 0010100_2
12. 011001_2
13. 0010111_2
14. 0011_2
15. 101001111_2
16. 100100100_2
17. 110011010_2
18. $001\ 001\ 111_2$
19. 10101010_2
20. 10111010_2

36–6 BINARY ADDITION

The addition of two quantities $a + b$, may, in binary devices, have only four possible values:

$$0 + 0 = 0 \qquad 0 + 1 = 1$$
$$1 + 0 = 1 \qquad 1 + 1 = 10$$

because of the dichotomous (two-state, on-off, open-closed, flipped-flopped, 1-0) nature of switching devices, and therefore the sum S of the addition $a + b$ will be limited to the four possible answers shown above. The first three forms present no difficulty, and we can add binary numbers which involve them very easily:

$$\begin{array}{r} 11001 \\ \underline{00100} \\ 11101 \end{array} \qquad \text{or} \qquad \begin{array}{r} 25 \\ \underline{4} \\ 29 \end{array}$$

But the addition of $1 + 1$ involves us in a two-part answer: 10. The 0 part of this answer is the *sum*, and the 1 part is the *carry*. This is similar to ordinary arithmetic. When the addition of two numbers requires it, say, $9 + 5$, we "put down 4 and carry 1."

EXAMPLE 7 Add 100110 and 110101.

SOLUTION

Set the two numbers down in traditional addition form, one above the other. Addition of $0 + 0$, $0 + 1$, and $1 + 0$ involves nothing new. When adding $1 + 1$, put down 0 and carry 1 over to the next stage of addition:

$$\begin{array}{r} 1 \\ 100110 \\ \underline{110101} \\ 1011011 \end{array} \qquad \text{or} \qquad \begin{array}{r} 38 \\ \underline{53} \\ 91 \end{array}$$

PROBLEMS 36–7

Add the following binary numbers:

1. $\begin{array}{r} 010101 \\ \underline{100010} \end{array}$
2. $\begin{array}{r} 100101 \\ \underline{010101} \end{array}$
3. $\begin{array}{r} 1101100 \\ \underline{0000100} \end{array}$
4. $\begin{array}{r} 0110110 \\ \underline{0100111} \end{array}$
5. $\begin{array}{r} 101110 \\ \underline{100100} \end{array}$
6. $\begin{array}{r} 101111 \\ \underline{010111} \end{array}$

7. 1011011
   1001010

8. 110010
   011010

9. 101010
   011011

10. 100101
    111011

11 to 20. Prove each of your answers by converting the individual parts into their decimal equivalents.

36–7 SUBTRACTION OF BINARY NUMBERS

Similarly to binary addition, binary subtraction is limited to four possibilities:

$0 - 0 = 0$
$1 - 0 = 1$
$1 - 1 = 0$
$0 - 1 = 1$ and carry 1 (or "borrow" 1)

When we are subtracting one ordinary number from another and come upon a step involving $5 - 8$, we borrow 1 from the digit to the left of the 5, subtract 8 from 15, and obtain 7. Binary subtraction is no different.

EXAMPLE 8 Subtract 0110 from 1011.

SOLUTION

Set the numbers in column form, the subtrahend below the minuend. When we must subtract 1 from 0, we borrow 1 from the number to the left of the 0 to make it 10. Then, $10 - 1 = 1$:

$$\begin{array}{r} 1011 \\ -0110 \\ \hline 0101 \end{array} \qquad \begin{array}{r} 11 \\ -6 \\ \hline 5 \end{array}$$

EXAMPLE 9

$$\begin{array}{r} 1000101 \\ -0110011 \\ \hline 0010010 \end{array}$$

You should convert these two binary numbers into their equivalent decimal numbers and test the solution.

PROBLEMS 36–8

Perform the following binary subtractions:

1. 011011
   −001001

2. 011011
   −010111

3. 010101
   −000100

4. 110111
   −011101

5. 100110
   −100101

6. 110100
   −101111

7. $\begin{array}{r} 1011011 \\ -1001010 \\ \hline \end{array}$

8. $\begin{array}{r} 100111 \\ -100011 \\ \hline \end{array}$

9. $\begin{array}{r} 111111 \\ -110110 \\ \hline \end{array}$

10. $\begin{array}{r} 100110 \\ -100101 \\ \hline \end{array}$

11 to 20. Prove each answer by converting all parts of each problem into their equivalent decimal forms.

36–8 SUBTRACTION BY ADDING COMPLEMENTS

One of the oldest rules in subtraction is ''change the sign and add.'' This policy makes binary subtraction extremely simple. Changing the sign of a binary number is like changing the condition of a switch. *On* becomes *off,* and *open* becomes *closed. Flipped* becomes *flopped,* 1 becomes 0, and 0 becomes 1.

EXAMPLE 10 Subtract 01101 from 11001 by means of complementation.

SOLUTION

Rewrite the problem, changing the subtrahend to its 1's complement; then add:

$$\begin{array}{r} 11001 \\ -01101 \\ \hline \end{array} \quad \text{becomes} \quad \begin{array}{r} 11001 \\ +10010 \\ \hline 101011 \end{array} \quad \text{that is,} \quad \begin{array}{r} 25 \\ -13 \\ \hline 43 \end{array}$$

43?! Well, when the answer to such a process comes out with one more digit than the number of digits we had to start with, we transfer this extra digit as an ''end-carry''* and add it back in:

$$\begin{array}{r} 11001 \\ 10010 \\ \hline 101011 \\ + \quad \rightarrow 1 \\ \hline 01100 \end{array} \qquad \text{which is } 12_{10}$$

EXAMPLE 11 Perform the subtraction 11101101 − 01001011 by means of 1's complement.

SOLUTION

Rewrite the subtrahend into its 1's complement and add. Bring down the extra 1, if any, and add it as an end-carry:

$$\begin{array}{r} 11111 \\ 11101101 \\ +10110100 \\ \hline 110100001 \\ \rightarrow 1 \\ \hline 10100010 \end{array} \qquad \begin{array}{r} \\ 237 \\ -75 \\ \hline \\ \\ 162 \end{array}$$

*This function is built into the calculator/computer circuitry.

PROBLEMS 36–9

Perform the following subtractions by means of complementation:

1. 110100 − 100011
2. 101101 − 010010
3. 101011 − 010001
4. 001101 − 000110
5. 101010 − 010010
6. 101011 − 001010
7. 111011 − 101010
8. 101111 − 001100
9. 101100 − 010011
10. 001110 − 001001

11 to 20. Prove each answer by converting all the parts into their decimal equivalents.

36–9 BINARY MULTIPLICATION

Since 1 times anything is the thing itself and 0 times anything is 0, binary multiplication is very easy.

EXAMPLE 12

Multiply 1101 by 100.

SOLUTION

Set down the numbers as for ordinary multiplication and multiply in the usual way. Add the partial answer rows in binary form:

```
    1101       13
 ×   100      × 4
 -------      ---
    0000
   0000
  1101
 -------
  110100       52
```

EXAMPLE 13

Multiply 10011 by 101.

SOLUTION

As before, multiply by long multiplication methods. There is no need to write a complete line of 0's—just set down the right-hand 0 and shift the line for the following multiplier one step to the left:

```
   10011       19
 ×   101     ×  5
 -------     ----
   10011
  100110
 -------
 1011111       95
```

PROBLEMS 36–10

Multiply:

1. 100111 by 10
2. 110011 by 11
3. 101101 by 101
4. 010111 by 100

5. 1010010 by 110
6. 110011 by 110
7. 110101 by 1101
8. 11001110 by 1101
9. 101001101 by 1001
10. 111001111 by 1011

11 to 20. Prove each solution to Probs. 1 to 10 by converting all parts into their equivalent decimal forms.

36–10 BINARY DIVISION

Dividing by binary numbers is as easy as multiplying. Either the divisor is smaller than the dividend and the quotient is 1 or the divisor is larger than the dividend and the quotient is 0.

EXAMPLE 14 Divide 1000001 by 101.

SOLUTION

Write the numbers as for ordinary long division. Will the three-bit divisor go into the first three bits of the dividend or not? If it will, put down a 1 as the first item in the quotient, and carry on. If it will not, bring down the next digit in the dividend, and put down a 0 as the first item of the quotient:

```
       01101             13
  101)1000001          5)65
      000
      1000
       101
       0110
        101
        0010
        0000
         0101
          101
           xx
```

PROBLEMS 36–11

Perform the following divisions:

1. 100111 by 011
2. 011110 by 101
3. 101100101 by 10001
4. 10011010 by 111
5. 111010000 by 11101
6. 001001011 by 1111
7. 11100001 by 1001
8. 101000100 by 10010
9. 111100000 by 10000
10. 1101010011 by 10111

11 to 20. Prove each answer by converting all parts into their decimal equivalents.

36–11 FRACTIONAL BINARY NUMBERS

To this point, as in preceding editions, any binary arithmetic operations have involved whole-number quantities. But computers and pocket calculators would be seriously limited in their operation if they were able to function only with whole numbers. A computer may receive its *commands* in analog form or in any one of a host of binary codes, but internally it performs the desired arithmetic operations by using only the binary digits 1 and 0.

Obviously, in its internal workings, the computer must be able to translate both whole and fractional numbers in the denary system into their binary equivalents to permit proper operation of the arithmetic unit within the device.

From this point on, we shall use the word "denary" for a number to radix ten, in order to avoid any confusion between decimal numbers and decimal fractions, which are often referred to simply as "decimals."

EXAMPLE 15 Add \$1.00 + \$4.00 + \$0.25 and express the answer in binary form.

SOLUTION

$$1 + 4 + 0.25 = \$5.25$$

which may be written as

$$1 + 4 + \tfrac{1}{4} = \$5\tfrac{1}{4}$$

But in binary form,

$$\tfrac{1}{4} = \tfrac{1}{2^2} = 2^{-2} \qquad (\text{and } 2^2 = 4,\ 2^0 = 1)$$

Therefore, \$5.25 in binary form is shown as

$$1 \times 2^2 + 0 \times 2^1 + 1 \times 2^0 + 0 \times 2^{-1} + 1 \times 2^{-2}$$

Binary digits 1 0 1 • 0 1

(• — Decimal-binary point)

The binary solution is $5.25_{10} = 101.01_2$. Also, whether it be dollars, meters, or any unit with the denary number 5.25, the binary equivalent will always be 101.01. As with most numbering systems, the position of the integers is the most important point to note.

In the preceding example, note the use of the negative exponents; they are the key to expressing fractional denary numbers as fractional binary numbers by using the binary digits 1 and/or 0. The method of conversion presupposes that the fraction to be converted *has* a denominator with a base number of 2.

EXAMPLES

$$0.5 = \frac{1}{2} = 1 \times 2^{-1} = 2^{-1} = 0.1_2$$

$$0.25 = \frac{1}{4} = 1 \times 2^{-2} = 2^{-2} = 0.01_2$$

$$0.125 = \frac{1}{8} = 1 \times 2^{-3} = 2^{-3} = 0.001_2$$

$$0.875 = \frac{7}{8} = 7 \times 2^{-3} = 0.111_2$$

Therefore,

$$\frac{1}{2} + \frac{1}{4} + \frac{1}{8} = \frac{7}{8} = 0.111_2$$

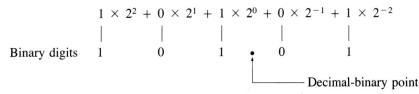

The positions of the binary digits provide the solution for the sum of $\frac{1}{2} + \frac{1}{4} + \frac{1}{8}$. However, there are many fractions with denominators that are not convenient numbers when converted to 2. The conversion of such fractions can be accomplished by successive multiplication by the desired base number, in this case 2, as illustrated below.

EXAMPLE 16 Convert 0.25 denary into its binary equivalent.

SOLUTION

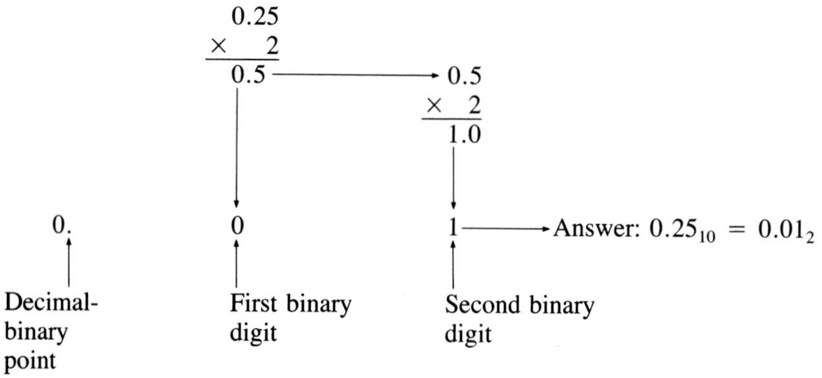

EXAMPLE 17 Convert $\frac{1}{5}$ denary into its binary equivalent.

SOLUTION

$$\frac{1}{5} = 0.2$$

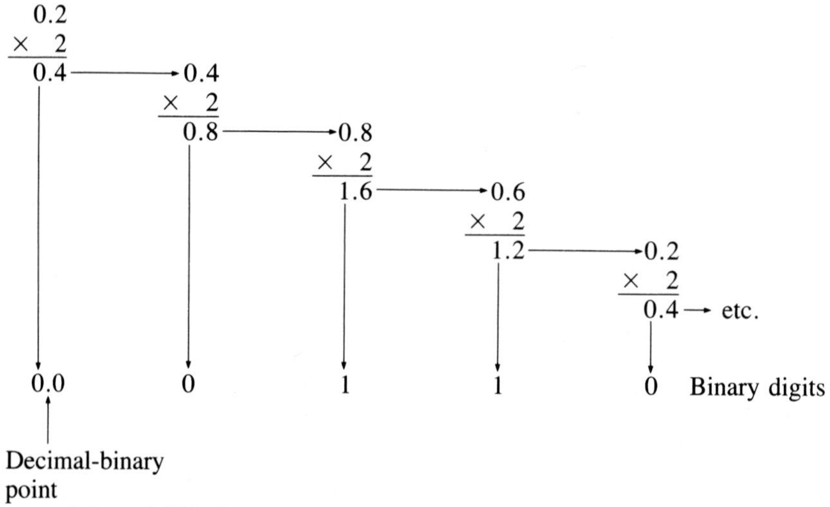

$$0.2 = 0.00110\ldots$$

Note that it is only the fractional part that is moved and then multiplied by 2. The whole-number portion is brought down to form part of the new binary number, whether it be 1 or 0.

The binary equivalent of 0.2 appears to be 0.00110, but the reconversion of the binary fraction shows:

$$\begin{aligned} 0 \times 2^{-1} &= 0 \\ 0 \times 2^{-2} &= 0 \\ 1 \times 2^{-3} &= 2^{-3} = 0.125 \\ 1 \times 2^{-4} &= 2^{-4} = \underline{0.0625} \\ & \qquad\quad\; 0.1875 \quad \text{Which is } \textit{not} \text{ } 0.2. \end{aligned}$$

This difference does provide some question about accuracy, since 0.1875 is 93.75% of 0.2, or −6.25%. In most pocket calculators each binary digit occupies at least one circuit, and 0.2 denary will use four circuits to provide an answer that is only 93.75% accurate. To provide greater accuracy, more binary digits must be used. To be within 1% accurate of 0.2, at least eight digits must be used; and that entails more circuits. If 0.2 denary is shown as 0.00110011 binary, the error is only −0.39%, but at the cost of using eight or nine circuits.

By now it should be obvious that 0.2 denary is a recurring binary number, and many calculators and most computers are programmed to recognize the binary sequence 0.001100110011 as denary 0.2. The 12-digit binary number could present a size problem when discrete components are used. The point is worth mentioning because 0.2 will appear often, even in other denary fractions. The *recognition circuit* in the calculator is only one solution. Another solution is to use a different *input code* to the calculator or computer, such as the octal or hexadecimal numbering system, for example.

The point being made is that less expensive calculators will probably have no more accuracy than a "good" slide rule would once have had, especially if the operator has been rounding off throughout the problem and then expressing the final answer to only three decimal places. The user of the calculator should be aware that any result given to the *n*th decimal place should not be assumed more accurate than three significant figures from a slide rule in the *final* answer. It is not our intent to discredit the calculator; we want only to make the operator aware that if the value of any constant, say π for example, is rounded off to three significant digits, the error introduced will be compounded throughout the complete calculation.

To prove the point, and the frequency with which 0.2 occurs in most calculations done with electronic calculators, complete the following problems. Express the binary answer to within 1% of the original denary fraction and reconvert the answer obtained to prove its accuracy.

PROBLEMS 36–12

Express the following fractions as binary numbers to within ±1%:

1. $\frac{9}{10}$
2. $\frac{4}{5}$
3. $\frac{2}{5}$
4. $\frac{3}{5}$
5. $\frac{5}{8}$
6. $\frac{1}{2}$
7. $\frac{5}{13}$
8. $\frac{9}{16}$
9. $\frac{3}{13}$
10. $\frac{5}{11}$

36–12 ARITHMETIC OPERATIONS WITH FRACTIONAL BINARY NUMBERS

ADDITION AND SUBTRACTION The rules for binary addition were outlined in Sec. 36–6; they are

$$0 + 0 = 0 \qquad 0 + 1 = 1$$
$$1 + 0 = 1 \qquad 1 + 1 = 10$$

The rules continue to apply; and as in addition with denary fractions and mixed numbers, the decimal-binary point is aligned.

EXAMPLE 18 Add 10001.01 binary to 111.111 binary and express the answer in both binary and denary form.

SOLUTION

$$\begin{array}{r} 10001.010 \\ +111.111 \\ \hline 11001.001_2 \end{array} \qquad \begin{array}{rl} 11001.001_2 = 1 \times 2^4 & = 16 \\ 1 \times 2^3 & = 8 \\ 0 \times 2^2 & = 0 \\ 0 \times 2^1 & = 0 \\ 1 \times 2^0 & = 1 \\ 0 \times 2^{-1} & = 0 \\ 0 \times 2^{-2} & = 0 \\ 1 \times 2^{-3} & = \underline{0.125} \end{array}$$

Therefore, $11001.0012 = 16 + 8 + 1 + \frac{1}{8} = 25.12510$

The rules for binary subtraction outlined in Sec. 36–7 are applied to direct subtraction. Again, remember to align the binary point.

EXAMPLE 19

$$\begin{array}{rcr} 1111.1110 & = & 15.875 \\ -\ 111.1111 & = & -\ 7.9375 \\ \hline 111.1111_2 & = & 7.9375_{10} \end{array}$$

The same result could have been found by using complement addition. First find the subtrahend's complement (111.1111) by making its number of digits the same as for the minuend (1111.111) without changing the value of either number.

EXAMPLE 20 Using the data from Example 19,

```
  1111.1110 ←———————— Zero added
− 0111.1111
  ↑
  Zero added
```

Complement of 0111.1111 = 1000.0000

Add complement,

$$\begin{array}{r} 1111.1110 \\ +\ 1000.0000 \\ \hline 10111.1110 \\ \llcorner\!\!\longrightarrow 1 \\ \hline 0111.1111_2 \end{array}$$

This agrees with the answer found by using *direct* subtraction. The point to note is that the end-carry is carried to the very last digit, whether it be part of the fraction, as in this case, or part of the whole number, as in Sec. 36–8.

PROBLEMS 36–13

Perform the indicated operations on the following fractions, and express the binary answer to within 1% accuracy:

1. $\frac{3}{16} + 0.011_2$
2. $\frac{11}{32} + \frac{4}{9}$
3. $\frac{115}{128} + \frac{20}{64}$
4. $\frac{11}{132} + \frac{17}{24}$
5. $\frac{132}{144} + \frac{86}{96}$
6. $\frac{17}{32} - \frac{21}{64}$
7. $1.0101011_2 - 0.375_{10}$
8. $\frac{59}{256} - \frac{987}{1024}$
9. $18.625_{10} - 110.11011_2$
10. $37.0875 - 27.125$

11 to 15. Perform the subtractions of Probs. 6 to 10 by using complementary addition. Express the answers in both denary and binary forms.

MULTIPLICATION The rules of binary arithmetic (Sec. 36–9) for multiplication still apply, along with the rules for denary arithmetic to place the decimal-binary point. That is,

$$1 \times 1 = 1$$

and all other combinations equal zero.

EXAMPLE 21 Multiply 1.101 by 1.111.

SOLUTION

```
     1.101
     1.111
     -----
     1 101
    11 01
   110 1
  1 101
 ---------
 11.000 011
```

The binary point is placed just as it would be in denary multiplication: the number of places after the point is the sum of the places in the multiplicand and the multiplier. Therefore, with six places of binary digits after the two binary points, 1.101 × 1.111 = 11.000 011. This can be easily proved:

1.101	=	1.625
1.111	=	1.875
11.000 011	=	3.046 875

Or
$$1.101 = 1\frac{5}{8}$$
$$1.111 = 1\frac{7}{8}$$

Which gives
$$\frac{13}{18} \times \frac{15}{8} = \frac{195}{64}$$
$$= 3 + (1 \times 2^{-5}) + (1 \times 2^{-6})$$
$$= 11 + 0.00001 + 0.000001$$
$$= 11.000011_2$$

By inspection, multiplication of whole binary numbers is a series of additions and shifts to the left and multiplication of fractional binary numbers is a series of additions and shifts to the right. In other words, in binary fractional multiplication the number gets smaller, as it does in denary fractional multiplication.

PROBLEMS 36–14

Perform the indicated multiplications. Show all binary working, and express the answers in both binary and denary form.

1. (1.1011)(10.11)
2. (111)(0.101)
3. (10101)(1.011)
4. (1000000)(0.000001)
5. (1101)(0.101)
6. (0.0101)(0.101)
7. (0.0011)(11010)
8. (111)(0.111)
9. (1111.100)(1.001)
10. (0.1111)(0.011)

DIVISION When in Sec. 36–10 if we had divided 1111_2, by 10_2, we would have produced the following result:

```
    0111
10)1111
   10
    11
    10
     11
     10
      1 ← Remainder
```

The accepted answer was 111 with $\frac{1}{10}$ remainder; the denary answer was 7.5. The denary equivalent of 7.5 can be shown from

$$1111_2 = 15 \qquad \text{and} \qquad 10_2 = 2$$
$$\frac{15}{2} = 7.5$$

The binary equivalent of 7.5 can be shown to be

$$7.5 = (7 \times 2^0) + (1 \times 2^{-1})$$
$$= 111 + 0.1$$
$$= 111.1_2$$

It is *not* necessary to convert the binary fractions into their denary equivalents before writing them as fractional binary numbers. Division with binary fractional numbers is no more complicated than it is with denary fractional numbers—add zeros where necessary without changing the value of the number.

EXAMPLE 22 Divide 1101_2 by 10_2.

SOLUTION

```
     110.1
  10)1101
     10
      10
      10
       010 ← Added zero places
        10   the binary point.
        00
```

The final answer is 110.1_2.

In Example 22 the division was accomplished in the same manner that division by a denary fraction would be. Division by a denary fraction is achieved by moving the decimal points in both dividend and divisor to convert the divisor into a whole number:

$$\frac{6.25}{0.5} = \frac{62.5}{5}$$

Similarly, division by a binary fraction is achieved by moving the binary points, as shown in the following example.

EXAMPLE 23 Divide 0.01_2 by 0.11_2.

SOLUTION

$$\frac{0.01}{0.11} = \frac{1.0}{11.0}$$

```
     0.0101
  11)1.0000
       11
        100
         11
          1    More zeros would be
               considered at this point.
```

The preceding examples have served their purpose, which is to illustrate that binary division is no more involved than denary division is. Division of whole binary numbers is no more than a series of subtractions and shifts of the binary point to the right. Fractional division is a series of subtractions; but depending upon the magnitude of the divisor, the result may be a shift of the binary point to the left or right.

If the divisor is of lesser magnitude than the dividend, the shift will be to the left, indicating the resulting binary number is greater in magnitude. A shift to the right would indicate that the quotient had a divisor greater than the dividend, and the result would be a smaller binary number, as in Example 23.

EXAMPLE 24 Divide 1.01_2 by 0.1_2.

SOLUTION

$$0.1\overline{)1.01}\quad \text{quotient } 10.1 \qquad \text{Or } 101_2 \text{ divided by } 10_2$$

The answer, 10.1_2, is a shift to the right; the quotient is larger than the divisor.

Note that the binary numbers have been *treated* in most respects in the same way as denary numbers. That is to say, if the binary point in the divisor was moved two places to the right, the corresponding move was made in the dividend.

PROBLEMS 36–15

Perform the indicated divisions, and express answers in both denary and binary forms to within 1% accuracy:

1. $101110_2 \div 111_2$
2. $101.01_2 \div 0.011_2$
3. $1000_2 \div 0.00101$
4. $100000_2 \div 11.11_2$
5. $1.001_2 \div 0.0011_2$
6. $10101_2 \div 0.011101_2$
7. $0.0101_2 \div 0.000101_2$
8. $0.110001_2 \div 101_2$
9. $0.0001_2 \div 0.0000011_2$
10. $1000000_2 \div 11.111_2$

36–13 FRACTIONAL OCTAL NUMBERS

In earlier sections in this chapter we have shown how a denary number could be converted into a binary or octal number or, in fact, a number to any radix. Throughout these conversions only whole numbers were considered. The three-bit code was developed to illustrate how to convert from binary to octal systems, and vice versa, without having to return through the denary system. This code uses a series of short steps, and it also illustrates the influence of octal numbers and their fractional parts compared to binary fractional numbers.

Certain binary fractions encountered have presented an accuracy problem unless a considerable number of digits were used. The main problem seemed to come from a denary fraction of 0.2; to present that number in its binary equivalent to within 1% accuracy, 12 or more binary digits were required. The same denary fraction can be represented by its octal equivalent by using only four octal digits, and the accuracy is better than 1%.

EXAMPLE 25 Convert 0.2 denary into its binary and octal equivalents to within 1% accuracy.

SOLUTION

From Ex. 17, 0.2_{10}		$= 0.001100110011_2$
2^0	$0 = 0$	$= 0.00000$
2^{-1}	$0 = 0$	$= 0.00000$

2^{-2}	$0 = 0$		$= 0.00000$
2^{-3}	$1 = 1 \times 2^{-3}$		$= 0.12500$
2^{-4}	$1 = 1 \times 2^{-4}$		$= 0.06250$
2^{-5}	$0 = 0$		$= 0.00000$
2^{-6}	$0 = 0$		$= 0.00000$
2^{-7}	$1 = 1 \times 2^{-7}$		$= 0.007813$
2^{-8}	$1 = 1 \times 2^{-8}$		$= 0.003906$
2^{-9}	$0 = 0$		$= 0.00000$
2^{-10}	$0 = 0$		$= 0.00000$
2^{-11}	$1 = 1 \times 2^{-11}$		$= 0.000488$
2^{-12}	$1 = 1 \times 2^{-12}$		$= 0.000244$
			0.199951 Which is 99.9756% of 0.2

To convert denary 0.2 to its octal equivalent, either of two methods may be used: (1) repeated multiplication by 8 or (2) applying the three-bit code. By the first method,

0.2	→ 0.6	→ 0.8	→ 0.4	→ 0.2	
× 8	× 8	× 8	× 8	× 8	
1.6	4.8	6.4	3.2	1.6	etc.
↓	↓	↓	↓	↓	
0.1	4	6	3	1	Octal digits

This shows that 0.2_{10} equals 0.14631_8. The proof follows.

8^0	0	$= 0.0000$
8^{-1}	$1 = 1 \times 8^{-1}$	$= 0.12500$
8^{-2}	$4 = 4 \times 8^{-2}$	$= 0.06250$
8^{-3}	$6 = 6 \times 8^{-3}$	$= 0.011719$
8^{-4}	$3 = 3 \times 8^{-4}$	$= 0.00073$
8^{-5}	$1 = 1 \times 8^{-5}$	$= 0.00003$
		0.199979 Which is 99.989% of 0.2.

The second method, using the three-bit code, saves considerable time, especially as we already have the binary equivalent. That is,

$$
\begin{array}{cccccc}
0.2 = 0. & 001 & 100 & 110 & 011 \\
\downarrow & \downarrow & \downarrow & \downarrow & \downarrow \\
0. & 1 & 4 & 6 & 3
\end{array}
$$

These are the first four digits obtained by using method 1, and 0.1463_8 is within the required 1% limit, with an accuracy of 99.99975%.

Obviously, since each octal digit represents three binary bits, the octal system must give greater accuracy for fewer digit places. Even greater accuracy can be obtained by using base 16, or the hexadecimal system; see Sec. 36–14. The point to note is that octal-binary conversions are made relatively simple by use of the three-bit code.

RULE

To use the three-bit code to convert fractional binary numbers into their octal equivalents, group the binary digits into groups of three, starting at the binary point and working to the right.

EXAMPLE 26 Convert 0.100001 to its octal equivalent.

SOLUTION

$$
\begin{array}{l}
0.\ 100\ 001 = \frac{1}{2} + \frac{1}{64} = \frac{33}{64} = 0.516\ (0.515625)\ \text{(binary solution)} \\
\quad\ \ \downarrow \quad\ \downarrow \\
0.\ \ \ 4 \quad\ 1 \ \ = \frac{4}{8} + \frac{1}{64} = \frac{33}{64} = 0.516\ (0.515625)\ \text{(octal solution)}
\end{array}
$$

CHECK

Convert denary to octal and then to binary.

$$
\begin{array}{r}
0.515625 \\
\times \qquad\quad 8 \\
\hline
4.125000
\end{array}
\longrightarrow
\begin{array}{r}
0.125 \\
\times \quad 8 \\
\hline
1.000
\end{array}
$$

0.4 ⟶ 0.100

1 ← Octal answer

001 ← Binary answer

Therefore, $0.100001_2 = 0.41_8 = 0.515625_{10}$, which was the original number.

PROBLEMS 36–16

Use the three-bit code to perform the binary to octal conversions of the following:

1. 100101.11_2
2. 110.0011_2
3. 1011.111_2
4. 1011.1_2
5. 1101.101101_2
6. 1001.011_2
7. 1111.1111_2
8. 1.001101_2
9. 0.010101_2
10. 1000.101101_2

36–14 HEXADECIMAL NUMBERING SYSTEM

In the preceding sections of this chapter, denary numbers have been converted to numbers with any radix, although particular attention was given to base 2 (binary) and base 8 (octal) systems. Now that striking advances have been made in computers, microprocessors, and word processors, one other system should be given more than a cursory treatment. It is the hexadecimal system, which has base 16.

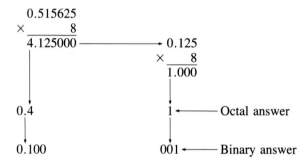

The hexadecimal system is an *alphanumeric code* used to solve the data entry problems of microprocessor units, but it is related to the binary numbering system needed by the computer sections that perform the arithmetic functions. The system requires that each of the numbers from 0 through 16 in the denary system be expressed in no more than four binary bits, generally termed a *nibble*.

This presents one minor problem: if each denary integer is to be represented by four bits, any number greater than 9 will require eight or more bits. The system is already used in what is known as a weighted 8421 binary-coded decimal (BCD). That is, 9 = 1001 in BCD, and 12 = 0001 0010 in BCD. However, in their binary equivalents:

$$\begin{aligned} 9_{10} &= 1001_2 & 13_{10} &= 1101_2 \\ 10_{10} &= 1010_2 & 14_{10} &= 1110_2 \\ 11_{10} &= 1011_2 & 15_{10} &= 1111_2 \\ 12_{10} &= 1100_2 & 16_{10} &= 10000_2 \end{aligned}$$

Obviously, the binary equivalent of denary 16 requires more than four bits. But as in all numbering systems, the greatest allowable number is less than the base number by one. For that reason 16 cannot appear in hexadecimal; the largest number is 15.

To differentiate between existing codes using four binary bits and the hexadecimal system, the following convention was established: 10 is represented by A, 11 by B, 12 by C, 13 by D, 14 by E, and 15 by F. Note that 16 is *not* represented by a letter for the reason given in the preceding paragraph. It follows that there must be some rule for numbers that are 16 or greater.

RULE

(for denary numbers greater than 16)

$$\begin{aligned} 1_{10} &= 0001_{16} \\ 5_{10} &= 0005_{16} \\ 15_{10} &= 000F_{16} \\ 16_{10} &= 0010_{16} \\ 17_{10} &= 0011_{16} \\ 27_{10} &= 001B_{16} \end{aligned}$$

Note that the value of 27 denary is the sum of 16_{10} and 11_{10}. But since 11_{10} is represented by B and 16_{10} by 0010_{16},

$$\begin{aligned} 27_{10} &= 000B_{16} + 0010_{16} \\ &= 001B_{16} \end{aligned}$$

Hexadecimal equivalents of denary numbers greater than 16 can be derived in the same way, but for the conversion of very large numbers hexadecimal arithmetic must be used. It is not discussed in this text; instead it is left to more advanced studies in computer systems.

The hexadecimal system is related to the binary system; Table 36–1 shows how. The table also includes octal numbers; they are supplied for use with one of the methods for converting denary numbers into hexadecimal numbers. The method for deriving binary and octal equivalents of denary numbers is repeated division by the new base number. It can be applied to get hexadecimal equivalents also, but there are certain rules that must be observed when dealing with remainders.

TABLE 36–1 RELATIONS AMONG FOUR NUMBER SYSTEMS

Denary	Hexadecimal	Binary	Octal
0	00	0000	0
1	01	0001	1
2	02	0010	2
3	03	0011	3
4	04	0100	4
5	05	0101	5
6	06	0110	6
7	07	0111	7
8	08	1000	10
9	09	1001	11
10	0A	1010	12
11	0B	1011	13
12	0C	1100	14
13	0D	1101	15
14	0E	1110	16
15	0F	1111	17

EXAMPLE 27 Convert the denary number 934 into its hexadecimal equivalent.

SOLUTION

Divide 934 by 16 and obtain 58 with remainder 6,
Divide 58 by 16 and obtain 3 with remainder 10,
3 is not divisible by 16, and the answer is
$934_{10} = 3\ 10\ 06_{16}$.

16)934(6
16) 58(10
3

But 10_{10} is represented by A in base 16; therefore, the correct answer is $934_{10} = 3A6_{16}$

Example 27 illustrates that care must be taken with remainders, especially 10, 11, 12, 13, 14, and 15. Those remainders must be written into final answers by substituting the alphabetic designations for them. That may appear to be clumsy and open to error, but the operator of a word- or microprocessor can supply information to the main terminal with an ordinary typewriter keyboard. The letters A, B, C, D, E, and F can be interpreted as base 10 integers, or 10, 11, 12, 13, 14, and 15 can be entered in their equivalent alphabetic forms. Usually the latter course is taken. But many minicomputer terminals and pocket calculators, with the exception of some liquid-crystal devices, do not have the capability for an alphanumeric display. The operator of such a machine must recognize that 031006 on an LED display is 3A6 in the hexadecimal system.

As an exercise, show that $187_{10} = 0BB_{16}$.

Did you show that

$$\begin{array}{r} 16\overline{)187}(11 \\ 16\overline{)11}(11 \\ 0 \end{array}$$

Reading up, $187_{10} = 0\ 11\ 11_{16}$

But $11_{10} = B_{16}$; therefore, $187_{10} = 0BB_{16}$.

The division method can become cumbersome for very large, very small, or fractional numbers. An alternative method is to use octal numbers coupled with the three-bit-code conversion into binary number equivalents.

EXAMPLE 28 Convert 187_{10} into its binary, octal, and hexadecimal equivalents.

SOLUTION

Conversion to octal,

$$\begin{array}{r} 8\overline{)187}(3 \\ 8\overline{)23}(7 \\ 2 \end{array} \qquad 187_{10} = 273_8$$

Using the three-bit-code,

$$273_8 = 010\ 111\ 011_2 \qquad \text{and} \qquad 187_{10} = 010111011_2$$

The next step is to divide the binary number 010111011_2 into four-bit nibbles starting at the least significant bit (LSB): $1011\ 1011_2$. From Table 36–1,

$$1011_2 = 11_{10} \quad \text{and} \quad 11_{10} = B_{16}$$

Therefore, the hexadecimal answer is

$$1011\ 1011_2 = BB_{16} = 187_{10}$$

This answer confirms the solution by the division method, which is what we set out to do.

The major advantage of using the three-bit code is in the reconversion from hexadecimal to denary.

EXAMPLE 29 Convert BB_{16} to its denary equivalent.

SOLUTION

$B_{16} = 1011_2$; therefore, $BB_{16} = 10111011_2$. Using the three-bit code on 10111011_2

$$\begin{array}{ccc} 010 & 111 & 011_2 \\ | & | & | \\ 2 & 7 & 3_8 \end{array} \begin{aligned} &= 2_{10} \times 8^2 + 7_{10} \times 8^1 + 3_{10} \times 8^0 \\ &= 128_{10} + 56_{10} + 3_{10} \\ &= 187_{10} \end{aligned}$$

Alternatively, the conversion could have been made from the hexadecimal number without the intermediate steps:

$$\begin{aligned} BB_{16} = (B \times 16^1) + (B \times 16^0) &= (11 \times 16) + (11 \times 1) \\ &= 176_{10} + 11_{10} \\ &= 187_{10} \end{aligned}$$

By either method, however, conversion from hexadecimal to denary numbers, although straightforward, is time-consuming. For this reason tables have been constructed to provide fast and easy conversion in either direction.

Since the principal use of the hexadecimal system is in connection with byte-organized machines, computer operators usually become quite adept with either or both the octal and hexadecimal systems. We suggest that you learn to recognize the binary, octal, and hexadecimal equivalents for the denary integers from 0 through 15. You will find the ability useful not only for study of this chapter but also for the study of any other BCD systems such as 8421 and excess-3 weighted codes.

Before we examine the conversion tables, a summary of denary to hexadecimal conversion will help you to speed up the process without using the division method.

1. Convert denary numbers greater than 16 to their octal equivalents (easy mental arithmetic).
2. Convert octal to binary three-bit code; one digit octal = three digits binary.
3. Rewrite the binary triads into one binary "word." Start at LSB and group into binary tetrads (four-bit bytes). Maximum denary value per tetrad = 15.
4. Convert the binary tetrads into hexadecimal equivalents. Remember

10	11	12	13	14	15
A	B	C	D	E	F

EXAMPLE 30 Convert 144 into its hexadecimal equivalent.

SOLUTION

$144_{10} = 220_8$; then

$$\begin{matrix} 2 & 2 & 0_8 \\ | & | & | \\ 010 & 010 & 000_2 \end{matrix} = 010010000_2$$

Regrouping,

$$\begin{matrix} 1001 & 0000_2 \\ | & | \\ 9 & 0 \end{matrix} = 90_{16} \qquad \text{From Table 36–1, } 1001_2 = 9_{16}$$

CHECK

$$(9 \times 16^1) + (0 \times 16^0) = 144_{10} = 90_{16}$$

Do *not* drop the zero; it places the integer 9 in the correct position.

Having once mastered this technique, you will find it a very simple task to convert denary to hexadecimal—and, by reversing the process, hexadecimal to denary—provided that only whole-number integers are involved. Fractional hex numbers will be considered later.

PROBLEMS 36–17

Use the octal-triad-binary to hexadecimal-tetrad-binary method to convert the following denary numbers to hexadecimal numbers:

1. 285_{10}
2. 396_{10}
3. 1515_{10}
4. 512_{10}
5. 82_{10}

Reverse the triad-tetrad-binary process to convert the following hexadecimal numbers into denary numbers:

6. $1C_{16}$
7. $1E8_{16}$
8. $53E_{16}$
9. 260_{16}
10. 400_{16}

The usefulness of Table 36–2 is best seen in the conversion of hexadecimal integers into denary integers. Given, say, 700_{16}, we first locate 7_{16} in the position 3 of the table and read across for the denary equivalent, which is 1792_{10}. Since both position 1 and position 2 are equal to 0, $700_{16} = 1792_{10}$.

> Note the vertical number patterns that run through the denary ($x \times 16^3$) column of Table 36–2.

TABLE 36–2 DENARY HEXADECIMAL INTERCONVERSIONS

Hex Pos 4	Denary $x \times 16^3$	Hex Pos 3	Denary $x \times 16^2$	Hex Pos 2	Denary $x \times 16^1$	Hex Pos 1	Denary $x \times 16^0$
1	4096	1	256	1	16	1	1
2	8192	2	512	2	32	2	2
3	12288	3	768	3	48	3	3
4	16384	4	1024	4	64	4	4
5	20480	5	1280	5	80	5	5
6	24576	6	1536	6	96	6	6
7	28672	7	1792	7	112	7	7
8	32768	8	2048	8	128	8	8
9	36864	9	2304	9	144	9	9
A	40960	A	2560	A	160	A	10
B	45056	B	2816	B	176	B	11
C	49152	C	3072	C	192	C	12
D	53248	D	3328	D	208	D	13
E	57344	E	3584	E	224	E	14
F	61440	F	3840	F	240	F	15

The application of Table 36–2 for hexadecimal to denary conversion is as follows:

1. Identify the position of each hexadecimal integer.
2. Look up the denary equivalent for the hexadecimal integer position located in step 1.
3. Repeat the procedure in step 2 for *all* the integers for the four hexadecimal positions. Read from left to right for the complete hexadecimal number.
4. Add all the denary values found from the four positions. The sum is the solution to the problem.

EXAMPLE 31 Use Table 36–2 to convert 2234_{16} to its denary equivalent.

SOLUTION 1

In the fourth position is a 2 corresponding to	8192_{10}
In the third position is a 2 corresponding to	512_{10}
In the second position is a 3 corresponding to	48_{10}
In the first position is a 4 corresponding to	4_{10}
Adding the four positions of the denary values,	8756_{10}

SOLUTION 2

2234_{16}

$$= 2_{10} \times 16_{10}^3 + 2_{10} \times 16_{10}^2 + 3_{10} \times 16_{10}^1 + 4_{10} \times 16_{10}^0 = 8756_{10}$$
$$= 8192_{10} + 512_{10} + 48_{10} + 4_{10} = 8756_{10}$$

Solution 2 of Example 31 confirms Solution 1, which was obtained by using Table 36–2. It does, however, involve raising 16_{10} to several different powers, which the table does for you. Once the hexadecimal positions have been located, the arithmetic is reduced to just addition. Now what remains is to show that Table 36–2 can be used for denary to hexadecimal conversions also.

EXAMPLE 32 Convert 5003_{10} into its hexadecimal equivalent by using Table 36–2.

SOLUTION

In the table, locate the denary number closest to but not greater than 5003_{10}. It is found in position 4. The hex 1 value is 4096_{10}, which is 907_{10} below 5003_{10}.

Write down in position 4 hexadecimal integer 1_{16}. Now locate the number closest to but not greater than 907_{10} and record the hexadecimal integer. It is 768_{10} in hex position 3. Record the value, 3_{16}. The difference between 907_{10} and 768_{10} is 139_{10}. The number closest to but not greater than 139_{10} is in hex position 2. The value is 128_{10}, and the difference is 11_{10}. Record the position 2 hexadecimal integer 8_{16}.

The first hexadecimal position is very straightforward, since $11_{10} = B_{16}$. We now have the complete set of hexadecimal integers for the denary number: $5003_{10} = 138B_{16}$.

Proof of the hexadecimal equivalent can be given in the same manner as in Solution 2 of Example 31. It is left as an exercise for you to show that

$$5003_{10} = 4096_{10} + 768_{10} + 128_{10} + 11_{10}$$

or that

$$138B_{16} = (1_{10} \times 16^3_{10}) + (3_{10} \times 16^2_{10}) + (8_{10} \times 16^1_{10}) + (11_{10} \times 16^0_{10})$$

A quick check of the calculations will show that the values are equal and that denary to hexadecimal conversion with the aid of Table 36–2 is possible.

Once you identify the position and hex location, simple addition and subtraction yield the denary number equivalent to the given hexadecimal number or, if the denary is given, the equivalent hexadecimal number. Table 36–2 is only part of a composite table which includes powers of 16 greater than the third power and fractional conversion tables out to at least 12 decimal places. For this study, however, we shall confine Table 36–2 to an upper limit of the third power. The fractional conversions are dealt with in Sec. 36–15.

PROBLEMS 36–18

Use Table 36–2 to express the following as denary numbers:

1. $2E7_{16}$
2. 801_{16}
3. $3F5_{16}$
4. $AC1_{16}$
5. $5F1_{16}$

Use the table to express the following as hexadecimal numbers:

6. 288_{10}
7. 684_{10}
8. 560_{10}
9. 8264_{10}
10. 1056_{10}

36–15 FRACTIONAL HEXADECIMAL NUMBERS

In the preceding section, Table 36–2 was used for interconversion of hexadecimal and denary numbers. The construction of a table for fractional conversions requires more careful treatment, especially when the conversion is from hexadecimal to denary, than that of a table for whole-number conversions.

We have already discussed the accuracy aspect, the number of binary bits required for $\pm 1\%$ representation of the equivalent denary fraction. Each hexadecimal integer represents four binary bits, so a four-integer hexadecimal requires the accuracy of 16 binary bits. The one denary fraction that produced some problems when converted to binary was the denary quantity 0.2_{10}; in binary it was 0.001100110011_2 for an accuracy of better than 99%. By using the four-bit-nibble conversion from binary to hexadecimal, we can show that $0.2_{10} = 0.333_{16}$ for an accuracy of 99.976% (rounded off to three decimal places). This is an error of only -0.024%.

Although accuracy is a major point, the space required to enter 0.333_{16} is obviously smaller than that for writing 0.001100110011_2 with less accuracy. That may appear to be splitting hairs, but the point to be stressed is that a calculator, computer, or microprocessor with an accuracy as good as -0.024% using binary numbers is not very common. Unless in a laboratory situation, most of the more common calculating devices are no better than $\pm 1\%$.

One last word on the denary quantity 0.2_{10}: If your calculator is able to convert different base numbers into one another and it does not have an alphanumeric display, it will represent the value 0.2_{10} in digital form as 0.030303_{16}. That is true of the Hewlett-Packard model 29C. Using the HP-29C FIXED to four decimal places, $0.2_{10} = 0.0303_{16}$. Note that 0.0303 displayed on the eight-segment LED readout of the HP-29C corresponds to 0.33_{16}; see Table 36–1. Converting 0.33_{16} back to denary in the conventional manner,

$$\begin{aligned} 0.33_{16} &= 3_{10} \times 16_{10}{}^{-1} + 3_{10} \times 16_{10}{}^{-2} \\ &= 3_{10}(0.0625_{10}) + 3_{10}(0.003906_{10}) \\ &= 0.1875_{10} + 0.011719_{10} \\ &= 0.199219_{10} \qquad \text{99.6\% accurate} \end{aligned}$$

In Sec. 36–14 we demonstrated different ways to convert denary whole numbers into their hexadecimal equivalents. Earlier we showed how to convert the fractional portion of a denary number to either its binary or octal equivalent. The method, successive multiplication by the desired base number, is valid, although time-consuming, for fractional hexadecimal numbers also. The following examples illustrate fractional denary and hexadecimal number conversion by using the values set by Table 36–3.

TABLE 36–3 FRACTIONAL HEXADECIMAL CONVERSIONS

x	Denary $x \times 16^{-1}$	Hex Pos 1	Denary $x \times 16^{-2}$	Hex Pos 2	Denary $x \times 16^{-3}$	Hex Pos 3
1	0.0625	0.1	0.003 906 250	0.01	0.000 244 141	0.001
2	0.1250	0.2	0.007 812 500	0.02	0.000 488 281	0.002
3	0.1875	0.3	0.011 718 750	0.03	0.000 732 422	0.003
4	0.2500	0.4	0.015 625 000	0.04	0.000 976 563	0.004
5	0.3125	0.5	0.019 531 250	0.05	0.001 220 703	0.005
6	0.3750	0.6	0.023 437 500	0.06	0.001 464 844	0.006
7	0.4375	0.7	0.027 343 750	0.07	0.001 708 984	0.007
8	0.5000	0.8	0.031 250 000	0.08	0.001 953 125	0.008
9	0.5625	0.9	0.035 156 250	0.09	0.002 197 266	0.009
A	0.6250	0.A	0.039 062 500	0.0A	0.002 441 406	0.00A
B	0.6875	0.B	0.042 968 750	0.0B	0.002 685 547	0.00B
C	0.7500	0.C	0.046 875 000	0.0C	0.002 929 688	0.00C
D	0.8125	0.D	0.050 781 250	0.0D	0.003 173 828	0.00D
E	0.8750	0.E	0.054 687 500	0.0E	0.003 417 969	0.00E
F	0.9375	0.F	0.058 593 750	0.0F	0.003 662 109	0.00F

Example 33 shows that the multiplication method has validity for hexadecimal numbers. It also shows, by using several denary fractions as examples, how the method was used to construct the table. One denary fraction is that used earlier, 0.2_{10}.

Table 36–3 has been rounded off to nine places of decimals for convenience of table size, and most calculators have only nine-place readouts.

EXAMPLE 33 Convert the following denary fractions to their hexadecimal equivalents: (*a*) 0.2_{10} and (*b*) 0.071044922_{10}.

SOLUTION

(*a*) Multiply 0.2_{10} by 16_{10}.

$0.2_{10} \times 16_{10} = 3.2_{10}$ → $0.2_{10} \times 16_{10} = 3.2_{10}$ → $0.2_{10} \times 16_{10} = 3.2_{10}$ etc.

0.3_{16} 3_{16} 3_{16}

Hexadecimal numbers (16^{-1}, 16^{-2}, 16^{-3})

$$0.2_{10} = 0.333_{16}$$

(*b*) Multiply 0.071044922_{10} by 16_{10}.

$0.071044922_{10} \times 16_{10}$:
0.426269532
0.710449220
1.136718752 → $0.136718752_{10} \times 16_{10}$:
0.820312512
1.367187520
2.187500032 → $0.187500032_{10} \times 16_{10}$:
1.125000192
1.875000320
3.000000412

0.1_{16} 2_{16} 3_{16}

Hexadecimal number, 0.123_{16}

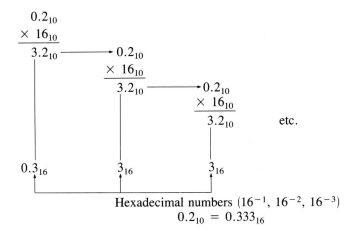

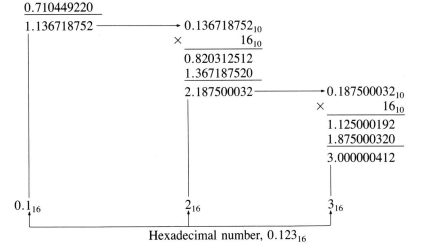

Example 33 illustrates that the multiplication *method* is quite lengthy and therefore time-consuming. The reconversion is somewhat more compact although still cumbersome.

EXAMPLE 34 Convert 0.123_{16} into its denary equivalent.

SOLUTION

As with all numbering systems, the *position* of the integers is all-important:

$$\begin{aligned} 0.123_{16} &= (1_{10} \times 16_{10}{}^{-1}) + (2_{10} \times 16_{10}{}^{-2}) + (3_{10} \times 16_{10}{}^{-3}) \\ &= 0.06250_{10} + 0.0078125_{10} + 0.000732422_{10} \\ &= 0.071044922_{10} \end{aligned}$$

Because of the sometimes involved fractional multiplications, Table 36–3 was constructed. Examples 33 and 34 illustrate the need for a better method, especially for dealing with fractional quantities. Table 36–3 can, with practice, provide a quick conversion from hexadecimal to denary fractions and, with a little more care, from denary fractions to their hexadecimal equivalents. For reason of their simplicity, Tables 36–2 and 36–3 should be used to interconvert from hexadecimal to denary and denary to hexadecimal. Particular attention should be paid, as always, to the position of the integers in either system.

EXAMPLE 35 Positions to the right of the decimal point are shown as

$$\begin{aligned} \text{Position } 1_{16} &= 16_{10}{}^{-1} \quad \text{or} \quad 0.062\ 50_{10} \\ \text{Position } 2_{16} &= 16_{10}{}^{-2} \quad \text{or} \quad 0.003\ 906\ 250_{10} \\ \text{Position } 3_{16} &= 16_{10}{}^{-3} \quad \text{or} \quad 0.000\ 244\ 141_{10} \end{aligned}$$

Succeeding positions to the right will be the successive negative powers of 16. This progression, along with the increasing powers for the whole numbers in Table 36–2, lets computer programmers convert readily in either direction from the decimal point. Although ''commercial'' conversion tables would be expansions of Tables 36–2 and 36–3, in this text we will not go beyond 16^3 to the left and 16^{-3} to the right of the decimal point.

Table 36–3 was constructed by using the multiplication method as follows:

$$1_{10} \times 16_{10}{}^{-1} = 0.06250_{10} \qquad \begin{array}{r} 0.06250_{10} \\ \times \quad 16_{10} \\ \hline 0.37500 \\ 0.62500 \\ \hline 1.00000_{10} \\ \downarrow \quad\quad \\ 0.1_{16} \quad\quad \end{array}$$

It is then reasonable to say that $1_{10} \times 16_{10}{}^{-1} = 0.06250_{10} = 0.1_{16}$. By using the same procedure, it can be shown that the following statements are correct. It is left to the student as an exercise to validate the statements:

$$1_{10} \times 16_{10}{}^{-2} = 0.003\ 906\ 250_{10} = 0.01_{16}$$
$$1_{10} \times 16_{10}{}^{-3} = 0.000\ 244\ 141_{10} = 0.001_{16}$$

The general formula for constructing Table 36–3 can be stated as

$$x \times 16^{-1}, \qquad x \times 16^{-2}, \qquad x \times 16^{-3},$$

where x takes on any integral value from 1 to 15 denary. To reduce the size of the table, zero has been omitted. The table headings, reading from left to right, provide the value of x, the corresponding denary value, and the hexadecimal equivalent and its respective position.

EXAMPLE 36 What is the denary equivalent of 0.07_{16}?

SOLUTION

Locate $x = 7$; then read across to column hex position $2 = 0.07_{16}$. To the left of hex position 2, read the denary number $0.027\ 343\ 750_{10}$.

EXAMPLE 37 Use Table 36–3 to perform the following conversions: (*a*) $0.35B_{16}$ to denary and (*b*) $0.431\ 396\ 484_{10}$ to hexadecimal.

SOLUTION

(*a*) Locate 0.3_{16}, hex position 1, and read denary number	$0.187\ 5_{10}$
Locate 0.05_{16}, hex position 2, and read denary number	$0.019\ 531\ 250_{10}$
Locate $0.00B_{16}$, hex position 3, and read denary number	$0.002\ 685\ 547_{10}$
Add the three denary numbers	$0.209\ 716\ 797_{10}$

This sum is the solution: $0.35B_{16} = 0.209\ 716\ 797_{10}$

Therefore, the hexadecimal to denary conversion is achieved by locating the denary equivalents for the corresponding hex positions and then adding them. Without the table, the solution could have involved some nine-place decimal multiplication.

(*b*) Converting $0.431\ 396\ 484_{10}$ to hexadecimal will require more careful thought. The first step is to locate the denary number closest to but not higher in value than the given number. If the denary number were an *exact* equivalent of a hexadecimal number, the problem would be solved, but examination of the table shows this *not* to be the case.

By inspection, the first digit to the right of the decimal point of the denary number is 4. The denary number closest to but not greater than 0.4 is located where x is equal to 6. It has the value 0.3750, which makes 6_{16} the first hexadecimal digit to the right of the hexadecimal point.

Now 0.3750_{10} is subtracted from the original denary number to provide a difference of $0.056\ 396\ 484_{10}$. It too is not an exact hexadecimal equivalent. In position 2, where x is E, we find the hex value closest to but not greater than 0.056. It is $0.054\ 687\ 500_{10}$, and, subtracted from the preceding remainder, it yields $0.001\ 708\ 984_{10}$. Therefore, we record E_{16} as the second hexadecimal number and look in the table for a denary value that is closest to but not greater than $0.001\ 708\ 984_{10}$. This time we find the exact number; it is the denary equivalent corresponding to x equal to 7.

In the third position, therefore, we put 7_{16}.

With this last value our hexadecimal equivalent is complete, and $0.431\ 396\ 484_{10} = 0.6E7_{16}$. It is left as an exercise for you to confirm this answer by using the table or any other conversion method.

From the examples and the following problems you will see that conversion from fractional hexadecimal numbers to the denary equivalents is easier than conversion from the fractional denary number to the equivalent hexadecimal fraction. Practice and use of the conversion tables will remove this difficulty, and it will soon be apparent that the tables are speedier and more accurate to at least nine places of decimals.

However, if the tables are not available or you have the time to perform the successive multiplications, fractional denary-hexadecimal conversions are, in principle, no more complex than fractional binary or octal conversions.

PROBLEMS 36–19

Convert the following hexadecimal numbers to their denary equivalents by using the values from Table 36–2 or 36–3:

1. $13.8B_{16}$
2. $0.B7C_{16}$
3. $12.6A_{16}$
4. $0.56B_{16}$
5. $BE.2D_{16}$

Convert the following denary numbers into their hexadecimal equivalents by using Tables 36–2 and 36–3:
(If your calculator allows you to retain the fraction portion of a display (see your instruction manual for INT and FRAC), repeated multiplication of the decimal portion by a radix, such as 16 (as in Example 33), may be performed with ease. Set up and perform the first multiplication, record the integer portion of the first multiplication, call for FRAC, multiply by the radix, record the integer portion, etc.)

6. 5002_{10}
7. $0.835\ 693\ 359_{10}$
8. 1342_{10}
9. 63.9_{10}
10. $15.796\ 142\ 578_{10}$

SELF TEST

1. Write the decimal form of 101011_2.
2. What is the binary form of 213_{10}?
3. What is the decimal equivalent of 00517_8?
4. What is the octal equivalent of 2417_{10}?
5. Write 429_{10} in the system of radix 7.
6. Write the decimal equivalent of 132_5.
7. What is the binary equivalent of 517_8?
8. What is the octal equivalent of 1001011001_2?
9. Add 010011 and 101101.
10. Subtract (011010) from (100101).
11. Multiply 110101 by 1101.
12. Divide 10100101110 by 1101.
13. Express $\frac{7}{16}$ as a binary number to five binary places.
14. Perform the indicated operations, and express your answer in denary form:

$$0.1111 + 0.01 - 0.011$$

15. Multiply, express your answer in denary form:

$$0.0101 \times 0.1$$

16. Divide 101.010101 by 10.11.
17. What is the octal equivalent of 0.101001_2?
18. Convert $A7_{16}$ to its binary equivalent.
19. Use Table 36–2 to convert $6A37_{16}$ to its denary equivalent.
20. Use Table 36–3 to convert $0.7A6_{16}$ to its denary equivalent.

CHAPTER 37

BOOLEAN ALGEBRA

More and more, electronic devices are being put to work in computing machines and controlling machines. First, electronic tubes superseded relays, and then transistors took the place of tubes. Now Very Large Scale Integrated circuits have been added to the list of computer and control components.

And with these applications of electronic devices, there is a growing need for technologists to know at least something about the logic operations of computers. The subject, generally, is known as *Boolean algebra* in honor of George Boole (1815–1864), who developed the work upon which the subject is now based. It is also often referred to as propositional calculus, mathematical logic, and truth-functional logic.

Here we are going to explore the basic ideas of Boolean algebra to see how we can put *logic* to work for us in two ways: (1) to describe circuits mathematically after they have been designed or assembled and (2) to design circuits mathematically before they are assembled. We are not going to do any work in the *philosophical* field, where logic and its algebra are extremely useful. Several excellent books have been written from that point of view, whereas there has been little introductory work from the point of view of switching or logic circuits.

37–1 SYMBOLS OF LOGIC CIRCUITRY

Different associations, different manufacturers, different authors, and different publishers have their own ideas as to what symbols should be used in logic circuits. Table 37–1 shows the ANSI Y32.14 standard symbols which will be used in this book. However, you must be prepared to recognize others in other technical publications.

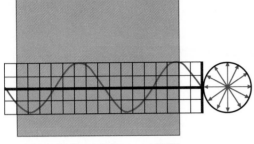

TABLE 37–1

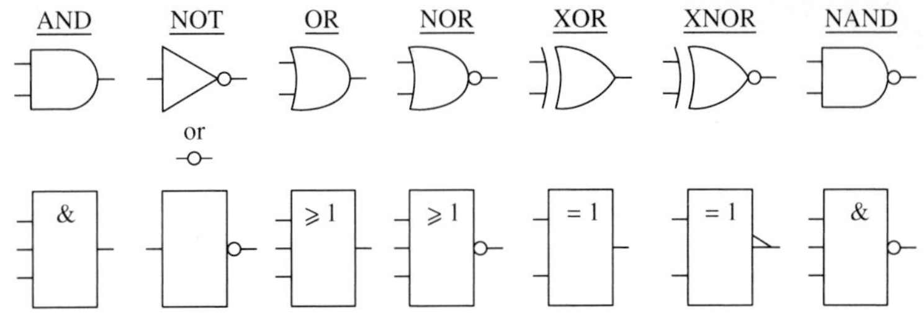

37–2 SYMBOLS OF MATHEMATICAL LOGIC

Just as the symbol for resistance appearing in circuit diagrams is replaced in the electronics mathematics by the symbol *R*, so the symbols of logic circuitry shown in Table 37–1 are replaced in the logic mathematics by their own special mathematical symbols which are shown in Table 37–2. Let us look further into the meanings of the circuit symbols and see what mathematical expressions are required.

TABLE 37–2 LOGICAL MATHEMATICAL SYMBOLS

	AND	OR	NOT
Symbols used in this text	•	+	—
Other symbols sometimes used	juxtaposition &	v	

AND The AND symbol means that an output signal will be produced by the particular device, regardless of the total amount of circuitry involved, only when both the *a* and *b* input signals are applied. Our mathematical counterpart must carry this meaning of AND.

OR The OR symbol means that an output will be produced by the device when either the *a* input or the *b* input signal is applied or when *both* are applied. Our mathematical replacement must give this meaning of "either . . . OR . . ., or both."

NOT The NOT (inverter) symbol means that either (1) there will *not* be an output when the input signal *is* applied or (2) there *is* an output when the input signal *is not* applied. Our mathematical symbol must carry the meaning of "not" or "reversed."

Now we must develop mathematical operators, sometimes referred to as *truth functors,* which will simply and effectively describe these circuit requirements. Table 37–2 shows the variety of symbols used in the literature, and, again, the symbol at the head of each column is the one to be used throughout this book. As well as knowing the appearance of the symbols and their general purpose, we must take particular pains to be able to pronounce the symbols.

AND	$a \cdot b$	May be pronounced: *a* and *b* both *a* and *b* the logical product of *a* and *b* *a* conjunct *b* the conjunction of *a* and *b* *a* in series with *b* if, and only if, *a* as well as *b*
OR	$a + b$	May be pronounced: *a* or *b* or both either *a* or *b* (or both) the inclusive OR of *a* and *b* the disjunction of *a* and *b* the alternation of *a* and *b* the logical sum of *a* and *b* *a* in parallel with *b* at least one of *a* and *b* if, and only if, *a* or *b* or both true if, and only if, *a* or *b* or both
NOT	$\bar{a}$	May be pronounced: not *a* the complement of *a* the inverse of *a* the negation of *a* the rejection of *a* it is false that *a* *a* is not assertable "not *a*" is true the valence of *a* is false

These pronunciations are the ones often met with in dealing with logic statements. Those appearing at the end of each group are the ones more usually found in philosophical statements, and they are included as a general-interest addition to our main study. At the same time, special symbols are often used for the *exclusive* OR operator, when we want to say "either *a* OR *b*, but not both together." Note that our definition of OR does not suit this requirement. However, we will say this in symbol form later *without* using any other special symbol.

AGGREGATE SYMBOLS: ( ), [ ] In addition to the operator symbols are the symbols of aggregation, already met with in Sec. 3–9. Everything inside an aggregate symbol is subject to any operator symbol which may be applied to the aggregate: $\overline{(a + b)}$ means "when input signal a or input signal b or both are applied, there will be no output signal." (Can you see that this could be said, "not a and not b"?)

TRUTH SYMBOLS: 1, 0 In addition to the operators and aggregates, we require "truth symbols" to say whether a signal is *true* or *false*, whether there is a signal or there is not a signal, whether a switch is closed or open. Sometimes the letters T and F are used for these designations, but more frequently 1 and 0 are used. (See how these *two possible states* lead us into applications of *binary* arithmetic.)

Thus, if switch . . . a . . . is closed, its value is 1. When switch . . . c . . . is open, its value is 0.

EXAMPLE 1 Express in logical mathematical symbols the statement, "It is raining and the wind is blowing."

SOLUTION

First of all, select identification symbols to stand for the two propositions which make up the statement, say r for "it is raining" and b for "the wind is blowing." Second, since these two propositions are connected, we must choose the operational symbol which will represent AND, using the • or mere juxtaposition of the identification symbols.

"It is raining and the wind is blowing" $= r \cdot b$
or $= rb$

EXAMPLE 2 Express in logical symbols the statement, "Either switch p is open when switch q is closed or switch p is closed when switch q is open."

SOLUTION

Select identification symbols:

p = switch p closed
$\bar{p}$ = switch p open
q = switch q closed
$\bar{q}$ = switch q open

Then select the operational symbols to represent the conditions:

1. The requirements of "either . . . or . . ." are met by the use of + = OR.
2. The requirements of "when" = "at the same time" = AND is met with • or juxtaposition.

"Either switch p is open when switch q is closed or switch p is closed when switch q is open" $= \bar{p}q + p\bar{q}$.

PROBLEMS 37–1

By using s to represent ''We are going to school'' and l to represent ''We are learning something new,'' write in symbolic form the following statements:

1. We are going to school, and we are learning something new.
2. We are going to school, but we are not learning something new.
3. Either we are going to school or we are learning something new, or both.
4. We are not going to school, but we are learning something new.
5. When we are going to school, then we are learning something new.
6. We are not going to school; therefore, we are not learning anything new.
7. Either we are not going to school or we are learning something new, or both.
8. We are neither going to school nor learning something new.
9. We are (*a*) both going to school and learning something new or else (*b*) we are not going to school and we are not learning something new.
10. Either we are going to school or we are learning something new, but not both.

37–3 THE AXIOMATIC TAUTOLOGIES

In Sec. 5–2 we have already learned that an axiom is a statement which is so self-evident that it need not be formally proved. And a tautology is nothing more than a statement or equation which shows two different ways of saying the same thing. This is a specific mathematician's version of the dictionary definition. For example, $\sin\theta = \dfrac{\text{opp}}{\text{hyp}}\,\theta$ is a tautology. Sometimes it is convenient to use one relationship; sometimes the other.

While philosophical logic introduces many tautologies and develops them with great care, the following brief introduction will serve the purposes of most students working in this text. Some, who go on to computer or control engineering, will want to study further to broaden their scope in the subject.

T.1 $$a \cdot a = a$$

This is the *redundancy law of multiplication*. It means that whenever a circuit design calls for a contact on relay a to be closed and later calls for another contact on the same relay a to be closed in series with the first, we really need only a single contact on relay a.

T.2 $$a + a = a$$

This is the *redundancy law of addition*. It means that when a circuit calls for a contact on relay a to be closed and later for another contact on the same relay to be closed in parallel with the first, we need only a single contact on relay a.

These first two tautologies, or laws, really say, ''Saying the same thing over and over again does not make it any more true.''

T.3 $$a \cdot b = b \cdot a$$

This is the *commutative law of multiplication*. In the mathematics of logic, as in many other systems (but not all), it does not matter what the order of the multiplication is or, in switching algebra, what the physical order of the switches in series is.

T.4 $$a + b = b + a$$

This is the *addition law of commutation*. It does not matter whether a is in parallel with b or b is in parallel with a.

T.5 $$(a \cdot b)c = a \cdot (b \cdot c)$$

This is the *associative law of multiplication* and means, again, that the order of switches in series or the order of factors in multiplication does not matter.

T.6 $$(a + b) + c = a + (b + c)$$

This is the *associative law of addition,* which is applied in the same way as T.5 and in ordinary algebra.

T.7 $$\overline{\overline{a}} = a$$

This is the *law of double complementation,* and it means that an inverted inversion has the same effect as the original proposition. (A switch, which can only be open or closed, if changed in position twice, is back in its original position.)

NOTE *Ordinary English grammar does not follow this definition because we do not always understand that two negatives make a positive in an ordinary English statement.*

T.8 $$a + \overline{a} = 1$$

This is the *first law of complementation*. Since the circuit will always give an output signal if one contact is normally closed and the other, in parallel, is normally open, a *true* indication will always appear.

T.9 $$a \cdot \overline{a} = 0$$

This is the *second law of complementation*. It is impossible to achieve an output signal with one contact open in series with another that is closed.

T.10 $$a(b + \overline{b}) = a$$

This tautology says that a contact a in series with a circuit that is always operating (T.8) will have the same effect as if that contact were alone.

T.11 $$a + (b \cdot \overline{b}) = a$$

Any contact a in parallel with a permanent open circuit (T.9) will have the same effect as a alone.

T.12 $$\overline{a \cdot b} = \bar{a} + \bar{b}$$

This is the first of De Morgan's *laws of negation*. Some serious thought, coupled with the work which will follow, will prove the truth of this and the next tautology.

T.13 $$\overline{a + b} = \bar{a} \cdot \bar{b}$$

The second of De Morgan's laws of negation.

Some additional tautologies will be found on page 685, and they will be referred to in the text below.

37–4 TRUTH TABLES

Analysis of circuits by mathematical logic may be carried out by purely algebraic means, using the tautologies, and this method will be investigated shortly. But another useful method of analyzing circuits is the method of *truth tables*. These are a fairly systematic mechanical method of examining the possible combinations of truths (or circuit conditions) existing in a particular problem. For instance, consider the circuit in Fig. 37–1, which may be described mathematically as $a \cdot b + c$. We may set up the truth table for this circuit to determine which combinations of closed (1) or open (0) conditions of the switches will produce an output signal.

The first step is to list the three possible contacts, a, b, and c, and the possibilities appearing in the formula. This step gives us the row of headings across the top of Table 37–3. Under these headings there will appear eight rows of data and calculations: 2^3, where the 2 represents the two possible states 1 and 0 and the 3 represents the three different switches, or contacts, a, b, and c. Note the mechanical method of establishing the possible combinations: each of the contacts will be open for half of the possibilities, and each will be closed for half. By making the first half of the eight possibilities for a 1 and the second half 0, half of a's 1 conditions will see b 1 and half will see b 0, and so on.

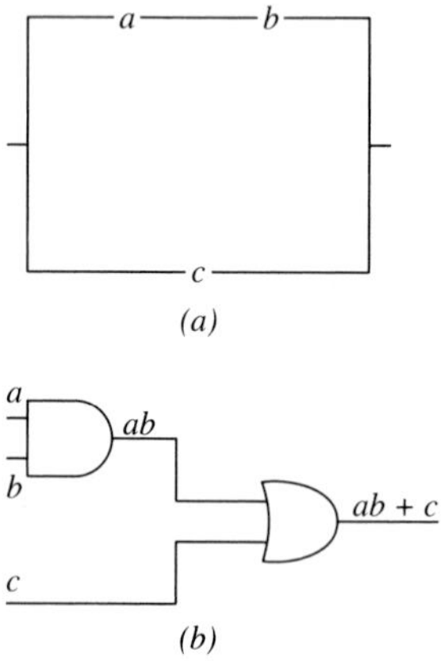

FIG. 37–1 Switching circuit for $a \cdot b + c$ in Table 37–3.

TABLE 37–3 TRUTH TABLE FOR ab + c

Combination	a	b	c	$a \cdot b$	$a \cdot b + c$
1	1	1	1	1	1
2	1	1	0	1	1
3	1	0	1	0	1
4	1	0	0	0	0
5	0	1	1	0	1
6	0	1	0	0	0
7	0	0	1	0	1
8	0	0	0	0	0

Now, referring to the circuit of Fig. 37–1 and Table 37–3, check the circuit for each row of combinations:

- Combination 1. When switches *a, b,* and *c* are closed (1), there is a complete circuit through the series leg (*ab*) and a complete circuit through the parallel switch *c*. Then the two closed parallel circuits will give a true (1) result, and there will be an output signal.
- Combination 2. When both *a* and *b* are closed, then even with *c* open, there will be an output signal, and again the last, or *total circuit,* column reads 1.
- Combination 3. Here *a* and *c* are closed and *b* is open. Hence, even when the series leg is an open circuit (0), the closed switch *c* in parallel yields an output signal.
- Combination 4. When *a* is the only closed switch, open *b* prevents a signal getting through the series leg and open *c* in parallel means that there will be *no* output signal from the circuit. The final column reads 0.
- Combination 5. In combinations 5 through 8, since *a* is open, the condition of *b* has no effect, since the series leg is of necessity open. (See column *ab*.) Switch *c,* in parallel with this open circuit, determines that there will be an output signal when *c* is closed and no output signal when *c* is open.

You must satisfy yourself that there are no other possible switch combinations and that there will be a complete circuit, or an output signal, only for combinations 1, 2, 3, 5, and 7 and no output signal for combinations 4, 6, and 8. The formula for the circuit, $ab + c$, is sometimes said to be a tautology for the five closed combinations, although this is a loose use of the word.

PROBLEMS 37–2

Prepare the truth tables for the following expressions:

1. $ab + ac$
2. $c(a + b)$
3. $\bar{a} \cdot \bar{b} \cdot \bar{c} = \overline{a + b + c}$
4. $\overline{abc} = \bar{a} + \bar{b} + \bar{c}$
5. $a + \bar{a} = 1$
6. $a(b + c) = ab + ac$
7. $a(a + b) = a$
8. $a + ab = a$
9. $p + \bar{p}q = p + q$
10. $\overline{a \cdot b} = \bar{a} + \bar{b}$
11. $a\bar{c} + b\bar{c} = \bar{c}(a + b)$
12. $a + bc = (a + b)(a + c)$
13. $\overline{x + y} + \overline{x + z} = \overline{x + yz}$
14. $(p + q)(\bar{q} + r)(q + 1) = p\bar{q} + qr$
15. $(a + b)(\bar{a} + c)(b + c) = \bar{a}b + ac$
16. $(a + c)(a + d)(b + c)(b + d) = ab + cd$

37–5 PROPOSITIONAL INVESTIGATIONS

Sometimes it happens that a proposed circuit is described in Boolean algebra in a rather complicated manner and it is possible to use the tautologies to simplify it.

EXAMPLE 3 A designer asks for a circuit which will perform the following switching function:

$$\overline{a + b} + \overline{a + c}$$

Can we simplify the circuit requirements before drawing modules from stock and putting them together as requisitioned?

SOLUTION

Choosing the appropriate tautologies (and here practice is the only cure), we alter the appearance of the original problem formula and see what might be done. (In the example, each step below has been identified with the number of tautology applied. Refer to the simple Boolean relationships section that follows.)

- Given $\overline{a + b} + \overline{a + c}$
- T.13 $\overline{a + b}$ may be written $\overline{a} \cdot \overline{b}$
- T.13 $\overline{a + c}$ may be written $\overline{a} \cdot \overline{c}$

and the formula becomes $\overline{a} \cdot \overline{b} + \overline{a} \cdot \overline{c}$

- T.14 $\overline{a}(\overline{b} + \overline{c})$
- T.12 $\overline{a}(\overline{b \cdot c})$
- T.13 $\overline{a + bc}$

Compare the original circuit, as requested, with the simplified version (Fig. 37–2*a* versus *b*). You should prepare a truth table for the two circuits and prove

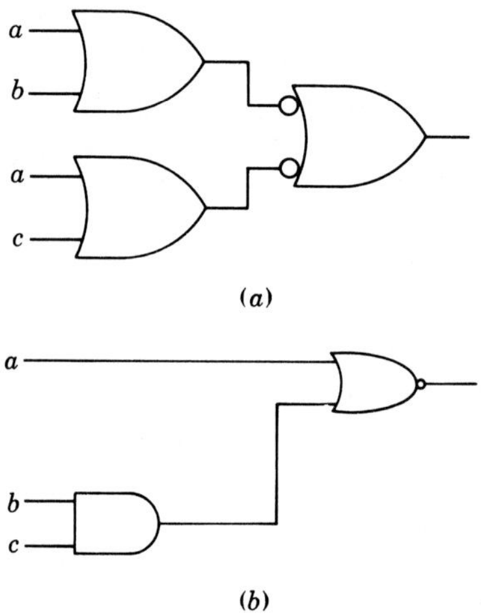

FIG. 37–2 Equivalent switching combinations of Example 3.

that the two forms are tautological, that is, when one set of switches is true, then the other also is true for all possible identical combinations. Check also to satisfy yourself that there are no combinations other than 2^3.

SIMPLE BOOLEAN RELATIONSHIPS

$$a \cdot a = a$$
$$a + a = a$$
$$a \cdot b = b \cdot a$$
$$a + b = b + a$$
$$(a \cdot b) \cdot c = a \cdot (b \cdot c)$$
$$(a + b) + c = a + (b + c)$$
$$(\bar{\bar{a}}) = a$$
$$a + \bar{a} = 1$$
$$a \cdot \bar{a} = 0$$
$$a(b + \bar{b}) = a$$
$$a + (b \cdot \bar{b}) = a$$
$$\overline{(a \cdot b)} = \bar{a} + \bar{b}$$
$$\overline{(a + b)} = \bar{a} \cdot \bar{b}$$
$$a(b + c) = ab + ac$$
$$a + bc = (a + b)(a + c)$$
$$a + \bar{a}b = a + b$$
$$a(a + b) = a$$
$$a + ab = a$$

$$a + 0 = a$$
$$a + 1 = 1$$
$$a \cdot 0 = 0$$
$$a \cdot 1 = a$$

Refer to your computer user's manual. If your computer is programmed to work with Boolean algebra, it probably uses multiplication (*) to indicate AND and addition (+) to indicate OR. Don't forget to insert an * between each of the elements to be ANDed.

PROBLEMS 37–3

Use truth tables to prove the following statements:

1. $\bar{a} + ab(ab + \bar{a}) = \bar{a} + b$
2. $(a + b)(\bar{a} + c)(b + c) = \bar{a}b + ac$
3. $(\bar{a}b + a)(\bar{a}b + c) = (a + b)(\bar{a} + c)(b + c)$
4. $a(\bar{a} + b)(\bar{a} + b + c) = ab$
5. $abc(a + b + c) = abc(ab + bc + ac) + abc(abc + ab)$
6. $(a + b)\overline{(ab)}$
7. $st + vw = (s + v)(s + w)(t + v)(t + w)$

8. $ABC + A\bar{B}C + AB\bar{C} + A\bar{B}\,\bar{C} + \bar{A}BC + \bar{A}\,\bar{B}C + \bar{A}B\bar{C} =$
$A + B + C$
9. $(\alpha + \beta)(\alpha + \gamma) = \alpha + \beta\gamma$
10. $\overline{(a \cdot b + bc + ac)} = \bar{a} \cdot \bar{b} + \bar{b} \cdot \bar{c} + \bar{a} \cdot \bar{c}$

37–6 SWITCHING NETWORKS

Although actual switches may be so adjusted that some contacts *make* before others *break,* or vice versa, or some close or open in a special sequence, in general, every individual switch is either open or closed, off or on, flipped or flopped. This two-state condition lends itself to binary operation (1 or 0), and to Boolean analysis. When a switch is closed, it provides, theoretically, perfect permittance to a current flow, and when it is open, perfect hindrance. It is convenient to define Y_{pq} as the permittance of a circuit between the points p and q and Z_{pq} as the hindrance of the circuit between the same points. Obviously, $Y_{pq} = \bar{Z}_{pq}$.

EXAMPLE 4 Write the expressions for the permittance and the hindrance of the circuit of Fig. 37–3.

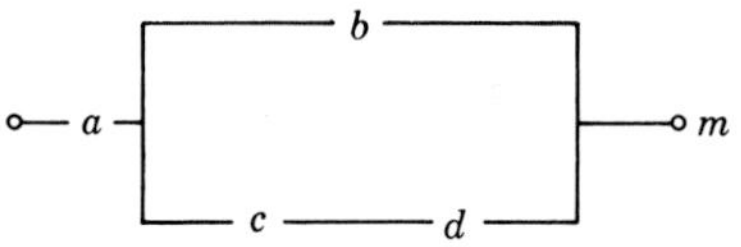

FIG. 37–3 Switching circuit of Example 4.

SOLUTION

To write the expression for the permittance of the circuit Y_{lm}, we agree that

$$Y_{lm} = Y_a(Y_b + Y_cY_d)$$

where Y_a is the permittance of switch a, and so on. We may write this simply as

$$Y_{lm} = a(b + cd)$$

and we understand that the letter designation for a switch without an overbar indicates that the switch is closed, that is, offers perfect permittance. Studying the circuit, you can see that when contact a is closed and then either b or c and d in series is closed, the circuit will offer permittance—there will be an output signal.

Similarly, the hindrance of a contact, that is, an open switch, is indicated by the letter designation with an overbar, so that Z_{lm} must be written:

$$Z_{lm} = \bar{a} + (\bar{b})(\bar{c} + \bar{d})$$

When contact a is open, or else when both b is open and either c or d is open, then there will be no output signal—or perfect hindrance.

You should prepare a set of truth tables to show that $Y_{lm} = \bar{Z}_{lm}$.

PROBLEMS 37–4

1. Write the expressions for (*a*) the hindrance and (*b*) the permittance of the circuit of Fig. 37–4.

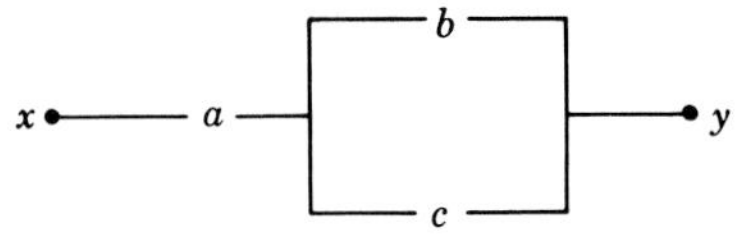

FIG. 37–4 Switching circuit of Prob. 1.

2. Write the expressions for (*a*) the hindrance and (*b*) the permittance of the circuit of Fig. 37–5.

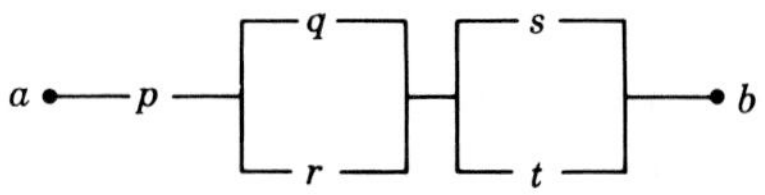

FIG. 37–5 Switching circuit of Prob. 2.

3. Write the expressions for (*a*) the hindrance and (*b*) the permittance of the circuit of Fig. 37–6.

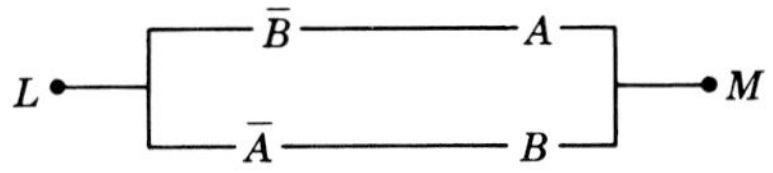

FIG. 37–6 Switching circuit of Prob. 3.

4. Write the expressions for (*a*) the hindrance and (*b*) the permittance of the circuit of Fig. 37–7.

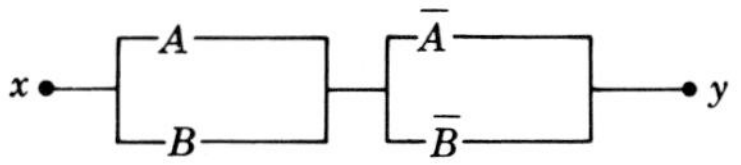

FIG. 37–7 Switching circuit of Prob. 4.

5. Write the expressions for (*a*) the hindrance and (*b*) the permittance of the circuit of Fig. 37–8.

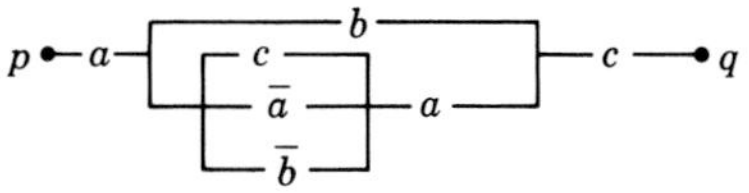

FIG. 37–8 Switching circuit of Prob. 5.

Draw the circuits for the following expressions:

6. $Y_{pq} = a(b + c)(ad)$
7. $Y_{lm} = xy(\bar{y}z + \bar{x})a$
8. $Y_{ab} = [\alpha(\beta + \bar{\gamma}) + \beta]\gamma$

Determine equivalent expressions for the following expressions using the least number of gates:

9. $Z_{\text{out}} = ABC + AC(A + B) + B + C$
10. $Y_{pq} = \bar{A}\bar{B}(C + D)\bar{B} + D$

Equivalent switching networks may be developed mathematically by using the tautologies of Boolean algebra, whereby somewhat complicated circuits may be reduced to circuits which will perform identical services with less hardware or, alternatively, to circuits which will perform identical services with readily available, although not simpler, hardware.

EXAMPLE 5 Given the switching network of Fig. 37–9, develop a simpler circuit which will provide an identical switching service.

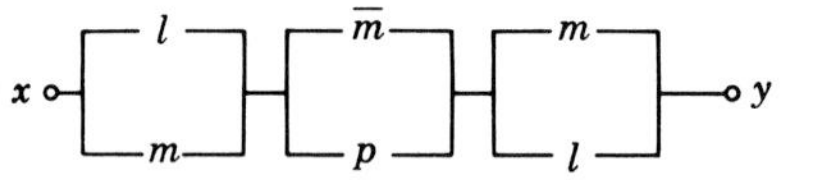

FIG. 37–9 Switching circuit of Example 5.

SOLUTION

Write either the permittance or hindrance function of the circuit:

$$Y_{xy} = (l + m)(\bar{m} + p)(m + l)$$
$$\text{T.4:}(l + m)(\bar{m} + p)(l + m)$$
$$\text{T.2:}(l + m)(\bar{m} + p)$$

That is, the network shown in Fig. 37–9 may be replaced by that shown in Fig. 37–10. You should prepare a truth table to prove that the two circuits are tautological.

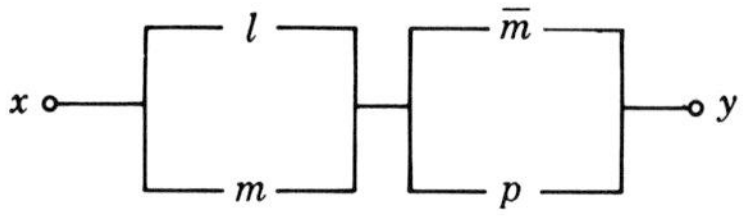

FIG. 37–10 Simpler circuit equivalent of Fig. 37–9.

PROBLEMS 37–5

1. By using the appropriate tautologies, develop a simpler circuit to replace that of Fig. 37–11.

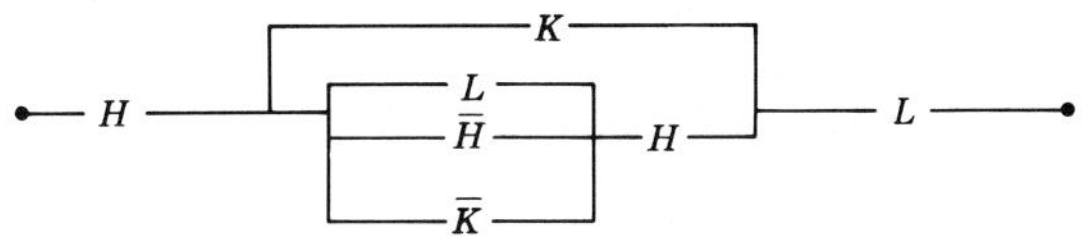

FIG. 37–11 Switching circuit of Prob. 1.

2. Develop a simpler circuit to replace that of Fig. 37–12.

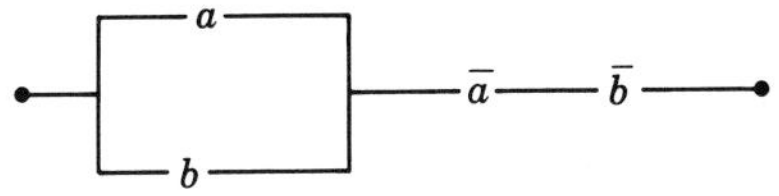

FIG. 37–12 Switching circuit of Prob. 2.

3. Develop a simpler circuit to replace that of Fig. 37–13.

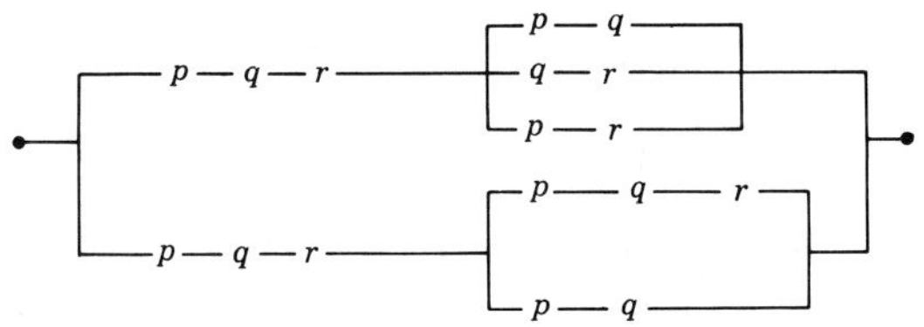

FIG. 37–13 Switching circuit of Prob. 3.

4. Develop a simpler circuit to replace that of Fig. 37–14.

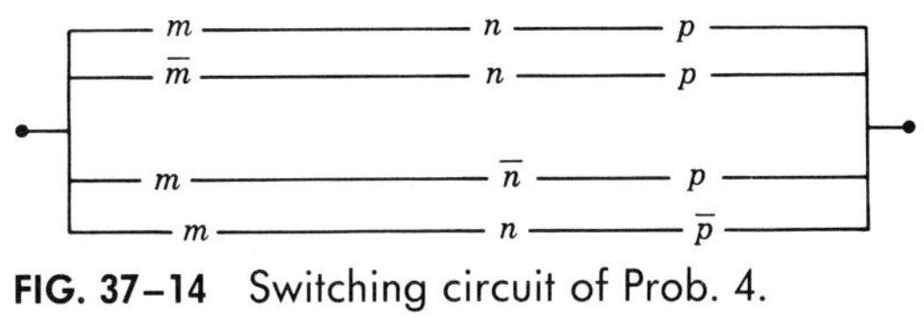

FIG. 37–14 Switching circuit of Prob. 4.

5. Develop a simpler circuit to replace that of Fig. 37–15.

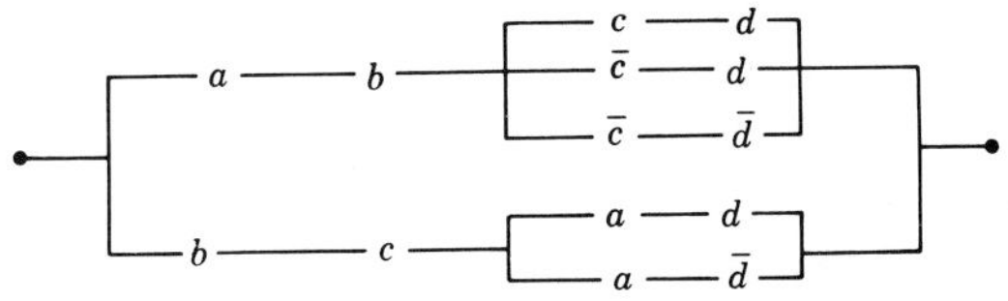

FIG. 37–15 Switching circuit of Prob. 5.

6 to 10. Check each of your solutions above by means of truth tables.

37–7 COMPUTER GATING APPLICATIONS

The standard computer gating symbols are shown in Table 37–1. These simple symbols (and the circuits for which they stand) may be combined into *adders* or *half-adders* or other more complex components. Let us look at a few of the simple tautologies as they would appear in gating configurations.

EXAMPLE 6 Tautology T.14 states that $a(b + c) = ab + ac$. The two circuit configurations are shown in Fig. 37–16.

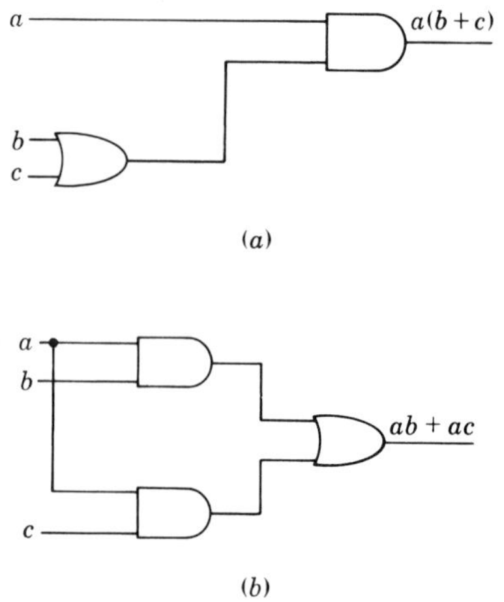

FIG. 37–16 Equivalent circuits of inclusive OR.

INVESTIGATION

You should check the two parts of Fig. 37–16 and satisfy yourself that the two circuits do perform the same functions. Then, by preparing a truth table for the two statements, you will see that when $a(b + c)$ is 1, so also is $ab + ac$, and when $a(b + c)$ is 0, so also is $ab + ac$. Then, since the two forms have been proved by tracing and by truth table to be tautological, the end results of using one will be identical with those of using the other. There may be times when availability of circuit wiring boards or parts may make it more desirable to use one circuit rather than the other, but the results will be the same regardless of the circuit configuration chosen.

You can see, then, that it may often be convenient to spend time exploring the possibilities mathematically, before even breadboarding a circuit, in order to reduce the total number of components or the number of different components required.

PROBLEMS 37–6

1. Write the output expression for the circuit of Fig. 37–17 and develop an alternate circuit. Test your answer by means of a truth table.
2. Write the output expression for the circuit of Fig. 37–18 and develop an alternate circuit. Test your answer by means of a truth table.

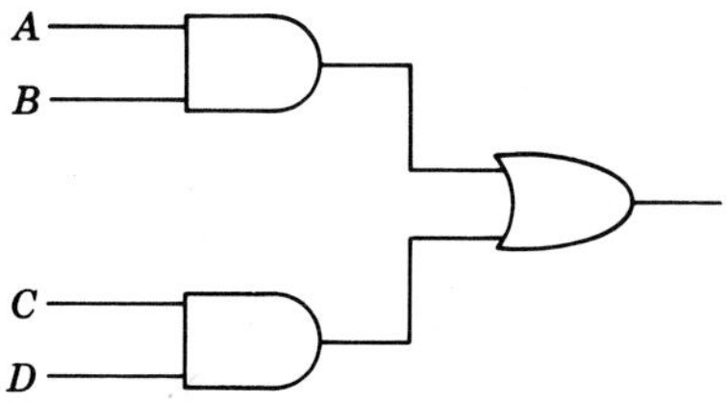

FIG. 37–17 Switching circuit of Prob. 1.

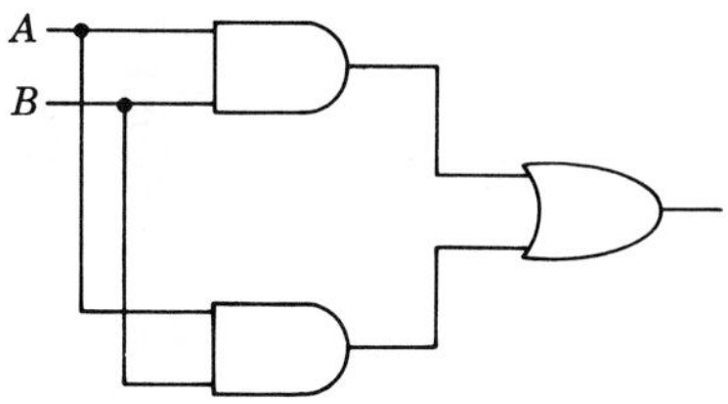

FIG. 37–18 Switching circuit of Prob. 2.

3. The *half-adder* circuit produces two outputs, a sum S and a carry C. The circuit is shown in Fig. 37–19. Show that the same result can be achieved by using three AND gates, one OR gate, and one INVERTER.

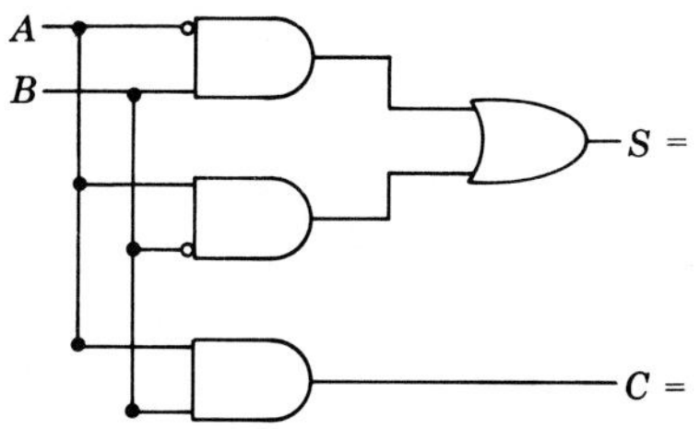

FIG. 37–19 Half-adder circuit of Prob. 3.

4. The classic *full adder,* shown in Fig. 37–20, involves the two quantities to be added (a and b) by a digital computer, plus the carry from the preceding step (c_p). The circuit requires eight AND gates, two OR gates, and nine INVERTERS. Show that the carry portion of the output may be simplified with a saving of one AND gate and three INVERTERS.

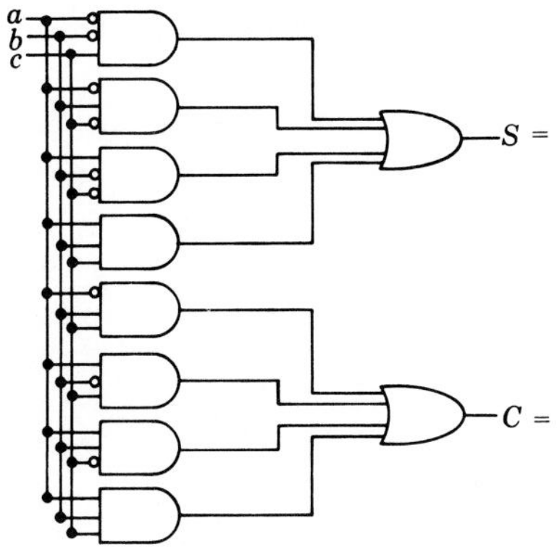

FIG. 37–20 Full-adder circuit of Prob. 4.

37–8 NOR COMBINATIONS

Over the last few years, many manufacturers have found it convenient for a number of reasons to build their logic circuits as multiples of a single type of gate. Often NOR gates (Fig. 37–21) are used because of the simplicity of circuit elements and design. In Problems 37–7 you will be asked to determine the gating equivalents of various combinations of NOR gates.

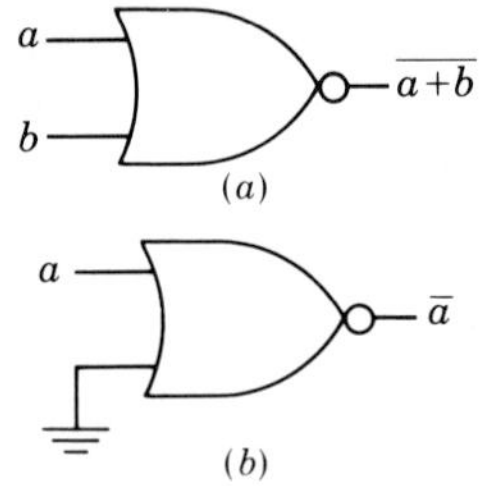

FIG. 37–21 (a) The NOR gate delivers a negation of the OR function:

$$\overline{a+b} = \overline{a} \cdot \overline{b}.$$

(b) When an input is grounded, it represents a 0:

$$\overline{a+0} = \overline{a} \cdot 1 = \overline{a},$$

and we have a NOT gate.

PROBLEMS 37–7

1. Write the output expression for the gating circuit of Fig. 37–22, and determine the simplest equivalent function.

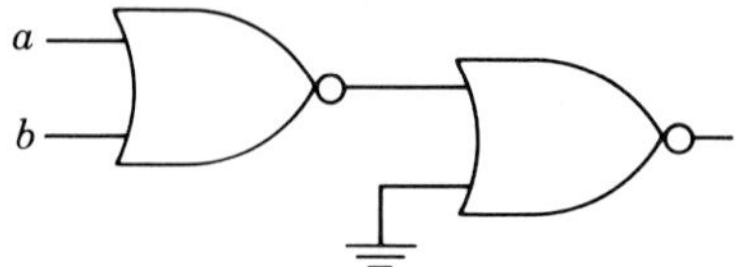

FIG. 37–22 NOR gate combination for Prob. 1.

2. Write the output expression for the gating circuit of Fig. 37–23, and determine the simplest equivalent function.

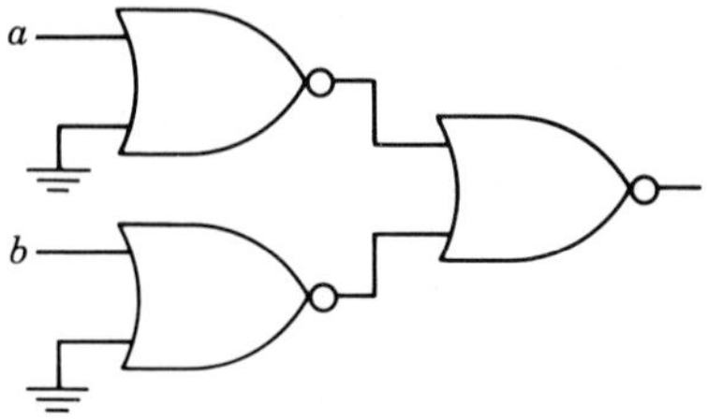

FIG. 37–23 NOR gate combination for Prob. 2.

3. Write the output expression for the gating circuit of Fig. 37–24, and determine the simplest equivalent function.

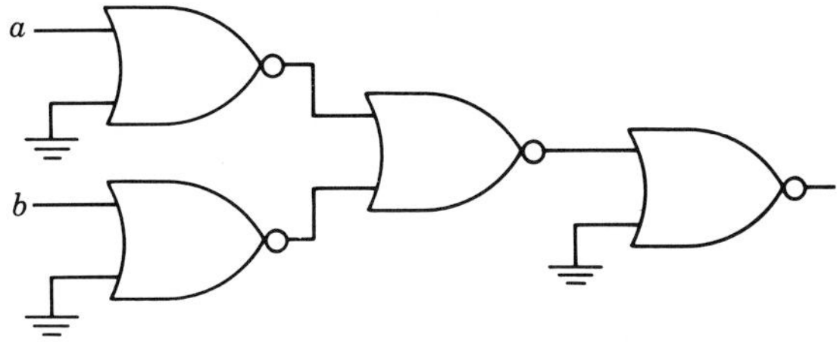

FIG. 37–24 NOR gate combination for Prob. 3.

4. Write the output expression for the gating circuit of Fig. 37–25, and determine the simplest equivalent function.

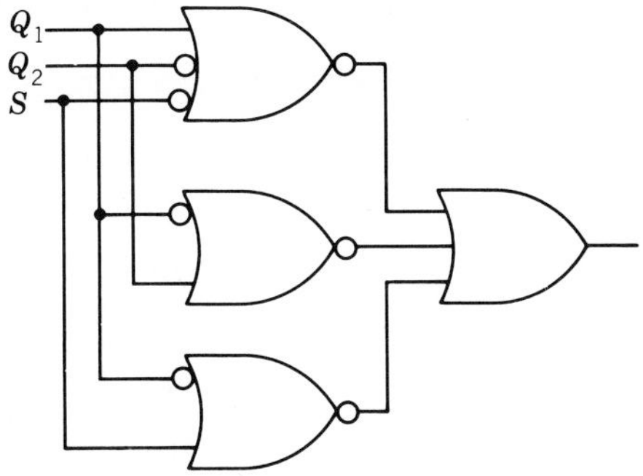

FIG. 37–25 NOR gate combination for Prob. 4.

SELF TEST

1. Use t to represent "watching television," r to represent "reading a book," and p to represent "eating popcorn" to write in symbolic form: "Whenever I see you, you are either watching television or reading a book, and you are always eating popcorn."
2. Prepare a truth table for the following expression:

$$pt + r$$

3. Use a truth table to test the truth or falsity of the following expression:

$$(x + y)(\bar{x} + z) = \bar{x}y + z(x + y)$$

4. Use appropriate tautologies to develop a simpler expression for:

$$a + ab + abc + \bar{a}cd$$

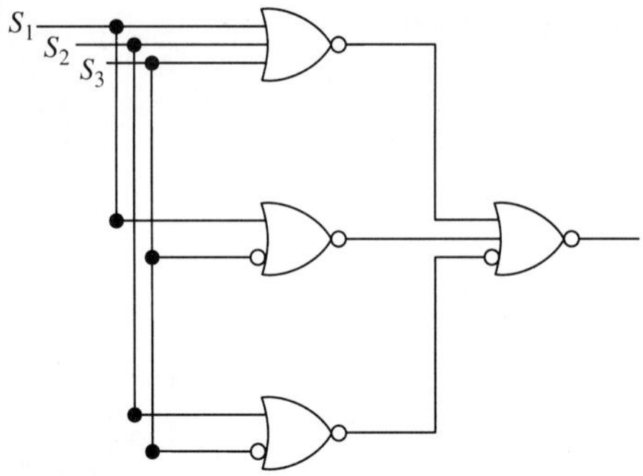

FIG. 37–26 Gating circuit of self test Prob. 5.

5. Write the output expression for the gating circuit of Fig. 37–26.
6. Use tautologies to simplify the following expression:

$$\overline{\overline{S_1S_2C} + \overline{S_1\overline{\overline{C}}} + \overline{S_2\overline{\overline{C}}}}$$

CHAPTER 38

KARNAUGH MAPS

In Sec. 37–4, we discovered the simplicity and convenience of *truth tables* in the investigation of Boolean expressions. In this chapter we will enlarge on this mechanical, tabular method of displaying data, and we will discover automatic methods of simplifying complicated expressions. If your understanding of truth tables is not complete, please review their development and application in Chap. 37.

The Karnaugh map was named for Professor Maurice Karnaugh, who developed and established the concept in 1953. At first glance the map may appear similar to a truth table, but unlike the truth table, it provides a means to simplify data. A one (1) is used, but for zero the absence of one is used. Some other map variations will describe zero as NOT-one. The argument stands, since the only two possibilities in a truth table are one (1) and zero (0); if it is NOT-one, it *must* be zero.

The map is constructed of ''boxes'' representing areas, each area indicating one unique AND function combination of variables at the input of some logic diagram. Although the map may be used for any number of variables, it is seldom used for more than six. For this analysis four variables will be considered as a maximum.

38–1 KARNAUGH MAPS FOR A SINGLE VARIABLE

A complete truth table of x variables, each of which may have two possible states (true or false, 1 or 0, etc.), requires 2^x lines to display all the possible combinations. Thus in Table 37–3, which displays three variables, there are 2^3 ($=$ 8) lines, of which the first four examine combinations in which a has the value 1 and the last four examine the combinations in which a has the value zero (0), or NOT-one ($\overline{a}$).

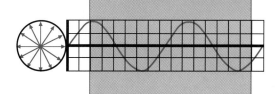

In the Karnaugh map, we provide boxes instead of lines. Thus, Fig. 38–1 illustrates the Karnaugh map for a single variable A. Since A may exist in two possible states, A or $\bar{A}$, we will require 2^1 (= 2) boxes, suitably labeled. Then, having considered all the *possible* combinations (in the case of Fig. 38–1, two possible combinations) and provided a box for each possibility (two boxes), we could enter a 1 in whichever box represents the data. If A were true, we would put a 1 in the A box. If $\bar{A}$ were true, we would put a 1 in the $\bar{A}$ box.

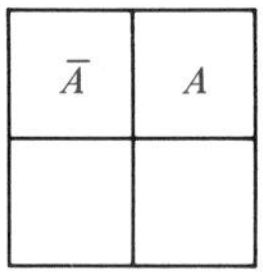

FIG. 38–1 Karnaugh map for a single variable which may have two states requires two boxes ($2^1 = 2$). A may be true or false. Total function = $\bar{A} + A$.

38–2 THE GENERAL KARNAUGH MAP

If an expression involves two variables, the Karnaugh map requires 2^2 (= 4) boxes (Fig. 38–2). An expression involving three variables requires 2^3 (= 8) boxes (Fig. 38–3). An expression involving four variables requires 2^4 (= 16) boxes, as shown in Fig. 38–4.

Because we shall be referring repeatedly to particular boxes within various maps, we need a simple method of identifying the boxes without repetitive wordiness. By numbering the boxes horizontally from left to right, and by rows from top to bottom (Fig. 38–5), we can easily refer to "box 10" rather than "the box in the second column of row three." You may find it convenient to number the boxes in your own maps, using this sequence, in order to keep track of your investigations. By keeping the box numbers in the upper corners, much like the dates on an appointment calendar, you can readily identify the boxes without interfering with the Boolean 1 notations that will belong in some of the boxes.

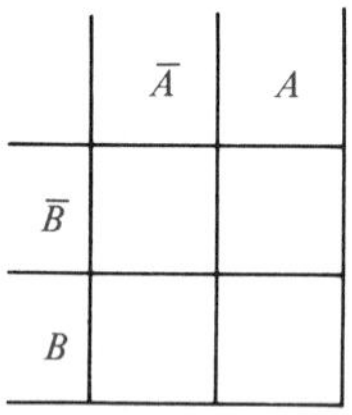

FIG. 38–2 Karnaugh map for two variables. Four boxes are required to describe two variables, each of which may have two states ($2^2 = 4$).

Refer now to Fig. 38–2, the Karnaugh map for two variables. It consists of four boxes, 1 and 2 in the $\bar{B}$ row and 3 and 4 in the B row. Each box describes a unique combination of AND functions of the variables A and B. You should note that it is not possible to represent an OR function within a single box in a Karnaugh map; two boxes must be used. If a 1 is placed in box 1, it represents the Boolean expression $\bar{A}\,\bar{B}$. If a 1 is now placed in box 2, the Boolean expression represented is $A\bar{B}$, and the total Boolean expression for boxes 1 and 2 would be $\bar{A}\,\bar{B}$ OR $A\bar{B}$, which, as we saw in Chap. 37, is usually written as $\bar{A}\,\bar{B} + A\bar{B}$.

You should confirm the following tabulation possibilities for Fig. 38–2:

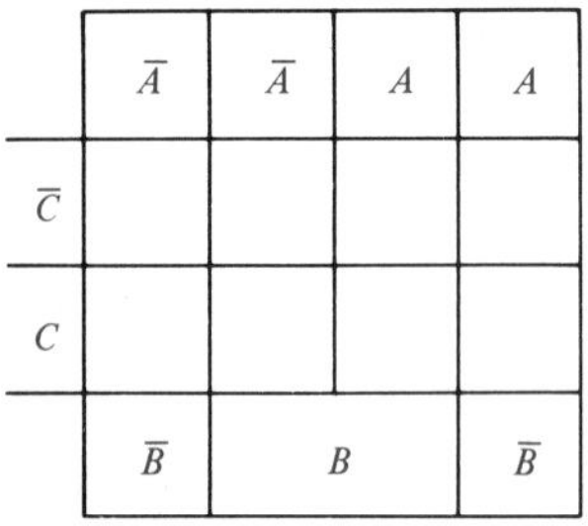

FIG. 38–3 Karnaugh map for three variables, each of which may have two states, requires eight boxes ($2^3 = 8$).

Box	Boolean Expression
1	$\bar{A}\,\bar{B}$
2	$A\bar{B}$
3	$\bar{A}B$
4	AB

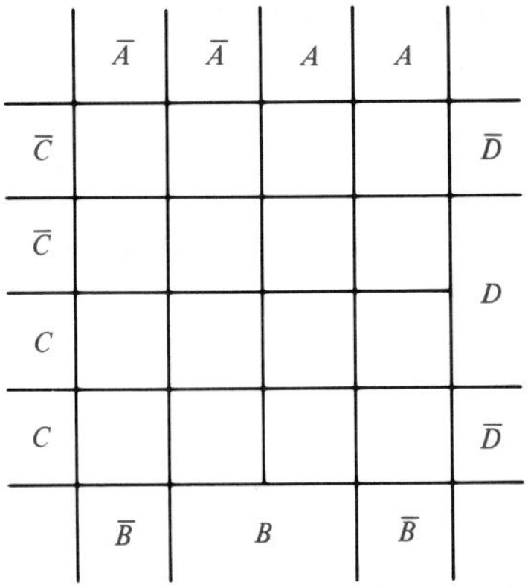

FIG. 38–4 Karnaugh map for four variables requires 16 boxes ($2^4 = 16$).

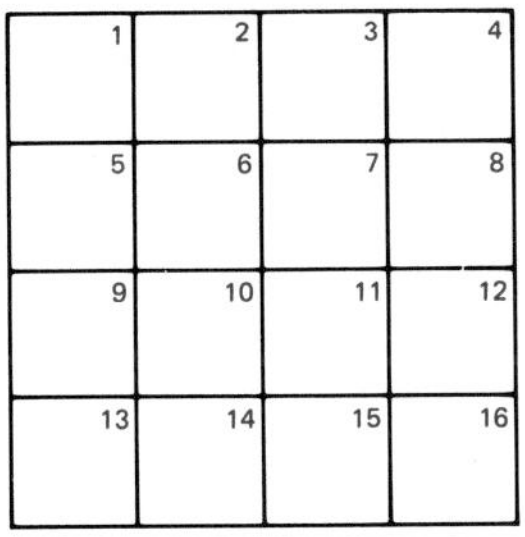

FIG. 38–5 Numbering system showing sequence of readout from a Karnaugh map.

Thus, if all four boxes of the Karnaugh map in Fig. 38–2 contained 1s, the map would illustrate the Boolean expression

$$f = \bar{A}\bar{B} + A\bar{B} + \bar{A}B + AB \ (= 1)$$

This type of expression was simplified in Chap. 37 using Boolean identities and DeMorgan's theorem. In Sec. 38–3 we will see how to use map grouping to simplify such an expression automatically. If there were 1s in only boxes 1 and 3, the Karnaugh map would represent the function

$$f = \bar{A}\bar{B} + \bar{A}B$$

In Fig. 38–3, a 1 in box 3 would represent the Boolean expression $AB\bar{C}$. You should confirm the following tabulation of possibilities for the Karnaugh map of Fig. 38–3:

Boolean expression:	$\bar{A}\bar{B}\bar{C}$	$\bar{A}B\bar{C}$	$AB\bar{C}$	$A\bar{B}\bar{C}$	$\bar{A}\bar{B}C$	$\bar{A}BC$	ABC	$A\bar{B}C$
Box number:	1	2	3	4	5	6	7	8

In Fig. 38–3, the expression $\bar{A}BC + ABC + A\bar{B}C$ would be represented by 1s in boxes 6, 7, and 8.

Figure 38–5 shows the box numbering sequence for a Boolean expression containing four terms; since $2^4 = 16$, 16 boxes will be required. Figure 38–4 shows a Karnaugh map for the variables *A*, *B*, *C*, and *D*. You should again confirm that the following tabulation covers all the possibilities of the Karnaugh map of Fig. 38–4:

$\bar{A}\bar{B}\bar{C}\bar{D}$	$\bar{A}B\bar{C}\bar{D}$	$AB\bar{C}\bar{D}$	$A\bar{B}\bar{C}\bar{D}$
1	2	3	4
$\bar{A}\bar{B}\bar{C}D$	$\bar{A}B\bar{C}D$	$AB\bar{C}D$	$A\bar{B}\bar{C}D$
5	6	7	8
$\bar{A}\bar{B}CD$	$\bar{A}BCD$	$ABCD$	$A\bar{B}CD$
9	10	11	12
$\bar{A}\bar{B}C\bar{D}$	$\bar{A}BC\bar{D}$	$ABC\bar{D}$	$A\bar{B}C\bar{D}$
13	14	15	16

Thus, the expression $\bar{A}\bar{B}\bar{C}D + \bar{A}B\bar{C}D + A\bar{B}\bar{C}D + A\bar{B}CD + \bar{A}BC\bar{D}$ would be represented by 1s in boxes 5, 6, 8, 12, and 14 of the Karnaugh map of Fig. 38–4.

Any four-variable expression of AND combinations containing the variables A, B, C, D, and $\bar{A}, \bar{B}, \bar{C}, \bar{D}$ can be represented on this four-variable map. Their box location and row number will be dependent upon the location of the labels for $A, \bar{A}, B, \bar{B}, C, \bar{C}$, and $D, \bar{D}$, all of which are quite arbitrary. This means that the box locations for the variables A, B, C, D and $\bar{A}, \bar{B}, \bar{C}, \bar{D}$ could be different from that shown in Fig. 38–4, but the 16 AND expressions would be identical to those in Fig. 38–4. Examination of the 16 terms will show that there are no two identical terms. This can be verified by constructing a truth table for the four variables A, B, C, D and producing 16 terms identical to those of Fig. 38–4.

To illustrate the map labeling as being a matter of choice, we would have been perfectly correct if we had reversed the labels in Fig. 38–1, so that box 1 represented A and box 2 represented $\bar{A}$. Similarly, the labels in any of the Karnaugh maps illustrated could have been relocated, so that, in the map of Fig. 38–4, the A's could have been interchanged with the $\bar{A}$'s, all the B labels switched with the C's and so on. While the location of the boxes containing the 1s would be different, by reading the labels for each such box and separating the boxes with *OR* signs (+), the Boolean expression could be written for any given map.

EXAMPLE 1 Draw the Karnaugh map for each of the following expressions: (*a*) $\bar{A}\bar{B} + AB$, (*b*) $AB + \overline{A + B}$.

SOLUTION

(*a*) Figure 38–6 illustrates one possible Karnaugh map for this expression. If A and $\bar{A}$ were interchanged, the 1s would appear in boxes 2 and 3. If the labels were changed so that $\bar{B}$ and B appeared across the top and $\bar{A}$ and A appeared down the side, then the 1s would appear in boxes 1 and 4, and so on.

(*b*) For $f = AB + \overline{A + B}$, since an OR function cannot be represented in a single box on a Karnaugh map, the second term $\overline{A + B}$ must be converted to an AND function, using the tautological relationships of Chap. 37. From Sec. 37–3, T.13 (also T.12), $\bar{A}\bar{B} = \overline{A + B}$, which is DeMorgan's second law of negation. Then it appears that the Boolean expression for (*b*) becomes $AB + \bar{A}\bar{B}$, which is tautological with the expression for (*a*). The only reason for preferring one expression over the other (outside of Karnaugh mapping) would merely be the choice of available circuitry and/or hardware.

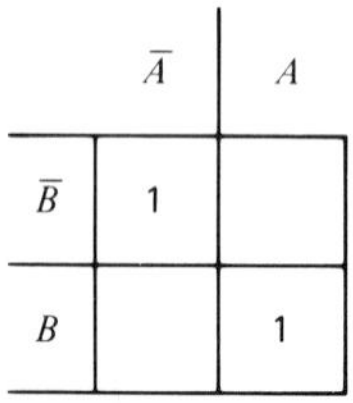

FIG. 38–6 Function (*a*) of Example 1. This uses only two boxes because the function contains only two terms. In the map, box 1 is $\bar{A}\bar{B}$ and box 4 is AB.

Analysis of the algebraic expressions within the algebraic statements for the two functions would allow us to list the hardware necessary to build the circuits. The function in (*a*) requires two ANDS, two INVERTERS, and one OR, while the function in (*b*) requires one AND, one OR, and one NOR gate to achieve the same result.

The map labeling and layout as mentioned, and shown, is quite arbitrary. Various books and journals have proposed a total of 13 different styles, none of which contradicts any other—all produce the same end result from different routines. In the following set of problems we will read the Boolean function from a Karnaugh map, and then construct a Karnaugh map from the given Boolean function.

PROBLEMS 38–1

1. Write the Boolean algebra expression for the Karnaugh map of Fig. 38–7.
2. Write the Boolean algebra expression for the Karnaugh map of Fig. 38–8.

	$\overline{A}$	A
$\overline{B}$	1	
B		1

FIG. 38–7 Karnaugh map for Prob. 1.

	$\overline{A}$	A
$\overline{B}$		1
B	1	

FIG. 38–8 Karnaugh map for Prob. 2.

3. Write the Boolean algebra expression for the Karnaugh map of Fig. 38–9.
4. Write the Boolean algebra expression for the Karnaugh map of Fig. 38–10.

	$\overline{A}$		A	
$\overline{B}$	1	1		
B		1	1	
	$\overline{C}$	C		$\overline{C}$

FIG. 38–9 Karnaugh map for Prob. 3.

	$\overline{A}$		A	
$\overline{B}$	1		1	1
B			1	
	$\overline{C}$	C		$\overline{C}$

FIG. 38–10 Karnaugh map for Prob. 4.

In the following questions, draw the Karnaugh map for the given functions:

5. $f = \overline{A}BC + A\overline{B}C + A\overline{B}\,\overline{C} + \overline{A}B\overline{C} + AB\overline{C}$
6. $f = \overline{A}\,\overline{B}\,\overline{C} + \overline{A}\,\overline{B}C + \overline{A}BC + ABC + AB\overline{C}$
7. $f = \overline{A}\,\overline{B}\,\overline{C}\,\overline{D} + \overline{A}\,\overline{B}C\overline{D} + \overline{A}\,\overline{B}CD + \overline{A}BCD + ABCD + \overline{A}B\overline{C}\,\overline{D} + \overline{A}BC\overline{D} + AB\overline{C}\,\overline{D}$
8. $f = \overline{A}\,\overline{B}\,\overline{C}\,\overline{D} + A\overline{B}C\overline{D} + A\overline{B}\,\overline{C}\,\overline{D} + \overline{A}\,\overline{B}CD + \overline{A}BCD + \overline{A}B\overline{C}\,\overline{D} + \overline{A}BC\overline{D} + AB\overline{C}\,\overline{D} + ABC\overline{D}$

38–3 GROUPING WITHIN KARNAUGH MAPS

If, in the Karnaugh map of Fig. 38–2, there were 1s in boxes 1 and 2, they would represent the Boolean expression $\overline{A}\,\overline{B} + A\overline{B}$. From the tautologies and relationships of Secs. 37–3 and 37–5, we could simplify this expression:

$$\begin{aligned}\overline{A}\,\overline{B} + A\overline{B} &= \overline{B}(\overline{A} + A)\\ &= \overline{B}(1)\\ &= \overline{B}\end{aligned}$$

No matter how we redraw the Karnaugh map, interchanging the A's and B's or the A and $\overline{A}$, B and $\overline{B}$, etc., the 1s representing the given expression will always be located in adjacent boxes—either adjacent horizontally or adjacent vertically. Try a few variations of this map, by interchanging the A's and B's, and prove this statement for yourself.

Now when two boxes on a properly labeled Karnaugh map are adjacent, either vertically or horizontally, the Boolean expression that they represent can be simplified by eliminating the matching complementary variables that are part of the labels for those boxes. Thus, in the example above, since boxes 1 and 2 are adjacent horizontally, we can eliminate the A and $\overline{A}$ components of their labels and write the simplification $\overline{B}$, which is the label of the horizontal row containing boxes 1 and 2.

EXAMPLE 2 Redraw Fig. 38–2, and insert 1s in boxes 2 and 4. Simplify the Boolean expression represented by this Karnaugh map.

SOLUTION

The Boolean expression is $A\overline{B} + AB$. Using the tautologies and relationships, this simplifies

$$\begin{aligned} A\overline{B} + AB &= A(\overline{B} + B) \\ &= A(1) \\ &= A \end{aligned}$$

Using the Karnaugh map, we have two boxes adjacent vertically. The grouping of these boxes allows us to eliminate the vertical labels $\overline{B}$ and B, and write the simplified answer A, which is the label of the vertical column containing boxes 2 and 4.

Study the Karnaugh map of Fig. 38–4. Any grouping of two adjacent boxes, horizontally or vertically, will allow us to eliminate one set of labels. The group consisting of boxes 2 and 3 allows us to eliminate A and $\overline{A}$. Boxes 1 and 2 combine to eliminate B and $\overline{B}$. Boxes 5 and 9 combine to eliminate C and $\overline{C}$. Boxes 11 and 15 combine to eliminate D and $\overline{D}$. You can easily identify other pairs of boxes that eliminate a pair of complementary labels.

Figure 38–11a and b illustrates the conditions for grouping:

Any pair of boxes that are adjacent, either horizontally or vertically, in a Karnaugh map can be grouped to eliminate a complementary pair of labels.

Diagonal pairs may *not* be grouped for such elimination.

There is no elimination possible for diagonal pairs, because the pair group does not contain any complementary pair relating to a single remaining label. In Fig. 38–11c boxes 1 and 4 involve all four possible conditions, $\overline{B}$, B, $\overline{A}$ and A, but not in any pair combination that puts two adjacent boxes under either $\overline{A}$ and A or $\overline{B}$ and B. Similarly, in Fig. 38–11d there are no *adjacent* horizontal or vertical pairs.

However, we can *always* combine adjacent pairs. The format we have arbitrarily chosen for the various Karnaugh maps for two, three, and four variables always permits the elimination of a complementary pair of labels by pair grouping. At this point we should consider the rules for map grouping for expressions containing more than two variables, but no more than four variables.

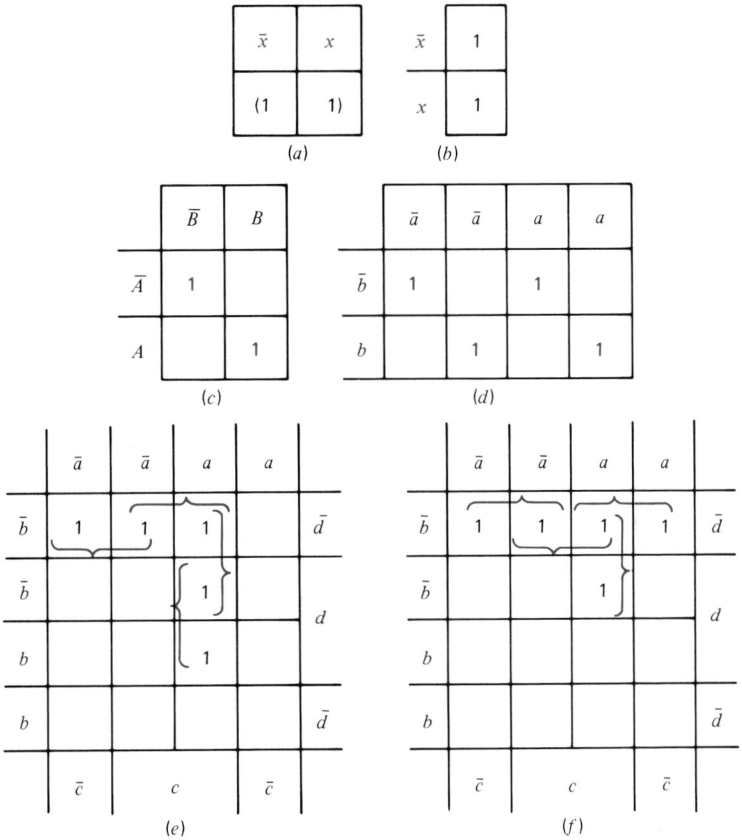

FIG. 38–11 Illustration of the conditions for grouping. (*a*) Both terms combine to become an allowed group (adjacent horizontally). (*b*) Both terms combine to become an allowed group (adjacent vertically). (*c*) Diagonally adjacent terms may not be grouped. (*d*) Diagonally adjacent terms may not be grouped. (*e*) Three adjacent boxes may be grouped in pairs. (*f*) The variable in box 3 may be grouped in three possible combinations: with box 2, box 4, or box 7.

Grouping of boxes in the Karnaugh map has resulted in simplification of the original expression by elimination of AND combinations. The rules for grouping any function including up to four variables are quite straightforward, and simply allow a more complex function (compared with the previous examples) to be reduced to a more simplified form. The following three rules will reduce the amount of Boolean algebra otherwise required to perform the same simplification:

1. Any group of an even number of boxes (2, 4, 8, or 16) may be combined in either a horizontal or a vertical direction, provided the boxes are adjacent (Fig. 38–11*a* and *b*). However, no grouping may be made in a diagonal direction.

2. Three adjacent boxes in a row or column may be grouped as two sets of two, but *not* as one group of three (Fig. 38–11*e*).
3. Any variable may be used in two different pair groups in the same row (or column) and could, under certain map conditions, be grouped a third time with an adjacent box in a column (or row). (Box 3 in Fig. 38–11*f* pairs with box 2, pairs with box 4, and pairs with box 7.)

Four adjacent boxes on a Karnaugh map might appear as

(*a*) Three sets of two horizontally,
(*b*) Three sets of two vertically, or
(*c*) A single set of four in a row, column, or block,

A set of eight adjacent boxes on a Karnaugh map might appear as

(*a*) Two sets of four,
(*b*) Four vertical pairs,
(*c*) Four horizontal pairs, or
(*d*) One block of eight boxes

A set of 16 adjacent boxes could have pair groups made both horizontally and vertically. Or there could be four sets of four boxes. Obviously there could be two sets of eight boxes, and finally one set of 16 boxes.

All of the above possibilities must obey the rules for grouping in adjacent boxes. The above conditions and the rules for grouping are best shown in the following example, which will (*a*) follow the rules for mapping the expression, (*b*) set out the conditions for grouping, and (*c*) apply these rules to simplify a Boolean expression.

EXAMPLE 3 Using the Karnaugh map in Fig. 38–11*e*, write the Boolean expression for the map, and simplify by grouping, using rules 1 and 2.

SOLUTION

From Fig. 38–11*e* the Boolean expression may be written as

$$f = \underset{1}{\bar{a}\bar{b}\bar{c}\bar{d}} + \underset{2}{\bar{a}\bar{b}c\bar{d}} + \underset{3}{a\bar{b}c\bar{d}} + \underset{7}{a\bar{b}cd} + \underset{11}{abcd}$$

Inspection of the map shows that there are four pair groupings that can be made: box 1 with box 2, box 2 with box 3, box 3 with box 7, and finally box 7 with box 11. This set of groupings is from rule 3, and applies rules 1 and 2 for two adjacent boxes. Once the groupings are complete, we can use any of the tautologies from Chap. 37 to reduce the expression further.

Now consider the pair groupings listed above. Applying rules 1 and 2, we get

$$\underset{1}{\bar{a}\bar{b}\bar{c}\bar{d}} + \underset{2}{\bar{a}\bar{b}c\bar{d}} = \bar{a}\bar{b}\bar{d}(\bar{c} + c) = \bar{a}\bar{b}\bar{d}(1) = \bar{a}\bar{b}\bar{d}$$

and

$$\underset{2}{\bar{a}\bar{b}c\bar{d}} + \underset{3}{a\bar{b}c\bar{d}} = \bar{b}c\bar{d}(\bar{a} + a) = \bar{b}c\bar{d}(1) = \bar{b}c\bar{d}$$

and $$\underset{3}{a\overline{b}c\overline{d}} + \underset{7}{a\overline{b}cd} = a\overline{b}c(\overline{d} + d) = a\overline{b}c(1) = a\overline{b}c$$

and $$\underset{7}{a\overline{b}cd} + \underset{11}{abcd} = acd(\overline{b} + b) = acd(1) = acd$$

The original function for Fig. 38–11(e) has been simplified by means of pair grouping to

$$f = \overline{a}\,\overline{b}\,\overline{d} + \overline{b}c\overline{d} + a\overline{b}c + acd$$

Obviously each pair grouping quickly eliminated one of the indexing variables and reduced four-variable AND expressions to three-variable AND expressions. The final expression also has one fewer set of variables than the original expression. In the final comparison we see that we have reduced the number of inverters by four, and the number of AND gates by one (in modern-day integrated circuits, a considerable saving).

Before continuing, you should satisfy yourself that by using the same approach that was used in Example 3, the Boolean expression for Fig. 38–11f can be simplified to $f = \overline{a}\,\overline{b}\,\overline{d} + \overline{b}c\overline{d} + a\overline{b}\,\overline{d} + a\overline{b}c$. Repeat Example 3 if this is not apparent, or follow the rationale below:

Boxes 1 and 2 produce $\overline{a}\,\overline{b}\,\overline{d}$ by eliminating $\overline{c}$ and c.
Boxes 2 and 3 produce $\overline{b}c\overline{d}$ by eliminating $\overline{a}$ and a.
Boxes 3 and 4 produce $a\overline{b}\,\overline{d}$ by eliminating $\overline{c}$ and c.

But further inspection shows that as well as being grouped with boxes 2 and 4, box 3 may also be grouped with box 7, eliminating $\overline{d}$ and d to produce $a\overline{b}c$. In this way we can see that pair grouping enables us to quickly reduce, and easily write the simplified Boolean expressions for, the function:

$$f = \overline{a}\,\overline{b}\,\overline{d} + \overline{b}c\overline{d} + a\overline{b}\,\overline{d} + a\overline{b}c$$

EXAMPLE 4 Draw the Karnaugh map for the following Boolean expression, and use pair groupings to achieve a simpler expression for the function:

$$f = \overline{a}bcd + abcd + a\overline{b}cd + abc\overline{d}$$

SOLUTION
The Karnaugh map is shown in Fig. 38–12a. There are three possible pair groupings, shown in Fig. 38–12(b):

1. Boxes 10 and 11 combine to eliminate a and $\overline{a}$: bcd.
2. Boxes 11 and 12 combine to eliminate b and $\overline{b}$: acd.
3. Boxes 11 and 15 combine to eliminate d and $\overline{d}$: abc.

Thus the simplified form of the given expression is:

$$f = bcd + acd + abc$$

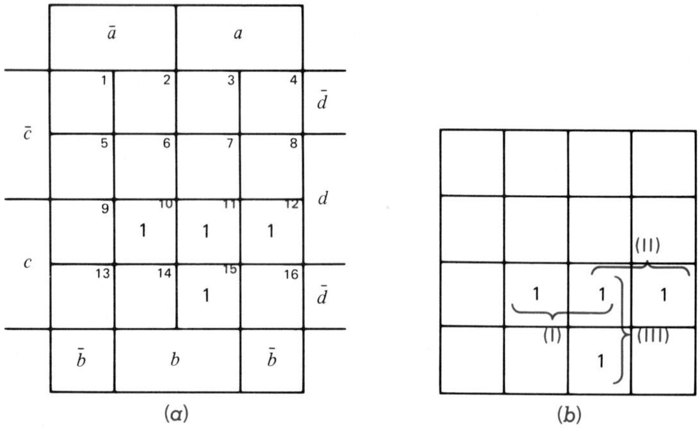

FIG. 38–12 (*a*) Karnaugh map for Example 4. (*b*) The four terms shown in (*a*) may be combined into three possible pair groupings.

Note that there are no inverted terms in the simplified form of the expression, thus removing the need for inverters from the circuit.

Before leaving Example 4, let us examine further the expression derived from the Karnaugh map:

$$f = \underset{\text{(i)}}{bcd} + \underset{\text{(ii)}}{acd} + \underset{\text{(iii)}}{abc}$$

With the aid of basic identities, the expression may be rewritten as

$f = bcd + ac(d + b)$, or $c(bd + ad + ab)$, or $f = abc + cd(b + a)$.

If there is any advantage in doing so, it would involve the final selection of circuitry desired and the circuitry available.

It is left as an exercise for you to construct the logic diagram using the least number of gates.

PROBLEMS 38–2

1. Use Karnaugh maps to simplify the following expressions, and verify the simplified form using Boolean identities. Draw the logic gating required for the *simplified* function.
   (*a*) $\overline{A}\,\overline{B}\,\overline{C} + \overline{A}B\overline{C} + \overline{A}BC + A\overline{B}C + AB\overline{C} + ABC =$ function
   (*b*) $\overline{X}Y\overline{Z} + X\overline{Y}Z + \overline{X}YZ + XY\overline{Z} + XYZ =$ function
   (*c*) $AB\overline{C} + A\overline{B}\,\overline{C} + A\overline{B}C + ABC =$ function
2. Refer to Prob. 3 of Problems 38–1. Your answer was $\overline{A}\,\overline{B}\,\overline{C} + \overline{A}\,\overline{B}C + \overline{A}BC + ABC$. Using the tautologies of Chap. 37, reduce this answer to $\overline{A}\,\overline{B} + BC$. Use pair groupings of the boxes of Fig. 38–9 to produce *three* AND groups. Can you see that since boxes 2 and 6 have both already been used in the horizontal pairs 1 and 2 and 6 and 7, the vertical pair grouping of boxes 2 and 6 does not introduce any new information? If you omit the redundant data from boxes 2 and 6, do you achieve the simplified form that resulted from the tautological simplification?

3. Refer to the Boolean expression of Prob. 5 in Problems 38–1. Use tautological analysis to show that the expression may be represented by $\overline{A}B + A\overline{B} + A\overline{C}$. Further analysis would produce another possible simplification, given by $f = \overline{A}B + A\overline{B} + B\overline{C}$.
   (*a*) Draw the Karnaugh map of the original expression using the map layout as shown in Fig. 38–3. Use pair grouping to write the simplified expression read from the map.
   (*b*) Draw a truth table for the three variables, and verify that the two given solutions for simplification are tautological with each other, and with the simplified expression read from the Karnaugh map. (All of these should be tautological with the original expression in Prob. 5 of Problems 38–1.)
4. Apply the tautological simplifications required to eliminate any redundant boxes from the Karnaugh map answer of Prob. 7 of Problems 38–1. Provide a simpler answer for the expression if possible.
5. Refer to Prob. 8 of Problems 38–1. Apply tautological simplification to eliminate any redundant boxes from the Karnaugh map answer. Can you convert the answer into a simpler expression?
6. Refer to Prob. 8 of Problems 37–3. Draw the Karnaugh map and apply pair grouping with tautological simplification to show that

$$ABC + A\overline{B}C + AB\overline{C} + A\overline{B}\,\overline{C} + \overline{A}BC + \overline{A}\,\overline{B}C + \overline{A}B\overline{C} = A + B + C$$

7. Refer to Fig. 37–20. Write the Boolean expression for the sum output S, draw the Karnaugh map for this expression, and use grouping techniques to simplify the expression $\overline{a}\,\overline{b}c + \overline{a}b\overline{c} + a\overline{b}\,\overline{c} + abc$.
8. Refer to Fig. 37–20. Write the Boolean expression for the carry output C, draw the Karnaugh map for this expression, and use grouping techniques to simplify the expression

$$\overline{a}bc + a\overline{b}c + ab\overline{c} + abc$$

9. Refer back to Prob. 3. You should note that when using the Karnaugh map, you *cannot* convert four pair groups into a three-pair answer (which *can* be obtained by using the tautologies from Chap. 37). The truth table verification will show that the two given expressions are tautologies, as is the expression read from the map with the original expression. The two forms of the simplified expression merely state:

   If $A\overline{C}$ is used with $\overline{A}B + A\overline{B}$ to perform the function, $B\overline{C}$ is a redundant term; if $B\overline{C}$ is used with $\overline{A}B + A\overline{B}$, then $A\overline{C}$ is the redundant term.

   It is left as an exercise for you to identify any redundant terms from Probs. 6, 7, and 8 of Problems 38–1.

38–4 ROLLAROUND OF KARNAUGH MAPS

The rollaround feature of Karnaugh maps allows several additional pair-grouping possibilities. To fully understand this aspect of the Karnaugh map, we should first refer back to Secs. 38–1 and 38–2. In these earlier sections we commented about the possible interchanging of indexing labels on the Karnaugh map, without changing the final simplification of the expression represented by the 1's on the map.

Since the labeling is quite arbitrary, let us examine the Karnaugh map of Fig. 38–11*f*. The labeling shown for the boxes with 1's indicates that boxes 1, 2, 3, 4, and 7 represent the Boolean expression for the function as

$$\begin{array}{lccccc} f = & \bar{a}\bar{b}\bar{c}\bar{d} + & \bar{a}\bar{b}c\bar{d} + & a\bar{b}c\bar{d} + & a\bar{b}\bar{c}\bar{d} + & a\bar{b}cd \\ & 1 & 2 & 3 & 4 & 7 \end{array}$$

This expression is reduced, using pair groupings of boxes 1 and 2, 2 and 3, 3 and 4, and 3 and 7, to $\bar{a}\bar{b}\bar{d} + \bar{b}c\bar{d} + a\bar{b}\bar{d} + a\bar{b}c$. It was left as an exercise for you to show that the expression could be further reduced with the aid of basic Boolean tautologies to $\bar{b}\bar{d} + a\bar{b}c$.

Now suppose the indexing variables $\bar{a}$–$\bar{a}$–a–a and $\bar{c}$–c–c–$\bar{c}$ had been written as a–a–$\bar{a}$–$\bar{a}$ and c–$\bar{c}$–$\bar{c}$–c. The variable AND combination $(\bar{a}\bar{b}\bar{c}\bar{d})$ that appeared in box 1 of the original map is box 3 on the relabeled map. You should satisfy yourself that the variable AND combination that is represented by box 4 $(a\bar{b}\bar{c}\bar{d})$ on the original map has become box 2 on the revised map. By drawing the revised map you should verify that the earlier statement regarding the simplification by pair groupings will produce the same Boolean expression as the original map of Fig. 38–11*f*.

By reading the AND combinations for the top row of boxes in the revised map, we obtain

$$\begin{array}{lcccc} f = & a\bar{b}c\bar{d} + & a\bar{b}\bar{c}\bar{d} + & \bar{a}\bar{b}\bar{c}\bar{d} + & \bar{a}\bar{b}c\bar{d} \\ \text{Box No.} \quad 1 & & 2 & 3 & 4 \end{array}$$

You should be able to see that grouping boxes 1, 2, 3, and 4 reduces this to $\bar{b}\bar{d}$, since the four AND terms are *identical* to those produced by the original map of Fig. 38–11*f*. It is apparent, as was stated in Sec. 38–2, that the only difference is in the box numbers, or, in other words, the location of the AND combinations is different: boxes 1, 2, 3, and 4 of the original map have become boxes 3, 4, 1, and 2, respectively.

Turning our attention to the original map, box 7, which represented the AND combination $a\bar{b}cd$, is now box 5 of the revised map. This location places it directly below box 1, as it was in the original, where it combined with box 3. You should verify that the combination of the revised box labeling for boxes 1 and 5 produces the pair grouping

$$a\bar{b}c\bar{d} \text{ with } a\bar{b}cd$$

to eliminate $\bar{d} + d$ and produce $a\bar{b}c$, which gives the same result as pairing boxes 3 and 7 in the original map. We can now complete the pair grouping for the revised map and produce the simplified function $\bar{b}\bar{d} + a\bar{b}c$. However, consider the following pair groupings:

Boxes 1 and 2 of the revised map eliminate $\bar{c} + c$: $a\bar{b}\bar{d}$.
Boxes 2 and 3 of the revised map eliminate $a + \bar{a}$: $\bar{b}\bar{c}\bar{d}$.
Boxes 3 and 4 of the revised map eliminate $\bar{c} + c$: $\bar{a}\bar{b}\bar{d}$.
Boxes 1 and 5 of the revised map eliminate $\bar{d} + d$: $a\bar{b}c$.

The pair grouping of the revised map produces the Boolean expression $f = a\bar{b}\,\bar{d} + \bar{b}\,\bar{c}\,\bar{d} + \bar{a}\,\bar{b}\,\bar{d} + a\bar{b}c$. Obviously this expression is *not* identical to the original simplification. However, a further reduction is possible using the basic Boolean tautologies:

$$\begin{aligned} a\bar{b}\,\bar{d} + \bar{b}\,\bar{c}\,\bar{d} + \bar{a}\,\bar{b}\,\bar{d} + a\bar{b}c &= \bar{b}\,\bar{d}(a + \bar{a}) + \bar{b}\,\bar{c}\,\bar{d} + a\bar{b}c \\ &= \bar{b}\,\bar{d}(1 + \bar{c}) + a\bar{b}c \\ &= \bar{b}\,\bar{d} + a\bar{b}c \end{aligned}$$

which is what was previously obtained for the simplification of the original map. The differences occur with the pair grouping of boxes 2 and 3. In the original map this pair group produced $\bar{b}c\bar{d}$. This group does not appear in the revised map. But boxes 2 and 3 of the original map are now boxes 4 and 1 of the revised map. It is also apparent that boxes 2 and 3 of the revised map were boxes 4 and 1 of the original map. By interchanging the $\bar{a}$–$\bar{a}$–a–a and $\bar{c}$–c–c–$\bar{c}$ variable labels, we have grouped boxes 1 and 4 of the original map.

It should be obvious that we can pair-group boxes 1 and 4 of the original map without interchanging labels. Closer examination would show that on the revised map, the column of boxes numbered 1, 5, 8, and 13 has been made adjacent to the column of boxes numbered 4, 8, 12, and 16. This gives the map the appearance of having the right and left columns "rolled around" to produce the possibility of another four pair groups, as shown in Fig. 38–13*b*, which depicts the left and right column *rollaround*. Study Fig. 38–13*b* closely, then add the pair grouping of boxes 1 and 4 to the original four pair groupings of the map in Fig. 38–11*f*:

$$\begin{array}{ccccccccccc} f = & \bar{a}\,\bar{b}\,\bar{c}\,\bar{d} & + & \bar{a}\,\bar{b}c\bar{d} & + & a\bar{b}\,c\bar{d} & + & a\bar{b}\bar{c}\,\bar{d} & + & a\bar{b}cd \\ & 1 & & 2 & & 3 & & 4 & & 7 \end{array}$$

This will reduce to

$$\begin{array}{ccccccccccc} f = & \bar{a}\,\bar{b}\,\bar{d} & + & \bar{b}c\bar{d} & + & a\bar{b}\,\bar{d} & + & a\bar{b}c & + & \bar{b}\,\bar{c}\,\bar{d} \\ & 1-2 & & 2-3 & & 3-4 & & 3-7 & & 1-4 \end{array}$$

which simplifies to

$$\bar{b}\,\bar{d} + a\bar{b}c$$

In this instance the extra pair grouping did not help to simplify the Boolean expression further. Later examples will show how the left and right rollaround will provide simplifications not possible without this feature. But first, let us consider what would have happened if, instead of interchanging the $\bar{a}$–$\bar{a}$–a–a and $\bar{c}$–c–c–$\bar{c}$ variable labels, we had interchanged the $\bar{b}$–$\bar{b}$–b–b and $\bar{d}$–d–d–$\bar{d}$ labels and written them as b–b–$\bar{b}$–$\bar{b}$ and d–$\bar{d}$–$\bar{d}$–d in the original Karnaugh map of Fig. 38–11*f*.

This relabeling would have put the row of boxes containing boxes 1, 2, 3, and 4 in the original map into row 3 of the revised Karnaugh map, and boxes 9, 10, 11, and 12 would now contain 1s to represent the function of the original

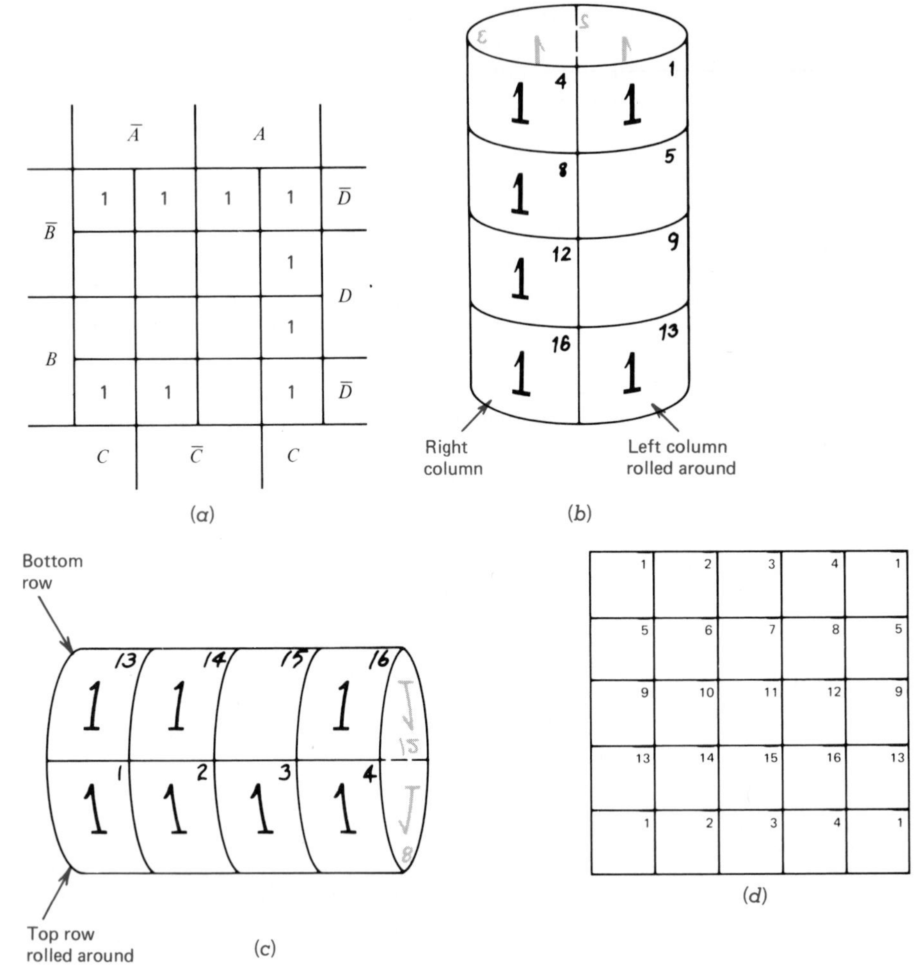

FIG. 38–13 (a) Typical Karnaugh map of a logic function involving combinations of four variables. (b) "Rollaround" of Karnaugh map of (a) makes right and left columns adjacent for grouping of boxes. (c) Rollaround of Karnaugh map makes top and bottom rows adjacent for grouping of boxes. (d) Simplified view of Karnaugh map, with rollaround groupings taken into consideration.

map. This revised labeling would also put the variable AND combinations represented by boxes 13, 14, 15, and 16 into row 2 and boxes 5, 6, 7, and 8 of the revised map. This would make boxes 1 and 13, 2 and 14, 3 and 15, and 4 and 16 a further four pair groups. These additional pair groups are more easily seen if we study the Karnaugh map shown in Fig. 38–13*c*, which depicts top and bottom rollaround. This Karnaugh rollaround shows that the pair groupings of boxes 1 and 13, 2 and 14, 3 and 15, and 4 and 16 can be achieved *without* relabeling the boxes of the original Karnaugh map.

The full rollaround feature has revealed an additional eight pair groups that can be added to the already available pair groups of the four-variable Karnaugh map. These additional pair groups, shown by the rollaround depicted in

Fig. 38–13*b* and *c*, can be seen immediately when we consider the expanded Karnaugh map shown in Fig. 38–13*d*. You should notice the "extra" row and column—these are the result of *full rollaround.*

In Fig. 38–11*d*, did you notice that boxes 1, 4, 13, and 16 now form a block of four adjacent boxes? You should study the Karnaugh map in Fig. 38–13*d* and confirm that boxes 1, 4, 8, and 5 will also form a block of four adjacent boxes. The top and bottom left-hand corners of the map, containing boxes 1, 2, 5, and 6 along with boxes 13, 14, 1, and 2, will also provide blocks-of-four groupings.

The full advantages of the complete rollaround are best seen when we see *how* the function illustrated by a Karnaugh map is affected. To do this we will consider the Boolean function represented by the Karnaugh map of Fig. 38–13*a*.

EXAMPLE 5 Use the full rollaround feature of Karnaugh mapping techniques to develop pair groupings to simplify the Boolean expression in the Karnaugh map of Fig. 38–13*a*.

SOLUTION

From the map, the Boolean expression of the function is seen to be

$$F = \underset{1}{\overline{A}\,\overline{B}C\overline{D}} + \underset{2}{\overline{A}\,\overline{B}\,\overline{C}\,\overline{D}} + \underset{3}{A\overline{B}\,\overline{C}\,\overline{D}} + \underset{4}{A\overline{B}C\overline{D}} + \underset{8}{A\overline{B}CD} + \underset{12}{ABCD} + \underset{13}{\overline{A}BC\overline{D}} + \underset{14}{\overline{A}B\overline{C}\,\overline{D}} + \underset{16}{ABC\overline{D}}$$

See how the function is affected when we just consider the full rollaround for horizontal and vertical pair groupings: the top and bottom rows roll together (Fig. 38–13*c*), producing three pairs of adjacent boxes, 1 and 13, 2 and 14, and 4 and 16. By rolling the extreme left and right columns together (Fig. 38–13*b*), we produce further adjacent pairs (1 and 4 and 13 and 16). You should note that relabeling the indexing variables produces the same results. Now what appeared in Fig. 38–13*a* as a function that involved nine of the boxes on a four-variable Karnaugh map and permitted pair groupings that involved boxes 1 and 2, 2 and 3, 3 and 4, and 13 and 14 horizontally and boxes 4 and 8, 8 and 12, and 12 and 16 vertically is seen to permit five additional pair groupings. Each pair grouping allows the elimination of one variable from the total expression for the function. With the rollaround feature considered, the pair groupings are:

Box Numbers:

1–2:	eliminates	$C + \overline{C}$: $\overline{A}\,\overline{B}\,\overline{D}$
2–3:	eliminates	$\overline{A} + A$: $\overline{B}\,\overline{C}\,\overline{D}$
3–4:	eliminates	$\overline{C} + C$: $A\overline{B}\,\overline{D}$
13–14:	eliminates	$C + \overline{C}$: $\overline{A}B\overline{D}$
4–8:	eliminates	$\overline{D} + D$: $A\overline{B}C$
8–12:	eliminates	$\overline{B} + B$: ACD
12–16:	eliminates	$D + \overline{D}$: ABC
4–1:	eliminates	$A + \overline{A}$: $\overline{B}C\overline{D}$
16–13:	eliminates	$A + \overline{A}$: $BC\overline{D}$
13–1:	eliminates	$B + \overline{B}$: $\overline{A}C\overline{D}$
14–2:	eliminates	$B + \overline{B}$: $\overline{A}\,\overline{C}\,\overline{D}$
16–4:	eliminates	$B + \overline{B}$: $AC\overline{D}$

With just a preliminary inspection of the three-variable AND functions, you should see that there are further reductions to be made using the tautologies from Chap. 37. It is left as an exercise for you to show that the final expression for the 12 three-variable AND functions of Example 5 will reduce to

$$F = AC + C\overline{D} + \overline{A}\,\overline{D} + \overline{B}\,\overline{D}$$

Variations of this final expression will merely be a choice of available circuitry. For example,

$$F = AC + \overline{D}(\overline{A} + \overline{B} + C)$$

would reduce the number of AND gates and inverters, without changing the requirements for output.

In Example 5, we could have read the final function directly from the Karnaugh map. However, to do this we must look more closely at larger groupings of the variables on the Karnaugh map.

With the possibility of larger groupings, we should make ourselves aware of the *extra* sets of grouping made possible by the rollaround feature of the Karnaugh map. Re-examine Fig. 38–13*b, c,* and *d,* and consider the total pair groups in a Karnaugh map numbered as in Fig. 38–5, but with the top and bottom rows and the right and left columns rolled together, as previously described (Fig. 38–13*d*).

The extra sets of grouping become as many as the original 16 boxes, so we have a possible 32 *pair* groups. The combinations would be as follows:

1–2, 2–3, 3–4, 4–1
5–6, 6–7, 7–8, 8–5
9–10, 10–11, 11–12, 12–9
13–14, 14–15, 15–16, 16–13

1–5, 5–9, 9–13, 13–1
2–6, 6–10, 10–14, 14–2
3–7, 7–11, 11–15, 15–3
4–8, 8–12, 12–16, 16–4

38–5 LARGER GROUPINGS WITHIN KARNAUGH MAPS

Refer again to Fig. 38–11*f*. In Sec. 38–3, we saw that four pair groupings could be achieved, producing the simplified Boolean expression

$$\overline{a}\,\overline{b}\,\overline{d} + \overline{b}c\overline{d} + a\overline{b}\,\overline{d} + a\overline{b}c$$

It presented no problem to apply the tautological relationships to this ''simplified'' form in order to simplify it further to produce $\overline{b}\,\overline{d} + a\overline{b}c$. The question you should *ask* is: can we achieve the same result directly from the Karnaugh map?

In Sec. 38–3 we stated that grouping within the Karnaugh map could consist of:

1. Any group of an even number of boxes (2, 4, 8, or 16) that may be combined in either a horizontal or a vertical direction, provided that the boxes are adjacent.

2. Four adjacent boxes on a Karnaugh map might appear as a single set of four, in a row, column, or block.

If these statements are not familiar to you, refer back to Sec. 38–3 and reread the three *rules* regarding map grouping.

Up to this point we have thoroughly examined and used pair grouping to simplify Boolean functions illustrated by a Karnaugh map. If we reexamine the steps taken to simplify the top row of the Karnaugh map of Fig. 38–11*f,* we see that we paired boxes 1 and 2 to eliminate $\bar{c}$ and c, and we paired boxes 3 and 4 to eliminate $\bar{c}$ and c again. By pairing boxes 2 and 3, both of which have already been used in the previous pair groupings, we eliminated $\bar{a}$ and a. That is, the *four-box* grouping of boxes 1, 2, 3, and 4 combines in a *single step* to eliminate the variables $\bar{a}$, a, $\bar{c}$, *and* c, leaving only $\bar{b}\,\bar{d}$ to represent the entire top row of the map. The additional grouping of the vertical boxes 3 and 7 gives us the second term of the simplest statement of the Karnaugh map: $a\bar{b}c$. As a final check, you should prepare a truth table to show that

$$\bar{a}\,\bar{b}\,\bar{c}\,\bar{d} + \bar{a}\bar{b}c\bar{d} + a\bar{b}c\bar{d} + a\bar{b}\,\bar{c}\,\bar{d} + a\bar{b}cd = \bar{a}\,\bar{b}\,\bar{d} + \bar{b}c\bar{d} + a\bar{b}\,\bar{d} + a\bar{b}c$$
$$= \bar{b}\,\bar{d} + a\bar{b}c$$

We can make the following statement about groups of four adjacent boxes:

RULE

Either a row or a column of four adjacent boxes or a block of four adjacent boxes on a four-variable map will eliminate two of the four variables.

We will remember the result of combining the top row of boxes 1, 2, 3, and 4, and apply the four-grouping method to the following example.

EXAMPLE 6 Consider the pair-grouping result of Example 5. For Fig. 38–13*a,* the total pairings of the top row result in

Top row: $$\underset{1-2}{\bar{A}\,\bar{B}\,\bar{D}} + \underset{2-3}{\bar{B}\,\bar{C}\,\bar{D}} + \underset{3-4}{A\bar{B}\,\bar{D}} + \underset{4-1}{\bar{B}C\bar{D}}$$

From the preceding examination of the top row for Fig. 38–11*f,* we can see that the top row of Fig. 38–13*a* will simplify to just $\bar{B}\,\bar{D}$; pairs 1–2 and 3–4 eliminated $\bar{C}$ and C, while pairs 2–3 and 4–1 eliminated $\bar{A}$ and A.

With variables $\bar{A}$, A, $\bar{C}$, and C eliminated, the top row becomes just $\bar{B}\,\bar{D}$. You should satisfy yourself that boxes 1–2 and 3–4 eliminate C and $\bar{C}$ to produce $\bar{A}\,\bar{B}\,\bar{D} + A\bar{B}\,\bar{D}$, which with the aid of tautologies can be reduced to $\bar{B}\,\bar{D}(\bar{A} + A) = \bar{B}\,\bar{D}$. Since the tautology removed the A variable, and boxes 2–3 and 4–1 also eliminate the A variable, the total grouping of these four boxes 1–2–3–4 (including rollaround 4–1) is $\bar{B}\,\bar{D}$.

Similarly, if we look at the vertical column of boxes 4–8–12–16 (including rollaround 16–4) and group them as a column of four, we eliminate the vertical indexing variables $\bar{B}$, B, $\bar{D}$, and D to produce simply AC.

Now consider a block of four adjacent boxes 1–13–2–14: Pairs 13–1 and 14–2 eliminate B and $\bar{B}$, to yield $\bar{A}C\bar{D} + \bar{A}\,\bar{C}\,\bar{D}$, which simplifies to $\bar{A}\,\bar{D}$. Pairs

1–2 and 13–14 eliminate $\overline{C}$ and C, to yield $\overline{A}\,\overline{B}\,\overline{D} + \overline{A}B\overline{D}$, which simplifies to $\overline{A}\,\overline{D}$. Similarly, pairs 4–1 and 16–13 eliminate $\overline{A}$ and A, and pairs 13–1 and 16–4 eliminate $\overline{B}$ and B, which leaves $C\overline{D}$. This would appear to be the proof that a block of four adjacent boxes (4–1, 16–13, 13–1, and 16–4) eliminates two of the variables on this four-variable map. Thus we confirm that four adjacent boxes, whether they be in rows, columns, or blocks, will eliminate two of the four variables. Combining the results of groupings of two and four adjacent boxes enables us to simplify the results of Example 5 without referring to the tautologies: From the simplification noted above, and the following tabulation of Example 5,

Boxes 1–2–3–4 yield $\overline{B}\,\overline{D}$.
Boxes 1–2–13–14 yield $\overline{A}\,\overline{D}$.
Boxes 4–8–12–16 yield AC.
Boxes 1–13–4–16 yield $C\overline{D}$.

All combinations of boxes in Fig. 38–13*a* for Examples 5 and 6 have been considered (four of them twice), and the function originally given as

$$\underset{1}{\overline{A}\,\overline{B}C\overline{D}} + \underset{2}{\overline{A}\,\overline{B}\,\overline{C}\,\overline{D}} + \underset{3}{A\overline{B}\,\overline{C}\,\overline{D}} + \underset{4}{A\overline{B}C\overline{D}} + \underset{8}{A\overline{B}CD} + \underset{12}{ABCD} + \underset{13}{\overline{A}BC\overline{D}} + \underset{14}{\overline{A}B\overline{C}\,\overline{D}} + \underset{16}{ABC\overline{D}}$$

in the Karnaugh map of Fig. 38–13*a* is now simplified by Karnaugh grouping techniques to

$$f = AC + \overline{A}\,\overline{D} + \overline{B}\,\overline{D} + C\overline{D}$$

which we *may* choose to write as

$$f = AC + \overline{D}(\overline{A} + \overline{B} + C)$$

If we now consider the numbering system for all boxes, as illustrated in Fig. 38–5, and combine all possible blocks of four adjacent boxes on a four-variable map, without rollaround, there is the possibility of:

1. Four horizontal rows
2. Four vertical columns
3. Nine blocks of four

If we include rollarounds (Fig. 38–13*d*), we should find that the possible total number of blocks of four adjacent boxes has increased from the groupings previously considered without rollaround to a possible total of 24 four-variable groups. These rows, columns, and blocks of four adjacent boxes are as follows:

Horizontal rows of four adjacent boxes: 1–2–3–4, 5–6–7–8, 9–10–11–12, 13–14–15–16.
Vertical columns of four adjacent boxes: 1–5–9–13, 2–6–10–14, 3–7–11–15, 4–8–12–16.

Blocks of four adjacent boxes without rollaround: 1–2–5–6, 2–3–6–7, 3–4–7–8, 5–6–9–10, 6–7–10–11, 7–8–11–12, 9–10–13–14, 10–11–14–15, 11–12–15–16.

With rollaround, the *extra* blocks of four adjacent boxes are: 13–14–1–2, 14–15–2–3, 15–16–3–4, 4–1–8–5, 8–5–12–9, 12–9–16–13, and 16–13–4–1.

In Sec. 38–3 we developed pair groupings, which we later combined into groupings of four adjacent boxes in rows, columns, and blocks. It only remains for us to agree that on a four-variable Karnaugh map, there may also be possible groups of eight adjacent boxes, including the side-to-side and top-to-bottom rollarounds. These groups are as follows:

Without rollaround: 1–2–3–4–5–6–7–8, or 5–6–7–8–9–10–11–12, or 9–10–11–12–13–14–15–16, or 1–5–9–13–2–6–10–14, or 2–6–10–14–3–7–11–15, or 3–7–11–15–4–8–12–16.

With rollaround: 13–14–15–16–1–2–3–4 or 4–8–12–16–1–5–9–13.

The eight-adjacent-box groupings should present little or no difficulty, since eight adjacent boxes are only two sets of four adjacent boxes (block 5–6–9–10 and block 7–8–11–12, for example).

PROBLEMS 38–3

1. In Fig. 38–11*f*, relabel the $\bar{a}$–$\bar{a}$–a–a and $\bar{c}$–c–c–$\bar{c}$ variables to read a–a–$\bar{a}$–$\bar{a}$ and c-$\bar{c}$-$\bar{c}$-c. Show that, in this revised map, boxes 1–4, 5–8, 9–12, and 13–16 are adjacent.
2. In Fig. 38–11*f*, relabel the $\bar{b}$–$\bar{b}$–b–b and $\bar{d}$–d–d–$\bar{d}$ variables to read b–b–$\bar{b}$–$\bar{b}$ *and* d–$\bar{d}$–$\bar{d}$–d. Show that, in this revised map, original boxes 1–13, 2–14, 3–15, and 4–16 become adjacent.
3. Use Karnaugh map groupings to simplify Problems 1 through 10 of Problems 37–3, and show that the same, or equivalent, answers are obtained.
4. In the Karnaugh map of Fig. 38–11*f*, show that, when only the $\bar{a}$–a variable labels are interchanged to read a–a–$\bar{a}$–$\bar{a}$, box 2 of the original map becomes box 3 and box 3 becomes box 2, and they are still adjacent boxes.
5. From Prob. 4, identify the boxes that will describe the AND combination of variables given in the original map of Fig. 38–11*f* as: (*a*) $a\bar{b}c\bar{d}$, (*b*) $a\bar{b}cd$.
6. From the revised Karnaugh map of Prob. 4, write the simplified Boolean expression for the least number of AND combinations, using only map grouping techniques.
7. Use Karnaugh maps to confirm your previous simplifications of problems in Chap. 37.

SELF TEST

1. Write the Boolean algebra expression for the Karnaugh map of Fig. 38–14.
2. What is the simplest equivalent of the expression for the K-map of Prob. 1?
3. Write the Boolean expression for the K-map of Fig. 38–15.

	\bar{A}	A
\bar{B}	1	1
B		

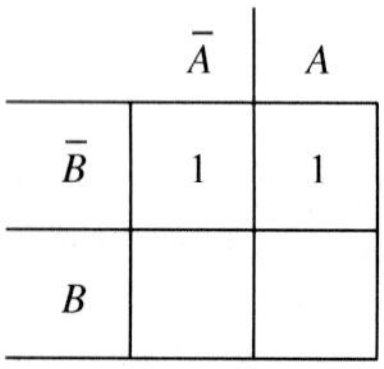

FIG. 38–14 Karnaugh map for self test Prob. 1.

	\bar{A}		A		\bar{A}
\bar{B}	1	1		1	
B		1	1		
\bar{B}					
	\bar{C}		C		\bar{C}

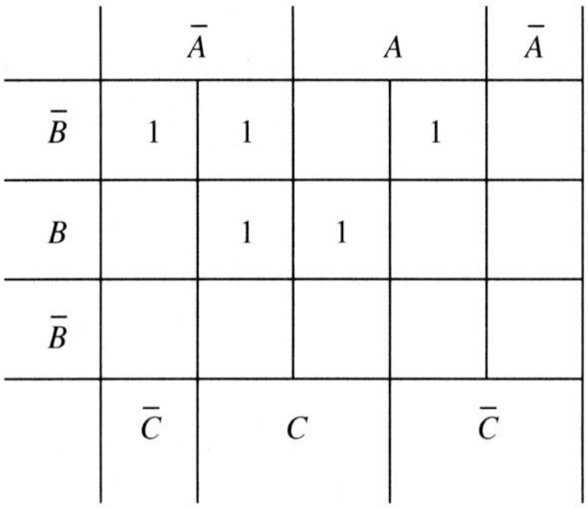

FIG. 38–15 Karnaugh map for self test Prob. 3.

4. Redraw the K-map of Fig. 38–15 to complete the rollaround.
5. Use pairing to simplify the Boolean expression of Probs. 3 and 4 to produce four terms.
6. Draw the K-map for the following function:

$$\bar{A}\,\bar{B}\,\bar{C}\,\bar{D} + A\bar{B}\,\bar{C}\,\bar{D} + \bar{A}BCD + AB\bar{C}D + ABCD + AB\bar{C}\,\bar{D}$$

7. Redraw the K-map of Prob. 6, completing the rollaround; apply pairing, and simplify the expression to six terms.

APPENDIX A

TABLES

TABLE 1 MATHEMATICAL SYMBOLS

× or ·	multiplied by, AND	≅	is approximately equal to	∴	therefore
÷ or /	divided by	≠	does not equal	∠	angle
+	positive, plus, add, OR	>	is greater than	⊥	perpendicular to
−	negative, minus, subtract	≫	is much greater than	‖	parallel to
±	positive or negative, plus or minus	<	is less than	$\|n\|$	absolute value of n
∓	negative or positive, minus or plus	≪	is much less than	△	increment of
= or ::	equals	≧	greater than or equal to	%	percent
≡	identity	≦	less than or equal to	∝	is proportional to

TABLE 2 LETTER SYMBOLS

Term	Symbol	Term	Symbol	Term	Symbol	Term	Symbol
Altitude	a	Drain	D	Number of turns	n	Speed of light	c
Area	A	Electromotive force	V, v	Ohm	Ω	Temperature	t
Base	B	Emitter	E	Period (of time)	T	Time	t
Capacitance	C	Frequency	f	Power	P	Transistor	Q
Cathode	K	Gate	G	Reactance	X	Tube (valve)	V
Collector	C	Impedance	Z	Resistance	R, r	Voltage	V, v
Current	I, i	Inductance	L	Resonant frequency	f_r	Wavelength	λ
Diode	D	Length	l	Rise time	t_r	Width	w

TABLE 3 ABBREVIATIONS AND UNIT SYMBOLS

Term	Abbreviation	Term	Abbreviation	Term	Abbreviation
Alternating current	ac	Cotangent	cot	Degree (interval or change)	deg
Ampere	A	Coulomb	C	Degrees Celsius	°C
Ampere-hour	Ah	Counterclockwise	ccw	Degrees Fahrenheit	°F
Amplitude modulation	AM	Counter electromotive force	cemf	Diameter	diam
Antilogarithm	antilog	Cubic	$\ldots^3$	Direct current	dc
Audio frequency	AF	Cubic centimeter	cm^3		
Bel	B	Cubic foot	ft^3	Dozen	spell
Candela	cd	Cubic inch	in^3	Efficiency	spell
Centimeter	cm	Cubic meter	m^3	Electromotive force	emf
Circular	cir	Cubic yard	yd^3	Equation	Eq.
Clockwise	cw	Cycles per second	Hz	Farad	F
Cologarithm	colog	Decibel	dB	Foot, feet	ft
Continuous wave	CW	Decibels referred to a level of one milliwatt	dBm	Feet per minute	ft/min
Cosecant	csc			Feet per second	ft/s
Cosine	cos				

ABBREVIATIONS AND UNIT SYMBOLS (*Continued*)

Term	Abbreviation	Term	Abbreviation	Term	Abbreviation
Feet per second squared	ft/s^2	Mega	M	Ohms	Ω
Figure	Fig.	(prefix, = 1×10^6)		Ohms per kilometer	Ω/km
Frequency	spell	Megacycles per second	MHz	Ounce	oz
Frequency modulation	FM	Megahertz	MHz	Peak-to-peak	p-p
Giga (prefix, =		Megavolt	MV	Pico (prefix, =	p
1×10^9)	G	Megawatt	MW	1×10^{-12})	
Gigacycles per second	GHz	Megohm	MΩ	Picoampere	pA
Gigahertz	GHz	Meter	m	Picofarad	pF
Gram	g	Meter-kilogram-second		Picosecond	ps
Henry	H	system	MKS	Picowatt	pW
Hertz	Hz	Meters per second	m/s	Pound	lb
High frequency	HF	Mho	S	Power factor	PF
Highest common factor	HCF	Micro (prefix, =	μ	Problem	Prob.
Hour	h	1×10^{-6})		Radian	$...^r$
Hundred	spell, or	Microampere	μA	Radians per second	$^r/s$
	× 10	Microfarad	μF		
Inch	in	Microhenry	μH	Radio frequency	RF
Inches per second	in/s	Micromho	μS	Radius	r, R
Intermediate frequency	IF	Micromicro (prefix, =	p	Range (distance)	R
Joule	J	1×10^{-12})		Revolutions per minute	rev/min
Kelvin	K	Micromicrofarad	pF	Revolutions per second	rev/s
Kilo (prefix, = 1×10^3)	k	Microsecond	μs	Root mean square	rms
Kilocycles per second	kHz	Microsiemens	μS	Secant	sec
Kilogram	kg	Microvolt	μV		
Kilohertz	kHz	Microwatt	μW	Second	s
Kilohm	kΩ	Mile	mi	Siemens	S
		Miles per hour	mi/h		
Kilometer	km	Miles per minute	mi/min	Sine	sin
Kilometers per hour	km/h	Miles per second	mi/s	Square centimeter	cm^2
Kilovars	kvar	Milli (prefix, =	m	Square foot	ft^2
Kilovolt	kV	1×10^{-3})		Square inch	in^2
Kilovoltampere	kVA	Milliampere	mA	Square meter	m^2
Kilowatt	kW	Millihenry	mH	Square yard	yd^2
Kilowatthour	kWh	Millimeter	mm	Tangent	tan
Knot	kn	Millisecond	ms		
Logarithm (common,		Millivolt	mV	Ultrahigh frequency	UHF
base 10)	log	Milliwatt	mW	Var (reactive	var
Logarithm (any base)	log_a	Minimum	min	voltampere)	
Logarithm (natural		Minute	min	Very high frequency	VHF
base *e*)	log_e, ln	Most significant bit	MSB	Volt	V
Low frequency	LF	Nano (prefix, =	n	Voltampere	VA
Lowest common		1×10^{-9})		Watt	W
denominator	LCD	Nanoampere	nA	Watthour	Wh
Lowest common		Nanofarad	nF	Wattsecond	Ws
multiple	LCM	Nanosecond	ns	Webers per square	Wb/m^2
Lowest significant bit	LSB	Nanowatt	nW	meter	
Lumen	lm	Neper	Np	Yard	yd
Lux	lx	Number	No. or		
Maximum	max		spell		

TABLE 4 GREEK ALPHABET

Name	Capital	Lowercase	Commonly Used to Designate
Alpha	A	α	Angles, area, coefficients
Beta	B	β	Angles, flux density, coefficients
Gamma	Γ	γ	Conductivity, specific gravity
Delta	Δ	δ	Variation, density
Epsilon	E	ε	Base of natural logarithms (*e*)
Zeta	Z	ζ	Impedance, coefficients, coordinates
Eta	H	η	Hysteresis coefficient, efficiency
Theta	Θ	θ	Temperature, phase angle
Iota	I	ι	
Kappa	K	κ	Dielectric constant, susceptibility
Lambda	Λ	λ	Wavelength
Mu	M	μ	Micro, amplification factor, permeability
Nu	N	ν	Reluctivity
Xi	Ξ	ξ	
Omicron	O	ο	
Pi	Π	π	Ratio of circumference to diameter = 3.1416
Rho	P	ρ	Resistivity
Sigma	Σ	σ	Summation
Tau	T	τ	Time constant, time phase displacement
Upsilon	Υ	υ	
Phi	Φ	ϕ	Magnetic flux, angles
Chi	X	χ	
Psi	Ψ	ψ	Dielectric flux, phase difference
Omega	Ω	ω	Capital, ohms; lowercase, angular velocity

TABLE 5 CONVERSION FACTORS*

Multiply	By	To Obtain
Avoirdupois pounds	0.4536	Kilograms
Circular mils	5.067×10^{-4}	Square millimeters
Coulombs	6.242×10^{18}	Electric charges
Feet	0.3048	Meters
Gallons (imperial)	4.546	Liters
Gallons (U.S. dry)	4.405	Liters
Gallons (U.S. liquid)	3.785	Liters
Horsepower	0.746	Kilowatts
Inches	25.4	Millimeters
Kilograms	2.205	Avoirdupois pounds
Kilometers	3.28×10^{3}	Feet
Meters	3.28	Feet
Meters	39.37	Inches
Mils	2.54×10^{-2}	Millimeters
Statute miles	1.609	Kilometers
Yards	0.9144	Meters

*Selected from H. F. R Adams, *SI Metric Units: An Introduction,* McGraw-Hill Ryerson Ltd., 1974, by permission.

TABLE 6 STANDARD ANNEALED COPPER WIRE SOLID* AMERICAN WIRE GAGE (BROWN AND SHARPE) (20°C)

Gage	Diameter, mm	Cross Section, mm^2	Ohms per Kilometer	Meters per Ohm	Kilograms per Kilometer
0000	11.68	107.2	0.160 8	6 219	953.2
000	10.40	85.01	0.202 8	4 931	755.8
00	9.266	67.43	0.255 7	3 911	599.5
0	8.252	53.49	0.322 3	3 102	475.5
1	7.348	42.41	0.406 5	2 460	377.0
2	6.543	33.62	0.512 8	1 950	298.9
3	5.827	26.67	0.646 6	1 547	237.1
4	5.189	21.15	0.815 2	1 227	188.0
5	4.620	16.77	1.028	972.4	149.0
6	4.115	13.30	1.297	771.3	118.2
7	3.665	10.55	1.634	612.0	93.80
8	3.264	8.367	2.061	485.3	74.38
9	2.906	6.631	2.600	384.6	58.95
10	2.588	5.261	3.277	305.2	46.77
11	2.30	4.17	4.14	242	37.1
12	2.05	3.31	5.21	192	29.4
13	1.83	2.63	6.56	152	23.4
14	1.63	2.08	8.28	121	18.5
15	1.45	1.65	10.4	95.8	14.7
16	1.29	1.31	13.2	75.8	11.6
17	1.15	1.04	16.6	60.3	9.24
18	1.02	0.823	21.0	47.7	7.32
19	0.912	.653	26.4	37.9	5.81
20	.813	.519	33.2	30.1	4.61
21	.724	.412	41.9	23.9	3.66
22	.643	.324	53.2	18.8	2.88
23	.574	.259	66.6	15.0	2.30
24	.511	.205	84.2	11.9	1.82
25	.455	.162	106	9.42	1.44
26	.404	.128	135	7.43	1.14
27	.361	.102	169	5.93	0.908
28	.320	.080 4	214	4.67	.715
29	.287	.064 7	266	3.75	.575
30	.254	.050 7	340	2.94	.450
31	.226	.040 1	430	2.33	.357
32	.203	.032 4	532	1.88	.288
33	.180	.025 5	675	1.48	.227
34	.160	.020 1	857	1.17	.179
35	.142	.015 9	1 090	0.922	.141

TABLE 6 STANDARD ANNEALED COPPER WIRE SOLID* AMERICAN WIRE GAGE (BROWN AND SHARPE) (20°C) (*Continued*)

Gage	Diameter, mm	Cross Section, mm^2	Ohms per Kilometer	Meters per Ohm	Kilograms per Kilometer
36	.127	.012 7	1 360	.735	.113
37	.114	.010 3	1 680	.595	.091 2
38	.102	.008 11	2 130	.470	.072 1
39	.089	.006 21	2 780	.360	.055 2
40	.079	.004 87	3 540	.282	.043 3
41	.071	.003 97	4 340	.230	.035 3
42	.064	.003 17	5 440	.184	.028 2
43	.056	.002 45	7 030	.142	.021 8
44	.051	.002 03	8 510	.118	.018 0
45	.0447	.001 57	11 000	.0910	.014 0
46	.0399	.001 25	13 800	.0724	.011 1
47	.0356	.000 993	17 400	.0576	.008 83
48	.0315	.000 779	22 100	.0452	.006 93
49	.0282	.000 624	27 600	.0362	.005 55
50	.0251	.000 497	34 700	.0288	.004 41
51	.0224	.000 392	43 900	.0228	.003 49
52	.0198	.000 308	55 900	.0179	.002 74
53	.0178	.000 248	69 400	.0144	.002 21
54	.0157	.000 195	88 500	.0113	.001 73
55	.0140	.000 153	112 000	.008 89	.001 36
56	.0124	.000 122	142 000	.007 06	.001 08

* Bureau of Standards Handbook 100, reproduced by permission.

TABLE 7 DECIMAL MULTIPLIERS

0.000 000 000 000 000 001	= 10^{-18}	= ten to the negative *eighteenth* power	= atto a
0.000 000 000 000 001	= 10^{-15}	= ten to the negative *fifteenth* power	= femto f
0.000 000 000 001	= 10^{-12}	= ten to the negative *twelfth* power	= pico p
0.000 000 001	= 10^{-9}	= ten to the negative *ninth* power	= nano n
0.000 001	= 10^{-6}	= ten to the negative *sixth* power	= micro μ
0.001	= 10^{-3}	= ten to the negative *third* power	= milli m
1	= 10^{0}	= ten to the *zero* power	= unit
1 000	= 10^{3}	= ten to the *third* power	= kilo k
1 000 000	= 10^{6}	= ten to the *sixth* power	= Mega M
1 000 000 000	= 10^{9}	= ten to the *ninth* power	= Giga G
1 000 000 000 000	= 10^{12}	= ten to the *twelfth* power	= Tera T
1 000 000 000 000 000	= 10^{15}	= ten to the *fifteenth* power	= Peta P
1 000 000 000 000 000 000	= 10^{18}	= ten to the *eighteenth* power	= Exa E

TABLE 8 ROUNDED VALUES OF PREFERRED NUMBERS*

Series	*R*5	*R*10	*R*20	*R*40	*R*6†	*R*12†	*R*24†
Approximate Ratio	1.6	1.25	1.12	1.06	1.46	1.21	1.1
	1	1	1	1	1	1	1
				1.06			1.1
			1.12	1.12		1.2	1.2
				1.18			1.3
		1.25	1.25	1.25	1.5	1.5	1.5
				1.32			1.6
			1.40	1.40		1.8	1.8
				1.50			2.0
	1.60	1.60	1.60	1.60	2.2	2.2	2.2
				1.70			2.4
			1.80	1.80		2.7	2.7
				1.90			3.0
		2.0	2.0	2.0	3.3	3.3	3.3
				2.12			3.6
			2.24	2.24		3.9	3.9
				2.36			4.3
	2.50	2.50	2.50	2.50	4.7	4.7	4.7
				2.65			5.1
			2.80	2.80		5.6	5.6
				3.0			6.2
		3.15	3.15	3.15	6.8	6.8	6.8
				3.35			7.5
			3.55	3.55		8.2	8.2
				3.75			9.1
	4.00	4.00	4.00	4.00	10	10	10
				4.25			
			4.50	4.50			
				4.75			
		5.00	5.00	5.00			
				5.30			
			5.60	5.60			
				6.00			
	6.30	6.30	6.30	6.30			
				6.70			
			7.10	7.10			
				7.50			
		8.00	8.00	8.00			
				8.50			
			9.00	9.00			
				9.50			
	10.00	10.00	10.00	10.00			

*These tables have been adapted from various international, American, and British standards.
†The *R*6, *R*12, and *R*24 tables are sometimes referred to as *E*6, *E*12, and *E*24, respectively.

APPENDIX B

ANSWERS TO ODD-NUMBERED PROBLEMS

NOTE *The accuracy of answers to numerical computations is, in general, to three significant figures from a five-figure readout of a hand-held electronic calculator.*

PROBLEMS 2–1

1. (*a*) 15 times x
   (*b*) 12 times i
   (*c*) 0.05 times R
3. (*a*) \$496.80
   (*b*) \$3.45n
5. 12.5I A
7. $\frac{2}{3}C$ pF, 4C pF, 48C pF
9. (*a*) $16 + R$ Ω
   (*b*) $v + 220$ V
   (*c*) $i - I$ A
11. $L_2 = L_1 - 115$ mH
13. (*a*) 44 A
    (*b*) 0.25 A
15. (*a*) 2.99 s
    (*b*) 0.685 s
17. 0.125 m

PROBLEMS 2–2

1. (*a*) 80
   (*b*) 175
   (*c*) 1600
   (*d*) 1192
   (*e*) 16
   (*f*) 198
3. (*a*) Monomial
   (*b*) Monomial
   (*c*) Monomial
   (*d*) Binomial
   (*e*) Trinomial
   (*f*) Binomial
   (*g*) Trinomial
   (*h*) Trinomial
   (*i*) Monomial
   (*j*) Trinomial
5. (*a*) $I = \frac{V}{R}$
   (*b*) $V = IR$
   (*c*) $P = RI^2$
   (*d*) $R_1 = R_2 + R_3$
   (*e*) $K = \frac{M}{\sqrt{L_1 L_2}}$
   (*f*) $R_p = \frac{R_1 R_2}{R_1 + R_2}$
   (*g*) $N = \frac{R_m}{R_s} + 1$

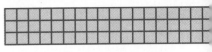

7. 78 606 μH Note that, all other factors remaining equal, if the number of turns is tripled, the inductance is multiplied by a factor of nine (3^2).
9. (*a*) Increased by a factor of 4
   (*b*) Increased by a factor of 16
   (*c*) Reduced to a value one-ninth the original

PROBLEMS 3–1

1. 68
3. -34
5. 28
7. -736
9. 429.18
11. $4\frac{1}{2}$
13. $-11\frac{11}{32}$
15. $\frac{2}{15}$

PROBLEMS 3–2

1. 63
3. 179
5. 1245
7. 2.16
9. $-4\frac{15}{16}$
11. (*a*) 67°
    (*b*) 26°
    (*c*) 159°
13. $573.45
15. 8 V

PROBLEMS 3–3

1. $3i$
3. $112IZ$
5. $-p - 6P$
7. $-12I + 18\frac{V}{R}$
9. $58\beta - 43\alpha$
11. $27i^2r + 10W - 3vi + 49w$
13. $11V_1 - 3V_2 - 14V_3$
15. $-\frac{11}{48}\pi ft - 2\frac{1}{8}\pi Z$
17. $13\phi + 11\theta$
19. $-4.7I^2R + 5.8VI + 3\frac{V^2}{R}$
21. $3.90IZ - 1.31IR - 0.41IX$
23. $10.93\mu - 14.1\lambda$

PROBLEMS 3–4

1. $2 - 7y$
3. $7R - 3X + 7$
5. $5W - 7w - 4VI$
7. $x + 2y$
9. $18x - 5y + 9z$

PROBLEMS 3–5

1. (*a*) $2L_1 + (5L_2 - 4L_3 + L_T)$
   (*b*) $2w + (12x - 3y + z)$
   (*c*) $8\theta + (2\phi - 6\psi - 10\Omega)$
   (*d*) $2R + \left(-11\frac{V}{I} + 5\frac{P}{I^2} - 3X\right)$
   (*e*) $10R_T - 3R_1 + (6R_2 - 5R_3 + 4R_4)$
3. $\sqrt{R^2 + X^2} - Z$
5. $38.2 + v$ V
7. $Z - \sqrt{r^2 + x^2}$
9. $V - \left(IR + \frac{P}{I}\right)$
11. $X_L - 2\pi f L$

PROBLEMS 4–1

1. 24
3. 22.14
5. $-\frac{15}{512}$
7. 0.000 000 833 76
9. $ac\omega^2$
11. $2\pi f L_1 L_2$
13. $\frac{1}{2\pi f C_p}$
15. $-\frac{\alpha\varepsilon}{\mu\psi}$

PROBLEMS 4–2

1. a^8
3. $-x^9$
5. $12y^6$
7. $-72p^4x^3$
9. abm^{n+p}
11. $8p^3$
13. $6a^3b^4c^3d^7$
15. $-\frac{\pi M X_L}{4}$
17. $-0.0288a^3dx^3y^2$
19. r^{12}

PROBLEMS 4–3

1. $20x + 15y$
3. $4Y_1Z^2 + 8Y_2Z^2$
5. $6\mu^3\pi + 12\lambda\pi - 18\pi\phi$

7. $12\alpha^2\beta^2\phi - 30\beta\theta\phi^2 - 24\alpha\beta\theta\phi + 36\beta\phi$
9. $36m^3p^2 + 15m^2p^3 - 12\,mp^4$
11. $\dfrac{iI^3RZ}{3} - \dfrac{iI^3R^2Z}{6} - \dfrac{2i^2IZ^2}{9}$
13. $3I^3PR - 6Ii^2Pr + 2IP^2$
15. $0.157V^3IZ^3 + 0.314VIZ^5 - 10.5IZ^6$
17. $-25x - 31y$
19. $\theta^3 - \phi^3$
21. $1.5IR_1 + 2.5R_1{}^2R_2 - 6.5IR_1R^2 + 3I^2R_2$
23. $\dfrac{3L^2M}{4} + \dfrac{3LIM}{8} - \dfrac{I^2M}{4}$
25. $0.125IR - 0.025IR_1 - 0.4125IR_2$
27. 0
29. $15s$

PROBLEMS 4–4

1. $\alpha^2 + 2\alpha + 1$
3. $\alpha^2 - 2\alpha + 1$
5. $\beta^2 - 9$
7. $x^2 + 7x + 12$
9. $p^2 - 8p - 48$
11. $y^2 + 7y + 10$
13. $3x^2 + 11xy - 20\,y^2$
15. $6N^2 - 13NP - 5P^2$
17. $6\theta^2 + 13\theta\phi + 6\phi^2$
19. $2V^2 - 9VI + 10I^2$
21. $6a^3 + 17a^2 + a - 10$
23. $2R^3 - 2R^2r - 2Rr^2 + 2r^3$
25. $a^3 - a^2b - ab^2 + b^3$
27. $\alpha^3 - \alpha^2\beta - \alpha\beta^2 + \beta^3$
29. $a^3 + 3a^2b + 3ab^2 + b^3$
31. $x^2 + 2xy + y^2$
33. $M^2 - 2MN + N^2$
35. $27x^3 + 81x^2y + 81xy^2 + 27y^3$
37. $44V^2I^2 - 21VI - 9$
39. $10a^3 + 4a^2b - 16a^2 - 5ab^2 - 14ab + a$

PROBLEMS 4–5

1. 9
3. 18
5. $-\dfrac{4}{3}$
7. $-2\pi fC$

9. $\dfrac{V \times 10^8}{L_v}$

11. -99

13. 320

15. $-\dfrac{5}{2}$

PROBLEMS 4–6

1. $4\alpha^2\beta^4$

3. $-3\theta\phi^3\psi$

5. $-\dfrac{5X_LZ}{3}$

7. $-9x^4y^3z$

9. $\dfrac{3mn^2p}{4}$

11. $-9c$

13. $-6\delta^3\varepsilon^2\eta^3$

15. $\dfrac{\beta^7\delta^4}{8\alpha\lambda^6}$

17. $-\dfrac{\phi^6}{4\theta^{12}\psi\Omega^3}$

19. $\dfrac{5000\alpha^2\beta^7}{\gamma^2}$

PROBLEMS 4–7

1. $6x + 4y$

3. $11m^2 - 5n^2$

5. $4R_1 + 6R_2 - 10R_3$

7. $\dfrac{0.005\mu^3}{\pi} + 10\mu\pi$

9. $\dfrac{3m^3}{10} - \dfrac{7m}{5} - \dfrac{6}{5m}$

11. $5 + 7xz - 3x^2z^2 - 6x^4y^4$

13. $2(\theta + \phi) - 4(\theta + \phi)^3 + 3(\theta + \phi)^5$

15. $\dfrac{(VI + P)^2}{2} - 2 + \dfrac{6}{VI + P}$

17. $\dfrac{1}{I\left(\omega L - \dfrac{1}{\omega C}\right)} - 2I\left(\omega L - \dfrac{1}{\omega C}\right) - 5I^3\left(\omega L - \dfrac{1}{\omega C}\right)^3$

19. $3(\theta - \phi)^2 - 6(\theta + \phi)(\theta - \phi)^3 - \dfrac{9(\theta - \phi)}{\theta + \phi}$

PROBLEMS 4–8

1. $a + 1$

3. $3x + 4$

5. $4E - 3$

7. $3R^2 - 4Z - 7$

9. $E^2 + 6E - 5$

11. $E + e$

13. $V^3 + V^2v + Vv^2 + v^3$

15. $V^2 + I^2R^2$

17. $X^5 - X^4Y + X^3Y^2 - X^2Y^3 + XY^4 - Y^5$

19. $x^2 + 2xy + y^2$

21. $2R_2 - 3$

23. $12x^2 - 32x + 17 + \dfrac{170x}{2x^2 + 3 - 5x}$

25. $3R + \dfrac{1}{3}$

27. $6x - \dfrac{y}{3} - \dfrac{1}{2}$

29. $\dfrac{3L_1^2}{8} - \dfrac{L_1}{4} - \dfrac{2}{3}$

PROBLEMS 5–1

1. $x = 4$
3. $a = 6$
5. $\theta = 6$
7. $\pi = 5$
9. $\omega L = 4$
11. $\alpha = -10$
13. $V = 8$
15. $Q = -2$
17. $I = -1.4$
19. $\beta = 1$

PROBLEMS 5–2

1. $V = p - 120$ V
3. $d = rt$ km
5. $x - u$ years
7. $\dfrac{D}{m}$ km/min
9. 220 V
11. $I = \dfrac{V}{R}$
13. 16 m by 32 m
15. 40, 60, 80 m
17. $h^2 = a^2 + b^2$
19. 63, 64, 65

PROBLEMS 5–3

1. $Q = CV,\ V = \dfrac{Q}{C}$

3. $R^2 = Z^2 - X^2, \quad X^2 = Z^2 - R^2$

5. $L = \dfrac{Rm}{K}, \quad R = \dfrac{KL}{m}, \quad m = \dfrac{KL}{R}$

7. $f = \dfrac{\nu}{\lambda}, \ \nu = \lambda f$

9. $L = \dfrac{QR}{\omega}, \quad R = \dfrac{\omega L}{Q}, \quad \omega = \dfrac{QR}{L}$

11. $X_C = \dfrac{1}{2\pi fC}, \quad C = \dfrac{1}{2\pi fX_C}$

13. $\phi = HA,\ A = \dfrac{\phi}{H}$

15. $B = \dfrac{(V)(10^8)}{Lv},\quad L = \dfrac{(V)(10^8)}{Bv},\quad V = \dfrac{BLv}{10^8}$

17. $I_p = \dfrac{V_s I_s}{V_p}$

19. $R = \dfrac{V - v}{I},\quad V = IR + v,\quad v = V - IR$

21. $\theta = \omega t,\ \omega = \dfrac{\theta}{t}$

23. $V = \dfrac{V_0 + V_t}{2},\quad V_t = 2V - V_0$

25. $A = \dfrac{4}{3}\pi r^3$

27. $Z_t = \dfrac{F(R - r)}{C},\quad F = \dfrac{CZ_t}{R - r},\quad R = \dfrac{CZ_t + Fr}{F},\quad r = \dfrac{FR - CZ_t}{F}$

29. $V_b = iR_L + v_b,\quad v_b = V_b - iR_L,\quad i = \dfrac{V_b - v_b}{R_L}$

31. $l = \dfrac{Rd^2}{\rho},\quad \rho = \dfrac{Rd^2}{l},\quad d^2 = \dfrac{\rho l}{R}$

33. $A = \dfrac{Cd}{0.0884K(n - 1)},\quad n = \dfrac{Cd + 0.0884KA}{0.0884KA}$

35. $L = CRZ_r,\quad C = \dfrac{L}{RZ_r},\quad R = \dfrac{L}{CZ_r}$

37. $\omega = \dfrac{\eta\beta}{\gamma\alpha}$

39. $Q = \dfrac{\rho h v}{e}$

41. $C_2 = \dfrac{V_3 - V_2}{V_3\omega^2 L}$

43. $I_n = \dfrac{Q - I_P p}{n}$

45. $\omega_{01} = \dfrac{1}{C(R_1 + R_2)}$

47. 4000 Ω

49. 5 m

PROBLEMS 5–4

1. $\dfrac{1}{5}$

3. $\dfrac{6}{4}, \dfrac{3}{2}, \dfrac{1.5}{1}$

5. $\dfrac{4}{14}, \dfrac{6}{21}$, etc.

7. $\dfrac{1}{3}$

9. 30:1

PROBLEMS 5–5

1. 6
3. 80
5. $X = 5$
7. 4.5
9. $Q = 0.0014$

PROBLEMS 5–6

1. $I \propto V$
3. $C \propto A, \quad C = kA$
5. $X_C \propto \frac{1}{C}, \quad X_C = \frac{k}{C}$
7. $T \propto \sqrt{L}, \quad T = k\sqrt{L}$
9. $P \propto \frac{1}{V}, \quad P = \frac{k}{V}$
11. $L \propto \frac{1}{d^2}, \quad L = \frac{k}{d^2}$
13. 2.46 kV
15. 5400 kg

PROBLEMS 6–1

1. 6
3. 6
5. 3
7. 1
9. 4

PROBLEMS 6–2

1. 2.75×10^5
3. 7.13×10^3
5. 5.13×10^{-9}
7. 3.67×10^{-1}
9. 2.50×10^{-1}
11. 3.99×10^4
13. 2.59×10^{-2}
15. 2.76×10^5
17. 1.08×10^{-7}
19. 2.00
21. (1) 0.275×10^6 or 275×10^3
    (3) 7.13×10^3
    (5) 5.13×10^{-9}
    (7) 367×10^{-3} or 0.367
    (9) 250×10^{-3} or 0.250
    (11) 39.9×10^3
    (13) 25.9×10^{-3}
    (15) 276×10^3
    (17) 108×10^{-9}
    (19) 2.00

PROBLEMS 6–3

1. 1×10^{-5}
3. 2.16×10^{-1}
5. 7.14×10^{-12}
7. 4.00×10^{10}
9. 3.20×10
11. $5.65 \times 10^{0}\ \Omega$
13. $9.42 \times 10^{4}\ \Omega$
15. $1.67 \times 10^{1}\ \Omega$

PROBLEMS 6–4

1. 1.56×10^{-7}
3. 1.05×10^{-4}
5. 3.30×10^{-13}
7. 2.87×10^{10}
9. 2.55×10^{2}
11. 1.56×10^{-3}
13. $6.63 \times 10^{2}\ \Omega$
15. 4.82×10^{-5}

PROBLEMS 6–5

1. 10^{6}
3. 10^{14}
5. 6.25×10^{14}
7. 2.56
9. 5×10^{-3}
11. 42
13. 1.01×10^{4}
15. 1.50×10^{6} Hz
17. 1.01×10^{7} Hz
19. 1.20×10^{6} Hz

PROBLEMS 6–6

1. (*a*) 5300
   (*b*) 5.3×10^{3}
3. (*a*) 4 190 000 000 000
   (*b*) 4.19×10^{12}
5. (*a*) 185 999 999.972
   (*b*) 1.86×10^{8}

PROBLEMS 7–1

1. (*a*) 2.7×10^{6} mV
   (*b*) 2.7×10^{9} μV
   (*c*) 2.7 kV
3. (*a*) 5.25×10^{-3} kV
   (*b*) 5.25×10^{6} μV
   (*c*) 5.25×10^{3} mV

5. (*a*) 3.9 kΩ
   (*b*) 3.9×10^{-3} MΩ
   (*c*) 2.56×10^{-4} S
7. (*a*) 5×10^{-8} F
   (*b*) 5×10^{-2} μF
9. (*a*) 8.68×10^{-1} kW
   (*b*) 8.68×10^{5} mW
   (*c*) 8.68×10^{8} μW
11. (*a*) 1.32×10^{0} MHz
    (*b*) 1.32×10^{6} Hz
13. (*a*) 4.00×10^{-1} W
    (*b*) 4.00×10^{-4} kW
15. (*a*) 1.50×10^{-2} MHz
    (*b*) 1.50×10^{4} Hz
17. (*a*) 5.50×10^{4} μA
    (*b*) 5.50×10^{1} mA
19. (*a*) 5.6×10^{6} Ω
    (*b*) 5.6×10^{3} kΩ
21. (*a*) 3.35×10^{6} μH
    (*b*) 3.35 H
23. (*a*) 5.00×10^{2} pF
    (*b*) 5.00×10^{-10} F
25. (*a*) 2.50×10^{6} μS
    (*b*) 4.00×10^{-1} Ω
27. (*a*) 1.88×10^{0} mA
    (*b*) 1.88×10^{-3} A
29. (*a*) 2.8×10^{8} W
    (*b*) 2.8×10^{5} kW

PROBLEMS 7–2

1. (*a*) 84 in
   (*b*) 213 cm
   (*c*) 2.13×10^{3} mm
3. (*a*) 69.3 in
   (*b*) 176 cm
   (*c*) 1.92 yd
5. (*a*) 4.56 mi
   (*b*) 7.34×10^{3} m
   (*c*) 7.34 km
7. 1.63 mm
9. 1 Gs = 0.1 mT
11. 6×10^{-2} dB/100 m
13. 3.22×10^{-3} Ω/cm
15. 4.72 in/min

PROBLEMS 7–3

3. $X_L = 2\pi f L\ \Omega$
5. $f = \dfrac{159}{\sqrt{LC}}$ MHz
7. $\delta = \dfrac{6.63 \times 10^{-3}}{\sqrt{f}}$ cm
9. $R_{ac} = 83.2 \times 10^{-6} \dfrac{\sqrt{f}}{d}\ \Omega/\text{cm}$
13. 435 cm
15. 66.2 cm
17. (*a*) 72.4 cm
    (*b*) 75.9 cm
    (*c*) 30.7 cm
19. (*a*) 77 cm
    (*b*) 81 cm
    (*c*) 73.2 cm
    (*d*) 32.8 cm
    (*e*) 16.4 cm

PROBLEMS 8–1

1. 2.45 A
3. 19.6 A
5. 22 mA
7. 3.93 A
9. (*a*) 527 mA
   (*b*) 657 mA

PROBLEMS 8–2

1. (*a*) 8.95×10^3 W
   (*b*) 8.95 kW
3. 217 mA
5. 8.50 kW
7. 1641 kW
9. 0.164 W
11. (*a*) 0.1 mA
    (*b*) 0.47 mW
13. (*a*) 9.41 pW
    (*b*) 118 nA
15. (*a*) 90.5%
    (*b*) $24.72
17. 18.4 hp
19. 2.4 kW

PROBLEMS 8–3

1. (*a*) 0.108 A
   (*b*) 23.7 V
   (*c*) 5.47 W
3. (*a*) 300 Ω
   (*b*) 32.9 W
   (*c*) 43.9 W
5. 269 Ω
7. 3.1 Ω
9. (*a*) 1.5 Ω
   (*b*) 1.8 Ω
   (*c*) 1.25 kW

PROBLEMS 8–4

1. (*a*) 500 Ω
   (*b*) 19.5 V
3. (*a*) R_S = 3.3 kΩ
   (*b*) I_D = 1.8 mA
   (*c*) I_T = 1.82 mA
   (*d*) P = 45 mW
5. (*a*) I_C = 0.75 mA; I_E = 0.765 mA
   (*b*) R_B = 773 kΩ
   (*c*) V_{CE} = 3 V
   (*d*) P = 2.25 mW
7. (*a*) 60 kΩ
   (*b*) 600 μW
9. 2.5 kΩ
9. 2.34 kΩ

PROBLEMS 9–1

1. (*a*) 6.60 Ω
   (*b*) 0.264 Ω
3. 66.4 Ω
5. 2.36 Ω
7. 0.513 Ω
9. 1.07 m

PROBLEMS 9–2

1. 6.55 Ω
3. 37 Ω
5. 0.038 μΩ·m
7. 1.834 km
9. 444 m

PROBLEMS 9–3

1. 12.1 Ω
3. 13.9 Ω
5. No

PROBLEMS 9–4

1. (*a*) 10.5 Ω
   (*b*) 150 kg
3. (*a*) 1.78 km
   (*b*) 2.30 Ω
5. 2.13 kΩ
7. No. 1 wire
9. (*a*) No. 5 wire
   (*b*) 96.3%

PROBLEMS 10–1

1. a^2c^2
3. $i^6r^3X^3$
5. $9\theta^2\psi^2$
7. $4\pi^2f^2L^2$
9. $4\pi^2f^2C^2$
11. $-\dfrac{1}{4\pi^2f^2C^2}$
13. $-\dfrac{27M^6}{L^3C^3}$
15. $-\dfrac{V^6}{8g^3}$
17. $\dfrac{27P^3}{125\ V^3\ I^3}$
19. $-\frac{16}{9}\pi^2R^6$
21. $\dfrac{x^{12}y^{18}}{p^{15}}$

PROBLEMS 10–2

1. $\pm x$
3. $\pm 4\omega$
5. $\pm \omega$
7. $\pm 2\pi LC$
9. $2a^4$
11. 4
13. $\pm 13m^2np^3$
15. $3\pi^2\theta^3\phi^4$
17. $\pm \dfrac{16\pi rx^2}{17z^3\phi^2}$
19. $\pm \dfrac{25r^3s^2t^4}{4x^3z^5}$
21. $\dfrac{4a\omega^2}{5x^2z^4}$
23. $\pm \dfrac{5vt}{16a^4bx}$

PROBLEMS 10–3

1. $3(x + 2)$
3. $\alpha(6 + \varepsilon + 4\phi)$
5. $4I(4Z - 3R)$
7. $\dfrac{l^2m}{24}(8 + 3lm - 12m^2)$
9. $2a^2bc(ab + 4c^2 + 6bc)$
11. $36\alpha^2\beta^2\omega^2(\alpha^2\beta - 2\omega^3 + 5\beta^3)$
13. $\dfrac{1}{135}IR(5IR^2 + 9R - 45I^2)$
15. $120\eta\theta^2\phi\omega(6\eta^3\phi^2 + 9\eta\theta^2\omega + 5\eta^2\theta\omega - 4\theta^4)$

PROBLEMS 10–4

1. $x^2 + 10x + 25$
3. $C^2 - 2Cw + w^2$
5. $\phi^2 + 24\phi + 144$
7. $25V^2 - 10Vv + v^2$
9. $M^2 - 2Mm + m^2$
11. $25\theta^2 + 40\theta\phi + 16\phi^2$
13. $81r_1^2 - 54r_1r_2 + 9r_2^2$
15. $1 + 2X_L^2 + X_L^4$
17. $25a^4 - 20a^2f^3 + 4f^6$
19. $400 - 200 + 25 = 225$
21. $9\pi^2L^4 - 12\pi^2L^2C^2 + 4\pi^2C^4$
23. $2.25\theta^4 - 1.5\theta^2\alpha + 0.25\alpha^2$
25. $\dfrac{9}{16}X^4 - \dfrac{3}{4}X^2Z^2 + \dfrac{1}{4}Z^4$
27. $36\phi^4\omega^2 - 3\phi^2\omega\lambda^2 + \dfrac{1}{16}\lambda^4$
29. $y^2 + \dfrac{1}{2}y + \dfrac{1}{16}$
31. $\dfrac{1}{4} - E + E^2$
33. $1 + 2x^5 + x^{10}$
35. $L^4 - \dfrac{7}{4}L^2P + \dfrac{49}{64}P^2$
37. $\dfrac{a^2}{25} + \dfrac{2ab}{15} + \dfrac{b^2}{9}$
39. $R_1^2 - \dfrac{5}{4}R_1R_2 + \dfrac{25}{64}R_2$

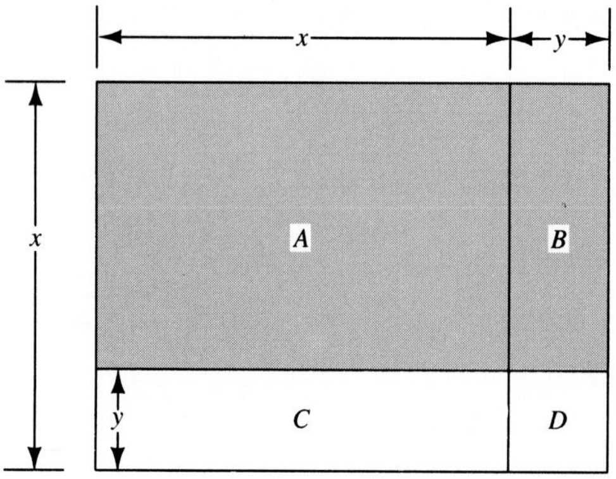

41. $(x + y)(x - y) = x^2 \quad + xy \quad - xy - y^2$
$A + B \quad = (A + C) + (B + D) - C - D$

PROBLEMS 10–5

1. $4a$
3. $4u$
5. $8lw$
7. $10lw$
9. $12mp$
11. b^2
13. $16p^2$
15. $\frac{1}{3}\theta\phi\omega$
17. $\frac{1}{9}\pi^2$
19. $\pm(E + 3)$
21. $3l + w$
23. $2\pi^2 r_1 + 7r_2$
25. $\pm\left(\frac{3}{5}\pi R^2 + \frac{2}{3}\right)$
27. $\pm\left(\frac{5}{6}\phi + \frac{2}{7}\lambda\right)$

PROBLEMS 10–6

1. $2y(x + 3z)$
3. $5A(C + D)^2$
5. $2i(3r + 5z)^2$
7. $\frac{24I}{V}(R_1 - R_2)^2$
9. $\frac{5r}{16e}(\lambda - 4f^2)^2$ or $\left(\frac{5}{e}\right)\left(\frac{r}{2^4}\right)(\lambda - 4f^2)^2$

PROBLEMS 10–7

1. $x^2 - 4$
3. $R^2 - Z^2$
5. $4M^2 - 9L^2$
7. $\frac{9}{16}R_1^2 - Q^2$
9. $\frac{4V^4}{R^2} - \frac{9I^4R^2}{P^2}$

PROBLEMS 10–8

1. $(I + Z)(I - Z)$
3. $(3x + 4y)(3x - 4y)$
5. $\left(\frac{1}{3} + x\right)\left(\frac{1}{3} - x\right)$
7. $(1 + 15\omega)(1 - 15\omega)$
9. $(8\pi\phi^2 + 1)(8\pi\phi^2 - 1)$
13. $3(xy - 2z - 2st)(xy - 2z + 2st)$
15. $(5a + 10cl + 12l)(5a + 10cl - 12l)$

PROBLEMS 10–9

1. $l^2 + 7l + 10$
3. $m^2 - m - 2$
5. $y^2 + 12y + 35$
7. $4V^2 - 4V - 3$
9. $R_1^2 - 7R_1 + 10$
11. $\alpha^2 - \frac{5\alpha}{4} + \frac{1}{4}$
13. $I^2R^2 + \frac{IR}{6} - \frac{1}{6}$
15. $\alpha^2 + \alpha + \frac{2}{9}$
17. $\frac{1}{LC} - \frac{4f}{\sqrt{LC}} + 3f^2$
19. $\alpha^2\beta^4 + \frac{3\alpha\beta^2}{10} + \frac{1}{50}$

PROBLEMS 10–10

1. $(Q + 1)(Q + 2)$
3. $(L + 3)(L + 5)$
5. $(\beta + 6)(\beta - 4)$
7. $(\theta + 4)(\theta + 6)$
9. $(t + 11)(t - 2)$
11. $(q + 6)(q - 2)$
13. $(\psi + 4)(\psi - 3)$
15. $(\omega + 2f)(\omega - 3f)$
17. $\left(\frac{1}{x} + 5\right)\left(\frac{1}{x} - 2\right)$
19. $\left(I + \frac{6Z}{X}\right)\left(I - \frac{4Z}{X}\right)$

PROBLEMS 10–11

1. $\theta^2 - 5\theta - 14$
3. $10F^2 + 11F + 3$
5. $12k^2 + 2k - 4$
7. $10\omega^2 + 33\omega - 7$
9. $\frac{\omega^2}{4} + 2\omega - 32$
11. $6Y^2 + 20YIB + 6I^2B^2$
13. $15X^2 - 94X - 40$

15. $15\theta^2 - 77\theta + 10$
17. $54 - 57\lambda + 15\lambda^2$
19. $8a^2 + 34ab + 21b^2$
21. $4a^2 - 24at + 35t^2$
23. $m^2 + 0.6mp - 0.07p^2$
25. $\frac{x^2}{4} - \frac{3x\lambda}{2} - 4\lambda^2$
27. $10\alpha^2 + \frac{8\alpha}{3\beta\theta} + \frac{1}{6\beta^2\theta^2}$
29. $0.16p^2 - 0.62pq + 0.21q^2$

PROBLEMS 10–12

1. $(x + 2)(x - 7)$
3. $4(y + 2)(2y - 3)$
5. $(2L + 3)(5L - 7)$
7. $(7\mu + 3)(2\mu + 5)$
9. $(\alpha + 3\beta)(2\alpha - 7\beta)$
11. $(9y - 7)(5y + 3)$
13. $(5\alpha + 2\omega)(10\alpha - 3\omega)$
15. $(3Q - 4M)(5Q - 7M)$
17. $(9lm - w)(3lm + 2w)$
19. $6(\psi + 2\Omega)(\psi - 2\Omega)$
21. $(5x + \Delta)(3x - 2\Delta)$
23. $\left(8\theta + \frac{1}{2}\right)\left(6\theta + \frac{1}{4}\right)$
25. $(0.8x + 2)(0.4x - 1)$

PROBLEMS 10–13

1. $4\omega^2C^2$
3. $\frac{h^8k^{12}l^4m^8}{x^8y^8z^{12}}$
5. $\pm\frac{12IR}{13FX_C^2}$
7. $-\frac{4pr^2}{5x^3y^4z}$
9. $6\theta\phi^2\omega^2$
11. $f(L_1 + L_2)(L_1 - L_2)$
13. $\frac{v^2}{8}\left(\frac{8}{r_1} + \frac{5}{r_2} - \frac{7}{r_3}\right)$
15. $\frac{p}{12}(m - 5r^2 - 7t)$
17. $M^2 + 16M + 64$
19. $49V^4 + \frac{35V^2}{4} + \frac{25}{64}$
21. $\frac{25\beta^2}{81} - \frac{10\beta\lambda}{3} + 9\lambda^2$
23. $8a$
25. $36M^2$
27. $\frac{y^2}{9}$
29. $\pm(x + 3)$
31. $\pm(25L_1 + 14X)$
33. $\pm\left(\frac{I}{4} - \frac{i}{2}\right)$
35. $2Y(4b + 5B)$
37. $3i(r + 3)^2$
39. $8p(3V - IR)^2$
41. $I^2 - 25i^2$
43. $Y^2 - 81$
45. $\frac{576V^2}{I^2R^2} - 4P^2$
47. $(A + 1)(A - 1)$
49. $\left(3fR + \frac{1}{2\pi C}\right)\left(3fR - \frac{1}{2\pi C}\right)$

51. $(0.05\psi + 0.6\mu)(0.05\psi - 0.6\mu)$
53. $a + 2$
55. $\frac{1}{3}\alpha + \frac{2}{7}\beta$
57. $\frac{3}{5}e - \frac{4}{9}ir$
59. $t^2 - t - 30$
61. $0.3X_L^2 - 1.94X_L - 0.4$
63. $p^2 - \frac{3p}{28} - \frac{1}{28}$
65. $10a^2 + 19ab - 15b^2$
67. $0.6R^2 + 0.1Rr - 0.2r^2$
69. $24\phi^2 + 2\theta\phi - \frac{\theta^2}{3}$
71. $(5\alpha + 1)(3\alpha + 4)$
73. $(\theta - 2)(\theta - 3)$
75. $(y - 3)(y - 0.2)$
77. $(5Y + 2G)(4Y - 3G)$
79. $(4X - 0.5Z)(X + 0.2Z)$
81. $\left(\frac{X_C}{3} + Z\right)^2$
83. $(3l + w)(5l - 3w)$
85. $5(a + 3)(a - 3)$
87. $\frac{3}{2i}(V - 6v)^2$
89. $\frac{c}{144d}(8a - 9b)(9a - 8b)$

PROBLEMS 11–1

1. 14
3. $8\lambda\mu$
5. $0.5a^2bc$
7. $21VI$
9. $X_L + X_C$
11. $\alpha + 1$
13. $P + 2\frac{V^2}{R} + I^2R$
15. $5\left(2I + 3\frac{V}{R}\right)$

PROBLEMS 11–2

1. 180
3. 3630
5. $R^2L_1^3L_2^4M^2Y$
7. $180\ m^3n^2p^4$
9. $2(a + 2)(a - 2)(a + 3)$
11. $t(t + 4)(t - 4)^2$
13. $11(3\theta - 1)(2\theta + 1)(2\theta + 3)$
15. $\left(Q + \frac{\omega L}{R}\right)\left(Q - \frac{\omega L}{R}\right)\left(4Q - \frac{5\omega L}{R}\right)\left(2Q - \frac{7\omega L}{R}\right)$

PROBLEMS 11–3

1. 12
3. R_1R_2
5. $9abd$
7. $a^2 + 4a + 3$
9. $6i + 6\alpha$
11. $\frac{40}{64}$

13. $\dfrac{\omega LR^3 + 2\omega LR^2X + \omega LRX^2}{R^4 + 2R^3X + R^2X^2}$

15. $\dfrac{2VCQ - 3Q}{2V^2C^2 - VC - 3}$

PROBLEMS 11–4

1. $\dfrac{5}{24}$
3. $\dfrac{1}{36}$
5. $\dfrac{x}{y^2}$
7. $\dfrac{5I}{R}$
9. $\dfrac{\theta}{\theta^2 - \phi^2}$
11. $\dfrac{\alpha + \beta}{\alpha - \beta}$
13. $\dfrac{I - R}{5(I + R)}$
15. $\dfrac{\omega(\pi + 3\lambda)}{3\pi + \lambda}$

PROBLEMS 11–5

1. $\dfrac{5}{16}$
3. $\dfrac{2\pi fL}{X_C - X_L}$
5. $\dfrac{IR}{Z - X}$
7. $\dfrac{-GY}{R + X}$
9. $\dfrac{\pi R^2}{A_2 - A_1}$
11. -1
13. $\dfrac{-1}{\phi + \theta}$
15. $\dfrac{3 + \omega}{2 - 5\omega}$

PROBLEMS 11–6

1. $\dfrac{13}{8}$
3. $\dfrac{IR + V}{I}$
5. $\dfrac{Q - 3}{Q}$
7. $\dfrac{5\beta + 13}{\beta + 2}$
9. $\dfrac{\omega LR - 5R - L}{R}$
11. $\dfrac{(9 + 2x)(1 - x)}{x^2}$
13. $\dfrac{(R + 3)(R - 7)}{R^2}$
15. $\dfrac{9\lambda^2 - 4\lambda - 2}{(3\lambda + 1)(3\lambda - 1)}$
17. $\dfrac{-\theta^3 + 13\theta^2 + 31\theta - 45}{\theta^2(\theta - 1)}$
19. $\dfrac{2\alpha^4 - 7\alpha^2 - 1}{\alpha^2 - 3}$
21. $1\frac{17}{18}$
23. $3 + \dfrac{7}{16}$
25. $x^2 + 11x - 8 - \dfrac{1}{x + 1}$

27. $E^3 - E^2e + Ee^2 - e^3 - \dfrac{1}{E + e}$

29. $2x + 2 - \dfrac{x}{x^2 + 1}$

PROBLEMS 11–7

1. $\dfrac{35}{140}, \dfrac{40}{140}, \dfrac{84}{140}$

3. $\dfrac{16}{32}, \dfrac{28}{32}, \dfrac{9}{32}$

5. $\dfrac{VX}{IRX}, \dfrac{IRZ}{IRX}$

7. $\dfrac{vi}{ir}, \dfrac{1}{ir}, \dfrac{iprv}{ir}$

9. $\dfrac{a + b}{a^2 - b^2}, \dfrac{a - b}{a^2 - b^2}$

11. $\dfrac{7l + 7w}{l^2 - w^2}, \dfrac{5l - 5w}{l^2 - w^2}$

13. $\dfrac{\alpha\theta - \alpha\phi}{\theta^2 - \phi^2}, \dfrac{\beta\theta + \beta\phi}{\theta^2 - \phi^2}, \dfrac{\beta\theta + \beta\phi - \alpha\theta - \alpha\phi}{\theta^2 - \phi^2}$

15. $\dfrac{\pi^2 - \phi^2}{\pi\phi}, \dfrac{\phi^2}{-\pi\phi}$

PROBLEMS 11–8

1. $\dfrac{11}{60}$

3. $\dfrac{11}{48}$

5. $\dfrac{83V}{48}$

7. $\dfrac{pt - qs}{qt}$

9. $\dfrac{\alpha NP - \beta MP - \gamma MN}{MNP}$

11. $\dfrac{8IR - 2V + 5}{IRV}$

13. $\dfrac{19I + i}{6}$

15. $\dfrac{2(2L_1 + L_2)}{L_1^2 - L_2^2}$

17. $\dfrac{3L_1 + 34}{L_1^2 + 4L_1 - 12}$

19. $-\dfrac{19\theta - 7}{6(\theta^2 - 1)}$

21. $\dfrac{2(Y^2 + 16Y + 20)}{Y(Y^2 - 25)}$

23. $\dfrac{2\pi + 7}{3 + \pi}$

25. $\dfrac{2x^2y + 6xy^2}{x^3 + x^2y - xy^2 - y^3}$

27. $\dfrac{10 - 4E}{(E - 4)(E - 5)(E - 6)}$

29. $\dfrac{12\omega^2}{\omega^3 + 27}$

PROBLEMS 11–9

1. $\frac{24}{65}$
3. $-\frac{1}{320}$
5. $3\frac{1}{3}$
7. $\frac{5\theta\phi}{3}$
9. $\frac{2\alpha}{\beta^2}$
11. $\frac{\omega}{2\pi fR}$
13. $\frac{4x + 4y}{(x - y)^2}$
15. $\frac{3B - Y}{7(2B + 5)}$
17. $\frac{\rho + 1}{\rho(\rho - 2)}$
19. $-\frac{3b}{2}$
21. $10V^2$
23. $4c$
25. ϕ^2
27. $\frac{1}{3\omega L + R}$
29. $3y$

PROBLEMS 11–10

1. $-1\frac{1}{2}$
3. $\frac{3}{4P}$
5. $\frac{fC_1C_2I}{C_1 + C_2}$
7. $\frac{BY}{BY - G}$
9. $\frac{2E(E - e)}{e(E + e)}$
11. $\frac{l - w}{l + w}$
13. $-\frac{y}{x}$
15. $\frac{v^2 - v + 1}{2v - 1}$

PROBLEMS 12–1

1. $\theta = 21$
3. $\gamma = 12$
5. $V = 3$
7. $\phi = 5$
9. $\lambda = 5$
11. $\theta = 35$
13. $Z = -3$
15. $\omega = \frac{2}{3}$

PROBLEMS 12–2

1. $I = 0.5$
3. $\theta = 5$
5. $r = 30$
7. $R = 2.5$
9. $b = \frac{1}{17}$
11. $\alpha = 40$
13. $Y = 1.5$
15. $\alpha = 3$

PROBLEMS 12–3

1. $a = 2$
3. $m = 4$
5. $L = 3$
7. $w = 3$
9. $\pi = 5$
11. $v_O = 5$

13. $x = 3$

15. $R = -\frac{4}{5}$

17. $\omega = 5$

19. $\theta = -2$

23. 5.09 h

25. $x = \frac{abc}{ab + ac + bc}$ days

27. 2:43 P.M.

31. 40 kg

33. 187.5 L

35. 70,350

37. $\frac{1}{2}, 1\frac{1}{2}, 2\frac{1}{2}$

39. 20×60 m

PROBLEMS 12–4

1. $h = \frac{2A}{a+b}$, $b = \frac{2A - ah}{h}$

3. $V_b = v + IR$

$I = \frac{V_b - v}{R}$

5. $C_2 = \frac{V_3 - V_2}{\omega^2 L V_3}$

$V_2 = V_3(1 - \omega^2 L C_2)$

7. $\alpha = \frac{R_t - R_o}{R_o \Delta t}$

$\Delta t = \frac{R_t - R_o}{\alpha R_o}$

9. $v = \frac{Vr}{R + r}$

$R = \frac{r(V - v)}{v}$

11. $C_1 = \frac{1}{\omega^2 C_2 R_1 R_3}$

13. $V_0 = \frac{I_0 R_0}{1 - \mu\beta}$

15. $E_t = E_g - IR$

$E_g = E_t + IR$

17. $\frac{E}{I} = \frac{Z_1Z_2 + Z_2Z_3 + Z_3Z_1}{Z_3}$

$Z_2 = \frac{Z_3(E - IZ_1)}{I(Z_1 + Z_3)}$

$Z_3 = \frac{IZ_1Z_2}{E - I(Z_1 + Z_2)}$

19. $R_x = \frac{R_y(AV_1 - V)}{V(A + 1)}$

$R_y = \frac{VR_x(A + 1)}{AV_1 - V}$

$\frac{V}{V_1} = \frac{AR_y}{(A + 1)R_x + R_y}$

21. $R = \frac{Z_{ab}^2(X_p - X_s)}{X_s^2 + Z_{ab}^2}$

$Z_{ab}^2 = \frac{X_s^2 R}{X_p - X_s - R}$

23. $V_n = \frac{I_2R(R_1 + R_2)}{2R_1 + R_2}$

$I_2R = \frac{V_n(2R_1 + R_2)}{R_1 + R_2}$

25. $F = \frac{9}{5}C + 32$

27. $R_0 = \frac{V_1 R}{BI_0 - V_1}$

$R = \frac{R_0(BI_0 - V_1)}{V_1}$

29. $R_H = \frac{\mu mN}{Kg} - r$

$K = \frac{\mu mN}{g(R_H + r)}$

31. $R = (1 - \alpha)Z_2 + (1 - \alpha + k\alpha)Z_1$

$Z_1 = \dfrac{R - Z_2(1 - \alpha)}{1 - \alpha + k\alpha}$

$k = \dfrac{R + (\alpha - 1)(Z_1 + Z_2)}{\alpha Z_1}$

33. $F_2 = \dfrac{F_s(\alpha F_{12} - 2f)}{2f}$

$F_{12} = \dfrac{2fF_2}{\alpha F_S} + \dfrac{2f}{\alpha}$

35. $\alpha = \dfrac{1}{H_2R_1} - S$

$S = \dfrac{1 - \alpha H_2R_1}{H_2R_1}$

37. $r_1 = \dfrac{r_2r_3}{r_4}$

$r_4 = \dfrac{r_2r_3}{r_1}$

39. $X = \left(1 - \dfrac{C_v}{C_0}\right)\left(\dfrac{f_c}{f}\right)$

$C_0 = \dfrac{C_vf_c}{f_c - fX}$

41. $I_2 = \dfrac{ER_0}{R_1R_0 + R_1R_2 + R_2R_0}$

43. $v_0 = \dfrac{Vi_1R_1}{R_2(i_1 + i_2) + i_1R_1}$

$i_1 = \dfrac{Rv_0i_2}{VR_1 - v_0(R_1 + R_2)}$

45. $\mu_1 = \dfrac{G(\mu_2\beta_2 - 1)}{\beta_1G(\mu_2\beta_2 - 1) - \mu_2}$

$\beta_1 = \dfrac{\mu_1\mu_2 + G(\mu_2\beta_2 - 1)}{G\mu_1(\mu_2\beta_2 - 1)}$

47. $I_f = \dfrac{\sigma_0(2I_1 + 1)(\gamma_1 + \gamma_f)}{4\pi\lambda^2\gamma_1} - \dfrac{1}{2}$

$\sigma_0 = \dfrac{2\pi\lambda^2\gamma_f(2I_f + 1)}{(2I_1 + 1)(\gamma_1 + \gamma_f)}$

$= 2\pi\lambda^2\left(\dfrac{\gamma_1}{\gamma_1 + \gamma_f}\right)\left(\dfrac{2I_f + 1}{2I_1 + 1}\right)$

49. $d_1 = \dfrac{\lambda + \pi n' d_0^2}{\lambda + \pi n' d_0}$

$n' = \dfrac{\lambda}{\pi d_0}\left(\dfrac{1 - d_1}{d_1 - d_0}\right)$

51. $R_2 = \dfrac{-Z_2(Z - Z\alpha + Rk\alpha)}{Z_1 + Z_2}$

$Z = \dfrac{RkZ_2\alpha + R_2(Z_1 + Z_2)}{Z_2(\alpha - 1)}$

$R = -\left(\dfrac{1}{k}\right)\left(\dfrac{Z_1R_2}{Z_2\alpha} + \dfrac{R_2}{\alpha} + \dfrac{Z(1 - \alpha)}{\alpha}\right)$

53. $r_1 = \dfrac{r_2r_3}{r_4}$

$r_3 = \dfrac{r_1r_4}{r_2}$

$r_4 = \dfrac{r_2r_3}{r_1}$

55. $V_{out} = \dfrac{GC_{fg}Q}{C_f[G(C_{fg} + C_d) + C_{fg}]}$

57. $p_2 = \dfrac{CNP_Lp_1}{p\omega\varepsilon_2 (\tan \delta) - CNP_L}$

59. $R_p = \dfrac{R_1R_2}{R_1 + R_2}$

$R_1 = \dfrac{R_2R_p}{R_2 - R_p}$

$R_2 = \dfrac{R_1R_p}{R_1 - R_p}$

61. $R_3 = \dfrac{V_0R_a}{\mu V - V_0(\mu + 1)}$

$R_a = \dfrac{R_3}{V_0}[\mu V - V_0(\mu + 1)]$

$\dfrac{V_0}{V} = \dfrac{\mu}{\mu + 1 + \dfrac{R_a}{R_3}}$

63. $\pi = \dfrac{MNk}{4(kH_0 - M)}$

$k = \dfrac{4\pi M}{4\pi H_0 - MN}$

65. $d = \dfrac{b(X + X')^2}{X^2 + X'^2} = b + \dfrac{2b}{\dfrac{X}{X'} + \dfrac{X'}{X}}$

67. $R_a = \dfrac{\mu R_1 R_3(V - V_0) - V_0 R_3(R_s + R_1)}{V_0(R_1 + R_3 + R_s) - VR_3}$

$R_s = \dfrac{\mu R_1 R_3(V - V_0) - V_0(R_1 R_3 + R_a R_3 + R_a R_1) + VR_a R_3}{V_0(R_a + R_3)}$

$\mu = \dfrac{R_a R_3(V - V_0) - V_0(R_a R_s + R_a R_1 + R_s R_3 + R_1 R_3)}{R_1 R_3(V_0 - V)}$

69. $R_a = \dfrac{\mu R_0 R_1(R_s + R_1 + R_2)}{R_2(R_s + R_1) - R_0(R_s + R_1 + R_2)}$

$R_2 = \dfrac{R_0(R_s + R_1)(R_a + \mu R_1)}{R_a(R_s + R_1) - R_0(R_a + \mu R_1)}$

$\mu = \dfrac{R_a[R_2(R_s + R_1) - R_0(R_s + R_1 + R_2)]}{R_0 R_1(R_s + R_1 + R_2)}$

71. $R_a = \dfrac{R_1 R_2 R_3(\mu R_1 - R_1)}{(R_i - R_1)(R_1 R_2 + R_2 R_3 + R_1 R_3)}$

73. $\pi = \dfrac{\alpha^2(\beta - \alpha)}{\alpha + 2\beta}$

$\beta = \dfrac{\alpha(\alpha^2 + \pi)}{\alpha^2 - 2\pi}$

75. $1.4 \times 10^{-13} m^2$

77. $Z_2 = 8.57\ \Omega^2$

79. $R_2 = 68.6\ \Omega$

81. $F = C$ at $-40°$

83. $R_0 = 37.7\ \Omega$

85. $C_1 = 5.56$ pF

87. $p = 120$

89. (*a*) Increased by a factor of 4 (*b*) Halved

91. $R = 593$ mΩ

93. $R = \dfrac{n(V - Ir)}{I}$

$n = \dfrac{IR}{V - Ir}$ cells

95. $g_m = \dfrac{v_d}{v_g i_d r_o} - \dfrac{1}{r_s}$

99. $V_0 = \dfrac{S}{t_s} - \dfrac{1}{2}gt$

101. $v_0 = 41.1$ m/s

103. $\Delta i_d = g_m v_{gs}$

105. $V_{CC} = -34.5$ V; PNP

107. $V_1 - V_2 = 300$ V

109. $\alpha = \dfrac{\beta}{1 + \beta}$

PROBLEMS 13–1

1. 235 Ω
3. 15 kΩ
5. (*a*) 600 Ω (*b*) 4.1 kΩ (*c*) 9 Ω
7. 120 V
9. 112 kΩ
11. 4.47 W
13. 14.7 W
15. 1 kV

PROBLEMS 13–2

1. 8.78 Ω
3. 4.63 Ω
5. 71.9 Ω
7. (*a*) 600 Ω
   (*b*) 3 kΩ
9. $R_p = \frac{R}{n}\ \Omega$
11. 1.5 kW
13. (*a*) 6.56 Ω
    (*b*) 2.75 kW
15. 499 Ω

PROBLEMS 13–3

1. 250 mA
3. 30.6 W
5. (*a*) $V_G = 230$ V
   (*b*) $R_3 = 20$ kΩ
   (*c*) $R_1 = 70.6$ kΩ
   (*d*) $I_2 = 1.86$ mA
   (*e*) $I_3 = 1.4$ mA
7. (*a*) $V_1 = 442$ V
   (*b*) $V_2 = 558$ V
   (*c*) $R_2 = 6.8$ kΩ
   (*d*) $R_3 = 4.7$ kΩ
7. (*e*) $I_t = 201$ mA
   (*f*) $I_3 = 119$ mA
   (*g*) $P_t = 201$ W
9. 354 Ω
11. 1 kΩ
13. (*a*) 4.2 kΩ
    (*b*) 10 kΩ
    (*c*) 48 W
15. 730 W
17. 2.47 A

PROBLEMS 14–1

1. 0.959 Ω
3. 2.82 m
5. (*a*) 0–10 mA: 6.11 Ω
   (*b*) 0–100 mA: 0.556 Ω
   (*c*) 0–1 A: 0.0551 Ω
   (*d*) 0–10 A: 0.005 51 Ω
7. $R_1 = 150$ Ω
   $R_2 = 15$ Ω
   $R_3 = 1.5$ Ω
   $R_4 = 0.167$ Ω

PROBLEMS 14–2

1. (*a*) 37.5 V
   (*b*) 25 V
3. $R_1 = 9.6\ \text{k}\Omega$
   $R_2 = 99.6\ \text{k}\Omega$
   $R_3 \cong 1\ \text{M}\Omega$
   $R_4 \cong 10\ \text{M}\Omega$

PROBLEMS 15–1

1. 1 = 60 V
   2 = 6 V
   3 = 0.6 V
   4 = 0.06 V
3. 27 kΩ: 0.114 W
   68 kΩ: 0.288 W
   75 kΩ: 0.318 W
5. $R_1 = 171\ \Omega$
   $R_2 = 150\ \Omega$
   $R_3 = 600\ \Omega$
7. $P_1 = 13.75$ W
   $P_2 = 7.75$ W
   $P_3 = 4.50$ W
9. $R_1 = 3\ \text{k}\Omega$
   $R_2 = 3\ \text{k}\Omega$
   $R_3 = 10\ \text{k}\Omega$
   $R_4 = 500\ \Omega$
11. 42 W

PROBLEMS 15–2

1. (*a*) $I_1 = 20.83$ mA
   (*b*) $I_2 = 6.94$ mA
3. (*a*) $I_1 = 179$ mA
   (*b*) $I_2 = 133$ mA
5. $I_2 = 4.94$ A

PROBLEMS 15–3

1. 1.94 Ω
3. 0.520 Ω
5. 35.9 km
7. $X = \dfrac{R_2L - R_1R_3}{R_1 + R_2}$

PROBLEMS 16–1

1. Current varies directly as the applied voltage. (Graph of current is a straight line.)
3. With velocity constant, distance varies directly as time. (Graph of distance is a straight line.)

5. (*a*) 2 P.M.
   (*b*) 480 km
   (*c*) 80 km
7. Third, sixth, ninth, and fifteenth
9. (*b*) $F = 50°$, $C = 10°$
   (*c*) above 3.2°C, 26.3°F (to 260°C, 500°F)

PROBLEMS 16–2

1. Latitude

PROBLEMS 16–5

1. $y = \frac{2}{5}x - 2$
3. $y = 0.08x + 0.4$
5. $R = -0.000\,667T + 0.4$
7. (*a*) 47:1
   (*b*) 0.021 25:1
   (*c*) 47 Ω

PROBLEMS 17–1

1. $x = 2.5$, $y = 4$
3. $\alpha = 6$, $\beta = -4$
5. $E = 6$, $I = -10$
7. $\alpha = -2$, $\beta = -3$
9. $L = 2$, $M = 3.5$

PROBLEMS 17–2

1. $a = 3$, $b = 2$
3. $x = -2$, $z = 3$
5. $R_1 = 4$, $R_2 = 1$
7. $p = 5$, $q = 5$
9. $L = 3$, $M = 2$
11. $I_1 = 5$, $I_2 = -3$
13. $i = 2$, $I = -3$
15. $\theta = 2$, $\phi = 3.5$
17. $V = -11$, $v = 12$
19. $I = 12$, $i = 9$

PROBLEMS 17–3

1. $v = 6$, $I = 2$
3. $R_1 = 5$, $R_2 = 3$
5. $B = 9$, $G = 4$
7. $X_L = 6$, $X_C = 1$
9. $I_1 = 15$, $I_2 = 12$
11. $L_1 = 3$, $L_2 = 4$
13. $Y = 12$, $Z = -6$
15. $R_1 = 12$, $R_2 = -3$
17. $\theta = 3$, $\phi = 1$
19. $a = 1.5$, $b = 0.4$

PROBLEMS 17–4

1. $G = Y = 1$
3. $\varepsilon = 2, \eta = 3$
5. $\alpha = 2, \beta = 8$
7. $X = 15, Z = 2$
9. $m = 3, p = 8$

PROBLEMS 17–5

1. $x = 12, y = 8$
3. $\lambda = 12, \mu = -2$
5. $V_1 = 120, V_2 = -95$
7. $B = -7, G = 4$
9. $\theta = 1.6, \phi = 0.8$

PROBLEMS 17–6

1. $R_1 = \frac{1}{3}, R_2 = \frac{1}{5}$
3. $G = 9, Y = -5$
5. $\theta = 5, \phi = 12$
7. $R_1 = -2, R_2 = 1$
9. $I_1 = \frac{3}{5}, I_2 = \frac{5}{16}$

PROBLEMS 17–7

1. $a = \dfrac{\theta + 5\pi}{2}, \quad b = \dfrac{\theta - 5\pi}{2}$
3. $a = \dfrac{3(G + Y)}{GY}, \quad b = \dfrac{5(G + Y)}{GY}$
5. $R_1 = \dfrac{iZ_1 + Z_2}{I + i}, \quad R_2 = \dfrac{IZ_1 - Z_2}{I + i}$
7. $X_1 = 60Z_2 - 3.2Z_1$
   $X_2 = 4.8Z_1 - 40Z_2$
9. $R_x = \dfrac{14R_PR_t}{3R_P - R_t}$
   $R_y = \dfrac{14R_PR_t}{5R_t - R_P}$

PROBLEMS 17–8

1. $I_t = 1, \quad I_1 = 3, \quad I_2 = 2$
3. $I_1 = 6, \quad I_2 = -5, \quad I_3 = 12$
5. $R_L = 2, \quad R_p = 7, \quad R_1 = 9$
7. $r = 1, \quad R = 3, \quad R_L = 7$
9. $a = 12, \quad b = 4, \quad c = 8$

PROBLEMS 17–9

1. $\dfrac{V_s + V_d}{2}$ V, $\dfrac{V_s - V_d}{2}$ V
3. $\dfrac{9}{15}$

5. $\frac{90 + \alpha°}{2}, \frac{90 - \alpha°}{2}$
7. Resistors, 15¢ each, capacitors, 35¢ each
9. $J = 80$ km/h, $A = 90$ km/h

11. $W = \frac{Q^2}{2C}$
13. $s = ut + \frac{1}{2}at^2$
15. Gain $= G_m r_d$
17. $R = \frac{L}{Cr}$
19. $Q = CV$
21. 10.67 J
23. $H = Pt$ J

25. $R_X = \frac{R_A}{R_A + R_B}(V_3 - V_2)$

$R_Y = \frac{R_A}{R_A + R_B}(V_2 - V_1)$

$R_T = \frac{R_A}{R_A + R_B}(V_3 - V_1)$

27. $R_1 = \frac{R_aR_b + R_bR_c + R_aR_c}{R_c}$

$R_2 = \frac{R_aR_b + R_bR_c + R_aR_c}{R_a}$

$R_3 = \frac{R_aR_b + R_bR_c + R_aR_c}{R_b}$

PROBLEMS 18–1

1. -7
3. 69
5. zero
7. -66
9. 0.25
11. 0
13. $bx - ay$
15. $bx - ay$

PROBLEMS 18–2

1. $x = 2.5, y = 4$
3. $I_1 = 3, I_2 = -2$
5. $I = 5, i = -3$
7. $\alpha = 4, \beta = -1$
9. $R_1 = 150, R_2 = 1200$

PROBLEMS 18–3

1. 21
3. -103
5. -163
7. $I_1 = 2,\ I_2 = 5,\ I_3 = 4$
9. $x = 2,\ y = 7,\ z = -1$

11. $V = 0.25,\ v = 0.375,\ IR = 0.625$

PROBLEMS 18–4

1. 32
3. 42
5. 102
7. zero

PROBLEMS 18–5

1. −1141
3. −206
5. 22.1
7. 1440
9. $I_1 = 1, \quad I_2 = 3, \quad I_3 = 5$
11. $\alpha = 2.6, \quad \beta = 5.7, \quad \gamma = 1.8$
13. $I_1 = 2, \quad I_2 = -3, \quad I_3 = 5, \quad I_4 = -2$
15. $\alpha = 3, \quad \beta = 1, \quad \gamma = 6, \quad \delta = -4$

PROBLEMS 19–1

1. 0.059 Ω
3. 0.169 Ω
5. (*a*) 3.48 W
   (*b*) 98.3%

PROBLEMS 19–2

1. (*a*) 1.15 V
   (*b*) 625 mA
   (*c*) 1.84 Ω
3. (*a*) 0.127 Ω
   (*b*) 3.16 mW
   (*c*) 10 Ω
   (*d*) 250 mW
   (*e*) 99%
5. 11.8 A
7. (*a*) 9 Ω
   (*b*) 600 mW
   (*c*) 16 A
9. (*a*) 385 mA
   (*b*) 12.7 V
   (*c*) 4.88 W
11. (*a*) 4.08 V
    (*b*) 1.6 V
    (*c*) 288 mW
13. (*a*) 0.173 Ω
    (*b*) 80 A
17. $V = 1.62$ V, $r = 0.169$ Ω
19. $V = 2.1$ V, $R = 0.665$ Ω

PROBLEMS 20–1

1. x^7
3. α^3
5. p^{l+q}
7. $V^{\alpha+\beta}$
9. x^5
11. X^{5y-2}

13. $\theta^{2\beta}$

15. I^{10}

17. x^6y^9

19. a^{mn}

21. $x^{4l}y^{4m}z^{4p}$

23. $\dfrac{V^2}{R^2}$

25. $\dfrac{4\pi^2f^2L^2}{R^2}$

27. $\dfrac{-X_C{}^6}{X_L{}^3}$

29. γ^{8x-4}

31. $\dfrac{1}{2\pi fC}$ or $\dfrac{0.159}{fC}$

33. $\dfrac{p^{2q}}{m^n}$

35. $\dfrac{\theta^4}{\phi^3\lambda^{2x}}$

37. $\dfrac{bc}{a^3}$

39. $\dfrac{Ir^3}{4R^2}$

PROBLEMS 20–2

1. ± 5

3. ± 3

5. $-3xy^4b^3c$

7. $\pm\omega^3L^3$

9. $\dfrac{4x^4}{\pi^6}$

11. $\sqrt{12}$

13. $3\sqrt[3]{Z}$

15. $\sqrt[4]{(\theta\lambda)^3}$

17. $x^{\frac{5}{2}}$

19. $3^{\frac{4}{3}}R^{\frac{2}{3}}$

21. $G^{\frac{2}{3}}B^2$

23. $(\alpha\beta)^{\frac{2}{5}}$

25. $2\pi f^{\frac{5}{3}}\,20^{\frac{1}{3}}$

PROBLEMS 20–3

1. $\pm 3\sqrt{3}$

3. $\pm 2\sqrt{3}$

5. $\pm 5\sqrt{3}$

7. $\pm 6\sqrt{5}$

9. $\pm 12\sqrt{5}$

11. $\pm 3\alpha\beta^2\sqrt{2}$

13. $\pm 30\omega\sqrt{L}$

15. $\pm 18\omega f^2FT^2\sqrt{7FT}$

17. $\pm 18A^4B^2C^3\sqrt{2AB}$

19. $\pm 26\pi^4C^2L^2\sqrt{3}$

PROBLEMS 20–4

1. $\dfrac{\sqrt{5}}{5}$

3. $\dfrac{\sqrt{10}}{5}$

5. $\pm\dfrac{\sqrt{6}}{4}$

7. $8\sqrt{2}$

9. $\dfrac{\sqrt{\lambda}}{\lambda}$

11. $\pm\dfrac{3\sqrt{\theta}}{4\theta}$

13. $\sqrt{\alpha\omega}$

15. $\dfrac{\beta\sqrt{\theta}}{\theta}$

17. $\dfrac{R^3\sqrt{\pi A}}{4A}$

19. $\pm\dfrac{X_L\sqrt{15}}{4}$

21. $\pm\dfrac{4Q^2\sqrt{5}}{9}$

PROBLEMS 20–5

1. $3\sqrt{2}$
3. $3\sqrt{3}$
5. $(a - b + c)\sqrt{2}$
7. $21\sqrt{3}$
9. 0
11. $7\sqrt{2}$
13. $\dfrac{\sqrt{7}}{7}$
15. $\dfrac{2\sqrt{3\alpha} - \sqrt{6\alpha}}{18}$

PROBLEMS 20–6

1. $\pm\sqrt{10}$
3. $\pm 9\sqrt{5}$
5. $\pm 12\sqrt{10}$
7. 4
9. $x^2 + R^2$
11. $2x - 5 - 2\sqrt{X^2 - 5X}$
13. $Z - Y$
15. $\pm 10\pi\varepsilon$
17. $19 + 7\sqrt{5}$
19. 9

PROBLEMS 20–7

1. $\pm\sqrt{3}$
3. $2(5 - \sqrt{7})$
5. $-(5 + 5\sqrt{2})$
7. $\dfrac{x^2 - 2x\sqrt{y} + y}{x^2 - y}$
9. $\dfrac{8 - 3\sqrt{2}}{5}$
11. $6 - 3\sqrt{3} + 2\sqrt{5} - \sqrt{15}$

PROBLEMS 20–8

1. j4
3. j25
5. $-\mathrm{j}Z$
7. $\mathrm{j}RZ^2$
9. −j15
11. $\mathrm{j}\dfrac{5}{9}$
13. $\mathrm{j}\dfrac{5}{14}\sqrt{21}$
15. $-\mathrm{j}\dfrac{V\sqrt{P}}{P}$

PROBLEMS 20–9

1. 7 + j16
3. 48 + j3
5. 143 − j6
7. 20 − j2
9. 1 + j2
11. 12 + j11
13. −87 − j16
15. 20 + j8

PROBLEMS 20–10

1. $10 - j30$
3. $a^2 + b^2$
5. $\theta^2 - \phi^2 + j2\theta\phi$
7. $\frac{1 - j1}{2}$
9. $j1$
11. $\frac{1 - j1}{2}$
13. $\frac{6(6 + jx)}{36 + x^2}$
15. $\frac{R^2 + j2R\omega X - \omega^2 X^2}{R^2 + \omega^2 X^2}$
17. $\frac{-\phi^2 + j\theta\phi}{\theta^2 + \phi^2}$
19. $\frac{R^2 - jR\left(\omega L - \frac{1}{\omega C}\right)}{R^2 + \left(\omega L - \frac{1}{\omega C}\right)^2}$

PROBLEMS 20–11

1. $x = 25$
3. $\gamma = 625$
5. $B = 900$
7. $P = 12$
9. $\lambda = 1$
11. $\phi = 25$
13. $L_1 = \frac{I_x^2}{2L_2\varepsilon^2}$
15. $C = \frac{1}{4\pi^2 f^2 L}$
17. $\alpha = \frac{V^2 v_1 v_2}{C^2 - V^2(v_1 - v_2)^2}$
19. $Q_2 = \frac{n^2(Y_n^2 - G^2)}{G^2(n^2 - 1)^2}$
21. $g_m^2 = \frac{G_L(G_1 - G_a^2)}{R_{eq}(G_a^2 - G_1) - G_1}$
23. $C = 250$ pF
25. $C_a = \frac{C_b}{2\pi f^2(LC_b - 1)}$

PROBLEMS 21–1

1. $x = \pm 6$
3. $i = \pm\sqrt{158}$
5. $V = \pm\sqrt{33}$
7. $\lambda = \pm\frac{3}{11}$
9. $\mu = \pm\frac{4}{5}$
11. $y = \pm 5$
13. $\theta = \pm 6$
15. $X_C = \pm\frac{\sqrt{95}}{5}$

PROBLEMS 21–2

1. $y = -2$ or -3
3. $R = 3$ or 5
5. $\theta = 2$ or -5
7. $X = 3$ or 17
9. $Q = 2$ or 11
11. $\psi = -3$ or -14
13. $\theta = 5$ or $\frac{-7}{5}$
15. $i = 3$ or $-\frac{7}{4}$

PROBLEMS 21–3

1. $x = 4$ or 5
3. $\omega = 6$ or 12
5. $x = 7$ or 15
7. $\eta = 5$ or 1
9. $M = -2$ or 24
11. $v = 6$ or -11
13. $G = 7$ or 13
15. $R = 5$ or $5\frac{1}{3}$

PROBLEMS 21–4

1. $v = 7$ or -2
3. $R = 5$ or -3
5. $L = \frac{1}{4}$ or $-\frac{1}{6}$
7. $x = 3.217$ or -6.217
9. $x = \frac{1}{2}$ or $-\frac{1}{3}$
11. $d = 3$ or 11
13. $\beta = 5$ or $5\frac{1}{3}$
15. $i = 2$ or $-\frac{2}{15}$

No answers are provided for Problems 21–5 and 21–6.

PROBLEMS 21–7

1. (*a*) 9, roots real and unequal
   (*b*) 0, roots are equal
   (*c*) −251, roots are imaginary
3. 15 and 17
5. 180 × 210 m
7. 180, 160

9. (*a*) $V = \pm\sqrt{\dfrac{PnR}{k}}$
   (*b*) No change in V

11. $r = \dfrac{-PXx \pm x\sqrt{P^2X^2 + 4R^2(P - 1)}}{2R(P - 1)}$

    $x = \dfrac{PXr \pm r\sqrt{P^2X^2 + 4R^2(P - 1)}}{2R}$

13. $v = 837$ m/s

15. $v = \sqrt{2gs}$ m/s

17. $h = 0.0156v^2$ m

19. $R = 800\ \Omega$

25. (*a*) 2 A
    (*b*) 120 V
    (*c*) $R_1 = 10\ \Omega$
    $R_2 = 20\ \Omega$
    $R_3 = 30\ \Omega$

27. 20 V and 15 A or 60 V and 5 A

PROBLEMS 22–1

1. 8.06 mA
3. 1.46 kV
5. 466 V
7. (a) 0.5 A
   (*b*) 12.3 V
9. $2.22\ \Omega$

PROBLEMS 22–2

1. 110 V
3. 44.1 V
5. (*a*) 1.19 A
   (*b*) 53.2 mW
7. (*a*) 1.0 A
   (*b*) From *a* to *b*

PROBLEMS 22–3

1. (*a*) 656 mA
   (*b*) 109 W
3. (*a*) 1.64 A
   (*b*) 2.86 A from *a* to *b*
5. (*a*) 95.5 W
   (*b*) 3.18 V
7. (*a*) 86.3 W
   (*b*) 16 V
9. (*a*) 7.81 A
   (*b*) 243 W
11. (*a*) 64.8 V
    (*b*) 2.1 kW
13. (*a*) 220 V
    (*b*) 313 W

PROBLEMS 22–4

1. $R_a = 3.43\ \Omega$, $R_b = 2.74\ \Omega$, $R_c = 5.14\ \Omega$
3. $R_a = R_b = R_c = 66.7\ \Omega$
5. $R_1 = 12\ \Omega$, $R_2 = 15\ \Omega$, $R_3 = 18\ \Omega$
7. $I = 443$ mA approx.
9. $I_2 = 317$ mA
11. Zero A
13. $I = 5$ A
15. 3.1 A
17. 14.7 A

PROBLEMS 22–5

1. (*a*) Constant 100-V source in series with $0.556\ \Omega$
   (*b*) Constant 180-A source in parallel with $0.556\ \Omega$
3. 3.5 V in series with $3.21\ \Omega$; $I_2 = 265$ mA
5. (*a*) 4.35 A in parallel with $2.37\ \Omega$
   (*b*) $I_5 = 1.40$ A

PROBLEMS 23–1

1. (*a*) 33° (*b*) 72° (*c*) 44° (*d*) −62° (*e*) −150° (*f*) 115°
7. 36
9. 10 800°/s

PROBLEMS 23–2

1. (*a*) $\frac{\pi^r}{6}$, 0.524^r (*b*) $\frac{\pi^r}{3}$, 1.05^r (*c*) $\frac{2\pi^r}{3}$, 2.09^r (*d*) $\frac{7\pi^r}{6}$, 3.67^r (*e*) $\frac{5\pi^r}{4}$, 3.93^r (*f*) $\frac{\pi^r}{15}$, 0.209^r
3. $390\pi^r$
5. $\frac{11\pi^r}{2} = 17.28^r$
7. $20\pi^r/\text{s}$
9. 30 rev/min

PROBLEMS 23–3

1. (*a*) 33.3^g (*b*) 50^g (*c*) 66.7^g (*d*) 150^g (*e*) 250^g (*f*) 366.7^g
3. (*a*) 0.523^r (*b*) 0.7854^r (*c*) 1.05^r (*d*) 2.09^r (*e*) 3.93^r (*f*) 4.71^r

PROBLEMS 23–4

1. 4 cm and 5 cm
3. $a = 3, c = 5, B = 53.1°$
5. $a = 16.2, c = 18.4, B = 40°$
7. $b = 16.2, c = 6.39, A = 40°$
9. $c = 10, A = 48.9°, B = 101.8°$

PROBLEMS 23–5

1. $c = 49.5, B = 14°$
3. $a = 40, A = 20°$
5. 36 m
7. 76.8 m
9. 120 m

PROBLEMS 24–1

1. $\sin\theta = \frac{a}{c}$ $\sin\phi = \frac{b}{c}$
   $\cos\theta = \frac{b}{c}$ $\cos\phi = \frac{a}{c}$
   $\tan\theta = \frac{a}{b}$ $\tan\phi = \frac{b}{a}$
   $\cot\theta = \frac{b}{a}$ $\cot\phi = \frac{a}{b}$
   $\sec\theta = \frac{c}{b}$ $\sec\phi = \frac{c}{a}$
   $\csc\theta = \frac{c}{a}$ $\csc\phi = \frac{c}{b}$
3. (*a*) $\frac{OP}{OR} = \tan\beta$
   (*b*) $\frac{PR}{PO} = \sec\alpha$
   (*c*) $\frac{OR}{PR} = \cos\beta$
   (*d*) $\frac{OP}{RP} = \sin\beta$
   (*e*) $\frac{PR}{RO} = \csc\alpha$
5. $\sin\theta = 0.707$
   $\cos\theta = 0.707$
   $\tan\theta = 1.00$
   $\cot\theta = 1.00$
   $\sec\theta = 1.41$
   $\csc\theta = 1.41$
7. $\sin\theta = 0.894$
   $\cos\theta = 0.447$
   $\tan\theta = 2.000$
9. $\sin x = \frac{12}{13}$
   $\cos x = \frac{5}{13}$
   $\tan x = \frac{12}{5}$
   $\sec x = \frac{13}{5}$
   $\csc x = \frac{13}{12}$
   $\cot x = \frac{5}{12}$
11. $\sin B = \frac{4}{5}$
    $\cos B = \frac{3}{5}$
    $\tan B = \frac{4}{3}$
    $\cot B = \frac{3}{4}$
    $\sec B = \frac{5}{3}$
    $\csc B = \frac{5}{4}$

PROBLEMS 24–2

1. I or II
3. III or IV
5. II or III
7. I
9. IV
11. I or III
13. No

Q	sin	cos	tan
15.	+	+	+
17.	+	−	−
19.	−	−	+
21.	−	−	+
23.	+	+	+

Q	sin	cos	tan	sec	csc	cot
27.	$\frac{5}{13}$	$\frac{12}{13}$	$\frac{5}{12}$	$\frac{13}{12}$	$\frac{13}{5}$	$\frac{12}{5}$
29.	$\frac{-5\sqrt{41}}{41}$	$\frac{-4\sqrt{41}}{41}$	$\frac{5}{4}$	$\frac{-\sqrt{41}}{4}$	$\frac{-\sqrt{41}}{5}$	$\frac{4}{5}$
31.	$-\frac{3}{5}$	$\frac{4}{5}$	$-\frac{3}{4}$	$\frac{5}{4}$	$-\frac{5}{3}$	$-\frac{4}{3}$
33.	$\frac{-3\sqrt{34}}{34}$	$\frac{-5\sqrt{34}}{34}$	$\frac{3}{5}$	$\frac{-\sqrt{34}}{5}$	$\frac{-\sqrt{34}}{3}$	$\frac{5}{3}$

PROBLEMS 24–3

1. 0
3. ∞
5. No
7. (*a*) 1, (*b*) −1, (*c*) −1, (*d*) 1

PROBLEMS 25–1

	sin	cos	tan
1. (*a*)	0.061 05	0.998 13	0.061 16
(*b*)	0.083 68	0.996 49	0.083 97
(*c*)	0.406 74	0.913 55	0.445 23
(*d*)	0.788 01	0.615 66	1.279 94
(*e*)	0.974 37	0.224 95	4.331 48

	sin	cos	tan
3. (*a*)	0.033 85	0.999 43	0.033 87
(*b*)	0.842 08	0.539 36	1.561 25
(*c*)	0.628 10	0.778 13	0.807 19
(*d*)	0.646 52	0.762 89	0.847 46
(*e*)	0.822 84	0.568 27	1.447 96

PROBLEMS 25–2

1. (*a*) 27°, (*b*) 6.7°, (*c*) 61.5°, (*d*) 40.1°, (*e*) 2.14°
3. (*a*) 2.8°, (*b*) 12.7°, (*c*) 29.6°, (*d*) 68.4°, (*e*) 82.7°
5. (*a*) 13.16°, (*b*) 0.53°, (*c*) 74.38°, (*d*) 41.77°, (*e*) 47.11°

PROBLEMS 25–3

	sin	cos	tan
1. (*a*)	0.956 30	−0.292 37	−3.270 85
(*b*)	0.342 02	−0.939 69	−0.363 97
(*c*)	0.764 92	−0.644 12	−1.187 54
(*d*)	0.537 30	−0.843 39	−0.637 07
(*e*)	0.066 27	−0.997 80	−0.066 42

	sin	cos	tan
3. (*a*)	−0.984 81	0.173 65	−5.671 28
(*b*)	−0.669 13	0.743 14	−0.900 40
(*c*)	−0.175 37	0.984 50	−0.178 13
(*d*)	−0.865 15	0.501 51	−1.725 09
(*e*)	−0.008 73	0.999 96	−0.008 73

	sin	cos	tan
5. (*a*)	−0.087 16	−0.996 19	0.087 49
(*b*)	−0.294 04	0.955 79	−0.307 64
(*c*)	−0.652 10	−0.758 13	0.860 14
(*d*)	−0.045 36	0.998 97	−0.045 41
(*e*)	−0.003 49	−0.999 99	0.003 49

7. (*a*) $\phi = -47.1°$
   (*b*) $\phi = 91.6°$
   (*c*) $\phi = 51.3°$
   (*d*) $\phi = 167.5°$
   (*e*) $\phi = -69.9°$
9. 1.38 m
11. 90°
13. 859 lx
15. No
17. 40.7°

PROBLEMS 26–1

1. $Z = 43.3$, $X = 16.7$, $\phi = 67.4°$
3. $\theta = 55°$, $Z = 361$, $R = 207$
5. $Z = 424.7$, $X = 407.9$, $\phi = 16.2°$
7. $\phi = 55.4°$, $Z = 3.23 \times 10^3$, $X = 1.84 \times 10^3$
9. $Z = 653$, $R = 501$, $\phi = 50°$

11. $\phi = 61.5°$, $Z = 0.793$, $X = 0.378$
13. $Z = 1.28$, $X = 0.96$, $\phi = 41.5°$
15. $Z = 0.378$, $R = 0.0500$, $\phi = 7.6°$

PROBLEMS 26–2

1. $R = 11.0$, $X = 5.87$, $\theta = 28°$
3. $R = 17.0$, $X = 44.5$, $\phi = 20.9°$
5. $R = 2.58 \times 10^3$, $X = 812$, $\theta = 17.5°$
7. $R = 0.932$, $X = 0.171$, $\theta = 10.4°$
9. $R = 13.8$, $X = 8.86$, $\phi = 57.3°$

PROBLEMS 26–3

1. $\theta = 49.2°$, $\phi = 40.8°$, $R = 121$
3. $\theta = 67.7°$, $\phi = 22.3°$, $X = 34.1$
5. $\theta = 56.9°$, $\phi = 33.1°$, $X = 1.07$
7. $\theta = 20.05°$, $\phi = 69.95°$, $R = 1.32 \times 10^3$
9. $\theta = 60.8°$, $\phi = 29.2°$, $X = 372$

PROBLEMS 26–4

1. $\theta = 30.9°$, $\phi = 59.1°$, $Z = 24.1$
3. $\theta = 36.4°$, $\phi = 53.6°$, $Z = 10.3$
5. $\theta = 2.7°$, $\phi = 87.3°$, $Z = 430$
7. $\theta = 73.9°$, $\phi = 16.1°$, $Z = 46.1$
9. $\theta = 46°$, $\phi = 44°$, $Z = 0.403$

PROBLEMS 26–5

1. 57.7°
3. 6.20°
5. 67.8°
7. 8.04 m
9. 20.2 m
11. 77.6 m

PROBLEMS 26–6

1. (*a*) $\phi = 65.4°$ (*b*) 682 m^2 (*c*) 54.5 m (*d*) 682 m^2
3. 2564 mm^2

PROBLEMS 27–2

1. $b = 14.55$, $c = 14.75$, $\gamma = 70°$
3. $a = 28.9$, $c = 41.5$, $\gamma = 74°$
5. $a = 33$, $c = 91.7$, $\gamma = 108°$
7. $a = 0.312$, $b = 0.274$, $\alpha = 42°$
9. $a = 11.3$, $c = 63.6$, $\beta = 55.5°$
11. 1.009 km

PROBLEMS 27–3

1. $a = 18.1$, $\beta = 40.9°$, $\gamma = 71.1°$
3. $c = 1.046$, $\alpha = 8.40°$, $\beta = 27.6°$
5. $c = 4093$, $\alpha = 15.6°$, $\beta = 33.9°$
7. $\alpha = 20.7°$, $\beta = 32.1°$, $\gamma = 127.2°$
9. $\alpha = 6.7°$, $\beta = 15.6°$, $\gamma = 157.7°$
11. 55.7 by 146.8 mm

PROBLEMS 27–4

1. $0.366 \sin \theta + 1.366 \cos \theta$
3. $-0.366(\cos \theta + \sin \theta)$
5. 0.47725

PROBLEMS 28–1

1. 182.4 at 28.3°
3. 239 at 244.7°

PROBLEMS 28–2

1. $x = 27.7$
   $y = 35.5$
3. $x = 0.425$
   $y = 1.29$
5. $x = -51.3$, $y = 0$
7. $x = -80.6$, $y = -195$
9. $x = -28.4$
   $y = 11.9$
11. 1001 N, 503 N
13. 1273 km
15. 261 N

PROBLEMS 28–3

1. $193\angle 77.5^\circ$
3. $6.85\angle 38.5^\circ$
5. $315\angle 20.5^\circ$
7. $107\angle 0^\circ$
9. $125\angle 270^\circ$
11. $10.6\angle 241^\circ$
13. $27.1\angle 161^\circ$
15. $24.4\angle 216.5^\circ$

PROBLEMS 28–4

1. $311\angle 46.5^\circ$
3. $69.9\angle 16.7^\circ$
5. $17.6\angle 196.6^\circ$ or $17.6\angle -163.4^\circ$

PROBLEMS 29–2

1. (*a*) $\frac{\pi^r}{21\,600}$/s (*b*) $\frac{\pi^r}{1800}$/s (*c*) $\frac{\pi^r}{30}$/s
3. (*a*) 4.62°/min (*b*) $\frac{\pi^r}{2340}$/s
5. (*a*) $24\pi^r$ (*b*) $2.4\pi^r$ (*c*) $1.2\pi^r$

PROBLEMS 29–3

	(*a*)	(*b*)	(*c*)	(*d*)	(*e*)
1.	250	2π	1	1	20° lead
3.	2.55	628	100	0.01	10° lead
5.	184	157	25	0.04	22° lag

13. (*b*) $y = 60 \sin 40\pi t$ cm,
 (*c*) −35.3 cm,
 (*d*) 60 cm,
 (*e*) $10\pi^r$

PROBLEMS 30–1

1. (*a*) 29.5 mA
 (*b*) 147.2 mA
 (*c*) 120.2 mA
 (*d*) −120.2 mA
 (*e*) −109.3 mA
3. 439 mA
5. −91.7 V
7. 3.5 V
9. 210° and 330°

PROBLEMS 30–2

1. (a) $f = 480$ Hz
   (b) T = 2.08 ms
   (c) $311 \sin 960\pi t$ V
3. (a) 40 poles
   (b) $v = 250 \sin 800\pi t$ V
   (c) −238 V
5. 3600 rev/min
7. 500 kHz
9. $i = (3 \times 10^{-5}) \sin (1000\pi \times 10^{6})t$ A

PROBLEMS 30–3

1. 7.96 V
3. 16.5 V
5. 69.4 V
7. 10.6 A
9. 156 mA

PROBLEMS 30–4

1. (a) $i = 15 \sin(240\pi t + 21°)$ A
   (b) 14.8 A
3. −9.84 A
5. (a) $i = 40 \sin(800\pi t - 20°)$ A
   (b) 5.57 A
7. 9.2° lag
9. 49° lead or lag

PROBLEMS 31–1

1. 13.9 + j9.7
3. 110 + j40 = $117\underline{/\ 20°}$
5. −84 + j629
7. 21.3 − j42.4
9. 2500 + j400 = $2532\underline{/\ 9.1°}$
11. 2.2 + j10 = $10.2\underline{/\ 77.6°}$

PROBLEMS 31–2

1. 34 + j8 = $34.9\underline{/\ 13.2°}$
3. 46.86 + j2.82 = $46.9\underline{/\ 3.44°}$
5. 0.618 + j0.542 = $0.822\underline{/\ 41.3°}$
7. 1.47 − j0.10 = $1.48\underline{/\ -3.89°}$
9. 140 + j52 = $149\underline{/\ 20.4°}$

PROBLEMS 31–3

1. $28.1 + j22.4 = 36\angle 38.6^\circ$
3. $1057.11 + 390.63 = 1126.97\angle 20.3^\circ$
5. $-445 + j741 = 864\angle 121^\circ$
7. $5.61 - j26.39 = 26.98\angle -78^\circ$
9. $2478 + j798 = 2.6 \times 10^3\angle 17.9^\circ$
11. $-300 + j400 = 500\angle 126.9^\circ$

PROBLEMS 31–4

1. $27.05 - j7.25 = 28\angle -15^\circ$
3. $-25.20 + j48.40 = 54.57\angle 117.50^\circ$
5. $4.43 - j74.6 = 74.7\angle -86.6^\circ$
7. $1.92 + j0.565 = 2\angle 16.4^\circ$
9. $-0.0925 - j0.2348 = 0.2524\angle -111.5^\circ$
11. $13.9 - j2.69 = 14.1\angle -11^\circ$
13. $144\angle 60^\circ = 72 + j125$
15. $1.7\angle 22^\circ = 1.58 + j0.637$
17. $64\angle 270^\circ = 0 - j64$
19. $3\angle 11^\circ = 2.94 + j0.572$

PROBLEMS 32–1

1. (*a*) 423 mA
   (*b*) $v = 156 \sin 377t$ V
   (*c*) $i = 0.598 \sin 377t$ A
   (*d*) 19.9 V
   (*e*) 5.91 W
   (*f*) 100 mA
3. 22.9 V
5. $i = 51.6 \sin (2 \times 10^4 \pi t)$ mA

PROBLEMS 32–2

1. 30.2 Ω
3. $L = 253\ \mu$H
5. 3.43 kΩ
7. 297 mA
9. 995 MHz
11. −424 mA
13. (*a*) X_L is doubled (*b*) X_L is tripled (*c*) X_L is halved

PROBLEMS 32–3

1. 15.9 Ω
3. 72.3 mΩ
5. 138 Ω
7. 2.95 μA
9. 5.17 μF
11. $i = 0.639 \sin(377t + 90°)$ A
13. 153 pF
15. 137 μA
17. X_C varies inversely as C
19. 2.32 kΩ

PROBLEMS 32–4

1. (*a*) 679 Ω,
   (*b*) $788\angle 59.5°$ Ω,
   (*c*) $152\angle -59.5°$ mA,
   (*d*) $i = 215 \sin(377t - 59.5°)$ mA
   (*e*) 60.9 V,
   (*f*) 103 V,
   (*g*) not in phase
3. (*a*) $f\downarrow$, $X_L\downarrow$, $Z\downarrow$, $I\uparrow$, (*b*) 12.7 kΩ, 3.9 MHz, 75° lag
5. (*a*) 121 Ω, (*b*) 351 Ω, (*c*) 342 mA, (*d*) 113 V, (*e*) 41.2 V
7. 373 Ω
9. (*a*) $V_g = 600\angle -14°$ V, (*b*) $f = 3.19$ MHz, (*c*) $V_C = 145\angle -90°$ V, (*d*) $V_R = 582$ V

PROBLEMS 32–5

Q	Z	I	i	PF	P
1.	$677\angle 53.8°$ Ω	177 mA	$i = 251 \sin(377t - 53.8°)$ mA	59%	12.6 W
3.	$1769\angle 32°$ Ω	67.8 mA	$i = 95.9 \sin(200\pi t - 32°)$ mA	84.7%	6.9 W
5.	$6453\angle 81.1°$ Ω	232 nA	$i = 329 \sin(125.7 \times 10^6\pi t - 81.1°)$ nA	15.5%	53.4 pW
7.	$15\angle 2.4°$ Ω	7.79 A	$i = 11 \sin(377t - 2.4°)$ A	99.9%	911 W
9.	$118\angle -40.5°$ Ω	1.01 A	$i = 1.43 \sin(6\pi t \times 10^6 + 40.5°)$ A	76%	92 W

11. (*a*) $R_L = 200$ Ω
    (*b*) $X_L = 346$ Ω
    (*c*) $L = 919$ mH
13. (*a*) $55.8\angle -32.6°$ Ω
    (*b*) 505 μF
15. (*a*) 23.5 A
    (*b*) 8.95 kW

PROBLEMS 32–6

1. (*a*) 20.4 mA
   (*b*) 460 mW
   (*c*) 434 V across the capacitor, 511 V across the coil
3. 13.8 kHz
5. (*a*) 0.239 mH
   (*b*) 2.3 MHz
7. (*a*) $Z = 91.1\angle -80.5°\ \Omega$
   (*b*) $i = 3.29\angle 0°\ \mu A$

PROBLEMS 33–1

1. 850 pF
3. (*a*) $v = 311 \sin 377t$ V
   (*b*) $i = 138 \sin (377t + 90°)$ mA
   (*c*) 104 V
5. (*a*) 0.47 H
   (*b*) 1.41 H

PROBLEMS 33–2

1. $65.2\angle 38.9°\ \Omega$
3. $42.3\angle 6.5°\ \Omega$
5. $104\angle 55.6°\ \Omega$
7. $68\angle 74.2°\ \Omega$
9. $114\angle 188°\ \Omega$
11. $144\angle 1.52°\ \Omega$
13. $2.82\angle 66.6°$ A

PROBLEMS 33–3

1. (*a*) 5.92 MHz
   (*b*) 5.93 MHz
   (*c*) 16
3. (*a*) 113 kHz, (*b*) 5.66 MΩ, (*c*) 8.84 Ω
5. 381 pF
7. 7.08 mW
9. 0.6% leading
11. 32.2 pF
13. (*a*) 48.5 kΩ, 188.4 kΩ (*b*) 9.74 mH, 46.5 mH
    (*c*) $6.265\angle -53°$ mA (*d*) $5.12\angle -51.5°$ mA
15. $Z_2 = 78.13\angle 51.5°$ kΩ

PROBLEMS 33–4

1. $Z_a = 24.2\angle 8.51°\ \Omega$
   $Z_b = 17.5\angle 86°\ \Omega$
   $Z_c = 21.7\angle 18.8°\ \Omega$
3. $Z_1 = 87.4\angle 36.7°\ \Omega$
   $Z_2 = 104.9\angle -68.5°\ \Omega$
   $Z_3 = 111.1\angle 46.5°\ \Omega$

5. $Z_{ab} = 76.1\underline{/ 2.86°}\ \Omega$
7. $Z_{ab} = 187\underline{/ 27.1°}\ \Omega$
9. 21.9 W
11. $Z_{ab} = 89\underline{/ -25°}\ \Omega$
13. 84.8 W
15. $Z_{ab} = 279\underline{/ 58.1°}\ \Omega$
17. $Z_{ab} = 81.5\underline{/ 67.3°}\ \Omega$

PROBLEMS 34–1

1. $2 = \log_{10} 100$
3. $\log_2 8 = 3$
5. $0.5 = \log_4 2$
7. $1 = \log_a a$
9. $0 = \log_a 1$
11. $10^3 = 1000$
13. $5^2 = 25$
15. $6^0 = 1$
17. $4^5 = 1024$
19. $s^t = r$
21. $x = 4$
23. $x = 4$
25. $x = 7$
27. $5 - 3 = 2$
29. 1, 2, 3, 4, 5, 6, 7, 8, 9

PROBLEMS 34–2

1. 0.861 53
3. 2.861 53
5. 1.981 44
7. 1.000 87
9. 6.000 87

PROBLEMS 34–3

1. 93.046 49
3. 0.009 304 6
5. $7.570\,072 \times 10^6$
7. $757.007\,2 \times 10^{-9}$
9. 6.518 3

PROBLEMS 34–4

1. $x = 6.4$
3. $x = 100$
5. $x = 79.4$
7. $P_1 = 1992$
9. $x = 3.484 \times 10^4$
11. $x = 4.5$
13. $m = -0.086\,7$
15. $x = 0.702\,9$
17. $y = 11.093$
19. $x = 1.247$
21. $x = 3.88$
23. $x = 16.634$

25. $x = 7$

27. $Z_1 = \sqrt{Z_a Z_b}$

29. $S_c = \dfrac{2.302\ 59}{t} \log \dfrac{iL}{Vt}$ or $S_c = \dfrac{\log_e \dfrac{iL}{Vt}}{t}$

31. (*a*) $A = \dfrac{I_k}{T^2} e^{\frac{B}{T}}$

(*b*) $B = 2.302\ 59 \log \dfrac{AT^2}{I_k}$ or $B = T \log_e \dfrac{AT^2}{I_k}$

33. (*a*) $V = \dfrac{q}{C(1 - e^{-x})}$

(*b*) $R = \dfrac{0.434\ 29t}{C[\log CV - \log(CV - q)]}$

(*c*) $t = 2.302\ 59RC \log \dfrac{CV}{CV - q}$

35. worked example

37. (*a*) $i = 76.1$ mA
(*b*) $i = 799.9$ mA

39. Did you show?

41. 315 words/min.

43. (*a*) $i = 50$ mA
(*b*) $i = 48$ mA

45. (*a*) $i = 162$ mA
(*b*) $t = 13.6$ ms
(*c*) Yes

47. $R = 868.6\ \Omega$

49. $I_p + I_g = 58.8$ mA

51. $Bn = 466$ Hz

PROBLEMS 34–5

1. $1.908\ 36 \times 10^3$
3. $-3.700\ 54$
5. 4
7. -141.968×10^{-6}
9. 7.297 86
11. 1.728×10^3
13. 9.340 84
15. 53.261

PROBLEMS 35–1

1. *R*12 series of preferred values; 1.0, 1.2, 1.5, 1.8, 2.2, 2.7, 3.3, 3.9, 4.7, 5.6, 6.8, 8.2, 10.
Maximum % error: ±11.1%

3. *R*10 series of preferred values: 1.0, 1.25, 1.6, 2.0, 2.5, 3.2, 4.0, 5.0, 6.4, 8.0, 10.
Probable published tolerance: ±15%

5. (*a*) R_{24} (*c*) R_{24}
(*b*) R_{12} (*d*) R_6

PROBLEMS 35–2

1. (*a*) 12 dB
(*b*) 14 dB
(*c*) 16 dB
(*d*) −21 dB

3. 671 mV

5. 548 mV

7. 1.73 V

9. (*a*) 12 mW, 1.9 V
   (*b*) 60 mW, 4.24 V
   (*c*) 0.6 mW, 0.424 V (600 μW, 424 mV)
   (*d*) 6 μW, 42.4 mV
   (*e*) 60 pW, 134 μV

11. 3.16×10^7
13. 100 kW
15. 100 nW
17. 10^{-8}
19. 10^3
21. 100 dB
23. 2.82×10^6
25. 54 dB
27. (*a*) 13.3 mV (*b*) 33 dB (*c*) 83 dB
29. (*a*) 223 (*b*) 37 dB (*c*) 6.02 mW
31. 3.92 W
33. 1.1 μV
35. 0.20 dB/km
37. 250:1
39. 109 kW
41. 25.8 dB

PROBLEMS 35–3

1. 321 mH
3. (*a*) 562 mH
   (*b*) 0.029 μF = 29 nF
5. 356 mm
7. No. 2
9. 171.2 nF/km
11. 115 km
13. (*a*) 19 nH/cm (*b*) 0.0584 pF/cm (*c*) 0.295 mH (*d*) 906 pF

PROBLEMS 35–4

1. 571 Ω
3. 1.05%
5. No
7. No. 1 wire (7.348 mm)
9. 141 Ω
13. (*a*) 0.3 dB (*b*) 93.3%
15. (*a*) 0.7 dB (*b*) 85.1%
17. 604 pF
19. 5.6%

PROBLEMS 36–1

1. 7
3. 195
5. 17
7. 47
9. 35

PROBLEMS 36–2

1. 111
3. 1001011
5. 100110
7. 1100001
9. 1011001101

PROBLEMS 36–3

1. 5
3. 23
5. 63
7. 1223
9. 3256

PROBLEMS 36–4

1. 35
3. 115
5. 1061
7. 1624
9. 20211

PROBLEMS 36–5

1. 200_5
3. 112_5
5. 21110_3
7. 3443_8
9. 5276_8
11. 242_{10}
13. 63
15. 3
17. 70
19. 420_{10}

PROBLEMS 36–6

1. 001 110 011
3. 100101110
5. 001 110 000
7. 110 010 101 110
9. 111 111 111 111
11. 24
13. 27
15. 517_8
17. 632
19. 252

PROBLEMS 36–7

1. 110111
3. 1 110 000
5. 1010010
7. 1 000 001
9. 1000101

PROBLEMS 36–8

1. 010010
3. 010001
5. 01
7. 10001
9. 001001

PROBLEMS 36–9

1. 010001
3. 011010
5. 011000
7. 010001
9. 011001

PROBLEMS 36–10

1. 1001110
3. 11100001
5. 111101100
7. 1010110001
9. 101110110101

PROBLEMS 36–11

1. 1101
3. 10101
5. 10000
7. 11001
9. 11110

PROBLEMS 36–12

1. 0.1011001 (99.33% accurate)
3. 0.0110011 (99.6% accurate)
5. 0.101 (100% accurate)
7. 0.0110001 (99.53% accurate)
9. 0.00111011_2 (99.87% accurate)

PROBLEMS 36–13

1. 0.1001_2 (100% accurate)
3. 1.0011011_2 (100% accurate)
5. 1.1101_2 (100% accurate)
7. 0.111101_2 (100% accurate)
9. 1011.11001 (100% accurate)

PROBLEMS 36–14

1. $100.101001_2 = 4.640\ 625_{10}$
3. $11100.111_2 = 28.875_{10}$
5. $1000.001_2 = 8.125_{10}$
7. $100.1110_2 = 4.8750_{10}$
9. $10001.0111_2 = 17.4375_{10}$

PROBLEMS 36–15

	Binary	Denary	Accuracy, %
1.	110.1	6.5	99.9
3.	110011.001	51.2	99.998
5.	110	6	100
7.	100	4.0	100
9.	10.101010	2.65625 (2.6667)	99.6

PROBLEMS 36–16

1. 45.6_8
3. 13.7_8
5. 15.55_8
7. 17.74_8
9. 0.25_8

PROBLEMS 36–17

1. $11D_{16}$
3. $5EB_{16}$
5. 52_{16}
7. 488_{10}
9. 608_{10}

PROBLEMS 36–18

1. 743_{10}
3. 1013_{10}
5. 1521_{10}
7. $2AC_{16}$
9. 2048_{16}

PROBLEMS 36–19

1. 19.54_{10}
3. $18.414\ 062\ 5_{10}$
5. $190.175\ 781\ 250_{10}$
7. $0.D5F_{16}$
9. $3F.E661_{16}$

PROBLEMS 37–1

1. sl
3. $s + l$
5. sl
7. $\bar{s} + l$
9. $sl + \bar{s} \cdot \bar{l}$

PROBLEMS 37–2

1.

a	b	c	ab	ac	$ab + ac$
1	1	1	1	1	1
1	1	0	1	0	1
1	0	1	0	1	1
1	0	0	0	0	0
0	1	1	0	0	0
0	1	0	0	0	0
0	0	1	0	0	0
0	0	0	0	0	0

3.

a	b	c	$\bar{a}$	$\bar{b}$	$\bar{c}$	$\bar{a} \cdot \bar{b} \cdot \bar{c}$	$a + b + c$	$\overline{a + b + c}$
1	1	1	0	0	0	0	1	0
1	1	0	0	0	1	0	1	0
1	0	1	0	1	0	0	1	0
1	0	0	0	1	1	0	1	0
0	1	1	1	0	0	0	1	0
0	1	0	1	0	1	0	1	0
0	0	1	1	1	0	0	1	0
0	0	0	1	1	1	1	0	1

5.

a	$\overline{a}$	$\overline{a} + a$	Output
0	1	1	1
0	1	1	1
1	0	1	1
1	0	1	1

$\overline{a} + a = 1$

7.

a	b	$a + b$	$a(a + b)$
0	0	0	0
0	1	1	0
1	0	1	1
1	1	1	1

$a(a + b) = a$

9.

p	q	$\overline{p}$	$p + q$	$\overline{p}q$	$p + \overline{p}q$
0	0	1	0	0	0
0	1	1	1	1	1
1	0	0	1	0	1
1	1	0	1	0	1

$p + q = p + \overline{p}q$

11.

a	b	c	$\overline{a}$	$\overline{b}$	$\overline{c}$	$a\overline{c}$	$b\overline{c}$	$(a + b)$	$(a\overline{c} + b\overline{c})$	$\overline{c}(a + b)$
0	0	0	1	1	1	0	0	0	0	0
0	0	1	1	1	0	0	0	0	0	0
0	1	0	1	0	1	0	1	1	1	1
0	1	1	1	0	0	0	0	1	0	0
1	0	0	0	1	1	1	0	1	1	1
1	0	1	0	1	0	0	0	1	0	0
1	1	0	0	0	1	1	1	1	1	1
1	1	1	0	0	0	0	0	1	0	0

$a\overline{c} + b\overline{c} = \overline{c}(a + b)$

13.

x	y	z	$x + y$	$x + z$	yz	$x + yz$	$\overline{x + y}$	$\overline{x + z}$	$\overline{x + yz}$	$(\overline{x + y} + \overline{x + z})$
0	0	0	0	0	0	0	1	1	1	1
0	0	1	0	1	0	0	1	0	1	1
0	1	0	1	0	0	0	0	1	1	1
0	1	1	1	1	1	1	0	0	0	0
1	0	0	1	1	0	1	0	0	0	0
1	0	1	1	1	0	1	0	0	0	0
1	1	0	1	1	0	1	0	0	0	0
1	1	1	1	1	1	1	0	0	0	0

$\overline{x + y} + \overline{x + z} = \overline{x + yz}$

15.

a	b	c	$\bar{a}$	$\bar{b}$	$\bar{c}$	$a + b$	$\bar{a} + c$	$b + c$	$\bar{a}b$	ac	$(a + b)(\bar{a} + c)(b + c)$	$\bar{a}b + ac$
0	0	0	1	1	1	0	1	0	0	0	0	0
0	0	1	1	1	0	0	1	1	0	0	0	0
0	1	0	1	0	1	1	1	1	1	0	1	1
0	1	1	1	0	0	1	1	1	1	0	1	1
1	0	0	0	1	1	1	0	0	0	0	0	0
1	0	1	0	1	0	1	1	1	0	1	1	1
1	1	0	0	0	1	1	0	1	0	0	0	0
1	1	1	0	0	0	1	1	1	0	1	1	1

$(a + b)(\bar{a} + c)(b + c) = \bar{a}b + ac$

PROBLEMS 37–4

1. (a) $Z_{xy} = \bar{a} + \bar{b} \cdot \bar{c}$
   (b) $Y_{xy} = a(b + c)$
3. (a) $Z_{LM} = (\bar{A} + B)(A + \bar{B})$
   (b) $Y_{LM} = A\bar{B} + \bar{A}B$ **or** $(A + B)(\overline{AB})$
5. (a) $Z_{pq} = \bar{a} + \bar{b}(ab\bar{c} + \bar{a}) + \bar{c}$ **or** $\bar{a} + \bar{a}\bar{b} + \bar{c}$
   (b) $Y_{pq} = abc + a(\bar{a} + \bar{b} + c)c$ **or** ac
7. 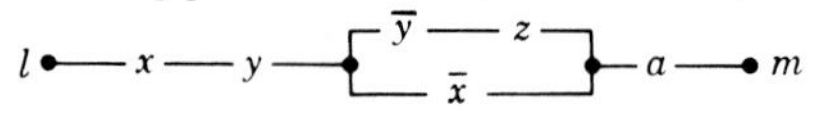

9. $Z_{\text{out}} = B + C$

PROBLEMS 37–7

1. $\overline{\overline{a + b}} = a + b$: OR gate
3. $\overline{\overline{\bar{a} + \bar{b}}} = \bar{a} + \bar{b} = \overline{ab}$: NAND gate

PROBLEMS 38–1

1. $\bar{A}\bar{B} + AB$
3. $\bar{A}\bar{B}\bar{C} + \bar{A}\bar{B}C + \bar{A}BC + ABC$

5.

	$\bar{A}\bar{C}$	$\bar{A}C$	AC	$A\bar{C}$
$\bar{B}$			1	1
B	1	1		1

7.

	$\bar{A}\bar{C}$	$\bar{A}C$	AC	$A\bar{C}$
$\bar{B}\bar{D}$	1	1		
$\bar{B}D$		1		
BD		1	1	
$B\bar{D}$	1	1		1

PROBLEMS 38–2

1. (*a*) $F = B + \bar{A}\bar{C} + AC$

	$\bar{A}\bar{C}$	$\bar{A}C$	AC	$A\bar{C}$
$\bar{B}$	1		1	
B	1	1	1	1

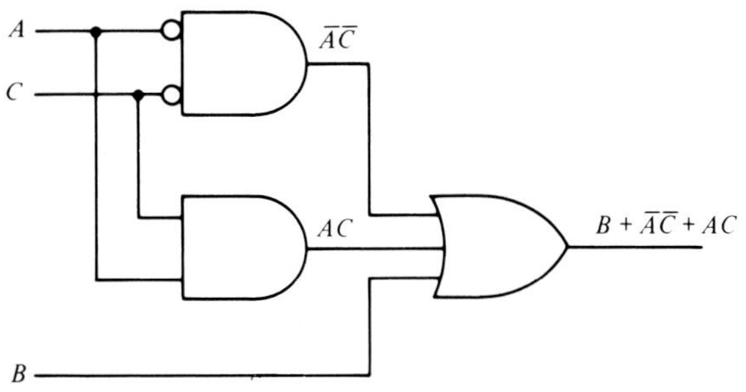

(*b*) $F = Y + XZ$

	$\bar{X}\bar{Z}$	$\bar{X}Z$	XZ	$X\bar{Z}$
$\bar{Y}$			1	
Y	1	1	1	1

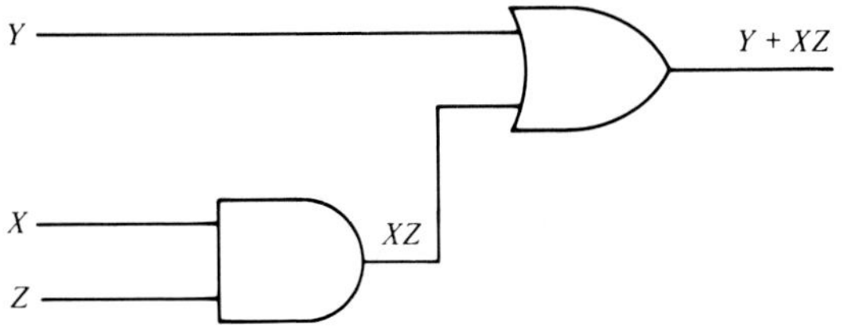

(c) $F = A$

	$\overline{A}$	$\overline{A}$	A	A
$\overline{B}$			1	1
B			1	1
	$\overline{C}$	C	C	$\overline{C}$

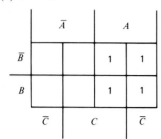

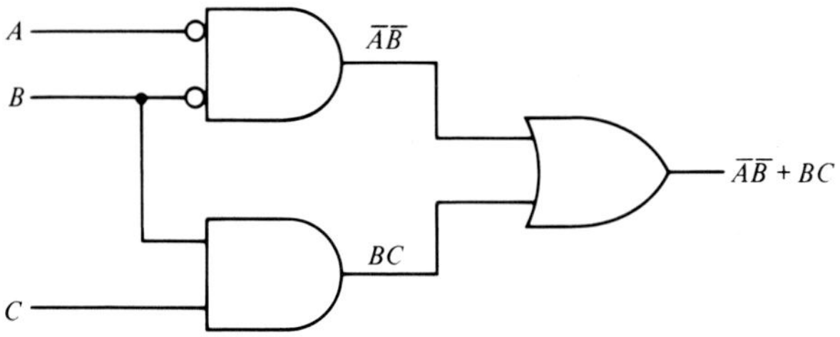

3. $\overline{B}C + A\overline{B}\,\overline{C} + B\overline{C} = \overline{B}C + A\overline{B} + B\overline{C}$ [$= \overline{B}C + A\overline{C} + \overline{A}\,\overline{B}C$
$= \overline{A}\,\overline{B}C + A\overline{B} + B\overline{C}$]

	$\overline{A}$	$\overline{A}$	A	A
$\overline{C}$		1	1	1
C		1		1
	$\overline{B}$	B	B	$\overline{B}$

Horizontal groupings;
Boxes 2 and 3 produce $B\overline{C}$,
Boxes 3 and 4 produce $A\overline{C}$.
Vertical groupings;
Boxes 2 and 6 produce $\overline{A}B$,
Boxes 4 and 8 produce $A\overline{B}$.

																			Map groupings				
A	B	C	$\overline{A}$	$\overline{B}$	$\overline{C}$	$\overline{A}B$	$A\overline{B}$	$A\overline{C}$	$B\overline{C}$	$\overline{A}B + A\overline{B} + A\overline{C}$			o/p	$\overline{A}B + A\overline{B} + B\overline{C}$			o/p	$\overline{A}B + A\overline{B} + A\overline{C} + B\overline{C}$				o/p
0	0	0	1	1	1	0	0	0	0	0	0	0	0	0	0	0	0	0	0	0	0	0
0	0	1	1	1	0	0	0	0	0	0	0	0	0	0	0	0	0	0	0	0	0	0
0	1	0	1	0	1	1	0	0	1	1	0	0	1	1	0	1	1	1	0	0	1	1
0	1	1	1	0	0	1	0	0	0	1	0	0	1	1	0	0	1	1	0	0	0	1
1	0	0	0	1	1	0	1	1	0	0	1	1	1	0	1	0	1	0	1	1	0	1
1	0	1	0	1	0	0	1	0	0	0	1	0	1	0	1	0	1	0	1	0	0	1
1	1	0	0	0	1	0	0	1	1	0	0	1	1	0	0	1	1	0	0	1	1	1
1	1	1	0	0	0	0	0	0	0	0	0	0	0	0	0	0	0	0	0	0	0	0

Original function

A	B	C	$\overline{A}$	$\overline{B}$	$\overline{C}$	$\overline{A}BC + A\overline{B}C + A\overline{B}\,\overline{C} + \overline{A}B\overline{C} + AB\overline{C}$					Output
0	0	0	1	1	1	0	0	0	0	0	0
0	0	1	1	1	0	0	0	0	0	0	0
0	1	0	1	0	1	0	0	0	1	1	1
0	1	1	1	0	0	1	0	0	1	0	1
1	0	0	0	1	1	0	0	1	0	0	1
1	0	1	0	1	0	0	1	0	0	0	1
1	1	0	0	0	1	0	0	0	0	1	1
1	1	1	0	0	0	0	0	0	0	0	0

All outputs are the same

5. $A\overline{B}\,\overline{D} + B\overline{D} + \overline{A}CD + \overline{C}\,\overline{D} + \overline{A}BC = \overline{D}(A + B + \overline{C}) + \overline{A}C(B + D)$

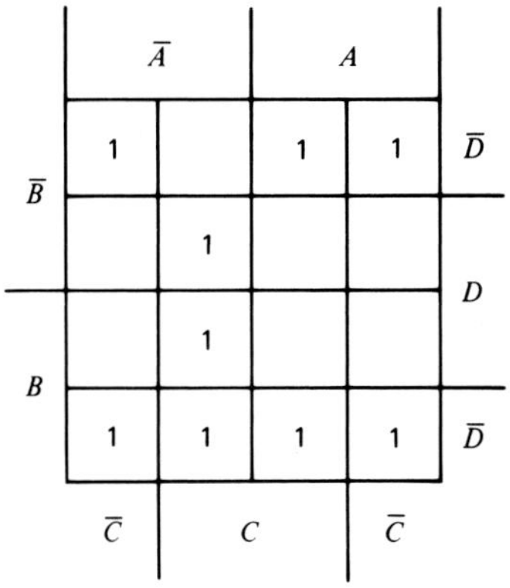

7. The Karnaugh map of the Sum function produces two sets of diagonal boxes with no grouping possible. Boolean algebra will provide some simplification:
$S = c(ab + \bar{a}\bar{b}) + \bar{c}(\bar{a}b + a\bar{b})$

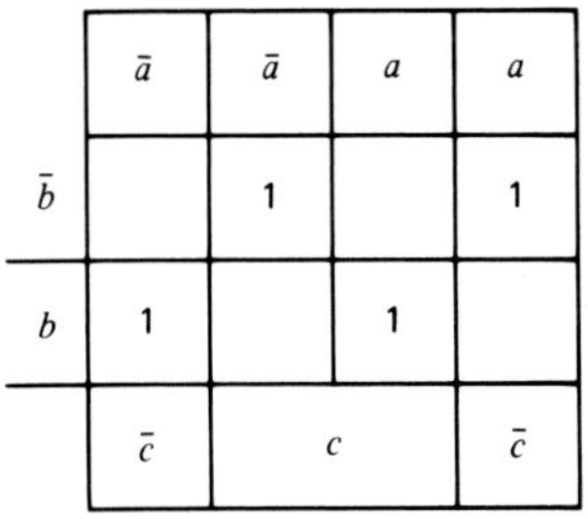

APPENDIX C

ANSWERS TO SELF TESTS

CHAPTER 2

1. $x + 12$ V
2. $\lambda = 0.600$ m
3. $C_3 = C_2 + 70\ \mu$F
4. $\frac{1}{L} = 10$
5. 0.036 W
6. $4D$ coins
7. $P = 0.000\ 000\ 750$ W
8. c
9. $L = 1.0$ H
10. $V_{av} = \frac{2\ V_{max}}{\pi}$

CHAPTER 3

1. 229
2. -434
3. -49
4. 2.4 m
5. $638.30
6. $4P + 5\frac{V^2}{R} + 7IR$
7. $9R + 3X_L + 15X_C$
8. $3M + P + 4R$
9. $3X^2 - 6Y^2 - 20Z^2$
10. $7a - 2b + 7c$

CHAPTER 4

1. -42
2. $2\pi fL$
3. $-\frac{1}{120}$
4. $15I^3RV - 10IPV^2$
5. $\theta^2 - 5\theta - 14$
6. -7
7. $\frac{-5\lambda}{\Omega}$
8. $3P^2 - 4P + 6$
9. $\theta^3 + \theta\phi^2 + 2\phi^3 + \frac{\phi^4}{\theta - \phi}$
10. $-2X^3 - 3X^2Y + 7XY^2 + 2Y^3$

CHAPTER 5

1. $m = 3$
2. $x = -5$
3. $t = 1.61$
4. $R = 1.13$

5. $R_2 = R_0(1 + \alpha\Delta t)$
6. $l = 20$ m, $w = 10$ m
7. 16 and 17
8. $t_2 = \frac{R_2}{R_1}(t_1 + 234.5) - 234.5$
9. $12a$
10. $l_B = 4l_A$

CHAPTER 6

1. 4
2. 1.7625×10^4
3. 27.9×10^3
4. 1.74×10^3
5. 243
6. 16
7. $R_X = 10.3 \times 10^3$
8. 1.14×10^6
9. $f = 5.81 \times 10^3$ Hz
10. $1.02 \times 10^3\ \Omega$

CHAPTER 7

1. 1.76 A
2. 18×10^6 mV
3. $277 \times 10^3\ \mu$A
4. 1.67×10^{-6} s
5. $5.76 \times 10^{-6}\ \mu$F/m
6. 25×10^{-3} m
7. 878 km/h
8. 2.01 mm/s
9. $2.63 \frac{\sqrt{f}}{d}\ \Omega$/m
10. 100 MHz

CHAPTER 8

1. 513 mA
2. 440 Ω
3. 5.94 V
4. 12 Ω
5. 2200 A
6. 66.7 Ω
7. 86.3%
8. Yes
9. 121 Ω 100 w (how about a 100-W 110-V light bulb?)
10. 117 V

CHAPTER 9

1. 8.82 Ω
2. 0.017 283 $\mu\Omega \cdot$ m
3. No. 6
4. 5.124 kg

5. $303\ \Omega$
6. $1.50\ \Omega$
7. No
8. No. 8
9. No. 2
10. 226 V

CHAPTER 10

1. $225P^6$
2. $\pm\frac{13R_1R_2^2}{9\alpha^2}$
3. $(5\pi R\lambda)(3 + 2\pi\lambda - 5R\lambda^2)$
4. $\omega^2 - \frac{7}{8}\omega + \frac{49}{256}$
5. $9x^2 - 42xy + 49y^2$
6. $2P(3X_L + 5X_C)^2$
7. $\frac{4\lambda^2}{9\pi^2} - \omega^2$
8. $(3a - 5b + c)(3a - 5b - c)$
9. $\theta^2 - \frac{11\theta\phi}{84} - \frac{5\phi^2}{42}$
10. $(3l^2m^2 - 2r)(2l^2m^2 + 3r)$
11. $15y^2 - 14yz - 8z^2$
12. $\frac{2y}{5\alpha}(3y + 2z)(y - 4z)$

CHAPTER 11

1. a
2. $720\theta\phi\lambda$
3. $? = 35 - 7x + 5L - Lx$
4. $\frac{x - y}{x + 3y}$
5. $\frac{m - l}{p - q}$ or $\frac{l - m}{q - p}$
6. $3R^2 + 2RV - 3V^2 + \frac{3}{R + V}$
7. $\frac{ab^2 + 4ab + 3a}{b^3 + 4b^2 + 3b}, \frac{ab^2 + 3ab + b^2 + 3b}{b^3 + 4b^2 + 3b}, \frac{ab^2 + ab - 2b^2 - 2b}{b^3 + 4b^2 + 3b}$
8. $\frac{2\theta + 5\theta^2 + 5\theta\lambda - 4\lambda}{\theta^2 - \lambda^2}$
9. $\frac{2(a - b)}{2a + 3b}$
10. $\frac{5x + 1}{5x - 1}$

CHAPTER 12

1. $\theta = 4$
2. $z = 1.2$
3. $\omega = 2$
4. $\phi = 8$
5. 9.26 h
6. 8.89 h
7. 22.3 L

8. $t_2 = \frac{R_2}{R_1}(234.5 + t_1) - 234.5$

9. $L_2 = \frac{L_p L_1}{L_1 - L_p}$

10. $d_2 = \frac{d_1}{1.5}[\mu_o(\alpha - 1) - 1]$

CHAPTER 13

1. 3.9 kΩ
2. 1.08 kΩ
3. .775 kΩ

CHAPTER 14

1. 0.9184 Ω
2. 219 955 Ω; use 220 kΩ
3. $R_1 = 43.2\ \Omega$, $R_2 = 1.62\ \Omega$, $R_3 = 0.18\ \Omega$
4. 110 V

CHAPTER 15

1. $R_1 = 150\ \Omega$, $P_1 = 2.16$ W
   $R_2 = 75\ \Omega$, $P_2 = 480$ mW
   $R_3 = 150\ \Omega$, $P_3 = 240$ mW
   $R_4 = 300\ \Omega$, $P_4 = 120$ mW
2. $I_1 = 75\%\ I_T$
3. 24 Ω
4. $P_3 = 314$ mW
5. 23.6 Ω

CHAPTER 16

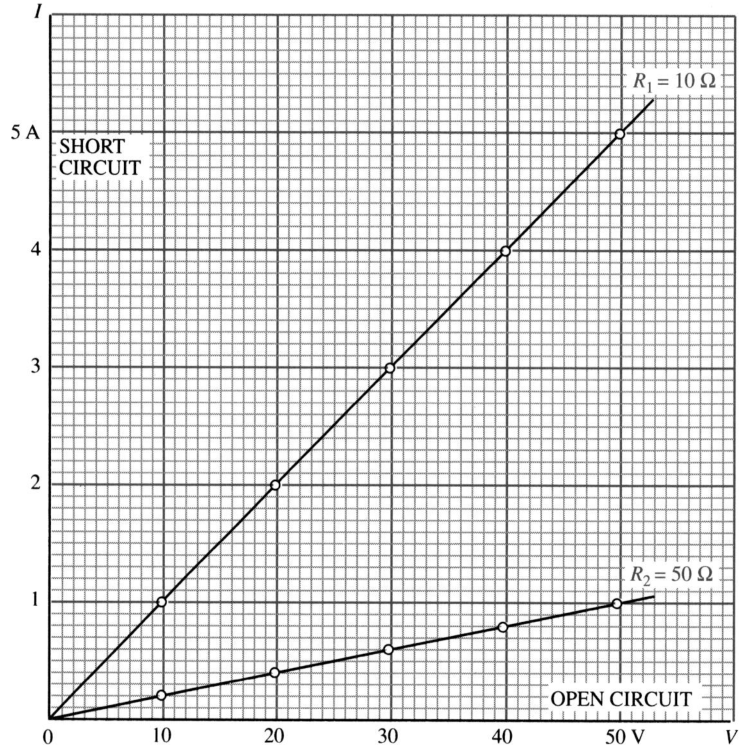
I
5 A
4
3
2
1
0
10
20
30
40
50 V
V
SHORT CIRCUIT
OPEN CIRCUIT
R1 = 10 Ω
R2 = 50 Ω

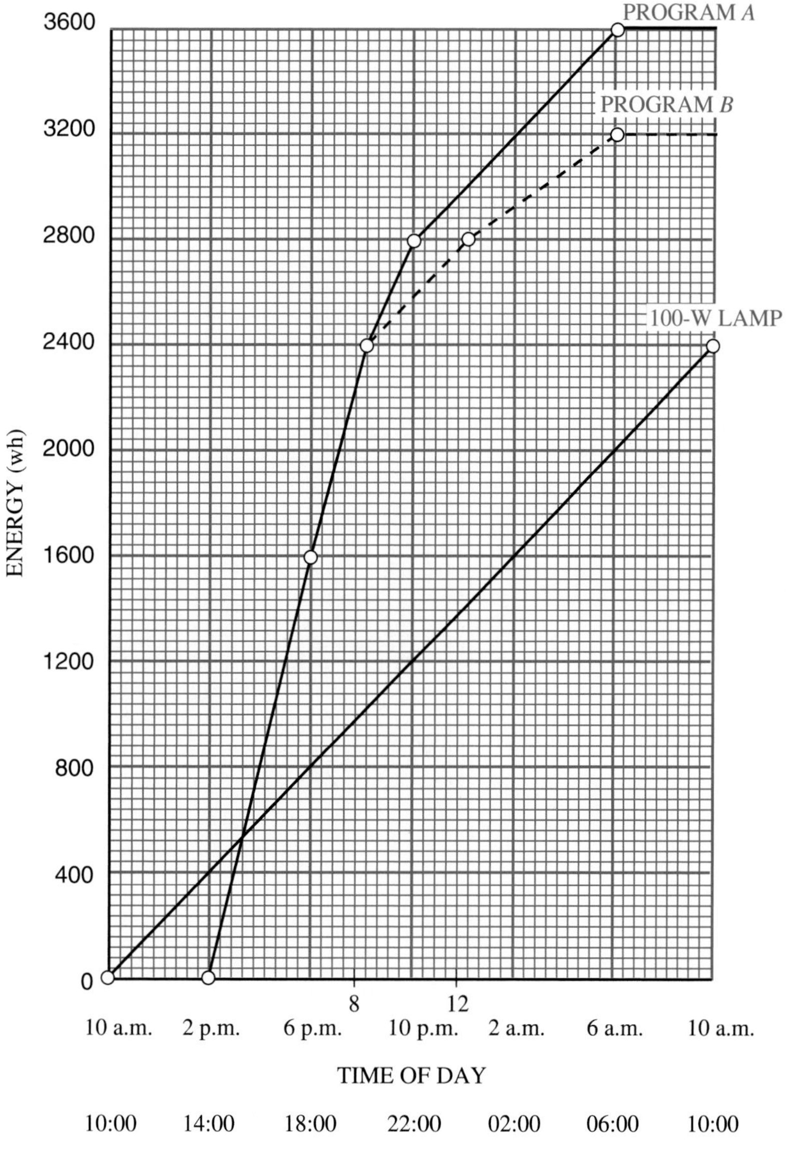

PROGRAM A
PROGRAM B
100-W LAMP
3600
3200
2800
2400
2000
1600
1200
800
400
0
ENERGY (wh)
8
12
10 a.m.
2 p.m.
6 p.m.
10 p.m.
2 a.m.
6 a.m.
10 a.m.
TIME OF DAY
10:00
14:00
18:00
22:00
02:00
06:00
10:00

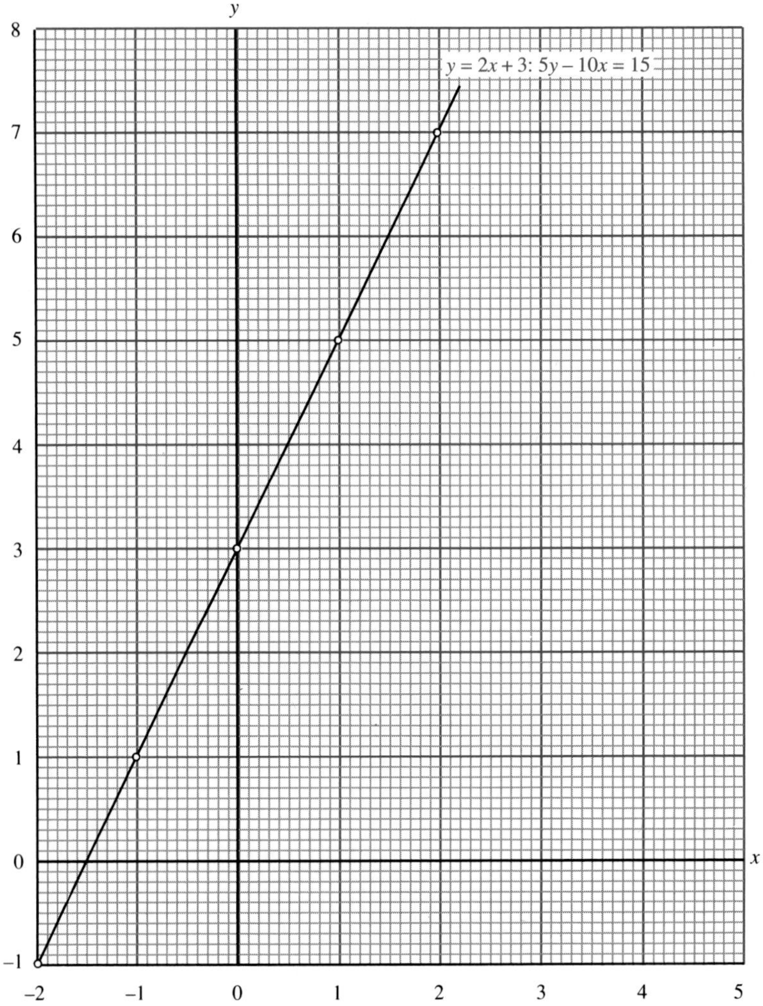
y
x
$y = 2x + 3$: $5y - 10x = 15$
8
7
6
5
4
3
2
1
0
–1
–2
–1
0
1
2
3
4
5

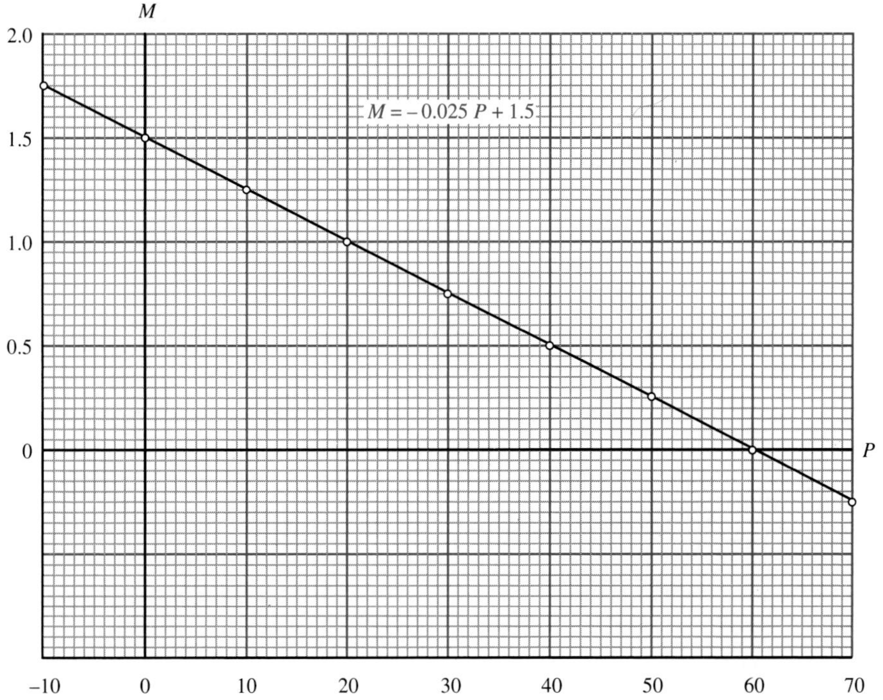

CHAPTER 17

1. $x = 1, y = 0.5$
2. $m = 6, n = 1$
3. $R = 3, Z = -2$
4. $x = 1.013, y = 0.4813$
5. $X = 2.4, R = 8.1$
6. $A = 5, B = 4$
7. $R_1 = \dfrac{3\beta + 4\alpha}{10}, R_2 = \dfrac{\beta - 2\alpha}{10}$
8. $I_1 = 7, I_2 = -3, I_3 = 5$
9. $R_a = \dfrac{R_1R_3}{R_1 + R_2 + R_3}$

   $R_b = \dfrac{R_1R_2}{R_1 + R_2 + R_3}$

   $R_c = \dfrac{R_2R_3}{R_1 + R_2 + R_3}$

CHAPTER 18

1. -25
2. $x = 1.013, y = 0.481$
3. -8
4. $\alpha = 7$
   $\beta = -3$
   $\gamma = 5$
5. -1141
6. $p = 3$

CHAPTER 19

1. 0.179 Ω
2. 110 mA
3. $V = 9.35$ V, $r = 0.195$ Ω

CHAPTER 20

1. a^{p+x}
2. x^{2z}
3. $\theta^6\phi^9$
4. R^9Z^6
5. $m^2n\sqrt{p}$
6. $\pm 95x^2y^4z\sqrt{xy}$
7. $\pm\dfrac{R}{2\pi fL}\sqrt{(2\pi fL)^2 - 1}$
8. $5\frac{1}{5}\sqrt{5a}$
9. $\dfrac{11 + 5\sqrt{5}}{4}$
10. $j\dfrac{6}{7}$
11. 5 + j12
12. 4 + j22
13. 27 + j24
14. −0.615 + j1.923
15. −39
16. $P = 88$

CHAPTER 21

1. $\theta = \pm 7$
2. $Q = 3$ or -8
3. $x = 3 \pm \sqrt{21}$
4. $p = 6$ or 3.5
5. (a) $y = -3$
   (b) $x = -3$ and $+1$
   (c) $x = -1$, $y = -4$
6. 236
7. 25 and 14
8. $R = 396$ Ω or 4.04 Ω
9. $R_2 = 6$ Ω and $R_3 = 20$ Ω; $R_2 = 16.67$ Ω and $R_3 = 33.33$ Ω
10. $I = 16.5$ A and $V = 27.5$ V; $I = 5.5$ A and $V = 82.5$ V

CHAPTER 22

1. 2.45 μA
2. 381 V
3. 544 mW

4. $R_a = 7.2\ \Omega$, $R_b = 4.91\ \Omega$, $R_c = 6\ \Omega$
5. 1.1 A
6. 12.6 V in series with 2.88 Ω
7. 357 mA in parallel with 5.93 Ω

CHAPTER 23

1. 0.0432°
2. 377 r/s
3. 108°
4. 5 mm
5. 16.3 mm

CHAPTER 24

1. $\frac{b}{c}$
2. $\frac{b}{c}$
3. θ
4. First quadrant
5. –
6. –
7. –
8. $\frac{\sqrt{5}}{3}$ (= 0.74536)
9. $\frac{\sqrt{207}}{7}$ (= 2.055)
10. −1

CHAPTER 25

1. 0.60460
2. −0.99705
3. −49.20° = 130.80°
4. 59.68° or 120.32°
5. −0.43130
6. 123.81° or 303.81°
7. 73.65° or 106.45°
8. −0.99229
9. 138.05°
10. 18.47° or 198.47°

CHAPTER 26

1. $X = 8.42$, $Z = 14.8$, $\phi = 55.4°$
2. $X = 0.313$, $R = 0.763$, $\theta = 22.3°$
3. $R = 591$, $\theta = 24.1°$, $\phi = 65.9°$
4. $Z = 260$, $\theta = 56.3°$, $\phi = 33.7°$
5. 14.8 m
6. 34.3×10^3 m²

CHAPTER 27

1. $\tan \theta$
2. $\lambda = 101°$, $P = 34.5$, $Q = 46.3$
3. $X = 396$, $\alpha = 45.2°$, $\beta = 37.8°$
4. $Q = 0.391\ 44 \sin \theta - 0.284\ 40 \cos \theta$

CHAPTER 28

1. $120.5\angle 45.1°$
2. 2.26 kN
3. $242\angle 150.3°$
4. $207\angle 40.5°$

CHAPTER 29

1. $2.25 \times 10^5\pi$ r/s
2. (*a*) 277
   (*b*) 120π r/s
   (*c*) 60 Hz
   (*d*) 0.016 67 s
   (*e*) 18° lead
   (*f*) 148.4

CHAPTER 30

1. 154 V
2. (*a*) 60 GHz
   (*b*) 102 μA
3. 140 μA
4. 156 μA
5. (*a*) $v = \dfrac{600}{0.707}\sin(880\pi t)$ V
   (*b*) $i = \dfrac{62}{0.707}\sin(880\pi t - 18°)$ A
   (*c*) 87.2 A

CHAPTER 31

1. $158.5 + j113.3$
2. $-19.4 + j19.6$
3. $60 + j45$
4. $-1.29 + j0.824$
5. $22.1\angle -3.06° = 22.1 - j1.18$
6. $8.91\angle 17.7° = 8.49 + j2.71$
7. $\pm 22.6\angle 20° = 21.3 + j7.74$ or $-21.3 - j7.74$
8. $25\angle 106.2° = -6.97 + j24.0$

CHAPTER 32

1. (*a*) $v = 311 \sin(880\pi t)$ V
   (*b*) $i = 8.30 \sin(880\pi t)$ A
   (*c*) 26.3 V
   (*d*) 1.14 kW
   (*e*) 2.19 A
   (*f*) 2.19 A
2. 723 kHz
3. 15.3 μF
4. (*a*) 4.52 kΩ
   (*b*) 4.68 kΩ
   (*c*) 6.94 mA
   (*d*) 8.33 V
   (*e*) 31.4 V
   (*f*) 39.8 V
   (*g*) 32.5 V
5. (*a*) 3.02 kΩ
   (*b*) 2.65 kΩ
   (*c*) $771\angle 28.1^\circ$ Ω
   (*d*) 130 mA
   (*e*) 392 V
   (*f*) 344 V
   (*g*) 88.2 V
   (*h*) 11.4 W
6. 56.3 Hz

CHAPTER 33

1. 72 μF
2. $24.6\angle -9.22^\circ$ Ω
3. $99\angle -8.2^\circ$ Ω = 98 − j14.1 Ω
4. (*a*) 398 kHz
   (*b*) 18.2
5. $Z_a = 155\angle 6.82^\circ \Omega$

CHAPTER 34 Test A

1. $\log_2 32 = 5$
2. $16^3 = 4096$
3. $x = 4$
4. $x = 4$
5. $2[\log 2 + 2\log x + \log 3 + 4\log y - (\log 5 + \log z)]$
6. 3
7. 0.243 04
8. −4
9. 2.158 36
10. 2.334 45
11. 3.500×10^3
12. 3
13. 2.083 75

CHAPTER 34 Test B

1. $x = 3.22$
2. $x = 5.5$
3. $x = 0.612\ 36$
4. $x = 4.477\ 69$
5. 5.176 15
6. $x = 2.163\ 49$
7. 2.461 23
8. $x = 2.517\ 68$
9. $Y = 8.055$
10. $V = 110$ V

CHAPTER 35

1.

1.00	39%
1.78	39%
3.16	39%
5.62	39%
10.0	39%
etc.	

2. 17 dB
3. 9.5 kW
4. 964 mV
5. 73 dB
6. 792 kW
7. 4.09 nF
8. 654 Ω
9. 15.8 mm
10. 0.421 dB

CHAPTER 36

1. 43
2. 11010101
3. 335_{10}
4. 4561_8
5. 1152_7
6. 42_{10}
7. 101001111_2
8. 1131_8
9. 1000000
10. 001011
11. 1010110001
12. 1100110
13. 0.01110
14. $\frac{13}{16}$
15. $\frac{5}{32}$
16. 1.1111
17. 0.51_8
18. 10100111_2
19. 27191_{10}
20. $0.478\ 027\ 344_{10}$

CHAPTER 37

1. $p(t + r)$
2.

p	t	pt	r	$pt + r$
1	1	1	1	1
1	1	1	0	1
1	0	0	1	1
1	0	0	0	0
0	1	0	1	1
0	1	0	0	0
0	0	0	1	1
0	0	0	0	0

3. True
4. $a + cd$
5. $\overline{\overline{S_1 + S_2 + C} + \overline{S_1 + \overline{C}} + \overline{\overline{S_2 + \overline{C}}}}$
6. 0

CHAPTER 38

1. $\overline{A}\,\overline{B} + A\overline{B}$
2. $\overline{B}$
3. $\overline{A}\,\overline{B}\,\overline{C} + \overline{A}\,\overline{B}C + A\overline{B}\,\overline{C} + \overline{A}BC + ABC$
4.

	$\overline{A}$		A		$\overline{A}$
$\overline{B}$	1	1		1	1
B		1	1		
$\overline{B}$					
	$\overline{C}$	C		$\overline{C}$	

5. $\overline{A}\,\overline{B} + \overline{B}\,\overline{C} + BC + \overline{A}C$
6.

	$\overline{A}$		A		$\overline{A}$	
$\overline{B}$		1	1			$\overline{D}$
$\overline{B}$						D
B	1		1	1		D
B			1			$\overline{D}$
$\overline{B}$						$\overline{D}$
	C	$\overline{C}$		C		

7. $\overline{B}\,\overline{C}\,\overline{D} + \overline{A}BCD + ABD + BCD + A\overline{C}\,\overline{D} + AB\overline{C}$

	$\overline{A}$		A		$\overline{A}$	
$\overline{B}$		1	1			$\overline{D}$
$\overline{B}$						D
B	1		1	1	1	D
B			1			$\overline{D}$
$\overline{B}$			1			$\overline{D}$
	C	$\overline{C}$		C		

INDEX

laws of logarithms

$$a^x = N \quad \log_a N = x \quad N = \text{antilog}_a\, x$$
$$\log_a a^b = b$$
$$\log_a (M \cdot N) = \log_a M + \log_a N$$
$$\log_a \frac{M}{N} = \log_a M - \log_a N$$
$$\log_a M^n = N \log_a M$$
$$\log_a M^{1/n} = \frac{\log_a M}{N}$$
$$\text{colog}_a N = \log_a \frac{1}{N}$$
$$\log_b a = \frac{1}{\log_a b}$$
$$\log_b N = \log_a N \cdot \log_b a = \frac{\log_a N}{\log_a b}$$

laws of exponents

$$a^m \cdot a^n = a^{m+n}$$
$$a^m \div a^n = a^{m-n}$$
$$(a^m)^n = a^{mn}$$
$$a^m = \frac{1}{a^{-m}}$$
$$(ab)^m = a^m b^m$$
$$\left(\frac{a}{b}\right)^m = \frac{a^m}{b^m}\ (b \neq 0)$$
$$a^0 = 1$$
$$a^{m/n} = \sqrt[n]{a^m} = (\sqrt[n]{a})^m$$